# STRUCTURAL BIOINFORMATICS

01235788

METHODS OF
BIOCHEMICAL ANALYSIS

Volume 44

# STRUCTURAL BIOINFORMATICS

**Edited By**

Philip E. Bourne
San Diego Supercomputer Center
Department of Pharmacology
University of California San Diego
La Jolla, CA

Helge Weissig
Structural Bioinformatics
San Diego SuperComputer Center
University of California San Diego
La Jolla, CA

A JOHN WILEY & SONS PUBLICATION

This book is printed on acid-free paper. ∞

Copyright © 2003 by John Wiley & Sons, Inc. All rights reserved.

Published by Wiley-Liss, Inc., Hoboken, New Jersey.
Published simultaneously in Canada.

No part of this publication may be reproduced, stored in a retrieval system, or transmitted in any form or by any means, electronic, mechanical, photocopying, recording, scanning, or otherwise, except as permitted under Section 107 or 108 of the 1976 United States Copyright Act, without either the prior written permission of the Publisher, or authorization through payment of the appropriate per-copy fee to the Copyright Clearance Center, Inc., 222 Rosewood Drive, Danvers, MA 01923, 978-750-8400, fax 978-750-4470, or on the web at www.copyright.com. Requests to the Publisher for permission should be addressed to the Permissions Department, John Wiley & Sons, Inc., 111 River Street, Hoboken, NJ 07030, (201) 748-6011, fax (201) 748-6008, e-mail: permreq@wiley.com.

Limit of Liability/Disclaimer of Warranty: While the publisher and author have used their best efforts in preparing this book, they make no representations or warranties with respect to the accuracy or completeness of the contents of this book and specifically disclaim any implied warranties of merchantability or fitness for a particular purpose. No warranty may be created or extended by sales representatives or written sales materials. The advice and strategies contained herein may not be suitable for your situation. You should consult with a professional where appropriate. Neither the publisher nor author shall be liable for any loss of profit or any other commercial damages, including but not limited to special, incidental, consequential, or other damages.

For general information on our other products and services please contact our Customer Care Department within the U.S. at 877-762-2974, outside the U.S. at 317-572-3993 or fax 317-572-4002.

Wiley also publishes its books in a variety of electronic formats. Some content that appears in print, however, may not be available in electronic format.

Cover Design: Jennifer Matthews

*Library of Congress Cataloging-in-Publication Data:*

Structural bioinformatics / edited by Philip E. Bourne, Helge
    Weissig.
        p. ; cm. – (Methods of biochemical analysis; v. 44)
    Includes bibliographical references and index.
    ISBN 0-471-20200-2 (cloth : alk. paper)
        1. Macromolecules–Structure–Mathematical models. 2. Mac-
romolecules–Structure–Computer simulation. 3. Bioinformatics.
    I. Bourne, Philip E. II. Weissig, Helge. III. Series.
        [DNLM: 1. Computational Biology. QH 506 S9266 2003]
    QP517.M3S776 2003
    572.8'733–dc21

                                                        2002011156

Printed in the United States of America.

10 9 8 7 6 5 4

# CONTENTS

## Section I  INTRODUCTION                                                    1

# FOREWORD

Both structural biology and bioinformatics have come of age as major arenas for productive research and are commanding a high level of attention. These fields are attracting the interest of traditionally "wet lab" biologists as well as of politicians, governments of the world, and industry. Job opportunities for young scientists in this domain exceed the pace of production, an atypical but wonderful circumstance. Indeed, the opportunities for those being trained today are virtually unlimited. This convergence—of independent assessments of merit—shows the extraordinary significance of the interface between structural biology and informatics, and clearly demonstrates just how timely this book is, as the initial foray into the new domain. In providing the very first textbook in structural bioinformatics, Philip E. Bourne and Helge Weissig have created a balanced set of contributions from the leading authorities in their respective research problems. This comprehensive collection fully captures the spirit of excitement at the "bleeding edge" of biodiscovery, or life at the frontier. The frontier between computing and biology itself reflects the decades of extraordinary progress and revolutionary advances in both domains. Indeed, for some time, the fields themselves have been mature enough to contribute significantly to each other's advancement. However, over the last decade, the science funding arms of the world's governments did not have the necessary mechanisms in place to build biological informatics; similarly, as a consequence of the lack of tools for data integration and biological modeling, the pace of biological understanding at an in-depth systems level has not kept pace with the technology advances that have powered experimental discovery and data collection, but could not power a higher-level integrative insight into biosystems. Today, with increased funding opportunities and recognition by the experimental community, we are in the midst of a radical shift, to which this collection speaks directly and powerfully, in the use of information technology and quantitative approaches in both the basic and the applied life sciences.

The early generation of practitioners of bioinformatics, who by and large came from the physical sciences and engineering, left academia when the large pharmaceutical firms recognized the importance of bioinformatics well before the sources of research funding did. As a consequence, we lost a generation of "seed corn": many of the individuals who would be faculty today never came to work in this domain—simply were never trained—since what opportunities there were existed largely within industry. Textbooks of the highest quality and clarity are essential today, as we all struggle to determine the right curriculum and the right content to train what will be the first generation of students who are truly bioinformaticians. To teach the vast numbers of young scientists looking to contribute in bioinformatics and, more generally, to enlarge the community who can exploit information technology in the era of high-throughput biology, requires authoritative treatments that provide the basis to pull in a new generation of scientists. Such works must also use the best treatments possible to reach a much larger audience, including mature scientists who wish to retrain themselves, and need to set a standard for training everywhere in the world. This collection admirably meets those goals and is a must read for all of us committed to understanding the interplay of structure and function. By assembling the best thinkers to address systematically all of the challenges at the next stage of the genome effort, the editors have created a book that will serve to educate the next generation, help us bring the best people to universities to ensure that in the future we will have the requisite seed corn to create the young faculty who will create the tools required for the new biology.

The role of modeling and theory (as predictive tools) have been brought home to experimental molecular biophysicists and biochemists through the Critical Assessment of Structure Prediction (CASP), a biannual contest to predict as accurately, routinely, and broadly as possible, structures of unknown proteins for which the actual structure will have been experimentally determined within a few months. Thus, the structures are then known in time for a comparative analysis of successes and failures in the ultimate science-product-on-the-line meeting, CASP, in each of its alternate year flavors. The fact that theoreticians *could* demonstrate high quality results and improvements each cycle in the power of a priori predictions of structure has had a profound influence on the community and the peer review process. No longer can it be said that knowledge of the actual experimental structure influenced the modeling. Reviewers and the community look to the accomplishments at CASP as setting a floor, as being the acid test of the hardness or robustness of individual approaches. (This effort has also removed some of the "not invented here" syndrome: it is better to use somebody else's method and be right than to be limited to one approach and fail.) CASP demonstrates better than any other activity that computational tools have come of age as tools for structural biology and structural genomics. (See the detailed and elegant description in the book.) We can only hope that this collection will inspire extensions of CASP; for example, to comparative analyses, from the perspective of sequence and of sequence and structure, on automated functional assignment approaches.

In an era based on the availability of ever more fully sequenced genomes and on a suite of high-throughput experimental biological methods, structure determination and the knowledge base-driven (annotated database-driven) exploration of the implications of structure have captured center stage in bringing the implicit information from the human genome to the service of society. The emphasis now is on obtaining a better understanding of biological processes and should extend in the future to providing improved health care delivery. How did we get here, given the past decades of excitement for a very different set of approaches in the life sciences, namely, those of molecular biology? Many metaphors capture today's challenge in bringing biology from a qualitative science (or at most a binary science of yes or no, a spot on a gel or its absence), to one in which the role of protein–protein interactions and the details of proteins working as macromolecular machines are characterized in quantitative terms. The accelerating pace of acquisition of experimental data and the challenges of data integration are driving the massive shift in biology of moving from a descriptive science to a predictive one. The data deluge following the technical advances in life sciences research suggests the need to drink from a firehose filled with rapidly flowing, incredibly complex data, and the need to swim in data to move forward in our understanding. This perception gives rise to the common expression that biology is becoming an information science. Mixing metaphors, biologists have left the twilight zone and fearlessly have taken up the decade old challenge originally proposed by Walter Gilbert: Use computational tools or become obsolete.

In somewhat less than five years, computational biology and bioinformatics have gone from uncertain stepchildren, often neglected, to the hottest job prospects and the buzzwords on everyone's lips. The reasons are obvious: We are already living in the future as biologists; we feel the impact of fully sequenced genomes; we are watching the transition to high-throughput biology and to a norm of asking global or systemic questions, rather than using individual macromolecule-specific probes. While it is amusing to read the phrase "classical bioinformatics" and more so to note its routine use to describe what was originally (and still should be termed) genome informatics, there is no doubt that the successes of the genome project and the impact of computer technology have, over just a few years, opened up new vistas in biology. We have only begun to see what will be and to learn how best to exploit computer and information technology.

The funding available for bioinformaticians around the world does not yet reflect the revolutionary opportunities. For biologists to live in the future, to use the metaphor *per jure*, all research funding organizations should recognize the level of substantial, sustained financial support needed to build a biological sciences informatics infrastructure and address the algorithmic, database, integration, and computational challenges. Let us hope that this book can contribute to informing such a decision. There are hopeful signs on the horizon for the larger nations and economies around the world. As biological science becomes more dependent on contributions from its interface with computing, the smaller nations (e.g., Belgium) will need to assert a level of contribution for an appropriate niche. For the past decade, a number of the larger economies, such as those of Germany, Japan, and the United Kingdom, initiated bioinformatics and specifically structural bioinformatics funding programs and that funding continues although it does not yet reflect the opportunities. However, the United States, a central contributor to the sea change, has had the insightful early support of private foundations. Following seminal awards, made by the Keck Foundation and by Burroughs-Welcome, to facilitate the development of teams and support education, the Howard Hughes Medical Institute instituted a series of awards in the form of new professorships in computational biology and bioinformatics. At the same time, the National Institutes of Health (NIH) began the process of implementing a major computational biology program that includes education and development of software tools within pilot projects. The NIH Protein Structure Initiative and the related biocomplexity program, as well as the Information Technology Research program at the National Science Foundation (NSF), provide a more stable basis for developing the infrastructure for this interface, and one can now expect extension to all biological subfields.

The excitement that began with recognizing how molecular sequence information could inform evolutionary studies, bring insights in cellular biology, identify genes involved in disease etiology, and provide drug targets captures what led researchers in the life sciences to adopt computational genomics or genome informatics as a routine part of their experimental arsenal. Today, the second phase of the genome projects are under way; this phase is known—depending on the cognizant federal agency—as the Protein Structure Initiative, structural genomics, Genomes to Life, systems biology, and/or quantitative biology. As such, the next step in bringing value of genomic information to the service of society is clearly to address structural informatics.

Those of us concerned with macromolecular structure have long spoken the mantra: form follows function, a given function requires a specific structure, or, as the inverse, structure in turn can be seen to determine function. That is, if we know structure, we can infer many aspects of biochemical and sometimes even cellular function, which can subsequently be experimentally tested. Backing up, given that we now *know* sequence, from the implicit information in genomes, we can build better and better tools to model structure explicitly and powerfully, and subsequently we can predict the details of probable function, thereby greatly reducing the search space and focus experimental work to test for only the most likely functions. We can also inform biological cartoons, the descriptive way in which we biologists think, by computational modeling of dynamic form and function—by pushing the boundaries beyond that which can be experimentally determined to fill in the details of the dynamics of macromolecules underlying living processes. To connect the abstract to the practical, this wonderful collection of insightful articles shows how we can build the infrastructure for an integrative approach to understanding biosystems through the power of knowing structure and its implications for the mechanics of function.

The basis for all structural bioinformatics, the central community database for structural biology, is the Protein Data Bank (PDB) and what it contains. Macromolecular structures are archived and represented in its master encyclopedia or community database. This book addresses the key points of the state of the art, beginning with definitions and scope and our knowledge of protein structure and the tools for its determination, and proceeds through what is needed from computer and information science and how these tools allow us to understand the complexity of biological systems. Overall, the individual chapters outline the suite of major basic life science questions such as the status of efforts to predict protein structure and how proteins carry out cellular functions, and also the applied life science questions such as how structural bioinformatics can improve health care through accelerating drug discovery. Dictated by the process of uncovering the mechanisms through which macromolecules act, this journey of discovery, into all the quirks and still-undiscovered mainstream events, will keep biologists entertained for centuries to come. This book is a great guidebook and a proud step toward this understanding, and I recommend it to all of you.

John C. Wooley
Associate Vice Chancellor, Research
University of California, San Diego
La Jolla, CA

# PREFACE

What is structural bioinformatics? As teachers of bioinformatics we are always surprised to encounter students who are committed to a career in bioinformatics, yet at the outset are not familiar with what bioinformatics encompasses. This situation is compounded by many biologists who perceive bioinformatics as simply the use of tools for sequence analysis. Here we define bioinformatics as the development and application of algorithms and methods to turn biological data into knowledge of biological systems, often requiring further experimentation suggested by initial data. More specifically, bioinformatics can be divided into three parts: (1) theory and methods: algorithms, statistical methods, machine learning, ontologies, and so on; (2) applications: for example, sequence analysis, whole genome assembly, protein structure prediction, and biological databases; and (3) kinds of data: for example, base reads from DNA microarrays, mass spectroscopy experiments, binding constants, and so forth. Structural bioinformatics then becomes all of (1) and (2) applied to biological structure data. Clearly, biological structure exists at different scales, from molecules to large complexes, organelles, cells, and finally to complete organisms and populations of organisms. A complete treatment of these biological scales would require many volumes. Here we restrict ourselves to molecules, primarily protein, DNA, RNA, ligands, and complexes thereof. Chapter 1 provides additional detail of the scope of this book and the history of the field.

The remainder of the introductory section familiarizes us with the data we are dealing with. For readers not already familiar with macromolecular structure data, Chapters 2 and 3 describe protein, DNA, and RNA structure, respectively. Understanding the nuances (scope, accuracy, completeness, etc.) of structural data is prerequisite to any effective use of that data. Effective data use in turn requires an understanding of the experiments that produce the data. The most popular methods for deriving macromolecular structure data are, in order (at this time), X-ray crystallography (Chapter 4), NMR spectroscopy (Chapter 5), and electron microscopy (Chapter 6). The raw data from these methods are most often a set of Cartesian coordinates representing the positions of the atoms in these structures. Browsing through tables of these data is clearly not useful for a human in interpreting structure. Thus, visualization of structure has evolved along with structural biology and this is discussed, with information about where to get free tools to visualize structure, in Chapter 7.

In the early days of structural biology (up to the late 1970s) those in the field could name all the structures that had been solved, many of which had Nobel prizes attached to them. As the field grew such a feat was no longer possible, and databases of structure data began to appear. Consistent use of structure data requires consistent data representation and Section II is devoted to this topic. These data formats, described in Chapter 8, are important because they contain information beyond the basic atomic coordinates. The field is very fortunate to have a single resource, the Protein Data Bank (PDB), that maintains the primary structure data for all publicly accessible structures worldwide (Chapter 9). Many resources take this primary data and derive other useful information (Chapters 10 and 11), at least a subset of which practitioners need to be aware of.

As the number of structures increased, comparative analysis, the subject of Section III, became possible. Chapters 12 and 13 describe structure classification schemes and Chapters 14 and 15 describe structure validation, which is important to understand the accuracy of data you are dealing with. The final chapter in this section (Chapter 16) describes methods for three-dimensional structure comparison and alignment.

The more we know from comparing structures, the more we can learn about structure and functional assignment. Secondary structure assignment can be made consistently and reliably for the majority of structures (Chapter 17). Proteins exist as one or more domains or compact structural and functional units. Hence, automated assignment of domains is important (Chapter 18). Through the structural genomics projects, structure determination is moving from a functional to a genomic initiative. That is, structure was traditionally determined in an effort to further elucidate a known function. We are now in a situation where structures are being determined with no elucidated functions, thus making functional assignment critical (Chapter 19).

Proteins do not function in isolation, but as the result of complex protein–protein, protein–ligand, and protein–solvent interactions. Section V describes these interactions. The majority of these interactions are not captured in a structure of a complex, but as a singleton with a signature that can be teased out to predict that interaction. Evolutionary information from sequence (Chapter 20) and electrostatics (Chapter 21) are important in this regard.

Section VI focuses on a special type of interaction where the protein and its signature structure are a potential drugable target. Chapter 22 considers the docking problem, predicting how two proteins or a protein and a ligand will interact, and Chapter 23 describes the commercial drug discovery process and how it is changing with the advent of more structures as potential drug targets and, of course, through structural bioinformatics!

While the number of structures may be increasing rapidly, so is the number of protein sequences, and so the idea of predicting a protein structure from its sequence remains an obsessive goal. Progress is being made, spurred by an unusual biannual competition, referred to as CASP—the Critical Assessment of Protein Structure Prediction (Chapter 24). Subcategories of efforts within CASP and the field in general are homology modeling (Chapter 25), fold recognition (Chapter 26) and *ab initio* structure prediction (Chapter 27). Other forms of prediction include secondary structure and membrane components of structure (Chapter 28).

Finally, Chapter 29 describes the paradigm shift in structural biology, which is much of the motivation for this book. Just as the human genome project could be held responsible for the emergence of bioinformatics as a field of study, so perhaps looking back in five years we will see structural bioinformatics having fully emerged as a result of high-throughput structure determination (i.e., structural genomics). Whatever the outcome, structural bioinformatics is an exciting place to be working right now. What makes it more so are the wonderful leaders in the field, many of whom have contributed (without much prodding) to the contents of the pages that follow. Their and our excitement is captured in what follows. Curious? Come share the fun.

*Philip E. Bourne*
*Helge Weissig*

# ACKNOWLEDGMENTS

We are grateful to the San Diego Supercomputer Center and the University of California San Diego for institutional support for all of our research and educational work, including this book. Likewise, the US Funding agencies, the National Science Foundation, the National Institutes of Health and the Department. As for many tasks, preparing this book was made easier by our administrative support, namely, Josie Alaoen, Dorothy Kegler, and Peggy Wagner. Finally, thanks to Luna Han at John Wiley & Sons for being so understanding in all the issues that arose in getting this book into print.

ACKNOWLEDGMENTS

# CONTRIBUTORS

**Paul D. Adams**, Lawrence Berkeley Laboratory, Berkeley, CA

**Russ B. Altman**, Stanford University, Stanford, CA

**Claus A. F. Andersen**, CUBIC, Department of Biochemistry and Molecular Biophysics, Columbia University, New York, NY; and Center for Biological Sequence Analysis, The Technical University of Denmark, Lyngby, Denmark

**Davis Baker**, Department of Biochemistry and Howard Hughes Medical Institute, University of Washington, Seattle, WA

**Nathan A. Baker**, University of California, San Diego, La Jolla, CA

**Gail J. Bartlett**, University College London, London, UK

**Helen M. Berman**, Department of Chemistry, Rutgers University, Piscataway, NJ

**Jeffrey B. Bonanno**, Howard Hughes Medical Institute, Laboratories of Molecular Biophysics, The Rockefeller University, New York, NY

**Richard Bonneau**, Department of Biochemistry and Howard Hughes Medical Institute, University of Washington, Seattle, WA

**Philip E. Bourne**, San Diego Supercomputer Center, Department of Pharmacology, University of California, San Diego, La Jolla, CA

**Natasja Brooijmans**, University of California, San Francisco, San Francisco, CA

**Axel T. Brunger**, The Howard Hughes Medical Institute and Departments of Molecular and Cellular Physiology, Neurology and Neurological Sciences, and Stanford Synchrotron Radiation Laboratory, Stanford University, Stanford, CA

**Stephen K. Burley**, Structural GenomiX Inc., San Diego, CA

**Dylan Chivian**, Department of Biochemistry, University of Washington, Seattle, WA

**Jonathan M. Dugan**, Section of Medical Informatics, Department of Medicine, Stanford University, Stanford, CA

**Eric B. Fauman**, Discovery Research Informatics, Ann Arbor, MI

**Zukang Feng**, Department of Chemistry, Rutgers University, Piscataway, NJ

**Jody Lynn Fink**, Biomedical Sciences Program, University of California, San Diego, La Jolla, CA

**Paula M. D. Fitzgerald**, Merck Research Laboratories, Whitehouse Station, NJ

**Adam Godzik**, Program in Bioinformatics and Biological Diversity, The Burnham Institute, La Jolla, CA

**Colin R. Groom**, Discovery Research Informatics, Ann Arbor, MI

**Ralf W. Grosse-Kunstleve**, Physical Biosciences Division, Lawrence Berkeley Laboratory, Berkeley, CA

**Dorit Hanein**, Program in Cell Adhesion & Extracellular Matrix Biology, The Burnham Institute, La Jolla, CA

**Andrew L. Hopkins**, Discovery Research Informatics, Ann Arbor, MI

**Lisa Iype**, Department of Chemistry, Rutgers University, Piscataway, NJ

**Elmar Krieger**, Center for Molecular and Biomolecular Informatics, University of Nijmegen, Nijmegen, The Netherlands

**Jennifer Krumrine**, Department of Pharmaceutical Chemistry, University of California, San Francisco, San Francisco, CA

**Irwin Kuntz**, Departmen to fPharmaceutical Chemistry, University of California, San Francisco, San Francisco, CA

**Roman A. Laskowski**, Department of Biochemistry and Molecular Biology, University College London, University of London, London, UK

**John L. Markley**, Department of Biochemistry, University of Wisconsin-Madison, Madison, WI

**J. Andrew McCammon**, Department of Chemistry and Biochemistry, University of California, San Diego, La Jolla, CA

**Sander B. Nabuurs**, Center for Molecular and Biomolecular Informatics, University of Nijmegen, Nijmegen, The Netherlands

**Stephen Neidle**, Structural Biology Section, CRC Biomolecular Structure Unit, Institute of Cancer Research, London, UK

**Christine A. Orengo**, Department of Biochemistry and Molecular Biology, University College London, University of London, London, UK

**Florencio Pazos**, Protein Design Group, National Center for Biotechnology, Madrid, Spain

**Frances M. G. Pearl**, Department of Biochemistry and Molecular Biology, University College London, University of London, London, UK

**Florian Raubacher**, Department of Pharmaceutical Chemistry, University of California, San Francisco, San Francisco, CA

**Boojala V. B. Reedy**, San Diego Supercomputer Center, University of California, San Diego, La Jolla, CA

**Jane S. Richardson**, Department of Biochemistry, Duke University, Durham, NC

**Timothy Robertson**, Department of Biochemistry, University of Washington, Seattle, WA

**Burkhard Rost**, Department of Biochemistry and Molecular Biophysics, Columbia University, New York, NY

**Eric D. Scheeff**, Biomedical Sciences Program, University of California, San Diego, La Jolla, CA

**Bohdan Schneider**, Department of Chemistry, Rutgers University, Piscataway, NJ

**Ilya Shindyalov**, San Diego Supercomputer Center, University of California, San Diego, La Jolla, CA

**John Tate**, European Bioinformatics Institute, Cambridge, UK

**Janet M. Thornton**, European Bioinformatics Institute, Cambridge, UK

**Annabel E. Todd**, University College London, London, UK

**Eldon L. Ulrich**, BioMagResBank, University of Wisconsin-Madison, Madison, WI

**Alfonso Valencia**, Protein Design Group, National Center for Biotechnology, Madrid, Spain

**Brian F. Volkman**, Department of Biochemistry, Medical College of Wisconsin, Milwaukee, WI

**Niels Volkmann**, Program in Bioinformatics and Systems Biology, The Burnham Institute, La Jolla, CA

**Gert Vriend**, Center for Molecular and Biomolecular Informatics, University of Nijmegen, Nijmegen, The Netherlands

**Helge Weissig**, ActivX Biosciences, Inc., La Jolla, CA

**Lorenz Wernisch**, European Molecular Biology Laboratory, European Bioinformatics Institute, Cambridge, UK

**John D. Westbrook**, Department of Chemistry, Rutgers University, Piscataway, NJ

**William M. Westler**, National Magnetic Resonance Facility at Madison, University of Wisconsin-Madison, Madison, WI

**Shoshana J. Wodak**, Unite de Conformation de Macromolecules Biologiques, Universite Libre de Bruxelles, Brussels, Belgium; and European Bioinformatics Institute, Cambridge, UK

**John C. Wooley**, Department of Biomedical Sciences, University of California, San Diego, La Jolla, CA

**Christine Zardecki**, Department of Chemistry, Rutgers University, Piscataway, NJ

# Section I

## INTRODUCTION

# 1

# DEFINING BIOINFORMATICS AND STRUCTURAL BIOINFORMATICS

Russ B. Altman and Jonathan M. Dugan

## WHAT IS BIOINFORMATICS?

The precise definition of bioinformatics is a matter of some debate. Some define it narrowly as the development of databases to store and manipulate genomic information. Others define it broadly as encompassing all of computational biology. Based on its current use in the scientific literature, bioinformatics can be defined as the study of two information flows in molecular biology (Altman, 1998). The first information flow is based on the central dogma of molecular biology: DNA sequences are transcribed into mRNA sequences, mRNA sequences are translated into protein sequences. Protein sequences fold into three-dimensional (3D) structures that have functions. These functions are selected for, in a Darwinian sense, by the environment of the organism, which drives the evolution of the DNA sequence within a population. The first class of bioinformatics applications, then, can address the transfer of information at any stage in the central dogma, including the organization and control of genes in the DNA sequence, the identification of transcriptional units in DNA, the prediction of protein structure from sequence, and the analysis of molecular function.

The second information flow is based on the scientific method: We create hypotheses regarding biological activity, design experiments to test these hypotheses, evaluate the resulting data for compatibility with the hypotheses, and extend or modify the hypotheses in response to the data. The second class of bioinformatics applications address the transfer of information within this protocol, including systems that generate hypotheses, design experiments, store and organize the data from these experiments in databases, test the compatibility of the data with models, and modify hypotheses.

*Structural Bioinformatics*
Edited by Philip E. Bourne and Helge Weissig
Copyright © 2003 by Wiley-Liss, Inc.

Although its use is still evolving, bioinformatics is not usually used to describe computational approaches to problems in biology above the cellular level. The explosion of interest in bioinformatics has been driven by the emergence of experimental techniques that generate data in a high throughput fashion—such as DNA sequencing, mass spectrometry, or microarray expression analysis (Miranker, 2000; Altman and Raychaudhuri, 2001; GISC, 2001; Venter et al., 2001). Bioinformatics depends on the availability of large data sets that are too complex to allow manual analysis. The rapid increase in the number of 3D macromolecular structures available in databases such as the Protein Data Bank (PDB,[1] Chapter 9; Berman et al., 2000), has driven the emergence of a subdiscipline of bioinformatics: *structural bioinformatics*. Structural bioinformatics is the subdiscipline of bioinformatics that focuses on the representation, storage, retrieval, analysis, and display of structural information at the atomic and subcellular spatial scales.

Structural bioinformatics, like many other subdisciplines within bioinformatics,[2] is characterized by two goals: the creation of general purpose methods for manipulating information about biological macromolecules, and the application of these methods to solving problems in biology and creating new knowledge. These two goals are intricately linked because part of the validation of new methods involves their successful use in solving real problems. At the same time, the current challenges in biology demand the development of new methods that can handle the volume of data now available, and the complexity of models that scientists must create to explain these data.

## Structural Bioinformatics Has Been Catalyzed by Large Amounts of Data

Biology has attracted computational scientists over the last 30 years in two distinct ways. First, the increasing availability of sequence data has been a magnet for those with an interest in string analysis, algorithms, and probabilistic models (Gusfield, 1997; Durbin et al., 1998). The major accomplishments have been the development of algorithms for pairwise sequence alignment, multiple alignment, the definition and discovery of sequence motifs, and the use of probabilistic models, such as Hidden Markov Models to find genes (Burge and Karlin, 1997), to align sequences (Hughey and Krogh, 1996), and to summarize protein families (Bateman et al., 2000). Second, the increasing availability of structural data has been a magnet for those with an interest in computational geometry, computer graphics, and algorithms for analyzing crystallographic data (Chapter 4) and NMR data (Chapter 5) and creating credible molecular models. Structural bioinformatics has its roots in this second group. The development of molecular graphics was one of the first applications of computer graphics (Langridge and Gomatos, 1963). The elucidation of the structure of DNA in the mid-1950s and the publication of the first protein crystal structures in the early 1960s created a demand for computerized methods for examining these complex molecules. At the same time, the need for computational algorithms to deconvolute X-ray crystallographic data, and to fit the resulting electron densities to the more manageable ball-and-stick models, created a cadre of structural biologists who were very well versed in computational

---

[1] See at http://www.rcsb.org.

[2] The International Society for Computational Biology (ISCB, http://www.iscb.org/) is the professional organization for bioinformatics, and many developments in structural bioinformatics are reported in the journals and conferences associated with this society.

technologies. The challenges of interpreting NMR-derived distance constraints into 3D structures further introduced computational technologies to biological structure. As the number of 3D structures increased, the need to create methods for storing and disseminating this data lead to the creation of the PDB, one of the earliest scientific databases.[3] It can be argued that we are currently seeing a third wave of interest in biological problems from a group that were not engaged by the availability of one-dimensional (1D) sequence data or 3D structural data. This third wave has arisen in response to the increased availability of RNA expression data, and has captured the interest of computational scientists with an interest in statistical analysis and machine learning, particularly in clustering methodologies and classification techniques. The problems posed by these data are different from those seen in both sequence and structural analysis data.

Structural bioinformatics is now in a renaissance with the success of the genome-sequencing projects, the emergence of high-throughput methods for expression analysis, and compound identification via mass spectrometry. There are now organized efforts in structural genomics (Chapter 29) to collect and analyze macromolecular structures in a high-throughput manner, (Teichmann, Chothia, and Gerstein, 1999; Teichmann, Murzin, and Chothia, 2001). These efforts include challenges in the selection of molecules to study, the robotic preparation and manipulation of samples to find crys-tallization conditions, the analysis of X-ray diffraction data, and the annotation of these structures as they are stored in databases (Chapter 4). In addition, the PDB now has a critical mass of structures that allows (indeed, requires!) statistical analysis of struc-tures in order to learn the rules for how active sites and binding sites are constructed, and that allows us to develop knowledge-based methods for the prediction of structure and function. Finally, the emergence of this structural information, when linked to the increasing amount of genomic information and expression data, provides opportunities for linking structural information to other data sources in order to understand how cellular pathways and processes work at a molecular level.

***Toward a High-Resolution Understanding of Biology.*** The great promise of structural bioinformatics is predicated on the belief that the availability of high-resolution structural information about biological systems will allow us to reason precisely about the function of these systems and the affects of modifications or per-turbations. Whereas genetic analyses can only associate genetic sequences with their functional consequences, structural biological analyses offer the additional promise of ultimate insight into the mechanisms of these consequences, and therefore a more pro-found understanding of how biological function follows from structure. The promise for structural bioinformatics lies in four areas: (1) creating an infrastructure for building up structural models from component parts; (2) gaining the ability to understand the design principles of proteins so that new functionalities can be created; (3) learning how to design drugs efficiently based on structural knowledge of their target; and (4) catalyzing the development of simulation models that can give insight into func-tion based on structural simulations. Each of these four areas has already seen success, and the structural genomics projects promise to create data sets sufficient to catalyze accelerated progress in all these areas.

With respect to creating an infrastructure for modeling larger structural ensembles, we are already seeing the emergence of a new generation of structures larger by an

---

[3]See at http://www.rcsb.org.

order of magnitude than the structures submitted to the PDB even a few years ago. The two main achievements in the last couple of years have been the elucidation of the structure of the bacterial ribosome (with more than 250,000 atoms) (Ban et al., 2000; Clemons et al., 2001; Yusupov et al., 2001), and the publication of the structure of the RNA polymerase structure (with about 500,000 atoms) (Cramer et al., 2000). These two accomplishments allow us to examine the principles for how a large number of component protein and nucleic acid structures can assemble to create macromolecular machines. With these successes, we can now target numerous other cellular ensembles for structural studies.

The design principles for proteins are now in reach because we have both a large "training set" of example proteins to study, and because methods for structure prediction are beginning to allow us to identify structures that are unlikely to be stable. There have been preliminary successes in the design of four-helix bundle proteins (DeGrado et al., 1987), and in the engineering of triose phosphate isomerase (TIM) barrels (Silverman et al., 2001). There has been interesting work in "reverse folding" in which a set of amino acid side chains is collected in order to stabilize a desired protein backbone conformation (Koehl and Levitt, 1999).

Rational drug design has not been the primary way for discovering major therapeutics (Chapter 23). However, recent successes in this area give reason to expect that drug discovery projects will increasingly be structure based. One of the most famous examples of rational drug design was the creation of HIV protease inhibitors based on the known 3D crystal structure (Kempf, 1994; Vacca, 1994). Methods for matching combinatorial libraries of chemicals against protein binding sites have matured and are in routine use at most pharmaceutical companies.

The simulation of biological macromolecular dynamics dates almost as far back as the elucidation of the first protein structure (Doniach and Eastman, 1999). These simulations are based on the integration of classical equations of motion and computation of electrostatic forces between atoms in a molecule. Methods for simulation now routinely include water molecules and are able to remain stable (the molecule does not fall apart) and reproduce experimental measurements with some fidelity. The simulation of larger ensembles and of structural variants (such as based on known genetic variations in sequence) should lead to a more profound understanding of how structural properties produce functional behavior.

## Special Challenges in Computing with Structural Data

Structural bioinformatics must overcome some special challenges that are either not present or not dominant in other types of bioinformatics domains (such as the analysis of sequence or microarray data). It is important to remember these challenges when assessing the opportunities in the field. They include:

- Structural data is not linear and therefore is not easily amenable to algorithms based on strings. In addition to this obvious nonlinearity, there are also nonlinear relationships between atoms (the forces are not linear), which means that most computations on structure need either to make approximations or to be very expensive.
- The search space for most structural problems is continuous. Structures are represented generally by atomic Cartesian coordinates (or internal angular coordinates) that are continuous variables. Thus, there are infinite search spaces for

algorithms attempting to assign atomic coordinate values. Many simplifications can be applied, such as lattice models for 3D structure (Hinds and Levitt, 1994), but these are attempts to manage the inherent continuous nature of these problems.

- There is a fundamental connection between molecular structure and physics. Although this statement seems obvious and trivial, it means that when reduced representations, such as pseudoatoms (Wuthrich et al., 1983) or lattice models are applied, they become more difficult to relate to the underlying physics that govern the interactions. The need to keep structural calculations physically reasonable is an important constraint.

- Reasoning about structure requires visualization. As mentioned earlier, the creation of computer graphics was driven, in part, by the need of structural biologists to look at molecules (Chapter 7). This visualization is both a benefit and a detriment: Structure is well defined and well-designed visualizations can provide insight into structural problems. However, graphic displays have a human user as a target and are not easily parsed or understood by computers, and thus represent something of a computational "dead end." The need to have expressive data structures underlying these visualizations allows the information to be understood and analyzed by computer programs, and thus opens the possibility of further downstream analysis.

- Structural data, like all biological data, can be noisy and imperfect. Despite some amazing successes in the elucidation of very high-resolution structures, the precision of our knowledge about many structures is likely to be limited by their flexibility, dynamics, or experimental noise. Understanding the protein structural disorder may be critical for understanding the protein's function. Thus, we must be comfortable reasoning about structures about which we have only partial knowledge.

- Protein and nucleic acid structures are generally conserved more than their associated sequence (Chapter 20). Thus, sequences will accumulate mutations over time that may make identification of their similarities more difficult, while their structures may remain essentially identical. However, sequence information is still much more abundant than structural information, and so for many molecules it is the sequence information that is readily available. Thus, the need to identify distant sequential similarities in order to gain structural insights can be a major challenge.

- Finally, we must recognize that there is a major gap in our knowledge of a large fraction of proteins that are not globular and water soluble. In particular, membrane-bound and fibrous proteins are simply not well understood and structures are not available in the numbers required to allow routine statistical and informatics approaches to their study. The importance of this shortcoming cannot be overemphasized, since these classes of proteins are among the most important for understanding a large number of cellular processes of great interest, including signal transduction, cytoskeletal dynamics, and cellular localizations and compartmentalization.

- Structural genomics will likely produce a large number of structures at the level of the domain—relatively well-defined modules that associate to form larger ensembles. The principles by which these domains associate and cooperatively function is a major challenge for structural biology.

## TECHNICAL CHALLENGES WITHIN STRUCTURAL BIOINFORMATICS

The scientific challenges within structural bioinformatics fall into two rough categories: (1) the creation of methods to support structural biology and structural genomics, and (2) the creation of methods to elucidate new biological knowledge. This distinction is not absolute, but is useful for dividing much structural bioinformatics work. The support of experimental structural biology is an area of particular interest currently with the emergence of efforts in high-throughput structural genomics. Informatics approaches are required for many aspects of this enterprise, and can be briefly reviewed here.

*Target Selection.* Structural genomics efforts with finite resources must carefully select proteins to study. Informatics methods are used to compare the database of existing structures and known sequences with potential targets in order to identify those that are most likely to add to our structural knowledge base. This selection can be informed by the expected novelty of the structure, and even its importance as reflected in the published literature (Linial and Yona, 2000). A critical part of target selection is the identification of domains within large proteins. Domains are often easier to study initially in isolation, and then to study in complexes. The definition of domains from sequence data alone is a challenging problem.

*Tracking Experimental Crystallization Trials.* One of the major bottlenecks in structural genomics is the discovery of crystallization conditions that work for proteins of interest. In addition to the obvious need for storing and tracking information on the proteins, the conditions attempted, and the results, there is also an opportunity to apply machine-learning methods to these data in order to extract rules that may help increase the yield of crystals based on previous experience (Hennessy et al., 2000). Until recently the results of failed crystallization experiments were not generally available, making it difficult to apply automated machine-learning methods to these data sets.

*Analysis of Crystallographic Data.* A long-standing area of computation within structural biology are the algorithms for deconvoluting the X-ray diffraction pattern, which involves computing an inverse Fourier transform with partial information (i.e., with missing phase information). There is interest in *ab initio* methods for automating these computations, and success in this area reduces the number of heavy atom derivatives that must be created for structures of interest (Gilmore et al., 1998). Multi-wavelength Anomalous Diffraction (MAD) (Hendrickson, 1991) is now the preferred method for solving the crystallographic phase problem. Over one-half of all structures are determined by MAD, a development in keeping with the availability of tunable synchrotron sources. Similarly, once the electron density is computed there is a challenge in fitting the density to a standard ball-and-stick model of the atoms. While this has been done manually (with graphic computer assistance), there is interest in finding methods for using image-processing techniques to automatically identify connected densities and matching them to the known shape of protein backbone and side-chain elements (Barr and Feigenbaum, 1982). Recent progress has been made on automated electron density map fitting and refinement (Chapter 4).

*Analysis of NMR Data.* NMR experiments provide complementary data to the crystallographic analyses. NMR experiments produce two-dimensional (or higher) spectra for which each individual peak must be assigned to an atomic interaction. The automated analysis and assignment of atoms in these spectra is a difficult search problem, but one in which progress has been made to accelerate the analysis of structure (Zimmerman and Montelione, 1995). Given a set of atomic proximities from NMR, we need methods to "embed" these distance measures into 3D structures that satisfy these constraints. Distance geometry (Moré and Wu, 1999), restrained molecular dynamics (Bassolino-Klimas et al., 1996) and other nonlinear optimization methods have been developed for this purpose (Altman, 1993; Williams et al., 2001).

*Assessment and Evaluation of Structures.* Given the results of a crystallographic or NMR structure determination effort, we must check the structures to be sure that they meet certain quality standards. Algorithms have been developed for assessing the basic chemistry of structural models, and also for identifying active sites and binding sites in these structures (Laskowski et al., 1993; Feng, Westbrook, and Berman, 1998; Vaguine, Richelle, and Wodak, 1999). Computational methods are still needed for automatically annotating 3D structures with functional information, based on an understanding of how molecular properties aggregate in three dimensions to produce function (such as binding, catalysis, motion, and signal transduction) (Wei, Huang, and Altman, 1999).

*Storing Molecular Structures in Databases.* The storage of the results of structural genomics efforts is an important task, requiring data structures and organizations that facilitate the most common queries. Ideally, databases of structure will store not only the resulting model, but also the raw data on which it is based. The PDB is the major repository for 3D structural information on proteins; the Nucleic Acids Database (NDB; Chapter 10) serves this function for nucleic acids. There is also an effort to store the raw data associated with crystallography in the PDB/NDB and the raw data associated with NMR in the BioMagResBank (BMRB).[4]

*Correlating Molecular Structural Information with Structural and Functional Information Gained from Other Types of Experimentation.* In the end, we perform structural studies in order to get an insight into how the molecules work. Structural studies with crystallography and NMR are but two methods that can be used to probe structure–function relationships. The integration of the results of these methods with other structural and functional data allows us to build comprehensive models of mechanism, specificity, and dynamics. A major bottleneck for using informatics methods for this integration is the lack of repositories of structural and functional data that can be accessed by computer programs doing systematic analyses. One exception is the noncrystallographic structural data about the 30S and 50S ribosomal subunits stored in the RiboWEB knowledge base (http://riboweb.stanford.edu/). RiboWEB is a knowledge base of ribosomal structural components that stores more than 8000 noncrystallographic structural and functional observations about the bacterial ribosome. It stores its information in structured "information templates" that are easily parsed by computer programs, thus making possible automated comparison and evaluation of structural models. For example, RiboWEB has been used to assess the compatibility of the published ribosomal crystal structures with over 1000 proximity measurements from cross-linking, chemical protection, and labeling experiments (collected during the last 25 years). Incompatibilities between these data and the crystal structures may suggest artifactual data or (more usefully) may suggest areas of important dynamic motion for the ribosome (Whirl-Carrillo et al., 2002).

---

[4]See at http://www.bmrb.wisc.edu.

## Understanding the Structural Basis for Biological Phenomenon

Given the structural information created by efforts in X-ray crystallography and NMR, there are a wide range of analytic and scientific challenges to informatics. It is not possible to cover the full scope of activities, but they can be reviewed briefly to show the richness of opportunities in the analysis of structural data.

*Visualization.* The creation of images of molecular structure remains a primary activity within structural biology (Chapter 7). The complexity of these molecules seems to demand novel display methods that are able to combine structural information with other information sources (such as electrostatic fields, the location of functional sites, and areas of structural or genetic variability). The issues for informatics include the creation of flexible software infrastructures for extending display capabilities, and the use of novel methods for rapidly rendering complex molecular structures (Huang et al., 1996; Sanner et al., 1999).

*Classification.* The database of known structures is already sufficiently large that it is necessary to cluster similar structures together, in order to form families of proteins. These families are often aggregated into superfamilies, and indeed entire structural hierarchies have been created. The Structural Classification of Proteins (SCOP; Chapter 12) is an example of a semiautomated classification of all protein structures (Murzin et al., 1995), and there have been numerous efforts to create automated classification—usually based on the pairwise comparison of all structures to create a matrix of distances (Chapter 13; Holm and Sander, 1996; Orengo et al., 1997).

*Prediction.* Despite the growth of the structural databases, the number of known 3D structures has lagged far behind the availability of sequence information. Thus, the prediction of 3D structure remains an area of keen interest. The Critical Assessment for Structure Prediction (CASP;[5] Chapter 24) meetings have provided a biennial forum for the comparison of methods for structure prediction. The main categories of prediction have been homology modeling (based on high sequence homology to a known structure (Chapter 25) (Sánchez and Sali, 1997), threading (based on remote sequence homology) (Bryant and Altschul, 1995), and *ab initio* prediction (based on no detectable homology (Chapter 27) (Osguthorpe, 2000). The diversity of methods invented and evaluated is quite inspiring, and the resulting lessons about how proteins are put together have been significant.

*Simulation.* The results of crystallographic studies (and to some extent, NMR studies) are primarily static structural models. However, the properties of these molecules that are of the greatest interest are often the result of their dynamic motions. The definition of energy functions that govern the folding of proteins and their subsequent stable dynamics has been an area of great interest since the first structure was determined. Unfortunately, the time scales on which macromolecular dynamics must be sampled (fractions of picoseconds) are much shorter than the time scale on which biologically important phenomena occur (microseconds to seconds). Nevertheless, the availability of increasingly powerful computers and the clever approximation and search methods are enabling molecular simulations of sufficient length and accuracy to

---

[5]See at http://predictioncenter.llnl.gov.

emerge, and are making contributions to our understanding of protein function.[6] The associated computation of electrostatic fields of macromolecular structures (Chapter 21) has emerged as an important component of understanding molecular function (Sheinerman et al., 1992).

## INTEGRATING STRUCTURAL DATA WITH OTHER DATA SOURCES

Structural bioinformatics has existed in some form or other ever since the determination of the first myoglobin structure. One could argue that the roots go back further to the time when small molecular structure determination was introduced. In any case, the challenges for the field are clearly abundant and significant. As we look to coming decades, it appears that a primary challenge in structural bioinformatics will be the integration of structural information with other biological information to yield a higher resolution understanding of biological function. The success of genome-sequencing projects has created information about all the structures that are present in individual organisms, as well as both the shared and unique features of these organisms. Even with the success of structural genomics projects, bioinformatics techniques will probably be used to create homology models of most of these genomic components. The resulting structures will be studied with respect to how they interact and perform their functions. Similarly, the emergence of microarray expression measurements provides an ability to consider how the expression of macromolecular structures is regulated at a structural level (including the key structural machinery associated with transcription, translation, and degradation). Mass spectroscopic methods that allow the identification of structural modifications and variations (such as genetic mutation or post-translational modifications) will need to be integrated with structural models to understand how they alter functional characteristics. Finally, cellular localization data will allow us to place 3D molecular structures into compartments within the cell as we build more complex models of how cells are organized structurally in order to optimize their function. This exciting activity will mark the next phase of structural bioinformatics—when the organization and physical structure of entire cells is understood and represented in computational models that provide insight into how thousands of structures within a cell work together to create the functions associated with life.

## REFERENCES

Altman RB (1993): *Probabilistic Structure Calculations: A Three-Dimensional tRNA Structure from Sequence Correlation Data*. First International Conference on Intelligent Systems for Molecular Biology, July 6–9, 1993, National Library of Medicine, Bethesda, MD.

Altman RB (1998): A curriculum for bioinformatics: the time is ripe. *Bioinformatics* 14:549–50. [The requirements for a graduate program in bioinformatics.]

Altman RB, Raychaudhuri S (2001): Whole-genome expression analysis: challenges beyond clustering. *Curr Opin Struct Biol* 11(3):340–7.

Ban N, Nissen P, Hansen J, Moore PB, Steitz TA (2000): The complete atomic structure of the large ribosomal subunit at 2.4 A resolution. *Science* 289:878–9.

---

[6]The IBM BlueGene project (http://www.research.ibm.com/bluegene) is focused on the creation of a very large supercomputer, with the theoretical capability of simulating the folding of a small protein in about one year. The computer is being designed to have $10^{15}$ floating point operations per second.

Barr A, Feigenbaum E (1982): Crysalis. In: Barr A, Feigenbaum EA, editors. *The Handbook of Artificial Intelligence*. Stanford, CA: HeurisTech Press, pp 124–33.

Bassolino-Klimas D, Tejero R, Krystek SR, Metzler WJ, Montelione GT, Bruccoleri RE (1996): Simulated annealing with restrained molecular dynamics using a flexible restraint potential: theory and evaluation with simulated NMR constraints. *Protein Sci* 5(4):593–603.

Bateman A, Birney E, Durbin R, Eddy SR, Finn RD, Sonnhammer EL (2000): The pfam protein families database. *Nucleic Acids Res* 28:263–6.

Berman HM, Westbrook J, Feng Z, Gilliland G, Bhat TN, Weissig H, Shindyalov I, Bourne PE (2000): The Protein Data Bank. *Nucleic Acids Res* 28:235–42.

Bryant SH, Altschul SF (1995): Statistics of sequence–structure threading. *Curr Opin Struct Biol* 5:236–44.

Burge C, Karlin S (1997): Prediction of complete gene structures in human genomic DNA. *J Mol Biol* 268:78–94.

Clemons WM Jr, Brodersen DE, McCutcheon JP, May JL, Carter AP, Morgan-Warren RJ, Wimberly BT, Ramakrishnan V (2001): Crystal structure of the 30S ribosomal subunit from thermus thermophilus: purification, crystallization and structure determination. *J. Mol. Biol.* 310:827–43.

Cramer P, Bushnell DA, Fu J, Gnatt AL, Maier-Davis B, Thompson NE, Burgess RR, Edwards AM, David PR, Kornberg RD (2000): Architecture of RNA polymerase II and implications for the transcription mechanism. *Science* 288:640–49.

DeGrado W, Regan L, Ho SP (1987): The design of a four-helix bundle protein. *Cold Spring Harb Symp Quant Biol* 52:521–6.

Doniach S, Eastman P (1999): Protein dynamics simulations from nanoseconds to microseconds. *Curr Opin Struct Biol* 9:157–63.

Durbin R, Eddy SR, Krogh A, Mitchison GJ (1998): *Biological Sequence Analysis: Probabilistic Models of Proteins and Nucleic Acids*. Cambridge: Cambridge University Press. [A good book for the study of sequence analysis.]

Feng Z, Westbrook J, Berman HM (1998): NUCheck. Rutgers publication NDB-407. Rutgers University, New Brunswick, New Jersey.

Gilmore CJ, Dong W, Bricogne G (1998): A multisolution method of phase determination by combined maximisation of entropy and likelihood. VI. The use of error-correcting codes as a source of phase permutation and their application to the phase problem in powder, electron and macromolecular crystallography. *Acta Crystallogr* A55:70–83.

[GISC] GENOME INTERNATIONAL SEQUENCING CONSORTIUM (2001): Initial sequencing and analysis of the human genome. *Nature* 409:860–921.

Gusfield D (1997): *Algorithms on Strings, Trees, and Sequences: Computer Science and Computational Biology*. Cambridge: Cambridge University Press.

Hendrickson WA (1991): Determination of macromolecular structures from anomalous diffraction of synchrotron radiation. *Science* 254:51–8.

Hennessy D, Buchanan B, Subramanian D, Wilkosz PA, Rosenberg JM (2000): Statistical methods for the objective design of screening procedures for macromolecular crystallizationĺ. *Acta Crystallogr* D56(Pt 7):817–27.

Hinds D, Levitt M (1994): Exploring conformational space with a simple lattice model for protein structure. *J Mol Biol* 243:668–82.

Holm L, Sander C (1996): Mapping the protein universe. *Science* 273:595–602.

Huang CC, Couch GS, Pettersen EF, Ferrin TE (1996): Chimera: An Extensible Molecular Modeling Application Constructed Using Standard Components. *Pacific Symp Biocomput.* 1:724.

Hughey R, Krogh A (1996): Hidden Markov models for sequence analysis: extension and analysis of the basic method. *CABIOS* 12:95–107.

Kempf D (1994): Design of symmetry-based, peptidomimetic inhibitors of human immunodeficiency virus protease. *Methods Enzymol* 241:334–54.

Koehl P, Levitt M (1999): Structure-based conformational preferences of amino acids. *Proc Natl Acad Sci USA* 96(22):12524–9.

Langridge R, Gomatos PJ (1963): The structure of RNA. *Science* 141:694–8.

Laskowski RA, McArthur MW, Moss DS, Thornton JM (1993): PROCHECK: a program to check the stereochemical quality of protein structures. *J Applied Crystallogr* 265:283–91.

Linial M, Yona G (2000): Methodologies for target selection in structural genomics. *Prog Biophys Mol Biol* 73(5):297–320.

Miranker AD (2000): Protein complexes and analysis of their assembly by mass spectrometry. *Curr Opin Struct Biol* 10(5):601–6.

Moré J., Wu Z (1999): Distance geometry optimization for protein structures. *J Global Optimiz* 15:219–34.

Murzin AG, Brenner SE, Hubbard T, Chothia C (1995): SCOP: a structural classification of proteins database for the investigation of sequences and structures. *J Mol Biol* 247:536–40.

Orengo CA, Michie AD, Jones S, Jones DT, Swindells MB, Thornton JM (1997): CATH—a hierarchic classification of protein domain structures. *Structure* 5(8):1093–108.

Osguthorpe DJ (2000): Ab initio protein folding. *Curr Opin Struct Biol* 10:146–52.

Sánchez R, Sali A (1997): Advances in comparative protein-structure modelling. *Curr Opin Struct Biol* 7:206–14.

Sanner MF, Duncan BS, Carrillo CJ, Olson AJ (1999): Integrating computation and visualization for biomolecular analysis: an example using PYTHON and AVS. *Pacific Symp Biocomput* 4:401–12.

Sheinerman FB, Norel R, Honig B (1992): Electrostatic aspects of protein–protein interactions. *Curr Opin Struct Biol* 10:153 59.

Silverman JA, Balakrishnan R, Harbury PB (2001): Reverse engineering the (beta/alpha)8 barrel fold. *Proc Natl Acad Sci USA* 98(6).3092–7.

Tcichmann SA, Chothia C, Gerstein M (1999): Advances in structural genomics. *Curr Opin Struct Biol* 9:390–9.

Teichmann SA, Murzin AG, Chothia C (2001): Determination of protein function, evolution and interactions by structural genomics. *Curr Opin Struct Biol* 11(3):354–63.

Vacca J (1994): Design of tight-binding human immunodeficiency virus type 1 protease inhibitors. *Methods Enzymol* 241:331–4.

Vaguine AA, Richelle J, Wodak SJ (1999): SFCheck: a unified set of procedures for evaluating the quality of macromolecular structure-factor data and their agreement with the atomic model. *Acta Crystalloger* D55:191–205.

Venter JC et al. (2001): The sequence of the human genome. *Science* 291:1304–51.

Wei L, Huang ES, Altman RB (1999): Are predicted structures good enough to preserve functional sites. *Structure* 7(6):643–50.

Whirl-Carrillo M, Gabashvili IS, Bada M, Banatao DR, Altman RB (2002): Mining biochemical information: lessons taught by the ribosome. *RNA* Mar; 8(3):279–89.

Williams GA, Dugan JM, Altman RB (2001): Constrained global optimization for estimating molecular structure from atomic distances. *J Comput Biol* 8(5):523–47.

Wuthrich K, Billeter M, Braun W (1983): Pseudo-structures for the 20 common amino acids for use in studies of protein conformations by measurements of intramolecular proton–proton distance constraints with nuclear magnetic resonance. *J Mol Biol* 169:949–61.

Yusupov MM, Yusupova GZ, Baucom A, Lieberman K, Earnest TN, Cate JH, Noller HF (2001): Crystal structure of the ribosome at 5.5Å resolution. *Science* 292:883–96.

Zimmerman DE, Montelione GT (1995): Automated analysis of nuclear magnetic resonance assignments for proteins. *Curr Opin Struct Biol* 5:664–73.

# 2

# FUNDAMENTALS OF PROTEIN STRUCTURE

Eric D. Scheeff and J. Lynn Fink

## THE IMPORTANCE OF PROTEIN STRUCTURE

Most of the essential structure and function of cells is mediated by proteins. These large, complex molecules exhibit a remarkable versatility that allows them to perform a myriad of activities that are fundamental to life. Indeed, no other type of biological macromolecule could possibly assume all of the functions that proteins have amassed over billions of years of evolution.

Any consideration of protein function must be grounded in an understanding of protein structure. A fundamental principle in all of protein science is that *protein structure leads to protein function*. The distinctive structures of proteins allow for the placement of particular chemical groups in specific places in three-dimensional space. It is this precision that allows proteins to act as catalysts (enzymes) for an impressive variety of chemical reactions. Precise placement of chemical groups also allows proteins to play important structural, transport, and regulatory functions in organisms. Since protein structure leads to function, and protein functions are diverse, it is no surprise that protein structure is similarly diverse. Further, the functional diversity of proteins is expanded through the interaction of proteins with small molecules, as well as other proteins.

For those who wish to study protein structure, this diversity represents a challenge. Upon their determination of the first three-dimensional globular protein structure (the oxygen-storage protein myoglobin) in 1958, John Kendrew and his co-workers registered their disappointment (Kendrew et al., 1958): "Perhaps the most remarkable features of the molecule are its complexity and its lack of symmetry. The arrangement

*Structural Bioinformatics*
Edited by Philip E. Bourne and Helge Weissig
Copyright © 2003 by Wiley-Liss, Inc.

seems to be almost totally lacking in the kind of regularities which one instinctively anticipates, and it is more complicated than has been predicated by any theory of protein structure." Despite these initial frustrations, subsequent studies of the myoglobin structure based on higher-quality data revealed that the protein did have *some* regularities; these regularities were also observed in other protein structures.

Decades of research have now yielded a coherent set of principles about the nature of protein structure and the way in which this structure is utilized to effect function. These principles have been organized into a four-tiered hierarchy that facilitates description and understanding of proteins: primary, secondary, tertiary, and quaternary structure. This hierarchy does not seek to describe precisely the physical laws that produce protein structure, but rather is an abstraction to make protein structural studies more tractable.

## THE PRIMARY STRUCTURE OF PROTEINS: THE AMINO ACID SEQUENCE

### Amino Acids

Proteins are linear polymers of amino acids,[1] and *it is the distinct sequence of component amino acids that determines the ultimate three-dimensional structure of the protein.* The sequence of a protein is often referred to as its primary structure. The concept of proteins as linear amino acid polymers was initially proposed by Fischer and Hofmeister in 1902 (Fruton, 1972). At that time, the prevailing theory in protein science was that proteins lacked a regular structure and consisted of loose associations of small molecules (colloids). This issue was hotly debated for over 20 years, until the linear polymer theory achieved general acceptance in the late 1920s (Fruton, 1972). In 1952, Fred Sanger made the important discovery that proteins could be distinguished by their amino acid sequences (Sanger, 1952). Indeed, he found that *proteins of exactly the same type have identical sequences.* Sanger's work helped to remove remaining doubts about the accuracy of the linear polymer theory.

Amino acids are small molecules that contain an amino group ($NH_2$), a carboxyl group (COOH), and a hydrogen atom attached to a central alpha ($\alpha$) carbon (Figure 2.1). In addition, amino acids also have a side chain (or R group) attached to the $\alpha$ carbon. It is this group, and this group alone, that distinguishes one amino acid

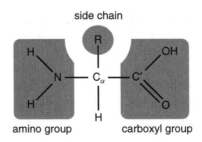

**Figure 2.1.** The structure of a prototypical amino acid. The chemical groups bound to the central alpha ($\alpha$) carbon are highlighted in gray. The R-group represents any of the possible 20 amino acid side chains.

[1]Specifically, the amino acids used in proteins are *alpha($\alpha$)* amino acids.

from another. Furthermore, the side chain confers the specific chemical properties of the amino acid.

Cellular genomes contain coded instructions for the production of multiple proteins, and there are 20 amino acids that can be incorporated into a protein via these instructions. The resulting sequence of a protein can contain any combination and number of the 20 amino acids, in any order. Though amino acids had been known to be the building blocks of proteins prior to the turn of the twentieth century, the exact set of amino acids used in proteins was not determined until 1940 (Fruton, 1972). This set of 20 amino acids is considered standard in that it is common to all observed organisms. Modified forms of these 20 amino acids do exist in proteins, but these are the product of modifications that occur subsequent to protein synthesis.[2]

The 20 standard amino acids can be loosely grouped into classes based on the chemical properties conferred by their side chains. Three classes are commonly accepted: hydrophobic, polar, and charged. Within these classes, additional subclassifications are possible; for example, aromatic or aliphatic, large or small, and so forth (Taylor, 1986). Figure 2.2 provides one possible amino acid classification.

A few amino acids have distinctive properties that merit closer attention. The side chain of proline forms a bond with its own amino group, causing it to be cyclic.[3] Though proline generally exhibits the properties of an aliphatic nonpolar amino acid, the cyclic construction limits its flexibility, and this impacts the overall structure of proteins that contain it.

Glycine is also of interest because its side chain consists of only a single hydrogen atom. In effect, glycine has no side chain, and this confers a unique property among the 20 amino acids: glycine is achiral. Any carbon bound to four distinct groups (as seen in the other 20 amino acids) is said to be chiral (Figure 2.3). Chiral molecules can exist in two distinct forms, which are in effect mirror images of each other. These two forms have been deemed the D and L forms.[4] In 1952, Fred Sanger discovered that proteins seem to be constructed entirely of L-amino acids (Sanger, 1952). Indeed, for unknown reasons, all known organisms have standardized on the L form of amino acids for the genetically directed production of proteins. D-amino acids *are* seen in polypeptides in rare cases, but they are a result of direct enzymatic synthesis (Krell, 1997).

## The Peptide Bond

Amino acids can form bonds with each other through a reaction of their respective carboxyl and amino groups. The resulting bond is called the peptide bond, and two or more amino acids linked by such a bond are referred to as a peptide (Figure 2.4). A protein is synthesized by the formation of a linear succession of peptide bonds between *many* amino acids (as directed by the genetic code) and can thus be referred to as a *poly*peptide. Once an amino acid is incorporated into a peptide, it is referred to as an amino acid residue, and the atoms involved in the peptide bond are referred to as the peptide backbone.

---

[2]There is one exception. In a few proteins, a selenium-containing residue, selenocysteine, can be incorporated during protein synthesis (Zinoni et al., 1986; Atkins and Gesteland, 2000).

[3]Because of the cyclic bond in proline, this molecule is technically an *imino* acid. However, proline is commonly referred to as one of the 20 amino acids.

[4]International chemical convention calls for the designations R and S, respectively, but D and L are the traditional, and currently dominant, terms.

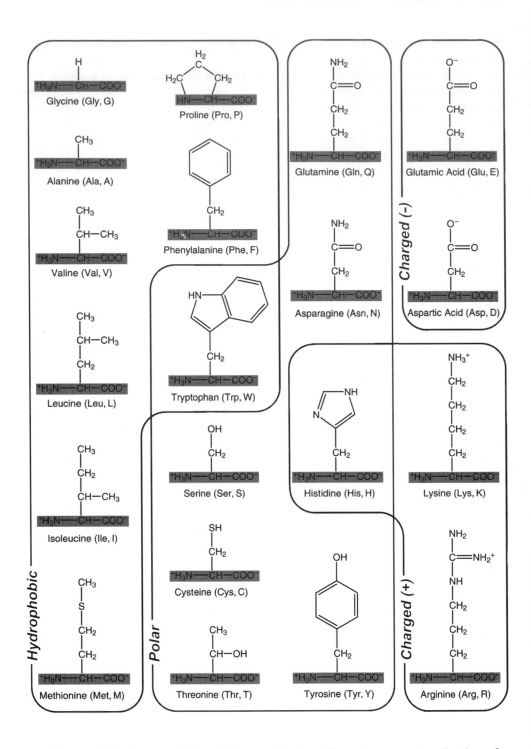

The specific characteristics of the peptide bond have important implications for the three-dimensional structures that can be formed by polypeptides. The peptide bond is planar and quite rigid. Therefore, the polypeptide chain has rotational freedom only about the bonds formed by the $\alpha$-carbons. These bonds have been termed the Phi ($\Phi$) and Psi ($\Psi$) angles (Figure 2.5). However, rotational freedom about the $\Phi$

**Figure 2.2.** The 20 standard amino acids used in proteins, grouped based on the properties of their side chains. The shared amino acid structure is shaded gray. Each amino acid is labeled with its full name, followed by its three-letter and one-letter abbreviations. This classification groups amino acids based on the form that predominates at physiological conditions (note that their amino and carboxyl groups are charged under these conditions). This classification is useful as a guideline, but does not convey the full complexity of side chain properties. For example, tryptophan and histidine do not fall clearly into a single grouping. Tryptophan is somewhat polar due to the nitrogen in its five-membered ring, but has a hydrophobic six-membered ring at the end of its side chain. Histidine can be neutral polar and/or positively charged under physiological conditions.

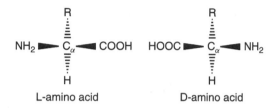

L-amino acid                              D-amino acid

**Figure 2.3.** The possible stereoisomers of a prototypical amino acid. Note that these structures are mirror images of each other. The L-form is the only type incorporated into proteins via the genetic machinery.

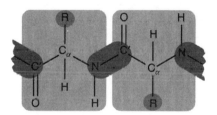

**Figure 2.4.** The peptide bond. Two peptide units (amino acid residues) are shown shaded in light gray. The peptide bond between them is shaded in dark gray. The R-group represents any of the possible 20 amino acid side chains.

($C_\alpha$–N) and $\Psi$ ($C_\alpha$–C') angles is limited by steric hindrance between the side chains of the residues and the peptide backbone. Consequently, the possible conformations of a given polypeptide chain are quite limited. A Ramachandran Plot (a plot of $\Phi$ vs. $\Psi$ angles) maps the entire conformational space of a polypeptide, and illuminates the allowed and disallowed conformations (Ramachandran and Sasisekharan, 1968) (Figure 2.6). These plots were developed by G. N. Ramachandran in the late 1960s based on studies of sterically allowed $\Phi$ and $\Psi$ angle combinations. See Chapter 14 for a more detailed discussion of these plots.

Some key exceptions to these conformational limitations can be attributed to glycine and proline. As noted previously, glycine's side chain (a single hydrogen atom) is very small. There is markedly reduced steric hindrance about the $\Phi$ and $\Psi$

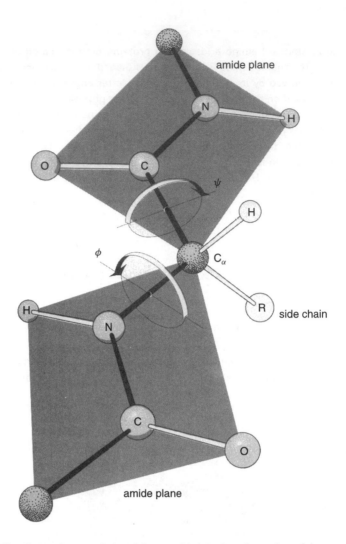

**Figure 2.5.** The planar characteristics of the peptide bond, and rotation of the peptide backbone about the $C_\alpha$ atom. Note the two planar peptide bonds about a central alpha carbon, shown here as a ball-and-stick model. Rotation is only possible about the $\Phi$ ($C_\alpha$–N) and $\Psi$ ($C_\alpha$–C') angles. Arrows about the two angles show the direction that is considered positive rotation. In this figure, both angles are approximately 180°. *From R.E. Dickerson and I. Geis. The Structure and Action of Proteins. New York: Harper & Row, 1969. Used with permission from Geis Archives.*

angles of this residue, thus, expanding the possible conformational space. Conversely, the cyclic bond present in proline residues reduces the conformational freedom beyond the limitations observed with other amino acids.

## THE SECONDARY STRUCTURE OF PROTEINS: THE LOCAL THREE-DIMENSIONAL STRUCTURE

The secondary structure of a protein can be thought of as the local conformation of the polypeptide chain, independent of the rest of the protein. The limitations imposed on the

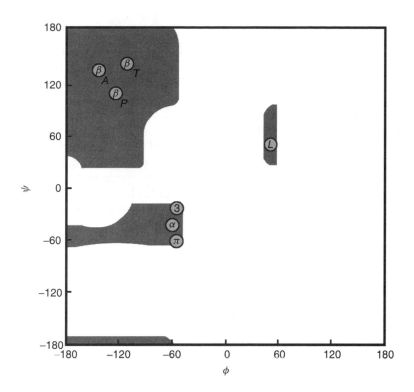

**Figure 2.6.** A schematic representation of a Ramachandran plot (a plot of $\Phi$ vs. $\Psi$ angles). Gray regions denote the allowed conformations of the polypeptide backbone. Circles indicate the paired angle values of the repetitive secondary structures. Definitions of symbols: $\beta_A$, antiparallel $\beta$ sheet; $\beta_P$, parallel $\beta$ sheet; $\beta_T$, twisted $\beta$ sheet (parallel or antiparallel); $\alpha$, right-handed $\alpha$ helix; $l$, left-handed helix; 3, $3_{10}$ helix; $\pi$, $\pi$ helix.

primary structure of a protein by the peptide bond and hydrogen bonding considerations dictate the secondary structure that is possible. During the course of protein structure research, two types of secondary structure have emerged as the dominant local conformations of polypeptide chains: alpha ($\alpha$) helix and beta ($\beta$) sheets. Interestingly, these structures were actually predicted by Linus Pauling, Robert Corey, and H. R. Branson, based on the known physical limitations of polypeptide chains, prior to the experimental determination of protein structures (Pauling, Corey, and Branson, 1951; Pauling and Corey, 1951a). Indeed, if the Ramachandran plot is examined, helices and sheets contain $\Phi$ and $\Psi$ angles that fall within the two largest regions of allowed conformation (Figure 2.6). These structures exhibit a high degree of regularity: the particular $\Phi$ and $\Psi$ angle combinations in the polypeptide chain are approximately repeated for the duration of the secondary structure.

Although helices and sheets satisfy the peptide bond constraints, this is not the only factor that explains their ubiquity. Both of these structural elements are stabilized by hydrogen bond interactions between the backbone atoms of the participating residues, making them a highly favorable conformation for the polypeptide chain. Helices and sheets are the only *regular* secondary structural elements present in proteins. However, *irregular* secondary structural elements are also observed in proteins, and are vital to both structure and function.

## α Helices

A helix is created by a curving of the polypeptide backbone such that a regular coil shape is produced. Because the polypeptide backbone can be coiled in two directions (left or right), helices exhibit *handedness*. A helix with a rightward coil is known as a right-handed helix. Almost all helices observed in proteins are right-handed, as steric restrictions limit the ability of left-handed helices to form. Among the right-handed helices, the α helix is by far the most prevalent.

An α helix is distinguished by having a period of 3.6 residues per turn of the backbone coil. The structure of this helix is stabilized by hydrogen-bonding interactions between the carbonyl oxygen of each residue and the amide proton of the residue 4 residues ahead in the helix (Figure 2.7). Consequently, all possible backbone hydrogen bonds are satisfied within the α helix, with the exception of a few at each end of the helix, where a partner is not available.

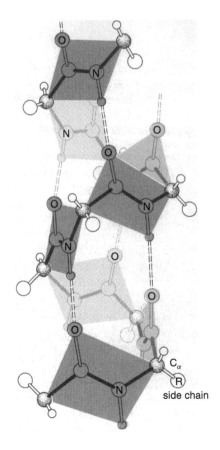

**Figure 2.7.** Diagram of an α helix using a ball-and-stick model. The bonds forming the backbone of the polypeptide are darkly shaded. The α helix is stabilized by internal hydrogen bonds formed between the carbonyl oxygen of each residue and the amide proton of the residue 4 residues ahead in the helix, shown here as dashed lines. Note that the polypeptide backbone curves towards the right, and as such the α helix is a right-handed helix. *From R.E. Dickerson and I. Geis. The Structure and Action of Proteins. New York: Harper & Row, 1969. Used with permission from Geis Archives.*

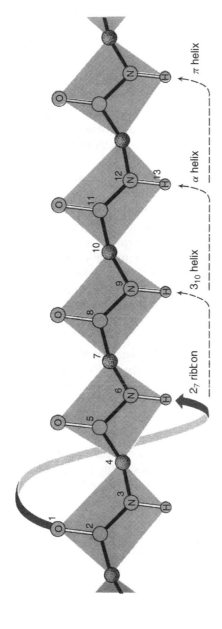

Figure 2.8. The hydrogen bonding patterns of different helical secondary structures. The peptide backbone is shown in an extended conformation, with an arrow denoting the hydrogen bonding pairings that would occur in each type of helix. The common α helix, depicted in Figure 2.7, forms hydrogen bonds between the carbonyl oxygen of each residue and the amide proton of the residue 4 residues ahead in the helix. The $3_{10}$ helix forms hydrogen bonds between the carbonyl oxygen of each residue and the amide proton of the residue 3 residues ahead, forming a more narrow and elongated helix. The π helix forms hydrogen bonds between the carbonyl oxygen of each residue and the amide proton of the residue five residues ahead, forming a wider helix. The $2_7$ ribbon is not a regular secondary structure, but is shown here to demonstrate all possible hydrogen bond pairings. *From R. E. Dickerson and I. Geis. The Structure and Action of Proteins. New York: Harper & Row, 1969. Used with permission from Geis Archives.*

Other helices have also been observed in proteins, though much less frequently due to their less favorable geometry. The $3_{10}$ helix has a period of 3 residues per turn, with hydrogen bonds between each residue and the residue 3 positions ahead (Taylor, 1941; Huggins, 1943) (Figure 2.8). This type of helix is usually seen only in short segments, often at the ends of an $\alpha$ helix. The very rare pi ($\pi$) helix has a period of 4.4 residues per turn, with hydrogen bonds between each residue and the residue 5 positions ahead, and has only been seen at the ends of $\alpha$ helices (Low and Baybutt, 1952).

## $\beta$ Sheets

Unlike helices, $\beta$ sheets are formed by hydrogen bonds between adjacent polypeptide chains rather than within a single chain. Sections of the polypeptide chain participating in the sheet are known as $\beta$ strands. $\beta$ strands represent an extended conformation of the polypeptide chain, where the $\Phi$ and $\Psi$ angles are rotated approximately 180° with respect to each other. This arrangement produces a sheet that is pleated, with the residue side chains alternating positions on opposite sides of the sheet (Figure 2.9).

Two configurations of $\beta$ sheet are possible: parallel and antiparallel. In parallel sheets, the strands are arranged in the same direction with respect to their amino-terminal (N) and carboxy-terminal (C) ends. In antiparallel sheets, the strands alternate their amino and carboxy terminal ends, such that a given strand interacts with strands in the opposite orientation. $\beta$ sheets can also form in a mixed configuration, with both parallel and antiparallel sections, but this configuration is less common than the uniform types mentioned above. Almost all $\beta$ sheets exhibit some degree of twist when the sheet is viewed edge on, along an axis perpendicular to the direction of the polypeptide chains. This twist is always right-handed.

An important variant of the classical $\beta$ sheet structure is the $\beta$ bulge. The $\beta$ bulge, most often observed in antiparallel $\beta$ sheets, is a hydrogen bond between two residues on one $\beta$ strand with one residue on the adjacent strand (Richardson, Getzoff, and Richardson, 1978; Chan et al., 1993). This structure can alter the direction of the polypeptide chain and augment the right-handed twist of the sheet (Richardson, 1981).

## Other Secondary Structure

$\alpha$ helices and $\beta$ sheets account for the majority of secondary structure seen in proteins. However, these regular structures are interspersed with regions of irregular structure that are referred to as *loop* or *coil*. Loop regions are usually present at the surface of the protein. These regions are often simply transitions between regular structures, but they also can possess structural significance, and can be the location of the functional portion, or *active site*, of the protein.

Most of these structures were predicted and observed in the 1960s and 1970s (Venkatachalam, 1968; Chandrasekaran et al., 1973; Lewis et al., 1973; Chou and Fasman, 1977). Because of their irregularity, these elements are difficult to classify. However, some types of structure are ubiquitous enough that they have been loosely categorized. Hairpin loops (or reverse turns) often occur between antiparallel beta strands, and involve the minimum number of residues (4–5) required to begin the next strand. Many other types of turns exist, and can be classified by the regular secondary structures that they connect and by the $\Phi$ and $\Psi$ angles of the residues involved in the turn. Transitions involving more residues (6–16) are often referred to as omega ($\Omega$) loops because they resemble the shape of the capital Greek letter omega (Richardson,

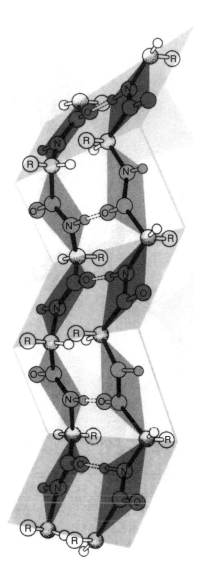

**Figure 2.9.** Diagram of an antiparallel $\beta$ sheet using a ball-and-stick model. The bonds forming the backbone of the polypeptide are darkly shaded. The $\beta$ sheet is stabilized by hydrogen bonds (shown here as dashed lines) formed between the carbonyl oxygen of a residue on one strand and the amide proton of a residue on the adjacent strand. Note that this arrangement produces a sheet that is pleated, with the residue side chains alternating positions on opposite sides of the sheet. *From R. E. Dickerson and I. Geis. Hemoglobin: Structure, Function, Evolution, and Pathology. Menlo Park, CA: The Benjamin/Cummings Publishing Company, 1983. Used with permission from Geis Archives.*

1981). These loops can involve complex interactions that include the side chains, in addition to the polypeptide backbone. Extended loop regions involving more than 16 residues have also been observed.

Although loop regions are irregular, these structures are generally as well ordered as the regular secondary structures. However, experimental determination of protein

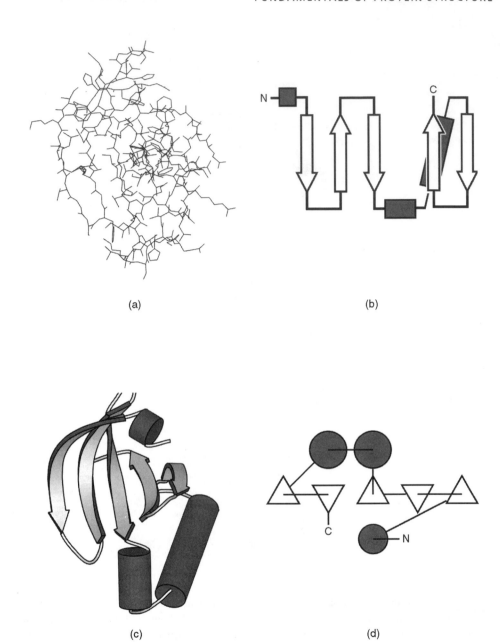

(a)

(b)

(c)

(d)

## Schematic Diagrams of Proteins

Figure 2.10. One of the earliest pioneers in protein visualization was Irving Geis (Dickerson and Geis, 1969; Kendrew, 1961), who created some of the most definitive representations of protein structure. His hand-drawn depictions were so enlightening that they appear in textbooks to this day (and appear in this chapter). Although hand drawings are extremely valuable, they are ultimately impractical because of the large number of protein structures that must be depicted (and the unusual level of talent required). Arthur Lesk and Karl Hardman (1982) were the first to popularize the use of computers to automatically generate schematic diagrams, given the (experimentally determined) spatial coordinates of the atoms in the protein. For more on molecular visualization, see Chapter 7.

The N-terminal domain of the catalytic core of eukaryotic protein kinase A (PKA, PDB id 1APM: residues 35–123) is depicted here using four different representations. This section of PKA contains a five-stranded antiparallel $\beta$ sheet and three helices. Figure 2.10a depicts this domain using an all-atom line representation. As can be seen, it is difficult to determine the overall structural characteristics of this protein using such a representation. Because proteins are often large and complex structures, views at the atomic level tend to obfuscate the important features. For this reason, a variety of schematic diagrams have been developed for the visual representation of protein structure. These diagrams replace the individual residues with shapes that represent the secondary structure they belong to, and facilitate recognition of motifs and domains.

Simple topology diagrams are two-dimensional projections of the protein structure that are particularly useful for comparing the tertiary structures of different proteins (Holbrook et al., 1970). In these diagrams, $\beta$ strands are represented by arrows that point from the N terminus to the C terminus; $\alpha$ helices are represented by cylinders. Connections between the secondary structural elements (loops) are simply represented as lines. These diagrams clearly illustrate the topology (connectivity) between the secondary structural elements and parallel or antiparallel nature of $\beta$ sheets (Figure 2.10b).

Cartoon diagrams illustrate the topology of the protein, as well as the spatial relationship between the structural components. These diagrams represent the three-dimensional structure of the protein as it actually occurs, with the atoms replaced by the same elements used in topology diagrams. Initially conceived by Jane Richardson, and presented as hand drawings (Richardson, 1981), this representation is now very commonly used in protein visualization software packages (Figure 2.10c). Figures 2.10a and 2.10c were generated with the MolScript package (Kraulis, 1991), and are shown in an identical spatial orientation. The cartoon images in Tables 2.1 and 2.2 were also generated with the MolScript package.

TOPS diagrams, developed by Michael Sternberg and Janet Thornton, allow both the topology and interaction of structural elements to be represented in a two-dimensional format (Sternberg and Thornton, 1977). Here, the secondary structural elements are viewed edge on, as if they were projecting outward in a direction perpendicular to the plane of the page. $\beta$ strands are represented by triangles; an upward-pointing triangle portrays a strand pointing out of the page while a downward-pointing triangle portrays a strand pointing into the page. Helices are represented by circles. Lines represent the loops connecting these elements and also help to portray chain direction (Figure 2.10d). Software is available to automatically generate these figures (Flores et al., 1994).

structures has shown that some loop regions are disordered, and thus do not achieve a stable structure. This type of loop region is referred to as *random* coil.

## THE TERTIARY STRUCTURE OF PROTEINS: THE GLOBAL THREE-DIMENSIONAL STRUCTURE

The tertiary structure of a protein is defined as the global three-dimensional structure of its polypeptide chain. In a striking display of foresight, Alfred Mirsky and Linus Pauling correctly described several important aspects of protein tertiary structure in 1936 (Mirsky and Pauling, 1936): "Our conception of a native protein molecule . . . is the following. The molecule consists of one polypeptide chain which continues without interruption throughout the molecule . . .; this chain is folded into a uniquely defined configuration, in which it is held by hydrogen bonds between the peptide nitrogen and oxygen atoms and also between the free amino and carboxyl groups of the diamino and dicarboxyl amino acid residues."

The prediction of Mirsky and Pauling is especially striking considering that very little structural data was available and the linear polymer theory was still unproven at that time. As described previously, hydrogen bonds are important in the stabilization of secondary structure, but Mirsky and Pauling correctly divined their importance in tertiary structure stabilization as well.

Mirsky and Pauling correctly predicted the role of hydrogen bonds in protein structure; however, it subsequently became apparent that other forces were important. With the determination of the structure of myoglobin by John Kendrew (Kendrew et al., 1958), protein scientists at last could begin to confirm many of their assumptions about various aspects of tertiary structure. Subsequent determination of the myoglobin structure using more accurate data (Kendrew et al., 1960), and the determination of the structure of hemoglobin by Max Perutz (Perutz et al., 1960) allowed scientists to begin to directly catalogue aspects of protein tertiary structure for the first time. Years of experimentation now make it possible to understand how secondary structural elements combine in three-dimensional space to yield the tertiary structure of a protein.

### Side Chains and Tertiary Structure

There are a wide variety of ways in which the various helix, sheet, and loop elements can combine to produce a complete structure. These combinations are brought about largely through interactions between the side chains of the constituent amino acid residues of the protein. Thus, *at the level of tertiary structure, the side chains play a much more active role in creating the final structure*. In contrast, backbone interactions are primarily responsible for the generation of secondary structure (particularly in the case of helices and sheets). Most proteins each have a distinctive tertiary structure; the secondary structural elements of these proteins will always fold in the same way to produce the same tertiary structure.[5] This consistency is vital to proper function of the protein.

---

[5]Some proteins do not have an intrinsically ordered structure, but rather spend much of their time in a disordered state. Often, the function of these proteins depends on achieving an ordered structure only under certain conditions or while engaged in specific interactions (Wright and Dyson, 1999).

## The Protein Fold

The final three-dimensional tertiary structure of a protein is commonly referred to as its fold. Appropriately, the process by which a linear polypeptide chain achieves its distinctive fold is known as protein folding. Protein folding is a complex process that is not yet completely understood, and will not be described in great detail here. In most (if not all) cases, the primary structure of the protein contains all of the information required for it to acquire the correct fold. Further, random polypeptide sequences will almost never fold into an ordered structure; an important property of protein sequences is that they have been selected by the evolutionary process to achieve a reproducible, stable structure (Richardson, 1992). Despite the deterministic nature of protein folding, it is not yet possible to accurately predict the final structure of a protein given only its sequence (see Section VII for a discussion of the methods currently used to tackle this problem).

## Domains and Motifs

Within the overall protein fold, distinct domains and motifs can be recognized. Domains are compact sections of the protein that represent structurally (and usually function- ally) independent regions. In other words, a domain is a subsection of the protein that would maintain its characteristic structure, even if separated from the overall protein. Motifs (also referred to as supersecondary structure) are small substructures that are not necessarily structurally independent: generally, they consist of only a few secondary structural elements. Specific motifs are seen repeatedly in many different protein struc- tures; they are integral elements of protein folds. Further, motifs often have a functional significance, and in these cases represent a minimal functional unit within a protein. Several motifs can combine to form specific domains.

## Molecular Interactions in Tertiary Structure

As with secondary structure, intramolecular forces are integral in defining and stabiliz- ing the tertiary structure. The molecular interactions between structural elements, and individual side chains within these elements, help determine the protein fold. Because of the variety of chemical characteristics of the 20 standard amino acids, many types of interactions between them are possible. Furthermore, it is important to note that many proteins exist in an aqueous environment; intermolecular interactions with water (solvent) must be considered in addition to interactions within the protein itself.

A dominant molecular interaction in the tertiary structure of proteins is the hydrophobic effect (Tanford, 1978). Residues with hydrophobic side chains are packed into the core of the protein, away from the solvent, while charged and polar residues form the surface of the protein and are able to interact with polar water molecules and solvated ions. Although residues with hydrophobic side chains fit naturally into the core of the protein, their *polar* polypeptide backbone does not. In order for the backbone to participate in the hydrophobic core, its hydrogen bonding groups must be satisfied such that their polarity is, in effect, neutralized. Ordered secondary structural elements (helices and sheets) provide this neutralization through their regular hydrogen bonding patterns. Thus, secondary structural elements are critical to the formation of the hydrophobic core.

Residues with polar side chains can also participate in the hydrophobic core. They are, however, subject to the same restrictions as the polypeptide backbone: they must be involved in an interaction that neutralizes their polarity. Buried polar residues can form hydrogen bonds with other polar residues, or with sections of the polypeptide backbone not participating in a regular secondary structure. Alternately, in some proteins small pockets exist in which buried polar residues satisfy their hydrogen bonds with water molecules. These water molecules are completely isolated from the solvent and are integral to the protein structure.

Charged residues can also occur within the hydrophobic core. This arrangement is possible only if the charged residue is paired with another residue of opposite charge, such that the net charge is zero. This interaction, initially proposed by Henry Eyring and Allen Stearn in 1939, is known as an ion pair or salt bridge (Eyring and Stearn, 1939).

Eyring and Stearn also surmised that covalent connections between residue side chains were important in maintaining tertiary structure (Eyring and Stearn, 1939). Indeed, covalent interactions have been observed in some proteins. However, only one of the standard 20 amino acids is capable of participating in a covalent linkage: cysteine. The disulfide bond ($-S-S-$) can occur between the thiol ($-SH$) groups of two cysteine residues.[6] This interaction exists in proteins in order to further stabilize the protein fold. Cysteine residues do not always participate in disulfide bonds; in proteins, the majority of these residues are not part of a disulfide linkage.

## Protein Modifications

Even though amino acids have a wide variety of chemical characteristics, chemical modifications and interaction with nonpolypeptide molecules can further extend the capabilities of proteins. At the level of tertiary structure, it is important to note these phenomena, as they can be critical to both the structure and function of proteins.

The chemical properties of the 20 standard amino acids can be extended through side chain modification. In some cases, such modification can be indispensable to the proper formation of tertiary structures. For example, the fibrous protein collagen contains a modified proline residue, hydroxyproline, which greatly stabilizes the protein fold (Pauling and Corey, 1951b). In other cases, residue modification has no effect on fold, but rather extends the functional repertoire of the protein.

Molecules, such as carbohydrates and lipids, can be attached to the protein via covalent bonds with specific residues. Carbohydrate modifications are often seen on proteins that function extracellularly and are known to play a role in intracellular protein localization. Lipid modifications can help anchor a protein in the cell membrane.

Proteins can also associate with small molecules or metal atoms (covalently or noncovalently) in order to diversify their functional and structural capabilities. Many enzymes employ such molecules as cofactors, which assist them in chemical catalysis (Karlin, 1993). Other proteins require these molecules or atoms for proper tertiary structure formation. For example, the zinc-finger motif, a structural motif important for the interaction of some proteins with DNA, cannot form without the covalent interaction of a zinc atom with specific cysteine, or cysteine and histidine, residues (Lee, 1989).

## Fold Space and Protein Evolution

One might imagine that the ways in which secondary structures can be combined to form a complete protein fold are almost limitless. Indeed, the variety of protein folds

---

[6]When cysteine is involved in a disulfide bond it is referred to as *cystine*.

and the chemical diversity of their residues lead to a wide array of functions. This universe of extant folds is often called fold space. Interestingly, currently available protein structure data suggest that fold space is in fact quite limited, relative to the range of folds that would seem possible a priori. Current estimates suggest that there may be about 1000 unique protein folds (Chothia, 1992; Govindarajan, Recabarren, and Goldstein, 1999; Wolf, Grishin, and Koonin, 2000). It is likely that two forces have played a role in the limitation of fold space: divergent evolution of protein function and convergent evolution of protein structure.

In the case of divergent evolution, *the number of extant protein folds is limited because they are derived from a relatively small group of shared common ancestor proteins.* These early ancestor proteins would have discovered a stable fold, which has then been duplicated and reused by organisms for many other functions over the course of evolution. Presumably, modification of an existing fold is more likely to occur than the spontaneous generation of a novel fold. There is clear evidence that this sort of modification has occurred over and over; it is possible because the link between protein structure and function is not direct (Todd, Orengo, and Thornton, 2001). Though protein structure leads to protein function, *similar protein structures will not always have similar functions.* Many cases exist where two proteins have similar sequences and structures, but differ by a few key amino acid residues in an active site and hence have very different functions. Thus, it is important to consider a protein's overall tertiary structure as a *guide* to the function of that protein, rather than a *definition* of the function (see Chapter 19). This functional versatility suggests the possibility that many protein folds will never be seen because organisms have simply not required or developed them.

In the case of convergent evolution, *the number of extant folds is limited because certain folds are much more biophysically favored, and so have been created independently in multiple cases.* Certain folds are clearly over-represented in the set of proteins of known structure, even when efforts are made to reduce the representational bias inherent in the Protein Data Bank (PDB) (Berman et al., 2000; see Chapter 9). In some cases, there is no discernable sequence similarity between proteins of similar fold, suggesting that they have converged on a similar structure independently and do not share a common ancestor (Holm and Sander, 1996). Further, some protein-folding models have suggested that a small subset of folds will be biophysically favored over all others (Li et al., 1996; Govindarajan and Goldstein, 1996). Hence, it is possible that many folds will never be seen because they are not structurally favorable, and other more favorable folds can be adapted to the needed functions.

The divergent and convergent protein evolution scenarios are not mutually exclusive, and both appear to have had a part in limiting the range of folds in fold space. Even though fold space appears limited, it is still complex enough to make classification and comprehension of protein folds difficult (see Chapters 12 and 13).

## Biochemical Classification of Folds

One method of protein classification partially sidesteps the issue of structural organization in favor of biochemical properties. Here, proteins are classified into three major groups: globular, membrane, and fibrous.

Globular proteins exist in an aqueous environment, and thus fold as compact structures with hydrophobic cores and polar surfaces. They exhibit the typical structural elements that have been discussed above. This class of proteins is well represented in the PDB, partly because these structures are the easiest to determine experimentally (see Chapters 4–6). Because of the variety of determined structures available, globular proteins have been used as the basis for most protein structure studies (Berman et al., 2000). Indeed, the first two experimentally determined structures, myoglobin and hemoglobin, are globular (Kendrew et al., 1958; Perutz et al., 1960) (Table 2.1).

Membrane proteins exhibit many of the same characteristics as globular proteins, but they are distinguished in two important ways. First, they exist in a completely

TABLE 2.1. Biochemical Folds

| Fold | Protein Example | Protein Schematic |
|------|-----------------|-------------------|
| Globular proteins | Myoglobin<br><br>(PDB id 1A6M) | |
| Membrane proteins | Rhodopsin<br><br>(PDB id 1AT9) | |
| Fibrous proteins | Collagen<br><br>(PDB id 1QSU) | |

different environment from typical globular proteins: the cell membrane. In the interior of the cell membrane, the protein is surrounded by a hydrophobic environment. Thus, the regions of the protein within the cell membrane must have a hydrophobic *surface* in order to be stable. Some proteins exist almost entirely within the cell membrane, whereas others have membrane-spanning or membrane-interacting domains. Second, experimental structure determination of membrane proteins is more difficult than that encountered with globular proteins (see Chapters 4–6). These proteins are very poorly represented in the PDB, thus, are poorly understood. However, the structures that have been determined suggest that they use the same secondary structural elements and follow the same general folding principles as globular proteins (Table 2.1).

Fibrous proteins differ markedly from both membrane and globular proteins. These proteins are often constructed of repetitive amino acid sequences that form simple, elongated fibers. Their repetitive design is well suited to the structural roles they often play in organisms. Some fibrous proteins consist of a single type of regular secondary structure that is repeated over very long sequences. Others are formed from repetitive atypical secondary structures, or have no discernable secondary structure whatsoever (Table 2.1).

## Structural Classification of Folds

As more and more protein structures have been determined, development of increasingly specific fold classifications has become possible. Cyrus Chothia and Michael Levitt derived one of the first such classifications, which grouped proteins based on their predominant secondary structural element (Levitt and Chothia, 1976). This classification consisted of four groups: all $\alpha$, all $\beta$, $\alpha/\beta$, and $\alpha + \beta$. All $\alpha$ proteins, as the name suggests, are based almost entirely on $\alpha$ helical structure, and all $\beta$ structures are based almost entirely on $\beta$ sheet. $\alpha/\beta$ structure is based on a mixture of $\alpha$ helix and $\beta$ sheet, often organized as parallel $\beta$ strands connected by $\alpha$ helices. Finally, $\alpha + \beta$ structures consist of discrete $\alpha$ helix and $\beta$ sheet motifs that are not interwoven (as they are in $\alpha/\beta$ structure). As known fold space has become more complex, these types of classifications have been adjusted and extended such that a complete hierarchy is created that places almost all known protein structures into specific subclassifications. Two approaches to this sort of classification (SCOP and CATH) are described in Chapters 12 and 13.

## THE QUATERNARY STRUCTURE OF PROTEINS: ASSOCIATIONS OF MULTIPLE POLYPEPTIDE CHAINS

Tertiary structure describes the structural organization of a single polypeptide chain. However, many proteins do not function as a single chain, or monomer. Rather, they exist as a noncovalent association of two or more independently folded polypeptides. These proteins are referred to as multisubunit, or multimeric, proteins and are said to have a quaternary structure. The subunits, or protomers, may be identical, resulting in a homomeric protein, or they can be comprised of different subunits resulting in a heteromeric protein (Figure 2.11).

The first observation of quaternary structure has been attributed to The Svedberg. His use of the ultracentrifuge to determine the molecular weights of proteins in 1926

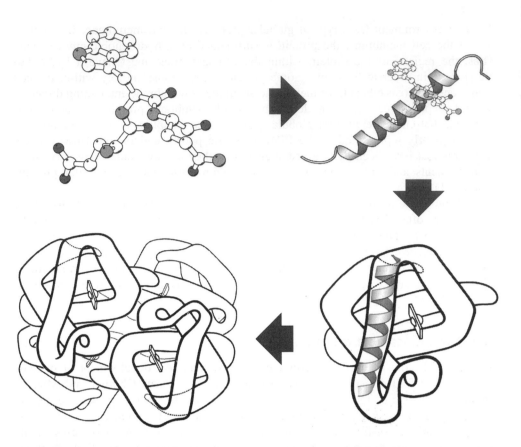

**Figure 2.11.** The four-tiered hierarchy of protein structure, depicted for the protein hemo-globin. Hemoglobin is a multisubunit, heteromeric protein consisting of four all-helical subunits. The figure begins with a depiction of primary structure in the upper left corner and proceeds to quaternary structure in a clockwise direction. The primary and secondary structure depictions were generated with the MolScript package (Kraulis, 1991). *Tertiary and quaternary structure depictions from R. E. Dickerson and I. Geis. Hemoglobin: Structure, Function, Evolution, and Pathology. Menlo Park, CA: The Benjamin/Cummings Publishing Company, 1983. Used with permission from Geis Archives.*

resulted in the separation of multisubunit proteins into their constituent protomers (Klotz, Langerman, and Darnall, 1970). The concept of quaternary structure was not of general biochemical interest until experiments on enzyme regulation in the early 1960s showed that protein subunits were crucial to understanding higher levels of cellular function (Gerhart and Pardee, 1962; Monod, Changeux, and Jacob, 1963; Klotz, Langerman, and Darrall, 1970).

Interestingly, most proteins are folded such that aggregation with other polypeptides is avoided (Richardson, 1992); the formation of multisubunit proteins is therefore a very specific interaction. Quaternary structures are stabilized by the same types of interactions employed in tertiary and secondary structure stabilization. The surface regions involved in subunit interactions generally resemble the cores of globular proteins; they are comprised of residues with nonpolar side chains, residues that can form hydrogen bonds, or residues that can participate in disulfide bonds. Table 2.2

TABLE 2.2. Functional Relevance of Quaternary Structure

| Functional Relevance | Protein Example | Protein Schematic |
| --- | --- | --- |
| *Cooperativity*<br>The association of subunits that bind the same substrate is often able to enhance binding capabilities of the multimer beyond what is possible with individual subunits. This cooperativity is realized through the ability of the subunits to influence each other based on their close proximity. | Hemoglobin<br><br>(PDB id 1A3N) | |
| *Co-localization of Function*<br><br>Different subunits can associate in order to confer multiple functions on a single protein. Often these functions involve distinct steps in the processing of a single substrate. Thus, the co-localization of function provided by a multisubunit complex can further enhance the abilities of a protein. | Tryptophan Synthase<br><br>(PDB id 1QOP) | |
| *Combinations of Subunits*<br><br>Combinatorial shifts in quaternary structure are able to bestow impressive versatility to protein function and regulation. Protein function can be altered by subunit swapping, and protein regulation can be achieved via interactions with different subunits. | Immunoglobulin<br><br>(PDB id 12E8) | |
| *Structural Assembly*<br><br>Very large structural proteins are made possible by the association of a large number of small subunits. This component-based assembly simplifies the construction of such structures and allows the information required to code these proteins to be more concise. | Actin<br><br>(PDB id 1ALM) | |

explains some of the functional advantages that are bestowed on proteins by their quaternary structure.

## CONCLUSION

Protein structure is complex, but it should be noted that by the 1970s protein scientists had determined most of the basic principles. These findings were confirmed in later years, as the pace of protein structure determination increased. These basic principles of protein structure now form the stable foundation needed for researching many of the remaining questions in protein science.

Computational tools now available, and described throughout this book, make management of the complexity of protein structure information more tractable. Protein structures can be more easily determined (see Chapter 29), and hypotheses as to the nature of protein structure can be more easily tested. As a result, it is now possible to pose more sophisticated questions, and in the coming years protein scientists can look forward to an understanding of protein structure and function on a much deeper level.

## INFORMATION ON THE INTERNET

Principles of Protein Structure Using the Internet, an on-line course at Birkbeck College, University of London: http://www.cryst.bbk.ac.uk/PPS2/. [A very useful source of information, it covers some areas of protein structure that were beyond the scope of this chapter.]

The RCSB Protein Data Bank (PDB): http://www.rcsb.org/pdb/. [The sole worldwide public repository for macromolecular structure data.]

Education Resources Listing at the PDB: http://www.rcsb.org/pdb/education.html. [A collection of links to educational resources covering protein structure.]

TOPS homepage and server: http://www.sander.embl-ebi.ac.uk/tops/. [Used to create one of the figures in this chapter.]

MolScript homepage: http://www.avatar.se/molscript/. [A popular package for generating images of protein structure, used to create many of the figures in this chapter.]

## FURTHER READING

Branden C, Tooze J (1999): *Introduction to Protein Structure* (2nd ed). New York: Garland Publishing. [Perhaps the finest book on the topic of protein structure. It features a well-written text and excellent hand-drawn illustrations.]

Mathews CK, van Holde KE (2000): *Biochemistry*. Redwood City, CA: The Benjamin/Cummings Publishing Company.

Voet D, Voet JG (1995): *Biochemistry* (2nd ed). New York: John Wiley & Sons.

## REFERENCES

Atkins JF, Gesteland RF (2000): The twenty-first amino acid. *Nature* 407:463–5.

Berman HM, Westbrook J, Feng Z, Gilliland G, Bhat TN, Weissig H, Shindyalov IN, Bourne PE (2000): The protein data bank. *Nucleic Acids Res* 28:235–42.

Chan AW, Hutchinson EG, Harris D, Thornton JM (1993): Identification, classification, and analysis of beta-bulges in proteins. *Protein Sci* 2:1574–90.

Chandrasekaran R, Lakshminarayanan AV, Pandya UV, Ramachandran GN (1973): Conformation of the LL and LD hairpin bends with internal hydrogen bonds in proteins and peptides. *Biochim Biophys Acta* 303:14–27.

Chothia C (1992): One thousand families for the molecular biologist. *Nature* 357:543–4.

Chou PY, Fasman GD (1977): $\beta$ turns in proteins. *J Mol Biol* 115:135–75.

Dickerson RE, Geis I (1969): *The Structure and Action of Proteins*. New York, Harper and Rowe. [A seminal book on protein structure, featuring the lucid illustrations of Irving Geis.]

Eyring H, Stearn AE (1939): The application of the theory of absolute reaction rates to proteins. *Chem Rev* 24:253–70.

Flores TP, Moss DM, Thornton JM (1994): An algorithm for automatically generating protein topology cartoons. *Protein Eng* 7:31–7.

Fruton JS (1972): *Molecules and Life: Historical Essays on the Interplay of Chemistry and Biology*. New York: Wiley-Interscience. [An informative perspective on the early history of protein science.]

Gerhart JC, Pardee AB (1962): The enzymology of control by feedback inhibition. *J Biol Chem* 237:891–6.

Govindarajan S, Goldstein RA (1996): Why are some protein structures so common? *Proc Natl Acad Sci USA* 93:3341–5.

Govindarajan S, Recabarren R, Goldstein RA (1999): Estimating the total number of protein folds. *Proteins* 35:408–14.

Holbrook JJ, Liljas A, Steindel SJ, Rossmann MG (1970): Oxidation-reduction. Part A. Dehydrogenases (I), Electron transfer (I). Volume 11. In: Boyer PD, editor. *The Enzymes*. 3rd ed. New York: Academic Press, p. 210.

Holm L, Sander C (1996): Mapping the protein universe. *Science* 273:595–603.

Huggins ML (1943): The structure of fibrous proteins. *Chem Rev* 32:195–218.

Karlin KD (1993): Metalloenzymes, structural motifs, and inorganic models. *Science* 261:701–8.

Kendrew JC (1961): The three-dimensional structure of a protein molecule. *Scientific American* 205:96–111. [An interesting follow-up to the determination of the myoglobin structure, this article also introduces the illustrations of Irving Geis.]

Kendrew JC, Bodo G, Dintzis HM, Parrish RG, Wyckoff H, Phillips DC (1958): A three-dimensional model of the myoglobin molecule obtained by x-ray analysis. *Nature* 181:662–6. [The publication announcing the experimental determination of the first protein structure. Worth reading for historical perspective.]

Kendrew JC, Dickerson RE, Strandberg BE, Hart RG, Davies DR, Phillips DC, Shore VC (1960): Structure of myoglobin. *Nature* 185:422–7.

Klotz IM, Langerman NR, Darnall DW (1970): Quaternary structure of proteins. *Ann Rev Biochem* 39:25–62.

Kraulis PJ (1991): MOLSCRIPT: A program to produce both detailed and schematic plots of protein structures. *J Applied Crystallogr* 24:946–50.

Kreil G (1997): D-Amino acids in animal peptides. *Ann Rev Biochem* 66:337–45.

Lee MS (1989): 3D structure of single zinc finger. *Science* 245:635–7.

Lesk AM, Hardman KD (1982): Computer-generated schematic diagrams of protein structures. *Science* 216:539–40.

Levitt M, Chothia C (1976): Structural patterns in globular proteins. *Nature* 261:552–8.

Lewis PN, Momany FA, Scheraga HA (1973): Chain reversal in proteins. *Biochim Biophys Acta* 303:211–29.

Li H, Helling R, Tang C, Wingreen N (1996): Emergence of preferred structures in a simple model of protein folding. *Science* 271:666–9.

Low BW, Baybutt RB (1952): The $\pi$-helix—A hydrogen bonded configuration of the polypeptide chain. *J Am Chem Soc* 74:5806–10.

Mirksy AE, Pauling L (1936): On the structure of native, denatured and coagulated proteins. *Proc Natl Acad Sci USA* 22:439–47. [A highly insightful and well-presented discussion of protein structure from two key pioneers in the field. Worth reading for historical perspective.]

Monod J, Changeux JP, Jacob F (1963): Allosteric proteins and cellular control systems. *J Mol Biol* 6:306–29.

Pauling L, Corcy RB (1951a): Configurations of polypeptide chains with favored orientations around single bonds: two new pleated sheets. *Proc Natl Acad Sci USA* 37:729–40. [This paper and the two that follow are part of a series of publications in which Linus Pauling and his colleagues discuss different aspects of secondary structure.]

Pauling L, Corey RB (1951b): The structure of fibrous proteins of the collagen-gelatin group. *Proc Natl Acad Sci USA* 37:272–81.

Pauling L, Corey RB, Branson HR (1951): The structure of proteins: two hydrogen bonded helical configurations of the polypeptide chain. *Proc Natl Acad Sci USA* 37:205–11.

Perutz MF, Rossmann MG, Cullis AF, Muirhead H, Will G, North ACT (1960): Structure of hemoglobin. *Nature* 185:416–22. [The publication announcing the experimental determination of the first multisubunit protein structure. Worth reading for historical perspective.]

Ramachandran GN, Sasisekharan V (1968): Conformation of polypeptides and proteins. *Adv Protein Chem* 23:283–437.

Richardson JS (1981): The anatomy and taxonomy of protein structure. *Adv Protein Chem* 34:167–339. [An exhaustive compendium of known protein structure, still useful to this day.]

Richardson JS (1992): Looking at proteins: representations, folding, packing, and design. *Biophys J* 63:1186–209. [An examination of protein structure and a discussion of efforts to engineer protein structures from designed amino-acid sequences.]

Richardson JS, Getzoff ED, Richardson DC (1978): The $\beta$ bulge: a common small unit of nonrepetitive protein structure. *Proc Natl Acad Sci USA* 75:2574–8.

Sanger F (1952): The arrangement of amino acids in proteins. *Adv Protein Chem* 7:1–67. [A presentation of the techniques and experimental data used to explore the peptide theory of proteins.]

Sternberg MJE, Thornton JM (1977): On the conformation of proteins: the handedness of the connection between parallel $\beta$ strands. *J Mol Biol* 110:269–83.

Tanford C (1978): The hydrophobic effect and the organization of living matter. *Science* 200:1012–8.

Taylor HS (1941): Large molecules through atomic spectacles. *Proc Am Phil Soc* 85:1–12.

Taylor WR (1986): The classification of amino acid conservation. *J Theor Biol* 119:205–18.

Todd AE, Orengo CA, Thornton JM (2001): Evolution of function in protein superfamilies, from a structural perspective. *J Mol Biol* 307:1113–43. [A comprehensive look at the relationship between protein structure and protein function in enzymes of known structure.]

Venkatachalam CM (1968): Stereochemical criteria for polypeptides and proteins. V. Conformation of a system of three linked peptide units. *Biopolymers* 6:1425–36.

Wolf YI, Grishin NV, Koonin EV (2000): Estimating the number of protein folds and families from complete genome data. *J Mol Biol* 299:897–905.

Wright PE, Dyson HJ (1999): Intrinsically unstructured proteins: re-assessing the protein structure-function paradigm. *J Mol Biol* 293:321–31. [A review discussing the current knowledge of proteins lacking an ordered structure, or achieving an ordered structure only under certain conditions or while engaged in specific interactions.]

Zinoni F, Birkmann A, Stadtman TC, Böck A (1986): Nucleotide sequence and expression of the selenocysteine-containing polypeptide of formate dehydrogenase (formate-hydrogen-lyase-linked) from *Escherichia coli*. *Proc Natl Acad Sci USA* 83:4650–4.

# 3

# FUNDAMENTALS OF DNA AND RNA STRUCTURE

Stephen Neidle, Bohdan Schneider, and Helen M. Berman

In 1946, Avery provided concrete experimental evidence that DNA was the main constituent of genes (Avery, MacLeod, and McCarthy, 1944); universal acceptance of this idea came with the publication of the Hershey–Chase experiments (Hershey and Chase, 1951). After the seminal discovery of the double helical nature of DNA in 1953 (Watson and Crick, 1953), the focus of nucleic acid structural research turned to fiber diffraction of natural and defined sequences (Arnott, 1970; Arnott, Campbell, Smith, and Chandrasekaran, 1976). Through these studies, we gained many insights into nucleic acid structure. We learned that hydration, ionic strength, and sequence all affect conformation type, and that nucleic acids can adopt a wide variety of structures, including single-stranded helices (Arnott, Chandrasekaran, and Leslie, 1976) and parallel helices (Rich et al., 1961), as well as triple and quadruple helices (Arnott, Chandrasekaran, and Marttila, 1974).

Once it was possible to synthesize and purify oligonucleotides (Khorana, Tener, and Moffatt, 1956), the use of crystallography and later NMR to determine nucleic acid structures became a reality. The first crystal structure that contained all the components that could have, in principle, allowed us to see a double helix was that of a very small piece of RNA-UpA (Seeman et al., 1971). Rather than forming a double helix, it displayed unusual conformations, which anticipated some of the many conformations that we now know exist in nucleic acids. In 1973, the double helix was visualized at atomic resolution with the determination of the crystal structures of two self-complementary RNA fragments, ApU and GpC (Rosenberg et al., 1973). The determination of the structures of dinucleoside phosphates complexed with drugs followed and laid the foundation for our understanding of nucleic acid recognition (Tsai, Jain, and Sobell, 1975).

*Structural Bioinformatics*
Edited by Philip E. Bourne and Helge Weissig
Copyright © 2003 by Wiley-Liss, Inc.

The publication in 1980 of a structure of more than a full turn of B-DNA (Wing et al., 1980) laid aside the doubts of even the most skeptical (Rodley et al., 1976) that DNA was a right-handed double helix. The structure is also a milestone in our understanding of the fine structure of DNA whereby it is possible to determine the effects of sequence on structure. Interestingly, it was at the same time that the structure of an unusual left-handed form of DNA—Z-DNA—was solved (Wang et al., 1979).

In parallel with the studies of these synthetic oligonucleotides, researchers were successful in purifying t-RNA. The publication of the structure of yeast phe t-RNA in 1974 (Kim et al., 1974; Robertus et al., 1974) represented the first, and until recently, the only structure of a natural intact nucleic acid. Now we are seeing an ever-increasing number of structures of RNA that are giving us insights into the RNA world.

In this chapter, we present the principles of nucleic acid structure. We then present a brief overview of the current state of our knowledge of nucleic acid structure determined using X-ray crystallographic methods. Discussion of nucleic acids in complex with proteins is reserved for another review.

## CHEMICAL STRUCTURE OF NUCLEIC ACIDS

In the early years of the twentieth century, chemical degradation studies on material extracted from cell nuclei established that the high molecular-weight "nucleic acid" was actually composed of individual acid units, termed nucleotides. Four distinct types were isolated: guanylic, adenylic, cytidylic, and thymidylic acids. These acids could be further cleaved to phosphate groups and four distinct nucleosides. The latter were subsequently identified as consisting of a deoxypentose sugar and one of four nitrogen-containing heterocyclic bases. Thus, each repeating unit in a nucleic acid polymer comprises these three units linked together: a phosphate group, a sugar, and one of the four bases.

The bases are planar aromatic heterocyclic molecules and are divided into two groups: the pyrimidine bases thymine and cytosine, and the purine bases adenine and guanine. Their major tautomeric forms are shown in Figure 3.1. Thymine is replaced by uracil in ribonucleic acids. RNA also has an extra hydroxyl group at the $2'$ position of their pentose sugar groups. The sugar present in RNA is ribose; in DNA, it is deoxyribose. The standard nomenclature for the atoms in nucleic acids is shown in Figures 3.1 and 3.2. Accurate bond length and angle geometries for all bases, nucleosides and nucleotides, have been well established by X-ray crystallographic analyses. The most recent surveys (Clowney et al., 1996; Gelbin et al., 1996) have calculated mean values for these parameters (which define their equilibrium values) from the most reliable structures in the Cambridge Structural Database (Allen et al., 1979) and the Nucleic Acid Database (Berman et al., 1992). These parameters have been incorporated in several implementations of the AMBER (Weiner and Kollman, 1981) and CHARMM (Brooks et al., 1983) force fields widely used in molecular mechanics and dynamics modeling, and in a number of computer packages for both crystallographic and NMR structural analyses (Parkinson et al., 1996). Accurate crystallographic analyses, at very high resolution, can also directly yield quantitative information on the electron-density distribution in a molecule, and, hence, on individual partial atomic charges. These charges for nucleosides have hitherto been obtained by *ab initio* quantum mechanical calculations, but are now available experimentally for all four DNA nucleosides (Pearlman and Kim, 1990).

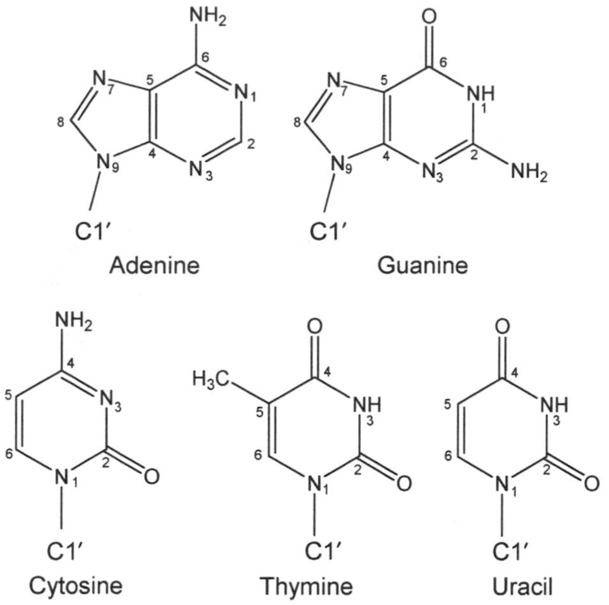

Figure 3.1. The five bases of DNA and RNA. The atoms are numbered according to standard nomenclature. From (Neidle, 2002). Reprinted by permission of Oxford University Press.

Individual nucleoside units are joined together in a nucleic acid in a linear manner, through phosphate groups that are attached to the 3′ and 5′ positions of the sugars (Figure 3.2). Hence, the full repeating unit in a nucleic acid is a 3′,5′-nucleotide.

Nucleic acid and oligonucleotide sequences use single-letter codes for the five-unit nucleotides—A, T, G, C, and U. The two classes of bases can be abbreviated as Y (pyrimidine) and R (purine). Phosphate groups are usually designated as "p". A single oligonucleotide chain is conventionally numbered from the 5′ end; for example, ApGpCpTpTpG has the 5′ terminal adenosine nucleoside, with a free hydroxyl at its 5′ position, and thus the 3′ end guanosine has a free 3′ terminal hydroxyl group. Intervening phosphate groups are sometimes omitted when a sequence is written. Chain direction is sometimes emphasized with 5′ and 3′ labels. Thus, an antiparallel double-helical sequence can be written as:

$$^{5'}CpGpCpGpApApTpTpCpGpCpG$$
$$^{3'}GpCpGpCpTpTpApApGpCpGpC$$

or simply as

$$(CGCGAATTCGCG)_2$$

Structural publications on DNA usually prefix a sequence with "d," as in d(CGAT), to emphasize that the oligonucleotide is a deoxyribose rather than an oligoribonucleotide.

The bond between sugar and base is known as the glycosidic bond. Its stereochemistry is important. In natural nucleic acids, the glycosidic bond is always $\beta$, which is to say that the base is above the plane of the sugar when viewed onto the plane and therefore on the same face of the plane as the 5′ hydroxyl substituent (Figure 3.3). The

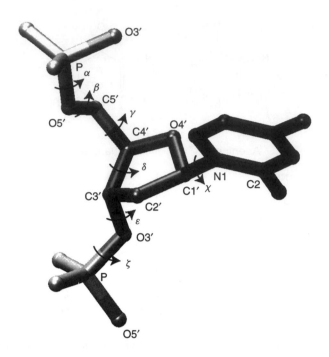

**Figure 3.2.** The organization and nomenclature of repeating units in a polynucleotide chain. The torsion angles for the sugar phosphate backbone are given as Greek letters. $\alpha$ = O3-P-O5'-C5'; $\beta$ = P-O5'-C5'-C4'; $\gamma$ = O5'-C5'-C4'-C3'; $\delta$ = C5'-C4'-C3'-O3'; $\varepsilon$ = C4'-C3'-O3'-P; $\zeta$ = C3'-O3'-P-O5'; $\chi$ = O4'-C1'-N1-C2. Reprinted with permission from John Wiley & Sons from (Berman, 1997).

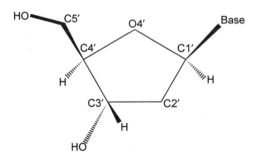

**Figure 3.3.** The stereochemistry of a natural $\beta$-nucleoside. Solid bonds are coming out of the plane of the page, toward the reader. Dashed bonds are going away from the reader. Reprinted from (Neidle, 2002). Reprinted by permission of Oxford University Press.

absolute stereochemistry of other substituent groups on the deoxyribose sugar ring of DNA is defined such that when viewed end-on with the sugar ring oxygen atom O4' at the rear, the hydroxyl group at the 3' position is below the ring and the hydroxymethyl group at the 4' position is above.

A unit nucleotide can have its phosphate group attached either at the 3' or 5' ends, and is thus termed either a 3' or a 5' nucleotide. It is chemically possible to construct $\alpha$-nucleosides, and from them $\alpha$-oligonucleosides that have their bases in the "below" configuration relative to the sugar rings and their other substituents. These $\alpha$-nucleotides are much more resistant to nuclease attack than standard natural $\beta$-oligomers and have been used as antisense oligomers to mRNAs on account of their superior intracellular stability.

## BASE PAIR GEOMETRY

The realization that the planar bases can associate in particular ways by means of hydrogen bonding was a crucial step in the elucidation of the structure of DNA. The important early experimental data of Chargaff (Zamenhof, Brawermann, and Chargaff, 1952) showed that the molar ratios of adenine : thymine and cytosine : guanine in DNA were both unity. This observation led to the proposal by Crick and Watson that in each of these pairs the purine and pyrimidine bases are held together by specific hydrogen bonds, to form planar base pairs. In native double-helical DNA, the two bases in a base pair necessarily arise from two separate strands of DNA (with intermolecular hydrogen bonds) and so hold the DNA double helix together (Watson and Crick, 1953).

The adenine : thymine (A·T) base pair has two hydrogen bonds compared to the three in a guanine : cytosine (G·C) one (Figure 3.4). Fundamental to the Watson–Crick arrangement is that the sugar groups are both attached to the bases asymmetrically on the same side of the base pair. This asymmetric arrangement defines the mutual positions of the two sugar–phosphate strands in DNA itself. Atoms at the surface of the sugar–phosphate backbone define two indentations with different dimensions called minor and major grooves. The major groove is by convention faced by C6/N7/C8 purine atoms and their substituents, and by C4/C5/C6 pyrimidine atoms and their substituents; the minor groove by C2/N3 purine and C2 pyrimidine atoms and their substituents.

The two base pairs are required to be almost identical in dimensions by the Watson–Crick model. High resolution (0.8–0.9 Å) X-ray crystallographic analyses of the ribodinucleoside monophosphate duplexes $(GpC)_2$ and $(ApU)_2$ by A. Rich and colleagues in the early 1970s has established accurate geometries for these A : T and G : C base pairs (Rosenberg et al., 1976; Seeman et al., 1976). These structure determinations showed that there are only small differences in size between the two types of base pairings, as indicated by the distance between glycosidic carbon atoms in a base pair. The $C1' \cdots C1'$ distance in the G·C base pair structure is 10.67 Å, and 10.48 Å in the A·U-containing dinucleoside.

The individual bases in a nucleic acid are flat aromatic rings, but base pairs bound together only by nonrigid hydrogen bonds can show considerable flexibility. The vertical arrangement of bases and base pairs is flexible and restrained mainly by stacking interactions of bases. This flexibility is to some extent dependent on the nature of the bases and base pairs themselves, but is more related to their base-stacking environments. Thus, descriptions of base morphology have become important in describing and understanding many sequence-dependent features and deformations of nucleic acids. The former features are often considered primarily at the dinucleoside local level, whereas longer-range effects, such as helix bending, can also be analyzed at a more global level.

A number of rotational and translational parameters have been devised to describe these geometric relations between bases and base pairs (Figure 3.5), which were originally defined in 1989 (the "Cambridge Accord") (Dickerson et al., 1989). These definitions, together with the Cambridge Accord sign conventions, are given for some key base parameters.

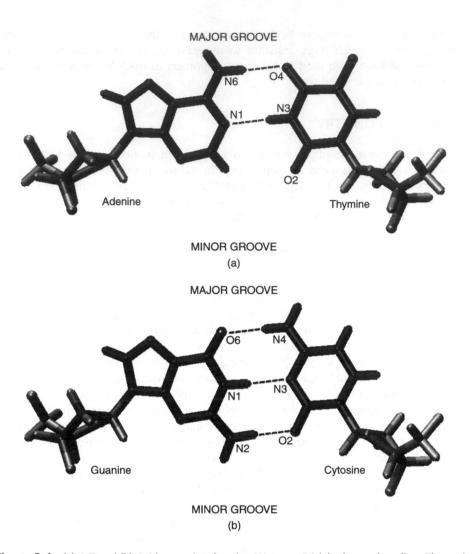

**Figure 3.4.** (a) A·T and (b) G·C base pairs, showing Watson–Crick hydrogen bonding. The major and minor sides of the bases are shown. Reprinted from with permission from John Wiley & Sons from (Berman, 1997).

*Propeller twist* ($\omega$) between bases is the dihedral angle between normals to the bases when viewed along the long axis of the base pair. The angle has a negative sign under normal circumstances, with a clockwise rotation of the nearer base when viewed down the long axis. The long axis for a purine–pyrimidine base pair is defined as the vector between the C8 atom of the purine and the C6 of a pyrimidine in a Watson–Crick base pair. Analogous definitions can be applied to other nonstandard base pairings in a duplex including purine–purine and pyrimidine–pyrimidine ones.

*Buckle* ($\kappa$) is the dihedral angle between bases, along their short axis, after propeller twist has been set to 0°. The sign of buckle is defined as positive if the distortion is convex in the direction $5' \to 3'$ of strand 1. The change in buckle for succeeding steps, termed *cup*, has been found to be a useful measure of changes along a sequence.

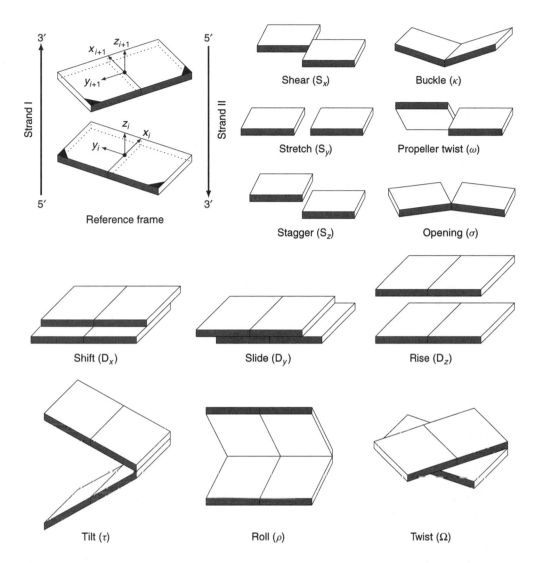

**Figure 3.5.** Pictorial definitions of parameters that relate complementary base pairs and sequential base-pair steps. The base-pair reference frame is constructed such that the $x$-axis points away from the (shaded) minor groove edge. Images illustrate positive values of the designated parameters. Reprinted with permission from Adenine Press from (Lu, Babcock, and Olson, 1999).

Cup is defined as the difference between the buckle at a given step, and that of the preceding one.

*Inclination* ($\eta$) is the angle between the long axis of a base pair and a plane perpendicular to the helix axis. This angle is defined as positive for right-handed rotation about a vector from the helix axis toward the major groove.

*X and Y displacements* define translations of a base pair within its mean plane in terms of the distance of the midpoint of the base pair long axis from the helix axis. X displacement is toward the major groove direction when it has a positive value.

Y displacement is orthogonal to this, and is positive if toward the first nucleic acid strand of the duplex.

The key parameters for base pair steps are:

*Helical twist* ($\Omega$) is the angle between successive base pairs, measured as the change in orientation of the C1'–C1' vectors on going from one base pair to the next, projected down the helix axis. For an exactly repetitious double helix, helical twist is 360°/n, where n is the unit repeat defined above.

*Roll* ($\rho$) is the dihedral angle for rotation of one base pair with respect to its neighbor, about the long axis of the base pair. A positive roll angle opens up a base pair step toward the minor groove. *Tilt* ($\tau$) is the corresponding dihedral angle along the short (i.e., *x*-axis) of the base pair.

*Slide* is the relative displacement of one base pair compared to another in the direction of nucleic acid strand one (i.e., the Y displacement), measured between the midpoints of each C6–C8 base pair long axis.

Unfortunately there is now some confusion in the literature regarding these parameters, in part because the Cambridge Accord did not define a single unambiguous convention for their calculation and two distinct types of approaches have been developed to calculate them (Lu, Babcock, and Olson, 1999). In one approach, the parameters are defined with respect to a global helical axis, which need not be linear. The other approach uses a set of local axes, one per dinucleotide step. Another ambiguity is that a variety of definitions of local and global axes have been used. Fortunately, the overall effect for most undistorted structures is that only a minority of parameters appear to have very different values depending on the method of calculation, using a number of the widely available programs: CEHS/SCHNAaP (El Hassan and Calladine, 1995; Lu, Hassan, and Hunter, 1997), CompDNA (Gorin, Zhurkin, and Olson, 1995; Kosikov et al., 1999), Curves (Lavery and Sklenar, 1988; Lavery and Sklenar, 1989) FREEHELIX (Dickerson, 1998), NGEOM (Soumpasis and Tung, 1988; Tung, Soumpasis, and Hummer, 1994) NUPARM (Bhattacharyya and Bansal, 1989; Bansal, Bhattacharyya, and Ravi, 1995) and RNA (Babcock, Pednault, and Olson, 1993; Babcock and Olson, 1994; Babcock, Pednault, and Olson, 1994).

To resolve the ambiguities in description of base morphology parameters, a standard coordinate reference frame for the calculation of these parameters has been proposed (Lu and Olson, 1999), and has been endorsed by the successor to the Cambridge Accord, the 1999 Tsukuba Accord described in detail in (Olson et al., 2001). The right-handed reference frame is shown in Figure 3.6. It has the *x*-axis directed toward the major groove along the pseudo two-fold axis (shown as •) of an idealized Watson–Crick base pair. The *y*-axis is along the long axis of the base pair, parallel to the C1'···C1' vector. The position of the origin is clearly dependent on the geometry of the bases and the base pair. The geometry of the bases has been taken from the published compilations (Clowney et al., 1996; Gelbin et al., 1996). The Tsukuba reference frame is unambiguous and has the advantage of being able to produce values for the majority of local base-pair and base-step parameters that are independent of the algorithm used.

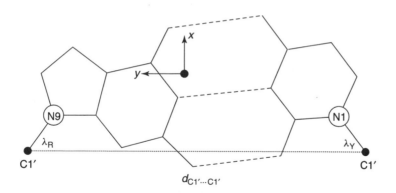

**Figure 3.6.** Illustration of idealized base-pair parameters, $d_{C1'\cdots C1'}$ and $\lambda$, used, respectively, to displace and pivot complementary bases in the optimization of the standard reference frame for right-handed A- and B-DNA, with the origin at • and the $x$- and $y$-axes pointing in the designated directions. Reprinted with permission from (Olson et al., 2001).

## CONFORMATION OF THE SUGAR PHOSPHATE BACKBONE

The five-member deoxyribose sugar ring in nucleic acids is inherently nonplanar. This nonplanarity is termed puckering. The precise conformation of a deoxyribose ring can be completely specified by the five endocyclic torsion angles within it (Figure 3.7a). The ring puckering arises from the effect of nonbonded interactions between substituents at the four ring carbon atoms—the energetically most stable conformation for the ring has all substituents as far apart as possible. Thus, different substituent atoms would be expected to produce differing types of puckering. The puckering can be described by either a simple qualitative description of the conformation in terms of atoms deviating from ring coplanarity, or precise descriptions in terms of the ring internal torsion angles.

In principle, there is a continuum of interconvertible puckers, separated by energy barriers. These various puckers are produced by systematic changes in the ring torsion angles. The puckers can be succinctly defined by the parameters P and $\tau_m$ (Altona and Sundaralingam, 1972). The value of P, the phase angle of pseudorotation, indicates the type of pucker since P is defined in terms of the five torsion angles $\tau_0 - \tau_4$:

$$\tan P = \frac{(\tau_4 + \tau_1) - (\tau_3 + \tau_0)}{2 * \tau_2 * (\sin 36° + \sin 72°)}$$

and the maximum degree of pucker, $\tau_m$, by

$$\tau_m = \tau_2/(\cos P)$$

The pseudorotation phase angle can take any value between 0° and 360°. If $\tau_2$ has a negative value, then 180° is added to the value of P. The pseudorotation phase angle is commonly represented by the pseudorotation wheel, which indicates the continuum of ring puckers (Figure 3.7b). Values of $\tau_m$ indicate the degree of puckering of the ring; typical experimental values from crystallographic studies on mononucleosides are in the range 25–45°. The five internal torsion angles are not independent of each other,

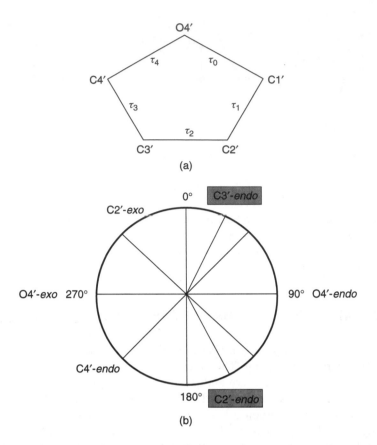

Figure 3.7. (a) The five internal torsion angles in a ribose ring. (b) The pseudorotation wheel for a deoxyribose sugar. The shaded areas indicate the preferred ranges of the pseudorotation angle for the two principal sugar conformations. From (Neidle, 2002). Reprinted by permission of Oxford University Press.

and so to a good approximation, any one angle $\tau_j$ can be represented in terms of just two variables:

$$\tau_j = \tau_m \cos[P + 0.8\pi(j - 2)]$$

A large number of distinct deoxyribose ring pucker geometries have been observed experimentally by X-ray crystallography and NMR techniques. When one ring atom is out of the plane of the other four, the pucker type is an envelope one. More commonly, two atoms deviate from the plane of the other three, with these two either side of the plane. It is usual for one of the two atoms to have a larger deviation from the plane than the other, resulting in a twist conformation. The direction of atomic displacement from the plane is important. If the major deviation is on the same side as the base and C4′–C5′ bond, then the atom involved is termed *endo*. If it is on the opposite side, it is called *exo*. The most commonly observed puckers in crystal structures of isolated nucleosides and nucleotides are either close to C2′-*endo* or C3′-*endo* types (Figure 3.8a, b). The C2′-*endo* family of puckers have P values in the range 140–185°; in view of their position on the pseudorotation wheel, they are sometimes termed S (south) conformations. The C3′-*endo* domain has P values in the range −10° to +40°,

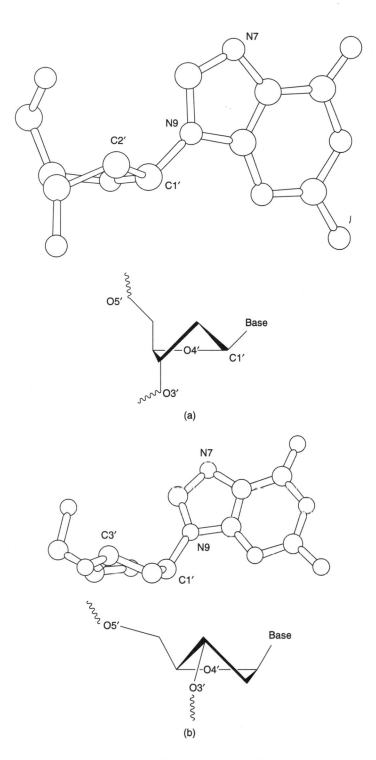

Figure 3.8. (a) C2′-*endo* sugar puckering for a guanosine nucleoside, viewed along the plane of the sugar ring. (b) C3′-*endo* sugar puckering for a guanosine nucleoside, with the sugar viewed in the same direction as in panel (a). From (Neidle, 2002). Reprinted by permission of Oxford University Press.

and its conformation is termed N (north). In practice, these pure envelope forms are rarely observed, largely because of the differing substituents on the ring. Consequently the puckers are then best described in terms of twist conformations. When the major out-of-plane deviation is on the *endo* side, there is a minor deviation on the opposite, *exo* side. The convention used for describing a twist deoxyribose conformation is that the major out-of-plane deviation is followed by the minor one, for example C2'-*endo*, C3'-*exo*.

The pseudorotation wheel implies that deoxyribose puckers are free to interconvert. In practice, there are energy barriers between major forms. The exact size of these barriers has been the subject of considerable study (Olson, 1982a; Olson and Sussman, 1982). The consensus is that the barrier height is dependent on the route around the pseudorotation wheel. For interconversion of C2'-*endo* to C3'-*endo*, the preferred pathway is via the O4'-*endo* state, with a barrier of 2–5 kcal/mole found from an analysis of a large body of experimental data (Olson & Sussman, 1982), and a somewhat smaller (potential energy) value of 1.5 kcal/mole from a molecular dynamics study (Harvey and Prabhakarn, 1986). The former value, being an experimental one, represents the total free energy for interconversion.

Relative populations of puckers can be monitored directly by NMR measurements of the ratio of coupling constants between H1'–H2' and H3'–H4' protons. These show that in contrast to the "frozen-out" puckers found in the solid-state structures of nucleosides and nucleotides, there is rapid interconversion in solution. Nonetheless, the relative populations of the major puckers are dependent on the type of base attached. Purines show a preference for the C2'-*endo* pucker conformational type, whereas pyrimidines favor C3'-*endo*. Deoxyribose nucleosides are primarily (>60%) in the C2'-*endo* form and ribonucleosides favor C3'-*endo*. Sugar pucker preferences have their origin in the nonbonded interactions between substituents on the sugar ring, and to some extent on their electronic characteristics. The C3'-*endo* pucker of ribose would have hydroxyl substituents at the 2' and 3' positions further apart than with C2'-*endo* pucker. Ribonucleosides are therefore significantly more restricted in their mobility than deoxyribonucleotides; this has significance for the structures of oligoribonucleotides. These differences in puckering equilibrium and hence in their relative populations in solution and in molecular dynamics simulations, are reflected in the patterns of puckers found in surveys of crystal structures (Murray-Rust and Motherwell, 1978).

Correlations have been found, from numerous crystallographic and NMR studies, between sugar pucker and several backbone conformational variables, both in isolated nucleosides/nucleotides and in oligonucleotide structures. These are discussed later in this chapter. Sugar pucker is thus an important determinant of oligo- and polynucleotide conformation because it can alter the orientation of C1', C3', and C4' substituents, resulting in major changes in backbone conformation and overall structure, as indeed is found.

The glycosidic bond links a deoxyribose sugar and a base, being the C1'–N9 bond for purines and the C1'–N1 bond for pyrimidines. The torsion angle $\chi$ around this single bond can in principle adopt a wide range of values, although as will be seen, structural constraints result in marked preferences being observed. Glycosidic torsion angles are defined in terms of the four atoms:

O4'–C1'–N9–C4 for purines

O4'–C1'–N1–C2 for pyrimidines

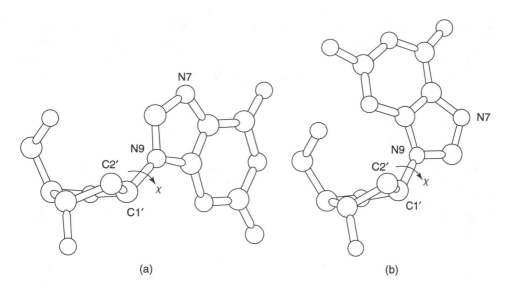

(a)                                                    (b)

**Figure 3.9.** (a) A guanosine nucleoside with the glycosidic angle $\chi$ set in an *anti* conformation. (b) Guanosine now in the *syn* conformation. From (Neidle, 2002). Reprinted by permission of Oxford University Press.

Theory has predicted two principal low-energy domains for the glycosidic angle, in accord with experimental findings for a large number of nucleosides and nucleotides. The *anti* conformation has the N1, C2 face of purines and the C2, N3 face of pyrimidines directed away from the sugar ring (Figure 3.9a) so that the hydrogen atoms attached to C8 of purines and C6 of pyrimidines are lying over the sugar ring. Thus, the Watson–Crick hydrogen-bonding groups of the bases are directed away from the sugar ring. These orientations are reversed for the *syn* conformation, with these hydrogen-bonding groups now oriented toward the sugar and especially its O5′ atom (Figure 3.9b). A number of crystal structures of *syn* purine nucleosides have found hydrogen bonding between the O5′ atom and the N3 base atom, which would stabilize this conformation. Otherwise, for purines, the *syn* conformation is slightly less preferred than the *anti*, on the basis of fewer nonbonded steric clashes in the latter case. The principal exceptions to this rule are guanosine-containing nucleotides, which have a small preference for the *syn* form because of favorable electrostatic interactions between the exocyclic N2 amino group of guanine and the 5′ phosphate group. For pyrimidine nucleotides, the *anti* conformation is preferred over the *syn*, because of unfavorable contacts between the O2 oxygen atom of the base and the 5′-phosphate group. The results of molecular-mechanics energy minimization's on all four DNA nucleotides in both *syn* and *anti* forms (using the AMBER all-atom force field) are fully in accord with these observations.

The sterically preferred ranges for the two domains of glycosidic angles are:

$$Anti: \quad -120 > \chi > 180°$$
$$Syn: \quad \quad 0 < \chi < 90°$$

Values of $\chi$ in the region of about $-90°$ are often described as "high *anti*." There are pronounced correlations between sugar pucker and glycosidic angle, which reflect the changes in nonbonded clashes produced by C2′-*endo* versus C3′-*endo* puckers. Thus, *syn* glycosidic angles are not found with C3′-*endo* puckers due to steric clashes between the base and the H3′ atom, which points toward the base in this pucker mode.

The phosphodiester backbone of an oligonucleotide has six variable torsion angles (Figure 3.2), designated $\alpha$, $\beta$, $\gamma$, $\delta$, $\varepsilon$, and $\zeta$, in addition to the five internal sugar torsions $\tau_0 \cdots \tau_4$ and the glycosidic angle $\chi$. As will be seen, a number of these have highly correlated values (and therefore correlated motions in a solution environment). Steric considerations alone dictate that the backbone angles are restricted to discrete ranges (Sundaralingam, 1969; Olson, 1982b), and are accordingly not free to adopt any value between 0° and 360°. Figure 3.10 uses a conformational wheel to show these preferred values, which are directly readable from their positions around the wheel. The fact that angles $\alpha$, $\beta$, $\gamma$, and $\zeta$ each have three allowed ranges, together with the broad range for angle $\varepsilon$ that includes two staggered regions, leads to a large number of possible low-energy conformations for the unit nucleotide, especially when glycosidic angle and sugar pucker flexibility is taken into account. In reality, only a small number of DNA oligonucleotide and polynucleotide structural classes have actually been observed out of this large range of possibilities; this is doubtless in large part due to the restraints imposed by Watson−Crick base pairing on the backbone conformations when two DNA strands are intertwined together. In contrast, crystallographic and NMR studies on a large number of standard and modified mononucleosides and nucleotides have shown their considerably greater conformational diversity. For mononucleotides backbone conformations in the solid state and in solution are not always in agreement;

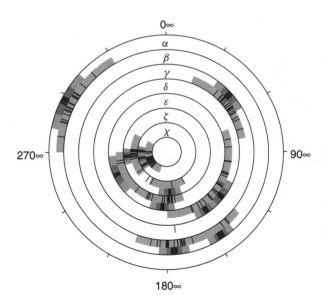

BDL001 over expected ranges for B-DNA.

**Figure 3.10.** Conformational wheel (Schneider, Neidle, and Berman, 1997) showing the torsion angles for BDL001 (Drew et al., 1981). Black lines show actual values of torsion angles, cyan background their allowed range in the B-type DNA conformation (Schneider, Neidle, and Berman, 1997). The grey shades in the outer rings show the average value(s) of the torsions in dark grey flanked by values of one and two estimated standard deviations in lighter grey. Figure also appears in Color Figure section.

the requirements for efficient packing in the crystal can often overcome the modest energy barriers between different values for a particular torsion angle. A large range of base... base interactions characterize many larger RNA molecules, which can therefore adopt a variety of backbone conformations.

A common convention for describing these backbone angles is to term values of $\sim60°$ as *gauche*$^+$ (g$^+$), $-60°$ as *gauche*$^-$ (g$^-$) and $\sim180°$ as *trans* (t). Thus, for example, angles $\alpha$ (about the P-O5$'$ bond) and $\gamma$ (the exocyclic angle about the C4$'$–C5$'$ bond), can be in the g$^+$, g$^-$ or t conformations. The two torsion angles around the phosphate group itself, $\alpha$ and $\zeta$, have been found to show a high degree of flexibility in various dinucleoside crystal structures, with the tg$^-$, g$^-$g$^-$ and g$^+$g$^+$ conformations all having been observed (Kim et al., 1973). A- and B-DNA adopt to g$^-$g$^-$ and g$^-$t conformations; Z-DNA adopts the g$^+$g$^+$, tg$^-$ and tg$^+$ conformations. The torsion angle $\beta$, about the O5$'$–C5$'$ bond, is usually *trans*. All three possibilities for the $\gamma$ angle have been observed in nucleoside crystal structures, although the g$^+$ conformation predominates in right-handed oligo- and polynucleotide double helices. The torsion angle $\delta$ around the C4$'$–C3$'$ bond adopts values that relate to the pucker of the sugar ring, since the internal ring torsion angle $\tau_3$, (also around this bond), has a value of about 35° for C2$'$-*endo* and about 40° for C3$'$-*endo* puckers; $\delta$ is about 75° for C3$'$-*endo* and about 150° for C2$'$-*endo* puckers.

There are a number of correlations involving pairs of these backbone torsion angles, as well as sugar pucker and glycosidic angle, that have been observed in mononucleosides and nucleotides (which are inherently more flexible in solution as well as being more subject to packing forces in the crystal), and more recently, in oligonucleotides (Schneider, Neidle, and Berman, 1997; Packer and Hunter, 1998). Some of the more significant correlations are:

- Between sugar pucker and glycosidic angle $\chi$, especially for pyrimidine nucleosides. C3$'$-*endo* pucker is usually associated with median-value *anti* glycosidic angles, whereas C2$'$-*endo* puckers are commonly found with high *anti* $\chi$ angles. *Syn* glycosidic angle conformations show a marked preference for C2$'$-*endo* sugar puckers.

- Scattergrams between $\alpha$ and $\zeta$ show clear distinctions for the A-, B-, and Z-DNA conformational classes. The same is true from scattergrams between $\chi$ and $\zeta$. Overall, values of torsion angles $\alpha, \zeta, \delta$, and $\chi$ form a "fingerprint" of a nucleotide or the whole structure (Schneider, Neidle, and Berman, 1997) with a predictive power sufficient to provide information for structural classification of the DNA double helix.

## STRUCTURES OF NUCLEIC ACIDS

### DNA Duplexes

The first structures of nucleic acids were of DNA duplexes derived using fiber diffraction methods. B-DNA, the classic structure first described by Watson and Crick (Watson and Crick, 1953), as refined using the linked-atom, least-squares procedure developed by Arnott and his group (Kim, et al., 1973). In canonical B-DNA (Figure 3.11), the backbone conformation has C2$'$-*endo* sugar puckers and high *anti* glycosidic angles. The right-handed double helix has 10 base pairs per complete turn, with the two polynucleotide chains antiparallel to each other and linked by Watson–Crick A·T and G·C

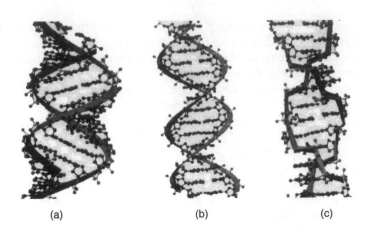

(a)                                    (b)                                    (c)

**Figure 3.11.** Canonical helical types of A-, B-, and Z-DNA from (Berman, Gelbin, and Westbrook, 1996) with permission from Elsevier Science. Figure also appears in Color Figure section.

base pairs. The paired bases are almost exactly perpendicular to the helix axis, and they are stacked over the axis itself. Consequently, the base-pair separation is the same as the helical rise—3.4 Å. An important consequence of the Watson–Crick base-pairing arrangement is that the two deoxyribose sugars linked to the bases of an individual pair are asymmetrically on the same side of it. So, when successive base pairs are stacked on each other in the helix, the gap between these sugars forms two continuous indentations with different dimensions in the surface that wind along, parallel to the sugar-phosphodiester chains. These indentations are termed grooves. The asymmetry in the base pairs results in two parallel types of groove, whose dimensions (especially their depths) are related to the distances of base pairs from the axis of the helix and their orientation with respect to the axis. The wide major groove is almost identical in depth to the much narrower minor groove, which has the hydrophobic hydrogen atoms of the sugar groups forming its walls. In general, the major groove is richer in base substituents—O6, N6 of purines and N4, O4 of pyrimidines—compared to the minor one (Figure 3.4). These differences in chemistry of the major and minor groove, together with the steric differences between the two, has important consequences for interaction with other molecules.

Over 230 single crystal structures with B-DNA conformations have been determined since the first such structure was published (Figure 3.12a) (Drew et al., 1981). The average values of the base morphology parameters are close to the canonical values derived from fiber studies. However, the individual structures have diverse shapes. Many DNA structures are bent by as much as 15° (Dickerson, Goodsell, and Kopka, 1996) and the groove widths are variable (Heinemann, Alings, and Hahn, 1994). The bending is a function of some of the base morphology parameters, in particular twist and roll. Attempts to relate the base morphology parameters to the sequence of the bases have shown some trends. Twist and roll appear to be correlated with regressions that depend on the nature of the bases in the step; purine–pyrimidine, purine–purine, and pyrimidine–purine steps show distinctive differences (Quintana et al., 1992; Gorin, Zhurkin, and Olson, 1995). A–T base pairs in AA steps have high propeller twist and bifurcated hydrogen bonds (Yanagi, Prive, and Dickerson, 1991). Stretches of A in sequences appear to be straight (Young et al., 1995). However, stretches of A in sequences appear to be straight, but certain steps such as TA and CA show a very

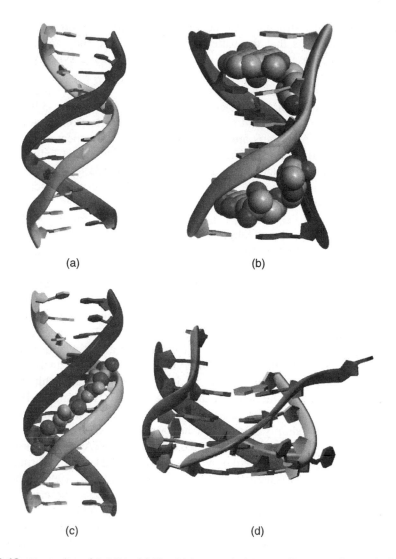

(a)                                             (b)

(c)                                             (d)

Figure 3.12. Examples of B-DNA. (a) The Dickerson dodecamer (Drew et al., 1981); (b) B-DNA daunomycin (Frederick et al., 1990). The drug is intercalated in between the CG base pairs; (c) Netropsin-DNA complex (Goodsell, Kopka, and Dickerson, 1995). The drug is bound in the minor groove; (d) DNA tetraplex (Phillips et al., 1997). Figure also appears in Color Figure section.

large variability due to either crystal packing, sequence context, or both (Lipanov et al., 1993).

Included among the many B-DNA structures are 25 with base mispairing. Interestingly, these mutations cause only local perturbations to the structures and the overall conformations of these structures remains the same as the parent structures. Hydration also plays an important role in the structure of B-DNA. The spine of hydration first seen in the dodecamer structure and described in some detail (Dickerson et al., 1983) has proven to be an enduring feature of these molecules. A detailed analysis of the hydration around the bases in DNA has demonstrated that the hydration is local (Schneider and Berman, 1995). Hydrated building blocks from decameric B-DNA were used to construct the dodecamer structure and the spine was faithfully reproduced. That same analysis strongly suggests that the patterns of hydration, be they spines or ribbons, are a function of the base morphology and that there is strong synergy between the hydration and base conformations.

Canonical A-DNA (Figure 3.11) has C3'-*endo* sugar puckers, which brings consecutive phosphate groups on the nucleotide chains closer together—5.9 Å compared to the 7.0 Å in B-form DNA—and alters the glycosidic angle from high *anti* to *anti*. As a consequence, the base pairs are twisted and tilted with respect to the helix axis, and are displaced nearly 5 Å from it, in striking contrast to the B helix. The helical rise is, as a consequence, much reduced to 2.56 Å, compared to 3.4 Å for canonical B-DNA. The helix is wider than the B one and has an 11 base-pair helical repeat. The combination of base pair tilt with respect to the helix axis and base pair displacement from the axis results in very different groove characteristics for the A double helix compared to the B form. This combination also results in the center of the A double helix being a hollow cylinder. The major groove is now deep and narrow, and the minor one is wide and very shallow.

Analysis of the A-DNA crystal structures has shown that there are variations in the helical parameters that appear to be related to crystal packing (Ramakrishnan and Sundaralingam, 1993). Two sequences each crystallize in two different space groups: GTGTACAC (Jain, Zon, and Sundaralingam, 1989; Jain, Zon, and Sundaralingam, 1991; Thota et al., 1993) and GGGCGCCC (Shakked et al., 1989). In both cases the helical parameters are different in different crystal forms. Another analysis of a series of A-DNA duplexes shows that there is an inverse linear relationship between the crystal packing density and the depth of the major groove (Heinemann, 1991).

In the earlier days of nucleic acid crystallography, it had been thought by some that the A conformation was simply a crystalline artifact and not likely to bear any relationship to biology. Now it has been demonstrated that in the TATA binding protein–DNA complex (Kim et al., 1993; Kim, Nikolov, and Burley, 1993), the conformation of the DNA is an A–B chimera (Guzikevich-Guerstein and Shakked, 1996). The apparent deformability of the base geometry seen in A-DNA oligonucleotide crystal structures may prove to be an advantage in forming protein-DNA complexes.

One of the earliest A-type structures determined demonstrated a unique arrangement of water molecules consisting of edge-linked pentagons (Shakked et al., 1983). This pattern, like the spine seen in B-DNA, provoked the continued pursuit of the role of hydration in the stabilization of the different forms of DNA. One theory is that the economy of hydration seen in A-type structures may provide a structural explanation for the B to A transition as the humidity is lowered (Saenger, Hunter, Kennard, 1986). In a study of the hydration of a series of A-DNA structures it was demonstrated that the bases have specific patterns and that both the direct and the water-mediated interactions could be related to recognition properties of DNA and proteins (Eisenstein and Shakked, 1995).

Around the same time that the single crystal structure of B-DNA was determined, a left-handed conformation called Z-DNA was discovered (Wang et al., 1979). The zigzag phosphate backbone defines a convex outer surface of the major groove and the deep central minor groove (Figure 3.11). Z-type structures have alternation of cytosine and guanine with a cytosine at the first position. Of the unmodified Z-DNA structures, there are very few examples in which there have been substitutions of A for G and T for C. Modifications of the cytosines with methyl groups at the five positions have allowed for more drastic sequence changes as exemplified for a structure with tandem

G's in the center (Schroth, Kagawa, and Ho, 1993), and another with an AT at that position (Wang et al., 1985).

The basic building block of Z-DNA is a dinucleotide with the twist of the CG steps 9° and of the GC steps 50°. The backbone conformation of Z-DNA is characterized by the alternation of the $\chi$ angle of the guanine and the cytosine where the former is *syn* and the latter *anti*. The values for backbone torsion angles do not resemble those found in either A- or B-DNA. In many Z-DNA crystals, the central step of one chain has a different conformation from the other steps. This conformational polymorphism is called ZI and ZII (Gessner et al., 1989). Analysis of the packing patterns in Z crystals has shown that it is possible to correlate the pattern of water bridges with the presence of the ZI–ZII pattern (Schneider et al., 1992).

The hydration characteristics of Z-DNA crystals are very uniform with a spine of hydration in the central minor groove and tightly bound water on the exterior major groove (Gessner, Quigley, and Egli, 1994). The effects of solvent reorganization as a result of subtle sequence changes has been offered as a very plausible explanation for the different stabilities of these sequences (Wang, et al., 1984; Kagawa et al., 1989).

## Drug Complexes

Three major types of drug interactions have been observed. The intercalation mode was first observed in a co-crystal between UpA and ethidium bromide (Tsai, Jain, and Sobell, 1975) and in a subsequent series of dinucleoside drug complexes (Berman and Young, 1981). The first intercalation complexes (Wang et al., 1987) with longer stretches contained daunorubicin sandwiched between the terminal CG base pair. Now more than 200 related structures have been determined, thus, shedding light on the effects of changes in the drug and the sequence. The determination of the structure of a complex between actinomycin D (Kamitori and Takusagawa, 1994) and an octamer in which the drug is bound to the central GC base pair showed how the double helix can accommodate the drug without itself fraying.

The second class of drug complexes have been those that bind the drug in the minor groove. Starting with the structure of netropsin bound to the Dickerson dodecamer (Kopka et al., 1985), there have been over 30 structures determined in which the drugs bind to the minor groove (Figure 3.12c). Several of these structures contain the Hoechst benzimidazole derivative, which is typical in having a planar heteroaromatic cross-section whose dimensions complement those of the A/T narrow minor groove in multiple binding geometries. In general, these groove binders show strong sequence specificity that is moderated by a combination of hydrophobic and hydrogen bonding interactions (Tabernero, Bella, and Alemán, 1996). There is an example in this class of complexes in which two drugs are bound in the minor groove of an octameric fragment of DNA (Chen et al., 1994).

Although covalent adducts are thought to be key in carcinogenesis as well as showing antitumor activity, in general, it has been very difficult to obtain crystalline samples. There are few examples: an anthramycin molecule bound to a dodecamer (Kopka et al., 1994). Other carcinogenic adducts that have been crystallized are simple modifications to the O6 methyl group of G. Two structures containing this type of modification crystallized in the same approximate unit cell as the parent dodecamer but are not isomorphous (Gao et al., 1993; Vojtechovsky et al., 1995). Such a variability of crystal packing may demonstrate how a local lesion may actually affect interaction and recognition properties. A pharmaceutically important structure is that of the antitumor agent cis platinum covalently bound to tandem guanine residues in a DNA duplex. The DNA is strongly bent as a result of this interaction (Takahara et al., 1995). In addition, several of the complexes containing daunomycin are actually covalent adducts that were formed from the presence of formaldehyde in the crystallization drop (Wang et al., 1991).

## DNA Quadruplexes

The existence of tetrameric arrangement of DNA and RNA helices was first shown in fiber of poly G and poly I (Arnott, Chandrasekaran, and Marttila, 1974). The biological relevance was discovered later when it was hypothesized that these types of conformations may occur in telomeres (Rhodes and Giraldo, 1995). There are now several examples of crystal structures of quadruplex DNA. Guanine-rich DNA sequences can form multistranded structures as a consequence of the ability of guanine bases to form hydrogen-bonded arrays involving two of its faces at once. The best-characterized such arrangement is the guanine quartet formed with four guanine bases. The resulting structures have four strands that can arise from a single strand folded back in a intramolecular arrangement. Four separate strands can also associate together, as can just two. The resulting structures, termed quadruplexes, have considerable diversity that depends on both the number of separate strands involved, and on the intervening nonguanine loop sequences. A number have been characterized by NMR, with rather less by X-ray crystallography. Figure 3.12d shows an example of one such crystal structure.

## RNA Duplexes

In the last few years, there has been a pronounced increase in the number of RNA structures determined. This is due in part to the improved ability to obtain pure material and crystallize it (Wyatt, Christain, and Puglisi, 1991; Wahl et al., 1996). Although RNA is generally single stranded, double-stranded RNA can be readily formed, analogous to duplex DNA. Uracil participates in U•A base pairs that are fully isomorphous with A•T ones in duplex DNA. Duplex RNA is conformationally rather rigid, and its behavior contrasts remarkably with the polymorphism of duplex DNA in that only one major polymorph of the RNA double helix has been observed. This double helix has many features in common with A-DNA, and accordingly is known as A-RNA. The conformational features of canonical RNA helices have been obtained from fiber-diffraction studies on duplex RNA polynucleotides from both viral and synthetic origins. A-RNA is an 11-fold helix, with a narrow and deep major groove and a wide, shallow minor groove, and base pairs inclined to and displaced from the helix axis. A-RNA helices have the C3′-endo sugar pucker. Another difference from duplex DNA is that RNA helices, though capable of a small degree of bending (up to ca. 15°), does not undergo the large-scale bending seen, for example, in A-tract DNA.

A number of structures of base-paired duplex RNA have been reported, the first being the structures of r(AU) and r(GC) (Rosenberg et al., 1976; Seeman et al., 1976). Crystallographic analyses of sequences, such as the octanucleotide r(CCCCGGGG), have shown helices of length sufficient for a full set of helical parameters to be extracted. This sequence crystallizes in two distinct crystal lattices, enabling the effects of crystal-packing factors on structure to be assessed (Portmann, Usman, and Egli, 1995). In each instance, rhombohedral and hexagonal, the RNA helices are very similar, and their features closely resemble those in fiber-diffraction canonical A-RNA. The

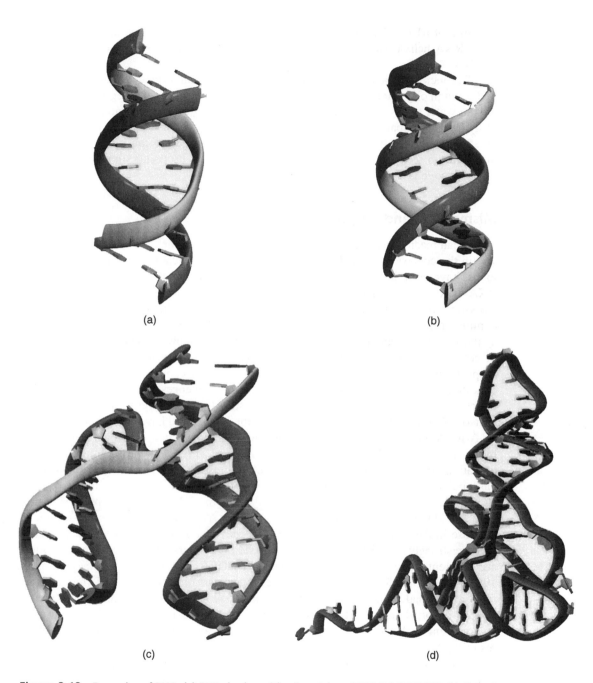

Figure 3.13. Examples of RNA. (a) RNA duplex with mismatches r(GGACUUCGGUCC) (Holbrook et al., 1991); (b) A-RNA duplex (Dock-Bregeon et al., 1989); (c) Hammerhead ribozyme (Pley, Flaherty, and McKay, 1994); (d) tRNA (Sussman et al., 1978). Figure also appears in Color Figure section.

structure of r(UUAUAUAUAUAUAA) (Figure 3.13b) was the first to show a full turn of an RNA helix (Dock-Bregeon et al., 1989). Again, the helix is essentially classical A-RNA.

The increasing availability of high-resolution crystal structures has enabled RNA hydration in oligonucleotides to be defined. Some general rules are beginning to emerge for the water arrangements beyond the obvious finding of an inherently greater degree of hydration compared to DNA oligonucleotides, on account of the 2′-OH group being an active hydrogen bond participant. There is no analogue of the minor-groove spine of hydration seen in B-DNA. Again, this finding is unsurprising since in A-RNA the minor groove is too wide for such an arrangement.

## Mismatched and Bulged RNA

RNAs can readily form stable base pair and triplet mismatches and extra-helical regions within the context of normal RNA double helices. Such features, notably extra-helical loops, have been the subject of intense structural study, since they are present in large RNAs (tRNAs, mRNAs, ribosomal RNA), and together with various types of base stacking, are responsible for maintaining their tertiary structures. It is paradoxically common for crystal structures of short sequences containing potential loop regions such as the UUCG "tetraloop" (which forms an especially stable extrahelical loop structure in solution), often not to show such features. This failure to produce extra helical loops is probably a consequence of the high ionic strength of many crystallization conditions, together with the preference of some sequences to pack in the crystal as helical arrays. So, instead of loops, these crystal structures tend to have runs of non-Watson–Crick mismatched base pairs where the loop would have formed. For example, an A-RNA double helix, albeit with G•U and U•U base pairs, is formed in the crystal by the sequence r(GCUUCGGC)d($^{Br}$U) (Cruse et al., 1994). There is evidence of some deviations from the exact canonical A-RNA duplexes formed by fully Watson–Crick base pairs, since this helix has <10 base pairs per turn. The dodecamer sequence r(GGACUUCGGUCC) (Figure 3.13a) similarly forms a base-paired duplex (Holbrook et al., 1991), with U•C and G•U base pairs. The A-RNA helix here has a significant increase in the width of its major groove, possibly on account of the water molecules that are strongly associated with the mismatched base pairs. Much greater perturbations are apparent in the structure of the dodecamer r(GGCCGAAAGGCC) (Baeyens et al., 1996), where the four non-Watson–Crick base pairs in the center of the sequence form an internal loop with sheared G•A and A•A base pairs. The resulting structure is very distorted from A-RNA ideality, with a compression of the major groove, an enlargement of the minor groove width to 13.5 Å, and a pronounced curvature of the resulting helix. This sequence forms a tetraloop in solution (Baeyens et al., 1996). The G•U base pair is a prominent and very important element of large RNA structures since it is especially stable (Varani and McClain, 2000), on account of its two hydrogen bonds. It is known as the "wobble" pair, since a G in the first position of a codon can accept either a C or a U in the third anticodon (wobble) position.

The RNA genome of the HIV-1 retrovirus contains many nonhelical features, some of which have been studied by structural methods. Its dimerization initiation site has features that act as signals for RNA packaging. The expected secondary structure (containing two loops in a "kissing-loop" arrangement) was not observed. Instead, the duplex contains two A•G base pairs, each adjacent to an extra-helical, bulged-out adenosine (Ennifar et al., 1999). This tendency of short RNA sequences to maximize helical features is also apparent in the crystal structure of a 29-nucleotide fragment from the signal recognition particle (Wild et al., 1999), which forms 28-mer heteroduplexes rather than a hairpin structure. Even so, this duplex has features of wider interest, since it has a number of non-Watson–Crick base pairs such as a 5′-GGAG/3′-GGAG purine bulge. Their overall effect is to produce backbone distortions so that the helix has non-A-RNA features such as a widening of the major groove, by ∼9 Å, and local undertwisting of base pairs adjacent to A•C and G•U mismatches.

## Transfer RNA

The crystal structure of yeast phenylalanine tRNA (Figure 3.13d) was determined in the early 1970s, simultaneously by two groups (Kim et al., 1974; Robertus et al., 1974). These tRNAs together with a few others remained the sole complex RNA molecular structures available for 20 years, until the first ribozyme structure. Both independent structures, one monoclinic (Robertus et al., 1974) and the other orthorhombic (Kim et al., 1974), are similar. They showed that the molecule is folded into an overall L-shape, with the two arms at right angles to each other. The original cloverleaf is still apparent, but with additional interactions between distant parts of the structure. The arms consist of short A-RNA helices, together with these extensive base–base interactions that hold the two arms together. The longer double helical anticodon arm has the short helix of the D stem stacked on it. The other arm is formed by the helix of the acceptor stem, on which is stacked the four base-pair T arm helix. This key feature of helix–helix stacking, has turned out to be of central importance for other complex RNAs. The nine additional base. . . base interactions that maintain the structural fold all tend to be in the elbow region, where the two arms are joined together. Almost all of these are of non-Watson–Crick type, and several are triplet interactions. Other subsequent crystal structures of tRNAs have shown that the overall L shape is invariant, as are many of the tertiary interactions.

## Ribozymes

The ability of certain RNA molecules to catalytically cleave either themselves or other RNAs is shown by five distinct categories of RNA to date, with undoubtedly more remaining to be discovered:

1. the RNA of self-splicing group I introns—that from *Tetrahymena* was the first ribozyme to be discovered (Cech, Zaug, and Grabowski, 1981). These introns contain four conserved sequence elements and form a characteristic secondary structure. The initial step of the cleavage involves nucleophilic attack by a conserved guanosine,

2. the RNA of self-splicing group II introns, which also have conserved sequence elements, but a very different secondary structure and a distinct mechanism of cleavage that involves the nucleophilic attack of a conserved adenosine, contained within the intron sequence,

3. the RNA subunit of the enzyme RNase P,

4. self-cleaving RNAs from viral and plant satellite RNAs. These are smaller ribozymes than the group I or II intron ones, and include the hammerhead ribozyme,

5. ribosomal RNA in the ribosome.

The crystal structures of hammerhead ribozymes were the first to be determined (Figure 3.13c). One is of a complex with a DNA strand containing the putative cleavage site (Pley, Flaherty, and McKay, 1994); since ribozymes do not cleave DNA, this is effectively an inhibitor complex. The structure shows three A-RNA stems connected to a central two-domain region containing the conserved residues. In the structure of the group I intron (Cate et al., 1996), the 160 residue single-stranded molecule is arranged such that the helices are aligned to maximize base stacking (Figure 3.14).

## The Ribosome

The ribosome is responsible for protein synthesis in all prokaryotic and eukaryotic cells. It consists of two subunits, each a complex of proteins and ribosomal RNA. The complete 70S prokaryotic ribosome has a total molecular weight of about 2.5 million Daltons. In prokaryotics the 30S subunit contains about 20 proteins and a single RNA molecule of around 1500 nucleotides in length. The larger 50S subunit contains half as many more proteins and a large RNA of about 3000 nucleotides, together with the small (120 nucleotide) 5S RNA. The primary function of the small subunit is to control tRNA interactions with messenger RNAs. The large subunit controls peptide transfer and undertakes the catalytic function of peptide bond formation.

Structural studies on bacterial ribosomes have been underway for almost 40 years, with the ultimate goal of achieving atomic resolution in order to understand the mechanics of ribosome function. In the last few years, several groups have successfully determined the structures of ribosomal subunits as well as the whole ribosome (Ban et al., 2000; Schluenzen et al., 2000; Wimberly et al., 2000; Yusupov et al., 2001).

The 3.0 Å crystal structure of the 30S subunit (Figure 3.15a) has located the complete 16S ribosomal RNA of 1511 nucleotides together with the ordered regions of 20 ribosomal proteins, altogether organized into four well-defined domains. The implication is that there is considerable flexibility between them, which is needed in order to ensure the movement of messenger and transfer RNAs. The overall shape of the 30S particle is dominated by the structure of the folded RNA. The secondary structure of the RNA shows over 50 helical regions. The numerous loops are mostly small and do not disrupt the runs of helix in which they are embedded. There are extensive interactions between helices, mostly involving co-axial stacking via the minor grooves. In one type of helix–helix interaction two minor grooves abut each other, with consequent distortions from A-type geometry. These distortions, which tend to involve runs of adenines, are facilitated by both extrahelical bulges and noncanonical base pairs, as have been observed in simple RNA structures. Less commonly, perpendicular packing of one helix against another (also via the minor groove) is mediated by an unpaired purine base. This mode is of especial importance since it involves the functionally significant helices in the 30S subunit. The motifs of RNA tertiary structure such as non-Watson–Crick base pairs, base triplets, and tetraloops all contribute to the overall structure.

The majority of the 20 ribosomal proteins in the structure each consist of a globular region and a long flexible arm. The latter have been too flexible to be observed in structural studies on the individual proteins, but have been located in the 30S subunit, where they play important roles in helping to stabilize the RNA folding, by essentially filling in the numerous spaces in the RNA folds.

The structure identifies the three sites where tRNA molecules bind and function, and where the essential proofreading checks for fidelity of code-reading and translation

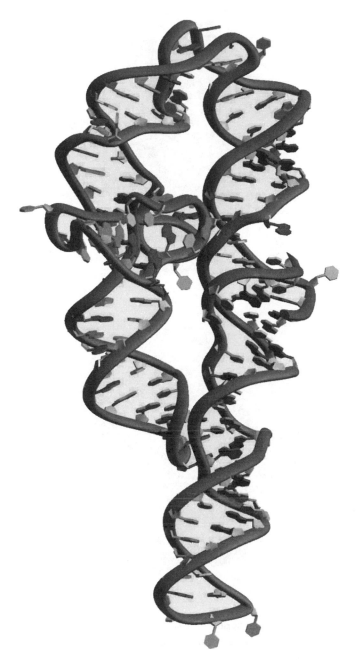

Figure 3.14. Group I intron ribozyme (Cate et al., 1996). Figure also appears in Color Figure section.

occur. These sites are: the P (peptidylation) site, when a tRNA anticodon base pairs with the appropriate codon in mRNA, and where the peptide chain is covalently linked to a tRNA; the A (acceptor) site, when peptide bonds are eventually formed (the actual peptidyl transferase steps occur in the 50S subunit); the E (exit) site, for tRNAs to be released from the subunit as part of the protein synthesis cycle.

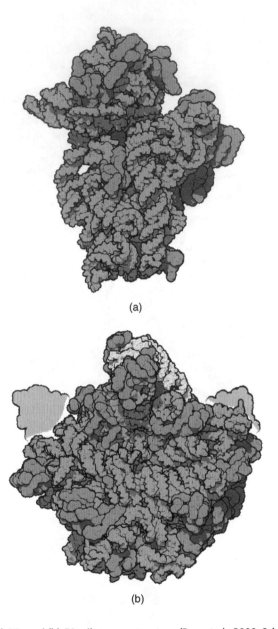

(a)

(b)

**Figure 3.15.** The (a) 30s and (b) 50s ribosome structure (Ban et al., 2000; Schluenzen et al., 2000; Wimberly et al., 2000). Image created by David Goodsell for the Protein Data Bank's Molecule of the Month series at http://www.pdb.org/. Figure also appears in Color Figure section.

tRNA itself is not present in this crystal structure, but the RNA from a symmetry-related 30S subunit effectively serves to mimic it as the anticodon stem-loop. Its interactions with the 30S RNA are also mediated (1) via minor groove surfaces, helped by some contacts with ribosomal proteins, and (2) via backbone contacts. Interestingly, the exit site of the tRNA is almost exclusively protein-associated, whereas the other functional sites are composed of RNA and not protein. It is thus the ribosomal RNA that mediates the functions of the 30S subunit, and not the ribosomal proteins.

In the 2.4 Å crystal structure of the 50S subunit (Figure 3.15b), 2711 out of the total of 2923 ribosomal RNA nucleotides have been observed, together with 27 ribosomal proteins, and 122 nucleotides of 5S RNA, with the subunit being about 250 Å in each dimension. Although the RNA itself can be arranged into six domains on the basis of its secondary structure, overall the 50S subunit is remarkably globular, reflecting its greater conformational rigidity compared to the 30S subunit, consistent with its functional need to be more flexible.

The structure of the complete 70S ribosome, although still at relatively low resolution, reveals much about the interactions between the subunits that are an essential element of the protein synthesis cycle (Yusupov et al., 2001). It is notable that even though the resolution as yet precludes any detailed study of the interactions involved, the three bound tRNA molecules are all extensively contacted by ribosomal RNA in addition to the necessary established functional interactions such as codon–anticodon recognition. The ribosome structures are rich mines of information on RNA tertiary folds. As they progress to increasingly higher resolution, we can anticipate the elucidation of some general rules governing RNA folding.

## CONCLUSION

Since the early 1980s, the number of structures of nucleic acids has grown exponentially. DNA crystallography has provided information about sequence effects, fine structures, and hydration. This information allows us to have a much better understanding of RNA crystallography, which has shown that RNA has a rich assortment of structural motifs. Continued efforts to uncover the underlying principles of nucleic acid structure will result in much greater insights into the complex functions of these molecules.

## ACKNOWLEDGMENTS

This chapter has made use of text materials from "Nucleic acid crystallography: a view from the Nucleic Acid Database" in *Prog. Biophys. Mol. Biol.* (Berman, et al., 1996) with permission from Elsevier Science and from Chapter 3 of *Nucleic Acid Structure and Recognition* by Stephen Neidle (2002). Reprinted by permission of Oxford University Press.

## FURTHER READING

Bartels H, Agmon I, Franceschi F, Yonath A (2000): Structure of functionally activated small ribosomal subunit at 3.3 Å resolution. *Cell* 102:615–623.

## REFERENCES

Allen FH, Bellard S, Brice MD, Cartright BA, Doubleday A, Higgs H, Hummelink T, Hummelink-Peters BG, Kennard O, Motherwell WDS, Rodgers JR, Watson DG (1979): The Cambridge Crystallographic Data Centre: computer-based search, retrieval, analysis and display of information. *Acta Crystallogr* B35:2331–9.

Altona C, Sundaralingam M (1972): Conformational analysis of the sugar ring in nucleosides and nucleotides: a new description using the concept of pseudorotation. *J Am Chem Soc* 94:8205–12.

Arnott S (1970): The geometry of nucleic acids. *Prog Biophys Mol Biol* 21:267–319.

Arnott S, Chandrasekaran R, Marttila CM (1974): Structures of polyinosinic acid and polyguanylic acid. *Biochem J* 141:537–8.

Arnott S, Campbell Smith PJ, Chandrasekaran R (1976): Atomic coordinates and molecular conformations for DNA–DNA, RNA–RNA, and DNA–RNA helices. In: Fasman GD, editor. *CRC Handbook of Biochemistry and Molecular Biology, Nucleic Acids*. Cleveland: CRC Press, 2:411–22.

Arnott S, Chandrasekaran R, Leslie AGW (1976): Structure of the single-stranded polyribonucleotide polycytidylic acid. *J Mol Biol* 106:735–8.

Avery OT, MacLeod CM, McCarthy M (1944): Studies on the chemical nature of the substance inducing transformation of pneumococcal types: induction of transformation by a desoxyribonucleic acid fraction isolated from Pneumococcus Type III. *J Exp Med* 79:137–58.

Babcock MS, Olson WK (1994): The effect of mathematics and coordinate system on comparability and "dependencies" of nucleic acid structure parameters. *J Mol Biol* 237:98–124.

Babcock MS, Pednault EPD, Olson WK (1993): Nucleic acid structure analysis: a users guide to a collection of new analysis programs. *J Biomol Struct Dyn* 11:597–628.

Babcock MS, Pednault EPD, Olson WK (1994): Nucleic acid structure analysis. Mathematics for local cartesian and helical structure parameters that are truly comparable between structures. *J Mol Biol* 237:125–56.

Baeyens KJ, De Bondt HL, Pardi A, Holbrook SR (1996): A curved RNA helix incorporating an internal loop with G–A and A–A non-Watson–Crick base pairing. *Proc Natl Acad Sci USA* 93:12851–5.

Ban N, Nissen P, Hansen J, Moore PB, Steitz TA (2000): The complete atomic structure of the large ribosomal subunit at a 2.4 Å resolution. *Science* 289:905–20. [This paper describes a key macro molecular complex involved in protein synthesis.]

Bansal M, Bhattacharyya D, Ravi B (1995): NUPARM and NUCGEN: software for analysis and generation of sequence dependent nucleic acid structures. *CABIOS* 11:281–7.

Berman HM (1997): Crystal studies of B-DNA: the answers and the questions. *Biopolymers* 44:23–44.

Berman HM, Young P (1981): The interaction of intercalating drugs with nucleic acids. *Ann Rev Biophys Bioeng* 10:87–114.

Berman HM, Olson WK, Beveridge DL, Westbrook J, Gelbin A, Demeny T, Hsieh SH, Srinivasan AR, Schneider B (1992): The Nucleic Acid Database—a comprehensive relational database of three-dimensional structures of nucleic acids. *Biophys J* 63:751–9.

Berman HM, Gelbin A, Westbrook J (1996): Nucleic acid crystallography: a view from the Nucleic Acid Database. *Prog Biophys Mol Biol* 66:255–88.

Bhattacharyya D, Bansal M (1989): A self-consistent formulation for analysis and generation of non-uniform DNA structures. *J Biomol Struct Dyn* 6:635–53.

Brooks BR, Bruccoleri RE, Olafson BD, States DJ, Swaminathan S, Karplus M (1983): CHARMM: a program for macromolecular energy, minimization, and dynamics calculations. *J Comput Chem* 4:187–217.

Cate JH, Gooding AR, Podell E, Zhou KH, Golden BL, Kundrot CE, Cech TR, Doudna JA (1996): Crystal structure of a group I ribozyme domain: principles of RNA packing. *Science* 273:1678–85. [One of the early structure determinations of a large ribozyme is described.]

Cech T, Zaug A, Grabowski P (1981): In vitro splicing of the ribosomal RNA precursor of Tetrahymena: involvement of a guanosine nucleotide in the excision of the intervening sequence. *Cell* 27:487–96.

Chen X, Ramakrishnan B, Rao ST, Sundaralingam M (1994): Side by side binding of two distamycin a drugs in the minor groove of an alternating B-DNA duplex. *Nat Struct Biol* 1:169–75.

Clowney L, Jain SC, Srinivasan AR, Westbrook J, Olson WK, Berman HM (1996): Geometric parameters in nucleic acids: nitrogenous bases. *J Am Chem Soc* 118:509–18.

Cruse WBT, Saludjian P, Biala E, Strazewski P, Prange T, Kennard O (1994): Structure of a mispaired RNA double helix at 1.6 Å resolution and implications for the prediction of RNA secondary structure. *Proc Natl Acad Sci USA* 91:4160–64.

Dickerson RE (1998): DNA bending: the prevalence of kinkiness and the virtues of normality. *Nucleic Acids Res* 26:1906–26.

Dickerson RE, Drew HR, Conner BN, Kopka ML, Pjura PE (1983): Helix geometry and hydration in A-DNA, B-DNA and Z-DNA. *Cold Spring Harb Symp Quant Biol* 47:13–24.

Dickerson RE, Bansal M, Calladine CR, Diekmann S, Hunter WN, Kennard O, von Kitzing E, Lavery R, Nelson HCM, Olson W, Saenger W, Shakked Z, Sklenar H, Soumpasis DM, Tung C-S, Wang AH-J, Zhurkin VB (1989): Definitions and nomenclature of nucleic acid structure parameters. *EMBO J* 8:1–4.

Dickerson RE, Goodsell D, Kopka ML (1996): MPD and DNA bending in crystals and in solution. *J Mol Biol* 256:108–25.

Dock-Bregeon AC, Chevrier B, Podjarny A, Johnson J, de Bear JS, Gough GR, Gilham PT, Moras D (1989): Crystallographic structure of an RNA helix: [U(UA)6A]2. *J Mol Biol* 209:459–74.

Drew HR, Wing RM, Takano T, Broka C, Tanaka S, Itakura K, Dickerson RE (1981): Structure of a B-DNA dodecamer: conformation and dynamics. *Proc Natl Acad Sci USA* 78:2179–83. [The first structure determination of a full turn of B-DNA is presented.]

Eisenstein M, Shakked Z (1995): Hydration patterns and intermolecular interactions in A-DNA crystal structures. Implications for DNA recognition. *J Mol Biol* 248:662–78.

El Hassan MA, Calladine CR (1995): The assessment of the geometry of dinucleotide steps in double-helical DNA: a new local calculation scheme with an appendix. *J Mol Biol* 251:648–64.

Ennifar E, Yusupov M, Walter P, Marquet R, Ehresmann B, Ehresmann C, Dumas P (1999): The crystal structure of the dimerization initiation site of genomic HIV-1 RNA reveals an extended duplex with two adenine bulges. *Structure* 7:1439–49.

Frederick CA, Williams LD, Ughetto G, van der Marel GA, van Boom JH, Rich A, Wang AH-J (1990): Structural comparison of anticancer drug-DNA complexes: adriamycin and daunomycin. *Biochemistry* 29:2538–49.

Gao Y-G, Sriram M, Denny WA, Wang AH-J (1993): Minor groove binding of SN6999 to an alkylated DNA: molecular structure of d(CGC[e⁶G]AATTCGCG)-SN6999 complex. *Biochemistry* 32:9639–48.

Gelbin A, Schneider B, Clowney L, Hsieh S-H, Olson WK, Berman HM (1996): Geometric parameters in nucleic acids: sugar and phosphate constituents. *J Am Chem Soc* 118:519–28.

Gessner RV, Frederick CA, Quigley GJ, Rich A, Wang AH-J (1989): The molecular structure of the left-handed Z-DNA double helix at 1.0-Å atomic resolution. *J Biol Chem* 264:7921–35.

Gessner RV, Quigley GJ, Egli M (1994): Comparative studies of high resolution Z-DNA crystal structures 1. Common hydration patterns of alternating DC–DG. *J Mol Biol* 236:1154–68.

Goodsell DS, Kopka ML, Dickerson RE (1995): Refinement of netropsin bound to DNA: bias and feedback in electron density map interpretation. *Biochemistry* 34:4983–93.

Gorin AA, Zhurkin VB, Olson WK (1995): B-DNA twisting correlates with base-pair morphology. *J Mol Biol* 247:34–48.

Guzikevich-Guerstein G, Shakked Z (1996): A novel form of the DNA double helix imposed on the TATA-box by the TATA-binding protein. *Nat Struct Biol* 3:32–7.

Harvey S, Prabhakaran M (1986): Ribose puckering: structure, dynamics, energetics, and the pseudorotation cycle. *J Am Chem Soc* 108:6128–36.

Heinemann U (1991): A note on crystal packing and global helix structure in short A-DNA duplexes. *J Biomol Struct Dyn* 8:801–11.

Heinemann U, Alings C, Hahn M (1994): Crystallographic studies of DNA helix structure. *Biophys Chem* 50:157–67.

Hershey AD, Chase M (1951): Genetic recombination and heterozygosis in bacteriophage. *Cold Spring Harb Symp Quant Biol* 16:471–9.

Holbrook SR, Cheong C, Tinoco I Jr, Kim S-H (1991): Crystal structure of an RNA double helix incorporating a track of non-Watson–Crick base pairs. *Nature* 353:579–81.

Jain SC, Zon G, Sundaralingam M (1989): Base only binding of spermine in the deep groove of the A-DNA octamer d(GTGTACAC). *Biochemistry* 28:2360–4.

Jain S, Zon G, Sundaralingam M (1991): The hexagonal crystal structure of the A-DNA octamer d(GTGTACAC) and its comparison with the tetragonal structure-correlated variations in helical parameters. *Biochemistry* 30:3567–76.

Kagawa TF, Stoddard D, Zhou G, Ho PS (1989): Quantitative analysis of DNA secondary structure from solvent accessible surfaces: the B- to Z-DNA transition as a model. *Biochemistry* 28:6642–51.

Kamitori S, Takusagawa F (1994): Multiple binding modes of anticancer drug actinomycin D: x-ray, molecular modeling, and spectroscopic studies of d(GAAGCTTC)(2)-actinomycine D complexes and its host DNA. *J Am Chem Soc* 116:4154–65.

Khorana HG, Tener GM, Moffatt JG, Pol EH (1956): A new approach to the synthesis of polynucleotides. *Chem Ind* 1523.

Kim JL, Nikolov DB, Burley SK (1993): Co-crystal structure of TBP recognizing the minor groove of a TATA element. *Nature* 365:520–7.

Kim S-H, Berman HM, Newton MD, Seeman NC (1973): Seven basic conformations of nucleic acid structural units. *Acta Crystallogr* B29:703–10.

Kim S-H, Suddath FL, Quigley GJ, McPherson A, Sussman JL, Wang AH-J, Seeman NC, Rich A (1974): Three-dimensional tertiary structure of yeast phenylalanine transfer RNA. *Science* 185:435–40.

Kim Y, Geiger JH, Hahn S, Sigler PB (1993): Crystal structure of a yeast TBP/TATA-box complex. *Nature* 365:512–20.

Kopka ML, Goodsell DS, Baikalov I, Grzeskowiak K, Cascio D, Dickerson RE (1994): Crystal structure of a covalent DNA-drug adduct: anthramycin bound to C-C-A-A-C-G-T-T-G-G and a molecular explanation of specificity. *Biochemistry* 33:13593–610.

Kopka ML, Yoon C, Goodsell D, Pjura P, Dickerson RE (1985): The molecular origin of DNA-drug specificity in netropsin and distamycin. *Proc Natl Acad Sci USA* 82:1376–80.

Kosikov KM, Gorin AA, Zhurkin VB, Olson WK (1999): DNA stretching and compression: large-scale simulations of double helical structures. *J Mol Biol* 289:1301–26.

Lavery R, Sklenar H (1988): The definition of generalized helicoidal parameters and of axis curvature for irregular nucleic acids. *J Biomol Struct Dyn* 6:63–91.

Lavery R, Sklenar H (1989): Defining the structure of irregular nucleic acids: conventions and principles. *J Biomol Struct Dyn* 6:655–67.

Lipanov A, Kopka ML, Kaczor-Grzeskowiak M, Quintana J, Dickerson RE (1993): Structure of the B-DNA decamer C-C-A-A-C-I-T-T-G-G in two different space groups: conformational flexibility of B-DNA. *Biochemistry* 32:1373–80.

Lu X-J, El Hassan MA, Hunter CA (1997): Structure and conformation of helical nucleic acids: analysis program (SCHNAaP). *J Mol Biol* 273:668–80.

Lu X-J, Babcock MS, Olson WK (1999): Mathematical overview of nucleic acid analysis programs. *J Biomol Struct Dynam* 16:833–43.

Lu X-J, Olson WK (1999): Resolving the discrepancies among nucleic acid conformation analyses. *J Mol Biol* 285:1563–75.

Murray-Rust P, Motherwell S (1978): Computer retrieval and analysis of molecular geometry. III. Geometry of the beta-1′-aminofuranoside fragment. *Acta Crystallogr* B34:2534–46.

Neidle S (2002): *Nucleic Acid Structure and Recognition*. Oxford: Oxford University Press.

Olson WK (1982a): How flexible is the furanose ring? II. An updated potential energy estimate. *J Am Chem Soc* 104:278–84.

Olson WK (1982b): Theoretical studies of nucleic acid conformation: potential energies, chain statistics, and model building. In: Neidle S, editor. *Topics in Nucleic Acid Structures, Part 2*. London: Macmillan Press, 1–79.

Olson WK, Sussman JL (1982): How flexible is the furanose ring? I. A comparison of experimental and theoretical studies. *J Am Chem Soc* 104:270–8.

Olson WK, Bansal M, Burley SK, Dickerson RE, Gerstein M, Harvey SC, Heinemann U, Lu X-J, Neidle S, Shakked Z, Sklenar H, Suzuki M, Tung C-S, Westhof E, Wolberger C, Berman HM (2001): A standard reference frame for the description of nucleic acid base-pair geometry. *J Mol Biol* 313:229–37.

Packer M, Hunter C (1998): Sequence-dependent DNA structure: the role of the sugar-phosphate backbone. *J Mol Biol* 280:407–20.

Parkinson G, Vojtechovsky J, Clowney L, Brünger AT, Berman HM (1996): New parameters for the refinement of nucleic acid containing structures. *Acta Crystallogr* D52:57–64.

Pearlman D, Kim S-H (1990): Atomic charges for DNA constituents derived from single-crystal X-ray diffraction data. *J Mol Biol* 211:171–87.

Phillips K, Dauter Z, Murchie AIH, Lilley DMJ, Luisi B (1997): The crystal structure of a parallel-stranded guanine tetraplex at 0.95 Å resolution. *J Mol Biol* 273:171–182.

Pley HW, Flaherty KM, McKay DB (1994): Three-dimensional structure of a hammerhead ribozyme. *Nature* 372:68–74.

Portmann S, Usman N, Egli M (1995): The crystal structure of r(CCCCGGGG) in two distinct lattices. *Biochemistry* 34:7569–75.

Quintana JR, Grzeskowiak K, Yanagi K, Dickerson RE (1992): The structure of a B-DNA decamer with a central T-A step: C-G-A-T-T-A-A-T-C-G. *J Mol Biol* 225:379–95.

Ramakrishnan B, Sundaralingam M (1993): Crystal packing effects on A-DNA helix parameters—a comparative study of the isoforms of the tetragonal and hexagonal family of octamers with differing base sequences. *J Biomol Struct Dyn* 11:11–26.

Rhodes D, Giraldo R (1995): Telomere structure and function. *Curr Opin Struct Biol* 5:311–22.

Rich A, Davies DR, Crick FHC, Watson JD (1961): The molecular structure of polyadenlyic acid. *J Mol Biol* 3:71–86.

Robertus JD, Ladner JE, Finch JT, Rhodes D, Brown RS, Clark BFC, Klug A (1974): Structure of yeast phenylalanine tRNA at 3 Å resolution. *Nature* 250:546–51. [One of the first structural determinations of tRNA is described.]

Rodley GA, Scobie RS, Bates RHT, Lewitt RM (1976): A possible conformation for double-stranded polynucleotides. *Proc Nat Acad Sci USA* 73:2959–63.

Rosenberg JM, Seeman NC, Day RO, Rich A (1976): RNA double helical fragments at atomic resolution: II. The structure of sodium guanylyl-3',5'-cytidine nonhydrate. *J Mol Biol* 104:145–67.

Rosenberg JM, Seeman NC, Kim JJP, Suddath FL, Nicholas HB, Rich A (1973): Double helix at atomic resolution. *Nature* 243:150–54.

Saenger W, Hunter WN, Kennard O (1986): DNA conformation is determined by economics in the hydration of phosphate groups. *Nature* 324:385–8.

Schluenzen F, Tocilj A, Zarivach R, Harms J, Gluehmann M, Janell D, Bashan A, Bartels H, Agmon I, Franceschi F, Yonath A (2000): Structure of functionally activated small ribosomal subunit at 3.3 Å resolution. *Cell* 102:615–23.

Schneider B, Ginell SL, Jones R, Gaffney B, Berman HM (1992): Crystal and molecular structure of a DNA fragment containing a 2-aminoadenine modification: the relationship between conformation, packing, and hydration in Z-DNA hexamers. *Biochemistry* 31:9622–8.

Schneider B, Berman HM (1995): Hydration of the DNA bases is local. *Biophys J* 69:2661–9.

Schneider B, Neidle S, Berman HM (1997): Conformations of the sugar-phosphate backbone in helical DNA crystal structures. *Biopolymers* 42:113–24.

Schroth GP, Kagawa TF, Ho PS (1993): Structure and thermodynamics of nonalternating C–G base pairs in Z-DNA: the 1.3 Å crystal structure of the asymmetric hexanucleotide d(m5CGGGm5CG).d(m5CGCCm5CG). *Biochemistry* 32:13381–92.

Seeman NC, Sussman JL, Berman HM, Kim S-H (1971): Nucleic acid conformation: crystal structure of a naturally occurring dinucleoside phosphate (UpA). *Nature New Biol* 233:90–2.

Seeman NC, Rosenberg JM, Suddath FL, Kim JJP, Rich A (1976): RNA double helical fragments at atomic resolution: I. The crystal and molecular structure of sodium adenylyl-3'-5'-uridine hexahydrate. *J Mol Biol* 104:109–44. [The first determination of the structure of the double helix is presented.]

Shakked Z, Rabinovich D, Kennard O, Cruse WBT, Salisbury SA, Viswamitra MA (1983): Sequence-dependent conformation of an A-DNA double helix. The crystal structure of the octamer d(G-G-T-A-T-A-C-C). *J Mol Biol* 166:183–201.

Shakked Z, Guerstein-Guzikevich G, Eisenstein M, Frolow F, Rabinovich D (1989): The conformation of the DNA double helix in the crystal Is dependent on its environment. *Nature* 342:456–60.

Soumpasis DM, Tung CS (1988): A rigorous basepair oriented description of DNA structures. *J Biomol Struct Dyn* 6:397–420.

Sundaralingam M (1969): Stereochemistry of nucleic acids and their constituents. IV. Allowed and preferred conformations of nucleosides, nucleoside mono-, di-, tri-, tetraphosphates, nucleic acids and polynucleotides. *Biopolymers* 7:821–60.

Sussman JL, Holbrook SR, Warrant RW, Church GM, Kim S-H (1978): Crystal structure of yeast phenylalanine transfer RNA. I. Crystallographic refinement. *J Mol Biol* 123:607–30.

Tabernero L, Bella J, Alemán C (1996): Hydrogen bond geometry in DNA-minor groove binding drug complexes. *Nucleic Acids Res* 24:3458–66.

Takahara PM, Rosenzweig AC, Frederick CA, Lippard SJ (1995): Crystal structure of double-stranded DNA containing the major adduct of the anticancer drug cisplatin. *Nature* 377:649–52.

Thota N, Li XH, Bingman C, Sundaralingam M (1993): High resolution refinement of the hexagonal A-DNA octamer d(GTGTACAC) at 1.4 Å resolution. *Acta Crystallogr* D49:282–91.

Tsai C-C, Jain SC, Sobell HM (1975): X-ray crystallographic visualizations of drug-nucleic acid intercalative binding: structure of an ethidium-Dinucleoside monophosphate crystalline complex, ethidium: 5-iodouridylyl(3'-5') adenosine. *Proc Natl Acad Sci USA* 72:628–32.

Tung C-S, Soumpasis DM, Hummer G (1994): An extension of the rigorous base-unit oriented description of nucleic acid structures. *J Biomol Struct Dynam* 11:1327–44.

Varani G, McClain W (2000): The G×U wobble base pair. A fundamental building block of RNA structure crucial to RNA function in diverse biological systems. *EMBO Rep* 1:18–23.

Vojtechovsky J, Eaton M, Gaffney B, Jones R, Berman H (1995): Structure of a new crystal form of a DNA dodecamer containing T•($O^6$Me)G base pairs. *Biochemistry* 34:16632–40.

Wahl MC, Ramakrishnan B, Ban CG, Chen X, Sundaralingam M (1996): RNA—synthesis, purification, and crystallization. *Acta Crystallogr* D52:668–75.

Wang AH-J, Quigley GJ, Kolpak FJ, Crawford JL, van Boom JH, van der Marel GA, Rich A (1979): Molecular structure of a left-handed double helical DNA fragment at atomic resolution. *Nature* 282:680–6.

Wang AH-J, Hakoshima T, van der Marel GA, van Boom JH, Rich A (1984): AT base pairs are less stable than GC base pairs in Z-DNA: the crystal structure of d($m^5$CGTAm$^5$CG). *Cell* 37:321–31.

Wang AH-J, Gessner RV, van der Marel GA, van Boom JH, Rich A (1985): Crystal structure of Z-DNA without an alternating purine–pyrimidine sequence. *Proc Natl Acad Sci USA* 82:3611–5.

Wang AH-J, Ughetto G, Quigley GJ, Rich A (1987): Interactions between an anthracycline antibiotic and DNA: molecular structure of daunomycin complexed to d(CpGpTpApCpG) at 1.2-Å resolution. *Biochemistry* 26:1152–63.

Wang AH-J, Gao YG, Liaw Y-C, Li Y-K (1991): Formaldehyde cross-links daunorubicin and DNA efficiently: HPLC and x-ray diffraction studies. *Biochemistry* 30:3812–15.

Watson JD, Crick FHC (1953): A structure for deoxyribose nucleic acid. *Nature* 171:737–8.

Weiner P, Kollman P (1981): Amber. *J Comput Chem* 2:287–303.

Wild K, Weichenrieder O, Leonard G, Cusack S (1999): The 2 A structure of helix 6 of the human signal recognition particle RNA. *Structure* 7:1345–52.

Wimberly BT, Brodersen DE, Clemons WM Jr, Morgan-Warren R, Carter AP, Vonrhein C, Hartsch T, Ramakrishnan V (2000): Structure of the 30S ribosomal subunit. *Nature* 407:327–39. [This paper describes the structure of the 30S subunit of the ribosome.]

Wing R, Drew HR, Takano T, Broka C, Tanaka S, Itakura K, Dickerson RE (1980): Crystal structure analysis of a complete turn of B-DNA. *Nature* 287:755–8.

Wyatt J, Christain M, Puglisi J (1991): Synthesis and purification of large amounts of RNA oligonucleotides. *BioTechniques* 11:764–9.

Yanagi K, Prive GG, Dickerson RE (1991): Analysis of local helix geometry in three B-DNA decamers and eight dodecamers. *J Mol Biol* 217:201–14.

Young MA, Ravishanker G, Beveridge DL, Berman HM (1995): Analysis of local helix bending in crystal structures of DNA oligonucleotides and DNA-protein complexes. *Biophys J* 68:2454–68.

Yusupov MM, Yusupova GZ, Baucom A, Lieberman K, Earnest TN, Cate JHD, Noller HF (2001): Crystal structure of the ribosome at 5.5 Å resolution. *Science* 282:883–96.

Zamenhof S, Brawermann G, Chargaff E (1952): On the desoxypentose nucleic acids from several microorganisms. *Biochim Biophys Acta* 9:402–5.

# 4

# COMPUTATIONAL ASPECTS OF HIGH-THROUGHPUT CRYSTALLOGRAPHIC MACROMOLECULAR STRUCTURE DETERMINATION

Paul D. Adams, Ralf W. Grosse-Kunstleve, and Axel T. Brunger

The desire to understand biological processes at a molecular level has led to the routine application of X-ray crystallography. However, significant time and effort usually are required to solve and complete a macromolecular crystal structure. Much of this effort is in the form of manual interpretation of complex numerical data using a diverse array of software packages, and the repeated use of interactive three-dimensional graphics. The need for extensive manual intervention leads to two major problems: significant bottlenecks that impede rapid structure solution (Burley et al., 1999), and the introduction of errors due to subjective interpretation of the data (Mowbray et al., 1999). These problems present a major impediment to the success of structural genomics efforts (Burley et al., 1999; Montelione and Anderson, 1999) that require the whole process of structure solution to be as streamlined as possible. See Chapter 29 for a detailed description of structural genomics. The automation of structure solution is thus necessary as it has the opportunity to produce minimally biased models in a short time. Recent technical advances are fundamental to achieving this automation and make high-throughput structure determination an obtainable goal.

*Structural Bioinformatics*
Edited by Philip E. Bourne and Helge Weissig
Copyright © 2003 by Wiley-Liss, Inc.

## HIGH-THROUGHPUT STRUCTURE DETERMINATION

Automation in macromolecular X-ray crystallography has been a goal for many researchers. The field of small-molecule crystallography, where atomic resolution data are routinely collected, is already highly automated. As a result, the current growth rate of the Cambridge Structural Database (CCSD) (Allen, Kennard, and Taylor, 1983) is more than 15,000 new structures per year. This growth rate is approximately 10 times the growth rate of the Protein Data Bank (PDB) (Berman et al., 2000). See Chapters 9, 10, and 11 for further details of structural databases. Automation of macromolecular crystallography could significantly improve the rate at which new structures are determined. Recently, the goal of automation has moved to a position of prime importance with the development of the concept of structural genomics (Burley et al., 1999; Montelione and Anderson, 1999). In order to exploit the information present in the rapidly expanding sequence databases, the structural database must also grow. Increased knowledge about the relationship between sequence, structure, and function will allow sequence information to be used to its full extent. For structural genomics to be successful, macromolecular structures will need to be solved at a rate significantly faster than at present. This high-throughput structure determination will require automation to reduce the bottlenecks related to human intervention. Automation will rely on: the development of algorithms that minimize or eliminate subjective input; the development of algorithms that automate procedures that were traditionally performed by hand; and, finally the development of software packages that allow a tight integration between these algorithms. Truly automated structure solution will require the computer to make decisions about how best to proceed in the light of the available data.

The automation of macromolecular structure solution applies to all of the procedures involved. There have been many technological advances that make macromolecular X-ray crystallography easier. In particular, cryoprotection to extend crystal life (Garman, 1999), the availability of tunable synchrotron sources (Walsh et al., 1999a), high-speed charge-coupled device (CCD) data collection devices (Walsh et al., 1999b), and the ability to incorporate anomalously scattering selenium atoms into proteins have all made structure solution much more efficient (Walsh et al., 1999b). The desire to make structure solution more efficient has led to investigations into the optimal data collection strategies for multiwavelength anomalous diffraction (Gonzalez et al., 1999) and phasing using single anomalous diffraction with sulfur or ions (Dauter et al., 1999; Dauter and Dauter, 1999). Gonzalez and her colleagues have shown that multiwavelength anomalous diffraction (MAD) phasing using only two wavelengths can be successful (Gonzalez et al., 1999). The optimum wavelengths for such an experiment are those that give a large contrast in the real part of the anomalous scattering factor (e.g., the inflection point and high-energy remote). However, Rice and his colleagues have also shown that, in general, a single wavelength collected at the anomalous peak is sufficient to solve a macromolecular structure (Rice, Earnest, and Brunger, 2000). Such an approach minimizes the amount of data to be collected and increases the efficiency of synchrotron beamlines, and is therefore likely to become an important and widely used technique in the future.

## DATA ANALYSIS

The first step of structure solution, once the raw images have been processed, is assessment of data quality. The intrinsic quality of the data must be quantified and the appropriate signal extracted. Observations that are in error must be rejected as outliers. Some observations will be rejected at the data-processing stage, where multiple observations are available. However, if redundancy is low, then probabilistic methods can be used (Read, 1999). The prior expectation, given either by a Wilson distribution of intensities or model-based structure-factor probability distributions, is used to detect outliers. This method is able to reject strong observations that are in error, which tend to dominate the features of electron-density and Patterson maps. This method could also be extended to the rejection of outliers during the model refinement process.

When using isomorphous substitution or anomalous diffraction methods for experimental phasing the relevant information lies in the differences between the multiple observations. In the case of anomalous diffraction, these differences are often very small, being of the same order as the noise in the data. In general the anomalous differences at the peak wavelength are sufficient to locate the heavy atoms, provided that a large enough anomalous signal is observed (Grosse-Kunstleve and Brunger, 1999). However, in less routine cases it can be very important to extract the maximum information from the data. One approach used in MAD phasing is to analyze the data sets to calculate $F_A$ structure factors, which correspond to the anomalously scattering substructure (Terwilliger, 1994). Several programs are available to estimate the $F_A$ structure factors: XPREP (Bruker, 2001), MADSYS (Hendrickson, 1991) and SOLVE (Terwilliger and Berendzen, 1999a). In another approach, a specialized procedure for the normalization of structure factor differences arising from either isomorphous or anomalous differences has been developed in order to facilitate the use of direct methods for heavy atom location (Blessing and Smith, 1999).

Merohedral twinning of the diffraction data can make structure solution difficult and in some cases impossible. The twinning occurs when a crystal contains multiple diffracting domains that are related by a simple transformation such as a twofold rotation about a crystallographic axis, a phenomenon that can only occur in certain space groups. As a result the observed diffraction intensities are the sum of the intensities from the two distinctly oriented domains. Fortunately, the presence of twinning can be detected at an early stage by the statistical analysis of structure factor distributions (Yeates, 1997). If the twinning is only partial, it is possible to detwin the data. Perfect twinning typically makes structure solution using experimental phasing methods difficult, but the molecular replacement method (see below) still can be successfully used.

## HEAVY ATOM LOCATION AND COMPUTATION OF EXPERIMENTAL PHASES

The location of heavy atoms in isomorphous replacement or the location of anomalous scatterers was traditionally performed by manual inspection of Patterson maps. However, in recent years labeling techniques such as seleno–methionyl incorporation have become widely used. Such labeling techniques lead to an increase in the number of atoms to be located, rendering manual interpretation of Patterson maps extremely difficult. As a result, automated heavy atom location methods have proliferated. The programs SOLVE (Terwilliger and Berendzen, 1999a) and CNS (Brunger et al., 1998; Grosse-Kunstleve and Brunger, 1999) use Patterson-based techniques to find a starting heavy atom configuration that is then completed using difference Fourier analyses. Both Shake-and-Bake (SnB) (Weeks and Miller, 1999) and SHELX-D (Sheldrick and Gould, 1995) use the direct methods reciprocal-space phase refinement combined with modifications in real-space. SnB refines phases derived from randomly positioned atoms, while SHELX-D derives starting phases by automatic inspection of the Patterson map. All methods have been used with great success to solve substructures with more than 60 selenium sites. SHELX-D and SnB have been used to find up to 150 and 160 selenium sites, respectively.

After the heavy atom or anomalously scattering substructure has been located, experimental phases can be calculated and the parameters of the substructure refined. A number of modern maximum-likelihood based methods for heavy atom refinement and phasing are readily available: MLPHARE (Otwinowski, 1991), CNS (Brunger et al., 1998), SHARP (La Fortelle and Bricogne, 1997), SOLVE (Terwilliger and Berendzen, 1999a). The SOLVE program has the advantage of fully integrating and automating heavy atom location, refinement, and phasing, and therefore is very easy to use. The SHARP program implements a more complex algorithm for phasing, making use of two-dimensional integration over both phases and amplitudes. This method is computationally expensive, rendering SHARP typically an order of magnitude slower than other phasing programs, but in the case of significant nonisomorphism between heavy atom derivative data sets the improvement in the phases can be worth the additional computing time.

## DENSITY MODIFICATION

Often the raw phases obtained from the experiment are not of sufficient quality to proceed with structure determination. However, there are many real space constraints, such as solvent flatness, that can be applied to electron density maps in an iterative fashion to improve initial phase estimates. This process of density modification is now routinely used to improve experimental phases prior to map interpretation and model building. However, due to the cyclic nature of the density modification process, where the original phases are combined with new phase estimates, introduction of bias is a serious problem. The $\gamma$ correction was developed to reduce the bias inherent in the process, and has been applied successfully in the method of solvent-flipping (Abrahams, 1997). The $\gamma$ correction has been generalized to the $\gamma$ perturbation method in the DM program, part of the CCP4 suite (Collaborative Computational Project 4, 1994), and can be applied to any arbitrary density modification procedure, including noncrystallographic symmetry averaging and histogram matching (Cowtan, 1999). After bias removal, histogram matching is significantly more powerful than solvent flattening for comparable volumes of protein and solvent (Cowtan, 1999). More recently a reciprocal-space, maximum-likelihood formulation of the density modification process has been devised and implemented in the program RESOLVE (Terwilliger, 2000). This method has the advantage that a likelihood function can be directly optimized with respect to the available parameters (phases and amplitudes), rather than indirectly through a weighted combination of starting parameters with those derived from flattened maps. In this way the problem of choices of weights for phase combination is avoided. The SOLVE and RESOLVE programs together provide a relatively automated way to go from experimental data to a map suitable for model building.

## MOLECULAR REPLACEMENT

The method of molecular replacement is commonly used to solve structures for which a homologous structure is already known. As the database of known structures expands as a result of structural genomics efforts, this technique will become more and more important. The method attempts to locate a molecule or fragments of a molecule, whose structure is known, in the unit cell of an unknown structure for which experimental data are available. In order to make the problem tractable, it has traditionally been broken down into two consecutive three-dimensional search problems: a search to determine the rotational orientation of the model followed by a search to determine the translational orientation for the rotated model (Rossmann and Blow, 1962). The method of Patterson Correlation (PC) refinement is often used to optimize the rotational orientation prior to the translation search, thus increasing the likelihood of finding the correct solution (Brunger, 1997). With currently available programs structure solution by molecular replacement usually involves significant manual input. Recently, however, methods have been developed to automate molecular replacement. One approach has used the exhaustive application of traditional rotation and translation methods to perform a complete six-dimensional search (Sheriff, Klei, and Davis, 1999). More recently, less time-consuming methods have been developed. The EPMR program implements an evolutionary algorithm to perform a very efficient six-dimensional search (Kissinger, Gehlhaar, and Fogel, 1999). A Monte-Carlo simulated annealing scheme is used in the program Queen of Spades to locate the positions of molecules in the asymmetric unit (Glykos and Kokkinidis, 2000).

To improve the sensitivity of any molecular replacement search algorithm, maximum likelihood methods have been developed (Read, 2001). The traditional scoring function of the search is replaced by a function that takes into account the errors in the model and the uncertainties at each stage. This approach is seen to greatly improve the chances of finding a correct solution using the traditional approach of rotation and translation searches. In addition, the method performs a statistically correct treatment of simultaneous information from multiple search models using multivariate statistical analysis (Read, 2001). This method will allow information from different structures to be used in highly automated procedures while minimizing the risk of introducing bias. In the future molecular replacement algorithms may permit experimental data to be exhaustively tested against all known structures to determine whether a homologous structure is already present in a database, which could then be used as an aid in structure determination.

## MAP INTERPRETATION

The interpretation of the initial electron density map, calculated using either experimental phasing or molecular replacement methods, is often performed in multiple stages (described below) with the final goal being the construction of an atomic model. If the interpretation cannot proceed to an atomic model, that is often an indication that the data collection must be repeated with improved crystals. Alternatively, repeating previous computational steps in data analysis or phasing may generate revised hypotheses about the crystal, such as a different space group symmetry or estimate of unit cell contents. Clearly, completely automating the process of structure solution will require that these eventualities are taken into consideration and dealt with in a rigorous manner.

The first stage of electron density map interpretation is an overall assessment of the information contained in a given map. The standard deviation of the local root-mean-square electron density can be calculated from the map. This variation is high when the electron-density map has well-defined protein and solvent regions and is low for maps calculated with random phases (Terwilliger and Berendzen, 1999b; Terwilliger, 1999). Terwilliger and Berendzen also have shown that the correlation of the local root-mean-square density in adjacent regions in the unit cell can be used as a measure of the presence of distinct, contiguous solvent and macromolecular regions in an electron density map (Terwilliger and Berendzen, 1999c).

Currently the process of analyzing an experimental electron density map to build the atomic model is a time-consuming, subjective process and almost entirely graphics based. Sophisticated programs such as O (Jones et al., 1991), XtalView (McRee, 1999), QUANTA (Oldfield, 2000), TurboFrodo (Jones, 1978), and MAIN (Turk, 2000) are commonly used for manual rebuilding. These greatly reduce the effort required to rebuild models by providing: libraries of side chain rotamers and peptide fragments (Kleywegt and Jones, 1998) and map interpretation tools and real space refinement of rebuilt fragments (Jones et al., 1991). However, Mowbray and her colleagues have shown that there are substantial differences in the models built manually by different people when presented with the same experimental data (Mowbray et al., 1999). The majority of time spent in completing a crystal structure is in the use of interactive graphics to manually modify the model. This manual modification is required either to correct parts of the model that are incorrectly placed or to add parts of the model that are currently missing. This process is prone to human error because of the large number of degrees of freedom of the model and the possible poor quality of regions of the electron density map.

Although interactive graphics systems for manual model building have made the process dramatically simpler, there have also been significant advances in making the process of map interpretation and model building truly automated. One route to automated analysis of the electron density map is the recognition of larger structural elements, such as $\alpha$-helices and $\beta$-strands. Location of these features can often be achieved even in electron density maps of low quality using exhaustive searches in either real space (Kleywegt and Jones, 1997) or reciprocal space (Cowtan, 1998; Cowtan, 2001), the latter having a significant advantage in speed because the translation search for each orientation can be calculated using a Fast Fourier Transform. The automatic location of secondary structure elements from skeletonized electron density maps can be combined with sequence information and databases of known structures to build an initial atomic model with little or no manual intervention from the user (Oldfield, 2000). This method has been seen to work even at relatively low resolution ($d_{min} \sim 3.0$Å). However, the implementation is still graphics based and requires user input. A related approach in the program MAID also uses a skeleton generated from the electron density map as the start point for locating secondary structure elements (Levitt, 2001). Trial points are extended in space by searching for connected electron density at $C_\alpha$ distance (approximately 3.7Å) with standard $\alpha$-helical or $\beta$-strand geometry. Real-space refinement of the fragments generated is used to improve the model. Both of these methods suffer from the limitation that they do not combine the model-building process with the generation of improved electron density maps derived from the starting phases and the partial models.

In order to completely automate the model-building process, a method has been developed that combines automated identification of potential atomic sites in the map (Perrakis et al., 1997) with model refinement (Murshudov, Vagin, and Dodson, 1997). An iterative procedure is used that describes the electron density map as a set of unconnected atoms from which proteinlike patterns, primarily the main-chain trace from peptide units, are extracted. From this information and knowledge of the protein sequence, a model can be automatically constructed (Perrakis, Morris, and Lamzin, 1999). This powerful procedure, known as warpNtrace in ARP/wARP, can gradually build a more complete model from the initial electron density map and in many cases is capable of building the majority of the protein structure in a completely automated way. Unfortunately, this method currently has the limitation of a need for relatively high-resolution data ($d_{min} < 2.0$Å). Data that extend to this resolution are available for less than 50% of the $\sim$16,500 X-ray structures in the PDB. To extend the applicability of automated map interpretation to lower resolution data, work has started using pattern recognition methods (Holton et al., 2000). The resulting program is called TEXTAL and shows great promise for the interpretation of maps, even at a data resolution as low as 3.0Å. Data of this quality are available for approximately 95% of the structures in the PDB. We anticipate that the combination of secondary structure fragment location, the pattern matching methods of the TEXTAL program, and iteration with structure refinement for map improvement will in the future provide a general solution to the problem of model building at resolutions better than 3.5Å.

## REFINEMENT

In general the atomic model obtained by automatic or manual methods contains some errors and must be optimized to best fit the experimental data and prior chemical information. In addition, the initial model is often incomplete and refinement is carried out to generate improved phases that can then be used to compute a more accurate electron density map. However, the refinement of macromolecular structures is often difficult for several reasons. First, the data-to-parameter ratio is low, creating the danger of overfitting the diffraction data. This method results in a good agreement of the model to the experimental data even when it contains significant errors. Therefore, the apparent ratio of data to parameters is often increased by incorporation of chemical information, that is, bond length and bond angle restraints obtained from ideal values seen in high-resolution structures (Hendrickson, 1985). Second, the initial model often has significant errors, often due to the limited quality of the experimental data, or a low level of homology between the search model and the true structure in molecular replacement. Third, local (false) minima exist in the target function. The more local minima and the deeper they are, the more likely refinement will fail. Fourth, model bias in the electron density maps complicates the process of manual rebuilding between cycles of automated refinement.

Methods have been devised to address these difficulties. Cross validation, in the form of the free $R$-value, can be used to detect overfitting (Brunger, 1992). The radius of convergence of refinement can be increased by the use of stochastic optimization methods such as molecular dynamics-based simulated annealing (Brunger, Kuriyan, and Karplus, 1987). Most recently, improved targets for refinement of incomplete, error-containing models have been obtained using the more general maximum likelihood formulation (Murshudov, Vagin, and Dodson, 1997; Pannu et al., 1998). The resulting maximum likelihood refinement targets have been successfully combined with the powerful optimization method of simulated annealing to provide a very robust and efficient refinement scheme (Adams et al., 1999). For many structures, some initial experimental phase information is available from either isomorphous heavy atom replacement or anomalous diffraction methods. These phases represent additional observations that can be incorporated in the refinement target. Tests have shown that the addition of experimental phase information greatly improves the results of refinement (Pannu et al., 1998;

Adams et al., 1999). We anticipate that the maximum likelihood refinement method will be extended further to incorporate multivariate statistical analysis, thus, allowing multiple models to be refined simultaneously against the experimental data without introducing bias (Read, 2001).

The refinement methods used in macromolecular structure determination work almost exclusively in reciprocal space. However, there has been renewed interest in the use of real-space refinement algorithms that can take advantage of high quality experimental phases from anomalous diffraction experiments or noncrystallographic symmetry averaging. Tests have shown that the method can be successfully combined with the technique of simulated annealing (Chen, Blanc, and Chapman, 1999).

The parameterization of the atomic model in refinement is of great importance. When the resolution of the experimental data is limited, then it is appropriate to use chemical constraints on bond lengths and angles. This torsion angle representation is seen to decrease overfitting and to improve the radius of convergence of refinement (Rice and Brunger, 1994). If data are available to high enough resolution, additional atomic displacement parameters can be used. Macromolecular structures often show anisotropic motion, which can be resolved at a broad spectrum of levels ranging from whole domains down to individual atoms. The use of the Fast Fourier Transform to refine anisotropic parameters in the program REFMAC has greatly improved the speed with which such models can be generated and tested (Murshudov et al., 1999). The method has been shown to improve the crystallographic R-value and free R-value as well as the fit to geometric targets for data with resolution higher than 2Å.

## VALIDATION

Validation of macromolecular models and their experimental data (Vaguine, Richelle, and Wodak, 1999) is an essential part of structure determination (Kleywegt, 2000). Validation is important both during the structure solution process and at the time of coordinate and data deposition at the Protein Data Bank, where extensive validation criteria are also applied (Berman et al., 2000). See Chapters 14 and 15 for descriptions of validation methods based on stereochemistry and atomic packing. In the future, the repeated application of validation criteria in automated structure solution will help avoid errors that currently occur as a result of subjective manual interpretation of data and models.

## CHALLENGES TO AUTOMATION

### Noncrystallographic Symmetry

It is not uncommon for macromolecules to crystallize with more than one copy in the asymmetric unit. This result leads to relationships between atoms in real space and diffraction intensities in reciprocal space. These relationships can be exploited in the structure solution process. However, the identification of noncrystallographic symmetry (NCS) is generally a manual process. A method for automatic location of proper NCS (i.e., a rotation axis) has been shown to be successful even at low resolution (Vonrhein and Schulz, 1999). A more general approach to finding NCS relationships uses skeletonization of electron density maps (Spraggon, 1999). A monomer envelope is calculated from the solvent mask generated by solvent flattening. The

NCS relationships between monomer envelopes can then be determined using standard molecular replacement methods.

These methods could be used in the future to automate the location of NCS operators and the determination of molecular masks. In the case of experimental phasing using heavy atoms or anomalous scatterers, it is possible to locate the NCS from the sites (Lu, 1999). The RESOLVE program automates this process such that NCS averaging can be automatically performed as part of the phase improvement procedure.

## Disorder

Except in the rare case of very well-ordered crystals of extremely rigid molecules, disorder of one form or another is a component of macromolecular structures. This disorder may take the form of discrete conformational substates for side chains (Wilson and Brunger, 2000) or surface loops, or small changes in the orientation of entire molecules throughout the crystal. The degree to which this disorder can be identified and interpreted typically depends on the quality of the diffraction data. With low-to-medium-resolution data, dual side chain conformations are occasionally observed. With high-resolution data (1.5Å or better) multiple side chain and main chain conformations are often seen. The challenge for automated structure solution is the identification of the disorder and its incorporation into the atomic model without the introduction of errors as a result of misinterpreting the data. Disorder of whole molecules within the crystal, as a result of small differences in packing between neighboring unit cells, cannot be visualized in electron density maps. However, the effect on refinement statistics such as the R and free-R value can be significant because no single atomic model can fit the observed diffraction data well. One approach to the problem is to simultaneously refine multiple models against the data (Burling and Brunger, 1994). An alternative approach is the refinement of Translation-Libration-Screw (TLS) parameters for whole molecules or subdomains of molecules (Winn, Isupov, and Murshudov, 2001). This introduces only a few additional parameters to be refined while still accounting for the majority of the disorder. However, it still remains a challenge to automatically identify subdomains.

## CONCLUSION

Over the last decade of the twentieth century there have been many significant advances toward automated structure determination. Programs such as SOLVE (Terwilliger and Berendzen, 1999a), RESOLVE (Terwilliger, 2000), and the warpNtrace suite (Perrakis et al., 1999) combine large functional blocks in an automated fashion. The program CNS (Brunger et al., 1998) provides a framework in which different algorithms can be combined and tested using a powerful scripting language. Progress toward full automation will be made in the short term by linking existing programs together using scripting languages or the World Wide Web. However, a long-term solution will require the construction of a fully integrated system that makes use of the latest advances in crystallographic algorithms and computer science. The software that truly automates the crystallographic process will need to be intimately associated with data collection and processing. We anticipate that the next generation of automated software will permit the heavy atom location and phasing steps of structure solution to be performed in a few minutes. This speed will enable real time assessment of diffraction data as it is collected at synchrotron beamlines. Map interpretation will be significantly faster than at present, with initial analysis of the electron density taking minutes rather than the hours or days required currently.

## REFERENCES

Abrahams JP (1997): Bias reduction in phase refinement by modified interference functions: introducing the gamma correction. *Acta Crystallogr* D53:371–76.

Adams PD, Pannu NS, Read RJ, Brunger AT (1999): Extending the limits of molecular replacement through combined simulated annealing and maximum likelihood refinement. *Acta Crystallogr* D55:181–90.

Allen FH, Kennard O, Taylor R (1983): Systematic analysis of structural data as a research technique in organic chemistry. *Acc Chem Res* 16:146–53.

Berman HM, Westbrook J, Feng Z, Gilliland G, Bhat TN, Weissig H, Shindyalov IN, Bourne PE (2000): The Protein Data Bank. *Nucleic Acids Res* 28:235–42.

Blessing RH, Smith GD (1999): Difference structure-factor normalization for heavy-atom or anomalous-scattering substructure determinations. *J Appl Crystallogr* 32:664–70.

Bruker AXS, Inc. (2001): *Analytical X-ray Solutions*. Madison, WI.

Brunger AT (1992): The Free R value: a novel statistical quantity for assessing the accuracy of crystal structures. *Nature* 355:472–4.

Brunger AT (1997): Patterson correlation searches and refinement. *Methods Enzymol* 276:558–80.

Brunger AT, Kuriyan J, Karplus M (1987): Crystallographic R factor refinement by molecular dynamics. *Science* 235:458–60.

Brunger AT, Adams PD, Clore GM, Gros P, Grosse–Kunstleve RW, Jiang J-S, Kuszewski J, Nilges M, Pannu NS, Read RJ, Rice LM, Simonson T, Warren GL (1998): Crystallography & NMR system (CNS) A new software system for macromolecular structure determination. *Acta Crystallogr* D54:905–21. [The design and implementation of the widely used CNS program is described. The use of a scripting language to develop, test, and implement new features is a powerful feature of the software.]

Burley SK, Almo SC, Bonanno JB, Capel M, Chance MR, Gaasterland T, Lin D, Sali A, Studier FW, Swaminathan S (1999): Structural genomics: beyond the human genome project. *Nat Genet* 23:151–7.

Burling FT, Brunger AT (1994): Thermal motion and conformational disorder in protein crystal structures: Comparison of multi-conformer and time-averaging models. *Israel J Chem* 34:165–75.

Chen Z, Blanc E, Chapman MS (1999): Real-space molecular-dynamics structure refinement. *Acta Crystallogr* D55:464–8.

Collaborative Computational Project, Number 4 (1994): The CCP4 suite: programs for protein crystallography. *Acta Crystallogr* D50:760–3.

Cowtan K (1998): Modified phased translation functions and their application to molecular-fragment location. *Acta Crystallogr* D54:750–6.

Cowtan K (1999): Error estimation and bias correction in phase-improvement calculations. *Acta Crystallogr* D55:1555–67.

Cowtan K (2001): Fast Fourier feature recognition. *Acta Crystallogr* D57:1435–44.

Dauter Z, Dauter M (1999): Anomalous signal of solvent bromides used for phasing of lysozyme. *J Mol Biol* 289:93–101.

Dauter Z, Dauter M, de La Fortelle E, Bricogne G, Sheldrick GM (1999): Can anomalous signal of sulfur become a tool for solving protein crystal structures? *J Mol Biol* 289:83–92. [If accurate diffraction data are collected at a wavelength close to that of Cu-K$\alpha$ radiation the anomalous signal present from sulfur atoms can be used to generate useful phase information. This technique will probably be used extensively in the future to solve protein structures.]

Garman E (1999): Cool data: quantity AND quality. *Acta Crystallogr* D55:1641–53.

Glykos NM, Kokkinidis M (2000): A stochastic approach to molecular replacement. *Acta Crystallogr* D56:169–74.

Gonzalez A, Pedelacq J-D, Sola M, Gomis-Rueth FX, Coll M, Samama J-P, Benini S (1999): Two-wavelength MAD phasing: in search of the optimal choice of wavelengths. *Acta Crystallogr* D55:1449–58.

Grosse-Kunstleve RW, Brunger AT (1999): A highly automated heavy-atom search procedure for macromolecular structures. *Acta Crystallogr* D55:1568–77.

Hendrickson WA (1985): Stereochemically restrained refinement of macromolecular structures. *Methods Enzymol* 115:252–70.

Hendrickson WA (1991): Determination of macromolecular structures from anomalous diffraction of synchrotron radiation. *Science* 254:51–8.

Holton T, Ioerger TR, Christopher JA, Sacchettini JC (2000): Determining protein structure from electron-density maps using pattern matching. *Acta Crystallogr* D56:722–34.

Jones TA (1978): A graphics model building and refinement system for macromolecules. *J Applied Crystallogr* 11:268–72.

Jones TA, Zou J-Y, Cowan SW, Kjeldgaard M (1991): Improved methods for the building of protein models in electron density maps and the location of errors in these models. *Acta Crystallogr* A47:110–9.

Kissinger CR, Gehlhaar DK, Fogel DB (1999): Rapid automated molecular replacement by evolutionary search. *Acta Crystallogr* D55:484–91.

Kleywegt GJ (2000): Validation of protein crystal structures. *Acta Crystallogr* D56:249–65.

Kleywegt GJ, Jones TA (1997): Template convolution to enhance or detect structural features in macromolecular electron-density maps. *Acta Crystallogr* D53:179–85.

Kleywegt GJ, Jones TA (1998): Databases in Protein Crystallography. *Acta Crystallogr* D54:1119–31.

La Fortelle E de, Bricogne G (1997): Maximum-likelihood heavy-atom parameter refinement in the MIR and MAD methods. *Methods Enzymol* 276:472–94.

Levitt DG (2001): A new software routine that automates the fitting of protein X-ray crystallographic electron-density maps. *Acta Crystallogr* D57:1013–9.

Lu G (1999): FINDNCS: A program to detect non-crystallographic symmetries in protein crystals from heavy atoms sites. *J Applied Crystallogr* 32:365–8.

McRee DE (1999): XtalView/Xfit—A versatile program for manipulating atomic coordinates and electron density. *J Struct Biol* 125:156–65.

Montelione GT, Anderson S (1999): Structural genomics: keystone for a human proteome project. *Nat Struct Biol* 6:11–2.

Mowbray SL, Helgstrand C, Sigrell JA, Cameron AD, Jones TA (1999): Errors and reproducibility in electron-density map interpretation. *Acta Crystallogr* D55:1309–19.

Murshudov GN, Vagin AA, Dodson EJ (1997): Refinement of macromolecular structures by the maximum-likelihood method. *Acta Crystallogr* D53:240–55.

Murshudov GN, Vagin AA, Lebedev A, Wilson KS, Dodson EJ (1999): Efficient anisotropic refinement of macromolecular structures using FFT. *Acta Crystallogr* D55:247–55.

Oldfield T (2002): A semi-automated map fitting procedure. In: Bourne PE, Watenpaugh K, editors. *Crystallographic Computing 7: Macromolecular Crystallographic Data (Crystallographic Computing)*. Oxford: Oxford University Press. In press.

Otwinowski Z (1991): Maximum likelihood refinement of heavy atom parameters. In: Wolf W, Evans PR and Leslie AGW, editors. *Isomorphous Replacement and Anomalous Scattering, Proc. Daresbury Study Weekend*. Warrington: SERC Daresbury Laboratory, p. 80–5.

Pannu NS, Murshudov GM, Dodson EJ, Read RJ (1998): Incorporation of prior phase information strengthens maximum-likelihood structure refinement. *Acta Crystallogr* D54:1285–94.

Perrakis A, Sixma TK, Wilson KS, Lamzin VS (1997): wARP: improvement and extension of crystallographic phases by weighted averaging of multiple-refined dummy atomic models. *Acta Crystallogr* D53:448–55.

Perrakis A, Morris R, Lamzin VS (1999): Automated protein model building combined with iterative structure refinement. *Nat Struct Biol* 6:458–63. [An automated method for building and refining a protein model is described. An iterative procedure is used that describes the electron density map as a set of unconnected atoms from which proteinlike patterns are extracted. This method is currently used by crystallographers to automate model building when high resolution data are available (approximately 2.0Å or better).]

Read RJ (1999): Detecting outliers in non-redundant diffraction data. *Acta Crystallogr* D55:1759–64.

Read RJ (2001): Pushing the boundaries of molecular replacement with maximum likelihood. *Acta Crystallogr* D57:1373–82.

Rice LM, Brunger AT (1994): Torsion angle dynamics: reduced variable conformational sampling enhances crystallographic structure refinement. *Proteins* 19:277–90.

Rice LM, Earnest TN, Brunger AT (2000): Single wavelength anomalous diffraction phasing revisited: a general phasing method? *Acta Crystallogr* D56:1413–20.

Rossmann MG, Blow DM (1962): The detection of sub-units within the crystallographic asymmetric unit. *Acta Crystallogr* 15:24–31.

Sheldrick GM, Gould RO (1995): Structure solution by iterative peaklist optimization and tangent expansion in space group P1. *Acta Crystallogr* B51:423–31.

Sheriff S, Klei HE, Davis ME (1999): Implementation of a six-dimensional search using the AMoRe translation function for difficult molecular-replacement problems. *J Applied Crystallogr* 32:98–101.

Spraggon G (1999): Envelope skeletonization as a means to determine monomer masks and non-crystallographic symmetry relationships: application in the solution of the structure of fibrinogen fragment D. *Acta Crystallogr* D55:458–63.

Terwilliger TC (1994): MAD phasing: Bayesian estimates of $F_A$. *Acta Crystallogr* D50:11–6.

Terwilliger TC (1999): $\sigma_R^2$, a reciprocal-space measure of the quality of macromolecular electron-density maps. *Acta Crystallogr* D55:1174–1178.

Terwilliger TC (2000): Maximum-likelihood density modification. *Acta Crystallogr* D56:965–72. [A procedure is described for reciprocal-space maximization of a likelihood function based on experimental phases and characteristics of the electron-density map. This powerful approach to phase improvement is able to generate minimally biased phase estimates and will be a valuable tool in the future for all aspects of phase improvement and phase combination.]

Terwilliger TC, Berendzen J (1999a): Automated MAD and MIR structure solution. *Acta Crystallogr* D55:849–61. [This important paper describes in detail the SOLVE program and the underlying algorithms that are used to automate heavy atom location and phasing. It is a good example of the considerations that go into automating a complex crystallographic process.]

Terwilliger TC, Berendzen J (1999b): Discrimination of solvent from protein regions in native Fouriers as a means of evaluating heavy-atom solutions in the MIR and MAD methods. *Acta Crystallogr* D55:501–5.

Terwilliger TC, Berendzen J (1999c): Evaluation of macromolecular electron-density map quality using the correlation of local r.m.s. density. *Acta Crystallogr* D55:1872–7.

Turk D (2000): MAIN 96: An interactive software for density modifications, model building, structure refinement and analysis. In: Bourne PE, Watenpaugh K, editors. *Crystallographic Computing 7: Macromolecular Crystallographic Data (Crystallographic Computing)*. Oxford: Oxford University Press.

Vaguine AA, Richelle J, Wodak SJ (1999): SFCHECK: A unified set of procedures for evaluating the quality of macromolecular structure-factor data and their agreement with the atomic model. *Acta Crystallogr* D55:191–205.

Vonrhein C, Schulz GE (1999): Locating proper non-crystallographic symmetry in low-resolution electron-density maps with the program GETAX. *Acta Crystallogr* D55:225–9.

Walsh MA, Evans G, Sanishvili R, Dementieva I, Joachimiak A (1999a): MAD data collection—current trends. *Acta Crystallogr* D55:1726–32.

Walsh MA, Dementieva I, Evans G, Sanishvili R, Joachimiak A (1999b): Taking MAD to the extreme: ultrafast protein structure determination. *Acta Crystallogr* D55:1168–73.

Weeks CM, Miller R (1999): The design and implementation of SnB v2.0. *J Applied Crystallogr* 32:120–4.

Wilson MA, Brunger AT (2000): The 1.0Å crystal structure of Ca2+ bound calmodulin: an analysis of disorder and implications for functionally relevant plasticity. *J Mol Biol* 301:1237–56.

Winn MD, Isupov MN, Murshudov GN (2001): Use of TLS parameters to model anisotropic displacements in macromolecular refinement. *Acta Crystallogr* D57:122–33.

Yeates TO (1997): Detecting and overcoming crystal twinning. *Methods Enzymol* 276:344–58.

# 5

# MACROMOLECULAR STRUCTURE DETERMINATION BY NMR SPECTROSCOPY

John L. Markley, Eldon L. Ulrich, William M. Westler, and Brian F. Volkman

The fundamental hypothesis of structural genomics is that a more complete mapping of peptide sequence space onto conformational space will lead to efficiencies in determining structure–function relationships (Burley, 2000; Heinemann, 2000; Terwilliger, 2000; Yokoyama et al., 2000). Longer-range scientific goals are the prediction of structure and function from sequence and simulations of the functions of a living cell. Structural genomics is part of a wider functional genomics effort, which promises to assign functions to proteins within complex biological pathways and to enlarge the understanding and appreciation of complex biological phenomena (Thornton et al., 2000). Because of its potential to greatly broaden the targets for new pharmaceuticals, structural genomics is expected to join combinatorial chemistry and screening as an integral approach to modern drug discovery (Dry, McCarthy, and Harris, 2000). It is clear that much larger databases of structures, dynamic properties of biomolecules, biochemical mechanisms, and biological functions are needed to approach these goals. Because of its ability to provide atomic-resolution structural and chemical information about proteins, NMR spectroscopy is positioned to play an important role in this endeavor. Already, NMR spectroscopy contributes about 15% of the protein structures deposited at the Protein Data Bank. In addition, NMR spectroscopy is used routinely in high-throughput screens to determine protein:ligand interactions (Hajduk et al., 1999; Shuker et al., 1996). NMR is also a key tool in mechanistic enzymology and in studies of protein folding and stability. As discussed here, advances in key technologies promise rapid increases in the efficiency and scope of NMR applications to structural and functional genomics.

*Structural Bioinformatics*
Edited by Philip E. Bourne and Helge Weissig
Copyright © 2003 by Wiley-Liss, Inc.

## COMPARISON OF NMR SPECTROSCOPY AND X-RAY CRYSTALLOGRAPHY

Currently, the only methods worthy of serious consideration for high-throughput protein structure determination are single-crystal diffraction and solution-state NMR spectroscopy. The two methods have complementary features. X-ray crystallography represents a mature and rapid approach for proteins that form suitable crystals. NMR has advantages for structural studies of small proteins that are partially disordered, exist in multiple stable conformations in solution, show weak interactions with important cofactors or ligands, or do not crystallize readily. Low-resolution structures derived from NMR data can be used in phasing X-ray data. NMR spectroscopy is an incremental method that can rapidly provide useful information concerning overall protein folding, local dynamics, existence of multiple-folded conformations, or protein–ligand or protein–protein interactions. This information can be useful in designing strategies for structure determinations by either NMR or X-ray crystallography. Several ongoing structural genomics pilot projects are employing a combination of X-ray crystallography and NMR spectroscopy. This chapter discusses the current status and future prospects of macromolecular structure determination by solution state NMR spectroscopy. Although the focus is on proteins, the approaches can be generalized to other classes of biological macromolecules. Solid-state NMR spectroscopy shows great promise for structural studies of proteins that may not be amenable to investigation in solution, such as membrane proteins, and for functional investigations of proteins in the solid state. Solid-state NMR strategies for biomolecular sample preparation, data collection, and analysis are developing rapidly, and are expected to assume prominence in the next few years. Further discussion of this highly specialized field is beyond the scope of this chapter, and interested readers are directed to recent articles and reviews (Ramamoorthy, Wu, and Opella, 1999; Bertram et al., 2000; Bertram et al., 2001; Farrar et al., 2001; Gu and Opella, 1999; Jaroniec et al., 2001; Kim, Quine, and Cross, 2001; Luca et al., 2001; Marassi and Opella, 2000).

## PHYSICAL BASIS FOR BIOMOLECULAR NMR SPECTROSCOPY

NMR spectroscopy investigates transitions between spin states of magnetically active nuclei in a magnetic field. The most important magnetically active nuclei for proteins are the proton ($^1$H), carbon-13 ($^{13}$C), nitrogen-15 ($^{15}$N) and phosphorus-31 ($^{31}$P). All are nuclei of spin one-half, which in an external magnetic field have two spin states, one at lower energy with the magnetic spin paired with the external field, and one at higher energy, with the magnetic spin opposing the external field. The magnetic moment of each nucleus precesses about the external magnetic field and is influenced by other fields. Influences on a given spin are generated by neighboring spins in the molecule, giving rise to intrinsic NMR parameters, or by radio-frequency pulses and/or pulsed field gradients as programmed by the NMR spectroscopist. In an NMR experiment, a sequence of radio-frequency pulses and pulsed field gradients are applied to the spins present in the molecule studied. The excited spins are allowed to interact with one another and with the external magnetic field. Then, the state of the system is read out by detection of the current induced by the nuclear spins of the sample in the receiver coil of the NMR spectrometer. The physical basis for NMR is well understood, and the spectroscopic consequences of a given pulse sequence applied to a particular molecule can be simulated to a good level of precision (Ernst, Bodenhausen, and Wokaun, 1987; Cavanagh et al., 1996).

## NMR EXPERIMENTS

The NMR spectrometer can be programmed to operate on all nuclei of a given atom type, or selectively on nuclei with particular spectral features (chemical shift range, spin–spin coupling range, etc.). NMR spectroscopy is unique in that the Hamiltonian of the system under study can be manipulated easily by the application of radio-frequency pulses and/or pulsed field gradients. The versatility in creating sequences of complex "spin gymnastics" makes possible a myriad of NMR experiments by which particular parameters can be investigated. A huge variety of NMR experiments are at one's disposal when collecting data for a structure determination or functional investigation. The NMR field is still young and dynamic, and these experiments continue to evolve; thus, the optimal set of experiments for a structure determination or structure–function study is a matter of exploration and individual taste. The approaches described here are ones we have found to be effective.

The classical approach to structure determination is to first use multidimensional, multinuclear NMR methods to determine "sequence-specific assignments"; that is, resolve signals from the $^1$H, $^{15}$N, and $^{13}$C nuclei of a protein and assign them to specific nuclei in the covalent structure of the molecule. The assigned chemical shifts themselves provide reliable information about the secondary structure of the protein (Wishart, Sykes, and Richards, 1991; Wishart, Sykes, and Richards, 1992; Wishart and Sykes, 1994; Wishart and Nip, 1998), and the oxidation states of cysteine residues (Sharma and Rajarathnam, 2000), and can be used to test or validate structural models. Additional structural restraints are obtained from an interpretation of data from one or more different classes of NMR experiments: (1) NOE spectra, which provide $^1$H–$^1$H distance constraints; (2) three-bond spin–spin coupling experiments, which provide torsion angle constraints; (3) residual dipolar couplings from partially oriented proteins, which provide distance and spatial constraints (with respect to the orientation axes) for pairs of coupled nuclei. Additional hydrogen bond constraints are determined from hydrogen exchange experiments, chemical shifts, and/or trans-hydrogen-bond couplings (Cordier et al., 1999; Cordier and Grzesiek, 1999).

## NMR BIOINFORMATICS

Table 5.1 summarizes NMR data classes that are specific to structural and functional genomics. The major international repository for biomolecular NMR data is BioMag-ResBank (BMRB) http://www.bmrb.wisc.edu, which is affiliated with the Research Collaboratory for Structural Bioinformatics (RCSB) http://www.rcsb.org/. Coordinates for structural models derived from NMR data, as from crystallography, are archived at the Protein Data Bank (PDB) (Berman et al., 2000) http://www.rcsb.org/pdb/ (see Chapter 9).

The primary information to be archived in a protein NMR investigation include: (1) a complete description of the protein system studied (specification of each constituent and the stoichiometry of interacting constituents); (2) the solution conditions for each protein sample investigated (solvent, pH, temperature, and pressure, and concentration of each constituent); (3) a full description of each NMR experiment used

T A B L E  5.1. Summary of the NMR Data Classes That Are Specific to Structural and Functional Genomics

---

Raw NMR data

---

Time-domain data: free induction decays from a particular experiment with a particular sample under defined conditions. The multiplicity of the spectrum (1D, 2D, 3D, 4D) results from the number of time domains sampled.

---

Processed NMR data

---

Frequency-domain data: spectra derived from time-domain data by Fourier transformation and or other signal processing methods.

---

NMR parameters extracted from NMR data

---

Peak lists derived from individual data sets
Chemical shifts
$^1$H–$^1$H NOE
J-couplings
Residual dipolar couplings
NMR relaxation rates

---

Derived information

---

Percentage of expected peaks observed in a data set
NMR peak assignments
Percentage of theoretical peaks assigned
Covalent structure
Bond hybridizations
Secondary structural elements
Interatomic distances
Torsion angles
Hydrogen bonds
Order parameters and other dynamic information
Solvent exposure
Three-dimensional structure (coordinates of the family of conformers that best correspond to the experimental data; coordinates of the conformer that is designated as "representative")
Binding constants
pH titration parameters (p$K_a$ values, Hill coefficients, titration shifts)
Hydrogen exchange rates
Delocalization of unpaired electrons
Thermodynamics and kinetics of structural interconversions
Disordered regions

---

in collecting data for the sample; (4) the NMR parameters and derived information obtained in the investigation; and (5) the unprocessed, time-domain NMR spectra collected. The present goal of the data banks (PDB and BMRB) is to accommodate the inclusion of additional information on the preparation of the sample and the methods for structure determination at the level of detail provided in the methods section of a research journal. BMRB is now accepting raw (time-domain) data associated with NMR structure determinations contributed by authors.

In the interest of maintaining a common chemical shift standard for biomolecules, the international community has adopted the methyl signal of internal 2,2-dimethylsila-pentane-5-sulfonic acid (DSS) at low concentration as the $^1$H chemical shift standard. Chemical shifts of nuclei other than $^1$H are referenced indirectly to the $^1$H standard through the application of a conversion factor for each nucleus derived from ratios of NMR frequencies (Markley et al., 1998).

Structures are represented by a description of the input data used for structure determination and refinement, the methods used to determine the structure, a statistical analysis of the results, and the structure itself, which usually is represented by a family of conformers that best satisfy the input constraints along with geometric and energetic criteria. One of these conformers (or an additional one derived by averaging and energy minimization) is specified as being the single representative structure.

BMRB archives additional information about the chemical properties of proteins derived from NMR spectroscopy, including hydrogen exchange rates at specified sites and pH titration parameters ($pK_a$ values, Hill coefficients, and pH-dependent spectroscopic shifts) for titratable groups. Figure 5.1 illustrates the full range of information at BMRB.

NMR-STAR is the tag-value data format for biomolecular NMR developed by BMRB in collaboration with the PDB and a number of contributors from the NMR community. NMR-STAR is an implementation of the STAR format developed for small-molecule crystallography (Hall, 1991; Hall and Spadaccini, 1994). The mmCIF data format used in biomolecular crystallography (see Chapter 8) is a relational version of STAR. BMRB is developing software tools for interconverting NMR-STAR and mmCIF. NMR-STAR is used by BMRB as its data input format and by many biomolecular NMR software packages as a data exchange format. Specification of the NMR-STAR format and software tools for operating on NMR-STAR files are available from the BMRB web site http://www.bmrb.wisc.edu. BMRB data are exported in the ASCII NMR-STAR format and in a format that can be loaded easily into a relational database.

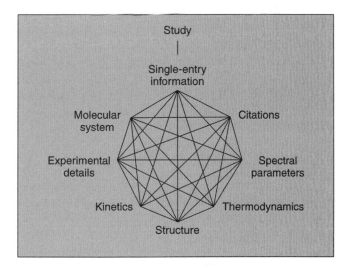

**Figure 5.1.** Scope of biomolecular NMR data archived at BMRB. More detailed views of the data types and the specifications for NMR-STAR data formats are available from the BMRB web site.

As the database of publicly accessible biomolecular NMR data at BMRB expands, the collection is becoming more and more valuable as a resource for data mining and for the development of statistical correlations between NMR observables and structural and chemical properties of biomolecules.

## NMR SCREENING METHODS

NMR spectroscopy can be used as a high-throughput method for screening the products of target protein production for protein folding and stability under various conditions (pH, ionic strength, buffer type, and exposure to libraries of common cofactors and metal ions) (Fejzo et al., 1999; Hajduk et al., 1999; Peng et al., 2001; Shuker et al., 1996). For proteins labeled with $^{15}$N, proton–nitrogen heteronuclear correlation methods, such as a fast $^1$H–$^{15}$N heteronuclear single-quantum correlation (HSQC) (Mori et al., 1995), provide a convenient approach to monitoring the state of the protein. With a cryogenic probe, $^1$H–$^{15}$N HSQC data can be collected in as little as 5 minutes. The spectrum yields a fingerprint for the protein that contains $^1$H–$^{15}$N cross peaks

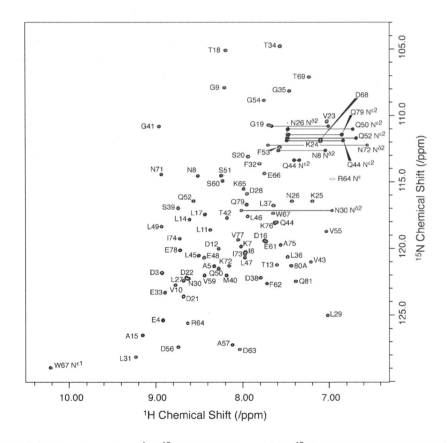

**Figure 5.2.** Two-dimensional $^1$H–$^{15}$N HSQC spectrum of [U–$^{15}$N]apo-D-alanyl carrier protein at 600 MHz and 25 °C. Backbone amide signals are labeled with the one-letter amino acid code and residue number. Side-chain amide NH$_2$ resonances are connected with horizontal lines and labeled.

from each residue in the sequence (except for proline, which has no NH group) plus signals from the side-chain amides of asparagine and glutamine. With transverse relaxation optimized spectroscopy (TROSY) approaches (Pervushin et al., 1997; Pervushin, 2000), this type of screening should be feasible with $^{15}$N-labeled proteins as large as 60 kDa or $^2$H, $^{15}$N-labeled proteins as large as 150 kDa.

Figure 5.2 shows the $^1$H–$^{15}$N HSQC spectrum of a typical small globular protein. By examination of the spectral dispersion, one can readily determine whether a protein is folded (well-dispersed pattern of peaks), unfolded (undispersed peak pattern), or in an aggregated or molten globule state (absence of most expected peaks).

For studies of ligand binding, one compares $^1$H–$^{15}$N HSQC spectra for the protein alone and in the presence of the ligand; the shifting and broadening of peaks indicate regions affected by ligand binding. The results of these screens can be used to decide which targets are worthy of further pursuit, which conditions are optimal for structural investigations by X-ray or NMR, or whether it may be profitable to pursue domain selection (paring down of the protein sequence into isolated folded domains by removal of unstructured intervening sequences). If ligand binding is detected, the identity of the ligand can be determined by subscreening or mass spectrometric analysis.

The longitudinal surveys of proteins studied by structural genomics projects are beginning to yield statistics on proteins that are soluble but largely unstructured. As more data accumulate on protein disorder, it should become possible to increase the success of predicted disorder in proteins (Li et al., 1999; Romero et al., 1998; Romero, Obradovic, and Dunker, 2001). Examples are already at hand for proteins that fail to fold in the absence of a partner protein or peptide or that fold only in the presence of a metal ion or cofactor.

Figure 5.3 shows how NMR spectroscopy can be used to quickly determine that an isolated protein is unfolded and that the addition of a ligand ($Ca^{2+}$ in this case) leads to a $1:2$ protein:$Ca^{2+}$ complex with concomitant folding of the protein (Lytle et al., 2000). The sequence of dockerin did not suggest initially that it binds calcium (the metal ligands fail to correspond to the standard EF hand topology). The ligand-binding screen, however, provided direct evidence, which was confirmed by

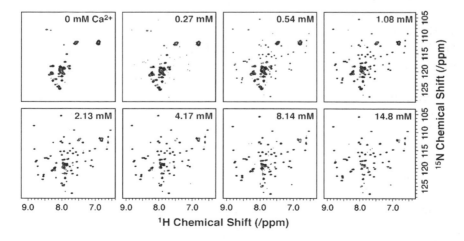

**Figure 5.3.** $Ca^{2+}$-induced folding of dockerin domain detected by 2D $^1$H–$^{15}$N HSQC NMR spectra. The lack of chemical shift dispersion without added $Ca^{2+}$ clearly indicates the absence of a unique folded tertiary structure for the dockerin.

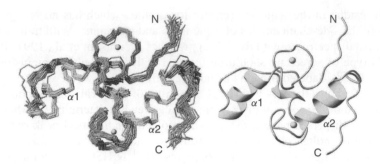

**Figure 5.4.** NMR structure of a type I dockerin domain from the cellulosome of *Clostridium thermocellum*. Backbone N, C$^\alpha$ and C' atoms of the 20 conformers with the lowest target function values out of 100 calculated structures are superimposed (left panel), and the minimized average structure is shown as a ribbon diagram (right panel). The two $\alpha$-helices are shown in cyan and the 3–10 helical turn in green. Calcium ions are shown as yellow spheres.

the subsequent solution structure (Fig. 5.4) (Lytle et al., 2001). A qualitative screen of this kind is possible with proteins 65 kDa or larger (with TROSY) and will be used for proteins destined for X-ray structure analysis that indicate ligand stabilization in prior screens.

## HOW PROTEIN NMR SAMPLES ARE PREPARED

### Protein Production and Labeling

When successful, bacterial expression provides a cost-effective, flexible, reliable, and scalable way to support structural characterizations. Protein production in *Escherichia coli* has an established record of being the most successful approach for providing protein targets for structural genomics. Suitable expression vectors are readily available, and the method is more economical than production in eukaryotic cells. Furthermore, promising proteins can be subsequently labeled metabolically with heavy-atom-labeled amino acids for X-ray or stable isotopes for NMR (Edwards et al., 2000). Metabolic labeling of biomolecules with stable isotopes ($^{15}$N, $^{13}$C and/or $^{2}$H) for NMR spectroscopy was pioneered with *E. coli* expression systems and have been extended successfully to only a few other systems (Markley and Kainosho, 1993). These efforts have paved the way for the development of automated downstream processes for larger-scale protein production and purification. Automation strategies are known for many of the steps in the bacterial growth and evaluation procedures (such as cloning, colony picking, cell growth, liquid handling, centrifugation, mixing, and fluorescence assay).

It is a general experience, however, that a certain percentage of proteins do not express abundantly in soluble form in bacteria. Insolubility arises either from an intrinsic property of the protein (for example, aggregation due to an exposed hydrophobic surface) or because inappropriate folding mechanisms in the expression host permit aggregation of folding intermediates (Chrunyk et al., 1993; Goff and Goldberg, 1987; Henrich, Lubitz, and Plapp, 1982). Expression can be particularly difficult for eukaryotic proteins that consist of multiple domains, that require cofactors or protein partners for proper folding, or that normally undergo extensive post-translational modification. The use of different expression systems can assist in the production of proteins that do not express well in *E. coli*. At present, eukaryotic options available for expressing these proteins, such as yeast, insect, or human cells, have disadvantages with respect to high-throughput operation. In their current forms, culture of insect and human cells is expensive and time-consuming, and current yeast systems, such as *Pichia*, lack the capacity to add some of the posttranslational modifications that are important for proteins from higher organisms and pose problems in lysing the cells. In addition, the development of metabolic labeling in these systems is in its infancy.

## Sample Concentration, Volume, and Stability

The higher the protein concentration, the faster the NMR data can be collected, provided that the protein does not aggregate. Practical lower limit protein concentrations are about 200 μM with ordinary probes and about 60 μM with cryogenic probes. The decreased concentration requirement of the cryogenic probe may make it possible to collect data from proteins that aggregate at higher concentrations. With ordinary NMR tubes, a sample volume of 300 to 500 μL is required, depending on the length of the detection coil in the probe. With a susceptibility matched NMR tube, it may be possible to use volumes as small as 200 μL.

The sample must remain stable over the data collection period. If the stability of the sample is a limiting factor, it is possible to use separate samples for different NMR experiments. Ideally, all data sets are collected with the same sample, because slight differences in solution conditions can lead to chemical shift differences that make it difficult to compare results from different experiments. One of the advantages of cryogenic probes and higher fields is that the overall data collection time for each experiment is shortened. This makes it possible to investigate systems that are less stable over time.

## INTRODUCTION TO PROTEIN STRUCTURE DETERMINATION BY NMR

Protein NMR methods have advanced to the point where small- to medium-sized protein domain structures can be determined in a routine manner. Key milestones in the development of this technique include NOE-based resonance assignment and structure refinement (Wüthrich, 1986), efficient stable isotope labeling methods (Markley and Kainosho, 1993), detection of multinuclear correlations (Oh et al., 1988; Oh, Westler, and Markley, 1989; Ortiz-Polo et al., 1986; Westler, Ortiz-Polo, and Markley, 1984; Westler et al., 1988a; Westler et al., 1988b) coupled with indirect detection and multinuclear NMR spectroscopy (Delaglio et al., 1995; Ikura, Kay, and Bax, 1990; Ikura, Kay, and Bax, 1991; Kay, Marion, and Bax, 1989; Kay et al., 1990). The last decade of the twentieth century saw refinements of these basic ideas, such as the use of pulsed field gradients for coherence selection and line-narrowing by TROSY (Pervushin et al., 1997; Czisch and Boelens, 1998; Pervushin et al., 1998a; Pervushin et al., 1998b; Salzmann et al., 1998; Salzmann et al., 1999a; Salzmann et al., 1999b; Xia, Sze, and Zhu, 2000; Zhu et al., 1999a; Zhu et al., 1999b), coupled with significant improvements in commercially available NMR hardware and the remarkable advances in desktop-computing capabilities.

Protein domains of up to roughly 25 kDa are generally amenable to NMR structure determination, provided that soluble, stable, U–$^{13}$C- and U–$^{15}$N-labeled samples can be obtained. This isotope enrichment enables the use of sensitive heteronuclear correlation experiments for the task of matching all chemical shifts in the protein with specific amino acid residues and atoms. Once all $^1$H, $^{15}$N, and $^{13}$C resonances have been assigned, the correlations observed in nuclear Overhauser enhancement (NOE) experiments can be interpreted in terms of short (<5 Å) interproton distances. NMR structures are obtained from constrained molecular dynamics calculations, with these NOE-derived $^1$H–$^1$H distances as the primary experimental constraints. As a consequence of chemical shift degeneracy, many NOE correlations may have multiple assignment possibilities, and the results of preliminary structure calculations are used to eliminate unlikely candidates on the basis of interproton distances. Refinement continues in an iterative manner until a self-consistent set of experimental constraints produces an ensemble of structures that also satisfies standard covalent geometry and steric overlap considerations. These structures are validated and reported in the PDB in a manner analogous to those obtained from X-ray crystallographic methods (see Chapters 14 and 15).

Structures of smaller proteins can be determined with samples at natural abundance. If the protein is reasonably soluble ($\geq 1$ mM), and particularly if a high-sensitivity cryogenic probe is available, it is possible with moderate effort to obtain $^1$H–$^{13}$C and $^1$H–$^{15}$N correlations at natural abundance. This additional information can be useful for signal assignments and for additional chemical shifts for secondary structure analysis. High-throughput strategies for protein structure determination generally are designed for proteins double-labeled with $^{13}$C and $^{15}$N ($M_r < \sim 20,000$) or triple-labeled with $^{13}$C, $^{15}$N, and $^2$H ($M_r > \sim 20,000$). NMR spectra of larger molecules are characterized by increasingly rapid decay of the signal to be detected, leading to increased line widths and lowered sensitivity (signal-to-noise ratio) and by greater overlap and signal degeneracy. Even with triple-labeled proteins, these problems make it increasingly difficult to solve NMR structures of proteins larger than 40 kDa with current methodology. Higher magnetic fields and TROSY methodology suggest that these limitations can be overcome to some extent.

Selective labeling approaches make it possible to obtain information about selected regions of proteins even as large as 150 kDa. These methods include labeling by single residue type (Markley and Kainosho, 1993) and segmental labeling by peptide semisynthesis, including the use of an autocatalytic cleaving element (intein technology) (Yamazaki et al., 1998; Xu et al., 1999).

Even with protein sample concentrations in the millimolar range (>10 mg/mL), co-addition of 8 or 16 transients may be required at each time point in some two- or three-dimensional experiments to achieve sufficient signal-to-noise (s/n). New cryogenic probe technology enables reduction of the data acquisition period by up to a factor of 10. This results from the three- to fourfold enhancement in sensitivity achieved by cooling the probe and preamplifier circuitry to very low temperatures. Because of the square-root relationship between time and sensitivity in signal averaging, an inherent sensitivity improvement of threefold is equivalent to a ninefold increase in signal averaging. In practical terms, the s/n of an experiment acquired with one or two transients at each time increment on a cryogenic probe should be equivalent (if not superior) to one acquired with 16 transients using a conventional (noncooled) probe. For this example, the cryogenic probe results in an eightfold reduction in experiment time with no negative consequences. The increased sensitivity afforded by a cryogenic probe is especially important for NOE experiments. The ultimate limit for rapid data collection is the longitudinal relaxation rate ($R_1$) of the nucleus to be detected.

# A PROTOCOL FOR PROTEIN STRUCTURE DETERMINATION BY NMR SPECTROSCOPY

As mentioned above, NMR methodology is evolving rapidly, and competing approaches are being explored in various structural genomics centers. We describe the NMR side of a protocol for semiautomated protein structure determination by a combination of NMR spectroscopy and X-ray crystallography. Proteins emerging from the protein production pipeline are divided by molecular weight: Those >60 kDa are diverted to the X-ray crystallization screens; those ≤60 kDa are labeled with $^{15}$N (from $^{15}$NH$_3$Cl in the minimal growth medium as the sole nitrogen source) and subjected to automated NMR screening to determine their state of folding and aggregation under a variety of solution conditions. Ideally, an NMR spectrometer equipped with a sample preparation robot, sample changer, and cryogenic probe will be used for these screens (a 600 MHz system should be adequate for this purpose). NMR screens are used to determine aggregation state (through measurement of transverse diffusion by gradient NMR), folding state (from analysis of the amide region of 1D $^1$H NMR spectra and the pattern of $^1$H–$^{15}$N correlation spectra), effect on structure of ligand binding, and to survey proteins for the presence of unstructured regions. The results of the NMR screens are used in deciding whether to proceed with NMR structure determination (for a smaller protein) or to transfer the protein to the crystallization trials pipeline, or to reengineer the protein for greater stability, solubility (by mutation or addition of a solubility tag) (Zhou, Lugovskoy, and Wagner, 2001), removal of unstructured regions, or domain separation. Proteins that pass these screens as candidates for NMR structure determination are double-labeled ($M_r < 20,000$) from $^{15}$NH$_3$Cl and [U–$^{13}$C]-glucose in the growth medium or triple-labeled ($M_r > 20,000$; high-throughput methods for triple labeling are still under development, prospects are discussed below in Future Prospects). The costs of the isotope in protein samples that produce at the level of a few mg/liter of culture are about \$100 for $^{15}$N-labeling and \$500 for the $^{15}$N,$^{13}$C-labeling.

Collection of residual dipolar coupling data for structure determination or refinement (Prestegard, 1998; Tian, Valafar, and Prestegard, 2001; Tjandra et al., 1997; Tjandra and Bax, 1997a; Tolman et al., 1995) requires that samples be oriented partially in the magnetic field. The orientation can be achieved in a number of ways; for example, by including in the sample bicelles (Ottiger and Bax, 1999; Sanders, Schaff, and Prestegard, 1993; Tjandra and Bax, 1997b) or phage particles (Hansen, Mueller, and Pardi, 1998). NMR spectrometers of 500 and 600 MHz equipped with cryogenic probes are adequate for the collection of triple-resonance data. If available, spectrometers with cryogenic probes operating at or greater than 750 MHz are preferably employed in collecting NOE and isotope-filtered/selected NOE data.

## Standard Data Collection Protocols for Proteins

The improved sensitivity from cryogenic probes permit data collection sufficient for a protein structure in seven days or less, rather than one or two months. If the protein concentration is sufficiently high that data can be recorded without signal averaging, the total data collection period can be as short as 30 hours. The two-dimensional (2D) and three-dimensional (3D) pulse schemes listed in Table 5.2 have been in routine use in many laboratories for several years.

Modifications of these pulse sequences are available for NMR instruments equipped with cryogenic probes (e.g. from the NMRFAM web site http://www.nmrfam.wisc.edu)

**T A B L E  5.2.** Typical Series of NMR Data Sets Collected for a Full Structure Determination

| Sample 1: [U–$^{15}$N]-protein, H$_2$O solution; 3 days | Sample 2: [U–$^{15}$N, U–$^{13}$C]-protein, H$_2$O solution; 4 days |
|---|---|
| 2D $^1$H–$^{15}$N HSQC | 2D $^1$H–$^{13}$C HSQC (aromatic) |
| 3D $^1$H–$^{15}$N NOESY-HSQC | 2D $^1$H–$^{13}$C CT-HSQC (aliphatic) |
| 3D $^1$H–$^{15}$N TOCSY-HSQC | 3D HNCO |
| 2D [$f_2$–$^{15}$N-filtered] $^1$H–$^1$H NOESY (for aromatics) | 3D HNCACO |
| 2D [$f_2$–$^{15}$N-filtered] $^1$H–$^1$H TOCSY (for aromatics) | 3D HNCA |
| | 3D HNCOCA |
| | 3D CBCACONH |
| | 3D CCONH |
| | 3D HNCACB |
| | 3D HCCH-TOCSY |
| | 3D $^1$H–$^{13}$C NOESY-HSQC (aliphatic) |
| | 3D $^1$H–$^{13}$C NOESY-HSQC (aromatic) |
| | Residual dipolar couplings |

and reduced-dimensionality versions also have been developed (Montelione et al., 2000). Protein samples with two labeling schemes are required: [U–$^{15}$N] and [U–$^{13}$C, U–$^{15}$N]. Protein labeled with $^{15}$N suffices for initial screening of conditions and is advantageous because of the low cost of the isotopically labeled precursor; material also can be used for collection of 3D $^{15}$N-edited Nuclear Overhauser Enhancement SpectroscopY (NOESY) and TOtal Correlated SpectroscopY (TOCSY) data. Two-dimensional homonuclear spectra often provide the most efficient means to obtain resonance assignments and NOE constraints for the side chains of aromatic residues. These spectra can be acquired on the [U–$^{15}$N] protein sample, using a $^{15}$N half-filter in the $f_2$ dimension to eliminate amide resonances, leaving only aromatic side-chain signals in the downfield half of the $^1$H spectrum. All other spectra require a sample of doubly labeled [U–$^{15}$N, U–$^{13}$C] protein.

## Processing and Analyzing NMR Data Sets: The Stepwise Approach

The usual approach in a biomolecular NMR study is to first convert time-domain data to frequency-domain spectra by Fourier transform. Then peaks are picked from each spectrum and analyzed. Methods have been developed for automated peak picking or global analysis of spectra to yield models consisting of peaks with known intensity, frequency, phase, and decay rate or line width (in each dimension) (Chylla and Markley, 1995). Our current protocols for processing, peak picking, and assignment of NMR spectra primarily use the programs NMRPipe (Fourier transformation) (Delaglio et al., 1995), SPARKY (peak picking and analysis) http://www.cgl.ucsf.edu/home/sparky/, XEASY (peak picking and semiautomated assignment) (Bartels et al., 1995), and GARANT (fully automated assignment) (Bartels et al., 1996; Bartels et al., 1997), on either Irix or Linux computing platforms. The iterative process of NOE assignment and structure calculations relies primarily on XEASY and DYANA (automated NOE assignment and torsion angle dynamics calculations) (Güntert, Mumenthaler, and Wüthrich, 1997).

***Convert and Process Raw Data.*** Before processing, time-domain data files acquired on Bruker spectrometers must be converted to the proper format for NMRPipe (Delaglio et al., 1995). This conversion is accomplished using the bruk2pipe program supplied with NMRPipe. The corresponding bruk2pipe and NMRPipe scripts for each experiment type are stored in the database at the time of the initial setup. Postconversion to XEASY format can also be performed.

*Peak Picking of All Spectra.* Standard methods for obtaining chemical shift assignments begin with triple-resonance experiments that correlate various combinations of backbone and side-chain $^{13}$C signals with the amide $^{15}$N and $^{1}$H signals. A 2D $^{1}$H–$^{15}$N HSQC serves as the initial reference spectrum for directing the identification of signals in all $^{1}$H$^{N}$-correlated 3D spectra (HNCO, HNCACO, HNCA, HNCOCA, CCONH, CBCACONH, HNCACB). In cases where significant degeneracy is apparent in the 2D HSQC, the 3D HNCO spectrum will be used to resolve overlapped spin systems. Currently, this task is performed in a semiautomated mode using the strip plot features of XEASY. On acceptance of the final peak list, each peak is assigned an arbitrary spin system identifier and stored in a database. This 2D $^{1}$H–$^{15}$N peak list is used to generate a strip plot representation of the first 3D (e.g., HNCO), and candidate signals along the orthogonal vector defining the strip are automatically picked, forming a preliminary peak list for the 3D spectrum. As in the 2D HSQC, this array of strips usually is inspected to determine the completeness and fidelity of the automatic peak picking.

## Sequence-Specific Assignments

At this stage, the set of carefully screened signals from seven 3D spectra are ready for analysis by semi- or fully automated methods for obtaining sequential assignments. The primary tools for semiautomated assignment at NMRFAM are XEASY and SPARKY (T. D. Goddard and D. G. Kneller, SPARKY 3, University of California, San Francisco). GARANT, which has been used at NMRFAM, shares the same file formats with XEASY and DYANA, simplifying the data pathway by eliminating file conversion problems, and it has the ability to provide side-chain assignments, along with backbone assignments.

A number of automated assignment strategies have been described in the literature. They include methods such as AUTOASSIGN (Zimmerman et al., 1997), which requires data from a particular set of experiments, and CONTRAST (Olson, 1995; Olson and Markley 1994), which handles data from a wide range of experiments. AUTOASSIGN can be accessed using a graphical interface developed for the program SPARKY, which may simplify the tasks of submitting data files to and evaluating results from AUTOASSIGN calculations.

## NOE Assignment and Structure Calculation

Efficient and accurate assignment of NOEs for structure determination is strongly dependent on the completeness and accuracy of the chemical shift assignments. Side-chain assignments normally are derived from 3D $^{15}$N TOCSY-HSQC, HCCH-TOCSY, CCONH, and CBCACONH data.

Semiautomated assignment of NOEs can be performed in XEASY using chemical shift filters to suggest possible assignment combinations. Peak lists in XEASY format serve as direct input for the torsion angle dynamics (TAD) program DYANA, which has useful features for automated assignment of NOEs, that use both chemical shift and distance filtering. Fully automated structure refinement using the NOAH module of

DYANA for iterative NOE assignment and TAD has proven successful in some cases at NMRFAM. A robust automated approach is the most desirable, and an enhanced version of DYANA with significantly improved automated NOE refinement capabilities has been described (Güntert, 2000; Güntert et al., 2000).

## Final Refinement and Validation

The final stages of refinement often require inspection of NOE assignments that result in consistently violated distance constraints. NOEs that have overestimated intensities due to peak overlap can produce too-restrictive constraints and require manual adjustment. A fully automated refinement scheme may be able to resolve many such conflicts, but manual intervention will probably be required at some point to achieve the desired goals for precision (lowest root-mean-square deviation [rmsd]), agreement with experimental and covalent geometry constraints (lowest energy/target function), and normal torsion angle geometry (no residues in disallowed regions of $\phi/\psi$ space). DYANA provides diagnostic output with concise summaries of all the necessary data for each ensemble of structures calculated, often simplifying the search for problematic restraints or assignments.

## Deposition of Completed NMR Structures

Normal deposition procedures require extensive data entry. The groups involved in structural genomics are developing software to harvest the information needed for data bank depositions in the course of the work on a particular target. The accumulated data can be checked by reference to software packages described below.

## VALIDATION OF STRUCTURAL MODELS

The optimal validation approach to be taken depends on the experimental data collected and how they were used in deriving the family of conformers that represents the structure. Two approaches are used for the validation of biomolecular NMR structures. In the first, the agreement between experimental data and coordinates is assessed. The second measures the match between many aspects of the geometry and the expected range of permissible values for each geometric parameter. Ranges of expected values can be derived theoretically, or can be obtained from databases of protein structures (PDB), nucleic acid structures (Nucleic Acids Database, NDB), and small molecules (Cambridge Structural Database, CSD). These two types of structure validation provide complementary checks on the quality of structures and are described below along with various existing software packages.

As with X-ray structures, the stereochemical quality of protein models can be checked with the software tools PROCHECK (Laskowski et al., 1993), and WHAT IF (Vriend, 1990). NMR-specific tools include AQUA and PROCHECK-NMR (Laskowski et al., 1996). Recently, WHAT IF has been extended to include a check on hydrogen geometry (Doreleijers et al., 1999b). Structural studies on peptides (Engh and Huber, 1991) and nucleic acids (Parkinson et al., 1996) solved at high resolution and present in the CSD resulted in a set of reference values and standard deviations of bond lengths and bond angles for heavy atoms that are taken as reference values for larger biomolecules. In most common force fields, the force constants for geometry constraints are chosen to reflect the variation in these geometric parameters. The check on bond lengths and bond angles is one of the many checks that should be performed before accepting a refined model. Severe deviations from the commonly accepted standard geometry signal problematic regions in a structure.

Software packages used to solve and refine NMR structures (DYANA, CNS, and others) provide a rich set of diagnostic tools and a summary of the violations between experimental data and the coordinates. In general, NMR structures of acceptable quality meet the following validation criteria: (1) an rmsd for NOE violations less than ~0.05 Å and no persistent NOE violation greater than 0.5 Å across the ensemble of structures (Doreleijers, Rullmann, and Kaptein, 1998); (2) an NOE completeness of at least 50% for all NMR observable proton contacts within 4 Å (Doreleijers et al., 1999a); (3) all torsion angles within the range of the restraints; (4) chemical shift values within acceptable ranges, unless verified independently. At the time the structures are released, full documentation of the structure and the experimental data as described in the International Union of Pure and Applied Chemistry (IUPAC) publication "Recommendations for the Presentation of NMR Structures of Proteins and Nucleic Acids," should be provided, as well as the reports generated by the software packages mentioned above.

A number of approaches to validating NMR structures against NOE restraints or peak volumes and against residual dipolar couplings have been proposed. H. R. Kalbitzer and co-workers have developed an improved software package for back-calculating NOESY spectra from structures that takes into account relaxation and scalar coupling effects; this approach and an associated "R" factor offer a promising way of validating NMR structures (Gronwald et al., 2000; Kalbitzer, 2001). Brünger and co-workers have developed a complete cross-validation technique, in analogy to the free R-factor used in X-ray crystallography, that provides an independent quality assessment of NMR structures based on the NOE violations (Brünger, 1992). The free R-factors of different NMR structures are only comparable if the NOE intensities have been translated into distance restraints in the exact same way. The Clore (Clore, Starich, and Gronenborn, 1998) and Bax (Ottiger and Bax, 1999) groups have developed quality factors for residual dipolar couplings. In cases where dipolar coupling data can be obtained easily, these factors provide an attractive approach to validation.

Validation of structural models against assigned chemical shifts is another promising technique for checking structures. The measurement of chemical shifts is easy and precise but the back-calculation of the expected chemical shift from a structure is less trivial (Case, 1998). Williamson and co-workers have related measured proton chemical shifts of proteins to values calculated on the basis of NMR and X-ray structures (Williamson, Kikuchi, and Asakura, 1995). The chemical shifts derived from NMR structures showed the same degree of agreement with the measured proton chemical shifts as X-ray structures with a resolution between 2 and 3 Å ($\sigma$ of 0.35 ppm). Methods for back-calculating chemical shifts from structure are improving rapidly (Case, 2000). The BioMagResBank (BMRB) (Seavey et al., 1991; Ulrich et al., 2000) contains over 900,000 experimental chemical shifts for proteins, nucleic acids, and small molecules and more than 7,000 experimental coupling constants. As of 2002, BMRB entries with related PDB coordinate entries totaled 965, and 342 of these BMRB entries were available with chemical shifts referenced consistently using the IUPAC recommendations. Workers are beginning to mine this wealth of information (Wishart et al., 1997), which is expected to grow rapidly in coming years (Cornilescu, Delaglio, Bax, 1999).

## NMR AND MOLECULAR DYNAMICS

Because signals observed in solution by NMR report on the chemical properties of nuclei, including their relative motions with respect to the laboratory and molecular frames, the method is uniquely suited for investigations of the kinetics and thermodynamics of molecular motions, segmental motions, dynamic conformational equilibria, and ligand-binding equilibria. Whereas regions of proteins that are statically or dynamically disordered may fail to produce resolvable electron density, NMR signals from such regions generally are observed and can be investigated.

## FUTURE PROSPECTS

In order to achieve a high-throughput mode of structure determination by NMR spectroscopy, it will be necessary to reduce, by a substantial fraction, the average cost of a structure in terms of both time and resources. Future improvements in the efficiency of the NMR approach are being achieved through: (1) reducing data acquisition times by capitalizing on the substantial sensitivity gains afforded by novel cryogenic probe technology; (2) streamlining project management and facilitating automation by storing all data in a relational database with customized interface modules for each analysis task; and (3) producing proteins by cell-free systems that offer reductions in sample cost, simplified spectral analysis, and higher-quality structures.

### Cell-Free Protein Production

The development of new systems and strategies capable of synthesizing any desired soluble, labeled protein or protein fragment on a preparative scale is one of the most important tasks in biotechnology today. Two strategies are currently employed: chemical synthesis and cell-free protein synthesis. Chemical synthesis is not suitable for long peptides, because the yields are low and the costs are high. In contrast, cell-free biological systems can synthesize proteins with high speed and accuracy, approaching in vivo rates (Kurland, 1982; Pavlov and Ehrenberg, 1996). Commercial implementations of this technology (although not yet sufficiently useful to routinely support structural proteomics efforts) are available, and recent research advances in this area have been documented in the literature (Baranov et al., 1989; Endo et al., 1992; Roberts and Paterson, 1973; Spirin et al., 1988). To date, an *E. coli* cell-free system yielding as much as 6 mg of protein per ml of reaction volume has given the best results (Kigawa et al., 1999). However, such high productivity can only be expected with relatively small proteins since with large proteins the increasing molecular weight of mRNAs results in increased probability of degradation by endogenous *E. coli* ribonucleases. Furthermore, it is likely that *E. coli* cell-free systems are not optimal for the synthesis of eukaryotic proteins. Recent important advances have been made in cell-free protein expression from wheat germ extracts: these include the removal of an inhibitor of protein synthesis (Madin et al., 2000), and improvements in coupled transcription–translation procedures (Endo, 2001), or capping and polyadenylation of messenger RNA used in direct cell-free translation (Kumar et al., 2000). Cell-free protein expression (both *E.coli* and wheat-germ cell-free systems) is being used for routine expression screening at RIKEN in Japan (Yokoyama, 2000). It appears that cell-free protein production methods will become increasingly important to the future of structural and functional genomics.

We envision cell-free protein production as a most promising long-term approach to protein production for structural proteomics. There are six reasons for this:

1. Volumes can be kept manageable. In favorable cases, sufficient protein for a structural investigation can be produced from a reaction volume of under 5 mL.
2. Proteins can be produced that would be toxic to cells.
3. The ratio of desired protein to unwanted protein is high, simplifying purification.
4. In a cell-free system, one has the potential for introducing a variety of agents to promote correct folding (chaperones, prolyl peptidyl isomerases, disulfide isomerases, etc.).
5. Enzymes and substrates can be added to promote post-translational modifications.
6. Labeled amino acids can be introduced without scrambling of the label.

The production of labeled proteins from cell-free systems requires that labeled amino acids be supplied in the reaction mixture. Whereas 1–2 g each of labeled ammonium chloride and glucose typically are required for the generation of a double-labeled NMR sample (at yields of 5–10 mg/L culture), considerably less amino acid mixture is needed for cell-free protein production. The optimal concentration of each amino acid in the cell-free reaction mixture is about 1 mM. At currently achievable protein yields (0.4 to 1.1 mg protein/mL of reaction mixture), a reaction volume of about 8 mL is needed to produce enough protein for 2–3 NMR samples. $U-^{15}N$- and $U-^{15}N,^{13}C$-amino acids are commercially available at prices that enable considerable price savings over production of labeled proteins from *E. coli* cells.

## Stereo Array Isotope Labeling (SAIL) and the Principle of Minimized Proton Density

The ideal labeling pattern for protein NMR experiments is one in which only one of the two prochiral groups, $-C-C(H_2)$ or $-CMe_2$, is visible by NMR. For aromatic rings, the principle is to minimize the proton density and indirect couplings (Kainosho, 2000; Kainosho et al., 2001). All methyls are replaced by $-CHD_2$. There are 10 advantages of this labeling pattern for NMR spectroscopy:

1. Defeat of the predominant dipolar relaxation mechanisms through the introduction of $-CHD-$ and $-CHD_2$ groups decreases $R_2$ relaxation; this makes signals sharper and increases signal-to-noise (s/n).
2. Increased s/n and resolution are obtained through reduction of losses through coherence transfer.
3. By reducing spin-diffusion pathways, it becomes possible to measure accurate NOEs and residual dipolar couplings at longer distances.
4. Prochiral assignments come directly from the labeling pattern and are known absolutely, which simplifies structure determinations and makes them more accurate.

5. The labeling pattern eliminates many longer-range couplings; this sharpens the signals and makes it easier to measure couplings more accurately ($J$ and residual dipolar).
6. Spectra are less crowded and are thus more amenable to automated assignments.
7. The labeling pattern will extend the range of high-throughput structure determinations to higher molecular weights (this issue has not been explored fully to date, but the cutoff certainly will be raised above 30 kDa).
8. The labeling pattern is compatible with all double- and triple-resonance NMR experiments, including TROSY.
9. Since both the methyl and methylene groups have the same (unit proton) intensity, internal mobility effects can be read from spectra by inspection of apparent intensity.
10. The labeling pattern supports the determination of NMR solution structures of proteins at high resolution.

Overall, the increase in s/n through the use of SAIL is about a factor of three (M. Kainosho, personal communication), similar to that achieved with cryogenic probes. The combination of SAIL and cryogenic probe data collection should lead to almost an increased s/n of 10. Structural studies will require less protein. Alternatively, at the same protein concentration used with conventional labeling, the data can be collected much faster. In either case, the quality of the data will be far superior to that achieved with conventional labeling. Still there is a limit imposed by the size of the molecule as imposed by $R_1$ relaxation.

Stereo-array-labeled amino acids must be produced by chemical synthesis. Kainosho and co-workers have worked out synthetic routes for all required amino acids, and it is hoped that this approach can be made commercially viable so that it will be available to all researchers in the field.

## Concerted Methods for Assignment and Structure Determination

Newly emerging methods indicate the possibility of determining structures without prior assignment of chemical shifts (Grishaev and Llinás, 2002; Madrid, Llinás, and Llinás, 1991; Tian, Valafar, and Prestegard, 2001; Zweckstetter and Bax, 2001). Such approaches bypass the tedious assignment process and ideally provide a probabilistic view of the structure determined by a given set of input data. These approaches are still at the early stage of development, but show great promise for accelerating the process of determining high-resolution NMR structures or for rapid determinations of protein folds.

## ACKNOWLEDGMENTS

The authors thank their colleagues Frits Abildgaard, David Aceti, Sam Butcher, Jurgen Doreleijers, Hamid Eghbalnia, Masatsune Kainosho, Steve Mading, Dmitriy Vinarov, Zsolt Zolnai, and others at the National Magnetic Resonance at Madison, BioMagResBank, and Center for Eukaryotic Structural Genomics who have contributed many of the ideas presented here and have participated in creating a most stimulating environment for biomolecular NMR spectroscopy.

## REFERENCES

Baranov VI, Morozov IY, Ortlepp SA, Spirin AS (1989): Gene-expression in a cell-free system on the preparative scale. *Gene* 84:463–6.

Bartels C, Xia T-H, Billeter M, Güntert P, Wüthrich K (1995): The Program XEASY for computer-supported NMR spectral-analysis of biological macromolecules. *J Biomol NMR* 5:1–10.

Bartels C, Billeter M, Güntert P, Wüthrich K (1996): Automated sequence-specific NMR assignment of homologous proteins using the program GARANT. *J Biomol NMR* 7:207–13.

Bartels C, Güntert P, Billeter M, Wüthrich K (1997): GARANT—a general algorithm for resonance assignment of multidimensional nuclear magnetic resonance spectra. *J Comput Chem* 18:139–49.

Berman HM, Westbrook J, Feng Z, Gilliland G, Bhat TN, Weissig H, Shindyalov IN, Bourne PE (2000): The Protein Data Bank. *Nucleic Acids Res* 28:235–42.

Bertram R, Quine JR, Chapman MS, Cross TA (2000): Atomic refinement using orientational restraints from solid-state NMR. *J Magn Reson* 147:9–16.

Bertram R, Kim S, Quine JR, Xu M, Chapman MS, Cross TA (2001): Molecular refinement and cross-validation with solid-state NMR orientational data. *Biophys J* 80:1548.

Brünger AT (1992): Free R-value—a novel statistical quantity for assessing the accuracy of crystal-structures. *Nature* 355:472–5.

Burley SK (2000): An overview of structural genomics. *Nat Struct Biol* 7:932–4.

Case DA (1998): The use of chemical shifts and their anisotropies in biomolecular structure determination. *Curr Opin Struct Biol* 8:624–30.

Case DA (2000): Interpretation of chemical shifts and coupling constants in macromolecules. *Curr Opin Struct Biol* 10:197–203.

Cavanagh J, Palmer AG 3rd, Fairbrother W, Skelton N (1996): *Protein NMR Spectroscopy: Principles and Practice*. San Diego: Academic Press.

Chrunyk BA, Evans J, Lillquist J, Young P, Wetzel R (1993): Inclusion-body formation and protein stability in sequence variants of interleukin-1-beta. *J Biol Chem* 268:18053–61.

Chylla RA, Markley JL (1995): Theory and application of the maximum likelihood principle to NMR parameter estimation of multidimensional NMR data. *J Biomol NMR* 5:245–58.

Clore GM, Starich MR, Gronenborn AM (1998): Measurements of residual dipolar couplings of macromolecules aligned in the nematic phase of a colloidal suspension of rod-shaped viruses. *J Am Chem Soc* 120:10571–2.

Cordier F, Grzesiek S (1999): Direct observation of hydrogen bonds in proteins by interresidue $^{3H}J_{NC'}$ scalar couplings. *J Am Chem Soc* 121:1601–2.

Cordier F, Rogowski M, Grzesiek S, Bax A (1999): Observation of through-hydrogen-bond (2h)J(HC') in a perdeuterated protein. *J Magn Reson* 140:510–2.

Cornilescu G, Delaglio F, Bax A (1999): Protein backbone angle restraints from searching a database for chemical shift and sequence homology. *J Biomol NMR* 13:289–302. [This is an excellent illustration of how databases of structures and NMR parameters can be combined to provide powerful tools for structure determination.]

Czisch M, Boelens R (1998): Sensitivity enhancement in the TROSY experiment. *J Magn Reson* 134:158–60.

Delaglio F, Grzesiek S, Vuister GW, Zhu G, Pfeifer J, Bax A (1995): NMRPIPE—a multidimensional spectral processing system based on UNIX pipes. *J Biomol NMR* 6:277–93.

Doreleijers JF, Rullmann JAC, Kaptein R (1998): Quality assessment of NMR structures: a statistical survey. *J Mol Biol* 281:149–64.

Doreleijers JF, Raves ML, Rullmann T, Kaptein R (1999a): Completeness of NOEs in protein structure: a statistical analysis of NMR data. *J Biomol NMR* 14:123–32.

Doreleijers JF, Vriend G, Raves ML, Kaptein R (1999b): Validation of nuclear magnetic resonance structures of proteins and nucleic acids: hydrogen geometry and nomenclature. *Proteins* 37:404–16.

Dry S, McCarthy S, Harris T (2000): Structural genomics in the biotechnology sector. *Nat Struct Biol* 7:946–9.

Edwards AE, Arrowsmith CH, Christendat D, Dharamsi A, Friesen JD, Greenblatt JF, Vedadi M (2000): Protein production: feeding the crystallographers and NMR spectroscopists. *Nat Struct Biol* 7:970–2.

Endo Y (2001): "Genomics to proteomics: a high-throughput cell-free protein synthesis system for practical use." The 3rd ORCS International Symposium on Ribosome Engineering; 2001 January 22–23; Tsukuba, Japan.

Endo Y, Otsuzuki S, Ito K, Miura K (1992): Production of an enzymatic active protein using a continuous-flow cell-free translation system. *J Biotechnology* 25:221–30.

Engh RA, Huber R (1991): Accurate bond and angle parameters for X-ray protein structure refinement. *Acta Crystallogr* A47:392–400.

Ernst RR, Bodenhausen G, Wokaun A (1987): *Principles of Nuclear Magnetic Resonance in One and Two Dimensions*. Oxford: Oxford University Press.

Farrar CT, Hall DA, Gerfen GJ, Inati SJ, Griffin RG (2001): Mechanism of dynamic nuclear polarization in high magnetic fields. *J Chem Phys* 114:4922–33.

Fejzo J, Lepre CA, Peng JW, Bemis GW, Ajay Murcko MA, Moore JM (1999): The SHAPES strategy: an NMR-based approach for lead generation in drug discovery. *Chem Biol* 6:755–69.

Goff SA, Goldberg AL (1987): An increased content of protease LA, the lon gene-product, increases protein-degradation and blocks growth in *Escherichia coli*. *J Biol Chem* 262:4508–15.

Grishaev A, Llinás M (2002): Protein structure elucidation from NMR proton densities. *Proc Natl Acad Sci USA* 99:6713–8; Grishaev A, Llinas M (2002): CLOUDS, a protocol for deriving a molecular proton density via NMR. *Proc Natl Acad Sci USA* 99:6707–12. [These two publications present a most promising implementation to date of an approach that uses NOESY data and prior information about protein structure and NMR chemical shifts to achieve coordinated structure determination and chemical shift assignments.]

Gronwald W, Kirchhofer R, Gorler A, Kremer W, Ganslmeier B, Neidig KP, Kalbitzer HR (2000): RFAC, a program for automated NMR R-factor estimation. *J Biomol NMR* 17:137–51.

Gu ZTT, Opella SJ (1999): Two- and three-dimensional H-1/C-13 PISEMA experiments and their application to backbone and side chain sites of amino acids and peptides. *J Magn Reson* 140:340–6.

Güntert P (2000): "Automated NMR structure calculation using network-anchored assignment and constraint combination." XIX International Conference on Magnetic Resonance in Biological Systems; 2000 August 20–25; Florence, Italy. Abs p 47.

Güntert P, Mumenthaler C, Wüthrich K (1997): Torsion angle dynamics for NMR structure calculation with the new program DYANA. *J Mol Biol* 273:283–98.

Güntert P, Salzmann M, Braun D, Wüthrich K (2000): Sequence-specific NMR assignment of proteins by global fragment mapping with the program MAPPER. *J Biomol NMR* 18:129–37.

Hajduk PJ, Gerfin T, Boehlen JM, Haberli M, Marek D, Fesik SW (1999): High-throughput nuclear magnetic resonance-based screening. *J Med Chem* 42:2315–7.

Hall SR (1991): The STAR File: a new format for electronic data transfer and archiving. *J Chem Inf Comput Sci* 31:326–33.

Hall SR, Spadaccini N (1994): The STAR file: detailed specifications. *J Chem Inf Comput Sci* 34:505–8.

Hansen MR, Mueller L, Pardi A (1998): Tunable alignment of macromolecules by filamentous phage yields dipolar coupling interactions. *Nat Struct Biol* 5:1065–74.

Heinemann U (2000): Structural genomics in Europe: slow start, strong finish? *Nat Struct Biol* 7:940–2.

Henrich B, Lubitz W, Plapp R (1982): Lysis of *Escherichia coli* by induction of cloned phi-X174 genes. *Molec General Genet* 185:493–7.

Ikura M, Kay LE, Bax A (1990): A novel approach for sequential assignment of $^1$H, $^{13}$C, and $^{15}$N spectra of larger proteins, heteronuclear triple-resonance 3-dimensional spectroscopy: application to calmodulin. *Biochemistry* 29:4659–67.

Ikura M, Kay LE, Bax A (1991): Improved three-dimensional $^1$H–$^{13}$C–$^1$H correlation spectroscopy of a $^{13}$C-labeled protein using constant-time evolution. *J Biomol NMR* 1:299–304.

Jaroniec CP, Tounge BA, Herzfeld J, Griffin RG (2001): Frequency selective heteronuclear dipolar recoupling in rotating solids: accurate C-13-N-15 distance measurements in uniformly C-13-N-15-labeled peptides. *J Am Chem Soc* 123:3507–19.

Kainosho M (2000): "Perspectives for isotope-assisted techniques in biological NMR spectroscopy." XIX International Conference on Magnetic Resonance in Biological Systems; 2000 August 20–25; Florence, Italy. Abs p 9.

Kainosho M, Terauchi T, Ohki S, Hayano T (2001): "Advanced isotope-labeling technology for structural studies of larger proteins." Keystone Symposia: Frontiers of NMR in Molecular Biology VII; 2001 January 20–26; Big Sky, MT. Abs p 48.

Kalbitzer HR (2001): "Automated structure calculation of proteins from NMR data." Keystone Symposia: Frontiers of NMR in Molecular Biology VII; 2001 January 20–26; Big Sky, MT.

Kay LE, Marion D, Bax A (1989): Practical aspects of 3D heteronuclear NMR of proteins. *J Magn Reson* 84:72–84. [Triple-resonance NMR experiments with $^{15}$N and $^{13}$C labeled proteins are now a standard approach for determining backbone shift assignments and secondary structure.]

Kay LE, Ikura M, Tschudin R, Bax A (1990): 3-dimensional triple-resonance NMR-spectroscopy of isotopically enriched proteins. *J Magn Reson* 89:496–514.

Kigawa T, Yabuki T, Yoshida Y, Tsutsui M, Ito Y, Shibata T, Yokoyama S (1999): Cell-free production and stable-isotope labeling of milligram quantities of proteins. *FEBS Lett* 442:15–9.

Kim S, Quine JR, Cross TA (2001): Complete cross-validation and R-factor calculation of a solid-state NMR derived structure. *J Am Chem Soc* 123:7292–8.

Kumar PKR, Hori H, Sawasaki T, Murthy MSRC, Kumaravel T, Nishikawa S, Katahira M, Mizuno H, Endo Y (2000): "Synthesis of fully deuterated proteins using wheat germ cell-free system." The 2nd International Symposium on Development of New Structural Biology Including Hydrogen and Hydration in Organized Research Combination System; 2000 October 27–29; Mito, Japan. Abs p 19.

Kurland CG (1982): Translational accuracy in vitro. *Cell* 28:201–2.

Laskowski RA, MacArthur MW, Moss DS, Thornton JM (1993): PROCHECK: a program to check the stereochemical quality of protein structures. *J Applied Crystallogr* 26:283–91.

Laskowski RA, Rullmann JAC, MacArthur MW, Kaptein R, Thornton JM (1996): AQUA and PROCHECK-NMR: programs for checking the quality of protein structures solved by NMR. *J Biomol NMR* 8:477–86.

Li X, Romero P, Rani M, Dunker AK, Obradovic Z (1999): Predicting protein disorder for N-, C-, and internal regions. *Genome Inform Ser Workshop Genome Inform* 10:30–40.

Luca S, Filippov DV, van Boom JH, Oschkinat H, De Groot HJM, Baldus M (2001): Secondary chemical shifts in immobilized peptides and proteins: a qualitative basis for structure refinement under magic angle spinning. *J Biomol NMR* 20:325–31.

Lytle BL, Volkman BF, Westler WM, Wu JH (2000): Secondary structure and calcium-induced folding of the *Clostridium thermocellum* dockerin domain determined by NMR spectroscopy. *Arch Biochem Biophys* 379:237–44.

Lytle BL, Volkman BF, Westler WM, Heckman MP, Wu JH (2001): Solution structure of a type I dockerin domain, a novel prokaryotic, extracellular calcium-binding domain. *J Mol Biol* 307:745–53.

Madin K, Sawasaki T, Ogasawara T, Endo Y (2000): A highly efficient and robust cell-free protein synthesis system prepared from wheat embryos: plants apparently contain a suicide system directed at ribosomes. *Proc Natl Acad Sci USA* 97:559–64.

Madrid M, Llinás E, Llinás M (1991): Model-independent refinement of interproton distances generated from $^1$H NMR Overhauser intensities. *J Magn Reson* 93:329–46.

Marassi FM, Opella SJ (2000): A solid-state NMR index of helical membrane protein structure and topology. *J Magn Reson* 144:150–5.

Markley JL, Kainosho M (1993): Stable isotope labeling and resonance assignments in larger proteins. In: Roberts GCK, editor. *NMR of Biological Macromolecules: A Practical Approach.* Oxford: Oxford University Press, pp 101–52.

Markley JL, Bax A, Arata Y, Hilbers CW, Kaptein R, Sykes BD, Wright PE, Wüthrich K (1998): Recommendations for the presentation of NMR structures of proteins and nucleic acids. *Pure Applied Chem* 70:117–42.

Montelione GT, Zheng D, Huang YJ, Gunsalus KC, Szyperski T (2000): Protein NMR spectroscopy in structural genomics. *Nat Struct Biol* 7:982–5.

Mori S, Abeygunawardana C, Johnson MO, van Zijl PCM (1995): Improved sensitivity of HSQC spectra of exchanging protons at short interscan delays using a new fast HSQC (FHSQC) detection scheme that avoids water saturation. *J Magn Reson* B108:94–8.

Oh B-H, Westler WM, Darba P, Markley JL (1988): Protein carbon-13 spin systems by a single two-dimensional nuclear magnetic resonance experiment. *Science* 240:908–11.

Oh B-H, Westler WM, Markley JL (1989): Carbon-13 spin system directed strategy for assigning cross peaks in the COSY fingerprint region of a protein. *J Am Chem Soc* 111:3083–5.

Olson JB Jr (1995): A general and adaptable method for the automated assignment of protein multidimensional nuclear magnetic resonance spectra [dissertation]. University of Wisconsin-Madison, 201 pp, available from ProQuest Information and Learning. Online Dissertation Services (Dissertation Express) http://www.umi.com/hp/Products/Dissertations.html under order number 9608152.

Olson JB Jr, Markley JL (1994): Evaluation of an algorithm for the automated sequential assignment of protein backbone resonances: a demonstration of the connectivity tracing assignment tools (CONTRAST) software package. *J Biomol NMR* 4:385–410.

Ortiz-Polo G, Krishnamoorthi R, Markley JL, Live DH, Davis DG, Cowburn D (1986): Natural-abundance $^{15}$N NMR studies of turkey ovomucoid third domain: assignment of peptide $^{15}$N resonances to the residues at the reactive site region via proton-detected multiple-quantum coherence. *J Magn Reson* 68:303–10.

Ottiger M, Bax A (1999): Bicelle-based liquid crystals for NMR-measurement of dipolar couplings at acidic and basic pH values. *J Biomol NMR* 13:187–91.

Parkinson G, Voitechovsky J, Clowney L, Brünger AT, Berman HM (1996): Bond lengths and angles, DNA/RNA. *Acta Crystallogr* D52:57–64.

Pavlov MY, Ehrenberg M (1996): Rate of translation of natural mRNAs in an optimized in vitro system. *Arch Biochem Biophys* 328:9–16.

Peng JW, Lepre CA, Fejzo J, Abdul-Manan N, Moore JM (2001): Nuclear magnetic resonance-based approaches for lead generation in drug discovery. *Methods Enzymol* 338:202–30.

Pervushin K (2000): Impact of Transverse Relaxation Optimized Spectroscopy (TROSY) on NMR as a technique in structural biology. *Quart Rev Biophys* 33:161–97.

Pervushin K, Riek R, Wider G, Wüthrich K (1997): Attenuated $T_2$ relaxation by mutual cancellation of dipole-dipole coupling and chemical shift anisotropy indicates an avenue to NMR structures of very large biological macromolecules in solution. *Proc Natl Acad Sci USA* 94:12366–71. [Transverse relaxation optimized spectroscopy (TROSY) provides the best NMR approach for investigating structure and function of larger proteins in solution.]

Pervushin KV, Ono A, Fernandez C, Szyperski T, Kainosho M, Wüthrich K (1998a): NMR scalar couplings across Watson–Crick base pair hydrogen bonds in DNA observed by transverse relaxation-optimized spectroscopy. *Proc Natl Acad Sci USA* 95:14147–51.

Pervushin KV, Riek R, Wider G, Wüthrich K (1998b): Transverse relaxation-optimized spectroscopy (TROSY) for NMR studies of aromatic spin systems in $^{13}$C-labeled proteins. *J Am Chem Soc* 120:6394–400.

Prestegard JH (1998): New techniques in structural NMR—anisotropic interactions. *Nat Struct Biol* 5:517–22.

Ramamoorthy A, Wu CH, Opella SJ (1999): Experimental aspects of multidimensional solid-state NMR correlation spectroscopy. *J Magn Reson* 140:131–40.

Roberts BE, Paterson BM (1973): Efficient translation of tobacco mosaic virus RNA and rabbit globin 9S RNA in a cell-free system from commercial wheat germ. *Proc Natl Acad Sci USA* 70:2330–4.

Romero P, Obradovic Z, Kissinger CR, Villafranca JE, Garner E, Guilliot S, Dunker AK (1998): Thousands of proteins likely to have long disordered regions. *Pacific Symp Biocomput* 1998:437–48.

Romero P, Obradovic Z, Dunker AK (2001): Intelligent data analysis for protein disorder prediction. *Artificial Intelligence Rev* 14:447–84.

Salzmann M, Pervushin KV, Wider G, Senn H, Wüthrich K (1998): TROSY in triple-resonance experiments: new perspectives for sequential NMR assignment of large proteins. *Proc Natl Acad Sci USA* 95:13585–90.

Salzmann M, Pervushin KV, Wider G, Senn H, Wüthrich K (1999a): [$^{13}$C]-Constant-time [$^{15}$N,$^1$H]-TROSY-HNCA for sequential assignments of large proteins. *J Biomol NMR* 14:85–8.

Salzmann M, Wider G, Pervushin KV, Senn H, Wüthrich K (1999b): TROSY-type triple-resonance experiments for sequential NMR assignments of large proteins. *J Am Chem Soc* 121:844–8.

Sanders II CR, Schaff JE, Prestegard JH (1993): Orientational behavior of phosphatidylcholine bilayers in the presence of aromatic amphiphiles and a magnetic field. *Biophys J* 64:1069–80.

Seavey BR, Farr EA, Westler WM, Markley JL (1991): A relational database for sequence-specific protein NMR data. *J Biomol NMR* 1:217–36.

Sharma D, Rajarathnam K (2000): NMR chemical shifts can predict disulfide bond formation. *J Biomol NMR* 18:165–171.

Shuker SB, Hajduk PJ, Meadows RP, Fesik SW (1996): Discovering high-affinity ligands for proteins—SAR by NMR. *Science* 274:1531–4.

Spirin AS, Baranov VI, Ryabova LA, Ovodov SY, Alakhov YB (1988): A continuous cell-free translation system capable of producing polypeptides in high-yield. *Science* 242:1162–4.

Terwilliger TC (2000): Structural genomics in North America. *Nat Struct Biol* 7:935–9.

Thornton JM, Todd AE, Milburn D, Borkakoti N, Orengo CA (2000): From structure to function: approaches and limitations. *Nat Struct Biol* 7:991–4.

Tian J, Valafar H, Prestegard JH (2001): A dipolar coupling based strategy for simultaneous resonance assignment and structure determination of protein backbones. *J Am Chem Soc* 123:11791–6.

Tjandra N, Bax A (1997a): Direct measurement of distances and angles in biomolecules by NMR in a dilute liquid crystalline medium [see comments]. *Science* 278:1111–4. [The measurement of residual dipolar couplings provides a powerful method for determining longer-range constraints for NMR structure determination.]

Tjandra N, Bax A (1997b): High-resolution heteronuclear NMR of human ubiquitin in an aqueous liquid crystalline medium. *J Biomol NMR* 10:289–92.

Tjandra N, Omichinski JG, Gronenborn AM, Clore GM, Bax A (1997): Use of dipolar $^1H$–$^{15}N$ and $^1H$–$^{13}C$ couplings in the structure determination of magnetically oriented macromolecules in solution. *Nat Struct Biol* 4:732–8.

Tolman JR, Flanagan JM, Kennedy MA, Prestegard JH (1995): Nuclear magnetic dipole interactions in field-oriented proteins: information for structure determination in solution. *Proc Natl Acad Sci USA* 92:9279–83.

Ulrich EL, Doreleijers JF, Chae S, Mading S, Muñoz A, Schnoes AM, Wenger RK, Zhao Q, Ioannidis YE, Livny M, Markley JL (2000): "BioMagResBank (BMRB)." XIX International Conference on Magnetic Resonance in Biological Systems; 2000 August 20–25; Florence, Italy. Abs 442.

Vriend G (1990): WHAT IF: a molecular modeling and drug design program. *J Molec Graphics* 8:52–6.

Westler WM, Ortiz-Polo G, Markley JL (1984): Two-dimensional $^1H$–$^{13}C$ chemical-shift correlated spectroscopy of a protein at natural abundance. *J Magn Reson* 58:354–7.

Westler WM, Kainosho M, Nagao H, Tomonaga N, Markley JL (1988a): Two-dimensional NMR strategies for carbon–carbon correlations and sequence-specific assignments in carbon-13 labeled proteins. *J Am Chem Soc* 110:4093–5.

Westler WM, Stockman BJ, Hosoya Y, Miyake Y, Kainosho M, Markley JL (1988b): Correlation of carbon-13 and nitrogen-15 chemical shifts in uniformly labeled proteins by heteronuclear two-dimensional NMR spectroscopy. *J Am Chem Soc* 110:6256–8.

Williamson MP, Kikuchi J, Asakura T (1995): Application of $^1H$ NMR chemical shifts to measure the quality of protein structures. *J Mol Biol* 247:541–6.

Wishart DS, Sykes BD, Richards FM (1991): Relationship between nuclear magnetic resonance chemical shift and protein secondary structure. *J Mol Biol* 222:311–33.

Wishart DS, Sykes BD, Richards FM (1992): The chemical shift index: a fast and simple method for the assignment of protein secondary structure through NMR spectroscopy. *Biochemistry* 31:1647–51.

Wishart DS, Sykes BD (1994): The $^{13}C$ chemical shift index: a simple method for the identification of protein secondary structure using $^{13}C$ chemical shifts. *J Biomol NMR* 4:171–80.

Wishart DS, Watson MS, Boyko RF, Sykes BD (1997): Automated $^1H$ and $^{13}C$ chemical shift prediction using the BioMagResBank. *J Biomol NMR* 10:329–36.

Wishart DS, Nip AM (1998): Protein chemical shift analysis: a practical guide. *Biochem Cell Biol* 76:153–63. [This work summarizes the wealth of information available from NMR chemical shifts about protein structure.]

Wüthrich K (1986): *NMR of Proteins and Nucleic Acids.* New York: John Wiley & Sons.

Xia YL, Sze KH, Zhu G (2000): Transverse relaxation optimized 3D and 4D N-15/N-15 separated NOESY experiments of N-15 labeled proteins. *J Biomol NMR* 18:261–8.

Xu R, Ayers B, Cowburn D, Muir TW (1999): Chemical ligation of folded recombinant proteins: segmental isotopic labeling of domains for NMR studies. *Proc Natl Acad Sci USA* 96:388–93.

Yamazaki T, Otomo T, Oda N, Kyogoku Y, Uegaki K, Ito N, Ishino Y, Nakamura H (1998): Segmental isotope labeling for protein NMR using peptide splicing. *J Am Chem Soc* 120:5591–2.

Yokoyama S (2000): "RIKEN Structural Genomics Initiative." International Conference on Structural Genomics 2000; 2000 November 2–5; Yokohama, Japan. Abs 40.

Yokoyama S, Hirota H, Kigawa T, Yabuki T, Shirouzu M, Terada T, Ito Y, Matsuo Y, Kuroda Y, Nishimura Y, Kyogoku Y, Miki K, Masui R, Kuramitsu S (2000): Structural genomics projects in Japan. *Nat Struct Biol* 7:943–5.

Zhou P, Lugovskoy AA, Wagner G (2001): A solubility-enhancement tag (SET) for NMR studies of poorly behaving proteins. *J Biomol NMR* 20:11–4.

Zhu G, Kong XM, Sze KH (1999a): Gradient and sensitivity enhancement of 2D TROSY with water flip-back, 3D NOESY-TROSY and TOCSY-TROSY experiments. *J Biomol NMR* 13:77–81.

Zhu G, Xia YL, Sze KH, Yan XZ (1999b): 2D and 3D TROSY-enhanced NOESY of N-15 labeled proteins. *J Biomol NMR* 14:377–81.

Zimmerman DE, Kulikowski CA, Huang Y, Feng W, Tashiro M, Shimotakahara S, Chien C-Y, Powers R, Montelione GT (1997): Automated analysis of protein NMR assignments using methods from artificial intelligence. *J Mol Biol* 269:592–610.

Zweckstetter M, Bax A (2001): Single-step determination of protein substructures using dipolar couplings: aid to structural genomics. *J Am Chem Soc* 123:9490–1. [This work demonstrates how it may be possible to determine low-resolution NMR structures of proteins routinely from minimal data sets.]

# 6

# ELECTRON MICROSCOPY

Niels Volkmann and Dorit Hanein

As modern molecular biology moves from single molecules toward more complex multimolecular machines, the need for structural information about these assemblies grows. Nuclear magnetic resonance (NMR) spectroscopy (see Chapter 5) and X-ray crystallography (see Chapter 4) are well-established approaches for obtaining atomic structures of biological macromolecules, but it has become increasingly clear that the structures of individual components of assemblies can be only a first step to understanding a biological phenomenon. Biological events are usually more than the sum of their parts.

Due to dramatic improvements in experimental methods and computational techniques, electron microscopy has matured into a powerful and diverse collection of methods that allow the visualization of the structure and dynamics of an extraordinary range of biological assemblies at resolutions spanning from molecular (about 2–3 nm) to near atomic (0.3 nm). Many of the restrictions of X-ray crystallography or NMR spectroscopy do not apply to electron microscopy. Crystalline order is helpful but not necessary; there is no upper size limit for the structures studied, the quantities of sample needed are relatively small, and cryomethods enable the observation of molecules in their native aqueous environment (Dubochet et al., 1988). All in all, imaging of large and multicomponent cellular machinery close to physiological conditions is possible using electron microscopy and image analysis.

In the early years of electron microscopy, electron micrographs of molecules in a thin film of heavy atom stain were used to produce structures that were interpreted directly. Later, the interpretation of the two-dimensional (2D) images as projected density summed along the direction of the electron beam led to the ability to reconstruct the three-dimensional (3D) object that was imaged (DeRosier and Klug, 1968). The 1980s marked the development of electron cryomicroscopy where macromolecules are examined without the use of heavy atom stains by embedding the specimens in a thin

*Structural Bioinformatics*
Edited by Philip E. Bourne and Helge Weissig
Copyright © 2003 by Wiley-Liss, Inc.

film of rapidly frozen water (Dubochet et al., 1988). This use of unstained specimens led to structure determination of the molecules themselves rather than the structure of a stain-excluding volume (negative stain). The staining procedures greatly enhance the signal-to-noise ratio for imaging of biological macromolecules but are severely limited by preservation artifacts. The signal-to-noise ratio in electron cryomicroscopy is much lower, but it allows imaging of biological specimens close to their native, fully hydrated state.

Up to November 2001, four atomic resolution structures had been obtained by electron cryomicroscopy of thin 2D crystalline arrays (Henderson et al., 1990; Kühlbrandt, Wang, and Fujiyoshi, 1994; Nogales, Wolf, and Downing, 1998; de Groot, Engel, and Grubmuller, 2001). For most biological macromolecules and assemblies, it has not yet been possible to determine their structure beyond 0.7–3 nm resolution using electron microscopy and image analysis. Although this resolution precludes atomic modeling directly from the data, near-atomic models often can be generated by combining high-resolution structures of individual components in a macromolecular complex with a low-resolution structure of the entire assembly. Combination of electron microscopy with bioinformatics-based technologies such as pattern recognition, database searches, or homology modeling are increasingly used to generate molecular models of large assemblies as well. This chapter gives an overview of key aspects of electron cryomicroscopy and puts them into context with structural bioinformatics. More information on the various aspects of electron cryomicroscopy can be obtained from several recent review articles (Baumeister, Grimm, and Walz, 1999; Baumeister and Steven, 2000; Chiu et al., 1999; Kühlbrandt and Williams, 1999; McEwen and Frank, 2001; Saibil, 2000).

## ELECTRON OPTICS AND IMAGE FORMATION

Electron cryomicroscopy provides 3D electron-density maps of macromolecules very similar to the electron-density maps determined by X-ray crystallography. In the imaging process of electron microscopy, the incident electron beam passes through the specimen and individual electrons are either unscattered or scattered by the specimen. Scattering occurs either elastically, with no loss of energy, or inelastically, with energy transfer from the scattering electrons to electrons in the specimen; thus, leading to radiation damage. The electrons emerging from the specimen are collected and focused by the imaging optics of the microscope (Fig. 6.1). In the viewing area either the electron diffraction pattern or the image can be seen directly by eye on the phosphor screen, detected by a change-coupled device (CCD) camera, or recorded on photographic film or imaging plate.

Structural information can only be obtained from coherent, elastic scattering of the electrons. The amplitudes and phases of the scattered electron beam are directly related to the Fourier components of the atomic distribution in the specimen. When the scattered beams are recombined with the unscattered beam in the image, they create an interference pattern that, for thin specimens, is directly related to density variations in the specimen. Thin samples of biological molecules fulfill the weak phase approximation, a theory of image formation that is used to describe the phase-contrast images of weakly scattering specimens. Although there is practically no contrast when the image is in focus, spherical aberration and defocus combine to give a phase-contrast image. The imaging characteristics are described by the contrast-transfer function (CTF),

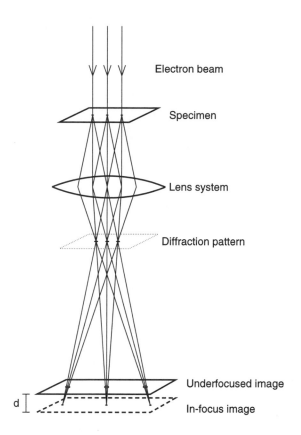

Figure 6.1. Simplified schematic diagram showing the principle of image formation in the electron microscope. The incident electron beam illuminates the specimen. Scattered and unscattered electrons are collected by the objective lens system and focused back to form first an electron diffraction pattern and then an image. In practice, an in-focus image has no contrast, so images are recorded with the objective lens system slightly defocused (d), taking advantage of the out-of-focus phase contrast mechanism.

which can be derived from the weak phase approximation. The CTF describes the contrast transfer as a function of spatial frequency. It has alternating bands of positive and negative contrast (Fig. 6.2), appearing in diffraction images as Thon rings. In order to restore the correct structural information, the images must be corrected for the CTF. For high-resolution studies, images must be collected at a range of defocus values to fill in missing data caused by zeros in the CTF whose positions vary with the actual defocus.

The most important consequence of inelastic scattering is the deposition of energy in the specimen, leading to radiation damage. Scattering events with X-rays are about 1000 times more damaging than those with electrons, but the cross-section for electron scattering is $10^5$ times greater. Therefore, radiation damage is a much more serious problem for electron microscopy and cooling is essential for imaging of high-resolution detail. Still, radiation damage is limiting even at low temperature. Therefore, the image exposures are chosen to be the weakest possible to obtain a measurable signal. Consequently, the signal-to-noise ratio of the recorded images is extremely low. This low signal-to-noise limits the amount of information that can be obtained from an image

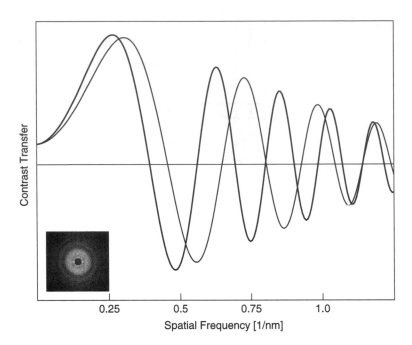

**Figure 6.2.** Graph of CTF plotted against spatial frequency. The lines represent underfocus levels of 2 μm (thin line) and 3 μm. The inset shows the typical Thon rings that are caused by the modulation effect of the CTF and are visible in diffraction patterns of electron micrographs recorded with defocused lens systems. The finite source size and nonmonochromatic electron beam both cause Gaussian decay of the CTF. In practice, contrast is much weaker at high spatial frequencies as a result of further decay caused by other factors such as inelastic scattering or specimen drift. When the CTF crosses the zero line (the straight line in the plot), the phases need to be flipped in order to correct for the contrast reversal at the corresponding spatial frequencies. At the spatial frequencies where the CTF actually approaches zero, no information is present. By collecting various data sets with different defocus, these information gaps can be filled.

of a single biological macromolecule. The high-resolution structure cannot be determined from a single molecule alone but requires the averaging of the information from at least 10,000 molecules in theory and even more in practice (Henderson, 1995). For example, five million molecules were used to determine the atomic structure for bacteriorhodopsin (Henderson et al., 1990).

The electron optical resolution of electron microscopy is on the order of 0.1 nm, much coarser than the diffraction limit imposed by the electron wavelength. The resolution is restricted by the small aperture size needed because of aberrations in the electromagnetic lenses. However, additional resolution restrictions for macromolecules come from radiation sensitivity, specimen movement in the electron beam, and low contrast. These effects have so far limited the resolution of macromolecular imaging to 0.3–0.4 nm in the best cases. Particularly for single particles, the loss of contrast beyond 2 nm resolution is a major limitation. This limit can be extended by the use of a field emission gun (FEG) electron source. The small apparent source size gives a highly coherent illumination that provides much better phase contrast at high resolution.

The microscope can be used either as an imaging or a diffraction instrument. Unlike diffraction experiments, in which phase information is lost, the image recorded by microscopy contains both amplitude and phase information. However, a much higher electron dose is needed to record the image than for the electron-diffraction pattern. Mechanical stability is particularly critical for obtaining phases. Movement does not affect the amplitudes in electron diffraction provided the crystalline area stays in the beam, but any movement during image recording distorts the phases and may easily make the image unusable.

## THREE-DIMENSIONAL RECONSTRUCTION

The determination of 3D structure by electron cryomicroscopy follows a common scheme for all macromolecules. Briefly, each sample must be prepared in a relatively homogeneous, aqueous form. This specimen is then rapidly frozen (vitrified) as a thin film, transferred to the electron microscope, and imaged under low-dose conditions (less than 5000 electrons/nm$^2$). Before image analysis, the best micrographs are selected in which the electron exposure is correct, there is no specimen movement, minimal astigmatism, and a reasonable amount of defocus. Interesting areas are boxed out for further use.

An electron micrograph consists of 2D projections of a 3D object. To retrieve its 3D structure, sufficiently sampled angular views of the object need to be aligned and combined. An object might possess crystalline, helical, icosahedral, or rotational symmetry, or no symmetry at all. The presence of symmetry means that redundant motifs are provided in the specimen, thereby enhancing the signal-to-noise ratio of the image, providing geometric constraints for the alignment of the objects, and reducing the number of images required to obtain a reconstruction. The exact steps of image analysis and image acquisition vary according to the symmetry and nature of the specimen. Three basic tasks are common to all samples (Fig. 6.3). First, images of the object must be obtained in a sufficient number of orientations. This task can be achieved using the natural design of the object (helical filaments or icosahedral viruses), using experimental design (by tilting the sample holder in the microscope to specified angles), or through a random distribution of orientations (single particles). Second, the orientation and center of the object needs to be determined. Iterative refinement of these parameters is usually carried out by cross-comparison between different images or by projection images of preliminary models. Third, image shifts must be applied computationally either in real space or Fourier space to bring all views of the object to a common origin. Only then can a 3D reconstruction be calculated. The different sample geometries require different data-collection schemes and image-processing approaches. These are described below.

### Crystalline Arrays

Ordered protein arrays, often only one molecule thick, are too insubstantial to be analyzed by X-ray crystallography. Due to the comparatively large cross-section of electron scattering and the resulting increase in scattering power, structure determination using electron beams is feasible for these samples. The strategy followed is to build up a 3D Fourier transform of the repeating unit by recording data from arrays tilted through various angles. The calculated Fourier transform of each image provides phases and amplitudes of a central plane through the 3D Fourier transform. For

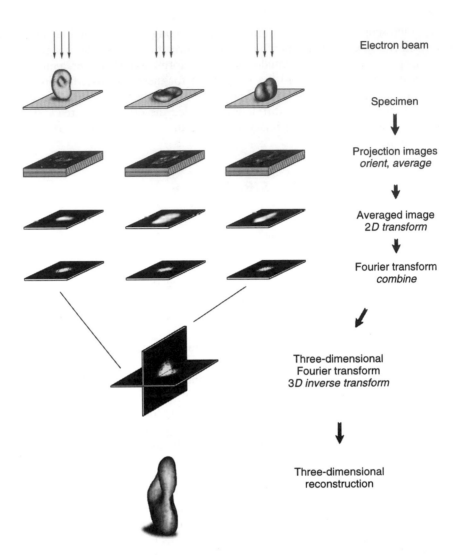

**Figure 6.3.** The principle of 3D reconstruction from 2D projections using the Arp2/3 complex (Volkmann et al., 2001a) as an example. The electron beam illuminates the molecule in different orientations. This illumination gives rise to sets of 2D projection images that need to be classified and oriented. In order to enhance the signal, as many projection images of the same molecule orientation as possible are averaged. The 2D Fourier transform of each projection (shown below the averaged projection images) is a section through the 3D Fourier transform of the underlying structure. The 3D Fourier transform is represented by two intersecting transform sections derived from the top and front view of the structure. Once enough sections are available, the full 3D Fourier transform can be interpolated and inverse transformed into a 3D density reconstruction.

well-ordered arrays, electron diffraction yields amplitudes superior to those calculated from the image and are consequently incorporated instead. In addition to the direct measurement of phases that distinguishes electron crystallography from X-ray crystallography, the images can be corrected for short-range lattice disorder such as bending or wrinkling of the arrays (Henderson et al., 1990; Kunji et al., 2000). This extends the resolution of the calculated transform from each image. There is a missing cone of data because the maximum tilt angle possible is 60–70°. This results in anisotropic resolution: features parallel to the plane of the array are better resolved than features perpendicular to the plane. Other possible complications with obtaining high-resolution structures from crystalline arrays include difficulties in finding proper crystallization conditions and lattice defects. Once images of sufficient quality are acquired, lattice reflections calculated from each image are assembled into the 3D Fourier transform, which is interpolated and inverse transformed to give the 3D density map.

Electron crystallography is well suited for intrinsic membrane proteins that are visualized in their natural environment embedded in a lipid bilayer. Bacteriorhodopsin, which naturally exists as planar arrays in the cell, was the first molecule with an atomic model provided by electron crystallography (Henderson et al., 1990). The plant light-harvesting complex is another example of an atomic model entirely based on electron crystallographic analysis (Kühlbrandt, Wang, and Fujiyoshi, 1994). Recently also electron-crystallography-based models of the membrane protein aquaporin became available (de Groot, Engel, and Grubmuller, 2001). In addition, the 3D structures of about a dozen or so membrane proteins were determined at a resolution of 0.5 nm to 1 nm (for a recent review see Stahlberg et al., 2001). An atomic model also has been obtained for tubulin, a major component of the cytoskeleton (Nogales, Wolf, and Downing, 1998), and several other proteins that are not integral membrane components have also been studied at somewhat lower resolution (see Stahlberg et al., 2001).

## Helical Assemblies

Many biological assemblies occur naturally in helical form, particularly cytoskeleton filaments. These filamentous structures are particularly attractive targets for helical reconstruction techniques as they are not usually amenable to crystallization due to their natural tendency to polymerize. Actin filaments (see, for example, De La Cruz et al., 2000; Orlova and Egelman, 2000; Steinmetz et al., 2000); actin filaments bound to capping proteins (McGough et al., 1997); actin complexed to domains of cytoskeletal proteins such as α-actinin (McGough, Way, and DeRosier, 1994), fimbrin (Hanein, Matsudaira, and DeRosier, 1997; Hanein et al., 1998), utrophin (Moores, Keep, and Kendrick-Jones, 2000), calponin (Hodgkinson et al., 1997) have all been studied by electron microscopy and helical reconstructions techniques. Conformational changes in motor proteins (myosin, kinesin, NCD) are being investigated by using electron cryomicroscopy techniques with actin- or microtubule-bound motors (for recent reviews, see Vale and Milligan, 2000; Volkmann and Hanein, 2000).

Helical crystallization has also been used for structure determination, particularly in the case of membrane proteins, which can be induced to form tubular crystals. Diffraction from a helix occurs on a set of layer lines related to the pitch repeat. An advantage of helical analysis over crystalline arrays is that the repeating unit is naturally presented over a range of angular views and tilting is not usually necessary for 3D reconstruction (DeRosier and Moore, 1970). However, it is more difficult to collect a large enough number of the repeating units and to achieve high resolution. Possible complications that arise during helical analysis include partial decoration of the helix under study, bending of the helix in the plane of the image and perpendicular to the plane, or imperfect helical symmetry. The majority of the helical reconstructions are in the 1.5–3 nm range, but 0.4–0.5 nm resolution also has been achieved using tubular crystals of the nicotinic acetylcholine receptor (Miyazawa et al., 1999).

## Single-Particle Analysis

Isolated single particles (macromolecules) offer certain practical advantages for electron cryomicroscopy. Because there is no requirement for crystallinity, virtually any particle is eligible. For more detailed information on single-particle reconstruction see recent reviews by Frank et al. (2000), van Heel et al. (2000) and the textbook by Frank (1996). In summary, single-particle reconstruction takes advantage of the fact that molecular complexes often exist as many copies in the specimen, visible as isolated particles, distinguished only by their orientations. Thus, a snapshot of a sufficiently large number of particles covers the complete angular range of possible orientations. A small number of micrographs, each of a different field of view, often contain enough particles to reconstruct the molecule in three dimensions. The averaging of many copies of the structure by this approach reduces the noise and carries potential for reaching high resolution, even for particles without symmetry. Computationally, the most challenging task is to determine and refine the particle orientations.

Orientation analysis is much more straightforward if the sample is biochemically homogeneous and is also facilitated by internal particle symmetry. In order to generate a high-resolution reconstruction, an initial starting model at moderate resolution (~3 nm) is acquired as a first step. Methods for acquiring an initial starting model include the random conical tilt method (Radermacher, 1988), angular reconstitution (van Heel, 1987), the use of models of related structures, or electron tomography (see below). The second stage involves cyclic model-based refinement. At each stage, more accurate values for the viewing angles of each particle are obtained by matching it against projections of the current model, each of which represents a particular view. Translational refinement is also fine-tuned. Then, a refined reconstruction is calculated and the procedure is iterated exhaustively until convergence.

The resolution attained depends on several factors, including the number of particles in the data set, the accuracy of the orientation parameters, and the quality of the original data. Even under the most favorable conditions macromolecules yield noisy, low-contrast images. For successful orientation determination, a sufficient signal for discrimination among projections must be generated. In practice, this requirement places a lower size limit on macromolecules that can be analyzed by this technique with the current limit being at 220 kDa for particles without internal symmetry (Volkmann et al., 2001a). Particles of 1–10 MDa are considered optimal for high-resolution single-particle analysis.

The first single-particle analysis at resolutions below 1 nm was of the capsid of hepatitis B virus (Böttcher, Wgnne, and Crowther, 1997; Conway et al., 1997) and papillomavirus (Trus et al., 1997). Icosahedral capsids have the advantage of 60-fold symmetry, which reduces the amount of particles required for averaging. As 1 nm is the spacing typical of close-packed $\alpha$-helices, density maps with resolutions higher than 1 nm are particularly informative for proteins with high $\alpha$-helix contents. In addition to a handful of icosahedral virus structures that were determined at resolutions better than 1 nm (Mancini et al., 2000; Zhou et al., 2000; Zhou et al., 2001), the structure of a ribosomal subunit (Matadeen et al., 1999), a particle without internal symmetry, was determined at a resolution of 0.75 nm. While these results are very encouraging, the vast majority of single-particle analyses today yield reconstructions with resolutions considerably lower than 1 nm, thus precluding direct analysis and modeling of helical arrangements.

## Electron Tomography

The most general method for obtaining 3D information by electron microscopy is tomography. The method is not only applicable to isolated particles but also to pleomorphous structures such as mitochondria, other organelles, or even whole cells (for a recent review, see Baumeister, Grimm, and Walz, 1999). A special issue of the *Journal of Structural Biology* edited by Koster and Agard (1997), a textbook edited by Frank (1992), and another upcoming special issue of the *Journal of Structural Biology* (in preparation) are all devoted to electron tomography. In this technique, a series of images is taken of a single specimen as the specimen is tilted over a wide range of angles. Sometimes, for better angular coverage, another tilt series is taken with the specimen rotated by 90°.

Tomography is the only method available for reconstruction of specimen with unique structure (no multiple copies). An entire cell for example would fall into this category, as it would be impossible to find two cells that are exactly identical. Today, with the use of computer-controlled microscopes and the availability of CCD cameras, it has become possible to image large-scale structures at a resolution of better than 5 nm, with data sets comprising up to 150 projections with a cumulative dose as low as 5000 electrons/nm$^2$. Tomography is undergoing considerable growth at the present time due to the realization that molecular information can be obtained from unstained, frozen-hydrated whole cells (Grimm et al., 1998) and isolated organelles (Nicastro et al., 2000). The main disadvantage of the tomographic approach is that radiation damage builds up during the multiple exposures as the specimen is being tilted. Although data collection with extremely low doses of radiation is under development, the experimental realization and image processing under such conditions still poses great challenges. Similar to electron crystallography, there is a missing wedge of data because of the maximum tilt angle, resulting in anisotropic resolution. Features perpendicular to the electron beam are better resolved than features parallel to the beam.

Although the main area of application for electron tomography is large, multicomponent objects, structures of single macromolecules also have been determined in negative stain (Rockel et al., 1999) as well as in vitrified ice (Nitsch et al., 1998). In this case, the 3D, noisy tomograms of the single macromolecules were aligned and averaged in three dimensions. To generate an averaged, high-quality 3D structure, only a few hundred particles are required using tomographic techniques (Koster et al., 1997), an advantage of using electron tomography instead of single-particle analysis that requires several thousand particles for structure determination. Resolution of up to 2.2 nm was achieved using this single-particle tomography approach.

## Hybrid Methods

A recent trend is the innovative combination of the more traditional methodologies mentioned above in order to push the limits of electron microscopy even further. For example, single-particle image-processing methods can be exploited for disorder correction of subunits or groups of subunits in ordered assemblies such as helical filaments (Egelman, 2000; Yu and Egelman, 1997) or crystalline arrays (Sherman et al., 1998; Verschoor, Tivol, and Mannella, 2001). Helical orientation parameters (that tend to be more accurate than single-particle orientation parameters) can be used to aid single-particle reconstructions of asymmetric particles attached to the end of helices (Yonekura et al., 2000). A combination of helical analysis and electron crystallographic techniques

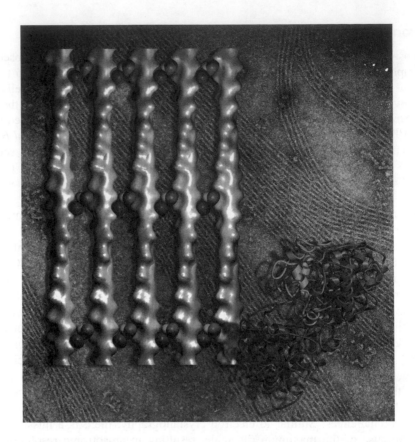

**Figure 6.4.** Example for a hybrid study that combines elements of electron crystallography and helical reconstruction with homology modeling and molecular docking approaches to elucidate the structure of an actin-fimbrin crosslink (Volkmann et al., 2001b). Fimbrin is a member of a large superfamily of actin-binding proteins and is responsible for cross-linking of actin filaments into ordered, tightly packed networks such as actin bundles in microvilli or stereocilia of the inner ear. The diffraction patterns of ordered paracrystalline actin-fimbrin arrays (background) were used to deduce the spatial relationship between the actin filaments and the various domains of the crosslinker. Combination of this data with homology modeling and data from docking the crystal structure of fimbrin's N-terminal actin-binding domain into helical reconstructions (Hanein et al., 1998) allowed us to build a complete atomic model of the cross-linking molecule. Figure also appears in Color Figure section.

is being used to interpret 2D paracrystalline arrays of filamentous structures (Sukow and DeRosier, 1998; Volkmann et al., 2001b and Fig. 6.4) and electron tomography is being used to generate starting models for single-particle analysis (Walz et al., 1999).

## Imaging Protein Dynamics

In electron cryomicroscopy, molecules can be imaged in their native, fully hydrated environment, unrestricted by a crystal lattice. This technique gives the opportunity to study macromolecules in their various functional states and to determine the associated conformational changes. Actin- and microtubule-based motor proteins are good examples where function-related motions can be detected by electron cryomicroscopy. Changes of the nucleotide state of several actin-bound myosin isoforms, for example, induce large conformational changes that are readily identified by eye (see for example Carragher et al., 1998; Jontes, Wilson-Kubalek, and Milligan, 1995; Whittaker et al., 1995). In addition, recent advances in computational methodology and image processing allowed more subtle changes to be identified and quantified in these systems (Volkmann et al., 2000). Another example for glimpses of a molecular machine in motion is provided by recent electron microscopy studies of the ribosome, which began to unravel the structural changes associated with different functional states in the translation process (Frank and Agrawal, 2000; Gabashvili et al., 2001; Spahn et al., 2001b; VanLoock et al., 2000). Electron microscopy of the chaperonin GroEL revealed a large repertoire of hinge rotations and allosteric movements in the chaperonin ATPase cycle (Roseman et al., 2001).

Because the freezing step in electron cryomicroscopy is very rapid (considerably less than 1 msec), it is possible to capture very short-lived structural states by spraying a ligand onto the specimen just before freezing. The spray method allowed capturing the acetylcholine-activated state of the nicotinic acetylcholine receptor, which has a lifetime of only 10 msec (Berriman and Unwin, 1994). A stopped-flow mixer system combined with an atomizer spray has been developed for actomyosin kinetic studies (Walker et al., 1999).

A possibility for detecting dynamics or mixed conformations directly in electron microscopic reconstructions is the analysis of the local variability using statistics on the single molecular units that contribute to the final average. Variants of this idea were developed for single-particle analysis (Liu, Boisset, and Frank, 1995) as well as for helical reconstructions (Rost, Hanein, and DeRosier, 1998).

## COMBINATION WITH OTHER APPROACHES

In the 3D structure determination of macromolecules, X-ray crystallography covers the full range from small molecules to very large assemblies such as viruses with molecular masses of megadaltons. The limiting factors are expression, crystallization, and the stability and homogeneity of the structure. In the case of NMR, structures can be determined from molecules in solution, but the size limit, although increasing, is presently of the order of 100 kDa. Dynamic aspects can be quantified, but again the structures of mixed conformational states cannot be determined. Electron microscopy provides complementary information to these other methods, being able to tackle very large assemblies as well as transient or mixed states, and usually requires small amounts of material. However, for the majority of the biological specimens studied, it has not yet been possible to determine their structures beyond 1.5 nm to 3 nm. By combining electron microscopy and image analysis with other sources of information, the gap between high-resolution information as obtained by X-ray crystallography and NMR and lower-resolution information such as that coming from light microscopy can be bridged.

### X-Ray Crystallography

Combining electron microscopy and crystallographic data can take two forms. When the whole specimen can be crystallized, the electron microscopy results can be used as an initial molecular-replacement phasing model. The principal difficulty here is the frequently poor overlap between the data from electron microscopy (highest resolution 2–3 nm) and the X-ray data (spots with resolution lower than 3 nm are often hidden in the beamstop). If successful phasing can be initiated with the electron microscopic reconstruction, it can be used to find heavy atoms, as was the case in the analysis of a ribosome subunit (Ban et al., 1998). Electron microscopic reconstructions can also be used to improve and extend crystallographic phases by noncrystallographic symmetry averaging (see, for example, Prasad et al., 1999) or as constraints in maximum-likelihood phasing procedures (Volkmann et al., 1995).

When the complete biomolecular assembly of interest can be imaged by electron microscopy but cannot be crystallized, lower-resolution information from electron microscopy can often be combined with the high-resolution information from atomic models of the assembly components. Manual fitting is widely used for this purpose (for a review, see Baker and Johnson, 1996). With this method, the fit of the atomic model into iso-surface envelopes calculated from the electron microscopic reconstruction is judged by eye and corrected manually until the fit "looks best." Objective scoring functions have been used occasionally to assess the quality and to refine the initial manual fit. If the components of the assembly under study are large molecules with distinctive shapes at the resolution of the reconstruction, manual fitting can often be performed with relatively little ambiguity (see, for example, Rayment et al., 1993; Smith et al., 1993). However, divergent models of the same complex docked by eye have also been reported (Hoenger et al., 1998; Kozielski, Arnal, and Wade, 1998). In addition to atomic models from X-ray crystallography, models derived by homology modeling are increasingly used for docking studies (see, for example, Spahn et al., 2001a; Volkmann et al., 2001b).

Recently, approaches aiming at automated, quantitative docking of atomic structures into lower-resolution reconstructions from electron microscopy have been developed. These methods employ global, exhaustive searches for the best fit, using various density correlation measures (Rossmann, 2000; Volkmann and Hanein, 1999), sometimes combined with density filtering operations (Roseman, 2000; Wriggers and Chacon, 2001), or by matching vector distributions derived by vector quantization of the atomic model and the reconstruction (Wriggers and Birmanns, 2001; Wriggers et al., 1999).

Open issues in this area include estimation of fitting quality, validation of results, estimation of fitting errors, and detection of ambiguities. A promising concept in this regard is that of solution sets (Volkmann and Hanein, 1999; Volkmann et al., 2000). In this approach, the global search is followed by a statistical analysis of the distribution of the fitting criterion. The analysis results in the definition of confidence intervals that lead eventually to solution sets. These sets contain all fits that satisfy the data within the error margin defined by the chosen confidence level. Structural parameters of interest can then be evaluated as properties of these sets. For example, the uncertainty of each atom position of the fitted structure can be approximated by calculating the root-mean-square deviation for each atom using all members of the solution set. Ambiguities in the fitting are clearly reflected in the shape of the solution set (Hanein et al., 1998). The size of the solution set can serve as a normalized goodness-of-fit criterion. The smaller the set, the better the data determines the position of the fitted atomic structure.

The statistical nature of the approach allows the use of standard statistical tests, such as Student's t-test, to evaluate differences between models in different functional states and to help model conformational changes (Volkmann et al., 2000). It also allows

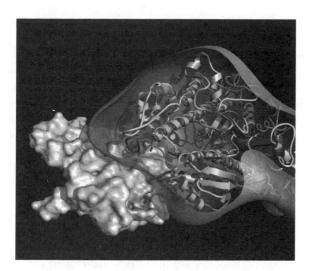

Figure 6.5. Example of a combination of high-resolution structural information from X-ray crystallography and medium-resolution information from electron cryomicroscopy (here 2.1 nm). Actin and myosin were docked into helical reconstructions of actin decorated with smooth-muscle myosin (Volkmann et al., 2000). Interaction of myosin with filamentous actin has been implicated in a variety of biological activities, including muscle contraction, cytokinesis, cell movement, membrane transport, and certain signal transduction pathways. Attempts to crystallize actomyosin failed due to the tendency of actin to polymerize. Docking was performed using a global search with a density correlation measure (Volkmann and Hanein, 1999). The estimated accuracy of the fit is 0.22 nm in the myosin portion and 0.18 nm in the actin portion. One actin molecule is shown on the left as a molecular surface representation. The fitted atomic model of myosin is shown on the right. The transparent envelope represents the density corresponding to myosin in the 3D reconstruction. The solution set concept (see text) was used to evaluate the results and to assign probabilities for residues to take part in the interaction. Figure also appear in Color Figure section.

estimating the probability that a certain residue is involved in the interaction between two components (Fig. 6.5). This probabilistic ranking of residues in terms of their involvement in binding gives a better starting point for the design of mutagenesis experiments.

Difference mapping between the density calculated from the fitted model and the reconstruction from electron microscopy (similar to $F_c$-$F_o$ maps in crystallography) is a powerful tool in locating portions of the structure that are not present in the crystal structure. For example, recent fitting of myosin crystal structures into reconstructions of actin-bound myosin revealed the location of a functionally important myosin loop of about 10 residues in reconstructions of about 2 nm resolution (Volkmann and Hanein, 1999; Volkmann et al., 2000). This loop was not resolved in any of the crystal structures due to structural flexibility. Presumably, actin binding stabilizes the loop so it becomes detectable in the reconstructions.

## Pattern Recognition

In the absence of atomic models, structural interpretation of large macromolecules at intermediate resolution is a difficult task, although structural information is still present.

Individual domains, components, and structural elements must be identified in the reconstruction to yield a tentative atomic model. If the resolution of the reconstruction under study is better than 1 nm, $\alpha$-helices become identifiable within the density by eye as well with pattern-recognition approaches. Very recently, a new method for helix recognition was developed (Jiang et al., 2001). This approach incorporates a multistep process that includes cross-correlation, density segmentation, segment quantification, and explicit description of the identified helices. The final helices are represented as cylinders, each specified by its length and six orientation parameters. The information encoded by these parameters as well as the relative position and orientation of the helices to each other can then be used to identify homologous structures based on spatial arrangement of secondary structural elements of proteins using a library based on protein structures in the Protein Data Bank (PDB) or to match helical regions from secondary structure prediction with the derived helical fragments in order to map sequence to structure. The latter method was recently used to derive the fold of rice dwarf virus that was reconstructed at 0.68 nm resolution (Zhou et al., 2001).

Pattern recognition approaches are also being developed for the interpretation of noisy, low-resolution tomograms of cells and large subcellular structures. The main experimental difficulty in interpreting these types of tomograms is the assignment of density to a particular molecular component. In this context, the pattern recognition can be divided in a feature extraction and a template-matching step. First, features of interest are extracted from the tomogram using segmentation algorithms (Frangakis and Hegerl, 1999; Volkmann, submitted). The second step consists of classification of the extracted features by template matching using a database of known atomic structures. Recently, a feasibility study of pattern recognition algorithms using calculated volumes and tomographic data sets containing isolated particles was conducted (Böhm et al., 2000). The tests demonstrate the feasibility of this strategy by showing that a distinction between the proteasome and the thermosome, molecules with similar shape but slightly different dimensions, can be made with reasonable confidence even in experimental tomograms at a resolution between 4 nm and 8 nm.

## FUTURE DIRECTIONS

Technical advances in data acquisition and in computational methods have made it possible to reconstruct biological macromolecular complexes at resolutions ranging from 3 nm to 0.35 nm. The development of more powerful computational methods coupled with the availability of faster computers with large storage capabilities will continue to have a major impact on the field. Parallel computing should further speed up the image analysis and 3D reconstruction process (Perkins et al., 1997). Automatization of data collection (see, for example, Carragher et al., 2000; Koster et al., 1992) and image analysis (see, for example, Ludtke, Baldwin, and Chiu, 1999) are important factors toward higher throughput. Electron microscopy provides complementary information to that from atomic-resolution techniques such as X-ray crystallography or NMR. Further development of methods for combining these data sources by docking is likely to play a major role in the future. An integration of these docking methods with structural database searches is also an attractive possibility for the future. Pattern recognition tools will make it possible to provide a bridge between cell biological function and molecular mechanism. Again, combination with database searches would add value. In summary, further technical progress with systematic integration of bioinformatics tools should allow electron microscopy to be a major player in the future of structural bioinformatics.

## ACKNOWLEDGMENTS

This work was supported by NIH research grants AR 47199 (DH) and GM/AR 64473 (NV). Dorit Hanein is a Pew Scholar in the Biomedical Sciences.

## FURTHER READING

Belmont LD, Orlova A, Drubin DG, Egelman EH (1999): A change in actin conformation associated with filament instability after Pi release. *Proc Natl Acad Sci USA* 96:29–34.

Egelman EH, Orlova A (1995): New insights into actin filament dynamics. *Curr Opin Struct Biol* 5:172–80.

## REFERENCES

Baker TS, Johnson JE (1996): Low resolution meets high: towards a resolution continuum from cells to atoms. *Curr Opin Struct Biol* 6:585–94.

Ban N, Freeborn B, Nissen P, Penczek P, Grassucci RA, Sweet R, Frank J, Moore PB, Steitz TA (1998): A 9 Å resolution X-ray crystallographic map of the large ribosomal subunit. *Cell* 93:1105–15.

Baumeister W, Grimm R, Walz J (1999): Electron tomography of molecules and cells. *Trends Cell Biol* 9:81–5.

Baumeister W, Steven AC (2000): Macromolecular electron microscopy in the era of structural genomics. *Trends Biochem Sci* 25:624–31.

Berriman J, Unwin N (1994): Analysis of transient structures by cryo-microscopy combined with rapid mixing of spray droplets. *Ultramicroscopy* 56:241–52.

Böhm J, Frangakis AS, Hegerl R, Nickell S, Typke D, Baumeister W (2000): Toward detecting and identifying macromolecules in a cellular context: template matching applied to electron tomograms. *Proc Natl Acad Sci USA* 97:14245–50.

Böttcher B, Wynne SA, Crowther RA (1997): Determination of the fold of the core protein of hepatitis B virus by electron cryomicroscopy. *Nature* 386:88–91.

Carragher BO, Cheng N, Wang ZY, Korn ED, Reilein A, Belnap DM, Hammer JAR, Steven AC (1998): Structural invariance of constitutively active and inactive mutants of acanthamoeba myosin IC bound to F-actin in the rigor and ADP-bound states. *Proc Natl Acad Sci USA* 95:15206–11.

Carragher B, Kisseberth N, Kriegman D, Milligan RA, Potter CS, Pulokas J, Reilein A (2000): Leginon: an automated system for acquisition of images from vitreous ice specimens. *J Struct Biol* 132:33–45.

Chiu W, McGough A, Sherman MB, Schmid MF (1999): High-resolution electron cryomicroscopy of macromolecular assemblies. *Trends Cell Biol* 9:154–9.

Conway JF, Cheng N, Zlotnick A, Wingfield PT, Stahl SJ, Steven AC (1997): Visualization of a 4-helix bundle in the hepatitis B virus capsid by cryo-electron microscopy. *Nature* 386:91–4.

DeRosier DJ, Klug A (1968): Reconstruction of three-dimensional structures from electron micrographs. *Nature* 217:130–4. [This seminal paper lays out the principles of 3D reconstruction from 2D projection images.]

DeRosier DJ, Moore PB (1970): Reconstruction of three-dimensional images from electron micrographs of structures with helical symmetry. *J Mol Biol* 52:355–69.

Dubochet J, Adrian M, Chang J-J, Homo J-C, Lepault J, McDowall AW, Schultz P (1988): Cryo-electron microscopy of vitrified specimens. *Quart Rev Biophys* 21:129–228. [This paper gives a detailed overview of the technical and experimental aspects of electron cryomicroscopy.]

Egelman EH (2000): A robust algorithm for the reconstruction of helical filaments using single-particle methods. *Ultramicroscopy* 85:225–34.

Frangakis A, Hegerl R (1999): Nonlinear anisotropic diffusion in three-dimensional electron microscopy. Lecture notes in computer science: scale space theories and computer vision 1682:386–97.

Frank J (1992): *Electron Tomography*. New York: Plenum Press.

Frank J, (1996): *Three-Dimensional Electron Microscopy of Macromolecular Assemblies*. San Diego: Academic Press. [This textbook on electron microscopy and 3D image reconstruction of macromolecules is an excellent starting point to gather more in-depth information on the field.]

Frank J, Agrawal RK (2000): A ratchet-like inter-subunit reorganization of the ribosome during translocation. *Nature* 406:318–22.

Frank J, Penczek P, Agrawal RK, Grassucci RA, Heagle AB (2000): Three-dimensional cryoelectron microscopy of ribosomes. *Methods Enzymol* 317:276–91.

Gabashvili IS, Gregory ST, Valle M, Grassucci R, Worbs M, Wahl MC, Dahlberg AE, Frank J (2001): The polypeptide tunnel system in the ribosome and its gating in erythromycin resistance mutants of L4 and L22. *Molec Cell* 8:181–8.

Grimm R, Singh H, Rachel R, Typke D, Zillig W, Baumeister W (1998): Electron tomography of ice-embedded prokaryotic cells. *Biophys J* 74:1031–42.

de Groot BL, Engel A, Grubmuller H (2001): A refined structure of human aquaporin-1. *FEBS Lett* 504:206–11.

Hanein D, Matsudaira P, DeRosier DJ (1997): Evidence for a conformational change in actin induced by fimbrin (N375) binding. *J Cell Biol* 139:387–96.

Hanein D, Volkmann N, Goldsmith S, Michon AM, Lehman W, Craig R, DeRosier D, Almo S, Matsudaira P (1998): An atomic model of fimbrin binding to F-actin and its implications for filament crosslinking and regulation. *Nat Struct Biol* 5:787–92.

Henderson R, (1995): The potential and limitations of neutrons, electrons and X-ray for atomic resolution microscopy of unstained biological molecules. *Quart Rev Biophys* 28:171–94. [In this paper, Henderson assesses the theoretical limitations of electron microscopy and puts it into perspective with neutron and X-ray diffraction.]

Henderson R, Baldwin JM, Ceska TA, Zemlin F, Beckmann E, Downing KH (1990): Model for the structure of bacteriorhodopsin based on high-resolution electron cryo-microscopy. *J Mol Biol* 213:899–929.

Hodgkinson JL, el-Mezgueldi M, Craig R, Vibert P, Marston SB, Lehman W (1997): 3-D image reconstruction of reconstituted smooth muscle thin filaments containing calponin: visualization of interactions between F-actin and calponin. *J Mol Biol* 273:150–9.

Hoenger A, Sack S, Thormahlen M, Marx A, Muller J, Gross H, Mandelkow E (1998): Image reconstructions of microtubules decorated with monomeric and dimeric kinesins: comparison with x-ray structure and implications for motility. *J Cell Biol* 141:419–30.

Jiang W, Baker ML, Ludtke SJ, Chiu W (2001): Bridging the information gap: computational tools for intermediate resolution structure interpretation. *J Mol Biol* 308:1033–44.

Jontes JD, Wilson-Kubalek EM, Milligan RA (1995): A 32 degree tail swing in brush border myosin I on ADP release. *Nature* 378:751–3.

Koster AJ, Chen H, Sedat JW, Agard DA (1992): Automated microscopy for electron tomography. *Ultramicroscopy* 46:207–27.

Koster AJ, Agard DA (1997): *J Struct Biol* 120:207–9. [This special issue of the *Journal of Structural Biology* is devoted to electron tomography. It is good starting point to learn more about this exciting technology. A follow up, edited by A. J. Koster and B. McEwen, is in preparation and is scheduled to appear early in 2002.]

Koster AJ, Grimm R, Typke D, Hegerl R, Stoschek A, Walz J, Baumeister W (1997): Perspectives of molecular and cellular electron tomography. *J Struct Biol* 120:276–308.

Kozielski F, Arnal I, Wade R (1998): A model of the microtubule-kinesin complex based on electron cryomicroscopy and X-ray crystallography. *Curr Biol* 8:191–8.

Kühlbrandt W, Wang DN, Fujiyoshi Y (1994): Atomic model of plant light-harvesting complex by electron crystallography. *Nature* 367:614–21.

Kühlbrandt W, Williams KA (1999): Analysis of macromolecular structure and dynamics by electron cryo-microscopy. *Curr Opin Chem Biol* 3:537–43.

Kunji ER, von Gronau S, Oesterhelt D, Henderson R (2000): The three-dimensional structure of halorhodopsin to 5 Å by electron crystallography: A new unbending procedure for two-dimensional crystals by using a global reference structure. *Proc Natl Acad Sci USA* 97:4637–42.

De La Cruz EM, Mandinova A, Steinmetz MO, Stoffler D, Aebi U, Pollard TD (2000): Polymerization and structure of nucleotide-free actin filaments. *J Mol Biol* 295:517–26.

Liu W, Boisset N, Frank J (1995): Estimation of variance distribution in three-dimensional reconstruction. II. Applications. *J Opt Soc Am* A12:2628–35.

Ludtke SJ, Baldwin PR, Chiu W (1999): EMAN: semiautomated software for high-resolution single-particle reconstructions. *J Struct Biol* 128:82–97.

Mancini EJ, Clarke M, Gowen BE, Rutten T, Fuller SD (2000): Cryo-electron microscopy reveals the functional organization of an enveloped virus, Semliki Forest virus. *Molec Cell* 5:255–66.

Matadeen R, Patwardhan A, Gowen B, Orlova EV, Pape T, Cuff M, Mueller F, Brimacombe R, van Heel M (1999): The *Escherichia coli* large ribosomal subunit at 7.5 Å resolution. *Structure Fold Des* 7:1575–83.

McEwen BF, Frank J (2001): Electron tomographic and other approaches for imaging molecular machines. *Curr Opin Neurobiol* 11:594–600.

McGough A, Way M, DeRosier D (1994): Determination of the alpha-actinin-binding site on actin filaments by cryoelectron microscopy and image analysis. *J Cell Biol* 126:433–43.

McGough A, Pope B, Chiu W, Weeds A (1997): Cofilin changes the twist of F-actin: implications for actin filament dynamics and cellular function. *J Cell Biol* 138:771–81.

Miyazawa A, Fujiyoshi Y, Stowell M, Unwin N (1999): Nicotinic acetylcholine receptor at 4.6 Å resolution: transverse tunnels in the channel wall. *J Mol Biol* 288:765–86.

Moores CA, Keep NH, Kendrick-Jones J (2000): Structure of the utrophin actin-binding domain bound to F-actin reveals binding by an induced fit mechanism. *J Mol Biol* 297:465–80.

Nicastro D, Frangakis AS, Typke D, Baumeister W (2000): Cryo-electron tomography of neurospora mitochondria. *J Struct Biol* 129:48–56.

Nitsch M, Walz J, Typke D, Klumpp M, Essen LO, Baumeister W (1998): Group II chaperonin in an open conformation examined by electron tomography. *Nat Struct Biol* 5:855–7.

Nogales E, Wolf SG, Downing KH (1998): Structure of the $\alpha\beta$-tubulin dimer by electron crystallography. *Nature* 391:199–203.

Orlova A, Egelman EH (2000): F-actin retains a memory of angular order. *Biophys J* 78:2180–5.

Perkins GA, Renken CW, Song JY, Frey TG, Young SJ, Lamont S, Martone ME, Lindsey S, Ellisman MH (1997): Electron tomography of large, multicomponent biological structures. *J Struct Biol* 120:219–27.

Prasad BV, Hardy ME, Dokland T, Bella J, Rossmann MG, Estes MK (1999): X-ray crystallographic structure of the Norwalk virus capsid. *Science* 286:287–90.

Radermacher M (1988): Three-dimensional reconstruction of single particles from random and nonrandom tilt series. *J Electron Microsc Tech* 9:359–94.

Rayment I, Holden HM, Whittaker M, Yohn CB, Lorenz M, Holmes KC, Milligan RA (1993): Structure of the actin-myosin complex and its implications for muscle contraction. *Science* 261:58–65.

Rockel B, Walz J, Hegerl R, Peters J, Typke D, Baumeister W (1999): Structure of VAT, a CDC48/p97 ATPase homologue from the archaeon Thermoplasma acidophilum as studied by electron tomography. *FEBS Lett* 451:27–32.

Roseman AM (2000): Docking structures of domains into maps from cryo-electron microscopy using local correlation. *Acta Crystallogr D* 56:1332–40.

Roseman AM, Ranson NA, Gowen B, Fuller SD, Saibil HR (2001): Structures of unliganded and ATP-bound states of the *Escherichia coli* chaperonin GroEL by cryoelectron microscopy. *J Struct Biol* 135:115–25.

Rossmann MG (2000): Fitting atomic models into electron-microscopy maps. *Acta Crystallogr D* 56:1341–9.

Rost LE, Hanein D, DeRosier DJ (1998): Reconstruction of symmetry deviations: a procedure to analyze partially decorated F-actin and other incomplete structures. *Ultramicroscopy* 72:187–97.

Saibil HR (2000): Conformational changes studied by cryo-electron microscopy. *Nat Struct Biol* 7:711–4.

Sherman MB, Soejima T, Chiu W, van Heel M (1998): Multivariate analysis of single unit cells in electron crystallography. *Ultramicroscopy* 74:179–99.

Smith TJ, Olson NH, Cheng RH, Liu H, Chase ES, Lee WM, Leippe DM, Mosser AG, Rueckert RR, Baker TS (1993): Structure of human rhinovirus complexed with Fab fragments from a neutralizing antibody. *J Virol* 67:1148–58.

Spahn CM, Beckmann R, Eswar N, Penczek PA, Sali A, Blobel G, Frank J (2001a): Structure of the 80S ribosome from Saccharomyces cerevisiae-tRNA-ribosome and subunit-subunit interactions. *Cell* 107:373–86.

Spahn CM, Kieft JS, Grassucci RA, Penczek PA, Zhou K, Doudna JA, Frank J (2001b): Hepatitis C virus IRES RNA-induced changes in the conformation of the 40s ribosomal subunit. *Science* 291:1959–62.

Stahlberg H, Fotiadis D, Scheuring S, Remigy H, Braun T, Mitsuoka K, Fujiyoshi Y, Engel A (2001): Two-dimensional crystals: a powerful approach to assess structure, function and dynamics of membrane proteins. *FEBS Lett* 504:166–72.

Steinmetz MO, Hoenger A, Stoffler D, Noegel AA, Aebi U, Schoenenberger CA (2000): Polymerization, three-dimensional structure and mechanical properties of dictyostelium versus rabbit muscle actin filaments. *J Mol Biol* 303:171–84.

Sukow C, DeRosier D (1998): How to analyze electron micrographs of rafts of actin filaments crosslinked by actin-binding proteins. *J Mol Biol* 284:1039–50.

Trus BL, Roden RB, Greenstone HL, Vrhel M, Schiller JT, Booy FP (1997): Novel structural features of bovine papillomavirus capsid revealed by a three-dimensional reconstruction to 9 Å resolution. *Nat Struct Biol* 4:413–20.

Vale RD, Milligan RA (2000): The way things move: looking under the hood of molecular motor proteins. *Science* 288:88–95.

van Heel M (1987): Angular reconstitution: a posteriori assignment of projection directions for 3D reconstruction. *Ultramicroscopy* 21:111–24.

van Heel M, Gowen B, Matadeen R, Orlova EV, Finn R, Pape T, Cohen D, Stark H, Schmidt R, Schatz M, Patwardhan A (2000): Single-particle electron cryo-microscopy: towards atomic resolution. *Quart Rev Biophys* 33:307–69.

VanLoock MS, Agrawal RK, Gabashvili IS, Qi L, Frank J, Harvey SC (2000): Movement of the decoding region of the 16 S ribosomal RNA accompanies tRNA translocation. *J Mol Biol* 304:507–15.

Verschoor A, Tivol WF, Mannella CA (2001): Single-particle approaches in the analysis of small 2d crystals of the mitochondrial channel vdac. *J Struct Biol* 133:254–65.

Volkmann N (2002): A novel three-dimensional variant of the watershed transform for segmentation of electron density maps. *J Struct Biol* 138:123–129.

Volkmann N, Schlünzen F, Vernoslava EA, Urzhumstev AG, Podjarny AD, Roth M, Pebay-Peyroula E, Berkovitch-Yellin Z, Zaytzev-Bashan A, Yonath A (1995): On ab initio Phasing of Ribosomal Particles at Very Low Resolution. *Joint CCP4 and ESF-EACBM Newsletters* 31:25–30.

Volkmann N, Hanein D (1999): Quantitative fitting of atomic models into observed densities derived by electron microscopy. *J Struct Biol* 125:176–84.

Volkmann N, Hanein D (2000): Actomyosin: law and order in motility. *Curr Opin Cell Biol* 12:26–34.

Volkmann N, Hanein D, Ouyang G, Trybus KM, DeRosier DJ, Lowey S (2000): Evidence for cleft closure in actomyosin upon ADP release. *Nat Struct Biol* 7:1147–55.

Volkmann N, Amann KJ, Stoilova-McPhie S, Egile C, Winter DC, Hazelwood L, Heuser JE, Li R, Pollard TD, Hanein D (2001a): Structure of Arp2/3 complex in its activated state and in actin filament branch junctions. *Science* 293:2456–9.

Volkmann N, DeRosier D, Matsudaira P, Hanein D (2001b): An atomic model of actin filaments cross-linked by fimbrin and its implications for bundle assembly and function. *J Cell Biol* 153:947–56.

Walker M, Zhang XZ, Jiang W, Trinick J, White HD (1999): Observation of transient disorder during myosin subfragment-1 binding to actin by stopped-flow fluorescence and millisecond time resolution electron cryomicroscopy: evidence that the start of the crossbridge power stroke in muscle has variable geometry. *Proc Natl Acad Sci USA* 96:465–70.

Walz J, Koster AJ, Tamura T, Baumeister W (1999): Capsids of tricorn protease studied by electron cryomicroscopy. *J Struct Biol* 128:65–8.

Whittaker M, Wilson-Kubalek EM, Smith JE, Faust L, Milligan RA, Sweeney HL (1995): A 35-A movement of smooth muscle myosin on ADP release. *Nature* 378:748–51.

Wriggers W, Milligan RA, McCammon JA (1999): Situs: A package for docking crystal structures into low-resolution maps from electron microscopy. *J Struct Biol* 125:185–95.

Wriggers W, Birmanns S (2001): Using situs for flexible and rigid-body fitting of multiresolution single-molecule data. *J Struct Biol* 133:193–202.

Wriggers W, Chacon P (2001): Modeling tricks and fitting techniques for multiresolution structures. *Structure* 9:779–88.

Yonekura K, Maki S, Morgan DG, DeRosier DJ, Vonderviszt F, Imada K, Namba K (2000): The bacterial flagellar cap as the rotary promoter of flagellin self-assembly. *Science* 290:2148–52.

Yu X, Egelman EH (1997): The RecA hexamer is a structural homologue of ring helicases. *Nat Struct Biol* 4:101–4.

Zhou ZH, Dougherty M, Jakana J, He J, Rixon FJ, Chiu W (2000): Seeing the herpes virus capsid at 8.5 Å. *Science* 288:877–80.

Zhou ZH, Baker ML, Jiang W, Dougherty M, Jakana J, Dong G, Lu G, Chiu W (2001): Electron cryomicroscopy and bioinformatics suggest protein fold models for rice dwarf virus. *Nat Struct Biol* 8:868–73.

# 7

# MOLECULAR VISUALIZATION

John Tate

A clear, concise visual representation of a macromolecular structure, be it a single image, a pair of stereo images, or a full-blown, interactive three-dimensional view, remains probably the most eloquent way to describe the very significant volume of data that is encapsulated in the atomic coordinates of a model. The goal of this chapter is to examine the different ways in which macromolecular structures may be represented, and to give brief overviews of a few of the macromolecular visualization packages that are currently available.

The programs and packages mentioned here are some of the most commonly used examples of macromolecular visualization software and they can be divided roughly into three classes: the first and possibly the least *visually* demanding visualization task surrounding macromolecular models is the construction of an atomic model. The starting point for model building may be a blank screen and a sequence, in which case the model must be built largely from scratch, or it may be an existing structure that can be modified and molded to fit the data for the target structure. Either way, the process is highly interactive, and packages that deal with this specialized area of visualization are generally large and complex but at the same time rather limited in the styles of representation that are available, the emphasis being on the data themselves rather than their representation.

Once a structure is built, refined, and available for wider use, the challenge becomes that of obtaining useful information from the structure. In many cases simply the shape and secondary structure composition of a structure can be invaluable, but an atomic resolution model contains dramatically more information than just this. Extracting detailed information from a model requires tools that allow interactive manipulation and query of atomic coordinates, from measuring distances and angles to displaying a series of overlaid multiple structures. Although there are more possibilities for using different styles of representation at this stage, the emphasis at this point is again on

*Structural Bioinformatics*
Edited by Philip E. Bourne and Helge Weissig
Copyright © 2003 by Wiley-Liss, Inc.

clarity and visual simplicity, and more importance is generally attached to tools that enable a user to interrogate a structure than to those that provide a range of visual styles.

Finally, once the structure is ready for publication, software is required to generate clear, informative, and attractive representations of atomic data, most often in the form of static images, but possibly also as two-dimensional animations or three-dimensional, interactive scenes. Displaying a macromolecular structure on a high-powered graphics workstation allows one or perhaps several users to interactively explore and investigate that structure, but unfortunately the mass of information that is easily gathered by a single user can be difficult to condense sufficiently to make it amenable to wider distribution. Presentation software must filter the large volumes of data into a form suitable for the low-detail mediums that are currently used to distribute structure information, generally still a two-dimensional figure in a standard printed journal.

## A BRIEF HISTORY

Those researchers trying to determine the first molecular structures were faced with not only the theoretical challenge of calculating electron density values, but also the daunting task of somehow interpreting this electron density and obtaining the atomic coordinates that constitute a theoretical model of a macromolecular structure. The earliest molecular models, such as that of myoglobin (Kendrew et al., 1958), were built from masses of rods, wires, and spheres, so complex that the molecule itself was often lost in the web of supporting metalwork that was required to maintain its structure (see Fig. 7.1). The technical challenge of actually constructing such models from electron

**Figure 7.1.** Sir John Kendrew with the model of insulin, one of the first protein structures to be determined by X-ray crystallography. Components of the actual model are just visible through the forest of vertical support rods.

density was eventually met by the Richards Box, a construction affectionately known as Fred's Folly (Richards, 1968). These elegant, if cumbersome, optical devices involved stacks of glass or Perspex sheets onto which electron density contours were traced out by hand. A half-silvered mirror was used to superimpose an image of the electron density contours on an image of the physical model, so that the operator could see the model overlaid with the experimentally derived map. Peering into the dim image the crystallographer could manually build the model of the protein structure by joining together small fragments of molecule and adjusting them to fit the faint outlines of electron density by eye.

As with many other areas of science, macromolecular structure determination truly took off with the advent of electronic computing, and, as graphics technologies developed, so did the field of macromolecular visualization. Until relatively cheap and plentiful computing power became available, any calculation of the kind required for molecular structure determination was a painful manual undertaking, while visualization still involved either a hand-built physical model or a computer-generated, two-dimensional representation, formed by plotting electron density values or atomic coordinates on paper. Gleaning any useful information from the model or plot required good spatial awareness and not a little imagination.

The earliest attempts at electronic representations of molecular models used a computer-controlled oscilloscope to display a rotating image of a protein structure, with the speed and direction of rotation being controlled by the user (Levinthal et al., 1968). The system had many shortcomings (the image had to be constantly rotated to give any impression of three-dimensionality, the model was fixed and could not be altered by the user, and the hardware required to drive the system was itself specialized and experimental), but it was an essential proof of concept and undoubtedly a herald of things to come.

Once Levinthal and colleagues had demonstrated the power of electronic visualization, several groups constructed graphics systems of their own, using whatever number-crunching and display systems they had available locally or could build for themselves. These early graphics systems were milestones in molecular visualization, as crystallographers were finally able to view their models as truly three-dimensional objects, but they were mostly ad hoc constructions, difficult to build and almost as difficult to use. At about the same time, in the late 1960s, the head of the computer science department of the University of Utah and a professor from Harvard founded a company with the goal of developing and advancing the new field of computer graphics. In 1969 Evans and Sutherland (E & S) (http://www.es.com/) produced one of the first commercial vector graphics systems, the crude and very expensive Line Drawing System (LDS1). Although it was far from a financial success, LDS1 proved that computer graphics was an invaluable way to display complex data, and gave the company the impetus to develop a range of graphics systems and workstations that were the workhorses of the field of macromolecular visualization for many years.

Vector graphics systems were limited in the range of styles of representation that were available, and they gradually gave way to far more flexible systems that used raster-based displays. Raster systems were capable of richer and more detailed depictions of molecules than simple lines, and, as they developed, the software that was used to display molecular structures was also improved to take advantage of the new possibilities. One of the earliest and most widely used programs for constructing and manipulating molecular structures was FRODO (Jones, 1978), written by Alwyn Jones and colleagues. FRODO provided high-quality, interactive, color images of electron density maps and structures, and gave the user the ability to fit a model into displayed density by moving fragments of the model or even individual atoms. The program also provided sophisticated tools such as a structure "regularizer" that could improve the geometry of a model by refining bond distances and angles against idealized values, helping to make the process of model building more accurate and less error prone. Originally written for the DEC PDP 11/40 with a Vector General 3404 display, FRODO was ported to the E & S Picture System 2 and other graphics platforms, including later E & S systems. The program remained in widespread use throughout the 1980s before eventually being superceded by O (Jones et al., 1991). O was intended to correct many of the architectural problems that had built up in FRODO, allowing more complex features to be added and giving more flexibility for the user. The program is still under development and remains one of the most popular crystallographic model-building packages.

At around the same time as the crystallographic community was making the switch from FRODO to O, at the other end of the spectrum in terms of complexity and features, lightweight structure viewers were also being developed, solely for the purpose of examining molecular structures. One such program was RasMol, developed by Roger Sayle as part of his graduate work in the early 1990s. The program started life as a test-bed for ideas about interactive rendering and gradually became what is now one of the most popular and widely used general-purpose molecular visualization programs. KINEMAGE (Richardson and Richardson, 1992) was even more lightweight than RasMol, but it was designed for a somewhat different purpose: the goal of a KINEMAGE scene was to encapsulate a visual description of the molecule that was created by the author of the structure, and provide the user with views and descriptions of the model that were written by the person who knew the structure best. Although some features of KINEMAGE were limited, even compared to RasMol, its ability to display an authored scene, with labels, annotation, and specific view orientations, was undoubtedly ahead of its time and most of these features are yet to be reproduced by any of the mainstream visualization programs currently available.

The next major sea change in molecular visualization was once again the result of new developments in computer technology. By the start of the 1980s the computer science department of the University of Utah had a reputation for computer graphics, having already spawned Evans and Sutherland and given a number of young researchers a head start in the computer graphics industry. Among the graduates of the Utah computer science department was Jim Clark, who, in 1982, founded Silicon Graphics, with the aim of making the most powerful graphics platform in the world and making it affordable. By 1987 Silicon Graphics had achieved these goals and had produced the de facto standard in computer graphics. Large Silicon Graphics systems powered flight simulators for the aviation industry, powerful rendering clusters were used to create stunning special effects for movies, and Silicon Graphics desktop workstations were *the* platform for molecular visualization.

Until very recently Silicon Graphics had a virtual monopoly on scientific computers, making powerful workstations with high quality three-dimensional-graphics capabilities, and selling them at a price that individual academic groups could readily afford. However, in late 1990s the revolution in Personal Computer (PC) hardware brought major changes to all aspects of structural biology, as cheap and extremely powerful desktop PCs largely outpaced all but the most expensive workstations in both raw power and graphics capability. Driven principally by the electronic games industry, PC graphics technology has rapidly caught up with the more established graphics platforms from established vendors such as Silicon Graphics, and at the time of writing it is possible to buy a $1,000 PC that can match or even outperform a dedicated graphics workstation costing an order of magnitude more. It is already possible for any user to generate views of macromolecular structures that are both visually appealing and at the same time almost bewilderingly detailed, and, as graphics technologies continue to develop, the possibilities for macromolecular visualization can only increase.

## VISUALIZATION STYLES AND SOFTWARE

The styles used to represent a given structure change radically according to the various uses of the representation. At the outset, when trying to solve a structure by crystallography, the goal of the crystallographer is to turn a blank screen and a sequence into a three-dimensional atomic model of a structure, and representation styles necessarily focus on the constituent atoms of the molecule. NMR structure solution is radically different from crystallographic structure solution, and although some interactive manipulation of models can be required at certain stages of the process, the tools used for this manipulation are often the same as those used for electron density interpretation. Although there are now refinement/model building packages, such as ARP/wARP (Lamzin and Wilson, 1993), that can reliably construct crystallographic atomic models largely automatically, they still rely on having relatively high-resolution electron density maps to work with, and in most cases crystallographers still pore over visual representations of electron density and construct models using complex, interactive software packages.

Several visualization packages are used for crystallographic model building, and common to all of them are the ability to display electron density, and tools for manipulating atomic coordinates to fit that density. Probably the most widely used of these packages is *O* (Jones et al., 1991). It incorporates a wide range of model-building tools, which, like those of its predecessor, FRODO, are aimed at making model construction and manipulation simpler and more accurate. With a similar set of features, *XFit*, part of the XtalView crystallography package (McRee, 1999), is an alternative model-building program. In contrast to the command-driven interface of O, XFit is entirely mouse-driven, and all functions can be accessed from a graphical user interface, making it slightly easier to learn than O and less intimidating for beginners. *QUANTA* (http://www.accelrys.com/quanta/) is a commercial package for macromolecular crystallography from MSI (now Accelrys) that provides a similar range of tools to the other model-building programs, but also integrates with other MSI products that can perform model refinement and simulations.

The representation styles used by all of these model-building packages are similar: electron density is invariably represented as the now-familiar three-dimensional "chicken-wire" contours (see Fig. 7.2), since this generally gives an observer the best impression of the three-dimensional shape of the node of density that is being interpreted without obscuring the atomic structure that is being manipulated. For the same reason, probably the most useful style for displaying atomic coordinates for manipulation is the simple wire-frame bonds representation (also visible in Fig. 7.2), with the lines being colored according to the type of the atoms being linked (red for oxygen, blue for nitrogen, etc.). Basic lines can usually be drawn very quickly by most

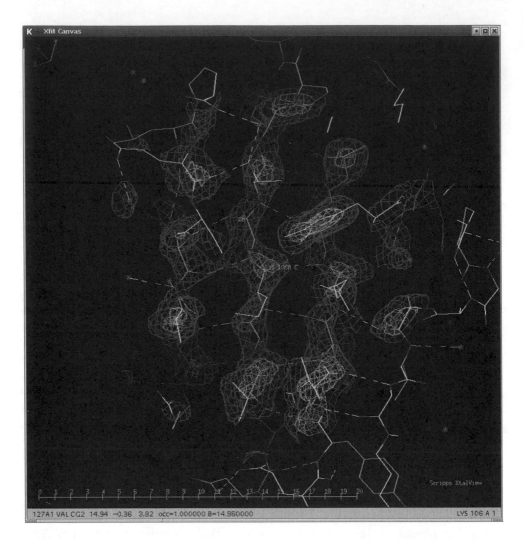

**Figure 7.2.** A typical fragment of electron density and a section of atomic model from the structure of the CuA domain from cytochrome BC3 (Williams et al., 1999) (PDB ID code 2CUA), displayed using *XFit* from the *XtalView* package. Bonds are colored according to the atoms that they join. Putative hydrogen bonds, are drawn as dashed white lines. Figure also appears in Color Figure section.

graphics systems, so even a large structural ensemble can generally be represented in this fashion without bringing the computer to a complete halt.

Other representation styles, although rarely used to represent an entire molecule, are often used to complement simple line drawings. In the case of ligands or heterogenous compounds that are bound to a larger structure, drawing the smaller molecule with atoms and bonds represented by solid spheres and rods (or "sticks") can be useful in helping the observer to pick it out from a complex scene. A common way to demonstrate the approximate van der Waal's radii of atoms is to draw them as large, solid spheres, also known as space filling or CPK representation, and this can also be useful for highlighting the position and interactions of smaller molecules bound to larger ones.

Figure 7.3 shows a representation of a region of the coat protein of human rhinovirus 1A, including a drug molecule that is bound in a small cavity within the virus protein. With the drug molecule represented as a space-filling sphere, it becomes significantly easier to locate it in the scene, and also to get an impression of the character of the structure surrounding it.

Many different programs can generate these relatively simple representations of molecules, but probably the most widely used is *RasMol* (http://www.umass.edu/ microbio/rasmol/). It is simple to use, yet flexible, and can be used for display and interrogation of atomic models. Thanks to the clever programming behind its graphical interface, and unlike most full-blown structure manipulation packages such as O and XFit, RasMol can comfortably display a full, all atom representation of even large molecules in CPK or ball-and-stick style and still allow the user to manipulate the view and query the scene interactively. RasMol runs well on practically

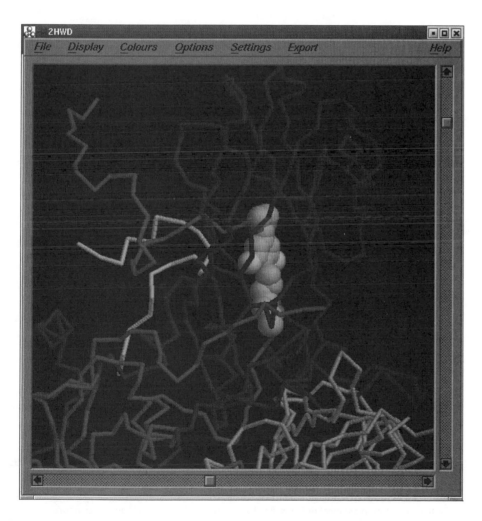

**Figure 7.3.** A region of human rhinovirus 1A (HRV-1A), including a bound drug molecule (Kim et al., 1993) (PDB ID code 2HWD). The virus proteins are shown as a simple backbone trace, with the drug represented as space-filling spheres. Figure also appears in Color Figure section.

every platform, from PC/Mac through all flavors of Unix and even VMS, and is probably the most useful general-purpose structure viewer currently available. As well as the original version, there are other versions of the program that build on the basic RasMol to add various new features. *RasTop* (http://www.bernstein-plus-sons.com/software/RasTop_1.3.1/), for example, is a Windows-only program that adds a more comprehensive user interface and extensions to the features of the original RasMol such as mouse-based selection of regions of a model, additional scripting commands, and improvements in the display options.

The next level of detail at which molecules are commonly represented utilizes somewhat abstract views of macromolecular (and in particular, protein) structures. The propensity of proteins to form well-defined secondary structural elements—$\alpha$-helices and $\beta$-strands—is a fundamental property of protein structure. Jane Richardson pioneered a style of representing $\alpha$-helices as simple cylinders or broad, spiral ribbons, and $\beta$-strands as broad, flat ribbons (Richardson, 1985), and this remains one of the

**Figure 7.4.** The structure of the reduced form of human thioredoxin (Weichsel et al., 1996) (PDB ID code 1ERT), drawn in the Richardson-style schematic secondary structure representation. The protein chain $\beta$-strands are represented by arrows pointing from the N- to the C-terminus, and $\alpha$-helices are drawn as spiral ribbons. Regions without defined secondary structure are shown as a simple, smooth tube. The four $\beta$-strands form a $\beta$-sheet at the center of the structure, which is easily visible in this kind of schematic representation. The image was generated using MolScript and render. Figure also appears in Color Figure section.

most enduring and appealing ways of representing protein secondary structure. By abstracting away the atomic coordinates and representing a structure according to secondary structure alone, this schematic style aptly describes both the arrangement of individual atoms in adjacent residues along the chain, as well as the interaction between more widely separated atoms through hydrogen bonding. Figure 7.4 shows the secondary structure of thioredoxin, illustrating the clarity that can be obtained with this simple representation style.

Practically all programs that can be used to view macromolecular structures can also generate interactive, Richardson-style, three-dimensional representations of protein structures. An important noninteractive program for generating attractive images of molecular structures is *MolScript* (Kraulis, 1991). MolScript scenes are described using a powerful scripting language and available representation styles for atomic coordinates include wire-frame, CPK, or ball-and-stick styles, while secondary structure elements may be drawn as solid spirals and arrows for $\alpha$-helices and $\beta$-strands, respectively. Output formats include PostScript and JPEG, but probably the most useful output format is a simple three-dimensional representation that can be fed in to *render*, a ray-tracing program from the *Raster3D* package (Merritt, 1997) that can be used to produce high-resolution ray-traced images of the MolScript scene (see Figure 7.4). Version 2 of MolScript adds a useful graphical front end, which allows scenes to be previewed in an interactive OpenGL viewer before they are written directly as an image or passed to an external rendering program. A widely used modification of the original version of MolScript, known as *bobscript* (Esnouf, 1997), adds enhanced coloring capabilities and, most significantly, the ability to display electron density maps as solid surfaces or meshes, although there is no facility for previewing a scene as in MolScript v2.

Along with electron density, many visualization programs can display various other kinds of three-dimensional data, such as electrostatic charge. In some cases it may be appropriate to display the data in the form of meshes, as for electron density, but another common style of representation is a projection of the data values onto molecular surfaces. The earliest visualization programs, such as FRODO, were able to give an idea of the van der Waal's surface of a molecule, usually using arrays of dots plotted at the van der Waal's radius for each atom, but as computing power and graphics capabilities have improved, many users now have access to graphics hardware that can readily display interactive representations of molecular surfaces as solid three-dimensional shapes. *Grasp* (Nicholls, 1993) was one of the earliest programs to be able to display surfaces interactively and although it is no longer in development and is available for only one computing platform (SGI), it remains one of the most commonly used programs for looking at the properties of macromolecules. Electrostatic potential is a property that lends itself well to being mapped onto a three-dimensional surface. Grasp is capable of coloring a solid surface according to the density and polarity of charge surrounding surface vertices, and the resulting patches of positive and negative charge dramatically illustrate the nature of the charge on the surface of a molecule.

Several newer programs can also display various three-dimensional data as solid and even translucent surfaces. One such package is *Chimera* (Huang et al., 1996), which can display both electron density and various kinds of surface as meshes and opaque or semitransparent solid surfaces. The core features of Chimera include a flexible interactive viewer for displaying molecular structures in a wide variety of styles, as well as basic structure editing tools, but the program also has a modular design that allows new features to be added easily. External modules, written in python, a powerful high-level scripting language, add the ability to render arbitrary volume data, to perform semiautomated docking of small molecules to larger structures, and to share modeling sessions across the network between users in different physical locations. Although designed to be extensible, Chimera is essentially closed-source, and is distributed only in binary form, at least at present. In contrast, *PyMol* (http://pymol.sourceforge.net/), another flexible, extensible package for molecular visualization, sports many of the same features as Chimera, but is freely available and open source. The program supports a wide range of representation styles, from simple line drawings through chicken wire or solid molecular surfaces (see Fig. 7.5) and density maps. Control is via a python-based scripting language and scenes can be manipulated interactively using a built-in viewer, output directly as images, or ray-traced with another built-in module to produce publication quality images.

Finally, one of the most powerful packages that can be used for visualizing molecular surfaces is *AVS*, from Advanced Visual Systems (http://www.avs.com/). AVS is a completely general-purpose visualization tool that uses a novel graphical editor to design networks that link together many separate modules, each of which performs a single task. It is notoriously difficult to master, but, with the right modules and careful design of networks, it can generate good results and provides an extremely flexible environment for investigation of molecular properties and structure and for all types of interactive visualization.

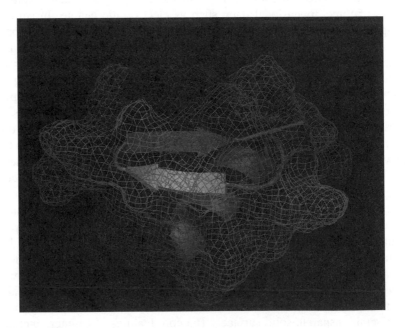

**Figure 7.5.** A molecular surface drawn as a mesh, overlaid on a secondary structure representation of the toxin LQ2 from *Leiurus Quinquestriatus* (Renisio et al., 1999) (PDB ID code 1LIR). The image was prepared entirely within PyMol. Figure also appears in Color Figure section.

## WEB-BASED VISUALIZATION SOFTWARE

Historically, in order to run a particular program, a user has been forced to obtain, compile, and install that program locally before he or she can even begin to use it. With the advent of the World Wide Web and the explosion of interest in server-based rather than locally installed software, this scenario is no longer always the case. Technologies such as Java, Active X, JavaScript, and so forth mean that users can run increasingly complex applications without having to explicitly download them, and rely instead on having the application delivered automatically along with a Web page, all for a single click of a mouse.

Java applets are probably the most commonly used form of Web-deliverable, platform-independent software, allowing developers to write a single program that can be deployed on many different kinds of computer, from PC to Macintosh to Unix, without any changes to the code or data it uses. Although Java has been slow to catch on with certain platforms, it is now widely available and is fairly well supported by all of the major Web browsers on most platforms. For simple, low-resolution visualization tasks, Java is a very capable solution and is becoming popular with developers because of its power and ease of use, and with users because of the ease of deployment of applications.

*WebMol* is a lightweight Java applet (http://www.cmpharm.ucsf.edu/~walther/webmol.html) (Fig. 7.6a) that can display and query a molecular structure. Since it is focused mainly on interrogation of a scene rather than visualization, it can display a structure in only a few different styles, but it does provide some sophisticated tools for querying atom-level data from a model. Tools such as live, interactive Ramachandran plots (see Fig. 7.6b) make WebMol a useful tool for assessing the quality of a model and for extracting detailed information from that model. Distance-matrix viewers make WebMol a useful tool for assessing the quality of a model and for extracting detailed information from that model. In the same vein, another Java applet, *QuickPDB* (http://cl.sdsc.cdu/QuickPDB.html) is one of very few visualization programs, Web-based or otherwise, that provides tools for interrogating both the structure and sequence of a protein structure simultaneously. Although the tools are rudimentary, QuickPDB is, as the name suggests, quick and easy to use, even over slow Internet connections, and is accessible from practically every reasonably recent Web browser. QuickPDB is one of the visualization options from the Research Collaboratory for Structural Bioinformatics (RCSB) Protein Data Bank (PDB) site (http://www.rcsb.org/pdb/). The *Molecular Interactive Collaborative Environment* (MICE) (Tate, Moreland J, Bourne PE, 2001) is another Java applet for viewing molecular structures, but it provides less support for querying the structure and places more emphasis on the representation of the structure. Figure 7.7 shows an example of a scene generated in MICE, in this case a molecule of reverse transcriptase (RT) from the human immunodeficiency virus (HIV). The key feature of MICE that differentiates it from other lightweight structure viewers, and indeed, from even the more complex molecular visualization packages, is that MICE allows users to generate interactive views of any structure in the PDB and to then share that view in real time with other users across a network. Multiple users in different physical locations can collaborate over a single shared scene, passing control between users as needed, and using tools such as a shared pointer to identify regions of a scene that are under discussion. Generation of the scene, creation of a collaboration, and control of an ongoing collaboration are all possible through simple form interfaces, and no configuration on the part of the user is required to make use of any features, making MICE simple to use, as well as powerful. Like QuickPDB, MICE is also available as one of the visualization options in the RCSB PDB site.

Web browsers began life as simple tools for displaying simple pages, but as the potential of the browser was realized, it became clear that there was no way that developers of the browser itself could hope to keep up with the mass of applications, technologies, and services that would soon be available on the Web. Netscape got around this problem by introducing a "plug-in" architecture to their browser that would allow third-party software developers to create modular applications that would be installed inside Netscape and used to cope with suitably enhanced Web pages and sites. Microsoft Internet Explorer soon followed suit and now most major Web browsers have some mechanism for making use of third-party modules. *Chime* is a commercial derivative of RasMol from MDL Information Systems (http://www.mdlchime.com/chime/), packaged in the form of a browser plug-in but having all of the same capabilities and flexibility as its stand-alone cousin. Once installed in a Web browser, Chime is invoked to handle PDB format files, creating a RasMol-style view of the structure. As well as

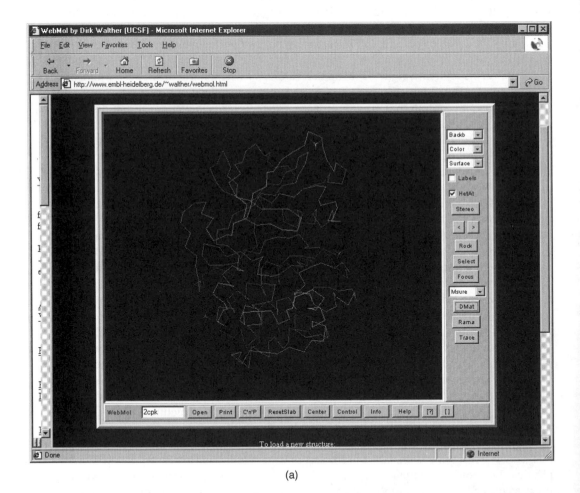

(a)

Figure 7.6. (*a*) The structure of c-AMP-Dependent protein kinase (Knighton et al., 1991) (PDB ID code 2CPK) displayed using *WebMol*. Although the representation styles are somewhat limited, WebMol does have some fairly advanced features, such as (*b*) an interactive Ramachandran plot that allows regions of a structure to be selected on the basis of the location of a residue in the plot. Figure 7.6(a) also appears in Color Figure section.

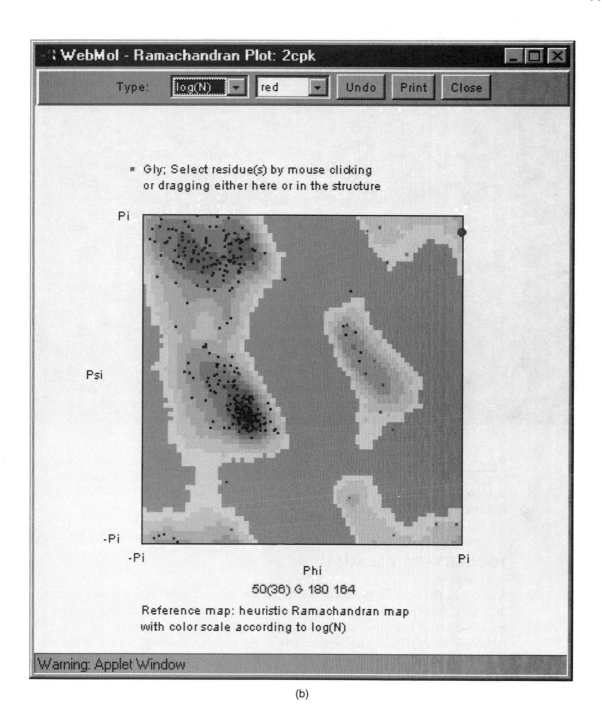

(b)

Figure 7.6. (*Continued*)

being useful in its own right as a means of viewing downloaded PDB files on the fly, Chime also forms the kernel of the *Protein Explorer* (http://proteinexplorer.org/), a Web application that provides a framework for examining and manipulating structures, using animations and pregenerated descriptions of structures of interest.

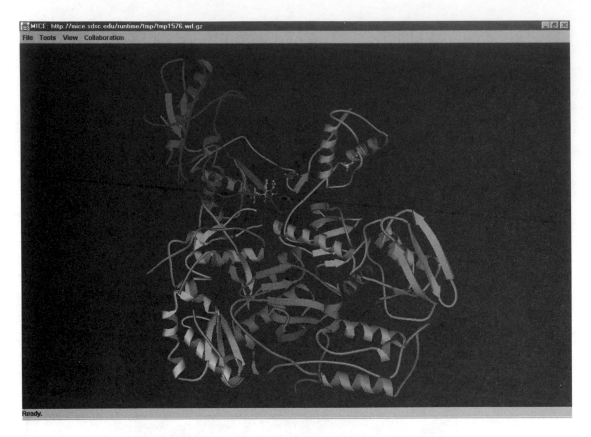

**Figure 7.7.** The structure of reverse transcriptase (RT) from the human immunodeficiency virus (HIV) (Hopkins et al., 1996) (PDB ID code 1RT1), displayed using *MICE*. This particular RT structure includes a drug molecule, just visible at the base of the cleft between the "finger" and "thumb" domains. Figure also appears in Color Figure section.

## TOOL KITS FOR VISUALIZATION

At least in the field of molecular visualization, most applications are still written as individual, stand-alone programs or as a collection interconnected applications that possess a set of features and requirements that are determined entirely by the author. Although the developer may be receptive to the requests and suggestions of the users, changes and enhancements are usually still under the immediate control of the author of a given package. Many visualization programs are distributed only as precompiled binaries for a given platform, and others may be available as source code that is designed to build and run on only a small range of platforms. Users do not expect packages to be significantly configurable, nor to be readily extensible by anyone but an experienced programmer.

    This model of software development is by far the most common, but it is also rather restrictive. The end users of a program must accept the shortcomings of the available software or they must generally start from scratch and write their own bespoke package to suit their needs, which inevitably leads to large numbers of small, unconnected programs being written by large numbers of developers working independently, each writing ad hoc, undocumented scripts or programs to solve precisely the same problems as their peers.

A better, more efficient solution is to provide users with high-level tool kits, from which they can quickly and relatively easily construct custom programs to fit any problem at hand. This model of development underlies several current visualization projects; the idea that users should be given high-level modules that each perform a specific task, and a framework within which to arrange them, rather than complete, monolithic programs with features and capabilities they will never use.

There are several of these tool-kit-style packages under development. One such is project is *Birdwash* (http://yoda.imsb.au.dk/birdwash/), a set of modules dealing with building, refinement, and analysis of macromolecular structures, written by one of the authors of O, and building on the experiences of designing and implementing a complex visualization and structure manipulation package. At time of writing the Birdwash tool kit includes modules for visualizing atomic structures and electron density, and a means of editing structures as is required for crystallographic model building. Similar but unconnected projects are the *Molecular Modeling Toolkit* (MMTK) (http://dirac.cnrs-orleans.fr/programs/mmtk.html), aimed more at molecular simulations rather than model construction, and the continuation of the MICE project, the *Molecular Biology Toolkit* (MBT). The goal of MBT is to construct a pure Java tool kit that will provide high-level tools for displaying and editing both macromolecular structures and sequences. Probably the most advanced tool kits and modules are those from the group of Art Olson (http://www.scripps.edu/pub/olson-web/), that specialize in creating graphical tools for structure visualization and manipulation. Having previously developed modules for manipulating molecular structures under AVS (Duncan, Macke TJ, Olson AJ, 1995), the group has considerable experience in the design of such tools, and their current tool kit includes a variety of python modules, including a structure loader, a surface generator, structure viewers, and molecule docking tools, all of which can be used to make purpose-built applications with only a minimum of high-level programming.

## CONCLUSION

Any list of software will inevitably be biased by the interests and experience of the author. Given the wealth of programs that are currently available for visualizing macromolecular structure and taking into account the ongoing development efforts in the field, it is also inevitable that such a list will be incomplete and out of date as soon as it is written. This chapter gives only the briefest of overviews of a few of the packages that are available to anyone wanting to look at the structures of macromolecules, but it is hoped that by illustrating just a few of the possibilities, the reader will be able to assess the usefulness and usability of these and other packages for themselves, and will be able to find the most appropriate one for any given task. Table 7.1 lists the programs discussed in this chapter along with their capabilities, comments, the pros and cons of each, on the platforms on which they run.

TABLE 7.1. Summaries of Some of the Currently Available Visualization Software

| Name | Capabilities | Comments | Pros | Cons | Platform | | | | | |
|---|---|---|---|---|---|---|---|---|---|---|
| | | | | | NT4.0 | W2000 | Mac | Linux | SGI | Unix |
| AVS | Structure query, presentation | Complex and extremely powerful framework for building custom applications using modular program fragments. Commercially available and supported. | Extremely flexible. Provides many powerful modules for all kinds of data manipulation and analysis. | Expensive and cumbersome. Difficult to learn. | | | Mac[a] | Linux[b] ✕ | SGI ✕ | Unix[c] ✕ |
| Bobscript | Presentation | An enhanced version of the original MolScript. | Powerful command language. Attractive representation styles. Improved coloring capabilities relative to Molscript Version 1. Support for electron density maps. | Command syntax can be unintuitive. Lacks the OpenGL preview mode of MolScript version 2. | | | | Linux ✕ | SGI ✕ | Unix ✕ |
| Chime | Browser-plug in version of RasMol | Commercially developed version of RasMol, designed to operate as a plug in for Netscape. Operates under some versions of Internet Explorer, via the Netscape-style plug-in architecture. | Same power and flexibility as RasMol. Web based | Available for only a limited number of (non-Unix) platforms. Commercial, closed source package. Apparently no longer under active development. | NT4.0 ✕ | W2000 ✕ | Mac ✕ | | | |

| Program | Application | Features | Notes | NT4.0 | W2000 | Mac | Linux | SGI | Unix |
|---|---|---|---|---|---|---|---|---|---|
| Chimera | Basic model building, model query, presentation | A modular package for all aspects of molecular visualization, from simple model building through production of high-quality images. | Flexible and extensible. Powerful tools, such as a docking module, are already available. Support for a wide range of data sets, including general volume data, such as, for example, cryo-EM data. | Closed source. Still in fairly early stages of development. | × | × |  | × | × | × |
| Dino | Presentation | Powerful and flexible general-purpose visualization program. Capable of displaying everything from protein structures to molecular surfaces to topography data from TEM. | Flexible representation styles. Visually very attractive representations. Output in a range of formats, including various image formats, Postscript and povray. | Still under development so bugs are quite common, although rapidly fixed. Quirky command language can be difficult to learn. |  |  |  | × | × | × |
| Grasp | Electrostatics calculation and representation | Widely used program for looking at molecular surfaces and electrostatics. Somewhat dated now. | Can generate and display molecular surfaces easily. Reads potential data calculated in standard electrostatic programs. | Complex command language. Unintuitive user interface. No longer under active development. Closed source and available only for SGI. |  |  |  |  | × |  |

(Continued overleaf)

152

TABLE 7.1. (*Continued*)

| Name | Capabilities | Comments | Pros | Cons | Platform |
|---|---|---|---|---|---|
| MICE | Java applet for structure display | Lightweight structure viewer with collaborative capabilities. | Simple to generate attractive and detailed views of structures via a form interface. Can access any structure in the PDB from just an ID code. Web deliverable. Collaborative. | Requires Java3D, a nonstandard Java extension that is available only on certain platforms. Limited tools for querying structures. | NT4.0 × W2000 × Mac Linux × SGI × Unix |
| MolScript | Presentation | Preparation of figures, as simple schematics or, in conjunction with Raster3D, ray-traced images. | Powerful command language. Attractive representation styles Wide range of output formats. An OpenGL front end allows interactive previewing of scenes. | Command syntax can be unintuitive. Apparently no longer under development. | NT4.0 W2000 Mac Linux × SGI × Unix × |
| O | Electron density map interpretation | Extremely powerful, if somewhat unintuitive, model-building package. | Very comprehensive model-building features. | Large, monolithic program. | NT4.0 × W2000 ? Mac Linux[d] |

| | Use | Description | Advantages | Disadvantages | NT4.0 | W2000 | Mac | Linux | SGI | Unix |
|---|---|---|---|---|---|---|---|---|---|---|
| | Model building | One of the most widely used model-building packages, certainly in academia. | Some presentation capabilities. Flexible, configurable interface. Large user base and good discussion groups available. | Can be difficult to learn and use well. Closed source. | | | | | × | ×$^e$ |
| PyMOL | Presentation | Powerful python-based visualization program. Input is via both a graphical user interface and a python-based command language. | Wide range of attractive representation styles. Flexible command language. Extensible. Open source. Runs (almost) anywhere that python runs. | Can be difficult to learn. Fairly heavy on memory and CPU resources, so a powerful computer is needed to do the program justice. | × | × | × | × | × | × |
| Quanta | Model building Structure interrogation | Commercial package for constructing, modifying, and interrogating atomic models. Widely used in industry but less common in academia because of the cost. | Wide range of tools for all aspects of structure work. | Commercial software, subject to license fees. Expensive, even for academic users. Older software, lacking many of the newer features of other comparable packages. | | | | | × | |

(Continued overleaf)

153

TABLE 7.1. (*Continued*)

| Name | Capabilities | Comments | Pros | Cons | Platform | | | | | |
|------|-------------|----------|------|------|------|------|------|------|------|------|
| | | | | | NT4.0 | W2000 | Mac | Linux | SGI | Unix |
| QuickPDB | Structure query | Simple structure and sequence display tool. | One of few tools that allows consideration of structure and sequence simultaneously. | Very basic. | × | × | × | × | × | × |
| | | Small and quick-to-load Java applet. | Web deliverable. | | | | | | | |
| | | | Uses an older version of Java that is embedded in many older versions of the popular Web browsers, making it largely platform independent. | | | | | | | |
| Rasmol | Structure interrogation | Lightweight but very powerful interactive structure viewer. | Quick, responsive interface. | Shows only one structure at a time. | × | × | × | × | × | × |
| | Presentation | The Swiss-Army knife of macromolecular visualization. | Powerful command language. | Ugly graphics—not really suitable for presentation or publication any more. | | | | | | |
| | | | Wide range of representation styles. | Unlikely to be adapted to make it truly Web deliverable (although see *Chime*) or to give it significantly more features. | | | | | | |
| | | | Open source and freely available—still developed (to a certain extent) and maintained by users. | | | | | | | |

| | | | NT4.0 | W2000 | Mac | Linux | SGI | Unix |
|---|---|---|---|---|---|---|---|---|
| RasTop | Enhanced RasMol | Improved version of RasMol, with many additional features such as an improved user interface, mouse-based selections, better support for animation. | X | X | | | | |
| | | Improved appearance over original RasMol. Windows only. | | | | | | |
| | | Corrects a number of serious deficiencies with original. | | | | | | |
| | | User interface removes the need to learn the command language. Runs well on virtually every platform, new or old. | | | | | | |
| | | Easy to use. | | | | | | |
| VMD | Structure analysis and presentation, with an emphasis on molecular dynamics | Includes numerous tools for dealing with molecular dynamics trajectories and models. Rather slow without a powerful machine with good graphics capabilities. | | | | X | X | X |
| | | General-purpose interactive visualization. Largely aimed at molecular dynamics. | | | | | | |
| | | Can be completely mouse driven, but also includes a powerful scripting language that can be used to extend the program and automate procedures. | | | | | | |
| | | Easy to use for simple visualizations. | | | | | | |
| | | Supported and documented and still under active development. | | | | | | |

# TABLE 7.1. (Continued)

| Name | Capabilities | Comments | Pros | Cons | Platform | |
|---|---|---|---|---|---|---|
| WebMol | Java applet for structure display | Lightweight and quick Java applet. Simple display styles only, but some sophisticated tools for structure interrogation. | Web deliverable—no installation required. Can load structures from URL or local files. Platform independent—runs wherever Java is available. | Limited display styles. Apparently no longer under active development. | NT4.0<br>W2000<br>Mac<br>Linux<br>SGI<br>Unix | ×<br>×<br>×<br>×<br>×<br>× |
| XFit | Electron density map interpretation<br><br>Model building<br><br>Presentation | Part of the XtalView crystallography package. Sports a comprehensive set of model building tools, most notably the ability to generate electron density from structure factors on the fly. | Completely mouse driven. Integrated into XtalView and therefore able to rapidly exchange data with other XtalView packages. Fairly lightweight compared to other model-building packages. | Can be difficult to find commands and features in the maze of panels and buttons until familiar with the program. Largely undocumented. | NT4.0<br>W2000<br>Mac<br>Linux<br>SGI<br>Unix | <br><br><br>×<br>×<br>× |

[a] "Mac" here refers to the classic Macintosh OS, rather than OSX.
[b] Notes here refer specifically to Linux running on the Intel platform, although it is likely that these programs will also run under Linux on other hardware, for example, LinuxPPC, where either source code or binaries for those platforms are available.
[c] "Unix" refers to various flavors of Unix, notably the various Compaq and Sun operating systems.
[d] Although there is a version of O for Linux, development seems to have stalled.
[e] O is supported on Compaq hardware running Compaq Unix.

156

# REFERENCES

Duncan BS, Macke TJ, Olson AJ (1995): Biomolecular visualization using AVS. *J Molec Graphics* 13(5):271–82, 299.

Esnouf RM (1997): An extensively modified version of MolScript that includes greatly enhanced coloring capabilities. *J Molec Graphics* 15(2):132–4, 112–3.

Hopkins AL, Ren J, Esnouf RM, Willcox BE, Jones EY, Ross C, Miyasaka T, Walker RT, Tanaka H, Stammers DK, Stuart DI (1996): Complexes of HIV-1 reverse transcriptase with inhibitors of the HEPT series reveal conformational changes relevant to the design of potent non-nucleoside inhibitors. *J Med Chem* 39:1589–1600.

Huang CC, Couch GS, Pettersen EF, Ferrin TE (1996): Chimera: An Extensible Molecular Modeling Application Constructed Using Standard Components. *Pacific Symposium on Biocomputing* 1:724.

Jones TA (1978): A graphics model building and refinement system for macromolecules. *J Applied Crystallogr* 11:268–72.

Jones TA, Zou J-Y, Cowan SW, Kjeldgaard M (1991): Improved methods for the building of protein models in electron density maps and the location of errors in these models. *Acta Crystallogr* A47:110–119.

Kendrew JC, Bodo G, Dintzis HM, Parrish RG, Wyckoff H, Phillips DC (1958): A three dimensional model of the myoglobin molecule obtained by x-ray analysis. *Nature* 181:662–6.

Kim KH, Willingmann P, Gong ZX, Kremer MJ, Chapman MS, Minor I, Oliveira MA, Rossmann MG, Andries K, Diana GD, Dutko FJ, McKinlay MA, Pevear DC (1993): A comparison of the anti-rhinoviral drug binding pocket in HRV14 and HRV1A. *J Mol Biol* 230:206–227.

Knighton DR, Zheng JH, Ten Eyck LF, Ashford VA, Xuong NH, Taylor SS, Sowadski JM (1991): Crystal structure of the catalytic subunit of cyclic adenosine monophosphate-dependent protein kinase. *Science* 253:407–14.

Kraulis PJ (1991): MOLSCRIPT: a program to produce both detailed and schematic plots of protein structures. *J Applied Crystallogr* 24:946–50.

Lamzin VS, Wilson KS (1993): Automated refinement of protein models. *Acta Crystallogr* D49:129–49.

Levinthal C, Barry CD, Ward SA, Zwick M (1968): Computer Graphics in Macromolecular Chemistry. *Emerging Concepts in Computer Graphics* (Nievergelt J, Secrest D, editors). New York W.A. Benjamin, pp. 231–53.

McRee DE (1999): XtalView/Xfit—A versatile program for manipulating atomic coordinates and electron density. *J Struct Biol* 125(2–3):156–65.

Merritt EA (1997): Raster3D: Photorealistic molecular graphics. *Methods Enzymol* 277:505–24.

Nicholls A (1993): GRASP: graphical representation and analysis of surface properties. *Biophys J* 64:A166.

Renisio JG, Lu Z, Blanc E, Jin W, Lewis JH, Bornet O, Darbon H (1999): Solution structure of potassium channel-inhibiting scorpion toxin Lq2. *Proteins* 34:417–26.

Richards FM (1968): The matching of physical models to three-dimensional electron-density maps: a simple optical device. *J Mol Biol* 37:225–30.

Richardson DC, Richardson JS (1992): The kinemage: a tool for scientific communication. *Protein Sci* 1(1):3–9.

Richardson JS (1985): Schematic drawings of protein structures. *Methods Enzymol* 115:359–380.

Tate JG, Moreland J, Bourne PE (2001): Design and Implementation of a Collaborative Molecular Graphics Environment. *J Molec Graphics* 19:280–7.

Weichsel A, Gasdaska JR, Powis G, Montfort WR (1996): Crystal structures of reduced, oxidized, and mutated human thioredoxins: evidence for a regulatory homodimer. *Structure* 4:735–51.

Williams PA, Blackburn NJ, Sanders D, Bellamy H, Stura EA, Fee JA, McKee DE (1999): The CuA domain of Thermus thermophilus ba3-type cytochrome c oxidase at 1.6 A resolution. *Nat Struct Biol* 6:509–519.

# Section II

## DATA REPRESENTATION AND DATABASES

# 8

# THE PDB FORMAT, mmCIF, AND OTHER DATA FORMATS

John D. Westbrook and P. M. D. Fitzgerald

In this chapter, the data formats and protocols used to represent primary macromolecular structure data are presented. The historical format used by the Protein Data Bank (PDB) is described first. Dictionary-based representations such as the macromolecular Crystallographic Information File (mmCIF) are presented. Finally, data structuring technologies employing markup languages are discussed along with protocols that provide data access through application program interfaces.

## THE PROTEIN DATA BANK FORMAT

The Protein Data Bank (PDB; http://www.pdb.org/; see also Chapter 3) (Bernstein et al., 1977; Berman, et al., 2000) was first established in 1971 by Walter Hamilton at Brookhaven National Laboratory in response to community requirements for a central repository for information about biological macromolecular structures. Seven structures were included in the PDB at its inception. The essential elements of the format used to encode these first entries are still the core of the PDB format used today. Because of the simplicity of the format and its consistency in representing three-dimensional structures, the PDB format remains the most widely supported means of exchanging macromolecular structure data.

    The PDB format consists of a collection of fixed format records that describe the atomic coordinates, chemical and biochemical features, experimental details of the structure determination, and some structural features such as secondary structure assignments, hydrogen bonding, and biological assemblies and active sites. The details

*Structural Bioinformatics*
Edited by Philip E. Bourne and Helge Weissig
Copyright © 2003 by Wiley-Liss, Inc.

```
          1         2         3         4         5         6         7
1234567890123456789012345678901234567890123456789012345678901234567890123456789

ATOM    145  N   VAL A  25      32.433  16.336  57.540  1.00 11.92      A1   N
ATOM    146  CA  VAL A  25      31.132  16.439  58.160  1.00 11.85      A1   C
ATOM    147  C   VAL A  25      30.447  15.105  58.363  1.00 12.34      A1   C
ATOM    148  O   VAL A  25      29.520  15.059  59.174  1.00 15.65      A1   O
ATOM    149  CB AVAL A  25      30.385  17.437  57.230  0.28 13.88      A1   C
ATOM    150  CB BVAL A  25      30.166  17.399  57.373  0.72 15.41      A1   C
ATOM    151  CG1AVAL A  25      28.870  17.401  57.336  0.28 12.64      A1   C
ATOM    152  CG1BVAL A  25      30.805  18.788  57.449  0.72 15.11      A1   C
ATOM    153  CG2AVAL A  25      30.835  18.826  57.661  0.28 13.58      A1   C
ATOM    154  CG2BVAL A  25      29.909  16.996  55.922  0.72 13.25      A1   C
```

Figure 8.1. An abbreviated example of the column-oriented data format for PDB ATOM records. The ATOM records in this example contain fields for the record name, atom serial number, an atom name, a residue name, a polymer chain identifier, a residue number, the x, y, z Cartesian coordinates, the isotropic thermal parameter and the occupancy. Atoms with serial numbers 149–154 also contain a label for alternative conformation in column 17.

of the format are described in the PDB Contents Guide (Callaway et al., 1996). This document enumerates the field formats for each PDB record and remark and describes the PDB conventions for naming atoms, residues, and nucleotides.

Each item of data in the PDB format is assigned to a range of character positions in one of many PDB record types (HEADER, SOURCE, REMARK, etc.). The ATOM records shown in Figure 8.1 encode the atomic coordinate data. ATOM records are among the more than 45 named data records in the PDB format. These named data records have strict column-formatting rules.

During its early history, the PDB served as a simple repository and a point of dissemination for structure data and was used primarily by crystallographers and later NMR spectroscopists. PDB entries during this early period resemble journal publications and contain lengthy descriptive text sections encoded in the REMARK records. An example of how refinement information was coded in pre-1994 PDB entries is shown in Figure 8.2a.

As the number of structures in the archive increased and the user base broadened, the PDB confronted changing requirements to enable comparative analysis of the data in the archive. Such studies minimally require a consistent representation of the data and an increase in the types of the data included in each entry. Accordingly, extensions in the PDB format were advanced in 1992 (Protein Data Bank, 1992) and 1996 (Callaway et al., 1996). Figure 8.2b is an example of how refinement information is coded in the current PDB format. Due to the archival nature of the prior entries, neither format change was backwardly propagated to earlier entries. Although format extensions significantly increased the encoding precision and level of detail for both the biochemical and experimental descriptions, the format of the PDB coordinate records has remained largely unchanged.[1]

Although the PDB format has served as the standard for representing macromolecular structure data for nearly three decades, both the underlying data and the user

---

[1]Coordinate records such as ATOM records were extended in column positions beyond 72 to include a segment identifier, element symbol, and atomic charge. In the earliest PDB format, column positions beyond 72 were reserved for punch card sequence numbers.

```
REMARK    3
REMARK    3 REFINEMENT. MOLECULAR DYNAMICS REFINEMENT BY THE METHOD OF
REMARK    3 A. BRUNGER, J. KURIYAN, AND M. KARPLUS (PROGRAM *XPLOR*).
REMARK    3 THE R VALUE IS 0.172 FOR ALL 32852 REFLECTIONS IN THE
REMARK    3 RESOLUTION RANGE 11 TO 2.1 ANGSTROMS.
```

(a)

```
REMARK    3
REMARK    3 DATA USED IN REFINEMENT.
REMARK    3  RESOLUTION RANGE HIGH (ANGSTROMS) : 2.1
REMARK    3  RESOLUTION RANGE LOW  (ANGSTROMS) : 11.0
REMARK    3  DATA CUTOFF              (SIGMA(F)) : 0.0
REMARK    3  DATA CUTOFF HIGH          (ABS(F)) : NULL
REMARK    3  DATA CUTOFF LOW           (ABS(F)) : NULL
REMARK    3  COMPLETENESS (WORKING+TEST)    (%) : NULL
REMARK    3  NUMBER OF REFLECTIONS          : 32852
```

(b)

**Figure 8.2.** (a) An example portion of PDB REMARK 3 given in the data format used prior to 1992. In this example information about the refinement is given as free text. (b) An example portion of PDB REMARK 3 given in the current data format. In this example the text is more structured. Figure reprinted from Bhat et al., 2001 with permission from Oxford University Press.

requirements for this data have changed dramatically. Together these considerations have posed informatics challenges that the current PDB format cannot fully address.

The macromolecular structure data represented in a PDB entry has increased in both the type and complexity. In addressing changes in experimental methodology, the PDB format has been extended with new REMARK records. For example, the organization and information content of REMARK 3 that encodes refinement information has been modified and extended for each new refinement program and program version. Although extending REMARK records in this way captures information in a manner that is easy for a human to read, the diversity of organization of this data makes it very difficult to design software that can automatically and reliably extract information from these records. Data in these records is also defined only in terms of the program that computed the information. Information between programs may not be directly comparable.

The PDB format uses fixed-width fields to represent data, and this restriction places absolute limits on the size of certain items of data. For instance, the maximum number of atom records that can be represented in a single structure model is limited to 99,999, and the field width of the identifier for polymer chains is limited to a single character. Although these restrictions were certainly reasonable when the format was first defined, this is no longer the case. Many large molecular systems, such as the ribosomal subunit structures, cannot be represented in a single PDB entry. These entries must be divided into multiple PDB files, seriously complicating their use.

As the size and diversity of structure data in the PDB archive has grown, it has become an increasingly important resource in structural biology. User requirements for PDB data have grown from accessing individual entries to analysis and comparison of experimental and structure data across the entire archive. The latter has been facilitated by the increased accessibility of database technologies. To support comparative analysis and database applications requires data uniformity and internal consistency that are

```
SEQRES   1    396   MET ASP GLU ASN ILE THR ALA ALA PRO ALA ASP PRO ILE
SEQRES   2    396   LEU GLY LEU ALA ASP LEU PHE ARG ALA ASP GLU ARG PRO

. . .

ATOM     1    N     MET       5       41.402  11.897  15.262  1.00 48.61
ATOM     2    CA    MET       5       40.919  13.262  15.600  1.00 47.70
ATOM     9    N     PHE       6       39.627  14.840  14.228  1.00 48.66
ATOM    10    CA    PHE       6       39.199  15.440  12.964  1.00 45.33

. . .
```

**Figure 8.3.** An abbreviated example illustrating sequence inconsistency between PDB records. The PDB SEQRES records describe the sequence of the polymer that was crystallized. The sequence labels in the PDB ATOM records report the coordinates and the residue sequence observed in the refined structure. These should be consistent; as shown in this example there have often been exceptions. This example highlights the sequence conflict between ASP in the chemical sequence (SEQRES) with PHE, residue number 6 in the ATOM records.

typically beyond the needs of software-accessing individual entries, such as molecular graphics applications.

The difficulty in the reliable extraction of experimental information from each entry has already been discussed in terms of the format variation of REMARK records used to encode experimental details (i.e., refinement information in REMARK 3). Internal consistency problems within PDB entries arise in cases in which the portions of structure or individual structural elements are referenced in different PDB records in a noncorresponding manner. Because the PDB format does not specify the precise relationship between the polymer sequence given in the SEQRES records and the observed residue sequence within the ATOM records, the sequence information that is presented in the PDB entry cannot be used directly. To use these data, additional sequence alignment must be performed to resolve gaps and possible conflicts. An example of this problem is shown in Figure 8.3.

Consistency problems can also arise in other records that reference structural features such as those records describing secondary structure, active sites, and biological assemblies. Although the relationships between these records and the coordinate data that they reference are obvious to the experienced user, they can only be understood by a careful reading of the format description document. Because these relationships are not electronically accessible, each such relationship must be coded as a special case by any software that needs to validate inter-record consistency.

## mmCIF—A DICTIONARY-BASED APPROACH TO DATA DESCRIPTION

The Crystallographic Information File (Hall, Allen, and Brown, 1991) was created to archive information about crystallographic experiments and results (Hall, 1991) and is the format in which all structures described in articles sent to *Acta Crystallographia C* are submitted. In 1990, the International Union of Crystallography (IUCr) formed a working group (Fitzgerald et al., 1992) to expand this dictionary so that it would be able to do the same for macromolecules.

The original short-term goal of the working group was to fulfill the mandate set by the IUCr: to define mmCIF data names that needed to be included in the CIF dictionary in order to adequately describe the macromolecular crystallographic experiment and its results. Implicitly this mandate included the need to describe all of the data items included in a PDB entry. Long-term goals were also determined: to provide sufficient data names so that the experimental section of a structure paper could be written automatically and to facilitate the development of tools so that computer programs could easily access and validate mmCIF data files.

In order to describe the progress of this project and to solicit community feedback, several informal and formal meetings were held. The first meeting, hosted by Eleanor Dodson, convened in April 1993 at the University of York. The attendees included the mmCIF working group, structural biologists, and computer scientists. A major focus of the discussion was whether the formal structure of the dictionary that was implemented using the then-current Dictionary Definition Language (DDL 1.0) (Westbrook and Hall, 1995) was adequate to deal with the complexity of the macromolecular data items. Criticisms included the idea that the data typing was not strong enough and that there were no formal links among the data items. A working group was formed to try to address these issues. The second workshop was hosted by Phil Bourne in Tarrytown, New York, in October 1993. The topics at that meeting focused on the development of software tools and the requirements of an enhanced DDL. In October 1994, a workshop hosted by Shoshana Wodak at the Free University of Brussels, resulted in the adoption of a new DDL that addressed the various problems that had been identified at the preceding workshops. The dictionary was cast in this new DDL 2 and was presented at the American Crystallographic Association (ACA) meeting in Montreal in July 1995. This dictionary was open for further community review. The dictionary was placed on a World Wide Web site and community comments were solicited via a list server. Lively discussions via this mmCIF list server ensued, resulting in the continuous correction and updating of the dictionary. Software was developed and was also presented on this web site. A workshop held at Rutgers in 1997 provided tutorials for using both the dictionary and the software tools that had been developed at that time.

In January 1997, the mmCIF dictionary containing 1700 definitions was completed and submitted to the IUCr committee that oversees dictionary development (COMCIFS) for review and in June 1997, Version 1.0 was released (Fitzgerald et al., 1996; Bourne et al., 1997). The method adopted for managing dictionary extensions uses a scientific journal as a model. Proposed extensions are sent to the editors of the mmCIF Dictionary (Fitzgerald et al., 1996, Editorial Board: Paula Fitzgerald, Editor, Helen Berman, Associate Editor) who send the new definitions to a member of the board of editors for scientific review. These editors have expertise in the various areas covered by the dictionary. Once the definitions are reviewed for their scientific content, they are sent to the technical editors. More than 100 new definitions have been proposed since the fall of 1997 and have been reviewed using the procedures outlined. Version 2 of the mmCIF dictionary contains many of these new definitions and was released the fall of 2000.

Software libraries to parse and access data CIF and mmCIF have been produced for a number of popular languages including: C/C++, JAVA, FORTRAN, PERL and Python (see http://deposit.pdb.org/mmcif/ for a list of programs).

## Dictionary and Data File Syntax

The syntax used in both mmCIF data files and dictionaries derived from the STAR (Self-defining Text Archive and Retrieval) (Hall, 1991) grammar and is similar in most respects to the syntax used by core CIF for describing small molecule crystallography.

```
_refine.ls_number_reflns_obs    32852
_refine.ndb_ls_sigma_F            0.0
_refine.ls_d_res_low             11.0
_refine.ls_d_res_high             2.1
_refine.ls_percent_reflns_obs   100.0
_refine.ls_R_Factor_obs         0.172
_refine.ls_R_Factor_all         0.172
```

**Figure 8.4.** An example portion of the mmCIF REFINE category containing the same informa-tion as in the PDB REMARKs in Figure 8.2a and 8.2b. This example illustrates keyword-value pair formatting that is characteristic of the mmCIF syntax.

```
_diffrn_measurement.diffrn_id          'Data set 1'
_diffrn_measurement.device             '3-circle camera'
_diffrn_measurement.method             'omega scan'
_diffrn_measurement.details
; 440 frames, 0.20 degrees, 150 sec, detector distance 12 cm,
detector angle 22.5 degrees
;
_diffrn_measurement.specimen_support      ?
```

**Figure 8.5.** An example illustrating the encoding of text strings using mmCIF. Short strings, such as *Data set 1* and *omega scan*, are surrounded by either single or double quotation marks. Multiline strings such as the value of _diffrn_measurement.diffrn_id are encapsulated by semicolons in the first column of the beginning and ending lines of the string.

In its simplest form, an mmCIF data file looks like a paired collection of data item names and values. Figure 8.4 illustrates the assignments of values to selected refinement parameters analogous to the PDB format data in Figures 8.2a and 8.2b. The syntax is described here.

The leading underscore character identifies data item names. The underscore char-acter is followed by a text string interpreted as containing both a category name and a keyword name separated by a period. The keyword portion of the name is the unique identifier of the data item within the category. In the examples shown in Figure 8.4, all of the data items belong to the REFINE category. This example also illustrates the one-to-one correspondence required between item names and item values. Data category and data item names are not case sensitive.

Figure 8.5 illustrates how text strings are expressed. Short text strings may be enclosed in single or double quotation marks. Text strings that span multiple lines are enclosed by semicolons that are placed at the first character position of the line. There are two special characters used as placeholders for item values, which for some reason cannot be explicitly assigned. The question mark (?) is used to mark an item value as missing. A period (.) may be used to identify that there is no appropriate value for the item or that a value has been intentionally omitted.

Vectors and tables of data may be encoded using a *loop_* directive. To build a table, the data item names corresponding to the table columns are preceded by the *loop_* directive, and followed by the corresponding rows of data. The mmCIF example in Figure 8.6 builds a table of atomic coordinates.

The use of the *loop_* directive has a few restrictions. First, it is required that all of the data items within the loop belong to the same data category. Second, the number

```
loop_
_atom_site.group_PDB
_atom_site.type_symbol
_atom_site.label_atom_id
_atom_site.label_comp_id
_atom_site.label_asym_id
_atom_site.label_seq_id
_atom_site.label_alt_id
_atom_site.cartn_x
_atom_site.cartn_y
_atom_site.cartn_z
_atom_site.occupancy
_atom_site.B_iso_or_equiv
_atom_site.footnote_id
_atom_site.auth_seq_id
_atom_site.id
ATOM N N   VAL A 11 . 25.369 30.691 11.795 1.00 17.93 . 11 1
ATOM C CA VAL A 11 . 25.970 31.965 12.332 1.00 17.75 . 11 2
ATOM C C   VAL A 11 . 25.569 32.010 13.881 1.00 17.83 . 11 3
# [data omitted]
```

**Figure 8.6.** An abbreviated example of the mmCIF category ATOM_SITE. This category is organized as a table and illustrates the use of the *loop_* directive, followed by the list of data items names as the simple means of declaring an mmCIF table. The data values that follow the list of data item names are assigned to each data item (column) in turn. Here, the value, ATOM, is assigned to the column for _atom_site.group_PDB in each of the three rows in this example.

of data values following the loop must be an exact multiple of the number of data item names. Finally, mmCIF does not support the nesting of *loop_* directives.

Data blocks are used to organize related information and data. A data block is a logical partition of a data file or dictionary created using a *data_* directive. A data block may be named by appending a text string after the *data_* directive, and a data block is terminated by either another *data_* directive or by the end of the file.

Figure 8.7 illustrates how data blocks can be used to separate similar information pertaining to different structures. This separation is required because the mmCIF syntax prohibits the repetition of the same category at multiple places within the same data block. As a result, the simple concatenation of the contents of the above two data blocks into a single data block would be syntactically incorrect.

Definitions in mmCIF data dictionaries are encapsulated in named save frames (Fig. 8.8). A save frame is a syntactical element that begins with the *save_* directive and is terminated by another *save_* directive, end of file, or new data block. Save frames are named by appending a text string to the *save_* directive. In the mmCIF dictionary, save frames are used to encapsulate item and category definitions. The mmCIF dictionary is composed of a data block containing thousands of save frames, where each save frame contains a different definition. Save frames appear in data dictionaries but they are not used in data files. Save frames may not be nested.

The content of this dictionary definition has the same item-value pair organization as in the previous data file examples. DDL2 dictionary definitions typically contain a small number of items that specify the essential features of the item. The example definition includes: a description or text definition, the name and category of the item, a code indicating that the item is optional (not mandatory), the name of a related

```
#
# --- Lines beginning with # are treated as comments
#
data_X987A
_entry.id                      X987A
_exptl_crystal.id              'Crystal A'
_exptl_crystal.colour          'pale yellow'
_exptl_crystal.density_diffrn  1.113
_exptl_crystal.density_Matthews 1.01

_cell.entry_id                 X987A
_cell.length_a                 95.39
_cell.length_a_esd             0.05
_cell.length_b                 48.80
_cell.length_b_esd             0.12
_cell.length_c                 56.27
_cell.length_c_esd             0.06

# Second data block
data_T100A

_entry.id                      T100A
_exptl_crystal.id              'Crystal B'
_exptl_crystal.colour          'orange'
_exptl_crystal.density_diffrn  1.156
_exptl_crystal.density_Matthews 1.06

_cell.entry_id                 T100A
_cell.length_a                 68.39
_cell.length_a_esd             0.05
_cell.length_b                 88.70
_cell.length_b_esd             0.12
_cell.length_c                 76.27
_cell.length_c_esd             0.06
```

**Figure 8.7.** An abbreviated example illustrating the organization of mmCIF files in data blocks. Data blocks are declared using the *data_* directive and optionally followed by a data block name. In this example, data blocks X987A and T100A are declared. The information in these data blocks is treated as logically distinct even if the data block exists within the same data file.

```
save__exptl.details
    item_description.description
;  Any special information about the experimental work prior to
   the intensity measurement. See also _exptl_crystal.preparation.
;
_item.name                 '_exptl.details'
_item.category_id           exptl
_item.mandatory_code        no
_item_aliases.alias_name   '_exptl_special_details'
_item_aliases.dictionary    cif_core.dic
_item_aliases.version       2.0.1
_item_type.code             text
save_
```

**Figure 8.8.** An example of the mmCIF data definition for data item _exptl.details. This definition contains a textual definition, name and category identity, a code indicating the item is optional, an alias name to a previous dictionary, and a data type.

definition in the core CIF dictionary, and code specifying that the data type is *text*. A further description of the elements of the dictionary definitions is presented in the next section.

## Semantic Elements of the mmCIF Data Dictionary

The elements of DDL provide the organizational framework for building data dictionaries like mmCIF. The role of the DDL is to define which data items may be used to construct the definitions in the data dictionary, and also to define the relationships between these defining data items.

The dictionary language contains no information about a particular discipline such as macromolecular crystallography; rather, it defines the data items that can be used to describe a discipline. The contents of the mmCIF dictionary are metadata, or data about data. The contents of the DDL are meta-metadata, the data defining the metadata. DDL defines data items that describe the general features of a data item like a textual description, a data type, a set of examples, a range of permissible values, or perhaps a discrete set of permitted values. Consequently, data modeling using DDL can be applied in many application areas not just macromolecular structure description.

The lowest level of organization provided by the DDL is the description of an individual data item. Collections of related data items are organized in categories. Categories are essentially tables in which each repetition of the group of related items adds a row. The terms *category* and *data item* are used here in order to conform with the previous use of these terms by STAR and CIF applications; these terms could be replaced by *relation* and *attribute* (or table and column) commonly used to describe the relational model that underlies the DDL.

Within a category, the set of data items determining the uniqueness of their group are designated as *key* items in the category. No data item group in a category is allowed to have a set of duplicate values of its key items. Each data item is assigned membership in one or more categories. Parent–child relationships may be specified for items belonging to multiple categories. These relationships permit the expression of the very complicated hierarchical data structures required to describe macromolecular structure.

Other levels of organization in addition to category are also supported. Related categories may be collected together in category groups, and parent relationships may be specified for these groups. This higher level of association provides a vehicle to organize a large, complicated collection of categories into smaller, more relevant, and potentially interrelated groups. This organization effectively provides a chaptering mechanism for large and complicated dictionaries like mmCIF. Within the level of a category, subcategories of data items may be defined among groups of related data items. The subcategory provides a mechanism to identify, for example, that the data items month, day, and year collectively define a date.

For categories, subcategories, and items *methods* may be specified. Methods are computational procedures that are defined and expressed in a programming language (e.g., C/C++, PERL, or JAVA) and stored within a dictionary. Among other things, these dictionary methods may be used to calculate a missing value or to check the validity of a particular value.

The highest levels of data organization provided by DDL2 are the data block and the dictionary. The dictionary level collects a set of related definitions into a single unit, and provides the attributes for a detailed revision history on the collection.

The detailed features of the DDL used to build the mmCIF data dictionary are described elsewhere (Westbrook and Bourne, 2000; Westbrook et al., forthcoming).

## mmCIF Dictionary Content

The mmCIF dictionary contains approximately 1700 definitions describing the macromolecular experiment and its structural results. This dictionary includes definitions describing all aspects of macromolecular structure; experimental details about crystallization, data collection, data processing, phasing, and refinement;

**Figure 8.9.** A schematic diagram illustrating the content and relationships among a portion of the mmCIF categories describing chemical entities, entity names, entity source organism, and the relationship between the entity and atomic level of description. In this figure, boxes enclose the data items within each mmCIF category, and arrows indicate the correspondence between data items that are common to multiple categories, with the arrowheads pointing in the direction of the parent data item. The category key data items are indicated with black dots.

and other supporting data categories describing citation and software. A complete discussion of the contents of the mmCIF dictionary has been previously described (Bourne et al., 1997; Fitzgerald et al., forthcoming).

mmCIF Molecular Entities provides a summary of a portion of mmCIF data categories describing chemical structure. In this discussion the data categories are presented in the form of a schematic diagram, a brief description, and a set of examples. In the diagrams, boxes enclose the data items within each mmCIF category, and arrows indicate the correspondence between data items that are common to multiple categories, with the arrowheads pointing in the direction of the parent data item. The category key data items are indicated with black dots.

***mmCIF Molecular Entities.*** An entity is a chemically distinct part of an mmCIF entry. There are three types of entities: polymer, nonpolymer, and water. A common name, systematic name, source information, and keyword description can be assigned to each mmCIF entity. The relationships between categories that describe these entity features are illustrated in Figure 8.9. Figure 8.10 is an example of an entity description taken from an HIV protease structure (PDB 5HVP [Fitzgerald et al., 1990]).

```
loop_
_entity.id
_entity.type
_entity.formula_weight
_entity.details
1   polymer        10916
;   The enzymatically competent form of HIV protease is a dimer. This
    entity corresponds to one monomer of an active dimer.

;
2   non-polymer  '647.2'  '.'
3   water           10        '.'
#
loop_
_entity_name_com.entity_id
_entity_name_com.name
1   'HIV-1 protease monomer'
1   'HIV-1 PR monomer'
2   'acetyl-pepstatin'
2   'acetyl-Ile-Val-Asp-Statine-Ala-Ile-Statine'
3   'water'
#
loop_
_entity_src_gen.entity_id
_entity_src_gen.gene_src_common_name
_entity_src_gen.gene_src_strain
_entity_src_gen.host_org_common_name
_entity_src_gen.host_org_genus
_entity_src_gen.host_org_species
_entity_src_gen.plasmid_name
1 'HIV-1' 'NY-5' 'bacteria'  'Escherichia'  'coli'  'pB322'
#
```

**Figure 8.10.** An abbreviated example of the mmCIF description of the chemical features of HIV protease. In this example, three entities are defined. One entity is a monomer of HIV protease, a second is the peptidic inhibitor, and the third is solvent. Common names for these entities are specified. The source organism from which the protein sequence was obtained is also specified in this example.

***mmCIF Polymer and Non-polymer Entities.*** Additional data categories are provided to describe polymeric entities. Polymer type, sequence length, information about nonstandard linkages, and chirality may be specified. The monomer sequence for each polymer entity is listed in category ENTITY_POLY_SEQ. This sequence information is directly linked to the sequence specified in the coordinate list. It is also linked to the full

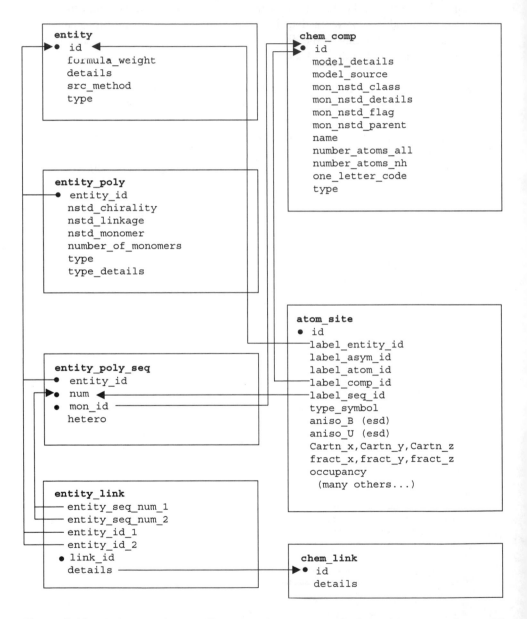

**Figure 8.11.** A schematic diagram illustrating the content and relationships among the mmCIF categories describing polymer entities. In this figure, boxes enclose the data items within each mmCIF category, and arrows indicate the correspondence between data items that are common to multiple categories, with the arrowheads pointing in the direction of the parent data item. The category key data items are indicated with black dots.

```
loop_
_entity_poly.entity_id
_entity_poly.type
_entity_poly.nstd_chirality
_entity_poly.nstd_linkage
_entity_poly.nstd_monomer
1   polypeptide(L)   no   no   no
      #
loop_
_entity_poly_seq.entity_id
_entity_poly_seq.num
_entity_poly_seq.mon_id
          1    1  PRO   1    2  GLN   1    3  ILE   1    4  THR   1    5  LEU
          1    6  TRP   1    7  GLN   1    8  ARG   1    9  PRO   1   10  LEU
          1   11  VAL   1   12  THR   1   13  ILE   1   14  LYS   1   15  ILE
          1   16  GLY   1   17  GLY   1   18  GLN   1   19  LEU   1   20  LYS
          1   21  GLU   1   22  ALA   1   23  LEU   1   24  LEU   1   25  ASP
        # - - - abbreviated - - -
```

**Figure 8.12.** An abbreviated mmCIF description of a polymer entity, including the enumeration of the three-letter residue codes of the entity polymer sequence.

chemical description of each monomer or nonstandard monomer in the CHEM_COMP category group. The relationships between categories describing polymer entities are illustrated in Figure 8.11 and Figure 8.12, which show an example of the description of a polymeric entity for a simple protein.

Non-polymeric entities are treated as individual chemical components. These entities may be fully described in the CHEM_COMP group of categories in the same manner as monomers within a polymeric entity. Like polymeric entities, each non-polymeric entity carries both an entity identifier and a component identifier. These identifiers form part of the label used to identify each atom (_atom_site.label_entity_id and _atom_site.label_comp_id). For polymeric entities the monomer identifier and the component identifier are the same; however, the atom label also includes an additional field for the sequence position (_atom_site.label_seq_id). An example for a drug–DNA complex illustrating both polymer and nonpolymer entity descriptions is shown in Figure 8.13.

***Atomic Positions.*** The refined coordinates are stored in the ATOM_SITE category. Atomic positions and their associated uncertainties may be stored in either Cartesian or fractional coordinates, and temperature factors and occupancies may be stored for each position.

Each atomic position must be uniquely identified by the data item _atom_site.id. Each position must also include a reference (_atom_site.type_symbol) to the table of elemental symbols in category ATOM_TYPE. All other data items in ATOM_SITE category are optional.

A typical atomic position for a macromolecule includes a variety of label information. The data items that label atomic positions, can be divided into two groups: those that are integrated into higher-level structural descriptions and those that are provided to hold alternative nomenclatures. The data items in the former group are prefixed by label_ and the latter carry an auth_ prefix.

```
#
loop_
_entity.id
_entity.type
_entity.src_method
    1   polymer        'man'
    2   non-polymer  'man'
    3   water           .
      #
loop_
_entity_keywords.entity_id
_entity_keywords.text
    1   'NUCLEIC ACID'
    2   'DRUG'
      #
loop_
_entity_name_com.entity_id
_entity_name_com.name
    2   ADRIAMYCIN
    3   WATER
      #
loop_
_entity_poly_seq.entity_id
_entity_poly_seq.mon_id
_entity_poly_seq.num
    1   T   1
    1   G   2
    1   G   3
    1   C   4
    1   C   5
    1   A   6
#
loop_
_entity_poly.entity_id
_entity_poly.number_of_monomers
_entity_poly.type
    1   6    'polydeoxyribonucleotide'
```

(a)

**Figure 8.13.** An abbreviated example of an mmCIF description of a drug–DNA complex. This example includes the description of the both the DNA strand and drug adriamycin. The mmCIF categories depicted in Figures 8.9 and 8.10 are populated in this example. In particular, this example illustrates how mmCIF describes the chemistry of both polymer and nonpolymer molecules. (a) The chemical entities in this drug complex are defined. The first entity is the DNA polymer strand and second entity is the drug adriamycin. The nucleotide sequence of the DNA is enumerated. (b) The list of chemical components in the complex is specified. This list includes the nucleotide monomers, the drug, and the solvent. The atomic coordinates a portion of the first nucleotide (thymine) are given. This coordinate data is explicitly related to the chemical description through identifiers for entity (_atom_site.label_entity_id =1), component (_atom_site.label_comp_id = T) and sequence order (_atom_site.label_seq_id =1).

```
loop_
_chem_comp.id
_chem_comp.name
        A     ADENINE
        T     THYMINE
        C     CYTOSINE
        G     GUANINE
        DM2   ADRIAMYCIN
        HOH   WATER
#
loop_
_atom_site.id
_atom_site.label_atom_id
_atom_site.label_comp_id
_atom_site.label_asym_id
_atom_site.label_seq_id
_atom_site.label_entity_id
_atom_site.Cartn_x
_atom_site.Cartn_y
_atom_site.Cartn_z
_atom_site.occupancy
_atom_site.B_iso_or_equiv
        1   O5*   T A   1   1   -18.744  20.195  22.722  1.00  36.68
        2   C5*   T A   1   1   -18.262  20.915  23.867  1.00   4.63
# - - - abbreviated - - -
```

(b)

Figure 8.13. (*Continued*)

## OTHER DATA DICTIONARIES

The methodology used to develop the mmCIF data dictionary has been applied to describe the content of a number of other content areas not completely covered in the mmCIF dictionary. All of these dictionaries have been developed to be consistent with the mmCIF data representation. These dictionaries, which are all available from the PDB mmCIF Resource Site (http://deposit.pdb.org/mmcif/), cover the following content areas:

- The imgCIF dictionary describes the details of crystallographic data collection, including data from image detectors in ASCII and binary data format.
- The BioSync dictionary describes the features and facilities provided by synchrotron beamlines (Kuller et al., 2002).
- The MDB dictionary is an extension of the mmCIF dictionary for homology models.
- The Symmetry extension supplements the mmCIF dictionary with detailed aspects of crystallographic symmetry.
- The Cryo-EM extension dictionary supplements the mmCIF dictionary for structure and volume data from three-dimensional electron microscopy experiments.
- PDB exchange dictionary supplements the mmCIF dictionary with data items used internally by PDB and data items required to describe high throughput structure determinations in structure genomics projects.

## OTHER FORMATS AND PROTOCOLS

Much attention in the area of data format has recently been focused on technologies related to the extensible markup language (XML). XML provides a framework for structuring complex information and documents. XML follows the syntax conventions of the simpler hypertext markup language (HTML); however, XML extends the functionality of HTML by clearly separating the description of information from its presentation. To achieve this, XML employs a strategy of customizable style sheets (http://www.w3c.org/Style/) that can be used to define how a particular set of data will be displayed. XML is used as a data exchange vehicle in a variety of commercial software platforms. Public domain software libraries for parsing, accessing, and traversing XML data are available for most popular programming languages.

An XML document may be described by a Document Type Definition (DTD). The DTD specifies the permissible syntax of an XML document. A DTD specification primarily addresses the particular data items that can appear in a document, and how the items must be ordered and/or organized in hierarchies. There is rather limited support within a DTD for defining data semantics at the level of the data dictionaries discussed previously.

An emerging alternative to the DTD is the XML schema. XML schemas extend the descriptive functionalities of DTDs by providing support for strong data typing, complex data types, enumerations, range restrictions, and parent–child and key relationships. At the time of this writing the first XML schema specification was still under review by the W3C (http://www.w3c.org/XML/Schema).

Both DTDs and XML schemas focus on those aspects of data that are interpretable by software. Neither document definition provides for the encoding of data definitions or the examples. To fully capture these important semantic elements of the mmCIF data dictionary a more robust approach is required.

The Object Management Group (OMG) has recently standardized an exchange protocol for metadata and metadata models. The XML Metadata Interchange (XMI; http://www.omg.org/technology/xmi/) standard is designed to provide for encoding and exchange of metadata models conforming to the OMG Meta Object Facility (MOF; http://cgi.omg.org/cgi-bin/doc?formal/00-04-03/) metadata model. MOF is a comprehensive framework for defining metadata models. The Unified Modeling Language (UML), the de facto standard for graphically modeling data and software, is based on MOF. The DDL meta-model underlying mmCIF also conforms to the OMG MOF specification and hence XMI could be used as a means of storing mmCIF metadata in an XML form.

A growing number of commercial and public domain software applications provide support for XML. Many of these applications provide sophisticated display and query features that can be applied to the XML data. Owing to the availability of off-the-shelf access and query tools, this format would appear to be a good choice for storing structure data.

Although it is straightforward to re-encode mmCIF data in XML, the trivial element-based encoding shown in Figure 8.14 results in a large overhead of XML tags. For a medium-sized protein structure of 200 residues this results in a 10-fold increase in data file size. Storage efficiency can be obtained by storing data column-wise. For instance, the entire list of _atom_site.id values could be stored between the single pair of XML tags, <id></id>. With such encodings, one loses many of the access, display, and query features that make XML desirable. It is also necessary

```
<atom_site_list>
   <atom_site>
      <id>1</id>
      <label_atom_id>O5*</label_atom_id>
      <label_comp_id>T</label_comp_id>
      <label_asym_id>A</label_asym_id>
      <label_seq_id>1</label_seq_id>
      <label_entity_id>1</label_entity_id>
      <Cartn_x>-18.744</Cartn_x>
      <Cartn_y>20.195</Cartn_y>
      <Cartn_z>22.772</Cartn_z>
      <occupancy>1.0</occupancy>
      <B_iso_or_equiv>36.68</B_iso_or_equiv>
   </atom_site>
   <atom_site>
      <id>18</id>
      <label_atom_id>C5*</label_atom_id>
      <label_comp_id>T</label_comp_id>
      <label_asym_id>A</label_asym_id>
      <label_seq_id>1</label_seq_id>
      <label_entity_id>1</label_entity_id>
      <Cartn_x>-18.262</Cartn_x>
      <Cartn_y>20.915</Cartn_y>
      <Cartn_z>23.867</Cartn_z>
      <occupancy>1.0</occupancy>
      <B_iso_or_equiv>4.63</B_iso_or_equiv>
   </atom_site>
<!--- abbreviated -->
</atom_site_list>
```

(a)

```
#
loop_
_atom_site.id
_atom_site.label_atom_id
_atom_site.label_comp_id
_atom_site.label_asym_id
_atom_site.label_seq_id
_atom_site.label_entity_id
_atom_site.cartn_x
_atom_site.cartn_y
_atom_site.cartn_z
_atom_site.occupancy
_atom_site.B_iso_or_equiv
      1  O5*   T A   1   1    -18.744   20.195   22.722   1.00   36.68

      2  C5*   T A   1   1    -18.262   20.915   23.867   1.00    4.63

   #
```

(b)

Figure 8.14. (a) An abbreviated example of XML element-based encoding of ATOM_SITE data. In this example, the coordinates of two atoms are included. Each coordinate record is enclosed within <ATOM_SITE> tags. The other element names are taken directly from the item names defined in the mmCIF dictionary less the category name. For instance, the value of _atom_site.id is stored with <id></id> tags. In this XML encoding, the XML tags are replicated for each item of data. (b) The data corresponding to Figure 8.14a expressed using mmCIF. In the mmCIF representation, the keywords defining the data items are specified once as part of the *loop_* declaration.

to customize parsing software to manage list-oriented data. It is hoped that as XML becomes more widely used by large data-oriented applications, these trade-offs between efficiency and functionality can be overcome.

No single file format is satisfactory for all users and applications. One way to avoid file format issues entirely is to provide access to data through an application program interface (API). Depending on the language implementation the API provides access to data through a collection of functions, procedures, or methods. The OMG standardizes API developed using its Common Object Request Broker Architecture (Corba). Corba supports an interface definition language (IDL) for defining programmable interfaces that are both language and platform independent. Corba has also been developed to support distributed cross-platform access. The Corba IDL for macromolecular structure (http://cgi.omg.org/cgi-bin/doc?lifesci/00-02-02) is based on the mmCIF data representation and provides efficient program access to all of the data in PDB entries (http://omg.sdsc.edu/; Greer, Westbrook, and Bourne, 2002).

## CONCLUSION

In this chapter, the syntactical features of a number of ways to represent macromolecular structure data have been discussed. Each of these has its strengths and weaknesses. The PDB format is simple and accessible with very simple software tools. XML provides great flexibility and is well supported by commercial and public domain software. Far more important than the details of syntax is the ability of a particular data representation to precisely define data in a manner that is completely electronically accessible. The particular strength of the mmCIF approach is that it is based on a data dictionary. This data dictionary provides the detailed ontology, including precise definitions and examples combined with a robust metadata model, that can be exploited by software to perform detailed checks on individual data items as well as checks of the internal consistency between data items.

## ACKNOWLEDGMENTS

The development of the mmCIF dictionary and the associated DDL was an enormous community task, and any list of contributors to the effort will certainly be incomplete. Much of the mmCIF dictionary development was done by the original working group including: Enrique Abola, Helen Berman, Phil Bourne, Eleanor Dodson, Art Olson, Wolfgang Steigemann, Lynn Ten Eyck, and Keith Watenpaugh. Evaluation and critique of the dictionary development process was greatly aided by the input from COMCIFS, the IUCr committee with oversight over this process (I. David Brown and Brian McMahon). Many members of the community provided valuable input during the public review of the mmCIF dictionary. Frances Bernstein, Herbert Bernstein, Dale Tronrud, and Peter Keller were particularly active in this review process. Sydney Hall, Michael Scharf, Peter Grey, Peter Murray-Rust, Dave Stampf, and Jan Zelinka contributed to defining the requirements for the mmCIF DDL.

# REFERENCES

Berman HM, Westbrook J, Feng Z, Gilliland G, Bhat TN, Weissig H, Shindyalov IN, Bourne PE (2000): The Protein Data Bank. *Nucleic Acids Res* 28:235–42.

Bernstein FC, Koetzle TF, Williams GJ, Meyer EE, Brice MD, Rodgers JR, Kennard O, Shimanouchi T, Tasumi M (1977): Protein Data Bank: a computer-based archival file for macromolecular structures. *J Mol Biol* 112:535–42.

Bhat TN, Bourne P, Feng Z, Gilliland G, Jain S, Ravichandran V, Schneider B, Schneider K, Thanki N, Weissig H, Westbrook J, Berman HM (2001): The PDB data uniformity project. *Nucleic Acids Res* 29:214–8.

Bourne PE, Berman HM, Watenpaugh K, Westbrook JD, Fitzgerald PMD (1997): The macromolecular Crystallographic Information File (mmCIF). *Methods Enzymol* 277:571–90.

Callaway J, Cummings M, Deroski B, Esposito P, Forman A, Langdon P, Libeson M, McCarthy J, Sikora J, Xue D, Abola E, Bernstein F, Manning N, Shea R, Stampf D, Sussman J (1996): Protein Data Bank contents guide: atomic coordinate entry format description. Brookhaven National Laboratory. http://www.rcsb.org/pdb/docs/format/pdbguide2.2/guide2.2-frame.html.

Fitzgerald PMD, McKeever BM, VanMiddlesworth JF, Springer JP, Heimbach JC, Leu C-T, Kerber WK, Dixon RAF, Darke PL (1990): Crystallographic analysis of a complex between Human Immunodeficiency Virus Type 1 protease and acetyl-pepstatin at 2.0 Å resolution. *J Biol Chem* 265:14209–19.

Fitzgerald PMD, Berman HM, Bourne PE, Watenpaugh K (1992): Macromolecular CIF working group, International Union of Crystallography.

Fitzgerald PMD, Berman HM, Bourne PE, McMahon B, Watenpaugh K, Westbrook J (1996): The mmCIF dictionary: community review and final approval. IUCr Congress and General Assembly. Seattle, WA, *Acta Crystallogr* A52 Supplement: MSWK.CF.06.

Fitzgerald PMD, Westbrook JD, Bourne PE, McMahon B, Watenpaugh KD, Berman HM, (forthcoming). The Macromolecular Crystallographic Information File (mmCIF). In: *International Tables for Crystallography vol. G*. Dordrecht, Kluwer Academic Publishers.

Hall SR (1991): The STAR file: A new format for electronic data transfer and archiving. *J Chem Inf Comput Sci* 31:326–31.

Hall SR, Allen AH, Brown ID (1991): The Crystallographic Information File (CIF): a new standard archive file for crystallography. *Acta Crystallogr* A47:655–85.

Kuller A, Fleri W, Bluhm WF, Smith JL, Westrook J, Bourne PE (2002): A biologist's guide to synchrotron facilities: the BioSync Web resource. *TIBS* 27:213–5.

Protein Data Bank (1992): Protein Data Bank atomic coordinate and bibliographic entry format description. Brookhaven National Laboratory. http://www.rcsb.org/pdb/docs/format/pdbguide2.2/guide2.2-frame.html.

Westbrook JD, Hall SR (1995): DDL. *A Dictionary Description Language for Structure Macromolecular, V. 2.1.1*. New Brunswick, NJ: Rutgers University, NDB-110.

Westbrook J, Bourne PE (2000): STAR/mmCIF: an extensive ontology for macromolecular structure and beyond. *Bioinformatics* 16:159–68.

Westbrook JD, Berman HM, Hall SR, (forth coming). Specification of a relational Dictionary Definition Language (DDL2). In: *International Tables for Crystallography vol. G*. Dordrecht, Kluwer Academic Publishers.

••• driven architecture for derived representations of macromolecular structure. *Bioinformatics* 18:1280–1.

# 9

# THE PROTEIN DATA BANK

The PDB Team[1]

The Protein Data Bank (PDB) was established at Brookhaven National Laboratory (BNL) (Bernstein et al., 1977) in 1971 as an archive for biological macromolecular crystal structures. Nobel prizes have been awarded for the determination and analysis of some of the structures in the PDB. It represents one of the earliest community-driven molecular biology data collections. In the beginning the archive held seven structures, and with each passing year a handful more were deposited. In the 1980s, the number of deposited structures began to increase dramatically. This increase was due to the improvements in technology for all aspects of the crystallographic process, the addition of structures determined by nuclear magnetic resonance (NMR) methods, and changes in community views about data sharing. By the early 1990s, the majority of journals required a PDBidentification (PDBid) for publication and at least one funding agency, the National Institute of General Medical Sciences (NIGMS), adopted the guidelines published by the International Union of Crystallography requiring data deposition for all structures determined using NIGMS funds. At the beginning of 2002 there were more than 17,000 entries in the archive.

Accompanying this rapid growth, the mode of access to PDB data has changed over the years as a result of improved technology. Data distribution is now primarily via the World Wide Web (www) rather than via magnetic media. Further, the need to analyze diverse subsets of the data has led to the development of modern data management systems.

Initial use of the PDB had been limited to a small group of experts involved in structural biology research. Today depositors to the PDB have expertise in the techniques of X-ray crystal structure determination, NMR, cryoelectron microscopy,

---

[1]For a current list of team members refer to http://www.rcsb.org/pdb/rcsb-group.html.

*Structural Bioinformatics*
Edited by Philip E. Bourne and Helge Weissig
Copyright © 2003 by Wiley-Liss, Inc.

and theoretical modeling. PDB users are a very diverse group of researchers in biology and chemistry, as well as educators and students at all levels. The tremendous influx of data soon to be fueled by the structural genomics initiative (see Chapter 29), and the increased recognition of the value of these data toward understanding biological function, continually demand new ways to collect, organize, and distribute the data (Berman et al., 2000a).

Since October 1998, the PDB has been managed by the Research Collaboratory for Structural Bioinformatics (RCSB), which is a consortium consisting of Rutgers, the State University of New Jersey; the San Diego Supercomputer Center at the University of California, San Diego; and the National Institute of Standards and Technology. In this chapter, we describe the current procedures for collecting, validating, annotating and distributing PDB data. Finally, we describe the plans for further automating and improving the PDB so it can meet emerging challenges posed by researchers and educators in the field of structural bioinformatics.

## DATA ACQUISITION AND PROCESSING

A key component of the PDB is the efficient capture and curation of the data—data processing. Data processing consists of data deposition, annotation, and validation. These steps are part of a fully documented and integrated data processing system shown in Figure 9.1.

In the present system, data (atomic coordinates, structure factors, and NMR restraints) may be submitted via e-mail or via the Web-based AutoDep Input Tool (ADIT) (Westbrook, Feng, and Berman, 1998; http://deposit.pdb.org/adit/) developed by the RCSB PDB. ADIT is built on top of the macromolecular Crystallographic Information File (mmCIF) dictionary that contains 1700 terms that define the macromolecular structure and the crystallographic experiment (Bourne et al., 1997; Westbrook and Bourne, 2000; see also Chapter 8). The mmCIF dictionary has been further extended to form the PDB exchange dictionary, which includes terms needed for tracking and other information management purposes. ADIT is complemented by MAXIT (MAcromolecular EXchange Input Tool; Feng, et al., 1998a), a program that performs many of the data-processing tasks and checks. This integrated system helps to ensure that the data submitted are consistent with the mmCIF dictionary, which defines data types, enumerates ranges of allowable values where possible, and describes allowable relationships between data values.

Each deposition to the PDB is represented by the PDBid—a four character code of the form nXYZ, where n is an integer and X, Y, and Z are alphanumeric characters, for example, 4HHB. The PDBid is assigned arbitrarily and is an immutable reference to the structure, and indeed is the only absolute way of retrieving a desired structure from the PDB, although this shortcoming is being addressed (refer to Data Uniformity below). PDBids are never reused and remain the link between the structure and the literature reference that describes that structure.

After a structure has been deposited using ADIT, a PDBid is automatically and immediately sent to the author (Fig. 9.1; Step 1). This procedure is the first stage in which information about the structure is loaded into the internal core database, validated, and annotated (see also Database Architecture and Validation and Annotation below). This step involves using ADIT to help diagnose errors or inconsistencies in the files. The completely annotated entry as it will appear in the PDB resource, together with the validation information, is sent back to the depositor (Fig. 9.1; Step 2).

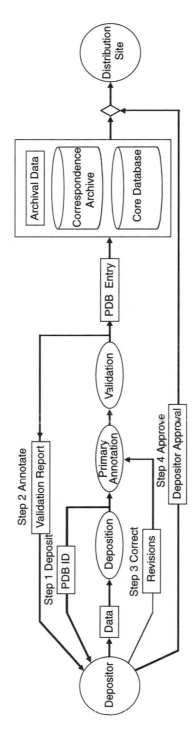

Figure 9.1. The steps involved in PDB data processing. Ellipses represent actions and rectangles define content. Figure reprinted from Berman et al. (2000b) with permission from the International Union of Crystallography.

After reviewing the processed file, the author sends any revisions (Fig. 9.1; Step 3). Depending on the nature of these revisions, steps 2 and 3 may be repeated. Once approval is received from the author (Fig. 9.1; Step 4), the entry and the tables in the internal core database are ready for distribution. The schema of this core database is a subset of the conceptual schema specified by the mmCIF dictionary.

All aspects of data processing, including communications with the author, are recorded and stored in the electronic correspondence archive. This record makes it possible for the PDB staff to retrieve information about any deposited entry. Current status information, including the entry's authors, title, and release status, is stored for each entry in the core database and is made accessible for query via the WWW interface (http://www.rcsb.org/pdb/status.html). Entries before release are categorized as "in processing" (PROC), "in depositor review" (WAIT), "to be held until publication" (HPUB) or "on hold until a depositor specified date" (HOLD).

## Content of the Data Collected by the PDB

All the data collected from depositors by the PDB are considered primary data. Primary data contain, in addition to the coordinates, general information required for all deposited structures and information specific to the method of structure determination. Table 9.1 contains the general information that the PDB collects for all structures as well as the information specific to X-ray and NMR experiments.

Historically, NMR data have been placed in a format defined around crystallographic information. The PDB is currently working with an NMR Task Force and the BioMagResBank (BMRB; Ulrich, Markley, and Kyogoku, 1989; see also Chapter 5) to develop an NMR data dictionary as an extension to the mmCIF. This dictionary includes descriptions of the solution components, the experimental conditions, enumerated lists of the instruments used, and information about structure refinement. This dictionary will be used as deposition and validation tools specifically for NMR structures. NMR coordinate data are currently deposited with the PDB and other NMR-specific experimental data are deposited with the BMRB. Plans are in place to have a single interface for the deposition of data to both the BMRB and the PDB.

The information content of data submitted by the depositor is likely to change as new methods for data collection, structure determination, and refinement evolve. A case in point is the need for structural genomics projects to collect all information that would be in the material and methods section of a paper describing a structure. The ways in which these data are captured are also changing as the software for structure determination and refinement evolve to produce the necessary data items as part of their output. The ontology-driven approach to software development used by the PDB makes it simple to collect new items of data once they are described in the mmCIF or extension dictionary.

## Validation and Annotation

Validation refers to the procedure for assessing the quality of deposited atomic models (structure validation) and for assessing how well these models fit the experimental data (experimental validation). Annotation refers to the process of adding information resulting from the validation to the entry. The PDB validates structures using accepted

## TABLE 9.1. Content of Data in the PDB

Content of All Depositions (X-ray and NMR)

Source: Specifications such as genus, species, strain, or variant of gene (cloned or synthetic); expression vector and host, or description of method of chemical synthesis
Sequence: Full sequence of all macromolecular components
Chemical structure of cofactors and prosthetic groups
Names of all components of the structure
Qualitative description of the characteristics of the structure
Literature citations for the structure submitted
Three-dimensional coordinates

Additional Items for X-ray Structure Determinations

Temperature factors and occupancies assigned to each atom
Crystallization conditions, including pH, temperature, solvents, salts, methods
Crystal data, including the unit cell dimensions and space group
Presence of noncrystallographic symmetry
Data collection information describing the methods used to collect the diffraction data including instrument, wavelength, temperature, and processing programs
Data collection statistics including data coverage, $R_{sym}$, data above 1, 2, 3 sigma levels and resolution limits
Refinement information including R factor, resolution limits, number of reflections, method of refinement, sigma cutoff, geometry rmsd, sigma
Structure factors: h, k, l, Fobs, $\sigma$ Fobs

Additional Items for NMR Structure Determinations

For an ensemble, the model number for each coordinate set that is deposited and an indication if one should be designated as a representative
Data collection information describing the types of methods used, instrumentation, magnetic field strength, console, probe head, sample tube
Sample conditions, including solvent, macromolecule concentration ranges, concentration ranges of buffers, salts, antibacterial agents, other components, isotopic composition
Experimental conditions, including temperature, pH, pressure, and oxidation state of structure determination and estimates of uncertainties in these values
Noncovalent heterogeneity of sample, including self-aggregation, partial isotope exchange, conformational heterogeneity resulting in slow chemical exchange
Chemical heterogeneity of the sample (e.g., evidence for deamidation or minor covalent species)
A list of NMR experiments used to determine the structure including those used to determine resonance assignments, NOE/ROE data, dynamical data, scalar coupling constants, and those used to infer hydrogen bonds and bound ligands. The relationship of these experiments to the constraint files are given explicitly
Constraint files used to derive the structure as described in Task Force recommendations

*Source*: Reprinted, by permission of the International Union of Crystallography, from Berman HM, Westbrook J, Feng Z, Gilliland G, Bhat TN, Weissig H, Shindyalov IN, Bourne PE (2001): The protein Data Bank, 1999. In: Rossman MG and Arnold E, editors. *International Tables for Crystallography*. F. Crystallography of Biological Macromolecules. Dordrecht: Kluwer Academic, p. 676

community standards as part of ADIT's integrated data-processing system. The following checks are run and are summarized in a letter that is communicated directly to the depositor:

*Covalent Bond Distances and Angles.* Proteins are compared against standard values from Engh and Huber (1991); nucleic acid bases are compared against standard values from Clowney et al. (1996); sugar and phosphates are compared against standard values from Gelbin et al. (1996).

*Stereochemical Validation.* All chiral centers of proteins and nucleic acids are checked for correct stereochemistry.

*Atom Nomenclature.* The nomenclature of all atoms is checked for compliance with International Union of Pure and Applied Chemistry (IUPAC) standards (IUPAC-IUB, 1983) and is adjusted if necessary.

*Close Contacts.* The distances between all atoms within the asymmetric unit of crystal structures and the unique molecule of NMR structures are calculated. For crystal structures, contacts between symmetry-related molecules are checked as well.

*Ligand and Atom Nomenclature.* Residue and atom nomenclature are compared against a standard dictionary (ftp://ftp.rcsb.org/pub/pdb/data/monomers/het_diction ary.txt) for all ligands as well as standard residues and bases. Unrecognized ligand groups are flagged and any discrepancies in known ligands are listed as extra or missing atoms. New ligands are added to the dictionary as they are deposited.

*Sequence Comparison.* The sequence provided by the depositor is compared against the sequence derived from the coordinate records. This information is displayed in a table where any differences or missing residues are annotated. During the annotation process the sequence database references provided by the author are checked for accuracy. If no reference is given, a BLAST (Zhang et al., 1991) search is used to find the best match. Any conflict between the depositor's sequence and the sequence derived from the coordinate records is further resolved and annotated by comparison with other sequence databases as needed.

*Distant Waters.* The distances between all water oxygen atoms and all polar atoms (oxygen and nitrogen) of the macromolecules, ligands, and solvent in the asymmetric unit are calculated. Distant solvent atoms are repositioned using crystallographic symmetry such that they fall within the solvation sphere of the macromolecule.

In almost all cases, serious errors detected by these checks have been corrected through annotation and correspondence with the authors. It is also possible to run these validation checks against structures before they are deposited. A validation server (http://deposit.pdb.org/validate/) has been made available for this purpose. In addition to the summary report letter, the server also provides output from PROCHECK (Laskowski et al., 1993), NUCheck (Feng, Westbrook, and Berman, 1998b), and SFCHECK (Vaguine, Richelle, and Wodak, 1999). A summary atlas page and molecular graphics images are also produced.

The PDB continuously reviews the validation methods used and will continue to integrate new procedures as they become available and are accepted as community standards.

## Data Deposition Sites

Data are deposited to the PDB to one of three sites. Because it is critical that the final archive is kept uniform, the content and format of the final files as well as the methods used to check them must be the same.

The RCSB–PDB deposition site (http://deposit.pdb.org/adit/) has developed software programs for data deposition, validation, and processing, including ADIT and the Validation Server. The ADIT system, as described above, is also used to process the data deposited.

The Institute for Protein Research at Osaka University in Japan has collaborated with the PDB to establish another deposition center (http://pdbdep.protein.osaka-u.ac.jp/adit/). All data deposited at this center (primarily depositors in Asia) are also processed by this Osaka group using the ADIT system.

The Macromolecular Structure Database group at the European Bioinformatics Institute (MSD–EBI) processes data that are submitted to them via AutoDep (http://autodep.ebi.ac.uk/). After processing, the data are sent to the RCSB in PDB format for inclusion in the central archive. A common mmCIF exchange dictionary has been developed with this group, which will help ensure a higher degree of data uniformity in the archival data in the future.

The PDB has also ported its data-processing software to a stand-alone system that does not require Internet access. This system is soon to be released for use by authors who wish to check data in their home laboratories.

## Data Processing Statistics

Production processing of PDB entries by the RCSB began on January 27, 1999. The median time from deposition to the completion of data processing including author interactions is less than two weeks. The number of structures with a HOLD release status remains at about 16% of all submissions; 63% are hold until publication; and 21% are released immediately after processing.

Figure 9.2 shows the growth of PDB data since the archive began. Figure 9.2a shows the total number of structures available in the archive per year. Figure 9.2b shows the number of residues released in the PDB each year, indicating how the complexity of structures released into the archive has increased over time.

The current breakdown of the types of structures in the PDB can be found at http://www.rcsb.org/pdb/statistics.html. Data at the end of September 2001 are shown in Table 9.2.

## Data Uniformity

A key goal of the PDB is to make the archive as consistent and error-free as possible. As indicated above, all new depositions are reviewed carefully by annotators before release. Errors found subsequent to release by authors and PDB users are addressed as rapidly as possible. Minor errors result in revisions to the entry that are annotated within the entry; major errors lead to a superceding entry or entry withdrawal. Corrections and updates to entries are sent to deposit@rcsb.rutgers.edu.

"Legacy data," that is, data submitted prior to October 1998, comply with several different PDB formats, and variation exists in how the same features are described for different structures within each format. The inconsistency of formats and nomenclature conventions makes it difficult to consistently parse these data and query across the

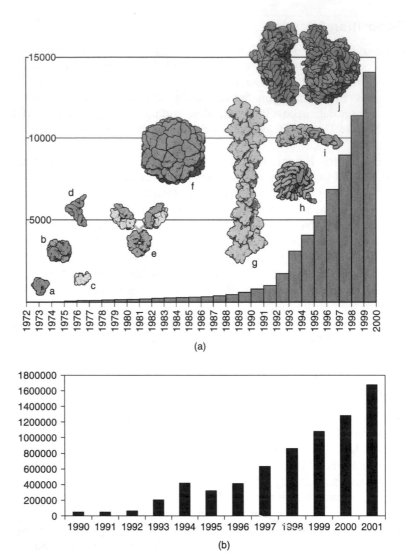

Figure 9.2. (A) Growth chart of the PDB showing the total number of structures available in the PDB archive per year and highlighting example structures from different time periods: a) myoglobin, b) hemoglobin, c) lysozyme, d) transfer RNA, e) antibodies, f) entire viruses, g) actin, h) the nucleosome, i) myosin, and j) 30s ribosomal subunits. Images were created by Dr. David Goodsell, who authors the PDB's Molecule of the Month series. Figure originally appeared in the International Union of Crystallography Newsletter (2001). Images, descriptions and the molecules, and links to related information can be found at http://www.rcsb.org/pdb/ molecules/molecule_list.html. (B) Number of residues released in the PDB per year. Figure also appears in Color Figure section.

archive. As an immediate solution to the query problem, particular records across all entries in the archive were corrected; these included citation, R factor, and resolution (Bhat et al., 2001). These corrections were loaded into the database and thus it was possible to query on these features and obtain accurate results (Table 9.3). However, these data were not available in the PDB files. To provide uniform data for each structure we

TABLE 9.2. Demographics of Data Released in the PDB (as of September 25, 2001)

| Experimental Technique | Molecule Type | | | | |
| --- | --- | --- | --- | --- | --- |
| | Proteins, Peptides, and Viruses | Protein–Nucleic Acid Complexes | Nucleic Acids | Carbohydrates and Other | Total |
| X-ray Diffraction and Other | 12,083 | 576 | 583 | 14 | 13,256 |
| NMR | 1997 | 73 | 399 | 4 | 2473 |
| Theoretical Modeling* | 297 | 21 | 23 | 0 | 341 |
| Total | 14,377 | 670 | 1005 | 18 | 16,070 |

*Theoretical models have subsequently been made available separately.

TABLE 9.3. Query Results on Uniform Versus Nonuniform Data (from August 29, 2000)

| Attribute | Query Term | Nonuniform | Uniform |
| --- | --- | --- | --- |
| Resolution | 2.1–2.5 Å | 3061 | 3492 |
| Primary Citation | J. Mol. Biol. | 1953 | 2331 |
| Journal Name | Biochemistry | 1919 | 2522 |
| | To be published | 2856 | 760 |
| EC Number | 3.2.1.17 | 264 | 570 |
| Source (Organism) | E. coli | 5 | 1278 |
| | Escherichia coli | 1103 | 1278 |
| | Mouse | 451 | 477 |
| | Mus musculus | 444 | 477 |
| | Human | 1988 | 2388 |
| | Homo sapiens | 2010 | 2388 |

*Source*: Reprinted, by permission of Oxford University Press, from Bhat TN, Bourne P, Feng Z, Gilliland G, Jain S, Ravichandran V, Schneider B, Schneider K, Thanki N, Weissig H, Westbrook J, Berman HM (2001): The PDB data uniformity project. *Nucleic Acids Research* **29**:217.
*Note*: The attributes listed can be searched by using the SearchFields interface. The numbers given are the result of entering the query term in the desired field on both nonuniform and uniform data. These data are currently available as database tables, but not available in the individual PDB data files. They are available in the mmCIF data files described in the Data Uniformity section of this chapter. Information about the data uniformity project is archived at http://www.rcsb.org/pdb/uniformity/.

used the software that was developed and tested for primary processing and revalidated all the data in the archive. Corrections were made to nomenclature and special attention was made to consistency of the chemical description of the macromolecule and the ligand. Examples of the types of errors that were found and corrected are shown in Table 9.4.

The corrected files were released in mmCIF format and can be found at (ftp:// beta.rcsb.org/pub/pdb/uniformity/data/mmCIF/) (Westbrook, et al., 2002). The original PDB files will continue to be available as they are a historical record and have been the basis of many research projects. Software is available from the PDB to transform the mmCIF files to PDB-formatted files. In the future, these mmCIF files will form the basis of the PDB databases accessible via the Web.

TABLE 9.4. Summary of Released Entries Containing Nomenclature and Chemical Representation Errors

|  | Incorrect Sequence | Sequence-Coordinate Mismatch | Atom Nomenclature Errors | Stereochemical Labeling Errors |
|---|---|---|---|---|
| Legacy data (8368 entries)[a] | 166 | 90 | 3311 | 294 |
| 1999 data (3150 entries)[b] | 0 | 5 | 162 | 3 |
| 2000 data (3569 entries)[b] | 0 | 0 | 31 | 3 |

[a] pre-October 1998 entries, excluding nucleic acid-containing crystal structures.
[b] Structures processed and released by the RCSB.

## DATA ACCESS

The PDB is presently incremented once per week with new data becoming available on Wednesday mornings in most parts of the world through a number of mirror sites. The following describes the database architecture used by the PDB, how users access these databases via the Web, and how data files can be accessed via the Web and via ftp.

### Database Architecture

The current PDB data management system consists of several heterogeneous data sources that are integrated through Perl CGI scripts (Fig. 9.3). Although this leads to some redundancy, since parts of the data are stored multiple times, it allows efficient access. The complete system is currently being reengineered to maintain this efficiency while providing more manageability with less redundancy. The new system will be based on the DB2 relational database management system. We consider here the five core components of the current system.

First, the core relational database (Sybase SQL server release 11.0, Emeryville, CA) stores the primary experimental and coordinate data as described in Table 9.1. These data are retrieved by the reporting options available through the Web interface. Second, the ftp archive (ftp://ftp.rcsb.org/pdb/) provides the data files in PDB and mmCIF formats as well as the data dictionaries to which they correspond. Third, the Property Object Model (POM) data management system (Shindyalov and Bourne, 1997) is used for more efficient access to certain structural features, such as sequence. POM consists of indexed objects containing native data (e.g., atomic coordinates) and derived properties (e.g., secondary structure calculated according to Kabsch and Sander (1983)). Fourth, the Netscape LDAP server is used to index the textual content of the PDB and provides support for keyword searches. Fifth, the Molecular Information Agent (MIA; http://mia.sdsc.edu) is used to collect and store hyperlinks and limited other information for approximately 60 external data resources in a separate Sybase database (see http://www.rcsb.org/pdb/mia.html). MIA formulates a query to each of these data sources based on the PDBid, and parses the results of the query to provide the information viewable through the "Other Sources" option of an entry's Structure Explorer page. MIA includes housekeeping software, for example, to coordinate the simultaneous access to these data sources and to timeout if a particular site is down.

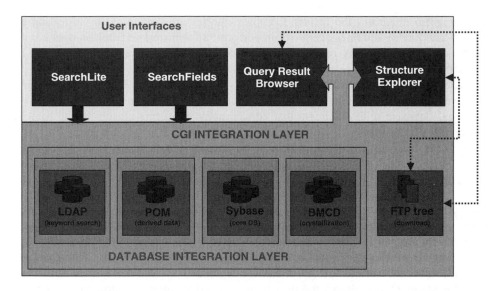

Figure 9.3. The integrated query interface to the PDB. Figure reprinted with permission from Berman et al. (2000).

These five components, associated software, and Web pages constitute the system that is mirrored to a number of sites worldwide (see below).

Finally, there is a close integration to three external resources (i.e., not mirrored as part of the PDB):

1. The Biological Macromolecule Crystallization Database (BMCD; (Gilliland, 1988)) is organized as a relational database within Sybase and contains three general categories of literature derived information: macromolecular, crystal, and summary data.
2. The Nucleic Acid Database (NDB; Berman, et al., 1992; see also Chapter 10) which contains information pertaining specifically to DNA and RNA.
3. The CE database (Shindyalov and Bourne, 1998; see also Chapter 16) of 3D protein structure alignments.

The latter raises an important point of PDB policy. As described in Chapter 16, alignment of structures depends to some degree on the assumptions of the method being used. Since there is no agreement in the community at present as to a de facto standard method for protein structure alignment, the PDB's policy is to provide access to a variety of alignments and classification schemes (Murzin et al., 1995; Gibrat, Madej, and Bryant, 1996; Orengo et al., 1997; Holm and Sander, 1998; Shindyalov and Bourne, 1998). In short, the PDB's policy is to provide a portal (entry point) to relevant information, but to not impose judgment on which methodology should be used.

In the current implementation, communication among the five PDB components and these databases has been accomplished using the Common Gateway Interface (CGI) in such a way as to hide the intricacies of the underlying databases from the user. An integrated Web interface dispatches a query to the appropriate database(s), which then execute the query. Each database returns the PDBids that satisfy the query, and the

CGI program integrates the results. Complex queries are performed by repeating the process and having the interface program perform the appropriate Boolean operation(s) on the collection of query results. A variety of output options are then available for use with the final list of selected structures.

The newly created and uniform mmCIFs will enable the PDB to substantially improve its underlying database architecture. The mmCIFs are loaded into a new relational database with a schema that conceptually conforms closely to the mmCIF dictionary. The results will provide access to data not currently available and do so in a way that is easier to maintain.

## User Web Access

Currently, three distinct query interfaces are available for the query of data within the PDB: Status Query (http://www.rcsb.org/pdb/status.html); SearchLite (http://www.rcsb.org/pdb/searchlite.html); and SearchFields (http://www.rcsb.org/pdb/queryForm.cgi). Table 9.5 summarizes the current query and analysis capabilities of the PDB. Figure 9.4 illustrates how the various query options are organized.

The Status Query allows the user to review information on structures deposited but not yet released. In addition to an author list, title, and release status, the depositor may opt to release sequence information for the unreleased entry. This provides a set of useful targets for structure prediction studies.

SearchLite provides a single form field for keyword searches. Textual information within the PDB file, such as dates and some experimental data, are searched. Boolean searching and restriction of keywords can be used to conform to specific attributes. For example, "green" can be attributed to an author name or a common name for a protein.

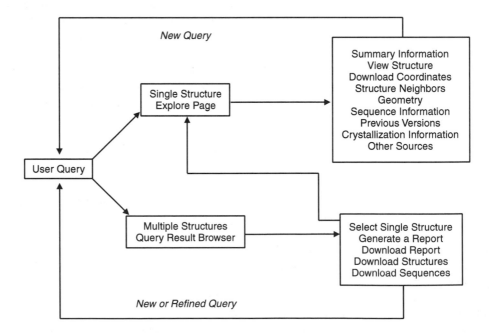

**Figure 9.4.** The layout of the PDB query system. Figure reprinted from Berman et al. (2001) with permission from the International Union of Crystallography.

TABLE 9.5. Current Query Capabilities of the PDB

| Query Options | |
| --- | --- |
| SearchLite | Any word or combination of words in the PDB |
| SearchFields | *General information*: PDB identifier, citation author, chain type (protein, DNA etc.), PDB HEADER, experimental technique, deposition/release date, citation, compound information, EC number, text search<br>*Sequence and secondary structure*: chain length, FASTA search, short sequence pattern, secondary structure content<br>*Crystallographic experimental information*: resolution, space group, unit cell dimensions, parameters |
| Status | PDB identifier, deposition author, title, holding status, deposition date, release date, prereleased sequences |

| Result Analysis | |
| --- | --- |

| Single Structure: Structure Explorer | |
| --- | --- |
| Summary | Compound name, authors, experimental method, classification, source, primary citation, deposition date, release date, resolution, R-value, space group, unit cell parameters, polymer chain identifiers, number of residues, HET groups, number of atoms |
| View Structure | VRML, RasMol, QuickPDB (Java Applet), Chime, still images |
| Download/Display file | HTML and text formats for display; PDB and mmCIF formats with different compression options for download |
| Structural Neighbors | List of sites for finding structural homologues |
| Geometry | Unusual dihedral angles, bond angles and bond lengths |
| Other Sources | Links to other sources of information |
| Sequence Details | Chain Ids, number of residues per chain, molecular weight, chain type, secondary structure assignment; download sequence only in FASTA format |
| Crystallization Information | Conditions under which the crystals were obtained |
| Previous versions | Versions of the structure replaced by the current version if applicable |
| Nucleic Acid Database Atlas Entry | Detailed information from the NDB (if applicable) |
| Quick Report | Nucleic acid geometry if applicable |
| Structure Factors | Experimental data if available |

| Multiple Structure: Results Browser | |
| --- | --- |
| Summary List | Deposition date, resolution, experimental method, classification, compound name |
| Download Structures or Sequences | mmCIF and PDB compressed files (gzip, tar, compressed); sequences in FASTA format |
| Query Refinement | Iterative query over result set using OR, AND or NOT Boolean logic |
| Tabular Report | Cell dimensions, primary citation, structure identifiers, sequence, experimental details, refinement details |
| Query Review | Summary of queries submitted thus far with the option to return to one of them |

*Source*: Reprinted, by permission of Oxford University Press, from Berman HM, Westbrook J, Feng Z, Gilliland G, Bhat TN, Weissig H, Shindyalov IN, Bourne PE (2000): The Protein Data Bank. *Nucleic Acids Research* **28**:239.

SearchFields is for more advanced searches and presents a customizable query form that can be used to search different data items, including macromolecule type, citation authors, sequence (via a FASTA search; Pearson and Lipman, 1988), and release dates. For enzymes it is possible to browse these structures using the Enzyme Commission hierarchy. The numbers of entries at each level are reported as the user traverses the hierarchy.

Search results are displayed in the "Query Result Browser," which can be used to generate reports, download data, and perform further searches. The "Structure Explorer" interface provides detailed information on a single structure. On-line tutorials for accessing PDB data via the Web are available at http://www.rcsb.org/pdb/info.html #PDB_Users_Guides.

## Application Web Access

Interfaces to both single and multiple structures are accessible to other Web resources and applications through the simple CGI application programmer interface (API) described at http://www.rcsb.org/pdb/linking.html. Stated another way, a URL can be constructed with either single or multiple embedded PDBids and used to return results on those structures. Many Web sites worldwide use this mechanism to reference PDB structures.

The PDB Web site is maintained on redundant load balanced servers and receives in excess of 100,000 page hits per day. On average a structure is downloaded every second 24 hours per day, 7 days per week.

## ftp Access

All structures, in PDB and mmCIF formats, are available for download from the PDB ftp site. Dictionaries, documentation, and PDB-provided software are also available. Instructions and software for mirroring the PDB ftp archive as a local copy are available at http://www.rcsb.org/pdb/ftpproc.final.html.

## Distribution

As stated, the PDB distributes coordinate data in PDB and CIF formats, structure factor files, and NMR constraint files. In addition, it provides derived data, documentation, and software. New data officially become available at 2:00 A.M. Pacific standard time each Wednesday. PDB mirrors have been established in Japan (Osaka University), Singapore (National University Hospital), Brazil (Universidade Federal de Minas Gerais), and in the UK (Cambridge Crystallographic Data Centre). Additionally, other sources of PDB data exist, but are provided through different interfaces. See http://www.rcsb.org/pdb/mirrors.html for a partial list. The PDB also distributes a quarterly CD-ROM set that is essentially a copy of the ftp site. Data are distributed as compressed files using the compression utility program gzip. Refer to http://www.rcsb.org/pdb/cdrom.html for details on how to order CD-ROM sets. There is no charge for this service.

## OUTREACH

Active outreach ensures that the community of PDB users is fully informed about our capabilities and activities and that the PDB receives feedback that allows it to improve its services. Outlined below are some of the key outreach activities.

*Help Desk.* The electronic help desk (info@rcsb.org) addresses questions about all aspects of the PDB and about general structural biology. Questions are generally addressed within one or two working days. The list receives an average of 130 inquiries per month. The PDB also maintains two other addresses: deposit@rcsb.rutgers.edu, for questions concerning data deposition and help@rcsb.rutgers.edu for questions concerning ADIT.

*pdb-l@rcsb.org.* A list server at pdb-l@rcsb.org is maintained for use by the community to make announcements and conduct discussions on activities relating to the PDB. It is a PDB policy that this list be reserved for open discussion by the community and not for use by the PDB itself. An archive of the discussions that take place on this list can be found at http://www.rcsb.org/pdb/lists/pdb-l/.

*PDB Web Site.* The Web site is updated weekly with news, recent developments, and improvements to existing documents. The site includes tutorials and user guides for query, deposition, and file formats.

*PDB Publications.* The PDB publishes a quarterly newsletter available via e-mail or postal mail (see http://www.rcsb.org/pdb/newsletter.html). PDF versions of the PDB newsletter dating back to September 1974 are available at this site. Flyers, tutorials, and an annual report are accessible from the PDB Web site and by sending mail to info@rcsb.org.

*Scientific Meetings.* PDB members attend a wide variety of meetings, presenting posters, talks, and exhibit booths. Among the meetings attended are the American Crystallographic Association's Annual Meeting, Protein Society's Symposiums, the Intelligent Systems in Molecular Biology's annual meeting, and the International Union of Crystallography's Congress and General Assembly.

TABLE 9.6. PDB Mirror Sites

| RCSB Partner Sites | |
| --- | --- |
| SDSC | http://www.pdb.org/ |
| La Jolla, CA (US) | ftp://ftp.rcsb.org/ |
| Rutgers University | http://rutgers.rcsb.org/ |
| Piscataway, NJ (US) | |
| NIST | http://nist.rcsb.org/ |
| Gaithersburg, MD (US) | |
| **Other RCSB Mirrors** | |
| CCDC | http://pdb.ccdc.cam.ac.uk/ |
| United Kingdom | ftp://pdb.ccdc.cam.ac.uk/rcsb/ |
| National University of | http://pdb.bic.nus.edu.sg/ |
| Singapore | ftp://pdb.bic.nus.edu.sg/pub/pdb/ |
| Singapore | |
| Osaka University | http://pdb.protein.osaka-u.ac.jp/ |
| Japan | ftp://ftp.protein.osaka-u.ac.jp/pub/pdb/ |
| Universidade Federal | http://www.pdb.ufmg.br/ |
| de Minas Gerais | ftp://vega.cenapad.ufmg.br/pub/pdb/ |
| Brazil | |

T A B L E  9.7. PDB Sites of Interest

| Source | Information Content |
| --- | --- |
| Deposition | |
| http://deposit.pdb.org/adit/ (RCSB-Rutgers) | ADIT Web site (deposit@rcsb.rutgers.edu) |
| http://pdbdep.protein.osaka-u.ac.jp/adit/ (Osaka University) | ADIT Web site (adit@adit.protein.osaka-u.ac.jp) |
| http://autodep.ebi.ac.uk/ (MSD-EBI) | AutoDep (pdbhelp@ebi.ac.uk) |
| http://deposit.pdb.org/validate/ | ADIT Validation Server |
| http://deposit.pdb.org/ | Deposition, Format, and ADIT FAQs |
| Query | |
| http://www.rcsb.org/pdb/status.html | PDB Status Search |
| http://www.rcsb.org/pdb/searchlite.html | SearchLite |
| http://www.rcsb.org/pdb/queryForm.cgi | SearchFields |
| http://www.rcsb.org/pdb/linking.html | Information on Linking to the PDB |
| http://mia.sdsc.edu/ | Molecular Information Agent |
| http://www.rcsb.org/pdb/mia.html | MIA at the PDB FAQ |
| http://www.rcsb.org/pdb/ftpproc.final.html | RCSB PDB Mirror Protocol |
| PDB Features | |
| http://www.rcsb.org/pdb/strucgen.html | Structural Genomics Resources |
| http://www.rcsb.org/pdb/uniformity/ | Information about the PDB Data Uniformity Project |
| http://www.rcsb.org/pdb/lists/pdb-l/ | PDB Listserv for community announcements |
| http://www.rcsb.org/pdb/cdrom.html | CD-ROM information |
| http://www.rcsb.org/pdb/info.html#PDB_Users_Guides | Tutorials for deposition and query |
| http://www.rcsb.org/pdb/newsletter/index.html | PDB Newsletters |
| http://www.rcsb.org/pdb/statistics.html | Statistics on the PDB archive |
| http://www.rcsb.org/pdb/ftpproc.final.html | FTP mirroring information |
| http://www.rcsb.org/pdb/cdrom.html | CD-ROM ordering information |
| info@rcsb.org | General help desk |

## FUTURE

Structural biology is a fast-evolving field that poses challenges to the collection, curation, and distribution of macromolecular structure data. Since 1999 the number of depositions has averaged approximately 50 per week. However, with the advent of a number of structure genomics initiatives worldwide this number is likely to increase. We estimate that the PDB could contain 35,000 structures by 2005. This growth presents a challenge to timely distribution while maintaining high quality. We believe our approach to information management should permit us to accommodate the anticipated large data influx. We are endeavoring to work closely with all structural genomics projects to automatically collect more data and are redesigning our database systems to provide a more scalable system. In terms of access, we have worked with the Object Management Group to define a Corba (Common Object Request Broker) standard for macromolecular structure, which is closely aligned with the mmCIF dictionary. Eventually this will provide a fine-grained access to items of PDB data by users and their applications. This will be achieved by providing a Corba server that is currently under development.

The maintenance and further development of the PDB are community efforts. The willingness of others to share ideas, software, and data provides a depth to the resource not obtainable otherwise. It is important to acknowledge the contribution of scientists and staff at the BNL, who maintained the archive for many years. New input is constantly being sought and the PDB invites you to make comments at any time by sending electronic mail to info@rcsb.org. A summary of all the URLs specified in this chapter is given in Table 9.6 and Table 9.7 for easy reference.

## ACKNOWLEDGMENTS

This work is supported by grants from the National Science Foundation, the Office of Biological and Environmental Research at the Department of Energy, and two units of the National Institutes of Health: the National Institute of General Medical Sciences and the National Institute of Medicine. This chapter is also appearing in *Acta Crystallographica* Special Issue of Sections B and D, entitled "Crystallographic Databases and their Applications" (Frank H. Allen and Jenny P. Glusker, Editors).

## REFERENCES

(2001): International Union of Crystallography Newsletter 9.

Berman HM, Olson WK, Beveridge DL, Westbrook J, Gelbin A, Demeny T, Hsieh SH, Srinivasan AR, Schneider B (1992): The Nucleic Acid Database—a comprehensive relational database of three-dimensional structures of nucleic acids. *Biophys J* 63:751–9.

Berman HM, Bhat TN, Bourne PE, Feng Z, Gilliland G, Weissig H, Westbrook J (2000a): The Protein Data Bank and the challenge of structural genomics. *Nat Struct Biol* 7:957–9.

Berman HM, Westbrook J, Feng Z, Gilliland G, Bhat TN, Weissig H, Shindyalov IN, Bourne PE (2000b): The Protein Data Bank. *Nucleic Acids Res* 28:235–42.

Berman HM, Westbrook J, Feng Z, Gilliland G, Bhat TN, Weissig H, Shindyalov IN, Bourne PE (2001): The Protein Data Bank, 1999. In: Rossman MG, Arnold E, editors. *International Tables for Crystallography. F. Crystallography of Biological Macromolecules.* Dordrecht: Kluwer Academic, pp. 675–62.

Bernstein FC, Koetzle TF, Williams GJ, Meyer EE, Brice MD, Rodgers JR, Kennard O, Shimanouchi T, Tasumi M (1977): Protein Data Bank: a computer-based archival file for macromolecular structures. *J Mol Biol* 112:535–42.

Bhat TN, Bourne P, Feng Z, Gilliland G, Jain S, Ravichandran V, Schneider B, Schneider K, Thanki N, Weissig H, Westbrook J, Berman HM (2001): The PDB data uniformity project. *Nucleic Acids Res* 29:214–8.

Bourne PE, Berman HM, Watenpaugh K, Westbrook JD, Fitzgerald PMD (1997): The macromolecular Crystallographic Information File (mmCIF). *Methods Enzymol* 277:571–90.

Clowney L, Jain SC, Srinivasan AR, Westbrook J, Olson WK, Berman HM (1996): Geometric parameters in nucleic acids: nitrogenous bases. *J Am Chem Soc* 118:509–18.

Engh RA, Huber R (1991): Accurate bond and angle parameters for X-ray protein structure refinement. *Acta Crystallogr* A47:392–400.

Feng Z, Hsieh S-H, Gelbin A, Westbrook J (1998a): *MAXIT: Macromolecular Exchange and Input Tool.* New Brunswick, NJ: Rutgers University, NDB-120.

Feng Z, Westbrook J, Berman HM (1998b): *NUCheck*. New Brunswick, NJ: Rutgers University, NDB-407.

Gelbin A, Schneider B, Clowney L, Hsieh S-H, Olson WK, Berman HM (1996): Geometric parameters in nucleic acids: sugar and phosphate constituents. *J Am Chem Soc* 118:519–28.

Gibrat J-F, Madej T, Bryant SH (1996): Surprising similarities in structure comparison. *Curr Opin Struct Biol* 6:377–85.

Gilliland GL (1988): A Biological Macromolecule Crystallization Database: a basis for a crystallization strategy. *J Cryst Growth* 90:51–9.

Holm L, Sander C (1998): Touring protein fold space with Dali/FSSP. *Nucleic Acids Res* 26:316–9.

[IUPAC-IUB] IUPAC-IUB Joint Commission on Biochemical Nomenclature (1983): Abbreviations and symbols for the description of conformations of polynucleotide chains. *Eur J Biochem* 131:9–15.

Kabsch W, Sander C (1983): Dictionary of protein secondary structure: pattern recognition of hydrogen-bonded and geometrical features. *Biopolymers* 22:2577–637.

Laskowski RA, McArthur MW, Moss DS, Thornton JM (1993): PROCHECK: a program to check the stereochemical quality of protein structures. *J Applied Crystallogr* 26:283–91.

Murzin AG, Brenner SE, Hubbard T, Chothia C (1995): SCOP: a structural classification of proteins database for the investigation of sequences and structures. *J Mol Biol* 247:536–40.

Orengo CA, Michie AD, Jones S, Jones DT, Swindells MB, Thornton JM (1997): CATH—a hierarchic classification of protein domain structures. *Structure* 5:1093–108.

Pearson WR, Lipman DJ (1988): Improved tools for biological sequence comparison. *Proc Natl Acad Sci USA* 24:2444–8.

Shindyalov IN, Bourne PE (1997): Protein data representation and query using optimized data decomposition. *CABIOS* 13:487–96.

Shindyalov IN, Bourne PE (1998): Protein structure alignment by incremental combinatory extension of the optimum path. *Protein Eng* 11:739–47.

Ulrich EL, Markley JL, Kyogoku Y (1989): Creation of a Nuclear Magnetic Resonance Data Repository and Literature Database. *Protein Seq Data Anal* 2:23–37.

Vaguine AA, Richelle J, Wodak SJ (1999): SFCHECK: a unified set of procedures for evaluating the quality of macromolecular structure-factor data and their agreement with the atomic model. *Acta Crystallogr* D55:191–205.

Westbrook J, Feng Z, Berman HM (1998): *ADIT—The AutoDep Input Tool*. Department of Chemistry, Rutgers, the State University of New Jersey, RCSB-99.

Westbrook J, Bourne PE (2000): STAR/mmCIF: An extensive ontology for macromolecular structure and beyond. *Bioinformatics* 16:159–68.

Westbrook J, Feng Z, Jain S, Bhat TN, Thanki N, Ravichandran V, Gilliland GL, Bluhm W, Weissig H, Greer DS, Bourne PE, Berman HM (2002): The Protein Data Bank: Unifying the archive. *Nucleic Acids Res* 30:245–48.

Zhang J, Cousens LS, Barr PJ, Sprang SR (1991): Three-dimensional structure of human basic fibroblast growth factor, a structural homolog of interleukin 1$\beta$. *Proc Natl Acad Sci USA* 88:3446–50.

# 10

# THE NUCLEIC ACID DATABASE

Helen M. Berman, John Westbrook, Zukang Feng, Lisa Iype,
Bohdan Schneider, and Christine Zardecki

The Nucleic Acid Database (NDB) (Berman et al., 1992) was established in 1991 as a resource for specialists in the field of nucleic acid structure. Over the years, the NDB has developed generalized software for processing, archiving, querying, and distributing structural data for nucleic acid-containing structures. The core of the NDB has been its relational database of nucleic acid-containing crystal structures. Recognizing the importance of a standard data representation in building a database, the NDB became an active participant in the macromolecular Crystallographic Information File (mmCIF) project and was the test-bed for this format. With a foundation of well-curated data, the NDB created a searchable relational database of primary and derivative data with very rich query and reporting capabilities. This robust database was unique in that it allowed researchers to do comparative analyses of nucleic acid-containing structures selected from the NDB according to the many attributes stored in the database.

In 1992, the NDB assumed responsibility for processing all nucleic acid crystal structures that were deposited into the Protein Data Bank (PDB); it became a direct deposit site for those structures in 1996. In order to meet data-processing requirements, the NDB created the first validation software for nucleic acids (Feng et al., 1998b). Until 1998, protein–nucleic acid crystal structures deposited into the PDB were post-processed and then incorporated into the NDB. When the Research Collaboratory for Structural Bioinformatics assumed the management of the PDB in 1998, the tools developed by the NDB were used to process all macromolecular structures (Berman et al., 2000; see also Chapter 9). The NDB continues to provide a high level of information about nucleic acids and serves as a specialty database for its community of researchers.

In this chapter, we describe the architecture and capabilities of the NDB and then present some of the research that has been enabled by this resource.

*Structural Bioinformatics*
Edited by Philip E. Bourne and Helge Weissig
Copyright © 2003 by Wiley-Liss, Inc.

## INFORMATION CONTENT OF THE NDB

Structures available in the NDB include RNA and DNA oligonucleotides with two or more bases either alone or complexed with ligands, natural nucleic acids such as tRNA, and protein–nucleic acid complexes. The archive stores both primary and derived information about the structures (Table 10.1). The primary data include: crystallographic coordinate data, structure factors, and information about the experiments used to determine the structures, such as crystallization information, data collection, and refinement statistics.

Derived information, such as valence geometry, torsion angles, and intermolecular contacts, are calculated and stored in the database. Database entries are further annotated to include information about the overall structural features, including conformational classes, special structural features, biological functions, and crystal-packing classifications.

### T A B L E  10.1. The Information Content of the NDB

Primary Experimental Information Stored in the NDB

Structure Summary: descriptor; NDB, PDB, and Cambridge Structural Database (CSD) names; coordinate availability; modifications, mismatches, and drug binding

Structural Description: sequence; structure type; descriptions about modifications, mismatches, and drugs; description of asymmetric and biological units

Citation: authors, title, journal, volume, pages, year

Crystal Data: cell dimensions; space group

Data Collection Description: radiation source and wavelength; data collection device; temperature; resolution range; total and unique number of reflections

Crystallization Description: method; temperature; pH value; solution composition

Refinement Information: method; program; number of reflections used for refinement; data cutoff; resolution range; R-factor; refinement of temperature factors and occupancies

Coordinate Information: atomic coordinates, occupancies, and temperature factors for asymmetric unit; coordinates for symmetry-related strands; coordinates for unit cell; symmetry-related coordinates; orthogonal or fractional coordinates

Derivative Information Stored in the NDB

Distances: chemical bond lengths; virtual bonds (involving phosphorus atoms)

Torsions: backbone and side chain torsion angles; pseudorotational parameters

Angles: valence bond angles, virtual angles (involving phosphorus atoms)

Base Morphology: parameters calculated by different algorithms

Nonbonded contacts

Valence geometry RMS deviations from small molecule standards

Sequence pattern statistics

*Sources*: Reprinted, by permission of the publisher, from H. M. Berman, Z. Feng, B. Schneider, J. Westbrook, and C. Zardecki (2001): The Nucleic Acid Database. In: M. G. Rossman and E. Arnold, editors. International Tables for Crystallography, F. Crystallography of Biological Macromolecules. Dordrecht: Kluwer Academic Publishers, p. 657.

Some features are derived by different algorithms, and it can be difficult to provide the most reliable values. Whenever possible, the NDB has tried to promote standards that allow structure comparison. An outstanding example of this lack of standards were the problems associated with different values for base morphology parameters produced by different programs (Lavery and Sklenar, 1988; Soumpasis and Tung, 1988; Bhattacharyya and Bansal, 1989; Lavery and Sklenar, 1989; Babcock, Pednault, and Olson, 1993; Babcock and Olson, 1994; Babcock, Pednault, and Olson, 1994; Tung, Soumpasis, and Hummer, 1994; Bansal, Bhattacharyya, and Ravi, 1995; El Hassan and Calladine, 1995; Gorin, Zhurkin, and Olson, 1995; Lu, El Hassan, and Hunter, 1997; Dickerson, 1998; Kosikov et al., 1999). These different values in base morphology meant that it was not possible to compare any two structures by using the numbers in the published literature and that it was necessary to recalculate these values for any analysis.

To help resolve this problem, the NDB cosponsored the Tsukuba Workshop on Nucleic Acid Structure and Interactions (January 12–14, 1999, AIST-NIBHT Structural Biology Centre, Tsukuba, Japan) to which all the key software developers in this field were invited. It was resolved that a single reference frame would be used to calculate these values and an agreement was reached about the definition of that reference frame (Olson et al., 2001). All the programs are being amended so that they will produce very similar values for the parameters. The NDB has recalculated these values for all the structures in the repository and will make them available as output from NDB Searches done over the Web (see The Database and Query Capabilities for more information).

## DATA VALIDATION AND PROCESSING

The NDB has created a robust data-processing system that produces high-quality data that is readily loaded into a database. The full capability of this system was recently demonstrated by the successful processing of ribosomal subunits, which are very large and complex structures.

Early on, the NDB adopted mmCIF (Bourne et al., 1997) as its data standard. This format has three advantages from the point of view of building a database: (1) the definitions for the data items are based on a comprehensive dictionary of crystallographic terminology and molecular structure description; (2) it is self-defining; and (3) the syntax contains explicit rules that further define the characteristics of the data items, particularly the relationships between data items (Westbrook and Bourne, 2000). The latter feature is important because it allows for rigorous checking of the data.

Structures are deposited via the Web with the AutoDep Input Tool (ADIT) (Westbrook, Feng, Berman, 1998) and then annotated using the same tool. ADIT operates on top of the mmCIF dictionary. In the next stage of data processing, a program called MAXIT (Macromolecular Exchange and Input Tool; Feng et al., 1998a) checks and corrects atom numbering and ordering as well as the correspondence between the SEQRES PDB record and the residue names in the coordinate files. Once these integrity checks are completed, the structures are validated.

NUCheck (Feng et al., 1998b) verifies valence geometry, torsion angles, intermolecular contacts, and the chiral centers of the sugars and phosphates. The dictionaries used for checking the structures were developed by the NDB Project from analyses (Clowney et al., 1996; Gelbin et al., 1996) of high-resolution, small-molecule structures from the CSD (Allen et al., 1979). The torsion angle ranges were derived from an analysis of well-resolved nucleic acid structures (Schneider, Neidle, and Berman, 1997).

One important outgrowth of these validation projects was the creation of the force constants and restraints that are now in common use for crystallographic refinement of nucleic acid structures (Parkinson et al., 1996a). The program SFCheck (Vaguine, Richelle, and Wodak, 1999) is used to validate the model against the structure factor data. The R factor and resolution are verified and the residue-based features are examined with this program. Once an entry has been processed satisfactorily, it is released based on its author-defined hold status.

## THE DATABASE AND QUERY CAPABILITIES

The core of the NDB project is a relational database in which all of the primary and derived data items are organized into tables. At present, there are over 90 tables in the NDB, with each table containing 5 to 20 data items. These tables contain both experimental and derived information. Example tables include: the citation table, which contains all the items that are contained in literature references; the cell_dimension table, which contains all items related to crystal data; and the refine_parameters table, which contains the items that describe the refinement statistics.

Interaction with the database is a two-step process (Fig. 10.1). In the first step, the user defines the selection criteria by combining different database items. As an example, the user could select all B-DNA structures whose resolution is better or equal to 2.0 Å, whose R-factor is better than 0.17, and that were determined by the authors Dickerson, Kennard, or Rich. Once the structures that meet the constraint criteria have

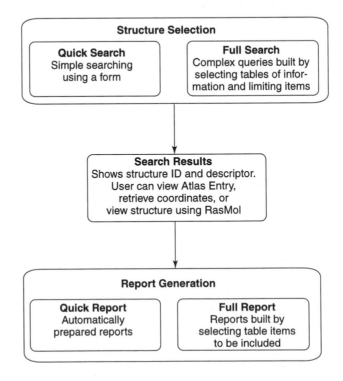

**Figure 10.1.** Flow chart demonstrating the two steps involved in searching the NDB: structure selection and report generation.

been selected, reports may be written using a combination of table items. For any set of chosen structures, a large variety of reports may be created. For the example set of structures given above, a crystal data report or a backbone torsion angle report can be easily generated, or the user could write a report that lists the twist values for all CG steps together with statistics, including mean, median, and range of values. The constraints used for the reports do not have to be the same as those used to select the structures. Some examples of the types of reports produced by the NDB are given in Fig. 10.2.

A Web interface was designed to make the query capabilities of the NDB as widely accessible as possible. In the Quick Search/Quick Report mode, several items, including structure ID, author, classification, and special features, can be limited either by entering text in a box or by selecting an option from the pull-down menu. Any

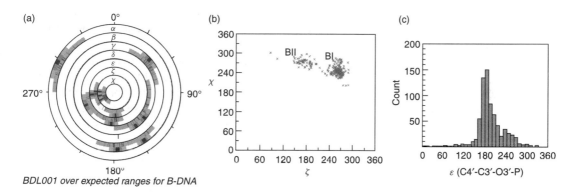

(a)

$BDL001$ over expected ranges for B-DNA

(b)

(c)

(d)    Torsion Angles alpha-chl

| NDB ID | Residue | Strand | o3_p_o5_c5 | p_o5_c5_c4 | o5_c5_c4_c3 | c5_o4_o3_o3 | c4_c3_o3_p | c3_o3_p_o5 | o4_c1_n1_9_c2_4 |
|--------|---------|--------|------------|------------|-------------|-------------|------------|------------|-----------------|
| BDL001 | C–1 | A | n.a. | n.a. | 174.23 | 156.78 | −141.35 | −143.95 | −104.97 |
| BDL001 | G–2 | A | −65.61 | 169.84 | 40.14 | 128.13 | 174.19 | −97.78 | −110.52 |
| BDL001 | C–3 | A | −62.60 | 171.80 | 58.83 | 98.34 | −176.69 | −87.62 | −135.14 |
| BDL001 | G–4 | A | −62.85 | −179.91 | 57.23 | 155.66 | −155.34 | −152.52 | −93.40 |
| BDL001 | A–5 | A | −43.02 | 142.81 | 52.42 | 119.60 | 179.94 | −92.20 | −126.25 |
| BDL001 | A–6 | A | −73.26 | 179.72 | 66.03 | 121.10 | 173.66 | −88.48 | −122.15 |
| BDL001 | T–7 | A | −56.62 | −179.17 | 52.22 | 98.91 | 173.58 | −85.86 | −127.34 |
| BDL001 | T–8 | A | −59.16 | 173.36 | 64.08 | 108.90 | 170.60 | −89.35 | −125.71 |
| BDL001 | C–9 | A | −58.54 | −179.50 | 60.50 | 128.69 | −156.91 | −94.01 | −119.51 |
| BDL001 | G–10 | A | −67.32 | 169.11 | 47.18 | 142.87 | −103.26 | 150.16 | −89.59 |
| BDL001 | C–11 | A | −73.90 | 139.26 | 56.30 | 135.66 | −161.81 | −89.57 | −125.10 |
| BDL001 | G–12 | A | −81.49 | 175.65 | 57.16 | 110.66 | n.a. | n.a. | −112.02 |
| BDL001 | C–13 | B | n.a. | n.a. | 55.92 | 136.69 | −158.65 | −124.92 | −127.60 |
| BDL001 | G–14 | B | −51.30 | 163.88 | 49.02 | 121.87 | 177.70 | −93.03 | −116.42 |
| BDL001 | C–15 | B | −62.96 | 168.85 | 60.43 | 85.68 | 174.83 | −85.52 | −133.77 |
| BDL001 | G–16 | B | −69.20 | 171.06 | 73.24 | 135.92 | 174.14 | −98.39 | −114.77 |
| BDL001 | A–17 | B | −56.62 | −169.54 | 53.84 | 146.60 | 176.95 | −97.12 | −106.44 |
| BDL001 | A–18 | B | −57.08 | −173.62 | 47.72 | 130.22 | 174.42 | −101.30 | −108.30 |
| BDL001 | T–19 | B | −58.29 | 173.59 | 60.04 | 109.21 | 178.77 | −88.28 | −131.27 |
| BDL001 | T–20 | B | −58.62 | 179.49 | 55.33 | 122.39 | 178.54 | −94.50 | −120.45 |
| BDL001 | C–21 | B | −59.10 | −175.42 | 44.95 | 110.35 | −176.68 | −86.46 | −114.28 |
| BDL001 | G–22 | B | −66.77 | 179.14 | 50.19 | 149.72 | −100.15 | 171.58 | −88.39 |
| BDL001 | C–23 | B | −72.24 | 138.53 | 44.64 | 112.75 | −174.38 | −96.75 | −125.35 |
| BDL001 | G–24 | B | −64.99 | 170.58 | 46.61 | 78.71 | n.a | n.a. | −135.15 |

Created by the Nucleic Acid Database Project on Mon Aug 20 16:02:17 2002

**Figure 10.2.** Examples of torsion angle reports generated from the NDB: (a) conformation wheel showing the torsion angles for structure BDL001 (Drew et al., 1981) over the average values for all B-DNA; (b) scattergram graph showing the relationship of $\chi$ vs. $\zeta$ for all B-DNA. Two clusters, BI and BII, are labeled; (c) histogram for $\varepsilon$ (C4′-C3′-O3′-P) for all B-DNA; (d) a torsion angle report for BDL001. Figure also appears in Color Figure section.

combination of these items may be used to constrain the structure selection. If none are used, the entire database will be selected. After selecting "Execute Selection," the user will be presented with a list of structure IDs and descriptors that match the desired conditions. Several viewing options for each structure in this list are possible. These include retrieving the coordinate files in either mmCIF or PDB format, retrieving the coordinates for the biological unit, viewing the structure with RasMol (Sayle and Milner-White, 1995), or viewing an NDB Atlas page.

Preformatted Quick Reports can then be generated for the structures in this result list. The user selects a report from a list of 13 report options (Table 10.2), and the report is created automatically. Multiple reports can be easily generated. These reports are particularly convenient for quickly producing reports based on derived features, such as torsion angles and base morphology (Fig. 10.3).

In the Full Search/Full Report mode, it is possible to access most of the tables in the NDB to build more complex queries. Instead of limiting items that are listed on a single page, the user builds a search by selecting the tables and then the items that contain the desired features. These queries can use Boolean and logical operators to make complex queries.

After selecting structures using the Full Search, a variety of reports can be written. The report columns are selected from a variety of database tables, and then the full report is automatically generated. Multiple reports can be generated for the same group of selected structures; for example, reports on crystallization, base modification, or a combination of these reports can be generated for a particular group of structures.

## TABLE 10.2. Quick Reports Available for the NDB

| Report Name | Contains |
| --- | --- |
| NDB Status | Processing status information |
| Cell Dimensions | Crystallographic cell constants |
| Primary Citation | Primary bibliographic citations |
| Structure Identifier | Identifiers, descriptor, coordinate availability |
| Sequence | Sequence |
| Nucleic Acid Sequence | Nucleic acid sequence only |
| Protein Sequence | Protein sequence only |
| Refinement Information | R-factor, resolution, and number of reflections used in refinement |
| NA Backbone Torsions (NDB) | Sugar-phosphate backbone torsion angles using NDB residue numbers |
| NA Backbone Torsions (PDB) | Sugar-phosphate backbone torsion angles using PDB residue numbers |
| Base Pair Parameters (global) | Global base pair parameters calculated using Curves 5.1 (Lavery, et al., 1989) |
| Base Pair Step Parameters (local) | Local base pair step parameters calculated using Curves 5.1 |
| Groove Dimensions | Groove dimensions using Stoffer & Lavery definitions from Curves 5.1 |

*Sources*: Reprinted, by permission of the publisher, from H. M. Berman, Z. Feng, B. Schneider, J. Westbrook, and C. Zardecki (2001): The Nucleic Acid Database. In: M. G. Rossman and E. Arnold, editors. International Tables for Crystallography, F. Crystallography of Biological Macromolecules. Dordrecht: Kluwer Academic Publishers, p. 659.

**Figure 10.3.** Examples of Quick Reports: (a) citation report for protein-RNA structures; (b) nucleic acid sequence report for protein-RNA structures; (c) refinement information for protein-RNA structures; (d) nucleic acid backbone torsions report for PR0001 (Rowsell et al., 1998).

## DATA DISTRIBUTION

Coordinate files, database reports, software programs, and other resources are available via the ftp server (ftp://ndbserver.rutgers.edu). In addition to links to information provided from the ftp server, the Web server (http://ndbserver.rutgers.edu/) provides a variety of methods for querying the NDB (described above). These sites are updated continually.

The NDB Archives, a section of the Web site, contain a large variety of information and tables useful for researchers. Prepared reports about the structure identifiers, citations, cell dimensions, and structure summaries are available and are sorted according to structure type. The dictionaries of standard geometries of nucleic acids as well as parameter files for X-PLOR (Brünger, 1992) are available. The NDB Archives section links to the ftp server, providing coordinates for the asymmetric unit and biological units in PDB and mmCIF formats, structure factor files, and coordinates for nucleic acid structures determined by NMR.

A very popular and useful report is the NDB Atlas report page (Fig. 10.4). An atlas page contains summary, crystallographic, and experimental information, a molecular view of the biological unit, and a crystal-packing picture for a particular structure. Atlas pages are created directly from the NDB database. The entries for all structures in the database are organized by structure type in the NDB Atlas.

## Mirrors

The NDB is based at Rutgers University (http://ndbserver.rutgers.edu/) and is currently mirrored at three other sites: the Institute of Cancer Research UK (http://www.ndb.icr.ac.uk/), the San Diego Supercomputer Center in San Diego, California (http://ndb.sdsc.edu/NDB/), and the Structural Biology Centre (http://ndbserver.nibh.go.jp/NDB/) in Tsukuba, Japan. These mirror sites are updated daily, are fully synchronous, and contain the ftp directories, the Web site, and the full database.

## Community Outreach

The NDB works closely with the research community to ensure that their needs are met. A newsletter is published electronically and provides information about the newest features of the system. To subscribe, send an email to ndbnews@ndbserver.rutgers.edu. Very complex queries will be done by the staff in response to user requests via e-mail to ndbadmin@ndbserver.rutgers.edu.

## APPLICATIONS OF THE NDB

The NDB has been used to analyze characteristics of nucleic acids alone and complexed with proteins. The ability to select structures according to many criteria has made it possible to create appropriate data sets for study. A few examples are given here.

The conformational characteristics of A-, B- and Z-DNA were examined (Schneider, Neidle, and Berman, 1997) by using carefully selected examples of well-resolved structures in these classes. Conformation wheels (Fig. 10.2a) for each conformation as well as scattergrams of selected torsion angles (Fig. 10.2b) were created. These diagrams can now be used to assess and classify new structures. Studies of B-DNA helices have shown that the base steps have characteristic values that depend on their sequence (Gorin, Zhurkin, and Olson, 1995). Plots of twist versus roll are different for purine–purine, purine–pyrimidine, and pyrimidine–purine steps. This particular analysis has been extended to derive energy parameters for B-DNA sequences (Olson et al., 1998).

In a series of systematic studies of the hydration patterns of DNA double helices, it was found that the hydration patterns around the bases are well defined and are local (Fig. 10.5a–c) (Schneider and Berman, 1995). That is, small changes in the

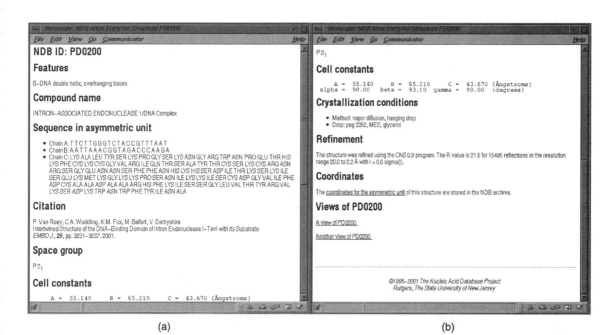

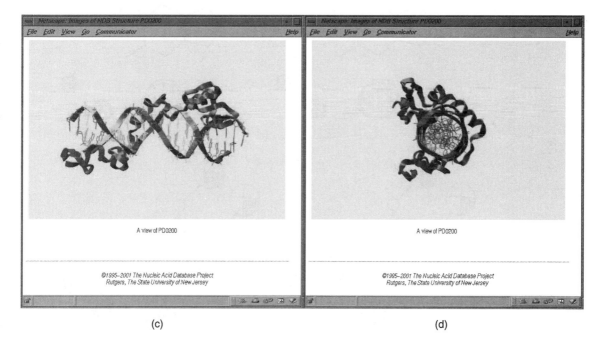

(c)                                                    (d)

Figure 10.4. NDB Atlas page for PD0200 (van Roey et al., 2001) that highlights structural information that is contained in the database. The NDB Atlas also includes images for biological and asymmetric units, and crystal-packing pictures for nucleic acids.

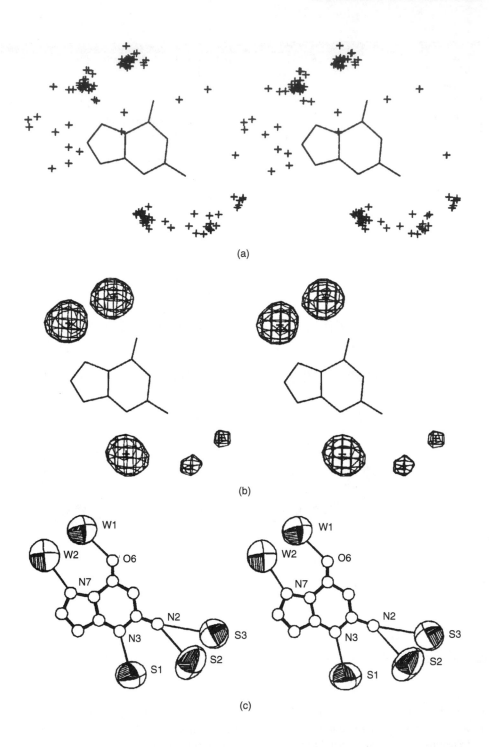

(a)

(b)

(c)

conformation of the backbone do not affect the hydration around the bases. It was also found that there are more diffuse patterns around the phosphate backbone that are dependent on the conformational class of the DNA. These analyses were used to attempt to predict the binding sites of protein side chains on the DNA. In a series of protein–nucleic acid complexes, the hydration sites of the DNA were calculated

**Figure 10.5.** Water environment of guanine residues in structures in the NDB. (a) Scattergram of 101 water molecules within 3.4 Å from any atom of 42 guanines found in 14 B-DNA decamer structures; (b) electron densities of the 101 water molecules plotted at the 4 $\sigma$ level. Each water is modeled as an oxygen atom with an occupancy of 1/42; (c) An ORTEP (Johnson, 1976) plot of the current guanine B-DNA hydration sites after refinement. Plotted are 50% probability thermal ellipsoids. The key guanine atoms and hydration sites are labeled. All plots are in stereo. Figure 10.5a–c are reprinted from (Schneider et al., 1995) with permission from the Biophysical society.

and then compared with the location of the amino acid side chain. The results were surprisingly good in that in most cases the side-chain site and hydration sites were very close. This result was true even in the case of a very bent DNA that is bound to Catabolite Activator Protein (Fig. 10.6) (Woda et al., 1998).

Systematic studies of the interface in protein–nucleic acid complexes have been done. In one analysis of protein-DNA complexes, 26 complexes were selected in which the proteins were nonhomologous (Jones et al., 1999). The results showed that there

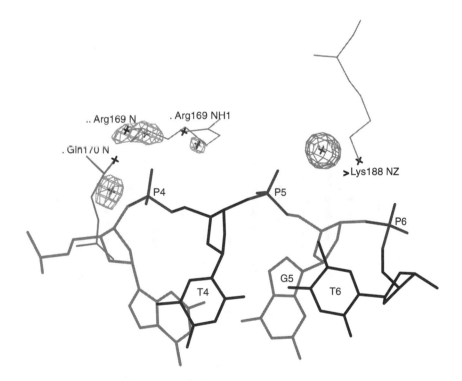

**Figure 10.6.** A view of the three residues in the consensus region for the high resolution CAP-DNA$_{GCE}$ complex (Parkinson et al., 1996b). The predicted phosphate hydration is drawn as pseudoelectron density in cyan, the interacting protein residues are shown in dark brown, and the phosphate groups are red. The protein atoms that contact the DNA shown as blue crosses. The predicted sites are the red crosses. Reprinted from (Woda et al., 1998) with permission from the Biophysical Society. Figure also appears in Color Figure section.

are amino acid propensities at the interface that are markedly different than in protein–protein complexes. It was also possible to place the complexes into three classes: double-headed, single-headed, and enveloping (Fig. 10.7–8). A similar analysis has also been done for protein-RNA complexes (Jones et al., 2001). There have also been detailed analyses of the hydrogen-bonding patterns at the protein DNA interface and it was found that CH•••O bonds are surprisingly common (Mandel-Gutfreund, et al., 1998).

Some analyses have been done on the relationships of crystal packing and conformation. Although there are more than 30 different crystal forms of B-DNA in the NDB, the actual number of packing motifs (Fig. 10.9) remains relatively small, with the most common motifs being minor groove–minor groove, stacking–lateral backbone, and major groove–backbone (Timsit and Moras, 1992).

Minor groove–minor groove interactions in which the guanines of one duplex form hydrogen bonds with the guanines of a neighboring duplex are seen not only in dodecamer structures but in an octamer sequence with three duplexes in the asymmetric unit (Urpi et al., 1996). The second motif contains duplexes stacked above one another with the adjoining phosphates forming lateral interactions. A large number of variations of this motif have been observed in decamer (Grzeskowiak et al., 1991) and hexamer structures (Cruse et al., 1986; Tari and Secco, 1995). The third type of packing involves the major groove of one helix interacting with the phosphate backbone of

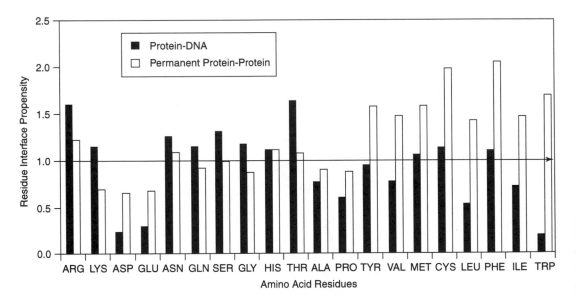

**Figure 10.7.** Histogram of the interface residue propensities calculated for 26 protein–DNA complexes and compared to those for permanent protein–protein complex (Jones & Thornton, 1996). "Permanent" complexes are those in which the components only exist as part of a complex; they do not exist in isolation. Generally, they have larger interfaces that are more hydrophobic and more complementary. A propensity of >1 indicates that a residue occurs more frequently in the interface than on the protein surface. The amino acid residues have been ordered using the Faucher & Pliska (Faucher & Pliska, 1983) hydrophobicity scale, with the most hydrophilic residues on the left-hand side and the most hydrophobic on the right-hand side of the graph. Reprinted with permission from (Jones et al., 1999).

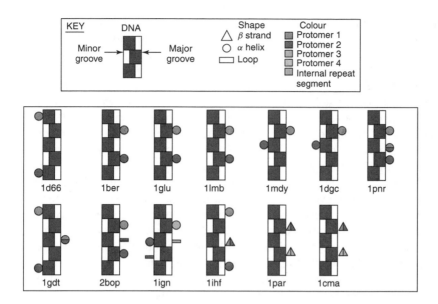

**Figure 10.8.** Simple model diagrams of protein-DNA complexes for double-headed binding proteins. The diagrams give an indication of the predominant secondary structure of the binding motif, protein symmetry and the type and relative position of the DNA groove bound. The secondary structure of the predominant binding motifs are indicated using different symbols analogous to those used in TOPS diagrams (Westhead & Thornton, 1998). Only one symbol of each type is indicated in any one groove, hence both a single sheet and two sheets are indicated by a single colored triangle. The symmetry of each protein is indicated by using a different color for each symmetry (or pseudo symmetry) related element. A single symbol shaded in two colors indicates that there are secondary structures of this type contributed by more than one symmetry-related element. Reprinted with permission from (Jones et al., 1999). Figure also appears in Color Figure section.

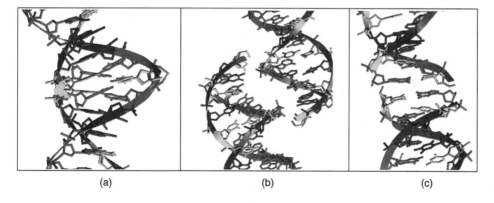

**Figure 10.9.** Examples of packing motifs in DNA duplexes in a B- and A-DNA. From left to right: (a) minor groove-minor groove interactions in BDL042 (Leonard & Hunter, 1993); (b) major groove-backbone interactions in BDJ060 (Goodsell et al., 1995); (c) stacking interactions in BDJ025 (Grzeskowiak et al., 1991); The bases are colored green for guanine, yellow for cytosine, red for adenine, and blue for thymine. Reprinted from (Berman et al., 1996) with permission from Elsevier Science. Figure also appears in Color Figure section.

another (Timsit et al., 1989). Sequence appears to be a large factor in determining these motifs, but it is not the only one. For example, the first structures exhibiting the major groove–phosphate interactions contained a cytosine that formed a hydrogen bond to the phosphate. However, not all structures that show this motif have this hydrogen bond (Wood et al., 1997). The particular sequence in this crystal is even more intriguing because it also crystallizes in another crystal form in which the terminal flips out to form a minor groove interaction with another duplex (Spink et al., 1995).

The task of trying to determine the relative effects of base sequence and crystal packing on the values of the base morphology parameters is hampered to some degree by the uneven distribution of the 16 different base steps among the different crystal types. Some steps such as CG are very well represented in B structures, whereas others such as AC have very few representatives in the data set. Nonetheless, there are a few steps that occur in crystals with different packing motifs. An analysis of the CG steps across all crystal types shows that its conformation is relatively insensitive to crystal packing and the distribution is similar to that found for all steps (see (Berman, Gelbin, and Westbrook, 1996)). However, the variability of the CA step appears to depend not simply on crystal type but on the packing motif. The values of twist for CA steps in minor groove–minor groove motifs are smaller than those in the major groove–backbone motif. Very high values are displayed for CA steps in the stacking–lateral backbone motif. Plots of twist versus roll for CG steps show the distribution noted by others (Gorin, Zhurkin, and Olson, 1995) and no clustering that depends on crystal type. However, the same plot for CA steps shows very distinctive differences that appear to depend on the packing motifs. It is important to note here that these motifs encompass several crystal types so that the structural variability observed is a function of a particular type of structural interaction rather than a particular crystal form. Before any definitive statements can be made about all the steps it will be necessary to have much more data.

With the recent increase in the number of RNA structures available there have been attempts to establish systems whereby it will be possible to systematically analyze these structures. The result of one of these studies has been the proposal of a classification scheme for the hydrogen bonds in the base pairs (Westhof and Fritsch, 2000). A new syntax (RNAML) has also been proposed for representing RNA structural features (http://www.smi.stanford.edu/projects/helix/rnaml/).

## THE CHANGING FACE OF THE NDB

When the NDB began, the world of nucleic acid structures consisted of DNA and RNA oligonucleotides, a few protein-DNA complexes, and some tRNA structures. Annotation of structural features was performed manually by visual inspection of molecular architectures. However, since the early 1990s a whole new universe of nucleic acid structures has emerged (Fig. 10.10; see also Chapter 3). There are many ribozyme structures and many different types of protein–nucleic acid complexes represented in over 500 structures. The newest additions to the archive—ribosomal structures—have increased the number of residues of RNA resident in the NDB several fold (Moore, 2001).

One outcome of the systematic studies that have been done with data from the NDB has been improved classification schemes for understanding nucleic acids. These classification schemes will be used to automatically annotate structures contained within

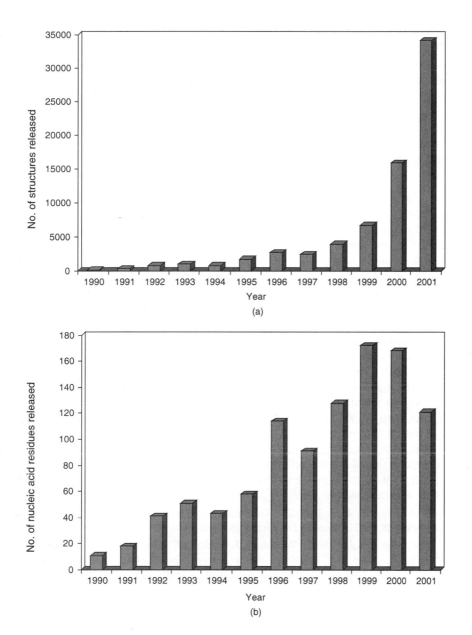

Figure 10.10. (a) the number of nucleic acid residues, and (b) the number of structures released in the NDB as of September 27, 2001.

the NDB, which will in turn improve the query capability of the NDB. This type of cycle shows the power of organizing information so that it is more accessible and can ultimately yield new knowledge.

## ACKNOWLEDGMENTS

The NDB Project is funded by the National Science Foundation and the Department of Energy. We would like to thank Wilma K. Olson and David Beveridge, co-founders of the NDB, for their collaborations on this project. This chapter also appears in *Acta Crystallographica*, 2002. D58:889–898. Special Issue of Sections B and D, entitled "Crystallographic Databases and their Applications" (Frank H. Allen and Jenny P. Glusker, Editors).

## REFERENCES

Allen FH, Bellard S, Brice MD, Cartright BA, Doubleday A, Higgs H, Hummelink T, Hummelink-Peters BG, Kennard O, Motherwell WDS, Rodgers JR, Watson DG (1979): The Cambridge Crystallographic Data Centre: computer-based search, retrieval, analysis and display of information. *Acta Crystallogr* B35:2331–9.

Babcock MS, Pednault EPD, Olson WK (1993): Nucleic acid structure analysis: a users guide to a collection of new analysis programs. *J Biomol Struct Dyn* 11:597–628.

Babcock MS, Olson WK (1994): The effect of mathematics and coordinate system on comparability and "dependencies" of nucleic acid structure parameters. *J Mol Biol* 237:98–124.

Babcock MS, Pednault EPD, Olson WK (1994): Nucleic acid structure analysis. Mathematics for local cartesian and helical structure parameters that are truly comparable between structures. *J Mol Biol* 237:125–56.

Bansal M, Bhattacharyya D, Ravi B (1995): NUPARM and NUCGEN: software for analysis and generation of sequence dependent nucleic acid structures. *CABIOS* 11:281–7.

Berman HM, Olson WK, Beveridge DL, Westbrook J, Gelbin A, Demeny T, Hsieh SH, Srinivasan AR, Schneider B (1992): The Nucleic Acid Database—a comprehensive relational database of three-dimensional structures of nucleic acids. *Biophys J* 63:751–9. [This paper gives the full description of the NDB system.]

Berman HM, Gelbin A, Westbrook J (1996): Nucleic acid crystallography: a view from the Nucleic Acid Database. *Prog Biophys Mol Biol* 66:255–88.

Berman HM, Westbrook J, Feng Z, Gilliland G, Bhat TN, Weissig H, Shindyalov IN, Bourne PE (2000): The Protein Data Bank. *Nucleic Acids Res* 28:235–42.

Berman HM, Feng Z, Schneider B, Westbrook J, Zardecki C (2001): The Nucleic Acid Database (NDB). In: Rossman MG, Arnold E, editors. *International Tables for Crystallography, F. Crystallography of Biological Macromolecules*. Dordrecht: Kluwer Academic Publishers. pp. 657–662.

Bhattacharyya D, Bansal M (1989): A self-consistent formulation for analysis and generation of non-uniform DNA structures. *J Biomol Struct Dyn* 6:635–53.

Bourne PE, Berman HM, Watenpaugh K, Westbrook JD, Fitzgerald PMD (1997): The macromolecular Crystallographic Information File (mmCIF). *Methods Enzymol* 277:571–90.

Brünger AT (1992): *X-PLOR, Version 3.1, A System for X-ray Crystallography and NMR*. New Haven, CT: Yale University Press.

Clowney L, Jain SC, Srinivasan AR, Westbrook J, Olson WK, Berman HM (1996): Geometric parameters in nucleic acids: nitrogenous bases. *J Am Chem Soc* 118:509–18.

Cruse WBT, Salisbury SA, Brown T, Cosstick R, Eckstein F, Kennard O (1986): Chiral phosphorothioate analogues of B-DNA. The crystal structure of Rp-d|Gp(S)CpGp(S)CpGp(S)C|. *J Mol Biol* 192:891–905.

Dickerson RE (1998): DNA bending: the prevalence of kinkiness and the virtues of normality. *Nucleic Acids Res* 26:1906–26.

Drew HR, Wing RM, Takano T, Broka C, Tanaka S, Itakura K, Dickerson RE (1981): Structure of a B-DNA dodecamer: conformation and dynamics. *Proc Natl Acad Sci USA* 78:2179–83.

El Hassan MA, Calladine CR (1995): The assessment of the geometry of dinucleotide steps in double-helical DNA: a new local calculation scheme with an appendix. *J Mol Biol* 251:648–64.

Faucher J, Pliska V (1983): Hydrophobic parameters pi of amino acid side-chains from the partitioning of *N*-acetyl-amino-acid amides. *Eur J Med Chem* 18:369–75.

Feng Z, Hsieh S-H, Gelbin A, Westbrook J (1998a): *MAXIT: Macromolecular Exchange and Input Tool*. New Brunswick, NJ: Rutgers University, NDB–120.

Feng Z, Westbrook J, Berman HM (1998b): *NUCheck*. New Brunswick, NJ: Rutgers University, NDB–407.

Gelbin A, Schneider B, Clowney L, Hsieh S-H, Olson WK, Berman HM (1996): Geometric parameters in nucleic acids: sugar and phosphate constituents. *J Am Chem Soc* 118:519–28.

Goodsell DS, Grzeskowiak K, Dickerson RE (1995): Crystal structure of C-T-C-T-C-G-A-G-A-G. Implications for the structure of the Holliday junction. *Biochemistry* 34:1022–9.

Gorin AA, Zhurkin VB, Olson WK (1995): B-DNA twisting correlates with base-pair morphology. *J Mol Biol* 247:34–48.

Grzeskowiak K, Yanagi K, Privé GG, Dickerson RE (1991): The structure of B-helical C-G-A-T-C-G-A-T-C-G and comparison with C-C-A-A-C-G-T-T-G-G: the effect of base pair reversal. *J Biol Chem* 266:8861–83.

Johnson CK (1976): *ORTEPII. Report ORNL-5138*. Oak Ridge, TN: Oak Ridge National Laboratory.

Jones S, Thornton JM (1996): Principles of protein–protein interactions. *Proc Natl Acad Sci USA* 93:13–20.

Jones S, van Heyningen P, Berman HM, Thornton JM (1999): Protein–DNA interactions: a structural analysis. *J Mol Biol* 287:877–96.

Jones S, Daley DTA, Luscombe NM, Berman HM, Thornton JM (2001): Protein–RNA interactions: A structural analysis. *Nucleic Acids Res* 29:934–54.

Kosikov KM, Gorin AA, Zhurkin VB, Olson WK (1999): DNA stretching and compression: large-scale simulations of double helical structures. *J Mol Biol* 289:1301–26.

Lavery R, Sklenar H (1988): The definition of generalized helicoidal parameters and of axis curvature for irregular nucleic acids. *J Biomol Struct Dyn* 6:63–91.

Lavery R, Sklenar H (1989): Defining the structure of irregular nucleic acids: conventions and principles. *J Biomol Struct Dyn* 6:655–67.

Leonard GA, Hunter WN (1993): Crystal and molecular structure of d(CGTAGATCTACG) at 2.25 Å resolution. *J Mol Biol* 234:198–208.

Lu X-J, El Hassan MA, Hunter CA (1997): Structure and conformation of helical nucleic acids: analysis program (SCHNAaP). *J Mol Biol* 273:668–80.

Mandel-Gutfreund Y, Margalit H, Jernigan R, Zhurkin V (1998): A role for CH•••O interactions in protein-DNA recognition. *J Mol Biol* 277:1129–40.

Moore P (2001): The ribosome at atomic resolution. *Biochemistry* 40:3243–50.

Olson WK, Gorin AA, Lu X-J, Hock LM, Zhurkin VB (1998): DNA sequence-dependent deformability deduced from protein-DNA crystal complexes. *Proc Natl Acad Sci USA* 95:11163–8.

Olson WK, Bansal M, Burley SK, Dickerson RE, Gerstein M, Harvey SC, Heinemann U, Lu X-J, Neidle S, Shakked Z, Sklenar H, Suzuki M, Tung C-S, Westhof E, Wolberger C, Berman HM (2001): A standard reference frame for the description of nucleic acid base-pair geometry. *J Mol Biol* 313:229–37. [This paper provides the standard reference frame for nucleic acid base-pair geometry.]

Parkinson G, Vojtechovsky J, Clowney L, Brünger AT, Berman HM (1996a): New parameters for the refinement of nucleic acid containing structures. *Acta Crystallogr* D52:57–64. [This paper describes the derivation of the parameters that are now widely used for refinement of structures containing nucleic acids.]

Parkinson G, Wilson C, Gunasekera A, Ebright YW, Ebright RE, Berman HM (1996b): Structure of the CAP-DNA complex at 2.5 Å resolution: a complete picture of the protein-DNA interface. *J Mol Biol* 260:395–408.

Rowsell S, Stonehouse NJ, Convery MA, Adams CJ, Ellington AD, Hirao I, Peabody DS, Stockley PG, Phillips SEV (1998): Crystal structures of a series of RNA aptamers complexed to the same protein target. *Nat Struct Biol* 5:970–5.

Sayle R, Milner-White EJ (1995): RasMol: biomolecular graphics for all. *Trends Biochem Sci* 20:374.

Schneider B, Berman HM (1995): Hydration of the DNA bases is local. *Biophys J* 69:2661–9.

Schneider B, Neidle S, Berman HM (1997): Conformations of the sugar-phosphate backbone in helical DNA crystal structures. *Biopolymers* 42:113–24.

Soumpasis DM, Tung CS (1988): A rigorous basepair oriented description of DNA structures. *J Biomol Struct Dyn* 6:397–420.

Spink N, Nunn C, Vojetchovsky J, Berman H, Neidle S (1995): Crystal structure of a DNA decamer showing a novel pseudo four-way helix-helix junction. *Proc Natl Acad Sci USA* 92:10767–71.

Tari LW, Secco AS (1995): Base-pair opening and spermine binding—B-DNA features displayed in the crystal structure of a gal operon fragment: implications for protein-DNA recognition. *Nucleic Acids Res* 23:2065–73.

Timsit Y, Westhof E, Fuchs RPP, Moras D (1989): Unusual helical packing in crystals of DNA bearing a mutation hot spot. *Nature* 341:459–62.

Timsit Y, Moras D (1992): Crystallization of DNA. *Methods Enzymol* 211:409–29.

Tung C-S, Soumpasis DM, Hummer G (1994): An extension of the rigorous base-unit oriented description of nucleic acid structures. *J Biomol Struct Dyn* 11:1327–44.

Urpi L, Tereshko V, Malinina L, Huynh-Dinh T, Subirana JA (1996): Structural comparison between the d(CTAG) sequence in oligonucleotides and trp and met repressor-operator complexes. *Nat Struct Biol* 3:325–8.

Vaguine AA, Richelle J, Wodak SJ (1999): SFCHECK: a unified set of procedures for evaluating the quality of macromolecular structure-factor data and their agreement with the atomic model. *Acta Crystallogr* D55:191–205.

van Roey P, Waddling CA, Fox KM, Belfort M, Derbyshire V (2001): Intertwined structure of the DNA-binding domain of intron endonuclease I-TEVI with its substrate. *EMBO J* 20:3631–7.

Westbrook J, Feng Z, Berman HM (1998): ADIT—The AutoDep Input Tool. Department of Chemistry, Rutgers, the State University of New Jersey, RCSB-99.

Westbrook J, Bourne PE (2000): STAR/mmCIF: an extensive ontology for macromolecular structure and beyond. *Bioinformatics* 16:159–68.

Westhof E, Fritsch V (2000): RNA folding: beyond Watson–Crick pairs. *Structure* 8(3):R55–65.

Woda J, Schneider B, Patel K, Mistry K, Berman HM (1998): An analysis of the relationship between hydration and protein-DNA interactions. *Biophys J* 75:2170–7.

Wood AA, Nunn CM, Trent JO, Neidle S (1997): Sequence-dependent crossed helix packing in the crystal structure of a B-DNA decamer yields a detailed model for the Holliday junction. *J Mol Biol* 269:827–41.

# 11

# OTHER STRUCTURE-BASED DATABASES

Helge Weissig and Philip E. Bourne

The single repository for experimentally derived macromolecular structures is the Protein Data Bank (PDB; Bernstein et al., 1977; Berman et al., 2000) described in Chapter 9. The *primary* data provided by the PDB are the Cartesian coordinates, occupancies, and temperature factors for the atoms in these structures. Additional information given includes literature references, author names, details of the experiment, links to the sequence in the sequence databases, and some limited annotation of the biological function (see Chapter 8). Collated into a single entry or, due to the PDB format restrictions, into multiple entries for very large X-ray structures and large NMR ensembles, these data constitute a concise description of the three-dimensional form of a molecule. The PDB currently releases the primary structure data once per week as requested by the depositor. Whereupon a number of sites worldwide acquire these data via the Internet, derive additional information, and constitute a set of *secondary* resources. Secondary resources cover features such as stereochemical quality (Table 11.1), protein structure classification (Table 11.2), protein–protein interaction data (Table 11.3), structure visualization (Table 11.4) and data on specific protein families. The secondary resources described in this chapter can be viewed as downstream of the PDB in an information flow diagram (Fig. 11.1). The number of these secondary resources is growing every year and no attempt is made at a complete overview, but rather to give a synopsis from several classes of resources (Figure 11.1) of what is available. A current compendium of secondary resources is maintained by the PDB at http://www.pdb.org/pdb/links.html. More detail on popular, well-established structure-based databases is available in other chapters. Chapter 5 includes a description of the NMR-specific BioMagResBank resource; the Nucleic Acid Database (NDB) is described in Chapter 4; the comparative fold classification databases SCOP and CATH are described in Chapters 12 and 13, respectively; Chapter 14 includes brief descriptions of stereochemical quality-oriented resources and additional resources are

*Structural Bioinformatics*
Edited by Philip E. Bourne and Helge Weissig
Copyright © 2003 by Wiley-Liss, Inc.

TABLE 11.1. Popular Software and Resources for Protein Structure Validation

| Resource | Details |
|---|---|
| PDBSum | Summaries for all protein structures including validation checks http://www.biochem.ucl.ac.uk/bsm/pdbsum/ |
| Procheck | Structure validation suite http://www.biochem.ucl.ac.uk/~roman/procheck/procheck.html Laskowski et al., 1993 |
| What_Check | Detailed stereochemical quality summaries for all protein structures. Part of the Whatif package. http://www.cmbi.kun.nl/gv/whatcheck/ |
| SFCheck | Validate the experimental structure factors associated with an X-ray diffraction experiment. Vaguine, Richelle, and Wodak, 1999 |
| PDB validation server | Validate the format and content of a PDB entry using the same software procedures as used by the PDB. Includes those listed above in this table http://pdb.rutgers.edu/validate/ |
| Protein–protein interaction server | http://www.biochem.ucl.ac.uk/bsm/PP/server/server_help.html Jones and Thornton, 1996 |
| Protein–DNA interaction server | http://www.biochem.ucl.ac.uk/bsm/DNA/server/ Jones et al., 1999 |

TABLE 11.2. Resources Classifying Protein Structure

| Resource | Details |
|---|---|
| SCOP | The Structure Classification of Proteins http://scop.mrc-lmb.cam.ac.uk/scop/ Murzin et al., 1995 |
| CATH | Class(C), Architecture(A), Topology(T) and Homologous superfamily (H). http://www.biochem.ucl.ac.uk/bsm/cath_new/index.html Orengo et al., 1997 |
| DALI | DALI Domain Dictionary http://www.embl-ebi.ac.uk/dali/domain/ Dietmann and Holm, 2001 |
| VAST | Vector Alignment Search Tool http://www.ncbi.nlm.nih.gov/Structure/VAST/vast.shtml Gibrat, Madej, and Bryant, 1997 |
| CE | Polypeptide chain comparison http://cl.sdsc.edu/ce.html Shindyalov and Bourne, 1998 |
| 3Dee | Protein Domain Definitions http://jura.ebi.ac.uk:8080/3Dee/help/help_intro.html Siddiqui and Barton, 1995 |
| CAMPASS | CAMbridge database of Protein Alignments organized as Structural Superfamilies http://www-cryst.bioc.cam.ac.uk/~campass/ Sowdhamini et al., 1998 |

TABLE 11.3. Popular Resources of Protein Interactions

| Resource | Details |
| --- | --- |
| DIP | Database of Interacting Proteins http://dip.doe-mbi.ucla.edu/ Xenarios et al., 2002 |
| BIND | The Biomolecular Interaction Network Database http://www.bind.ca/ Bader et al., 2001 |
| MINT | Molecular Interactions Database http://tweety.elm.eu.org/mint/index.html |

TABLE 11.4. Popular Resources Visualizing Macromolecular Structures

| Resource | Details |
| --- | --- |
| Jena Image Library | Images depicting biological function and useful links to other resources http://www.imb-jena.de/IMAGE.html Reichert and Sühnel, (2002). |
| PDBSum | Summaries for all protein structures including protein-ligand interaction http://www.biochem.ucl.ac.uk/bsm/pdbsum/ |
| NDB Atlas | Protein-DNA complexes http://ndbserver.rutgers.edu/NDB/NDBATLAS/ |
| STING | Sequence and property browser http://mirrors.rcsb.org/SMS/ |
| GRASS | Static GRASP images of electrostatic and surface properties http://trantor.bioc.columbia.edu/GRASS/surfserv_enter.cgi |
| General | World Index of Molecular Visualization Resources http://molvis.sdsc.edu/visres/ |

referenced throughout. The reader is also referred to the annual edition of Nucleic Acids Research dedicated to molecular biology databases that appears in January and that includes descriptions of many of the resources outlined here (volume 30(1) in 2002; 29(1) in 2001; 28(1) in 2000).

## THE ADDED-VALUE PHILOSOPHY

At the time of the its inception in the early 1970s, the PDB had only a few entries available and information technology to manage these data was in its infancy. However, as the number of entries in the PDB grew slowly during the 1980s, comparative analysis of these entries became possible, supported by new algorithms, faster computers to run these algorithms on this growing body of data, and the availability of databases to efficiently access these data.

Attempts at comparative analysis of PDB data revealed deficiencies in both the content and the format of the data. This subject is discussed in Chapter 8 and is not considered further here. Today the PDB is committed to provide consistent and complete information on the macromolecular structure and the experiment used to determine that structure. These rich and complex biological data provide many with the opportunity to add value to these data. As a consequence researchers are faced with a large array of resources from which to choose. This chapter introduces a subset of these resources that we consider to be important to a large audience.

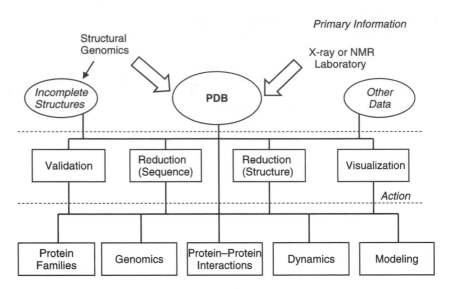

**Figure 11.1.** The flow of macromolecular structure data. *Primary Information* is derived directly from experiment. All completed macromolecular structures in the public domain are deposited with the PDB. It is anticipated that in the future incomplete structures will also be available from the structural genomics projects. Additional primary information, such as sequences, crystallization conditions, and the structure of small molecule ligands, are available in primary resources other than the PDB. A variety of *actions* are performed on these primary data and a set of *secondary resources* result.

## OTHER PRIMARY INFORMATION RESOURCES

Not all primary information on macromolecular structure is located in the PDB (Figure 11.1). Here we consider three additional sources of primary information. First, there is information of crystallization conditions which have been extracted from the literature. Second, there is information on small organic molecules, a number of which are covalently or non-covalently bound to large biological macromolecules. Third, there is the growing body of information derived from structural genomics projects.

### Biological Macromolecule Crystallization Database—http://wwwbmcd.nist.gov:8080/bmcd/bmcd.html

The Biological Macromolecule Crystallization Database (BMCD) contains crystal data and the crystallization conditions, which have been compiled by human annotators from the literature (Gilliland, Tung, and Ladner, 1996). Currently, BMCD includes 3547 crystal entries for 2526 biological macromolecules. These entries include proteins, protein–protein complexes, nucleic acids, nucleic acid–nucleic acid complexes, protein–nucleic acid complexes, and viruses. In addition to including crystallization data reported in the literature, the BMCD holds data from the NASA Protein Crystal Growth Archive, including data generated from studies carried out in NASA's microgravity environments as well as results from microgravity experiments conducted by other international space agencies.

BMCD addresses what is often the most difficult and time-consuming step in the determination of a macromolecular structure by X-ray crystallography—the crystallization of the macromolecule (see Chapter 4). The late Max Perutz once said "crystallization is a little like hunting, requiring knowledge of your prey and a certain cunning." Some structure models deposited in the PDB report the conditions used for the successful crystallization, but many do not. Hennessy and coworkers (Hennessy et al., 2000) have recently documented the usefulness of the information stored in BMCD for predicting crystallization conditions. The structural genomics projects (Chapter 29) are in addition now collecting and storing information on failed crystallization experiments, signaling a new era. These negative results can be considered as useful as those that led to successful diffraction experiments.

## Cambridge Structural Database: Small Molecule Organic Structures—http://www.ccdc.cam.ac.uk/prods/csd/csd.html

The Cambridge Structural Database (CSD) is the small molecule equivalent of the PDB, that is, a primary resource for crystal structure information of nearly a quarter million organic and metal organic compounds (Allen and Kennard, 1993). Crystal structures, deposited directly to the CSD or manually annotated from the literature, are derived from both X-ray and neutron diffraction. The CSD contains three distinct types of information for each entry that can be categorized according to their dimensionality:

1. One-dimensional (1D) information. This category includes all of the bibliographic information for the particular entry and a summary of the structural and experimental data. The text and numerical information include the names of the authors, compound names, and full journal references, as well as the crystallographic cell dimensions and space group. Where applicable, descriptions of absolute configuration, polymorphism form, and any drug or biological activity are also included.

2. Two-dimensional (2D) information. Data encoded as a chemical connection table including atom and bond properties and a chemical diagram of the molecule. Atom properties include the element symbol, the number of connected nonhydrogen atoms, the number of connected hydrogen atoms, and the net charge.

3. Three-dimensional (3D) information. Data used to generate a 3D representation of the molecule. These data include the atomic Cartesian coordinates, the space group symmetry, the covalent radii, and the crystallographic connectivity established by using those radii.

The data format used by the CSD, CIF or Crystallographic Information File (Hall, Allen, and Brown, 1991), is a small-molecule version of the macromolecular CIF (mmCIF, Chapter 8). Both CIF and mmCIF are endorsed and maintained by the International Union of Crystallography (IUCr).

CSD is distributed by the Cambridge Crystallographic Data Center (CCDC) as a commercial product for local installation. Network access to CSD information is currently made available free of charge to academic users in the United Kingdom and Europe. In addition, individual entries can be retrieved from the CSD using a simple form that requires at least a name, an e-mail address, and, for location of the entry, the CSD accession number and a complete journal reference of the CSD entry. On submission of the form, results are returned within three business days via e-mail. It is anticipated that the access to CSD will change in the near future and readers are referred to http://www.ccdc.cam.ac.uk/ for current information.

## Structures Not Yet Available

It is useful to know what macromolecular structures will likely be available at some point in the future. Two resources provide this information and both are maintained by the PDB. The first are those structures already solved and deposited in the PDB, but not yet available. These can be reviewed at http://www.rcsb.org/pdb/status.html. These structures are either on hold pending publication of the associated paper, or on hold for a longer period to permit the depositor to fully exploit his or her data. This period usually does not exceed one year. The second are those structures being determined by the structural genomic projects worldwide. These data range from a description of the target sequence under consideration, to a status of the structure determination, to a final 3D model. Details can be found at http://targetdb.pdb.org. A brief discussion of structural genomics concludes this chapter.

## SECONDARY RESOURCES

The resources described in this section are presented in no particular order and represent a cross section of what is available worldwide. Additional resources are listed in Tables 11.1 through 11.4. Where available, information is provided on current update frequencies, data formats and the underlying technology used. In most cases, users of these secondary resources can expect a delay between the release of a structure by the PDB and the availability of derivative information on the structure through a secondary resource. As the rate of deposition of structures increases (see Chapter 29), resources that rely on semiautomated update requiring human annotation will lag behind. As indicated throughout this book, the future will likely see a concerted effort in new and improved algorithms to automatically generate and statistically validate secondary information.

## Sequence and Structure Relationships to Provide Nonredundant Data: The ASTRAL compendium—http://astral.stanford.edu/

The PDB file format does not always provide an explicit relationship between the SEQRES records of biological sequence information and the ATOM/HETATM records that contain the Cartesian coordinates for each amino acid or nucleotide. While this shortcoming has been fully addressed in the mmCIF format, most structural bioinformatics software currently uses PDB files. The ASTRAL compendium (Brenner, Koehl, and Levitt, 2000) is a collection of data files and tools, providing a partially curated mapping of these records as produced by the program pdb2cif (Bernstein, Bernstein, and Bourne, 1998). The mapping is distributed in a text format named Rapid Access Format (RAF) that can easily be parsed by computer programs. The RAF file includes mappings for all PDB chains represented in the first seven classes of SCOP (see Chapter 12). It is used as the definitive sequence-mapping resource for ASTRAL and SCOP but is also intended as a useful resource for any PDB user.

Using RAF data the primary role of ASTRAL is to maintain nonredundant sequence sets corresponding to unique protein domains as defined by SCOP (Chapter 12). This information is helpful for the analysis of evolutionary relationships between domains based on sequence alignments. It also serves to reduce the redundancy in the PDB by filtering out protein sequences with varying degrees of sequence identity, leaving a representative conforming to the most accurate structure determination. The PDB has also recently begun to offer a similar capability.

Redundancy arises in the following ways, given a repository that accepts all structures solved by the worldwide community. First, different groups can determine the same or very similar structures independently. Second, a point mutation, occurring naturally or introduced post-translationally to analyze biological function or structure folding, leads to a very similar structure. For example, approximately 700 lysozyme structures can be found in the PDB where almost every position in the structure, and combinations thereof, have been modified. The Protein Mutation Resource (PMR; http://pmr.sdsc.edu) analyzes these data in more detail. Third, structures are often determined multiple times with different ligands bound to them (e.g., HIV-1 proteases with different inhibitors bound), without significant change in the protein itself.

PDBselect (Hobohm and Sander, 1994) was the first widely used reduced set of protein structure data. When performing such reduction an important question becomes, how to choose the representative? All approaches employ an initial ranking of structures based on the widely used quality parameters for X-ray structures: resolution and R-factor. ASTRAL uses in addition a Stereochemical Check Score (SCS), combining scores from PROCHECK (Laskowski et al., 1993) and WHATCHECK (Hooft et al., 1996), two well-known stereochemical quality assessment programs. Based on this combined overall "Summary PDB ASTRAL Check Index" (SPACI) score, structures are chosen as representatives for others at sequence identity cutoffs at set percentages. ASTRAL also provides access to nonredundant sets filtered by E-value or SCOP classification, that is, corresponding to class, fold, superfamily, and family.

## Providing Links to Literature, Sequence, and Genome Information: The MacroMolecular DataBase (MMDB)— http://www.ncbi.nlm.nih.gov/Structure/

The Macromolecular Database (MMDB) maintained by the National Center for Biotechnology Information (NCBI) contains all experimentally determined structures from the PDB (Ohkawa, Ostell, and Bryant, 1995). It is updated on a monthly basis and provides linkage to structural information from NCBI's integrated query interface Entrez. Entries in MMDB are specified using an Abstract Syntax Notation One (ASN.1; http://asn1.elibel.tm.fr/). MMDB provides access to coordinates, sequences, all bibliographic information and taxonomy data, as well as the authors and deposition dates together with the PDB assigned classification and compound information of a PDB entry.

The assignment of the correct species of origin of a specific PDB chain is based on a semiautomated procedure in which a human expert validates the automatically assigned taxonomy annotation based on sequence comparisons with GenBank and SwissProt. A set of rules ensures the consistency of this approach. Missing annotations are generated from literature information or using BLAST searches. Artificially generated protein and nucleic acid chains (excluding trivial modifications such as single amino acid substitutions or His-tags) are labeled as "synthetic."

Beyond enabling the query for structures based on the textual information described, MMDB also provides structural neighbor assignments produced by the Vector Alignment Search Tool (VAST; Madej et al., 1995; Gibrat et al., 1996). Each chain of each entry in MMDB is compared with every other chain to compile a list of structural neighbors. These are made available for individual chains as well as for domains. In addition, MMDB can be queried with user-supplied coordinate sets to find entries based on structural similarity.

The information stored in MMDB and Entrez allows for the seamless exploration and query of literature references, sequence information, taxonomical and genomic data associated with macromolecular structures. Other resources, including the PDB, provide links to some of these data, however, MMDB uniquely combines them into a single resource. NCBI also provides several graphics tools including the application CND3 for 3D structure visualization.

## Derived Secondary Structure of Proteins—http://www.sander.ebi.ac.uk/dssp/

The Derived Secondary Structure of Proteins (DSSP) resource provides secondary structure assignments computed from structure using an algorithm developed in the early 1980s by Wolfgang Kabsch and Chris Sander (Kabsch and Sander, 1983). This approach is discussed more fully in Chapter 17.

The DSSP resource consists of the dssp program itself (licensed at no cost to academic users and available for commercial licensing) and the dssp-generated flat files, one per PDB entry. Using a standardized representation, the DSSP file contains the secondary structure assignments, geometric structure, and solvent exposure for each residue. These data are also available from a variety of Web sites. In contrast, the PDB files provide annotator-validated secondary structure assignments based on the PROMOTIF program (Hutchinson and Thornton, 1996).

## Protein Quaternary Structure—http://pqs.ebi.ac.uk/

Structure determination does not always provide the functional form of a biological macromolecule. Rather, it provides the tertiary structure as found in the asymmetric unit of the crystal, whereas it is the quaternary structure—the macromolecular assembly of two or more copies of tertiary structure elements that form homo- or hetero-multimers—that infer biological function. Viral protein coats are beautiful examples of biological function inferred by the organization of tertiary structure into a quaternary biologically active assembly (see VIrus Particle ExploreR; VIPER; http://mmtsb.scripps.edu/viper/viper.html; Reddy et al., 2001 for examples specific to viruses). The Protein Quaternary Structure (PQS) resource maintained by the Macromolecular Structure Database (MSD) group at the European Bioinformatics Institute (EBI) provides an automatically derived assessment of the biological unit of a PDB entry determined by X-ray crystallography (Hendrick and Thornton, 1998).

The Cartesian coordinates found in a PDB entry generally correspond to the asymmetric unit of the molecule as found in the crystal and represent the unique atomic positions that are refined against the experimental data. However, these coordinates do not necessarily correspond to the biologically active molecule. The necessary crystallographic symmetry operations as defined by the space group and possibly noncrystallographic symmetry operations[1] must be applied to generate the biologically active quaternary structure. This is done through the application of rotation and translation to the individual atoms listed in the PDB entry.

Automating this procedure is nontrivial. The process must distinguish between an assembly that is a truly biologically active molecule and an assembly that is a number of discreet biologically active components associated through crystal packing, but having no physiological relevance. It should be noted that the PDB now seeks to capture the interpretation of the biologically active molecule from the structural biologists depositing the structure data rather than attempt to only determine it automatically.

The PQS procedure is well documented on the Web site and only a synopsis is given here. For nonvirus structure PQS performs two steps: generate the assembly and assess the assembly for the likelihood it is a quaternary structure. The first step involves applying any noncrystallographic symmetry and then recursively adding symmetry-related contents to the asymmetric unit. If close contacts are found this is considered a candidate quaternary structure. The second step determines the nature of the contacts using the solvent-accessible surface. The premise being that components forming a quaternary complex will have a lower solvent-accessible surface than those that exist as discreet globular proteins.

PQS provides its results in the form of PDB-formatted files that include the list of all symmetry operators and all calculated coordinates. In addition, PQS provides a description of the quaternary structure, for example, "homo dimer" or "hetero tetradecamer." Virus entries are treated differently in that several files are provided that include the complete virion, and separate, symmetry-related files as well as a file containing all chains needed to describe all the unique protein–protein interfaces.

Comparisons with literature derived information as well as information provided by individual researchers was used to determine a rough measure of accuracy for the PQS procedure (Hendrick and Thornton, 1998). Using 6739 entries available from the PDB in December 1997, 1398 were determined to be potential homo-dimers. Of these, 244 were assigned to have nonspecific (crystal-packing) contacts. This could be confirmed for 31 entries based on the available textual information. The remaining 1154 entries were assigned true homo-dimer status. This status could be confirmed for 385 entries, could not be confirmed for 386 entries, and was found to be false positive for 383 entries. Of those 383, 190 were lysozymes, which exhibit very strongly associated crystallographic packing, underscoring the difficulty in automatically determining the difference between specific and nonspecific macromolecular associations. Other examples of seemingly incorrect predictions include a prediction of a 24meric assembly of the transcription repressor protein rop (PDB identifier 1GTO). While the biologically active molecule in fact is a DNA-associated dimer, the authors of the crystal structure describe a "hyperstable helical bundle" in the crystal structure, possibly due to artificial solid-state interactions.

In summary, while caution needs to be exercised in using PQS-generated quaternary structure predictions, the resource nevertheless provides a valuable starting point to the determination of the biologically active molecules represented by the asymmetric units given in the PDB entries.

[1]If the molecule exhibits its own symmetry, then refinement of the structure may be undertaken only on the part considered unique, the tertiary structure is then generated from the coordinates present in the PDB and the application of noncrystallographic symmetry. See PDBids 2HHB and 3HHB for contrasting examples from the same molecule, deoxy hemoglobin.

## Protein–Ligand Interactions:
## ReliBase—http://relibase.ccdc.cam.ac.uk/

The biological function and regulation of proteins often involves the binding of smaller organic or inorganic molecules that commonly are grouped together under the term ligands. Metal ions, anions, solvate molecules (except water), cofactors, and inhibitors are generally all regarded as ligands. ReliBase, developed primarily by Dr. Manfred Hendlich and now maintained at the Cambridge Crystallographic Data Center (CCDC), contains experimental PDB structures with ligands and structures where only the ligand-binding partner was modeled into the structure (Hendlich, 1998). DNA and RNA strands are visualized in result sets as ligands but cannot be searched. ReliBase provides access to its entries via text queries over the HEADER, COMPOUND, and SOURCE records of the PDB files, as well as the names of authors, chemical names of ligands, and their PDB assigned three letter codes. In addition, ReliBase can be queried using a protein sequence or SMILES strings (string representations of 2D structural fragments or molecules; Siani, Weininger, and Blaney, 1994). Finally, it is possible to search ReliBase using 3D diagrams drawn using a Java applet.

ReliBase results are easy to browse and include 2D diagrams of ligands, bibliographic and some additional textual information from the PDB entries as well as convenient links to searches for similar ligands, binding sites, or protein chains. Query results can be stored as hitlists, which can be used in SMILES or 2D/3D searches. In addition, binding sites can be superimposed and visualized in different ways using static images, graphic applets, or client-side visualization tools such as Rasmol (Sayle and Milner-White, 1995). Third-party tools integrated into ReliBase include the sequence search package FASTA (Pearson, 1990) and the computational chemistry tool kit CACTVS (Ihlenfeldt et al., 2002), which is used to generate 2D diagrams in ReliBase.

ReliBase is the product of several industrial/academic partnerships and is written in C++ with a Perl CGI WWW front end. Stand-alone distributions for several platforms are available from the CCDC on request.

## Protein Families

Often macromolecular structure information is only a part of a larger study on a particular family of proteins that are functionally related. Resources capturing such comprehensive information are usually developed by individual research laboratories with interest in specific protein families. The general notion is to be narrow but deep versus resources such as the PDB, which are broad but shallow with respect to their information content. Stated in another way, the PDB contains a limited amount of information on all macromolecular structures; resources such as those described in this section integrate structure as part of additional information on a specific protein family. A couple of these resources are highlighted to indicate the kind of content that is available. Similar resources to those discussed here exist for chaperonins, the P450 family, cytokines, esterases, G protein-coupled receptors, glucoamylases, kinesins, thyroid hormone receptors, topoisomerases, and viruses. A more complete list and associated Web links can be found at the CMS Molecular Biology Resource at http://restools.sdsc.edu/biotools/biotools25.html.

*Protein Kinase Resource—http://pkr.sdsc.edu/.* The Protein Kinase Resource (PKR) includes detailed structure classifications, sequence and structure alignments, sequence classifications, motif recognition, and information on relationships to diseases with the overall goal of providing a comprehensive compendium for an important class of enzymes involved in cell signaling (Smith et al. 1997). PKR includes expert manual sequence alignments as well as automated alignments based on specific family profiles. Structure alignments are provided using a modified version of the Combinatorial Extension algorithm (CE; http://cl.sdsc.edu/ce.html; Shindyalov and Bourne, 1998).

PKR can be searched using a simple text form as well as advanced search forms and by queries based on protein sequences, isoelectric point, and molecular weight. Protein entries include bibliographic information, functional information, and sequence details, as well as links to other sequence and structure-based resources such as SwissProt, InterPro, PDB, and others. Other information includes a collection of experimental method descriptions such as activity assays or purification protocols, a directory of researchers, and a list of upcoming meetings, as well as a series of tutorials on kinases and their functions, including their involvement in cancer. PKR also maintains an active e-mail list for the discussion of protein kinase-related questions. Details on how to join the list are provided on the PKR web site.

PKR is maintained at the San Diego Supercomputer Center by Drs. Roland H. Nieder and Michael Gribskov using the relational database MySQL. PKR currently contains approximately 9100 protein kinase sequences.

***HIV Proteases—http://www.ncifcrf.gov/HIVdb/.*** The HIV Proteases resource (HIVpr) archives experimentally determined structures of Human Immunodeficiency Virus 1 (HIV1), Human Immunodeficiency Virus 2 (HIV2) and Simian Immunodeficiency Virus (SIV) proteases and their complexes (Vondrasek, Buskirk, and Wlodawer, 1997). The structures contained in HIVpr include 63 structures not currently available through the PDB that were made available by several pharmaceutical companies for exclusive use by the resource. An additional 120 structures are taken from the PDB.

The information provided by HIVpr includes tabular listings of ligand/enzyme complexes, enzyme inhibitors, and proteinase mutants. In addition, analytical information on volume analysis, interaction energy, surface analysis, subsite occupation, and structural superpositions are made available in graphic form. The resource is searchable through a simple text field and results are presented in tabular form including bibliographical information, PDB accession numbers, if applicable, and inhibitor information including graphic representations.

HIVpr was developed in the group of Dr. Alex Wlodawer and is maintained at the National Cancer Institute.

***Metalloproteins—http://metallo.scripps.edu/.*** The Metalloprotein Database and Browser (MDB) is part of the Metalloprotein Structure, Bioinformatics, and Design Program at The Scripps Research Institute (TSRI). MDB provides quantitative information and tools to visualize protein metal-binding sites from structures taken from the PDB (Castagnetto et al., 2002). Approximately one-third of all structures in the PDB contain a metal ion.

Entries are extracted from the PDB and added to MDB with a set of automatic tools that periodically scan newly released PDB structures for the occurrence of metal ions. An indexing tool extracts first- and second-shell data, recognizes multinuclear and cluster-containing sites, and classifies metal-binding sites according to criteria such as the number of metal ions in the site, the types of ions, and metal coordination. Noncovalent interactions are also determined within and among indexed shells.

MDB can be queried with a variety of methods ranging from simple text-based queries to fairly complex SQL queries that fully realize the power of the underlying, fully documented relational database schema. Real time 3D viewing of binding sites is provided through a Java applet that enables the user to inspect atom–atom distances, bond angles, and torsion angles. Structure superpositions, stereo viewing, and selection of atoms based on distance are also possible.

In addition to the interactive query and analysis interfaces provided to users, MDB offers noninteractive gateways for incorporation of MDB data into stand-alone programs. Most notably, MDB supports an XML-RPC-based interface, a remote procedure calling protocol that uses the hypertext transfer protocol (http) and Extensible Markup Language (XML) for the exchange of data. XML-RPC is a simple protocol that allows complex data structures to be transmitted, processed, and returned. The protocol would, for example, allow a metal-site design program to obtain an up-to-date list of observed ranges for a certain geometric feature (e.g., torsion angle) to compare a suggested model value with those found in known metalloproteins.

MDB is built on top of the relational database system MySQL and uses the powerful Web-scripting language PHP as a front end. The Java applet is also used by other sites such as the IMB Jena Image Library of Biological Macromolecules (Reichert and Suhnel, 2002) as a gateway to MDB.

### *Macromolecular Motions Database—http://molmovdb.mbb.yale.edu/ molmovdb/.*

The Macromolecular Motions Database (MolMovDB) describes and systematizes known motions that occur in proteins and other macromolecules. Associated with MolMovDB are a set of free software tools and servers for structural motion analysis (Gerstein and Krebs, 1998; Krebs and Gerstein, 2000).

MolMovDB addresses an important phenomenon in biochemistry, the precise movement of many atoms within a macromolecule that often plays a crucial role in its function. Macromolecular motions are essential in, for example, enzymatic reactions, allosteric regulation of activity, transporter functionality, and locomotion. Due to the involved timescales, which range from subnanosecond loop closures to refolding spanning several seconds, it is near impossible to study these motions with a single computational approach like molecular dynamics due to the computational intractability.

MolMovDB currently contains more than 4200 entries. Of these, 3800 have been automatically extracted from the PDB, 230 have been manually curated, and 200 have been submitted by users. Protein motions are categorized first by the information available on the motion, its size (distinguished are fragment, domain, and chain motions) and lastly by type of motion. Motions of proteins involving fragment or domain motions are primarily characterized as consisting of either a "shear" motion (sliding of a continuously maintained and tightly packed interface) or as a "hinge" motion (movement of two domains connected by a flexible linker without a continuously maintained interface). Motions of subunits are predominantly classified as "allosteric", "nonallosteric," or "complex." Each individual motion in the database is assigned a mnemonic accession code and a classification code. For example, the motion in calmodulin is accessible under the identifier "cm" and is classified as a "known domain motion, hinge mechanism" (D-h-2). A total of 29 such classifiers were established and are documented.

MolMovDB is searchable by keyword and/or by PDB identifier. Curated entries are also listed for easy access. Each entry is accompanied by its classification, links to PDB structures (via their PartsList entries, see Parts List: Dynamic Fold Comparisons below), a description of the motion, and particular values describing the motion. Movies are associated with each entry and available in several formats. The Morph Server software, which automatically generates these movies, was developed by Dr. Werner Krebs in collaboration with Prof. Mark Gerstein at Yale University. The Morph Server produces 2D and 3D animations of plausible pathways between two endpoints of a particular motion. A typical morph takes a few minutes to compute and results are stored for later access. Morphing involves an adiabatic-mapping algorithm to interpolate two PDB input files. A particular pathway is broken up into several equal length steps, at each step interpolated coordinates are subjected to an energy minimization "refinement" to correct for bond length, bond angle, and torsion angle aberrations. The Morph Server is accessible as a stand-alone tool for users wishing to generate their own movies based on two given structures.

MolMovDB is built on top of the freeware relational database MySQL and a Perl based CGI front end; some computationally intensive components of the site (Morph Server) are partially implemented in C/C++, FORTRAN, and Python/MMTK. The WWW front end is easy to navigate for any user but SQL dumps are also available for advanced users on request from the maintainers.

***Parts List: Dynamic Fold Comparisons—http://bioinfo.mbb.yale.edu/ partslist/.*** The number of structures in the PDB is expected to increase significantly in the next few years, specifically with the advent of structural genomics (see also Chapter 29 and a short perspective at the end of this chapter). However, the number of protein folds is quite limited and analyses and reanalyses of this finite parts list from an expanding number of perspectives will probably become more and more informative as the list reaches completeness. The resource described in this section, PartsList, allows users to dynamically compare this emerging and linked set of protein folds.

PartsList is based on the Structural Classification of Proteins (SCOP; see Chapter 12) fold classification and functions as supplemental annotation to SCOP. Folds in PartsList (represented by domains corresponding to specific folds and/or superfamilies in SCOP) are ranked on a growing number of currently more than 180 attributes. These attributes include the occurrence in completely sequenced genomes, the number of occurrences of a fold in the PDB, participation in protein–protein interactions, the number of known functions associated with a fold, the amino acid composition, participation in protein motions, and the level of similarity based on a comprehensive set of structural alignments using the Gerstein/Levitt algorithm. (Gerstein and Levitt, 1998; Quian et al., 2001).

Three ways of visualizing the fold rankings are provided by PartsList: first a profiler emphasizing the progression of high and low ranks across many preselected attributes, next a dynamic comparer for custom comparisons, and finally a numerical rankings correlator. Traditional single-structure reports are provided to summarize information related to genome occurrence, expression level, motion, function, and interaction with additional links to many other resources.

The ranking provided by PartsList allows for a comparison of folds using a unified approach. The numerical values associated with each rank can be used to compare the very different attributes of a fold; for example, expression levels and participation in protein–protein interaction. Access to tabular comparisons is made available for all individual fold rankings according to individual attributes. For example, users can readily switch between occurrence, interaction, motion, or alignment information for a fold identified with the Profiler, Comparer, or Correlator tool. In addition, PartsList is searchable by PDB or SCOP accession number and text files (summary tables and structural alignments) are made available for download.

PartsList is maintained in Prof. Mark Gerstein's group at Yale University. The resource provides "extrinsic" information on protein folds, that is, putting a fold into the context of all other folds according to specific criteria.

***Automated Comparative Modeling: Swiss-Model—http://www.expasy. org/swissmod/.*** Protein modeling involves the generation of a theoretical model of a protein structure based on its sequence and one or more known structures with more or less similar sequences. In recent years, many automated approaches have been reported in the literature and several servers are available for users to generate their own structural models (see Chapters 25–28). The Swiss-Model server (Guex and Peitsch, 1997) is one example of many structure-prediction and modeling resources and the reader is referred to a more comprehensive listing available at http://restools.sdsc.edu/biotools/biotools9.html.

Swiss-Model offers several modes in which users can generate and refine their models. In addition, the structure-viewing program Swiss-PDBViewer has been tightly integrated with the modeling resource. Swiss-PDBViewer enables the analysis of several proteins at the same time. Proteins can be superimposed to generate structural alignments in order to compare relevant parts, for example, their active sites. Amino acid mutations, hydrogen bonds, bond angles, and distances between atoms are displayed via graphic and menu interfaces. Swiss-PDBViewer can also read electron density maps for detailed interpretation of structures, various modeling tools are integrated, and command files for use in popular energy minimization packages can be generated. Although both Swiss-Model and Swiss-PDBViewer can be used independently, the combination of both can be used to generate structural models.

Swiss-Model uses structure templates extracted from the PDB, their sequences, and the ProModII modeling package to generate the actual models. Users are able to submit their own templates in PDB format for use in ProModII. The automatic template selection step involves a BLAST query of the Swiss-Model template database given user definable threshold values. The subsequent modeling procedure employed by Pro-ModII involves the following eight steps: (1) Superposition of related 3D structures; (2) Generation of a multiple alignment with the sequence to be modeled; (3) Generation of a framework for the new sequence; (4) A rebuild lacking loops; (5) Completion and correction of the structural backbone; (6) Correction and rebuilding of side chains; (7) Verification of the model structure's quality and a check of its packing; and (8) Refinement of the structure by energy minimization and molecular dynamics. Generated models are sent to users by e-mail and can be imported, analyzed, and manipulated in Swiss-PDBViewer.

Swiss-Model and Swiss-PDBViewer were developed in the group of Dr. Manuel Peitsch and are maintained at part of the Expert Protein Analysis System (ExPASy) server of the Swiss Institute of Bioinformatics.

***Sources of Targets and Prediction Methods.*** While the cost of structure determination is decreasing rapidly, it will probably never become as cheap as the cost of sequencing. Hence, the ratio of the number of structures to the number of sequences will remain at several orders of magnitude. Yet, as the number of structures continues to rise, they provide a rich source of template information for structure prediction using techniques such as homology modeling and threading. Progress in these areas is monitored by the Critical Assessment of Structure Prediction (CASP) experiments that are conducted every two years (Chapter 24). At the CASP meetings, prediction methods are compared, rated, and hotly debated (Venclovas et al., 2001). Predictions can be performed in 1D (secondary structure, solvent accessibility), 2D (inter-residue distances), and 3D (ab initio prediction, homology modeling such as implemented by Swiss-Model, and threading). Resources even exist to evaluate prediction servers (see, for example, EVA, http://cubic.bioc.columbia.edu/eva/ and LiveBench, http://bioinfo.pl/LiveBench/). To facilitate these prediction efforts, if the depositor permits, sequences of solved protein structures are now released ahead of the structures by the PDB to permit unbiased experiments from a continuous source of new targets (see http://www.rcsb.org/pdb/status.html). Another source of targets are the sequences registered by the structural genomics projects in a target database maintained by the PDB at http://targetdb.pdb.org/.

## STRUCTURAL DATABASES OF THE FUTURE

### Integration Over Multiple Resources

The world of on-line information available to structural biologists has become extremely balkanized as the number of resources available as well as the information content provided by these resources has increased exponentially in the last decade of the twentieth century (Williams, 1997). Most databases available today on the Web provide a good number of cross-links to other resources with relevant information. However, in almost all nontrivial cases (i.e., those cases where the link is not simply based on an obvious identifier in the remote resource), these cross-links have to be added and maintained by human curators. In order to create such links automatically, database maintainers have to first agree on a common nomenclature or provide a comprehensive ontology of the information available through their resources for interconnection with other ontologies. Much progress has been made in this area and the PDB curation efforts of the RCSB are a notable example.

The "following" of links provided to other (internal or external) information is a common action in browsing the content on any Web site. This process has been automated early on in the short history of the Web, leading to the creation of so-called "web crawlers," which retrieve a start page given by a Uniform Resource Locator (URL) and follow all links provided on this page. By recursively following all links on subsequently retrieved pages, this approach would theoretically lead to a comprehensive collection of all interlinked pages. A similar approach, refined for use with molecular biology-based information resources, is used by the Molecular Information Agent (MIA, http://mia.sdsc.edu/). MIA, which originates in the laboratory of Prof. Michael Gribskov at the University of California, San Diego, is an Application Programmer Interface (API) and program framework built on the notion of resource-specific reference templates and a set of keys or identifiers. A reference to a specific page can be generated from a template and a key. For example, taking the reference template http://www.rcsb.org/pdb/cgi/explore.cgi?pdbId= for the Structure Explorer page at the PDB together with the key 2CPK, a specific structure identifier, will result in the reference for the PDB Structure Explorer page summarizing data available for the structure of a cAMP-dependent protein kinase. Similarly, any Medline citations available for 2CPK can be assembled from the PubMed/Macromolecular Database citation template at the NCBI http://www.ncbi.nlm.nih.gov/htbin-post/Entrez/query?db=t&form= 6&Dopt=m&uid= and the same key 2CPK.

In a typical application, MIA would retrieve all specific references it can generate based on the templates it knows of (these are stored in a database) and a given key. By parsing the retrieved pages, all references to other resources or known keys can be extracted. The extracted references can then be used to extract additional keys or can be directly followed for further iterations. To expand on the example above, MIA parses the PDB Structure Explorer page for any occurrence of an enzyme classification (EC). Using that EC number (2.7.1.37 in the case of 2CPK) as a new key, MIA then assembles additional references for pages that are based on a qualified EC number, for example the EcoCyc metabolic pathway information (Karp and Paley, 1994).

Any reference generated by MIA is stored in a relational database for future retrieval together with information on the history of how the reference was obtained and when it was last checked for the availability of information at that URL. This specific check is important as not all databases are updated synchronously with each other and often information from a secondary resource will not become available for a while; for example, whereas the PDB is updated weekly on Tuesdays, the NCBI PubMed/MMDB references are updated only once a month. In this manner, MIA can be used to automatically assemble a large number of verified cross-links based on a single starting point. Since these links can be stored in a relational database, they can easily be retrieved "on the fly," a feature that is available on the PDB's Structure Explorer's "Other Resources" pages (Fig. 11.2).

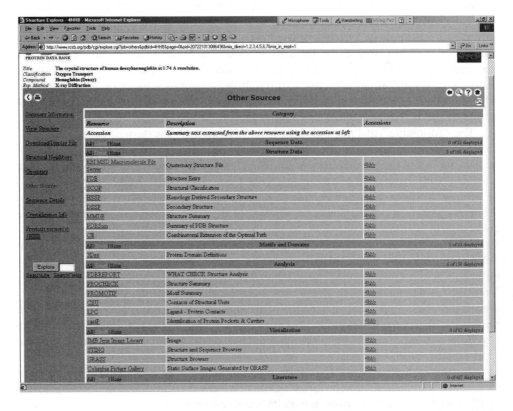

**Figure 11.2.** Example of the PDB's "Other Sources" Web page for deoxyhemoglobin (PDBid 4HHB).

In principle, the information retrieved by MIA could be stored in the same database and used for queries. This solution would provide a complement of local information, but raises issues of copyright.

## Database Interoperability Beyond Cross-Links

The approach taken by MIA is a simple mechanism for generating and maintaining a large number of specific cross-references. However, the approach has three drawbacks: (1) The assembly of a resource-specific URL based on a specific key relies on a published template for this URL, the key itself, and the parsing of the resulting page. If the maintainers of the remote resource decide on a design change, these assumptions may no longer be true and the system breaks down. (2) The repeated retrieval of information is very resource intensive both on the MIA end and on the remote resource end. A resource newly added to MIA that provides access to information based on a specific type of key would be queried for all known instances of this key within a short period of time using the Hypertext Transfer Protocol (http) protocol. (3) The MIA approach is somewhat wasteful in that very small pieces of information (often only the instance of a specific key) are extracted from a large body of data returned from a remote server. Several recent developments promise to overcome the limitations of approaches such as the one used by MIA through a much more closely connected database infrastructure. Most notably, Java based multitier architectures based on J2EE design patterns (www.java.sun.com) or programming-language independent Common Object Resource Broker (CORBA) provide standard interfaces not only to their specific data but also to the functionality available at a specific site. These interfaces allow programmers to incorporate remote procedure calls in their applications to efficiently retrieve and manipulate specific data items enabling the direct integration of multiple remote resources without the need for resource-specific parsers or the batch transfer of a lot of information not actually utilized in the particular application.

## The Impact of Structural Genomics

Structural genomics (Burley et al., 1999; see also Chapter 29) is an effort to develop and employ high-throughput structure determination for purposes including the filling in of protein fold space to facilitating comparative modeling, the determination of as many protein structures from a given genome as possible, or the furthering of our understanding of specific diseases or biochemical pathways. Although the goals may differ, the process is the same and will result in a large number of structures, estimated to reach 35,000 by 2005 (Bourne, 1999). Many of these structures will be incomplete, having been discarded in a partially completed state, since they were not deemed useful for the goals of a given project. Others will be complete, but for the first time functionally unclassified. Although efforts are under way to ensure the central deposition of all structural genomics results, many of these data might not be available centrally from the PDB, given the expected lack of annotation or their level of incompleteness. This situation will likely change, but it may be that the user will need to visit multiple sources of structure information for a complete coverage of all available macromolecular structures. The promise of what could come is given in part by the structural genomics target registration database maintained by the RCSB at http://targetdb.rcsb.org/. This database currently contains 14,000 entries, some of which will be solved and will further enrich the large variety of databases of derived information described in this chapter. While resource maintainers are faced with new challenges to judge and automatically handle the quality of the structure information available as well as to deal with the sheer amount of it, users will shortly have even richer information resources available from which to study structure function relationships. The fact that these resources already greatly increase our understanding of biological systems is a testament not only to those individuals who produce the primary structure data, but to all those who have developed and maintained the resources described herein.

## REFERENCES

Allen FH, Kennard O (1993): 3D Search and research using the Cambridge structural database. *Chem Des Autom News* 8:31–7.

Bader GD, Donaldson I, Wolting C, Ouellette BF, Pawson T, Hogue CW (2001): BIND—the biomolecular interaction network database. *Nucleic Acids Res* 29(1):242–5.

Berman HM, Westbrook J, Feng Z, Gilliland G, Bhat TN, Weissig H, Shindyalov IN, Bourne PE (2000): The Protein Data Bank. *Nucleic Acids Res* 28:235–42. [The standard reference for the PDB since it has been managed by the Research Collaboratory for Structural Bioinformatics.]

Bernstein FC, Koetzle TF, Williams GJ, Meyer EE, Brice MD, Rodgers JR, Kennard O, Shimanouchi T, Tasumi M (1977): Protein Data Bank: a computer-based archival file for macromolecular structures. *J Mol Biol* 112:535–42. [The original PDB reference.]

Bernstein H, Bernstein F, Bourne PE (1998): CIF Applications VIII. pdb2cif: translating PDB entries into mmCIF format. *J Applied Crystallogr* 31:282–95.

Bourne PE (1999): Editorial. *Bioinformatics* 15:715–6.

Brenner SE, Koehl P, Levitt M (2000): The ASTRAL Compendium for Protein Structure and Sequence Analysis. *Nucleic Acids Res* 28:254–6.

Burley SK, Almo SC, Bonanno JB, Capel M, Chance MR, Gaasterland T, Lin D, Sali A, Studier FW, Swaminathan S (1999): Structural genomics: beyond the human genome project. *Nat Genet* 23:151–7. [Original and highly cited article defining structural genomics.]

Castagnetto JM, Hennessy SW, Roberts VA, Getzoff ED, Tainer JA, Pique ME (2002): MDB: The Metalloprotein Database and Browser at the Scripps Research Institute. *Nucleic Acids Res* 30:379–82.

Dietmann S, Holm L (2001): Identification of homology in protein structure classification. *Nat Struct Biol* 8:953–7.

Gerstein M, Krebs W (1998): A database of macromolecular motions. *Nucleic Acids Res* 26:4280–90.

Gerstein M, Levitt M (1998): Comprehensive assessment of automatic structural alignment against a manual standard, the Scop classification of proteins. *Protein Sci* 7(2):445–56.

Gibrat J-F, Madej T, Bryant SH (1996): Surprising similarities in structure comparison. *Curr Opin Struct Biol* 6:377–85.

Gilliland GL, Tung M, Ladner J (1996): The Biological Macromolecule Crystallization Database and NASA Protein Crystal Growth Archive. *J Res Natl Inst Stand Technol* 101:309–20.

Guex N, Peitsch MC (1997): SWISS-MODEL and the Swiss-PDBViewer: an environment for comparative protein modeling. *Electrophoresis* 18:2714–23.

Hall SR, Allen FH, Brown ID (1991): The crystallographic information file (CIF): a new standard archive file for crystallography. *Acta Crystallogr* A47:655–85.

Hendlich M (1998): Databases for protein-ligand complexes. *Acta Crystallogr* D54:1178–82.

Hendrick K, Thornton JM (1998): PQS: A protein quaternary structure file server. *Trends Biochem Sci* 23:358–61.

Hennessy D, Buchanan B, Subramanian D, Wilkosz PA, Rosenberg JM (2000): Statistical methods for the objective design of screening procedures for macromolecular crystallization. *Acta Crystallogr* D56:817–27.

Hobohm U, Sander C (1994): Enlarged representative set of protein structures, *Protein Sci* 3:522–4. [A widely used set of structures in bioinformatics based on non-redundancy of sequence.]

Hooft RW, Vriend G, Sander C, Abola EE (1996): Errors in protein structures. *Nature* 381(6580):272.

Hutchinson EG, Thornton JM (1996): PROMOTIF—a program to identify and analyze structural motifs in proteins. *Protein Sci* 2:212–20.

Ihlenfeldt WD, Voigt JH, Bienfait B, Oellien F, Nicklaus MC (2002): Enhanced CACTVS browser of the open NCI database. *J Chem Inf Comput Sci* 42(1):46–57.

Jones S, Thornton JM (1996): Principles of protein–protein interactions. *Proc Natl Acad Sci USA* 93:13–20.

Jones S, van Heyningen P, Berman HM, Thornton JM (1999): Protein–DNA interactions: a structural analysis. *J Mol Biol* 287(5):877–96.

Kabsch W, Sander C (1983): Dictionary of protein secondary structure: pattern recognition of hydrogen-bonded and geometrical features. *Biopolymers* 22:2577–637.

Karp PD, Paley SM (1994): Representations of metabolic knowledge: pathways. *Proc Intel Sys Mol Biol* 2:203–311.

Krebs WG, Gerstein M (2000): The morph server: a standardized system for analyzing and visualizing macromolecular motions in a database framework. *Nucleic Acids Res* 28:1665–75.

Laskowski RA, MacArthur MW, Moss DS, Thornton JM (1993). PROCHECK: a program to check the stereochemical quality of protein structures. *J Applied Crystallogr* 26:283–91.

Madej T, Gibrat JF, Bryant SH (1995): Surprising similarities in structure comparison. *Curr Op Struct Biol* 6:377–85.

Murzin AG, Brenner SE, Hubbard T, Chothia C (1995): SCOP: a structural classification of proteins database for the investigation of sequences and structures. *J Mol Biol* 247:536–40. [Original paper describing the most widely used protein structure classification scheme.]

Ohkawa H, Ostell J, Bryant S (1995): MMDB: an ASN.1 specification for macromolecular structure. *Proc Intel Sys Mol Biol* 3:259–67.

Orengo CA, Michie AD, Jones S, Jones DT, Swindells MB, Thornton JM (1997): CATH—a hierarchic classification of protein domain structures. *Structure* 5(8):1093–108.

Pearson WR (1990): Rapid and sensitive sequence comparison with FASTP and FASTA. *Methods Enzymol* 183:63–98.

Qian J, Stenger B, Wilson CA, Lin J, Jansen R, Teichmann SA, Park J, Krebs WG, Yu H, Alexandrov V, Echols N, Gerstein M (2001): PartsList: a Web-based system for dynamically ranking protein folds based on disparate attributes, including whole-genome expression and interaction information. *Nucleic Acids Res* 29:1750–64.

Reddy VS, Natarajan P, Okerberg B, Li K, Damodaran KV, Morton RT, Brooks CL, Johnson JE (2001): Virus particle explorer (VIPER), a Website for virus capsid structures and their computational analyses. *J Virol* 24:11943–7.

Reichert J, Sühnel J (2002): The IMB Jena Image Library of Biological Macromolecules: 2002 update. *Nucleic Acids Res* 30:253–4.

Sayle RA, Milner-White EJ (1995): RASMOL: Biomolecular Graphics for All. *Trends Biochem Sci* 20(9):374.

Shindyalov IN, Bourne PE (1998): Protein structure alignment by incremental combinatorial extension (CE) of the optimal path. *Protein Eng* 11(9):739–47.

Siani MA, Weininger D, Blaney JM (1994): CHUCKLES: a method for representing and searching peptide and peptoid sequences on both monomer and atomic levels. *J Chem Inf Comput Sci* 34(3):588–93.

Siddiqui AS, Barton GJ (1995): Continuous and discontinuous domains: an algorithm for the automatic generation of reliable protein domain definitions. *Protein Sci* 4:872–84.

Smith CM, Shindyalov IN, Veretnik S, Gribskov M, Taylor SS, Ten Eyck LF, Bourne PE (1997): The protein kinase resource. *Trends Biochem Sci* 22:444–6.

Sowdhamini R, Burke DF, Huang J-F, Mizuguchi K, Nagarajaram HA, Srinivasan N, Steward RE, Blundell. TL (1998): CAMPASS: a database of structurally aligned protein superfamilies. *Structure* 6(9):1087–94.

Vaguine AA, Richelle J, Wodak SJ (1999): SFCHECK: a unified set of procedures for evaluating the quality of macromolecular structure factor data and their agreement with the atomic model. *Acta Crystallogr* D55(1):191–205.

Venclovas Č, Zemla A, Fidelis K, Moult J (2001): Comparison of performance in successive CASP experiments. *Proteins* 45(Sup. 5):163–70.

Vondrasek J, van, Buskirk CP, Wlodawer A (1997): Database of three-dimensional structures of HIV proteinases. *Nat Struct Biol* 4:8.

Williams N (1997): Bioinformatics: how to get databases talking the same language. *Science* 275:301–2.

Xenarios I, Salwinski L, Duan XJ, Higney P, Kim S, Eisenberg D (2002): DIP, the database of interacting proteins: a research tool for studying cellular networks of protein interactions. *Nucleic Acids Res* 30:303–5.

# Section III

## COMPARATIVE FEATURES

Section III

COMPARATIVE HEALTH CARE

# 12

# PROTEIN STRUCTURE EVOLUTION AND THE SCOP DATABASE

Boojala V. B. Reddy and Philip E. Bourne

The structure of a protein can reveal its function and its evolutionary history. Extracting this information requires knowledge of the structure and its relationships with other proteins. The structure and its relationships require a general knowledge of the folds that proteins adopt and detailed information about the structure of many proteins. Nearly all proteins have structural similarities with other proteins and, in many cases, share a common evolutionary origin. The knowledge of these relationships makes important contributions to structural bioinformatics and other related areas of science. Further, these relationships will play an important role in the interpretation of sequences produced by genome projects. To facilitate the understanding and access to available information for known protein structures, Murzin and co-workers (1995) have constructed a Structural Classification of Proteins (SCOP) database. The SCOP database is based on evolutionary relationships and on the principles that govern their three-dimensional structure. It provides for each entry links to coordinates, images of the structure, interactive viewers, sequence data, and literature references. The database is freely accessible on the World Wide Web (http://scop.mrc-lmb.cam.ac.uk/scop). To understand the rationale behind SCOP, we begin with a discussion of protein evolution from a sequence, structure, and functional perspective.

## THE EVOLUTION OF PROTEINS

Proteins that have descended from the same ancestor retain memory of that ancestor through the sequence, structure, and function. Murzin (1998) discussed various aspects of the evolution of sequence, structure, fold, and function with specific examples.

*Structural Bioinformatics*
Edited by Philip E. Bourne and Helge Weissig
Copyright © 2003 by Wiley-Liss, Inc.

Strong sequence similarity alone is considered to be sufficient evidence for common ancestry. Close structural and functional similarity together is also accepted as sufficient evidence for distant homology between proteins that lack significant sequence similarity. But neither structural nor functional similarity alone is considered to be strong evidence. Proteins of independent origin may well have similar structures due to physicochemical reasons, and they may also evolve similar functions due to functional selection referred to as divergent-evolved structures. Although in theory descendents from the same ancestor may have different functions and even different structures, they would be very difficult to detect.

## The Evolution of Fold

It is generally accepted that in distantly related proteins, structure is more conserved than sequence. Proteins that have diverged beyond detectable sequence similarity still retain the architecture and topology of their ancestral fold. The reasons for this structural conservation are not completely understood. In principle, a protein chain can have more than one stable fold. One theory states that protein evolution passed through a stage in which selective pressure from the physical demands on the protein chain required that it have just one stable and fast-folding conformation. Another theory states that half way toward the convergence of all protein structures into a set of small, mostly stable protein folds, this stage was aborted and replaced by the present state of affairs whereby the most selective pressure comes from functional constraints. Evolution requires that existing proteins continue to function, which would be interrupted if the protein's fold changed significantly. Notwithstanding, a protein fold can change in a minor way during evolution by keeping all the secondary structure elements the same, packed and connected exactly the same way, but some of the connecting loops may be organized differently. The structural similarity between seemingly unrelated proteins are often explained by *convergence* to a stable fold as opposed to divergence from a common ancestor. The convergence implies not only that given proteins are of independent origin but also that they had different original folds. With no evidence for their difference in original fold, they are referred as having undergone *parallel* evolution.

## The Evolution of Enzymatic Catalysis

Proven cases of distantly related enzymes with very different functions are very rare and therefore are of great value for the understanding of the origin of enzymatic activity. The precision and complexity of the active sites of modern enzymes did not happen by chance but evolved from primitive catalytic features of ancestral proteins. Primitive enzymes were probably less efficient, but they also could have had a broader range of activities. If this were the case, the devolution of these activities would allow improvements in both efficiency and specificity through customization of the active-site architectures with additional specificity-determining groups. Within the active site, the original catalytic features are likely to remain conserved and can be revealed by structural comparison. When a protein evolves to a new function, the protein fold can change before freezing again, having developed new functional constraints. A simple way of altering a protein fold without significant destabilization is to change its topology while maintaining its architecture. This can be done by the internal swapping of similar $\alpha$-helices and $\beta$-strands or by reversing the direction of some of its secondary structures. There are many proteins with similar secondary structures and architectures but different topologies that could be related in such a way; for example, an immunoglobulin domain and plastocyanin or an SH3 domain and GroES.

## The Comparison of Structures

In current methods for structure comparison the significance of structural similarity is measured by a score derived from the number of residues in the common structural core and their root mean square deviation. These values reflect the arrangement of regular secondary structures, whose packing and topology are determined by stereochemical rules rather than by evolutionary constraints. The common ancestry may manifest itself in more subtle features, such as the conservation of rare and unusual topological and packing details and other structural irregularities that are less likely to occur independently. Within the common structural core, there may also be conserved turns, $\alpha$-helical caps, $\beta$-bulges, and other small structural elements.

Early work on protein structures showed that there are striking regularities in the way in which secondary structures are assembled (Levitt and Chothia, 1976; Chothia, Levitt, and Rachardson, 1977) and in the topologies of the polypeptide chains (Richardson, 1976, 1977; Sternberg and Thornton, 1976). These regularities arise from the intrinsic physical and chemical properties of proteins (Chothia, 1984; Finkelstein and Ptitsyn, 1987) and provide the basis for the classification of protein folds (Levitt and Chothia, 1976; Richardson, 1981). The structure comparison methods (Holm and Sander, 1993; Shindyalov and Bourne, 1998) and structure classification (Orengo et al., 1993; Overington et al., 1993; Yee and Dill, 1993) provide a further systematic relationship among protein structures. Resources are now available for recognition of the relationships between protein structures and several are discussed in this book. The SCOP database hierarchically organizes proteins according to their structures and evolutionary origin (Murzin et al., 1995; Conte et al., 2000). The database forms a resource that allows researchers to study the nature of protein folds, to focus their investigation, and to rely on expert-defined relationships. SCOP is the most cited resource for classifying proteins. In the context of structural bioinformatics it provides a reductionism that facilitates many of the studies outlined in this book.

## SCOP HIERARCHY

The method used to construct the protein classification in SCOP is the visual inspection and comparison of structures first compared using automatic procedures. The SCOP database is organized on a number of hierarchical levels, with the principle ones being family, superfamily, fold, and class. Within this hierarchy, the unit of categorization is the protein domain since domains are typically the units of protein evolution, structure, and function. Small- and medium-sized proteins usually have a single domain and are treated as such. The domains in large proteins are classified individually. Thus, different regions of a single protein may appear in multiple places in the SCOP hierarchy under different folds or, in the case of repeated domains, several times under the same fold.

In SCOP, *families* contain protein domains that share a clear common evolutionary origin, as evidenced by sequence identity or extremely similar structure and function. *Superfamilies* consist of families whose proteins share very common structure and function, and therefore there is reason to believe that the different families are evolutionarily related. *Folds* consist of one or more superfamilies that share a common

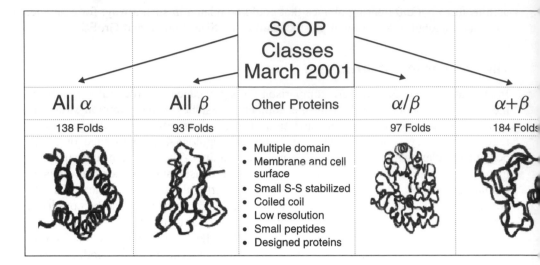

| All $\alpha$ | All $\beta$ | Other Proteins | $\alpha/\beta$ | $\alpha+\beta$ |
|---|---|---|---|---|
| 138 Folds | 93 Folds | | 97 Folds | 184 Folds |

SCOP Classes March 2001

Other Proteins:
- Multiple domain
- Membrane and cell surface
- Small S-S stabilized
- Coiled coil
- Low resolution
- Small peptides
- Designed proteins

Figure 12.1. Major SCOP classes and their fold content as of March 2001.

core structure (i.e., same secondary structure elements in the same arrangement with the same topological connections). Finally, depending on the type and organization of secondary structural elements, folds are grouped into four major *classes*. In addition, there are several other classes of proteins that are atypical and therefore difficult to classify (Fig. 12.1).

The following explanation, derived from the work of the authors of the SCOP database (Brenner et al., 1996), describes how the protein structures at each of the different levels have been classified.

## Classes

Initially a protein structure is classified into domains. A domain is a region of the protein that has its own hydrophobic core and has relatively little interaction with the rest of the protein, making it is structurally independent. Typically, domains are collinear in sequence, which aids in their identification, but occasionally one domain will involve two or more regions of sequence from one or more polypeptide chains that are not collinear. Hence, in some cases, automated, or for that matter, manual identification of domains by an expert is not straightforward (Chapter 18).

Assuming a domain structure, placing it in the appropriate class is a straightforward task. It should be readily apparent whether a domain consists exclusively of $\alpha$-helices, $\beta$-sheets, or some mixture thereof. It is possible that an all-$\beta$ protein can have small adornments of $\alpha$ or $3_{10}$ helix. Similarly, all-$\alpha$ structures may actually have several regions of $3_{10}$ helix, and in rare cases, small $\beta$-sheets outside the $\alpha$-helical core.

Domains with a mixture of helix and sheet structures are divided into two classes, $\alpha/\beta$ (alpha and beta), and $\alpha + \beta$ (alpha plus beta). The $\alpha/\beta$ domains consist principally of a single $\beta$-sheet, with $\alpha$-helices joining the C-terminus of one strand to the N-terminus of the next. Commonly in $\alpha/\beta$ proteins we see two subclasses: in one subclass, the $\beta$-sheet is wrapped to form a barrel surrounded by $\alpha$-helices; in the other, the central sheet is more planar and is flanked on either side by the helices. Domains that have the $\alpha$ and $\beta$ units largely separated in sequence fall into the $\alpha + \beta$ class.

The strands in these structures do not have the intervening helices; they are typically joined by hairpins, leading to antiparallel sheets such as are found in all-$\beta$ class folds. However, $\alpha + \beta$ structures may have one, and often a small cluster, of helices packing tightly and integrally against the sheet.

In addition to the four classes of globular protein structures, SCOP contains a few other classes, namely: multidomain proteins, membrane and cell surface proteins, small proteins, coiled coil proteins, low-resolution structures, peptides, and designed proteins (SCOP 1.55 release, March 2001). Multiple-domain proteins have different domains placed in different classes. However, the different domains of these proteins have never been seen independently of each other. The *small protein class* has structures stabilized by disulfide bridges or by metal ligands rather than by hydrophobic cores. Membrane proteins frequently have unique structures because of their unusual environment and are therefore placed in a separate class.

## Folds

Identification of the fold of a protein is the most difficult stage of classification. If the proteins have the same major secondary structures in the same arrangement with the same topological connections they are classified as one fold. A short description of the major structural features is used as the name for the fold. Different proteins with the same fold usually have peripheral elements of secondary structure and turn regions that differ in size and conformation and, in the more divergent cases, these differing regions may form half or more of each structure. For proteins placed together in the same fold category, the structural similarity probably arises from the physics and chemistry of proteins favoring certain packing arrangements and chain topologies. There may be cases where a common evolutionary origin is obscured by the extent of the divergence in sequence, structure, and function. In such cases it is possible that the discovery of new structures, with folds between those of the previously known structures, will make clear their common evolutionary relationship.

As of SCOP release 1.55 (March 2001) there are 138 classified folds in all $\alpha$ proteins, 93 folds in all $\beta$ proteins, 97 folds in $\alpha/\beta$ and 184 folds in $\alpha + \beta$ classes (Figure 12.1, Table 12.1). The best way to characterize fold is to look first at the major architectural features and then identify the more subtle characteristics.

TABLE 12.1. SCOP 1.55 Release Has 13,220 Protein Data Bank entries (March 2001), 31,474 Domains (Excluding Nucleic Acids and Theoretical Models)

| Class | Folds | Superfamilies | Families |
|---|---|---|---|
| All alpha proteins | 138 | 224 | 337 |
| All beta proteins | 93 | 171 | 276 |
| $\alpha/\beta$ proteins | 97 | 167 | 374 |
| $\alpha + \beta$ proteins | 184 | 263 | 391 |
| Multidomain proteins | 23 | 28 | 35 |
| Membrane and cell surface proteins | 11 | 17 | 28 |
| Small proteins | 54 | 77 | 116 |
| Total | 605 | 947 | 1557 |

## Superfamilies

Protein structures classified in the same superfamily are probably related evolutionarily and therefore they must share a common fold and usually perform similar functions. If the functional relationship is sufficiently strong, for example, the conserved interaction with substrate or cofactor molecules, the shared fold can be relatively small, provided it includes the active site(s). Proteins from the superfamily may have low sequence identity.

## Families

Proteins are clustered together into families on the basis of one of two criteria that imply their having a common evolutionary origin. First, all proteins that have residue identities of 30% and greater; second, proteins with lower sequence identity but whose functions and structures are very similar. A small number of SCOP families embrace a relationship that is above the standard family definition but below the superfamily level. It is suggested that proteins that have a similar domain organization and share a common fold in the catalytic domain, such as dihydrodipicolinate reductase and the glyceraldehyde-3-phosphate and glucose-6-phosphate dehydrogenases, are likely to be more closely related than those sharing a common fold in their coenzyme domain only.

Sequence comparison is a simple and reliable way of learning about the structural and evolutionary relationships of proteins. If a sequence has 30% identity to a protein of known structure, then an outline of its fold can be reliably deduced. If there is significant similarity between a sequence and a protein in SCOP, then that sequence can be put into the appropriate family, which then defines its superfamily, fold, and class.

The major limitation of sequence comparison is that it fails to identify many of the structural relationships in SCOP either because the sequence relationship has become too weak (for evolutionarily related proteins) or never existed (for evolutionarily unrelated proteins with similar folds). Structure-structure comparison programs use various methods to recognize similar arrangements of atomic coordinates and thus identify domains of similar structure. Although these methods lack complete accuracy, they can be used to suggest a shared fold between proteins of interest and others in SCOP. Manual inspection must then be used to verify the choice of fold and to select the appropriate superfamily. The selection of superfamily is the most challenging step of protein classification, for it ascribes a biological interpretation to chemical and physical data. Therefore, the assignment of all proteins of known structure to evolutionarily related superfamilies is perhaps the single most powerful and important feature of the SCOP database.

## ORGANIZATION AND CAPABILITIES OF THE SCOP RESOURCE

The SCOP database was originally created as a tool for understanding protein evolution through sequence–structure relationships and determining if new sequences and new structures are related to previously known protein structures. On a more general level, the highest levels of classification provide an overview of the diversity of protein structures. The specific lower levels are helpful for comparing individual structures with their evolutionary and structurally related counterparts.

TABLE 12.2. SCOP Mirror Sites at Different Locations Around the World

| Location | Site | URL |
|---|---|---|
| UK | SCOP home server in Cambridge | http://scop.mrc-lmb.cam.ac.uk/scop/ |
| USA | University of California, Berkeley | http://scop.berkeley.edu/ |
| Australia | Walter and Eliza Hall Institute Australian National Genomics Information Service (ANGIS) | http://scop.wehi.edu.au/scop/ |
| Japan | Biomolecular Engineering Research Institute (BERI) | http://www.beri.co.jp/scop/ |
| Taiwan | National Tsing Hua University | http://scop.life.nthu.edu.tw/ |
| Singapore | National University of Singapore | |
| India | Madurai Kamaraj University | http://gene.tn.nic.in/scop/ |
| | Indian Institute of Science | http://scop.physics.iisc.ernet.in/scop/ |
| | Center for DNA Finger Printing and Diagnostics (CDFD) | http://www.cdfd.org.in:5555/scop/ |
| | University of Pune | |
| Russia | Institute of Protein Research | http://scop.protres.ru/ |
| Israel | Wizmann Institute | http://pdb.weizmann.ac.il/scop/ |
| China | Peking University | http://mdl.ipc.pku.edu.cn/scop/ |
| Italy | Center for Biomedical Engineering, Politecnico of Turin | http://loki.polito.it/scop/ |

The SCOP database is available as a set of tightly coupled hypertext pages on the Web and can be accessed at the URL http://scop.mrc-lmb.cam.ac.uk/scop/

For rapid and effective access to SCOP, a number of mirrors have been established (Table 12.2). The facilities at various sites may differ with some sites providing sequence similarities and other sites providing sequence and structure based phylogenic relationships (Sujatha, Balaji, and Srinivasan, 2001). SCOP can be used for detailed searching of particular families and browsing of the whole database with a variety of techniques for navigation. Easy access to data and images makes SCOP a powerful general-purpose interface, providing a level of classification not present in the Protein Data Bank (PDB).

## Browsing Through the SCOP Hierarchy

SCOP is organized as a tree structure. Entering at the top of the hierarchy, the user can navigate through the levels of Class, Fold, Superfamily, Family, and Species to the leaves of the tree, which are the structural domains of individual PDB entries. The sequence similarity search facility allows any sequence of interest to be searched against databases of protein sequences classified in SCOP using the algorithms BLAST, FASTA, or SSEARCH. SCOP can be entered from the list of PDB chains found to be similar and the similarity can be displayed visually. The keyword search facility returns a list of SCOP pages containing the word entered or combinations of words separated by Boolean operators. Pages are provided that order folds, superfamilies, and families by date of entry into PDB.

In addition to the information on structural and evolutionary relationships contained within SCOP, each entry has links to images of the structure, interactive molecular viewers, the atomic co-ordinates, data on functional conformational changes, sequence data, homologs, and MEDLINE abstracts.

## SCOP Use

SCOP has broad utility with a wide range of users. Experimental structural biologists may wish to explore the region of *structure space* near their protein of current research. Molecular biologists may find the classification helpful because the categorization assists in locating proteins of interest and the links make exploration easy. As such, two important observations are made from this protein structure classification. First, strikingly skewed distributions occur at all levels relative to what exists in nature, based on analysis of complete genomes. This probably reflects the experimentalists' bias toward particular proteins and protein families, as well as to the bias of nature toward certain protein superfamilies and folds. Second, a retrospective analysis of the growth of structural data gives an estimate of the total numbers of protein folds and superfamilies that exist in nature and has shown them to be very limited indeed. It seems that we may have already seen the majority of folds and have determined at least one structure for some half of all superfamilies.

## SCOP FROM A USER'S PERSPECTIVE

At the time of writing, the SCOP database has received over 700 direct citations since it was made available in 1995 (Murzin et al., 1995). Here we summarize some areas of protein structural bioinformatics where the SCOP database has been used extensively. References cited in this paragraph are given in Table 12.3 broadly indicate how SCOP has been used in recent years. First, SCOP classified groups of proteins were used as a reference set of data to develop several automatic classification methods used in analyzing families, superfamilies, and folds. These classifications were then extensively used for integrative structural data mining to develop predictive methods and structure-comparison tools (Pasquier et al., 2001; Bertone and Gerstein, 2001; Stambuk and Konjevoda, 2001; Bukhman and Skolnick, 2001; Przytycka et al., 1999; Torshin, 2001; Lackner et al., 2000; Chou and Maggiora, 1998) Second, SCOP-classified proteins were extensively used in understanding evolution of protein enzymatic functions (Babbitt and Gerlt, 2001; Powlowski and Godzik, 2001; Todd et al., 1999; 2001; Konin et al., 1998; Murzin, 1998), evolutionary change of protein folds (Grishin, 2001; Lupas et al., 2001; Zhang and DeLisi, 2001; Thronton et al., 1999), and hierarchical structural evolution (Dokholyan and Shakhnovich, 2001; Paoli, 2001). Third, The SCOP classification of proteins at the superfamily and fold levels were used to study distantly related proteins with the same fold (Grigoriev et al., 2001; Teichmann et al., 2001b; Thornton, 2001; Koehl, 2001). Fourth, SCOP is used to study sequence and structure variability and its dependence in homologous proteins (D'Alfonso et al., 2001; Balaji and Srinivasan, 2001). Fifth, SCOP families are used to derive amino acid similarity matrices and substitution tables useful for sequence comparison and fold recognition studies (Dosztanyi and Torda, 2001; Shi et al., 2001). Sixth, SCOP is helpful in studying the structural anatomy of folds and domains, to extract structural principles for use in protein design experiments (Teichmann et al., 2001a; Helling et al., 2001; Taylor et al., 2001; Dengler, 2001). Seventh, SCOP domains have been used to study combinations of different domains and their decomposition in multidomain proteins

**T A B L E** 12.3. Partial List of References Where SCOP Classification Was Used as the Basis for Analysis

---

Apic G, et al. (2001) J Mol Biol 310:311–25
Babbitt PC, Gerlt JA. (2001) Adv Protein Chem 55:1–28
Balaji S, Srinivasan N. (2001) Protein Eng 14:219–26
Bertone P, Gerstein, M. (2001) IEEE Eng Med Mag 20:33–40
Bukhman YV, Skolnick J. (2001) Bioinformatics 17:468–78
Chou KC, Maggiora GM. (1998) Protein Eng 11:523–38
D'Alfonso G, et al. (2001) J Struct Biol 134:246–56
Dengler U, et al. (2001) Proteins 42:332–44
Dokholyan NV, Shakhnovich EI. (2001) J Mol Biol 312:289–307
Dosztanyi Z, Torda AE. (2001) Bioinformatics 17:686–99
Govindarajan S, et al. (1999) Proteins 35:408–14
Grigoriev IV, et al. (2001) Protein Eng 14:455–8
Grishin NV. (2001) J Struct Biol 134:167–85
Helling R, et al. (2001) J Mol Graphics & Modeling 19:157–67
Kinoshita K, et al. (1999) Protein Sci 8:1210–7
Koehl P. (2001) Curr Opin Struct Biol 11:348–53
Konin EV, et al. (1998) Curr Opin Struct Biol 8:355–63
Kuroda Y, et al. (2000) Protein Sci 9:2313–21
Lackner P, et al. (2000) Protein Eng 13:745–52
Lupas AN, et al. (2001) J Struct Biol 134:191–203
Mizuguchi K, et al. (1998) Protein Sci 7:2469–71
Murzin AG. (1998) Curr Opin Struct Biol 8:380–387
Paoli M. (2001) Prog Biophys Mol Biol 76:103–30
Pasquier C, et al. (2001) Proteins 44:361–9
Powlowski K, Godzik A. (2001) J Mol Biol 309:793–806
Przytycka T, et al. (1999) Nat Struct Biol 6:672–82
Shi JY, et al. (2001) J Mol Biol 310:243–57
Stambuk N, Konjevoda P. (2001) Int J Quantum Chem 84:13–22
Sowdhamini R, et al. (1998) Acta Crystallogr D 54:1168–77
Sujatha S, et al. (2001) Bioinformatics 17:375–6
Taylor WR, et al. (2001) Rep Prog Physics 64:517–90
Teichmann SA, et al. (2001a) J Mol Biol 311:693–708
Teichmann SA, et al. (2001b) Curr Opin Struct Biol 11:354–63
Thronton M, et al. (1999) J Mol Biol 293:333–42
Thornton M. (2001) Science 292:2095
Todd AE, et al. (2001) J Mol Biol 307:1113–43
Torshin IY. (2001) Frontiers Biosci 6:A1–A12
Vitkup D, et al. (2001) Nat Struct Biol 8:559–66
Wolf YI, et al. (2000) J Mol Biol 299:897–905
Xu Y, et al. (2000) Bioinformatics 16:1091–104
Zhang C, DeLisi C. (2001) Cell Molec Life Sci 58:72–9

---

(Apic et al., 2001; Xu et al., 2000; Kinoshita et al., 1999; Chou and Maggiora, 1998). Eighth, Recent structural genome projects have been using SCOP extensively in identifying new targets and for estimating the total number of protein folds (Vitkup et al., 2001; Kuroda et al., 2000; Wolf et al., 2000; Govindarajan et al., 1999). Ninth, SCOP families have been used for developing value-added and more specialized databases (Sujatha et al., 2001; Sowdhamini et al., 1998; Mizuguchi et al., 1998).

From this brief synopsis it should be apparent that we owe the SCOP authors a debt of gratitude for providing a resource that has had great impact on the field of structural bioinformatics.

## REFERENCES

Brenner SE, Chothia C, Hubbard TJP, Murzin AG (1996): Understanding protein structure: using SCOP for fold interpretation. *Methods Enzymol* 266:635–43.

Chothia C (1984): Principles that determine the structure of proteins. *Ann Rev Biochem* 53:537–72.

Chothia C, Levitt M, Rachardson D (1977): Structure of proteins: packing of α-helices and β-sheets. *Proc Nat Acad Sci USA* 74:4130–4.

Conte LL, Ailey B, Hubbard TJP, Brenner SE, Murzin AG, Chothia C (2000): SCOP: a structural classification of proteins database. *Nucleic Acids Res* 28:257–9.

Finkelstein AV, Ptitsyn OB (1987): Why do globular proteins fit the limited set of folding patterns? *Prog Biophys Mol Biol* 50:171–90.

Holm L, Sander C (1993): Protein structure comparison by alignment of distance matrices. *J Mol Biol* 233:123–38.

Hubbard TJP, Murzin AG, Brenner SE, Chothia C (1997): SCOP: a structural classification of protein database. *Nucleic Acids Res* 27:254–6.

Hubbard TJP, Ailey B, Brenner SE, Murzin AG, Chothia C (1999): SCOP: a structural classification of protein database. *Nucleic Acids Res* 25:236–9.

Levitt M, Chothia C (1976): Structural patterns in globular proteins. *Nature* 261:552–8.

Murzin AG (1998): How far divergent evolution goes in proteins. *Curr Opin Struct Biol* 8:380–7.

Murzin AG, Brenner SE, Hubbard TJP, Chothia C (1995): SCOP: a structural classification of proteins database for the investigation of sequences and structures. *J Mol Biol* 247:536–40. [The original SCOP paper from which all other references to SCOP are derived.]

Orengo C, Flores TP, Taylor WR, Thornton JM (1993): Identifying and classifying protein fold families. *Protein Eng* 6:485–500.

Overington JP, Zhu ZY, Sali A, Johnson MS, Sowdhamini R, Louie C, Blundell TL (1993): Molecular recognition in protein families: a database of three-dimensional structures of related proteins. *Biochem Soc Trans* 21:597–604.

Richardson JS (1976): Handedness of crossover connections in beta sheets. *Proc Natl Acad Sci USA* 73(8):2619–23.

Richardson JS (1977): Beta-sheet topology and the relatedness of proteins. *Nature* 268:495–500.

Richardson JS (1981): The anatomy and taxonomy of protein structure. *Adv Protein Chem* 34:167–339.

Sternberg MJ, Thornton JM (1976): On the conformation of proteins: the handedness of the beta-strand-alpha-helix-beta-strand unit. *J Mol Biol* 105:367–82.

Sujatha S, Balaji S, Srinivasan N (2001): PALI: A database of structural alignments and phylogeny of homologous protein structures. *Protein Eng* 17:375–6.

Shindyalov IN, Bourne PE (1998): Protein structure alignment by incremental combinatorial extension (CE) of the optimal path. *Protein Eng* 11:739–47.

Yee DP, Dill KA (1993): Families and the structural relatedness among globular proteins. *Protein Sci* 2:884–99.

# 13

# THE CATH DOMAIN STRUCTURE DATABASE

C. A. Orengo, F. M. G. Pearl, and J. M. Thornton

During evolution protein sequences change due to mutations in their residues and insertions and deletions of residues. These changes give rise to families of related proteins and the earliest protein family resources, based solely on sequence data, were first established in the 1970s by the pioneering work of Dayhoff. Since then many sequence databases have been established and relationships are often detected using alignment methods based on powerful dynamic programming algorithms adapted from the realm of computer science. Such methods handle the residue insertions and deletions occurring between distant evolutionary relatives very efficiently.

The structural data has always been more sparse than the sequence data due to the technical challenges of structure determination. There is currently over two orders of magnitude discrepancy between the sequence and structure resources. Thus, while the Protein Data Bank (PDB) contains about 16,000 structural entries, the nucleotide sequence databank at the National Centre for Biotechnology Information (NCBI) (GenBank) contains over 12 million entries.

Therefore, although the first crystal structures were solved in the early 1970s, it was not until the mid-1990s that structural classifications began to emerge, primarily with Structural Classification of Proteins (SCOP) (Murzin et al., 1995; Lo Conte, 2000), DALI (Holm and Sander, 1996), and CATH (Orengo et al., 1997; Pearl et al., 2001) databases and data resources (see Table 13.2). Several other classifications have arisen since (see, for example, DDBASE (Sowdhamini et al., 1998), 3Dee (Dengler, Siddiqui, and Barton, 2001), DaliDD (Holm and Sander, 1998; Dietmann and Holm, 2001), reviewed in Holm and Sander, 1994b, and Orengo, 1994. These databases use a variety of different algorithms for comparing three-dimensional (3D) structures (see Chapter 16). They also differ in methods for measuring similarity between the

*Structural Bioinformatics*
Edited by Philip E. Bourne and Helge Weissig
Copyright © 2003 by Wiley-Liss, Inc.

structures and for clustering them into fold groups or protein superfamilies and families. However, comparisons between three of the largest classifications (SCOP, DALI, CATH) recently revealed a reasonable degree of correspondence (more than 80%) between protein families generated using different protocols (Hadley and Jones, 1999).

Since structure is much more highly conserved than sequence during evolution, the discovery of structural alignment algorithms and the development of structural classifications have made a significant contribution to the understanding of evolutionary mechanisms as they have enabled much more distant evolutionary relatives to be identified. Furthermore, knowledge of a protein's structure can provide important clues to the functional mechanism and biological role of the protein, for example, protein–substrate and protein–protein interactions (see Chapter 20). Because a large proportion of the structural core of the protein (often more than 50%) is conserved even in very distant relatives, structure alignments are much more accurate than sequence alignments and this situation improves the identification of conserved structural features or sequence motifs, which are often associated with protein function.

The largest structure classifications (SCOP, CATH) currently contain between 950–1400 protein superfamilies. However, these superfamilies currently map to nearly one-third of the nonredundant sequences in the GenBank sequence database (~25% on the basis of equivalent residues). Furthermore, current structure genomics initiatives, described in Chapter 29, will significantly increase the number of structures determined over the next 10 years. Current estimates predict that there will be between 30,000 and 100,000 new structures before 2010. Because of the manner in which proteins are being selected for structure determination, these new structures will be predominantly from protein families currently unrepresented in the structure classifications or very distantly related to known structural families. Therefore, we may soon have structural representatives for most of the major protein families and for those of particular medical and biological interest, although some classes of structures such as transmembrane proteins may remain difficult to determine. It is also likely that methods for detecting distant relatives will improve in parallel and as a consequence of the growth in the sequence and structure databases. Links between the sequence and structure databases will also be promoted. The InterPro initiative has integrated several sequence databases (Pfam, PRINTS, PROSITE, SWISS-PROT). There are also plans to integrate the structure databases SCOP and CATH and the European Macromolecular Structure Database (EMSD) with the sequence databases InterPro and Pfam.

Structural classifications will therefore play an increasingly important role as representatives from more of the protein families are structurally determined and the mapping between structural families and genomic sequences improves. These resources will provide key data for understanding function at the molecular level. In this chapter we describe the CATH structural classification, its development, and the methods used to update and search the resource. Analysis of the classification has revealed that some protein families are very highly populated, a finding that has important implications for understanding evolutionary mechanisms.

## HISTORICAL DEVELOPMENT

The CATH domain structure database was established in 1993 when fewer than 3000 protein structures had been determined. Nearly a decade later the database has expanded considerably and contains ~13,000 protein structure entries from the PDB comprising 33,000 structural domains. CATH also contains over 200,000 sequence domains extracted from GenBank entries and assigned to one of the 1200 CATH homologous superfamilies using profile-based approaches. Since the domain was considered to be an important evolutionary unit and also because structural prediction and homology modeling methods are often more successful on a domain basis, CATH was initially established as a domain-based database. However, sequence- and structure-based relationships are also determined between multidomain proteins, and CATH now contains families and superfamilies of multidomain proteins with links to their constituent domains.

Most publicly available structure classifications are derived using sequence-based and/or structure-based protocols. These range from the completely automated approaches of DALI and the DALI Domain Database (Holm and Sander, 1998) through to the largely manual approach used in compiling the SCOP database. In the CATH database semiautomated protocols are used for clustering structures both phonetically, that is, purely on the basis of structural similarity, and phylogenetically on the basis of apparent evolutionary relatedness. Any ambiguities in the assignments from automated protocols are validated manually and major bottlenecks in the classification correspond to the detection of domain boundaries and the verification of homologous relationships.

CATH is a hierarchical classification comprising four major levels (see Fig. 13.1). In fact CATH is an acronym for these levels: Class, Architecture, Topology, and Homology. At the top, the protein class is determined by the secondary structure composition

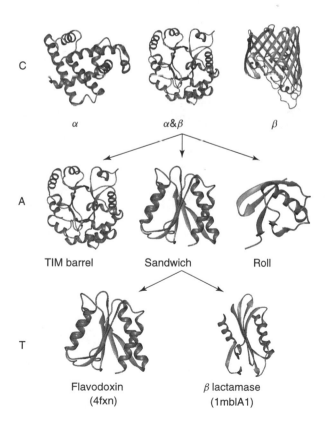

Figure 13.1. Schematic representation of the class, architecture, and topology/fold levels in the CATH database.

and packing using an automated approach. Architecture describes the orientation of the secondary structures in 3D space, regardless of their connectivity. For example, a large number of protein structures adopt alpha–beta barrel architectures, in which a central barrel of beta strands is enclosed within an outer barrel comprising a layer of alpha helices (see Figure 13.2). At the next level in the hierarchy, topology, both secondary

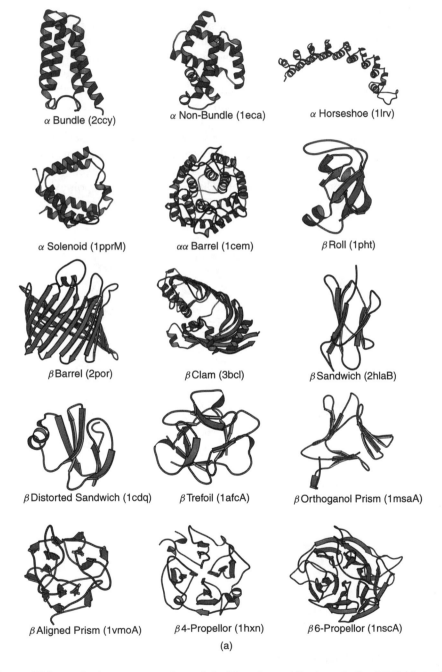

α Bundle (2ccy)            α Non-Bundle (1eca)            α Horseshoe (1lrv)

α Solenoid (1pprM)            αα Barrel (1cem)            β Roll (1pht)

β Barrel (2por)            β Clam (3bcl)            β Sandwich (2hlaB)

β Distorted Sandwich (1cdq)            β Trefoil (1afcA)            β Orthoganol Prism (1msaA)

β Aligned Prism (1vmoA)            β 4-Propellor (1hxn)            β 6-Propellor (1nscA)

(a)

Figure 13.2. Molscript representations of the 30 major architectures in the CATH hierarchy.

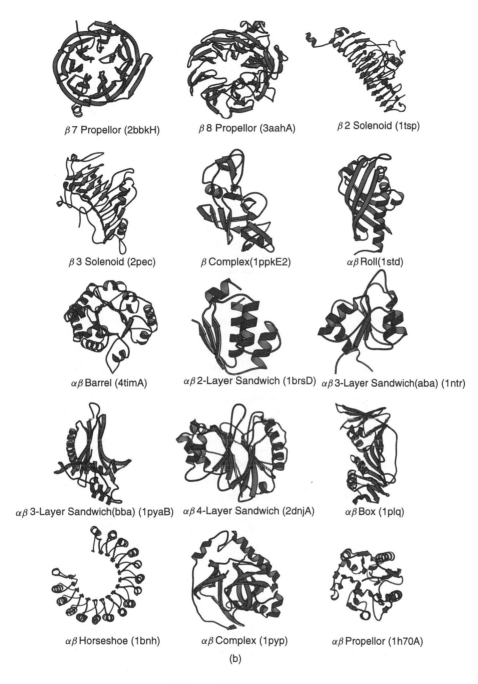

β 7 Propellor (2bbkH)   β 8 Propellor (3aahA)   β 2 Solenoid (1tsp)

β 3 Solenoid (2pec)   β Complex(1ppkE2)   αβ Roll(1std)

αβ Barrel (4timA)   αβ 2-Layer Sandwich (1brsD)   αβ 3-Layer Sandwich(aba) (1ntr)

αβ 3-Layer Sandwich(bba) (1pyaB)   αβ 4-Layer Sandwich (2dnjA)   αβ Box (1plq)

αβ Horseshoe (1bnh)   αβ Complex (1pyp)   αβ Propellor (1h70A)

(b)

Figure 13.2. (*Continued*)

structure orientation and connectivity between the secondary structures is taken into account in describing the fold of the protein. For the example shown in Figure 13.1, the three-layer alpha–beta sandwich architecture contains more than 100 different folds or topologies in which the secondary structures adopt a similar shape in 3D but the

TABLE 13.1. Description of the Levels in the Classification at the Architecture Level

| Primary Classification Number | Description of Level |
|---|---|
| 1 | Mainly $\alpha$ |
| 2 | Mainly $\beta$ |
| 3 | $\alpha\beta$ |
| 4 | Few secondary structures |
| 5 | Multidomain proteins |
| 6 | Single-domain proteins classified by sequence but not structure |
| 7 | Ambiguous multidomain proteins whose domain boundary assignment requires manual validation. Protein chains clustered at the sequence level |
| 8 | New proteins classified by sequence methods |
| 9 | Chains from multichain domains classified by sequence |

connectivities between them can differ considerably as shown by the schematic representations in the illustration. At the fourth and perhaps most biologically important level in the classification, homologous superfamily, proteins are grouped according to whether there is sufficient evidence (structural, sequence, and/or functional similarity) to support an evolutionary relationship. Within each homologous superfamily, proteins are clustered into sequence families at different levels of sequence identity (35%, 60%, 95%, 100%). More recently, protocols have been developed for identifying functional families within each superfamily.

There are currently three major classes within CATH, corresponding to mainly-alpha domains, mainly-beta domains, and alpha–beta domains. Other categories distinguished at the class level are multidomain proteins, domains comprising few secondary structures, and three groups corresponding to proteins at different stages in the classification and pending assignment to a particular fold group or superfamily (see Table 13.1). In the December 2001 release, CATH contained 36 architectures, the major architectures of which are shown in Figure 13.2, 780 fold groups, and 1390 homologous superfamilies. Further statistics on the database and discussion of the population of different levels are given later in this chapter.

## CURRENT METHODOLOGIES FOR IDENTIFYING STRUCTURAL AND PHYLOGENETIC RELATIONSHIPS IN CATH

Structural relationships in the CATH database were initially identified using the powerful structure-comparison algorithm, SSAP, devised by Taylor and Orengo in 1989 that incorporated a modified dynamic programming algorithm performing at two levels, thus enabling comparison of 3D information. More recently, fast graph-based methods for structure comparison have been implemented to enable the database to keep pace with the structure genomics initiatives. In fact sequence-based methods are first used to detect close relatives, as these are much faster than structure comparison and pairwise methods are reliable for relatives with 35% or more sequence identity. Sensitive profile-based sequence methods are used to detect more distant homologues that can then be verified by structural similarity.

The strategy used in classifying new structures into the database can thus be broken down into five major steps: (1) Close relatives are identified first using pairwise sequence methods. (2) Sequence profiles and structure comparison protocols are used to detect more distant relatives. (3) Structures unclassified at this stage are then examined using both automatic and manual procedures to determine domain boundaries. (4) Unclassified domain structures are recompared using the methods employed in steps 2 and 3. (5) Finally, any structures remaining unclassified are manually assigned to architectures within CATH or new architectures are described (see Fig. 13.3). The algorithms and manual validation protocols, used at different stages of the classification, are described below.

## Sequence-Based Protocols for Identifying Homologous Structures

Several studies have shown that when two proteins share more than 30% identities in their sequences, they have similar structures and can be assigned to the same super-family. Furthermore, pairwise sequence alignment methods are reasonably robust at these levels of sequence identity. In CATH, the global alignment method of Needle-man and Wunsch (1970) has been implemented and since Schneider and Sander (see Rost, 1999) have shown that there is a length dependence with small unrelated proteins, <100 residues, sometimes exhibiting sequence identities as high as 30%, a more cautious threshold of 35% identity is used to detect homologues. To reduce errors further, we check that at least 80% of the larger protein is aligned against the smaller protein. Any relatives missed at this stage of the classification are captured later using structure-comparison methods.

Relatives identified by these pairwise methods are clustered into their respective families using a single linkage clustering algorithm. Single linkage was chosen because the structural data is quite sparse in some families and it is known that sequences can diverge considerably, so it is not reasonable to expect a new relative to share significant sequence similarity with a large number of other relatives in the family.

Profile-based methods are used to detect more distant homologues. A number of protocols have been implemented. PSI-BLAST and the related IMPALA algorithms developed by the Altschul group (1997) have been shown to be among the most sensitive methods available. In these approaches, a sequence is scanned against a large sequence databank (i.e., GenBank at the NCBI) initially using pairwise methods to find close homologues, which are then multiply aligned to derive a sequence profile capturing the most specific residue preferences of that sequence and its relatives. Further iterations can result in highly specific profiles for the family to which the sequence belongs. Hidden Markov Models can also be applied and the SAMT method of Karplus and co-workers (Karplus and Hu, 2001) has been used to build profiles for each non-identical representative in CATH and also for structurally and functionally coherent families within the database. New structures are therefore scanned against both the IMPALA and SAMT profiles for representative structures already classified in CATH. For example, there are currently 7345 SAMT profiles including those built from individual single and multidomain proteins and those generated from multiple alignments of functionally related proteins.

Both methods (IMPALA and SAMT) have been benchmarked using known relatives from CATH and parameters and thresholds optimized to give the maximum

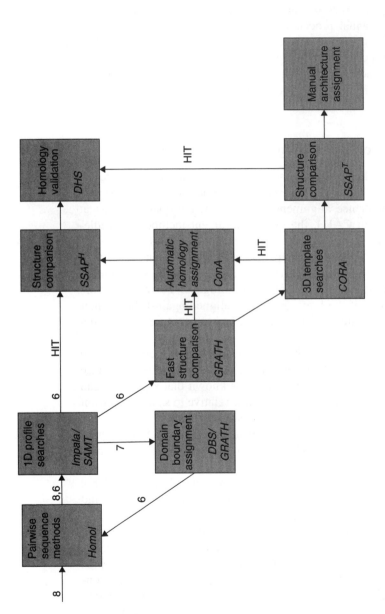

Figure 13.3. Flowchart describing the CATH classification procedure. New sequences are first compared pairwise, by sequence (HOMOL), with each other and all the entries in the database. Those that have not been identified as a sequence match are then compared using profile-based methods. If a homologue is found, the structure is compared structurally with all the members of the homologous superfamily (SSAP) and the DHS family data are updated. Those that are unmatched are assigned domain boundaries using DBS. The resulting single domains are again compared by pairwise and profile-based sequence methods. If no homologous relative is found, a fast structural comparison program (GRATH) is used to compare the domain with all sequence families within the CATH database. Structural templates (CORA) are then used to compare the structure with representatives from all the top scoring fold groups to assign homology using ConA. If GRATH does not find a significant hit, then the structure is compared against CORA templates from all superfamilies in the same class to identify a match. Finally, pairwise SSAPs against the database are run on any remaining unclassified structures as a final validation. If there is no significant fold match the architecture is assigned manually.

coverage for a small (<1%) error rate. Any relatives detected using these thresholds are subsequently validated by structure comparison (see below). The power of these sequence based approaches has been increased considerably by expanding CATH with sequence relatives from GenBank. These have been cautiously integrated into CATH superfamilies giving a 10-fold increase in the size of the database and effectively providing an intermediate sequence library for each family that broadens the scope of that family in sequence space, considerably enhancing the detection of distant homologues. Benchmarking trials using CATH (Pearl et al., 2001) demonstrated that although nearly 51% of distant structural homologues could be recognized by scanning against profiles derived from CATH structures, this percentage increased to 82% when profiles from intermediate sequence libraries for CATH were scanned.

Nearly three-quarters of all new structures are currently recognized using sequence-based methods, and the profile-based methods enable a further 10%, comprising very distant homologues, to be detected. This percentage will increase as the sequence databases and therefore the intermediate sequence libraries within CATH continue to grow with the international genome projects.

## Structure-Based Methods for Identifying Structural Homologues and Related Folds (SSAP and GRATH)

A significant proportion of distant relatives, currently about 15%, can only be recognized by comparing the structures directly. There are now many examples of evolutionary relatives that have diverged to an extent where no significant sequence similarity can be detected and yet the structural fold adopted by the protein remains highly similar. For example in the globins, sequence identities of relatives can fall below 10% but the structures and oxygen binding functions of the proteins remain highly conserved. As with sequence alignment, any method for comparing distant structural relatives must be able to cope with the extensive insertions or deletions occurring during evolution. Insertions and deletions are usually restricted to the loops between secondary structures where they are less likely to affect the fold and therefore the stability of the protein. However, in some families we have observed that secondary structures and sometimes quite large supersecondary motifs can be inserted or deleted.

In addition to residue insertions and deletions, shifts in the secondary structures can also occur to modulate the effects of volume changes caused by residue substitutions. Chothia and Lesk demonstrated quite significant movements of up to 40 degrees in some secondary structure pairs for two large protein families studied (the globins and the immunoglobulins). Analysis of families in CATH revealed that on average at least 50% of the secondary structures in the core of the protein are structurally well conserved across a family. However, although the fold of a protein family, comprising the core structural motif, may be well conserved currently, in about 10% of families there can be significant structural embellishments to this core and shifts in the orientations of the secondary structures. Sometimes these evolutionary changes give rise to modified or diverse functions within the family (see Todd, Orengo, and Thornton, 2001 for a review).

***Sequential Structure Alignment Program (SSAP).*** In order to address these problems, Taylor and Orengo adapted the dynamic programming methods used so successfully in sequence alignment to compare 3D structures (see Orengo, 1996 for a detailed description and review of applications). Instead of comparing residue identities, the method compares the structural environments of residues between two proteins. Structural environments can be simply encoded as the set of vectors from the C-beta atom of a particular residue to the C-beta atoms of all other residues within the same protein. Since there is no prior knowledge of which residues are equivalent between the proteins, dynamic programming must be applied at two levels: a lower level in which the structural environments of all residue pairs between the proteins are compared and an upper summary level in which information from putative equivalent pairs are accumulated. The method is therefore sometimes described as double dynamic programming.

SSAP has been benchmarked and optimized using sets of validated structural homologues. A logarithmic scoring scheme was implemented that was optimized to be largely independent of the size and class of the proteins, although proteins with large proportions of alpha helices tend to give slightly higher scores as the local similarity is so highly conserved. The score is normalized to be in the range of 0 to 100 for identical proteins, irrespective of size and structures, with similar folds tending to give scores above 70, while homologous proteins often give higher scores of 80 and above. When classifying new structures, scores of 70 and above are required before proteins are assigned to a particular fold group and again single-linkage clustering is used as some families are too small to ensure that a significant proportion of known relatives will have clear structural similarity (i.e., SSAP >70) to the new relative. Although high SSAP scores suggest that the proteins may be homologues, these proteins are not assigned at the superfamily level unless there is other evidence to support homology; for example, sequence similarity detectable by PSI-BLAST or functional similarity (discussed in more detail below).

Proteins are only classified into existing families of fold groups if the structural similarity detected extends for a significant proportion (more than 60%) of the larger structure. This preserves the domain-based nature of the classification. Thus, although new multidomain structures may contain one or more domains that match existing CATH families, these are not clustered into their families until a later stage in the classification when their domain boundaries have been reliably identified (see below).

***Graphical Method for Identifying Folds (GRATH).*** Although SSAP has proven to be reliable and accurate, it is computationally expensive and large structures such as TIM barrels with more than 300 residues can take several days to scan against the database, using the most powerful machines currently available. This has not proved to be a major bottleneck to date; however, structure genomics initiatives are expected to increase substantially the numbers of structures determined annually. Currently 50 to 100 new structures are determined weekly and this number may double or treble over the next decade. Therefore, to keep pace with structure genomics, a fast prefilter for SSAP has been designed, based on graph theory. Similar approaches have previously been implemented in the algorithm POSSUM designed by Artymiuk and co-workers (Mitchell et al., 1990).

The method implemented in CATH–GRATH (graphical representation of CATH structures) compares secondary structures between proteins, as there are an order of magnitude fewer of these than residues. These are represented as linear vectors and are associated with nodes in a protein graph. Edges between the nodes are characterized by the orientations of the secondary structures and the distances between their midpoints. Additional angles are used to describe the tilt and rotations of the vectors.

Ullmanns subgraph isomorphism algorithm is used to detect corresponding structural motifs between proteins from comparison of their graphs. Parameters for recognizing fold similarities have been optimized using validated relatives from CATH. GRATH recognizes the correct fold within the top 10 matches of a database search 98% of the time. Of importance, GRATH is 1000 times faster than SSAP for most proteins and the top 10 matches returned from a search can be validated using the more accurate SSAP method. Robust statistics (expectation values or E-values) have also been developed based on the extreme value distribution observed for a typical database scan. These statistics indicate the significance associated with a match and are important for determining the order in locating individual domain folds within multidomain proteins (see below).

Because GRATH is so fast, multidomain proteins can now be structurally compared to detect potential homologies that are then validated manually (see below).

## Methods for Generating Multiple Structure Alignments (CORA) and Protocols for Using 3D-Templates to Identify Distant Structural Relationships

Multiple structure alignments are generated for each superfamily in CATH using a modified version of the SSAP algorithm (consensus residue attributes, CORA). This algorithm is based on progressive alignment of relatives using a single-linkage tree derived from the pairwise SSAP similarity scores (see Orengo et al., 1997 for a more detailed description). The most similar structures are aligned first and the next relative, most similar to the aligned structures, is then iteratively selected and aligned in the order dictated by the tree. After addition of each relative, a consensus structure is derived consisting of average vectors between residue positions and information on the variability of these vectors. Further relatives are aligned against this consensus structure, weighting the alignment of conserved positions more highly.

Once all relatives have been aligned, information on the conservation of the residue structural environment, including residue contacts and various other attributes (e.g., accessibility, torsional angles), is compiled and encoded in a 3D template representing the set of structures. In a manner analogous to the improvements obtained using sequence profiles, structural templates have been shown to be far more effective at recognizing distant homologues as they capture the most conserved structural characteristics of the family or superfamily. New structures can be scanned against libraries of CORA templates, again using the double dynamic programming algorithm and the percentage of highly conserved contacts for the superfamily that are present in the putative relative is calculated (ConA), to determine homology. The pattern of conserved residue contacts has been shown to be a highly characteristic topological fingerprint for a particular superfamily.

CORA templates have been generated for each superfamily containing two or more nonidentical structures, and in the more highly populated superfamilies there are multiple templates corresponding to clusters of more closely related structures or families within the superfamily. Scanning against the template library is up to 100 times faster than performing pairwise searches with SSAP on representative structures from the superfamilies. The templates also show increased sensitivity and selectivity in recognising structural relatives over the pairwise SSAP and these attributes can be expected to increase as more in recognising structural relatives structures are determined and coverage within each superfamily increases.

## Identification of Domain Boundaries

Any proteins unclassified by the sequence-based methods are divided into their constituent domains where relevant, before resubmitting them to the sequence searches and to the slower structure-comparison protocols, which often facilitates the structure alignment as a smaller region of the comparison matrix is searched. Identification of domain boundaries is a difficult process and although numerous automatic algorithms have been devised (see Jones et al., 1998 for a review), most have an error rate of between 20% and 30% associated with them. The problem arises from the fact that no simple quantitative definition of a structural domain exists. Qualitatively, domains have been described by various authors as compact semi-independent folding units and many algorithms attempt to locate them by searching for large hydrophobic clusters indicative of domain cores and also by dividing structures so as to maximize internal residue contacts within putative domains and to minimize external contacts between them.

In the October 2001 release of CATH, nearly 40% of PDB entries are multidomain proteins, of which two-thirds comprise only two domains. Furthermore nearly one-quarter of domains in CATH are discontiguous; that is, the domain is formed from discontiguous segments of the polypeptide chain, and these cases are particularly difficult for the algorithms to detect. To improve accuracy, we employ a consensus-based protocol DBS (Domain Boundary Suite) that applies three independent algorithms PUU (Holm and Sander, 1994a), Domak (Siddiqui and Barton, 1995) and DETECTIVE (Swindells, 1995) (see Jones et al., 1998 for a description of methods). Where they agree within a tolerance of 10 residues, domains can be assigned completely automatically. Otherwise each of the individual assignments is manually checked.

In addition we use a more recently developed protocol, employing GRATH and based on the concept that domains are known to recur in different multidomain contexts. It is now well established that domain shuffling is a common evolutionary mechanism, often responsible for creating new or modified functions within an organism (see Todd, Orengo, and Thornton, 2001 and Teichmann et al., 2001 for reviews). Recent analysis of CATH revealed that at least 70% of the domains from multidomain proteins recurred in different multidomain families or also occurred as single domain proteins. Therefore, a simple domain-detection protocol is now used to search for known domains within new multidomain structures. This protocol uses GRATH to compare the secondary structure graph for each new putative multidomain protein with the library of graphs from representatives from all the CATH domain families. Domains found within the multidomain protein can then be extracted in order of decreasing statistical significance. This approach has led to significant improvements in domain assignment as the E-values returned by GRATH give a measure of the reliability in boundary prediction.

Any ambiguous assignments can be manually validated by coloring putative domains in the excellent molecular viewing package, RASMOL (Sayle and Milner-White, 1995). The majority of errors are found in this way, although in difficult cases boundary identification is a somewhat subjective process as witnessed by the fact that about 17% of boundary definitions in the SCOP and CATH databases disagree. This step is one of the most time-consuming but important stages in the classification. Although domain-boundary detection from structural data is difficult, it is considerably easier than assignment from sequence data and many sequence databases now incorporate the SCOP and CATH boundaries or use them to validate their own assignments.

## Structural and Functional Validation of Homologues—The Dictionary of Homologous Superfamilies

Sequence, structural, and functional data for each homologous superfamily in CATH are stored within a Dictionary of Homologous Superfamilies (DHS), which is also accessible over the Web. This dictionary was originally established in 1999 by Bray and now also contains all the sequence relatives for CATH superfamilies identified in GenBank. Information on pairwise relationships between all nonidentical structures and sequences is also stored, namely, sequence identities, expectation values from PSI-BLAST and SAM-T99; SSAP structural similarity scores and expectation values from GRATH, where appropriate.

Pairwise SSAP alignments between all nonidentical structures in the superfamily are also stored. Plots of sequence identities versus structural similarity scores can be used to illustrate the structural plasticity observed within a given superfamily (for example, see Fig. 13.4 for selected superfamilies), and any obvious outliers may indicate problems in the alignment. CORA multiple-structure alignments are generated for each superfamily and can be viewed over the Web. These alignments are annotated in various ways: by residue identities or physicochemical properties; by secondary structure; and by PROSITE motifs. RASMOL viewers for multiple superpositions of relatives allow the 3D location of conserved sequence motifs to be easily identified.

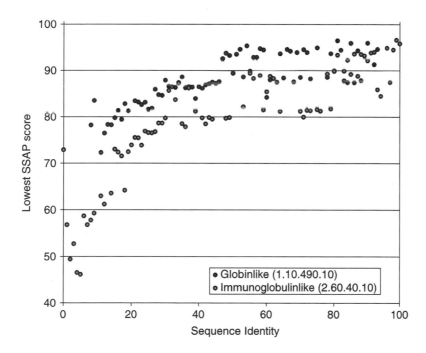

**Figure 13.4.** Structural plasticity plots showing the structural similarity for pairs of relatives in selected CATH superfamilies, as measured by the SSAP algorithm (Taylor and Orengo, 1989), as a function of sequence identity measured after structural alignment. To highlight the maximum deviation for each pairwise sequence identity the minimum SSAP value recorded is shown. Red dots are mainly-$\alpha$ globin superfamily and blue dots are the mainly-$\beta$ Immunoglobulin. Figure also appears in Color Figure Section.

Of importance, the DHS contains increasing amounts of functional data for each superfamily. These data are extracted from other public resources, for example, PDB header records, PDBsum; SWISS-PROT; the Enzyme database; GenProtEC, and PROSITE. In future releases the DHS will also provide functional annotations collected from InterPro, Pfam, Gene Ontology, and metabolic pathway data. Summary tables in the DHS can be used to peruse functional attributes for relatives across the family. This information can be used diagnostically to determine whether a new structural relative possesses a similar function to other members of the family and is therefore clearly a homologue. However, since recent analyses of CATH have revealed the extent to which function can be modified in some protein families (see Todd, Orengo, and Thornton, 2001 for a review), the DHS also contains all the sequence relatives detected for each superfamily, which considerably expands the functional information available for a particular superfamily, making it easier to find a relative with related function. It also reveals the extent to which functional change can occur within the family and this can be taken into account when assessing whether a protein with a similar fold is in fact a homologue.

Protocols are currently being devised for automatically comparing functional annotations between proteins in order to speed up the validation of homologues during classification. However, we do not expect this approach to be viable for more than about 30% to 50% of new relatives, a result suggested by the performance of similar protocols employed by other groups. However, because this validation stage is time consuming and another major bottleneck in the classification, we expect the new protocols to significantly improve the frequency of CATH releases.

## Recruiting Sequence Relatives into CATH Superfamilies

In order to gain as broad an understanding of function as possible within each protein family and superfamily, sequence relatives are regularly recruited into CATH from Gen-Bank. This recruitment is achieved using the profile-based methods described above, namely, the 1D-profiles (PSI-BLAST) and iterative Hidden Markov Models (SAMT). Nonidentical representatives from each class within CATH are scanned against Gen-Bank using both PSI-BLAST and SAM-T99. A consensus method, DomainFinder (Pearl et al., 2001), which uses conservative thresholds established by benchmarking against validated structural homologues, is then used to extract those domain regions from GenBank entries that can safely be assigned to CATH superfamilies. As with structural entries, pairwise sequence alignment methods are subsequently used to update sequence relationships within the superfamily and to enable clustering into sequence families at 35%, 60%, 95%, and 100% identities. CATH currently contains nearly 200,000 sequence relatives identified in this way.

## THE GENE3D RESOURCE

A related CATH-based Web resource, Gene3D, uses these assignments of gene sequences to CATH superfamilies to provide structural annotations for completed genomes. The June 2001 release of Gene3D contains 36 complete genomes obtained from GenBank. For each gene within the genome, the location of gene regions that match individual structural domains from CATH superfamilies are shown, together

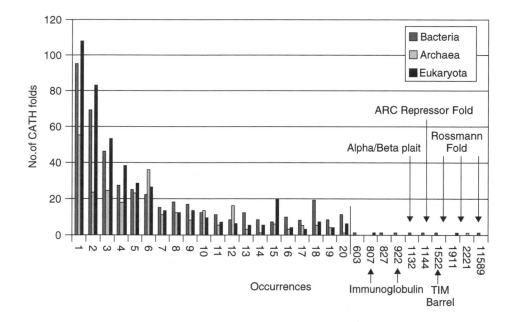

Figure 13.5. Comparison of fold usage in the eukaryotic, archaeal, and bacterial kingdoms using the November 2000 release of CATH. The majority of frequently recurring folds currently identified in all three kingdoms are alpha–beta proteins; particularly Rossman folds, TIM barrels. Structural annotation of 21 bacterial genomes, 8 archaeal genomes and 1 eukaryotic genome was performed using PSI-BLAST (Alschul et al., 1997) and the DomainFinder protocol (Pearl et al., 2002).

with a consensus domain region determined by the DomainFinder algorithm. Links to the relevant CATH superfamily and DHS entries are also provided. Statistics are given on the distributions of superfamilies and fold groups identified within each genome, enabling comparison of fold and superfamily usage between genomes (see, for example, Fig. 13.5).

## THE CATH WEB SITE AND SERVER

CATH is available over the Web at the address shown in Table 13.2. The site can be used for browsing the hierarchy, and there are representative MOLSCRIPT illustrations at each level together with links to other local resources (DHS, Gene3D, PDBsum (Laskowski et al., 2001). Each level has its own unique numeric identifier that is never changed, although some numbers may disappear; for example, if new evidence suggests two superfamilies should be merged.

CATH data are stored within an Oracle 8i relational database. Special Oracle features (e.g., associations, collections, inclusions) are utilized for storing relationships and family membership and another Oracle feature, materialized views, is used to create specialized views of the data for efficient searching. The database also contains genome information (e.g., taxonomy, gene locations where available) used by the Gene3D resource. The schema for CATH is part of a much larger schema for a

TABLE 13.2. URLs of the CATH Domain Structure Database and Related Resources

| Resource | URL |
| --- | --- |
| CATH Database: Classification of structural domains in the PDB. Domains are grouped by Class, Architecture, Topology (Fold) and Homologous superfamily. There are links to PDBsum. | http://www.biochem.ucl.ac.uk/bsm/ cath_new/ |
| SSAP Server: The SSAP server allows users to compare the structures of two proteins and view the subsequent structural alignment. | http://www.biochem.ucl.ac.uk/cgi-bin/ cath/GetSsapRasmol.pl |
| CATH Server: The CATH server allows users to compare a PDB or novel structure against a representative library of structures in CATH. | http://www.biochem.ucl.ac.uk/cgi-bin/ cath/CathServer.pl |
| Dictionary of Homologous Superfamilies: This resource displays the structural alignments for all members of a homologous superfamily classified in the CATH database. The alignments are augmented with ligand information and SWISS-PROT annotations. | http://www.biochem.ucl.ac.uk/ bsm/dhs |
| Gene3D: Database of precalculated structural assignments for genes and whole genomes. The data are derived using PSI-BLAST and IMPALA. | http://www.biochem.ucl.ac.uk/bsm/ Gene3D |
| IMPALA Server: This server allows the user to screen a sequence against the CATH set of IMPALA sequence profiles for protein structural domains. | http://www.biochem.ucl.ac.uk/bsm/ cath_new/Impala/ |

protein family database (PFDB), which also contains sequence-based families identified using other classification protocols and currently contains data for virus families, eye proteins, and other biological families being studied locally.

The database has a flexible design that allows proteins to participate in different structural and functional relationships, that is, alternative methods of clustering can easily be accommodated. The flexible design also has the benefit of allowing mapping between families or clusters generated using different protocols or based on different underlying philosophies, for example, functional versus structural.

## Regular Updates of the CATH Web Site

The protocols used for updating CATH have already been described above (see also Figure 13.3). To keep pace with structure genomics initiatives, sequence-based classification (see Figure 13.3) of newly determined structures will be run at weekly intervals and clearly identified relatives recruited into the database. Monthly releases of the database are planned to make this data publicly available on a more frequent basis. The assignment of domain boundaries is more time consuming as some validation is required where GRATH and DBS results are ambiguous. Structure classification is also slower as validation is required to confirm homologues. Therefore, both domain-boundary identification and structural classification are ongoing processes and those structures whose boundaries have been determined and classified by sequence- and structure-based methods will be added to the database to coincide with monthly updates. In future, we plan to display all newly determined protein structures that meet the criteria for integration in CATH (i.e., well resolved and not model or synthetic proteins) using additional protein classes to reflect the extent to which the structures have been classified in the database (see Table 13.1).

## THE CATH SERVER

Newly determined structures can be submitted to the CATH server, which scans them against representatives from the database in order to determine the putative superfamily or fold group to which the structure belongs or whether the protein comprises one or more novel folds. Domain boundaries can be supplied by the user or alternatively will be determined automatically using the consensus approach. GRATH is first used to identify a probable fold group. Subsequently, the new domain structure(s) is scanned against 3D templates (Cora, Orengo, 1999) for all the homologous superfamilies within the top three folds matched to identify the superfamily. Lists of structural neighbors are provided together with links to the appropriate superfamilies in CATH and the DHS. The user can also view superpositions of the structure with other relatives from the superfamily or fold group.

### Statistics on the Populations of Different Levels In the CATH Hierarchy

*Population of Folds Within Architectures.* Although there are currently 36 architectural groups in CATH only 28 can be well defined (see Figure 13.2), and the remaining groups can be thought of as bins comprising assorted irregular or complex folds or folds containing few secondary structures and often stabilized by a high proportion of disulphide bridges. Furthermore, among the well-defined architectures, some are much more highly populated than others with about half of the folds adopting one of six regular symmetric architectures; the mainly-$\alpha$ bundles, the two-layer $\beta$-sandwiches and $\beta$-barrels, and the two-layer and three-layer $\alpha\beta$-sandwiches and the $\alpha\beta$-barrels.

Recent analysis of structural relationships between fold groups in CATH using the GRATH algorithm has revealed that some fold groups are particularly "gregarious," that is, they have large motifs in common with many other folds within the database. For example, a high proportion of folds within the mainly-$\beta$ and the $\alpha\beta$-sandwich architectures can be categorized as gregarious in this way. This idea conforms with the idea of a structural continuum between fold groups, first suggested by an early analysis of CATH. However, the recent analysis also clearly showed that some 15% of folds, for example the beta trefoil interleukin fold, are much more distinct, representative more of discrete islands than progressions in a structural continuum.

The idea of a structural continuum is perhaps not surprising as it has long been known that certain structural motifs are favored and often recur within folds for a particular class; for example, the mainly-$\beta$ hairpins, the beta Greek keys, and the $\alpha\beta$ and split $\alpha\beta$ motifs (see Richardson, 1981). The TIM barrel fold in the $\alpha\beta$ class (see Figure 13.1 for a representative picture) comprises eight recurring $\alpha\beta$ motifs, which are also found to recur in the $\alpha\beta$ Rossmann folds (see also Figure 13.1), though the connectivity between two of the motifs differs, giving rise to a different topology for the fold.

***Population of Superfamilies and Families Within Folds.*** The population of the different levels in the CATH hierarchy, for the November 2000 release, are shown in Figures 13.6, 13.7, and 13.8 and it can be seen from Figure 13.7 that the number of new folds being determined each year is slowly decreasing. Early analysis of CATH revealed a small number of fold groups (~10) that were very highly populated containing many different homologous superfamilies and families. Reexamining these data six years later, following a nearly 10-fold expansion of the database from 3000 to 33,000 domain structures, it is clear that the trend still applies and it can be seen from Figure 13.8 that 5 large fold groups contain nearly one-fifth of all the homologous superfamilies in CATH. These highly populated fold groups have been

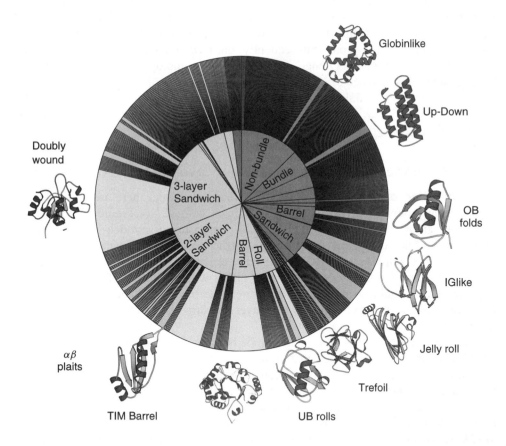

**Figure 13.6.** CATHerine wheel plot showing the distribution of nonhomologous structures (i.e., a single representative from each homologous superfamily [H-level in CATH]) among the different classes (C), architecture (A), and fold families (T) in the CATH database. Protein classes shown are mainly-alpha, mainly-beta, and alpha-beta. Within each class, the angle subtended for a given segment reflects the proportion of structures within the identified architectures (inner circle) or fold groups (outer circle). The superfolds are indicated and are illustrated with a Molscript drawing of a representative from the family.

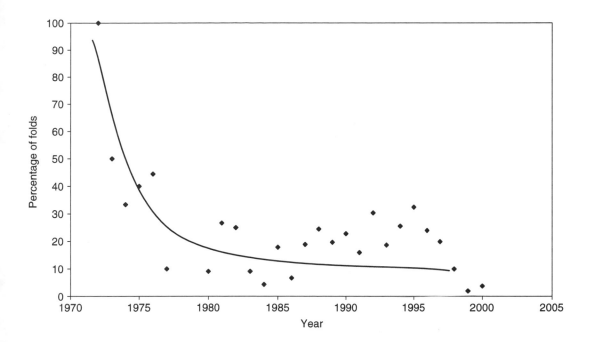

**Figure 13.7.** Number of unique folds identified annually as a percentage of the number of new structures determined.

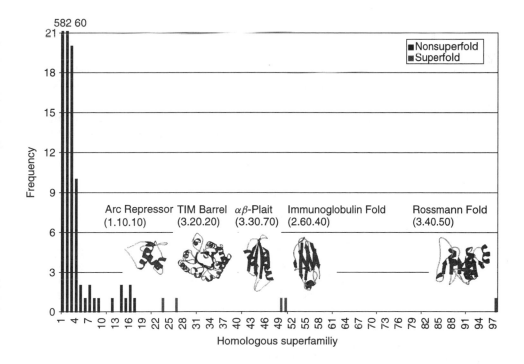

**Figure 13.8.** Populations of different levels in the CATH hierarchy; homologous superfamilies within fold groups. The October 2000 version of CATH was used to generate the histograms.

**TABLE 13.3. Names of the Researchers and Advisers Who Have Contributed to the CATH Database Since It Was Established in 1993**

List of Contributors (alphabetical)

Chris F. Bennett
James E. Bray
Daniel W.A. Buchan
Andrew Harrison
Gail Hutchinson
David T Jones
Susan Jones
Roman Laskowski
David Lee
Loredana Lo Conte
Andrew Martin
Alex D. Michie
Christine A. Orengo
Frances M.G. Pearl
Jane S. Richardson
Gabrielle A. Reeves
Adrian J. Shepherd
Ian Sillitoe
Mark B. Swindells
Willie R. Taylor
Janet M. Thornton
Annabel E. Todd

described as superfolds (Orengo, Jones, and Thornton, 1994) and other classifications (e.g., SCOP) have reported similar observations of frequently occurring domains within the database.

The popularity of these folds may be a result of divergent or convergent evolution. Divergent evolution gives rise to families of proteins in which the structure is generally well conserved but sequences may have changed to the extent that no significant similarity remains. In paralogues, which arise from duplication of the gene within an organism, the function of protein may also have been modified or changed. Thus, apparently diverse superfamilies within these superfolds may in fact be extremely distant relatives whose relationships cannot easily be verified from the available sequence or functional data.

Alternatively, Ptitsyn, Finkelstein, and others have suggested that there may be a limited number of folds in nature due to physical constraints on the packing of secondary structures. Chothia (1993) has suggested approximately 1000 folds and other similar estimates of a few thousand folds have also been made (Orengo, Jones, and Thornton, 1994). Structures sharing the same fold but arising from different ancestral proteins have been described as analogues.

The task of determining homology is often complicated by the lack of evolutionary clues. Murzin and others have shown several interesting examples of relatives possessing similar folds but disparate sequences and functions but where homology is suggested by the presence of some rare structural characteristic, for example, a conserved $\beta$-bulge. Detailed knowledge of the structural family can often provide insights that promote detection of evolutionary fingerprints. However, such practices are not readily amenable to automated protocols. The TIM barrel fold is one of the most highly populated, with 18 superfamilies in a recent CATH release. However, several detailed studies recently have provided evidence of common ligand-binding motifs or unusual structural characteristics, suggesting that many of these superfamilies should be merged. In this context, an analysis of functional properties observed within 167 enzyme superfamilies in CATH revealed that when all known relatives for the superfamily were considered, including sequence relatives identified in the genomes, more than 25% of the families exhibited functional diversity (see Todd, Orengo, and Thornton, 2001 for a review and Chapter 19).

## FURTHER READING

Apic G, Gough J, Teichmann SA (2001): Domain combinations in archaeal, eubacterial and eukaryotic proteomes. *J Mol Biol* 310:311–25. [Analysis of the domain superfamily pairwise combinations observed in different organisms. Proteins from 783 superfamilies are identified in 40 genomes, making 1307 pairwise combinations. Most superfamilies are observed in combination with one or two other superfamilies, one-quarter do not make combinations with any other superfamily, and a few combine with many other superfamilies.]

## REFERENCES

Altschul SF, Madden TL, Schaffer AA, Zhang J, Zhang Z, Miller W, Lipman DJ (1997): Gapped BLAST and PSI-BLAST: a new generation of protein database search programs. *Nucleic Acids Research* 25:3389–402.

Chothia C (1993): Proteins. One thousand families for the molecular biologist. *Nature* 357:543.

Chothia C, Lesk AM (1986): The relation between the divergence of sequence and structure in proteins. *EMBO J* 5:823–6.

Dengler U, Siddiqui AS, Barton GJ (2001): Protein structural domains: analysis of the 3Dee domains database. *Proteins* 42(3):332–44.

Dietmann S, Holm L (2001): Identification of homology in protein structure classification. *Nature Structural Biology* 8:953–7.

Hadley C, Jones DT (1999): A systematic comparison of protein structure classifications: SCOP, CATH and FSSP. *Structure Fold Des* 7:1099–112. [Systematic comparison between the most widely used and comprehensive databases SCOP, CATH, and FSSP, which represent three unique methods of classifying protein structures: purely manual, a combination of manual and automated, and purely automated, respectively. At the time of this analysis approximately two-thirds the protein chains in each database were common to all three databases and despite employing different classification methods, the majority of entries within these classifications were in agreement.]

Holm L, Sander C (1994a): Parser for protein folding units. *Proteins* 19:256–68. [Domain boundary prediction algorithm based on the premise that a domain will make more internal contacts (i.e., intradomain contacts) than external contacts (contact with residues in the remainder of the protein). The algorithm incorporates a harmonic model to approximate interdomain dynamics.]

Holm L, Sander C (1994b): Searching protein structure databases has come of age. *Proteins* 19:165–73.

Holm L, Sander C (1996): Mapping the protein universe. *Science* 273:595–603. [This paper describes structural comparison methods, the principles of structural alignment, and the clustering of similar proteins on the basis of their structure. It also describes highly populated regions in fold space, which are termed fold space attractors.]

Holm L, Sander C (1998): Dictionary of recurrent domains in protein structures. *Proteins* 33:88–96.

Jones S, Stewart M, Michie A, Swindells MB, Orengo CA, Thornton JM (1998): Domain assignment for protein structures using a consensus approach: characterisation and analysis. *Protein Science* 7:233–42. [A consensus approach for the assignment of structural domains is presented. The individual algorithms each use the 3D coordinates of the structures but measure different attributes or use a different decision algorithm to assign boundaries. By using the assignments from three different domain boundary assignment programs an increase in domain boundary prediction accuracy can be obtained.]

Karplus K, Hu B (2001): Evaluation of protein multiple alignments by SAM-T99 using the BAliBASE multiple alignment test set. *Bioinformatics* 8:713–20.

Laskowski RA (2001): PDBsum: summaries and analyses of PDB structures. *Nucleic Acids Res.* 29:221–2.

Lo Conte L, Ailey B, Hubbard TJ, Brenner SE, Murzin AG, Chothia C (2000): SCOP: a structural classification of proteins database. *Nucleic Acids Research* 28(1):257–9.

Mitchell EM, Artymiuk PJ, Rice DW, Willett P (1990): Use of techniques derived from graph theory to compare secondary structure motifs in proteins. *J. Mol. Biol.* 212:151–66.

Murzin AG, Brenner SE, Hubbard T, Chothia C (1995): SCOP: a structural classification of proteins database for the investigation of sequences and structures. *J Mol Biol* 247:536–40.

Needleman SB, Wunsch, CD (1970): A general method applicable to the search for similarities in the amino acid sequence of two proteins. *J. Mol. Biol.* 48:443.

Orengo CA (1994): Classification of protein folds. *Curr Opin Struct Biol* 4:429–40.

Orengo CA, Jones DT, Thornton JM (1994): Protein superfamilies and domain superfolds. *Nature* 372:631–4.

Orengo CA, Taylor WR (1996): SSAP: sequential structure alignment program for protein structure comparison. *Methods Enzymol* 266:617–35.

Orengo CA, Michie AD, Jones S, Jones DT, Swindells MB, Thornton JM (1997): CATH—A hierarchical classification of protein domain structures. *Structure* 5:1093–108.

Orengo CA (1999): CORA—topological fingerprints for protein structural families. *Protein Sci* 8:699–715.

Orengo CA, Sillitoe I, Reeves G, Pearl FM (2001): What can structural classifications reveal about protein evolution? *J Struct Biol* 134:145–65.

Pearl FM, Martin N, Bray JE, Buchan DW, Harrison AP, Lee D, Reeves GA, Shepherd AJ, Sillitoe I, Todd AE, Thornton JM, Orengo CA (2001): A rapid classification protocol for the CATH Domain Database to support structural genomics. *Nucleic Acids Research* 29:223–7.

Pearl FM, Lee D, Bray JE, Buchan DWA, Shepherd AJ, Orengo CA (2002): The CATH extended protein-family database: providing structural annotations for genome sequences. *Protein Sci* 11:233–44.

Richardson JS (1981): The anatomy and texonomy of protein structure. *Adv. Prot. Chem.* 34:167–339.

Rost B (1999): Twilight zone of protein sequence alignments. *Protein Eng.* 12:85–94.

Sayle RA, Milner-White EJ, (1995): RASMOL: biomolecular graphics for all. *Trends Biochem Sci* 20(9):374.

Siddiqui AS, Barton GJ (1995): Continuous and discontinuous domains: an algorithm for the automatic generation of reliable protein domain definitions. *Protein Sci* 4:872–84.

Sowdhamini R, Burke DF, Deane C, Huang JF, Mizuguchi K, Nagarajaram HA, Overington JP, Srinivasan N, Steward RE, Blundell TL (1998): Protein three-dimensional structural databases: domains, structurally aligned homologues and superfamilies. *Acta Crystallogr D* 54:1168–77. [This paper reports the availability of a database of protein structural domains (DDBASE), an alignment database of homologous proteins (HOMSTRAD), and a database of structurally aligned superfamilies (CAMPASS). The alignment of proteins in superfamilies has been performed on the basis of the structural features and relationships of individual residues using the program COMPARER.]

Swindells MB (1995): A procedure for the automatic determination of hydrophobic cores in protein structures. *Protein Sci* 4:93–102.

Teichmann SA, Rison SC, Thornton JM, Riley M, Gough J, Chothia C (2001): The evolution and structural anatomy of the small molecule metabolic pathways in *Escherichia coli*. *J Mol Biol* 311:693–708.

Todd AE, Orengo CA, Thornton JM (2001): Evolution of function in protein superfamilies, from a structural perspective. *J Mol Biol* 7:1113–43.

# 14

# STRUCTURAL QUALITY ASSURANCE

Roman A. Laskowski

The experimentally determined three-dimensional (3D) structures of proteins and nucleic acids represent the knowledge base from which so much understanding of biological processes has been derived over the last three decades of the twentieth century. Individual structures have provided explanations of specific biochemical functions and mechanisms, while comparisons of structures have given insights into general principles governing these complex molecules, the interactions they make, and their biological roles.

The 3D structures form the foundation of structural bioinformatics; all structural analyses depend on them and would be impossible without them. Therefore, it is crucial to bear in mind two important truths about these structures, both of which result from the fact that they have been determined experimentally. The first is that the result of any experiment is merely a *model* that aims to give as good an explanation for the experimental data as possible. The term *structure* is commonly used, but you should realize that this should be correctly read as *model*. As such the model may be an accurate and meaningful representation of the molecule, or it may be a poor one. The quality of the data and the care with which the experiment has been performed will determine which it is. Independently performed experiments can arrive at very similar models of the same molecule; this suggests that both are accurate representations, that they are good models.

The second important truth is that any experiment, however carefully performed, will have errors associated with it. These errors come in two distinct varieties: systematic and random. Systematic errors relate to the *accuracy* of the model—how well it corresponds to the true structure of the molecule in question. These often include errors of interpretation. In X-ray crystallography, for example, the molecule(s) need to be fitted to the electron density computed from the diffraction data. If the data are poor and the quality of the electron density map is low, it can be difficult to find the

*Structural Bioinformatics*
Edited by Philip E. Bourne and Helge Weissig
Copyright © 2003 by Wiley-Liss, Inc.

correct tracing of the molecule(s) through it. A degree of subjectivity is involved and errors of mistracing and frame-shift errors, described later, are not uncommon. In NMR spectroscopy, judgments must be made at the stage of spectral interpretation where the individual NMR signals are assigned to the atoms in the structure most likely to be responsible for them.

Random errors, on the other hand, depend on how precisely a given measurement can be made. All measurements contain errors at some degree of precision. If a model is essentially correct, the sizes of the random errors will determine how *precise* the model is. The distinction between accuracy and precision is an important one. It is of little use having a very precisely defined model if it is completely inaccurate.

The sizes of the systematic and random errors may limit the types of questions a given model can answer about the given biomolecule. If the model is essentially correct, but the data was of such poor quality that its level of precision is low, then it may be of use for studies of large scale properties—such as protein folds—but worthless for detailed studies requiring the atomic position to be precisely known; for example, to help understand a catalytic mechanism.

## STRUCTURES AS MODELS

To make the point about 3D structures being merely models it is instructive to consider the subtly different types of model obtained by the two principal experimental techniques: X-ray crystallography and NMR spectroscopy. Figure 14.1 shows the two different interpretations of the same protein that are given by the two methods, as explained below. The models are of the protein rubredoxin with a bound zinc ion held in place by four cysteines.

### Models from X-Ray Crystallography

Figure 14.1a is a representation of the protein model as obtained by X-ray crystallography. It is not a standard depiction of a protein structure; rather, its aim is to illustrate some of the components that go into the model. The components are: the $x$-, $y$-, $z$-coordinates, the $B$-factors, and occupancies of all the individual atoms in the structure. These parameters, together with the theory that explains how X-rays are scattered by the electron clouds of atoms, aim to account for the observed diffraction pattern. The $x$-, $y$-, $z$-coordinates define the mean position of each atom, whereas its $B$-factor and occupancy aim to model its apparent disorder about that mean. This disorder may be the result of variations in the atom's position in time, due to the dynamic motions of the molecule, or variations in space, corresponding to differences in conformation from one location in the crystal to another, or both. The higher the atom's disorder, the more "smeared out" its electron density. $B$-factors model this apparent smearing around the atom's mean location; at high resolution a better fit to the observations can often be obtained by assuming the $B$-factors to be anisotropic, as represented by the ellipsoids in Figure 14.1a. Occasionally, the data can be explained better by assuming that certain atoms can be in more than one place, due, say, to alternative conformations of a particular side chain (indicated by the arrows showing the two alternative positions of the glutamate sidechain in Figure 14.1a). The atom's occupancy defines how often it is found in one conformation and how often in another (for example, in the example

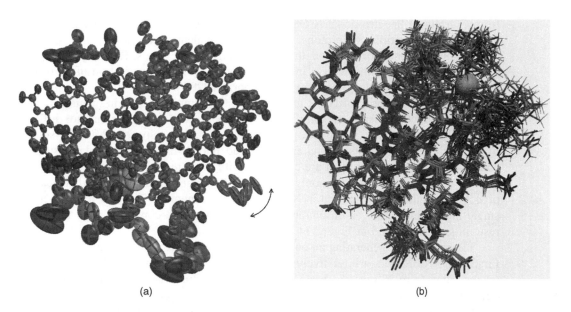

(a)                                                    (b)

**Figure 14.1.** The different types of model generated by X-ray crystallography and NMR spectroscopy. Both are representations of the same protein: rubredoxin. (a) In X-ray crystallography the model of a protein structure is given in terms of atomic coordinates, occupancies, and *B*-factors. The side chain of Glu50 has two alternative conformations, with the change from one conformation to the other identified by the double-headed arrow. The *B*-factors on all the atoms are illustrated by "thermal ellipsoids," which give an idea of each atom's anisotropic displacement about its mean position. The larger the ellipsoid, the more disordered the atom. Note that the main-chain atoms tend to be better defined than the side-chain atoms, some of which exhibit particularly large uncertainty of position. The region around the bound zinc ion appears well ordered. This is in stark contrast with the NMR case in (b). The coordinates and *B*-factors come from PDB entry 1irn, which was solved at 1.2Å and refined with anisotropic *B*-factors. (b) The result of an NMR structure determination is a whole ensemble of model structures, each of which is consistent with the experimental data. The ensemble shown here corresponds to 10 of the 20 structures deposited as PDB code 1bfy. In this case the metal ion, is iron. The more disordered regions represent either regions that are more mobile, or regions with a paucity of experimental data, or a combination of both. The region around the iron-binding site appears particularly disordered. Both diagrams were generated with the help of the Raster3D program (Merritt and Bacon, 1997). Figure also appears in Color Figure section.

given in Figure 14.1a the occupancies of the two alternative conformations are 56% and 44%).

## Models from NMR Spectroscopy

The data obtained from NMR experiments are very different, so the models obtained differ in their nature, too. The spectra measured by NMR provide a diversity of information on the average structure of the molecule, and its dynamics, in solution. The most numerous, but often least precise, data are from Nuclear Overhauser Effect SpectroscopY (NOESY) experiments where the intensities of particular signals correspond to the separations between spatially close protons ($\leq 6$Å) in the structure. The spectra from COrrelated SpectroscopY (COSY)-type experiments give more precise information on the separations of protons up to three covalent bonds apart, and in some cases on the presence, or even length, of specific hydrogen bonds. Recently developed dipolar-coupling experiments give information on the relative orientation of particular backbone covalent bonds (Clore and Gronenborn, 1998).

For the vast majority of NMR experiments, the sample of protein or nucleic acid is in solution, rather than in crystal form, which means that molecules that are difficult to crystallize, and hence impossible to solve by crystallography, can often be solved by NMR instead. The separations are converted into distance and angular restraints and models of the structure that are consistent with these restraints are generated using various techniques, most commonly molecular dynamics-based simulated annealing procedures similar to those used in X-ray structure refinement. The end result is not a single model, but rather an ensemble of models that are all consistent with the given restraints, as illustrated in Figure 14.1b.

The reasons for generating an ensemble of structures from NMR data are twofold. Firstly, the NMR data are relatively less precise and less numerous than experimental restraints from X rays so that a diversity of structures are consistent with them. Secondly, the biomolecules may genuinely possess heterogeneity in solution.

For general use, an ensemble of models is rather more difficult to handle than a single model. Ensembles deposited in the Protein Data Bank (PDB) can typically comprise 20 models. One of these is often designated as representative of the ensemble, or a separate file containing an average model may be deposited in addition to the ensemble. The separate average structure is energy minimized to counteract the unphysical bond lengths and angles that the averaging process introduces. Such a structure tends to have a separate PDB identifier from that of the ensemble—so the same structure, or rather the outcome of the same experiment, appears as two separate entries in the PDB. This is clearly potentially confusing and the use of separate files is now discouraged. The representative member of an ensemble is usually taken to be the structure that differs least from all other structures in the ensemble. An algorithmic Web-based tool called OLDERADO (http://neon.chem.le.ac.uk/olderado) allows you to select such a representative from an ensemble (Kelley, Gardner, Sutcliffe, 1996), but no single algorithm is universally agreed upon.

## AIM

The aim of this chapter is to demonstrate that not all structures are of equally high quality, usually because of the quality of the experimental data from which they were determined, and that care needs to be taken before using any structure to draw biological or other conclusions. When selecting data sets for deriving general principles about, say, protein structures it is important to filter out those that might give misleading results simply because they are unlikely to be sufficiently accurate or precise to contribute meaningful or correct data to the analysis.

It does seem slightly churlish to reject structures from consideration given the amount of time, care, and hard work the experimentalists have put into solving them. However, if you put unsound data into your analysis, you will get unsound conclusions out. This chapter hopes to explain the limitations of using 3D structures uncritically for structural bioinformatics purposes, and to provide some rules of thumb for weeding out the defective ones: what are the symptoms, what should you look for, and which structures should you reject?

## ERROR ESTIMATION AND PRECISION

All scientific measurements contain errors. No measurement can be made infinitely precisely; so, at some point, say after so many decimal places, the value quoted becomes unreliable. Scientists acknowledge this by estimating and quoting standard uncertainties on their results. For example, the latest value for Boltzmann's constant is $1.3806503(24) \times 10^{-23}$ J K$^{-1}$, where the two digits in brackets represent the standard uncertainty (or $s.u.$) in the last two digits quoted for the constant.

Compare this with the situation we have in relation to the 3D structures of biological macromolecules. Figure 14.2 shows a typical extract from the atom details section of a PDB file. It relates to a single amino acid residue (a lysine) and shows the information deposited about each atom in the protein's structure.

Looking at only the columns representing the $x$-, $y$-, $z$-coordinates you will notice that each value is quoted to three decimal places. This suggests a precision of 1 in $10^5$. Similarly, the $B$-factors (in the final column) are each quoted to two decimal places. Is it possible that the atomic positions and $B$-factors were really so precisely defined? What are the error bounds on these values? What are their $s.u.$s? Are the values accurate to the first place of decimals? The second? The third?

In fact, with the exception of a very few PDB structures, no error bounds are given. As at November 2001 there were 5 such exceptions out of 16,646 structures: one was a carbohydrate (cycloamylose, PDB code 1c58), three were marginally differing copies of the same 13-residue enterotoxin (1etl, 1etm and 1etn), and the fifth was the crystal structure of the 54-residue rubredoxin (4rxn). All had been solved at atomic resolution (ranging from 0.89Å to 1.2Å) and refined by the full-matrix least-squares method that is mentioned below.

Thus, in the overwhelming majority of cases one cannot tell how precisely defined the values are. Why is this so? What kind of scientific measurement is this? And how are we to judge how much reliance to place on the data given?

### Error Estimates in X-Ray Crystallography

*Estimation of Standard Uncertainties.* In X-ray crystallography it is, in theory, possible to calculate the standard uncertainties of the atomic coordinates and

| | Atom number | Atom name | Residue name | Residue number | Atomic coordinates | | | Occupancy | B-factor |
|---|---|---|---|---|---|---|---|---|---|
| | | | | | x | y | z | | |
| ATOM | 1 | N | LEU | 1 | -15.159 | 11.595 | 27.068 | 1.00 | 18.46 |
| ATOM | 2 | CA | LEU | 1 | -14.294 | 10.672 | 26.323 | 1.00 | 9.92 |
| ATOM | 3 | C | LEU | 1 | -14.694 | 9.210 | 26.499 | 1.00 | 12.20 |
| ATOM | 4 | O | LEU | 1 | -14.350 | 8.577 | 27.502 | 1.00 | 13.43 |
| ATOM | 5 | CB | LEU | 1 | -12.829 | 10.836 | 26.772 | 1.00 | 13.48 |
| ATOM | 6 | CG | LEU | 1 | -11.745 | 10.348 | 25.834 | 1.00 | 15.93 |
| ATOM | 7 | CD1 | LEU | 1 | -11.895 | 11.027 | 24.495 | 1.00 | 13.12 |
| ATOM | 8 | CD2 | LEU | 1 | -10.378 | 10.636 | 26.402 | 1.00 | 15.12 |

Figure 14.2. An extract from a PDB file of a protein structure showing how the atomic coordinates and other information on each atom are deposited. The atoms are of a single leucine residue in the protein. The contents of each column are labeled above the column. It can be seen that the $x$-, $y$-, $z$-coordinates of each atom are given to three decimal places.

*B*-factors. In fact, it is routinely done for the crystal structures of small molecules such as those deposited in the Cambridge Structural Database (CSD; Allen et al., 1979). The calculations of the *s.u.*s are performed during the refinement stage of the structure determination. As you learned in Chapter 4, refinement involves modifying the initial model to improve the match between the experimentally determined structure factors—as obtained from the observed X-ray diffraction pattern—and the calculated structure factors—as obtained from applying scattering theory to the current model of the structure. Figure 14.3 illustrates this principle.

In practice refinement is usually a long drawn-out procedure requiring many cycles of computation interspersed here and there with manual adjustments of the model using molecular graphics programs to nudge the refinement process out of any local

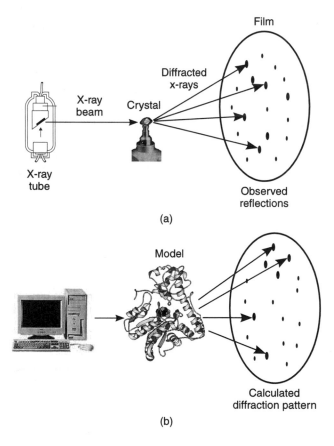

**Figure 14.3.** A schematic diagram illustrating the principle of structure refinement in X-ray crystallography. (a) X-rays are passed through a crystal of the molecule(s) of interest, generating a diffraction pattern from which, by one method or another (see Chapter 5), an initial model of the molecular structure is calculated. (b) Using the model, it is possible to apply scattering theory to calculate the diffraction pattern we would expect to observe. Usually this will differ from the experimental pattern. The process of structure refinement involves iteratively modifying the model of the structure until a better and better fit between the observed and calculated patterns is obtained. The goodness of fit of the two sets of data is measured by the reliability index, or *R*-factor.

minimum that it may have become trapped in. Furthermore, because in protein crystallography the data-to-parameter ratio is poor (the data being the reflections observed in the diffraction pattern and the parameters being those defining the model of the protein structure: the atomic $x$-, $y$-, $z$-coordinates, $B$-factors, and occupancies) the data need to be supplemented by additional information. This extra information is applied by way of geometrical restraints. These are target values for geometrical properties such as bond lengths and bond angles and are typically obtained from crystallographic studies of small molecules. The refinement process aims to prevent the bond lengths and angles in the model from drifting too far from these target values, which is achieved by applying additional terms to the function being minimized of the form:

$$\sum_{k=1}^{Distances} w_k(d_{k0} - d_k)^2,$$

where $d_k$ and $d_{k0}$ are the actual and target distance, and $w_k$ is the weight applied to each restraint.

If the structure is refined using full-matrix least-squares refinement, a by-product of this method is that the $s.u.$s of the refined parameters, such as the atomic coordinates and $B$-factors, can be obtained. However, their calculation involves inverting a matrix whose size depends on the number of parameters being refined. The larger the structure, the more atomic coordinates and $B$-factors, the larger the matrix. As matrix inversion is an order $n^3$ process, it has tended to be unfeasible for molecules the size of proteins and nucleic acids; these have several thousand parameters and consequently a matrix whose elements number several millions or tens of millions, which is why $s.u.$s have been routinely published for small-molecule crystal structures, but not for structures of biological macromolecules. It is purely a matter of size.

Recently, however, as faster workstations with larger memories have become available, the situation has started to change, and calculation of atomic errors has become more practicable (Tickle, Laskowski, and Moss, 1998). Indeed, $s.u.$s are now frequently calculated for small proteins using SHELX (Sheldrick and Schneider, 1997), the refinement package originally developed for small molecules, but sadly, the $s.u.$ data are still not commonly deposited in the PDB file. So this makes us none the wiser about the precision with which any given atom's location has been determined.

So what is to be done? What information is there on the reliability of an X-ray crystal structure? What should one look for?

First of all, there are several parameters relating to the overall quality of the structure commonly cited in the literature that can be found in the header records of the PDB file itself, as described in Global Parameters for X-ray Structures.

***Global Parameters for X-ray Structures.*** Figure 14.4 shows an extract from the header records of a PDB file showing some of the commonly cited global parameters.

RESOLUTION. The resolution at which a structure is determined provides a measure of the amount of detail that can be discerned in the computed electron density map. The reflections at larger scattering angles, $\theta$, in the diffraction pattern correspond to higher resolution information coming as they do from crystal planes with a smaller interplanar spacing. The high-angle reflections tend to be of a lower intensity and more difficult to measure and, the greater the disorder in the crystal, the more of these high-angle reflections will be lost. Resolution is related to how many of these

```
. . .
REMARK    2 RESOLUTION. 2.20 ANGSTROMS.

. . .
REMARK    3    R VALUE              (WORKING SET) : 0.198
REMARK    3    FREE R VALUE                      : 0.255
REMARK    3    FREE R VALUE TEST SET SIZE   (%) : 10.2

. . .
REMARK    3    ESTIMATED COORDINATE ERROR.
REMARK    3    ESD FROM LUZZATI PLOT         (A) : 0.23
REMARK    3    ESD FROM SIGMAA               (A) : 0.23
REMARK    3    LOW RESOLUTION CUTOFF         (A) : 5.00
REMARK    3
REMARK    3    CROSS-VALIDATED ESTIMATED COORDINATE ERROR.
REMARK    3    ESD FROM C-V LUZZATI PLOT     (A) : 0.30
REMARK    3    ESD FROM C-V SIGMAA           (A) : 0.27
. . .
```

Figure 14.4. Extracts from the header records of a PDB file (1ydv) showing some of the statistics pertaining to the quality of the structure as a whole. These include the resolution, $R$-factor, $R_{free}$, and various estimates of average positional errors (ranging from 0.23–0.30Å). The $R_{free}$ has been calculated on the basis of 10.2% of the reflections removed at the start of refinement and not used during it.

high-angle reflections can be observed, although the value actually quoted can vary from crystallographer to crystallographer as there is no clear definition of how it should be calculated. The higher resolution shells will tend to be less complete and some crystallographers will quote the highest resolution shell giving a 100% complete data set, whereas others may simply cite the resolution corresponding to the highest angle of scatter observed.

The higher the resolution the greater the level of detail, and hence the greater the accuracy of the final model. The resolution attainable for a given crystal depends on how well ordered the crystal is—that is, how close the unit cells throughout the crystal are to being identical copies of one another. A simple rule of thumb is that the larger the molecule the lower will be the resolution of data collected.

Figure 14.5 shows an example of how the electron density for a single side chain improves as resolution increases. In general, side chains are difficult to make out at very low resolution (4Å or lower), and the best that can be obtained is the overall shape of the molecule and the general locations of the regions of regular secondary structure. Models at such low resolution are clearly of no use for investigating side-chain conformations or interactions! At 3Å resolution, the path of a protein's chain can be traced through the density and at 2Å the side chains can be confidently fitted.

The most precise structures are the atomic resolution ones (from around 1.2Å resolution up to around 0.9Å). Here the electron density is so clear that many of the hydrogen atoms become visible, and alternate occupancies become more easily distinguishable. These structures require fewer geometrical constraints during refinement and hence give a better indication of the true geometry of protein structures.

The resolution of structures in the PDB varies from atomic resolution structures to very low resolution structures at around 4.0Å, with a definite peak at around 2.0Å. The lowest quoted resolution as of November 2001 was 30.0Å for PDB entry 1qgc—the structure of the capsid protein of the foot-and-mouth virus, complexed with antibody, and solved by a combination of cryoelectron microscopy and X-ray crystallography (Hewat et al., 1997).

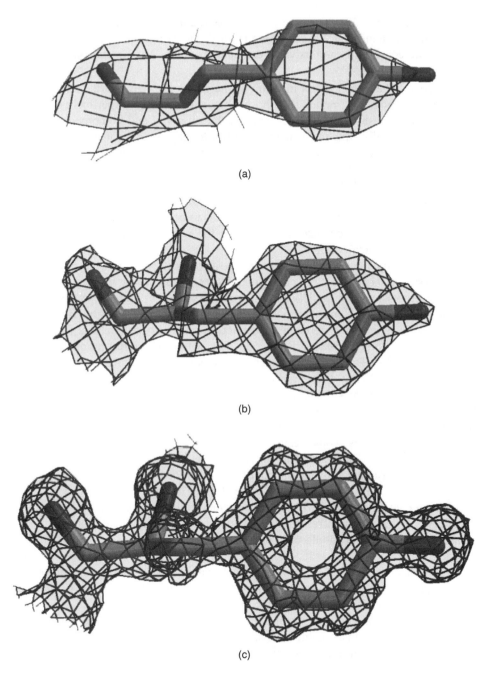

(a)

(b)

(c)

**Figure 14.5.** The effect of resolution on the quality of the electron density. The three plots show the electron density, as the wire-frame cage, surrounding a single tyrosine residue. The residue is Tyr100 from concanavalin A as found in three PDB structures solved at (a) 3.0Å resolution (PDB code 1val), (b) 2.0Å (1con), and (c) 1.2Å (1jbc). At the lowest resolution the electron density is merely a shapeless blob, but as the resolution improves the individual atoms come into clear focus. The electron density maps were taken from the Uppsala Electron Density Server (http://portray.bmc.uu.se/eds) and rendered using BobScript (Esnouf, 1997) and Raster3D (Merritt and Bacon, 1997).

Resolution is probably the clearest measure of the likely quality of the given model. However, bear in mind that, because there is no single definition of resolution, it tends not to be used consistently and its value can be overstated (Weissig and Bourne, 1999).

$R$-FACTOR. The $R$-factor is a measure of the difference between the structure factors calculated from the model and those obtained from the experimental data. In essence, it is a measure of the differences in the observed and computed diffraction patterns schematically illustrated in Figure 14.3. Higher values correspond to poorer agreement with the data, whereas lower values correspond to better agreement. Typically, for protein and nucleic acid structures, values quoted for the $R$-factor tend to be around 0.20 (or equivalently, 20%). Values in the range 0.40 to 0.60 can be obtained from a totally random structure, so structures with such values are unreliable and probably would never be published. Indeed, 0.20 seems to be something of a magical figure and many structures are deemed finished once the refinement process has taken the $R$-factor to this mystical value.

As a reliability measure, however, the $R$-factor is itself somewhat unreliable. It is quite easily susceptible to manipulation, either deliberate or unwary, during the refinement process, and so models with major errors can still have reasonable-looking $R$-factors. For example, one of the early incorrect structures, cited by Brändén and Jones (1990), was that of ferredoxin I, an electron transport protein. The fully refined structure was deposited in 1981 as PDB code 2fd1, with a quoted resolution of 2.0Å and an $R$-factor of 0.262. Due to the incorrect assignment of the crystal space group during the analysis of the X-ray diffraction data, this structure turned out to be completely wrong. The replacement structure, reanalyzed by the original authors and having the correct fold, was deposited as PDB entry 3df1 in 1988. Its resolution was given as 2.7Å and its $R$-factor as 0.35. On the face of it, therefore, mere comparison of the resolution and $R$-factor parameters would lead one to believe the first of the two structures to be the more reliable! The reason that an $R$-factor as low as 0.262 was achieved for a totally incorrect structure was that the coordinates included 344 water molecules, many extending far out from the protein molecule itself. This is a large number of waters for a protein containing only 107 residues. A rule of thumb suggested by Brändén and Jones (1990) is that, for high-resolution structures, one water molecule for each residue is reasonable, and waters should only be added to the structure if they make plausible hydrogen bonds.

Incidentally, the version of ferredoxin that was 3df1 was itself twice superseded, first by entry 4df1 in mid-1988 and then by entry 5df1 in 1993. The last of these had a quoted resolution of 1.9Å and $R$-factor of 0.215.

The ferredoxin example is one of overfitting; that is, having too many parameters for the experimental data available. It is always possible to fit a model, however wrong, to the data if there is an excess of parameters over observations.

$R_{FREE}$. A more reliable measure is Brünger's free $R$-factor, or $R_{free}$ (Brünger, 1992). This is less susceptible to manipulation during refinement. It is calculated in the same way as the standard $R$-factor and again measures the agreement between the structure factors as calculated from the model and as obtained from the experimental data. It differs in that its calculation uses only a small fraction of the experimental data, typically 5–10%, and, crucially, this fraction is excluded from the structure refinement procedure. The test set, as it is called, thus provides an independent measure of the goodness of fit of the model to the data while the refinement proceeds on the remaining data, the working set. Unless there are correlations between the data in the test set and those in the working set, the refinement process should not be able to influence the $R_{\text{free}}$ measure.

The value of $R_{\text{free}}$ will tend to be larger than the $R$-factor, although it is not clear what a good value might be. Brünger has suggested that any value above 0.40 should be treated with caution (Brünger, 1997). There were approximately 20 structures in this category in the PDB, as of November 2001. Not surprisingly, most are fairly low-resolution structures (3.0–4.0Å).

AVERAGE POSITIONAL ERROR. Even though atomic coordinate *s.u.*s are not commonly given, it is quite usual for an estimate of the *average* positional error of a structure's coordinates to be cited. There are two principal methods for estimating the average positional errors: the Luzzati plot (Luzzati, 1952) and the $\sigma_A$ plot (Read, 1986).

The Luzzati plot is obtained by partitioning the reflections from the diffraction pattern into bins according to their value of $\sin \theta$, where $\theta$ is the reflection's scattering angle, and then calculating the $R$-factor for each bin. The value calculated for each bin is plotted as a function of $\sin \theta/\lambda$, where $\lambda$ is the wavelength of radiation used. The resulting plot is compared against the theoretical curves of Luzzati (1952) to obtain an estimate of the average positional error. One problem with this method is that the actual curves do not usually resemble the theoretical ones at all well, and so the error estimate is somewhat crude and often merely provides an upper limit on the error. Better results are obtained if the $R_{\text{free}}$ is used instead of the traditional $R$-factor.

The $\sigma_A$ plot provides a better estimate still. It involves plotting $\ln \sigma_A$ against $(\sin \theta/\lambda)^2$, where $\sigma_A$ is a complicated function that has to be estimated for each $(\sin \theta/\lambda)^2$ bin, as described in Read (1986). The resultant plot should give a straight line whose slope provides an estimate of the average positional error.

Most refinement programs compute both error estimates from the Luzzati and Read methods, so these values are commonly cited in the PDB file. You will find them in the file's header records under the now unfashionable term "estimated standard deviation" (or ESD)—see Figure 14.4.

Bear in mind that an average *s.u.* is exactly what it says: an average over the whole structure. The *s.u.*s of the atoms in the core of the molecule, which tends to be more ordered, will be lower than the average, while those of the atoms in the more mobile and less well-determined surface—and often more biologically interesting—regions will be higher than the average.

ATOMIC $B$-FACTORS. A more direct, albeit merely qualitative, way of determining the precision of a given atom's coordinates is to look at its associated $B$-factor. $B$-factors are closely related to the positional errors of the atoms, although the relationship is not a simple one that can be easily formulated (Tickle, Laskowski, and Moss, 1998). It is safe to say, however, that atoms in a structure with the largest $B$-values will also be those having the largest positional uncertainty. So if high levels of precision are required in your analysis, leave out the atoms having the highest $B$-factors. As a rule of thumb, atoms with $B$-values in excess of 40.0 are often excluded as being too unreliable. Similarly, if atoms in your region of interest, such as an active site, are all cursed with high $B$-factors then your region of interest is not well determined and you will need to be careful about the conclusions you draw from it.

***Rules of Thumb for Selecting X-Ray Crystal Structures.*** Many analyses in structural bioinformatics require the selection of a dataset of 3D structures on which to perform one's analysis. A commonly used rule of thumb for selecting reliable structures for such analyses, where reasonably accurate models are required, is to choose those models that have a quoted resolution of 2.0Å or better, and an $R$-factor of 0.20 or lower. These criteria will give structures that are likely to be reasonably reliable down to the conformations of the side chains and local atom–atom interactions. One example that uses such a dataset is the Atlas of Protein Side-Chain Interactions (http://www.biochem.ucl.ac.uk/bsm/sidechains), which depicts how amino acid sidechains pack against one another within the known protein structures.

Of course, the selection criteria depend on the type of analysis required. For some analyses only atomic resolution structures (i.e., 1.2Å or better) will do, as in the accurate derivation geometrical properties of proteins—for example, side-chain torsional conformers and their standard deviations (EU 3-D Validation Network, 1998), or fine details of the peptide geometry in proteins that can reveal subtle information about their local electronic features (Esposito, et al., 2000). For other types of analysis, structures solved down to 3Å may be good enough, as in any comparison of protein folds. One interesting example is that of the lactose operon repressor. Three structures of this protein were solved to 4.8Å resolution, giving accurate position for only the protein's $C^\alpha$ atoms (Lewis et al., 1996). However, because the three structures were of the protein on its own, of the protein complexed with its inducer, and of the protein complexed with DNA, the global differences between the three structures showed how the protein's conformation changed between its induced and repressed states. Thus even very low resolution structures were able to help explain how this particular protein achieves its biological function (Lewis et al., 1996).

Often the above rule of thumb (resolution $\leq$2.0Å, and $R$-factor $\leq$0.20) is supplemented by a check on the year when the structure was determined. Structures are more likely to be less accurate the older they are simply because experimental techniques have improved markedly since the early pioneering days of the 1960s and 1970s. Indeed, many of the early structures have been replaced by more recent and accurate determinations.

## Error Estimates in NMR Spectroscopy

The theory of NMR spectroscopy does not provide a means of obtaining $s.u.$s for atomic coordinates directly from the experimental data, so estimates of a given structure's accuracy and precision have to be obtained by more indirect means.

*Global Parameters for NMR Structures.* As mentioned above, a number of models can be derived that are compatible with the NMR experimental data. It is difficult to distinguish whether this multiplicity of models reflects real motion within the molecules or simply results from insufficient experimentally derived restraints. (Compare how the most poorly defined regions of the X-ray model of rubredoxin in Figure 14.1a do not necessarily correspond to the most poorly defined regions of the NMR model in Figure 14.1b, although remembering that one structure was in crystal form and the other in solution). Generally, the agreement of NMR models with the NMR data is measured by the agreement between the distance and angular restraints applied during refinement of the models and the corresponding distances and angles in the final models. Large numbers of severe violations would indicate a serious problem of data interpretation and model building.

However, the errors associated with the original experimental data are sufficiently large that it is almost always possible to generate models that do not violate the restraints, or do so only slightly. Consequently, it is not possible to distinguish a merely adequate model from an excellent one by looking for restraint violations alone.

Traditionally, the quality of a structure solved by NMR has also been measured by the root-mean-squared deviation (rmsd) across the ensemble of solutions. Regions with high rmsd values are those that are less well defined by the data. In principle, such rmsd measures could provide a good indicator of uncertainty in the atomic coordinates; however, the values obtained are rather dependent on the procedure used to generate and select models for deposition. An experimentalist choosing the best few structures for deposition from a much larger draft ensemble can result in very misleading statistics for the PDB entry. For example, the best few structures may, in effect, be the same solution with minor variations—so the rmsd values will be small. Structures further down the original list may provide alternative solutions, which are slightly less consistent with the data, but that are radically different. The sizes of ensembles deposited in the PDB range from 1 to 85 models (as of November 2001).

The number of experimentally derived restraints per residue can give an indication of how effectively the NMR data define the structure in a manner analogous to the resolution of X-ray structures. Indeed, the number of restraints per residue correlates with the stereochemical quality of the structures to an extent, but some restraints may be completely redundant and no consistent method of counting is used by depositors.

None of these measures gives a true indication of the accuracy of the models, that is, how well they represent the true structure, and few are reported in the PDB file.

In recent years, NMR equivalents of the crystallographic $R$-factor have been introduced. One method involves the use of dipolar couplings. These provide long-range structural restraints that are independent of other NMR observables such as the NOEs, chemical shifts, and couplings constants that result from close spatial proximity of atoms. Because the expected dipolar couplings can be computed for a given model, they provide a means of comparing observed with expected, and obtaining an $R$-factor that is a measure of the difference between the two (Clore and Garrett, 1999). What is more, it is also possible to obtain a cross-validated $R$-factor, equivalent to the crystallographic $R_{free}$, wherein a subset of dipolar couplings are removed prior to the start of structure refinement and used only for computing the $R$-factor. This gives an unbiased measure of the quality of the fit to the experimental data. However, in the case of NMR, one cannot use a single test set of data; one has to perform a complete cross-validation. The reason for this is that, whereas in crystallography each reflection contains information about the whole molecule, in NMR each dipolar coupling does not. So a complete cross-validation is required, which means that a number of calculations have to be performed, each using a different selection of test sets and working data sets; the test set, which usually comprises 10% of the whole data set, being selected at random each time.

Another technique for calculating an NMR $R$-factor uses the NOEs and involves back-calculation of the NMR intensities from the models obtained and comparison with those observed in the experiment. This technique is implemented in the program RFAC (Gronwald et al., 2000), which calculates not only an overall $R$-factor for the entire structure, but also local $R$-factors, including residue-by-residue $R$-factors and individual $R$-factors for different groups of NOEs (e.g., medium-range NOEs, long-range NOEs, interresidue NOEs, etc.).

An additional back-calculation method for checking structure quality is to calculate the expected frequencies (positions) of spectral peaks from the structure and compare them to those observed. This comparison has the advantage that the frequencies are not usually a target of the structure refinement procedure (Williamson, Kikuchi, and Asakura, 1995).

However, the measures described here are not yet generally included in the deposited PDB files.

*Rules of Thumb for Selecting NMR Structures.* Historically, the rule of thumb for selecting NMR structures for inclusion in structural analyses has been the simple one of excluding them altogether! This early prejudice stems from the fact that they were viewed as being of generally lower quality than X-ray structures, there was no easy way of selecting them with a consistent rule as that used for selecting X-ray structures, and they represented only a minority of the PDB anyway. However, nowadays NMR structures provide much valuable information about protein and DNA structures not available from X-ray studies. Indeed, although only about one in eight PDB structures come from NMR experiments (as at November 2001), in data sets of representative structures (Hobohm and Sander, 1994) around one in four are NMR structures. This stems from the fact that many unique and important proteins can only be solved by NMR.

Nevertheless, it is still not possible to differentiate between reliable and unreliable NMR structures from the information given in the PDB files. There is no standardized information provided that is akin to the resolution, $R$-factor, and estimated $s.u.$s routinely quoted for X-ray crystal structures. The only way to get an idea of the quality of the structure is to read the paper describing it and judge from the statistics provided there or, more ambitiously, to carry out your own analysis of either the stereochemistry of the structure (using the programs that will be described later in this chapter) or the agreement between restraints and structures in those cases where the experimental data has been deposited along with the structure.

## ERRORS IN DEPOSITED STRUCTURES

### Serious Errors

There have been a number of serious errors in X-ray and NMR structures documented in the literature (for references see Brändén and Jones, 1990; Kleywegt, 2000). Many of the erroneous models have been retracted by their original authors, or replaced by improved versions. Structures are often re-refined, or solved with better data, and the models in the PDB are replaced by the improved versions.

The models that are replaced do not completely disappear, though. There is a growing graveyard of obsolete structures—some very, very incorrect, others merely slightly mistaken—available at the Archive of Obsolete PDB Entries (http://pdbobs.sdsc.edu). This Web site provides a graphic history of each structure, some of which have gone through several incarnations (e.g., 1atc, which has been replaced in turn by 3atc, 5atc, 7atc, and 5at1).

Of all errors, the most serious are those where the model is, essentially, completely wrong; for example, the trace of the protein chain follows the wrong path through the electron density and the resultant model has the wrong fold completely. Figures 14.6a and 14.6b give an example of such a case. There is practically no similarity between the correct and incorrect models.

The next most serious errors are where all, or most, of the secondary structural elements have been correctly traced, but the chain connectivity between them is wrong.

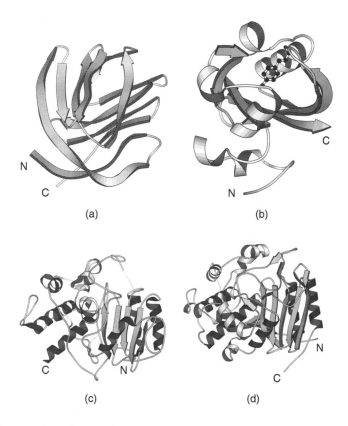

Figure 14.6. Examples of seriously wrong protein models and their corrected counterparts. (a) Incorrect model of photoactive yellow protein (PDB code, 1phy, an all-$C^\alpha$ atom model), and (b) the corrected model (2phy, all atoms plus bound ligand). Superposition of the two models gives an rmsd of 15Å between equivalent $C^\alpha$ atoms. Such a high value is hardly surprising given that the folds of the two models are so completely different. (c) Incorrect model of D-alanyl-D-alanine peptidase (1pte, an all-$C^\alpha$ atom model), and (d) corrected model (3pte, all atoms). The initial model had been solved at low resolution (2.8Å) at a time when the protein's sequence was unknown, so tracing the chain had been much more difficult than usual. Many of the secondary structure elements were correctly detected, but incorrectly connected. The matching secondary structures are shown as the darker shaded helices and strands. The connectivity between them is completely different in the two models, with the earlier model having completely wrong parts of the sequence threaded through the secondary structure elements. Indeed, you can see that the central strand of the $\beta$-sheet runs in the opposite direction in the two models. The N- and C-termini of all models are indicated. All plots were generated using the MOLSCRIPT program (Kraulis, 1991).

An example is given in Figures 14.6c and 14.6d. Here the erroneous model has most of the correct secondary structure elements, and has them arranged in the correct architecture. However, the protein sequence has been incorrectly traced through them (in one case going the wrong way down a $\beta$-strand). Thus most of the protein's residues are in the wrong place in the 3D structure. Such errors arise because the loop regions that connect the secondary structure elements tend to be more flexible, and more disordered, so their electron density tends to be more poorly defined and difficult to interpret correctly. This situation was particularly true in the case shown in Figure 14.6c as the *primary sequence* of the protein was unknown at the time the structure was being solved and had to be guessed from the limited clues in the electron density map.

Less serious are frame-shift errors, although they can often result in a significant part of the model being incorrect. These errors occur where a residue is fitted into the electron density that belongs to the next residue. The frame shift persists until a compensating error is made when two residues are fitted into the density belonging to a single residue. These mistakes often occur at turns in the structure, and almost exclusively at very low resolution (3Å or lower).

The least serious model-building errors involve the fitting of incorrect main-chain or side-chain conformations into the density. Of course, even such errors, depending on where they occur, can have an effect of the biological interpretation of what the structure does and how it does it.

## Typical Errors

Typically, the models deposited in the PDB will be essentially correct. The remaining errors will be the random errors associated with any experimental measurement. As mentioned above for X-ray structures, the average *s.u.*s—estimated on the basis of the Luzzati and $\sigma_A$ plots—can provide an idea of the magnitude of these errors. The values range from around 0.01Å to 1.27Å. Note that the latter value approaches the length of some covalent bonds! The median of the quoted *s.u.*s corresponds to estimated average coordinate errors of around 0.28Å. It has to be remembered that these values are estimates, and apply as an average over the whole model.

Figure 14.7 gives a feel of some typical uncertainties in atomic positions.

## Stereochemical Parameters

An alternative way of assessing a structure's quality, which complements the types of checks described so far, is to examine its geometry, stereochemistry, and other structural properties. A number of tests can be applied to a protein or nucleic acid structure that compare it against what is known about these molecules. This knowledge comes from systematic analyses of the existing structures in the PDB. In other words, the vast body of structures that have been solved to date provides a knowledge base of what is normal for proteins and nucleic acids.

The advantage of such tests of normality is that they do not require access to the original experimental data. Although it is possible to obtain the experimental data for many PDB entries—structure factors in the case of X-ray structures, and distance restraints for NMR ones—these entries are still the minority, and deposition of these data is still at the discretion of the depositors. Furthermore, to make use of the

**Figure 14.7.** Examples of typical uncertainties in atomic positions for (a) an *s.u.* of 0.2Å, (b) 0.3Å, and (c) 0.39Å. The protein is the same rubredoxin from Figure 14.1a. Of course, as shown in Figure 14.1a, the distribution of uncertainties would not normally be so uniform, with higher variability in the surface side-chain atoms than, say, the buried main-chain atoms. Figure also appears in Color Figure section.

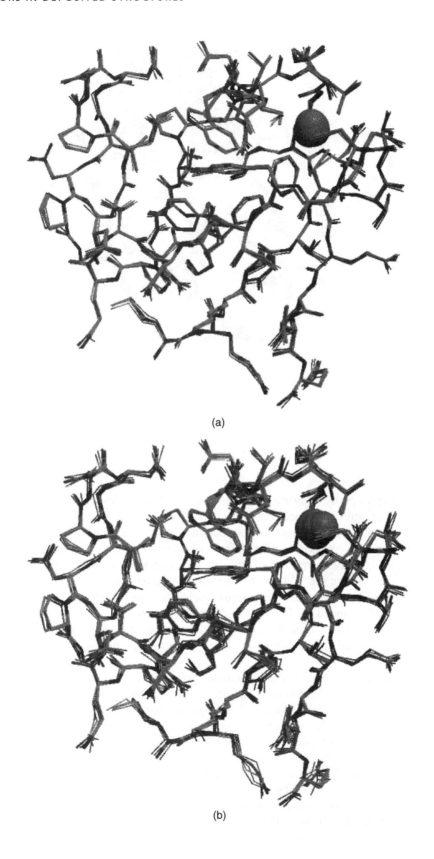

(a)

(b)

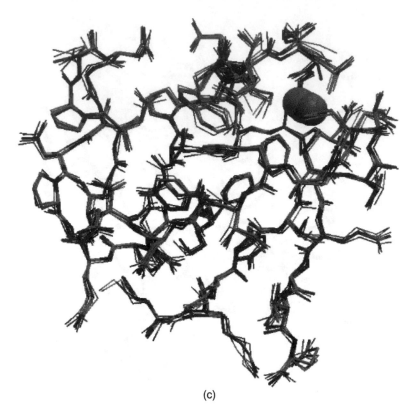

(c)

Figure 14.7. (*Continued*)

data requires appropriate software packages and expert know-how. The stereochemical tests, however, require no experimental data. So any structure, whether experimentally determined, or the result of homology modeling, molecular dynamics, threading, or blind guesswork can be checked. The software is freely available and easy to use and interpret. What is more, many of the results of such checks made on existing structures are readily available on the Web, as will be mentioned below.

Most of the tests described here apply exclusively to protein structures. Similar tests have been developed for DNA and for small molecules (hetero atom groups) that may be bound to protein or DNA. These will be mentioned later. The stereochemical tests include bond lengths, bond angles, torsion angles, hydrogen bond energies, and so on.

Before describing the checks, one crucial point needs to be stressed at the start. The majority of the checks compare a given structure's properties against what is the norm. Yet this norm has been derived from existing structures and could be the result of biases introduced by different refinement practices. Furthermore, outliers, such as an excessively long bond length or an unusual torsion angle, should not be construed as errors. They may be genuine—for example, as a result of strain in the conformation, say, at the active site. The only way of verifying whether oddities are errors or merely oddities is by referring back to the original experimental data. Indeed, the experimenters who solved the structure may already have done this, found the apparent oddity to be correct and commented to that effect in the literature.

Having said that, if a single structure exhibits a large number of outliers and oddities, then it probably does have problems and can safely be excluded from any analyses.

## Proteins

*The Ramachandran Plot.* Perhaps the best-known, and certainly the most powerful, check for the stereochemical quality of a protein structure is the Ramachandran plot (Ramachandran, Ramakrishnan, and Sasisekharan, 1963). This plot is of the $\psi$ main-chain torsion angle versus the $\phi$ main-chain torsion angle for every amino acid residue in the protein (except the two terminal residues, because the N-terminal residue has no $\phi$ and the C-terminus has no $\psi$). In the resulting scatter plot, the points tend to cluster in certain favorable regions, and tend to be excluded from certain disallowed regions due to steric hindrance of the side-chain atoms. Glycine and proline, which have no side chains as such, have slightly different distributions on the plot, although they too have regions from which they are excluded (Fig. 14.8).

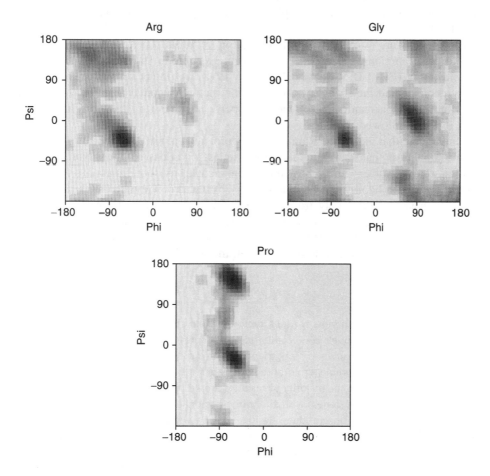

**Figure 14.8.** Differences in Ramachandran plots for (a) arginine, representing a fairly standard amino acid residue, (b) glycine which, due to its lack of a side chain, is able to reach the parts of the plot that other residue cannot reach, and (c) proline which, due to its restraints on the movement of the main chain, has a restricted range of $\phi$ values. The darker regions correspond to the more densely populated regions as observed in a representative sample of protein structures.

The favorable regions correspond to the regular secondary structures: right-handed helices, extended conformation (as found in $\beta$-strands), and left-handed helices. Even residues in loops tend to lie within these favored regions. Figure 14.9a shows a typical Ramachandran plot,. The residues show a tight clustering in the most favored regions with few or none in the disallowed regions. The regions themselves have been determined from an analysis of torsion angles in existing structures in the PDB (see, for example, Morris et al., 1992, or Kleywegt and Jones, 1996).

Figure 14.9b, shows a pathological Ramachandran plot. It comes from a structure that shall remain nameless. Here the majority of the residues lie in the disallowed regions, and it can be confidently concluded that the model has serious problems.

One caveat concerns proteins containing D-amino acids rather than the more common L-amino acids. These residues have the opposite chirality so their $\phi-\psi$ values will be negative with respect to their L-amino cousins. The Ramachandran plot for D-amino acids is the same as for L-amino acids, but with every point reflected through the origin. Thus, proteins such as gramicidin A (e.g., PDB code 1grm) that have many D-amino acids, give Ramachandran plots that look particularly troubling but that may be perfectly correct.

Few models are as extreme as the one in Figure 14.9b. The tightness of clustering tends to be a function of resolution, with atomic resolution structures exhibiting very tight clustering (EU 3-D Validation Network, 1998). At lower resolution, as the data quality declines and the model of the protein structure becomes less accurate, so the points on the Ramachandran plot tend to disperse and more of them are likely to be found in the disallowed regions.

One feature that makes the Ramachandran plot such a powerful indicator of protein structure quality is that it is difficult to fool (unless one does so intentionally by, say, restraining $\phi-\psi$ values during structure refinement as is sometimes done for NMR structures). This reliability was demonstrated by Gerard Kleywegt in Uppsala who once attempted to deliberately trace a protein chain *backwards* through its electron density to see whether it would refine and give the sorts of quality indicators that could fool people into believing it to be a reasonable model (Kleywegt and Jones, 1995). Of the parameters that he tried to fool, the two that seemed least gullible were

---

**Figure 14.9.** Ramachandran plots for (a) a typical protein structure, and (b) a poorly defined protein structure. Each residue's $\phi-\psi$ combination is represented as a black box, except for glycine residues, which are shown as black triangles. The most darkly shaded regions of the plot correspond to the most favorable, or core, regions (labeled A for $\alpha$-helix, B for $\beta$-sheet and L for left-handed helix) where the majority of residues should be found. The progressively lighter regions are the less-favored zones, with the white region corresponding to disallowed $\phi-\psi$ combinations for all but glycine residues. Residues falling within these disallowed regions are shown by the labeled boxes. The plot in a is for PDB code 1ubi, which is of the chromosomal protein ubiquitin. All but one of the protein's 66 nonglycine and nonproline residues are in the core regions of the Ramachandran plot (giving a core percentage of 98.5%). What is more, the points cluster reasonably well in the core regions. The structure was solved by X-ray crystallography at a resolution of 1.8Å. The plot in b exhibits many deviations from the core regions. The structure was solved by NMR, in the early days of the technique, and has a core percentage of 6.8%, while over a third of its residues lie in the disallowed regions. The plots were obtained using the PROCHECK program.

the $R_{\text{free}}$ factor mentioned above and the Ramachandran plot. The latter looked most unhealthy, with several residues in disallowed regions and no significant clustering in the most highly favored regions.

A simple measure of quality that can be derived from the plot is the percentage of residues in the most favorable or core regions. (Glycines and prolines are excluded from this percentage because of their unique distributions of available $\phi-\psi$ combinations).

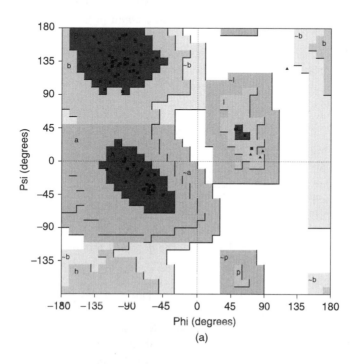

(a)

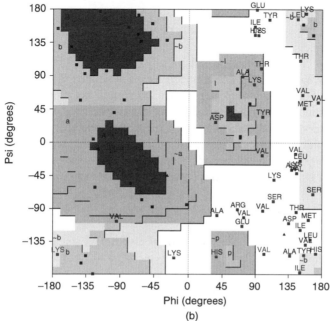

(b)

Using the core regions defined by the PROCHECK program (see below), one generally finds that atomic resolution structures have well over 90% of their residues in these most favorable regions. For lower and lower resolution structures this percentage drops, with structures solved to 3.0–4.0Å tending to have a core percentage around 70%. NMR structures also show increasing core percentage with increasing experimental information. However, NMR structures can have relatively good side-chain positions even with a poor core percentage as NMR data restrain side chains more strongly than the backbone because of the large number of side-chain protons.

***Side-Chain Torsion Angles.*** Protein side chains tend to have preferred conformations, known as rotamers, about their rotatable bonds, again as a result of steric hindrance. The rotamers are defined in terms of the side-chain torsion angles $\chi_1$, $\chi_2$, $\chi_3$, and so on. The first of these, $\chi_1$, is defined as the torsion angle about $N-C^{\alpha}-C^{\beta}-A^{\gamma}$, where $A^{\gamma}$ is the next atom along the side chain (for example, in lysine the $A^{\gamma}$ atom is $C^{\gamma}$). The next, $\chi_2$, is defined as $C^{\alpha}-C^{\beta}-A^{\gamma}-A^{\delta}$, and so on. The $\chi_1$ and $\chi_2$ distributions are both trimodal with the preferred torsion angle values being termed gauche-minus ($+60°$), trans ($+180°$), and gauche-plus ($-60°$). A plot of $\chi_2$ against $\chi_1$ for each residue has $3 \times 3$ preferred combinations, although the strength of each depends very much on the residue type. Figure 14.10 shows some examples of the distributions for different amino acid types.

Like the Ramachandran plot, a plot of the $\chi_1-\chi_2$ torsion angles can indicate problems with a protein model as these, like the $\phi$ and $\psi$ torsion angles, tend not to be restrained during refinement. What is more, these torsion angles tend to cluster more tightly toward their ideal rotameric values as resolution improves (EU 3-D Validation Network, 1998). For example, the standard deviation of the $\chi_1$ torsion angles about their ideal position tends to be around $8°$ for atomic resolution structures and can go as high as $25°$ for structures solved at 3.0Å. Similarly, the corresponding standard deviations for the $\chi_2$ torsion angles tend to be $10°$ and $30°$, respectively.

***Bad Contacts.*** Another good check for structures to be wary of is the count of bad and unfavorable atom–atom contacts that they possess. Too many and the model may be a poor one.

The simplest checks are those which merely count bad contacts, that is, those where the distance between any pair of nonbonded atoms is smaller than the sum of their van der Waals radii. Furthermore, the atoms checked should not merely be those involved in intraprotein contacts within the given protein structure; for X-ray crystal structures it is also necessary to consider atoms from molecules related by crystallographic and noncrystallographic symmetry.

More sophisticated checks consider each atom's environment and determine how happy that atom is likely to be in that environment. For example, the ERRAT program

**Figure 14.10.** Examples of $\chi_1-\chi_2$ distributions for six different amino acid residue types: Arg, Asn, Asp, His, Ile, and Leu. The darker regions correspond to the more densely populated regions as observed in a representative sample of protein structures. The dotted lines represent idealized rotameric torsion angles at $60°$, $180°$, and $300°$ (equivalent to $-60°$). It can be seen that the true rotameric conformations differ slightly from these values and that the different side-chain types have very different $\chi_1-\chi_2$ distribution preferences.

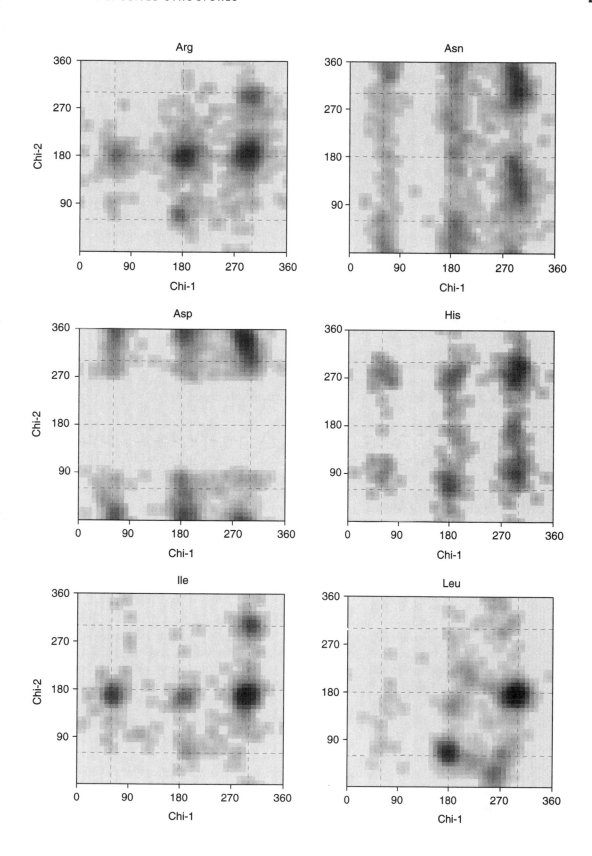

(Colovos and Yeates, 1993) counts the numbers of nonbonded contacts, within a cutoff distance of 3.5Å, between different pairs of atom types. The atoms are classified as carbon (C), nitrogen (N), and oxygen/sulfur (O), so there are six distinct interaction types: CC, CN, CO, NN, NO, OO. If the frequencies of these interaction types differ significantly from the norms (as obtained from well-refined high-resolution structures) the protein model may be somewhat suspect. A similar analysis can be used to locate local problem regions by using a nine-residue sliding window and obtaining the interaction frequencies at each window position.

One level up in sophistication is the DACA method (Vriend and Sander, 1993), which is implemented in the WHAT IF program (Vriend, 1990). DACA stands for Directional Atomic Contact Analysis and compares the 3D environment surrounding each residue fragment in the protein with normal environments computed from a high-quality data set of protein structures. There are 80 different fragment types, including main-chain fragments as well as side-chain fragments. The environment of each fragment is essentially the count of different nonbonded atoms in each $1\text{Å} \times 1\text{Å} \times 1\text{Å}$ cell of a $16\text{Å} \times 16\text{Å} \times 16\text{Å}$ cube surrounding the fragment.

A similar approach is that of the ANOLEA program (Atomic NOn-Local Environment Assessment), which calculates a nonlocal energy for atom–atom contacts based on an atomic mean force potential (Melo and Feytmans, 1998).

***Other Parameters.*** Other parameters that can be used to validate protein structures include counts of unsatisfied hydrogen bond donors and hydrogen-bonding energies as is done in the WHATCHECK program mentioned below (Hooft et el., 1996). See also Chapter 15.

***C-alpha Only Structures.*** As of November 2001, there were around 200 structures in the PDB (out of over 16,000) that contain one or more protein chains for which only the $C^\alpha$ coordinates have been deposited. The deposition of $C^\alpha$-only coordinate sets is usually done where the data quality has been too poor to resolve more of the structure. It was common in the early days of protein crystallography for only $C^\alpha$s to be deposited; nowadays it is still quite common for only $C^\alpha$s to be deposited for very large structures, such as the recently determined structure of the ribosome at 5.5Å (PDB codes 1gix and 1giy).

The standard validation checks are of no use for such models, lacking as they are in so much of their substance. However, there is an equivalent to the Ramachandran plot for these structures (Kleywegt, 1997). The parameters plotted are the $C^\alpha - C^\alpha - C^\alpha - C^\alpha$ torsion angle as a function of the $C^\alpha - C^\alpha - C^\alpha$ angle for every residue in the protein. As with the Ramachandran plot, there are regions of this plot that tend to be highly populated, and others that appear forbidden. So a structure with many outliers in the forbidden zones should be treated with caution. The checks are incorporated in the program MOLEMAN2 which can be run over the Web (see Table 14.1).

## Nucleic Acids

Finding validation tools for DNA and RNA is trickier than for proteins. The PDB's validation tool, ADIT (AutoDep Input Tool), incorporates a program called NuCheck (Feng, Westbrook, and Berman, 1998) for validating the geometry of DNA and RNA. Binary versions of the ADIT package can be downloaded for use on SGI and Linux machines (see Table 14.2).

TABLE 14.1. WWW Servers for Checking Structure Coordinates Online

| Program | Reference | Protein/DNA | URL |
|---|---|---|---|
| ANOLEA | Melo and Feytmans, 1998 | Protein | www.fundp.ac.be/sciences/ biologie/bms/CGI/test.htm |
| Biotech Validation: PROCHECK, PROVE, WHAT IF | EU 3-D Validation Network, 1998 | Protein | biotech.embl-ebi.ac.uk:8400 |
| DACA | Vriend and Sander, 1993 | Protein | www.cmbi.kun.nl/ gv/servers/WIWWWI/ oldqua.html |
| ERRAT | Colovos and Yeates, 1993 | Protein | www.doe-mbi.ucla.edu/ Services/ERRAT |
| MC-Annotate | Gendron, Lemieux, and Major, 2001 | RNA | www-lbit.iro.umontreal.ca/ mcannotate |
| MOLEMAN2 | Kleywegt, 1997 | Protein (C-alpha only) | xray.bmc.uu.se/cgi-bin/gerard/ rama_server.pl |
| Verify3D | Bowie, Lüthy and Eisenberg, 1991 | Protein | www.doe-mbi.ucla.edu/ Services/Verify_3D |

TABLE 14.2. Programs for Checking Structure Coordinates

| Program name | Reference | URL |
|---|---|---|
| ADIT | PDB | pdb.rutgers.edu/mmcif/ADIT |
| ERRAT | Colovos and Yeates, 1993 | www.doe-mbi.ucla.cdu/People/Yeates/ Gallery/Errat.html |
| PROCHECK | Laskowski et al., 1993 | www.biochem.ucl.ac.uk/~roman/ procheck/procheck.html |
| PROVE | Pontius, Richelle, and Wodak, 1996 | www.ucmb.ulb.ac.be/SCMBB/PROVE |
| SQUID | Oldfield, 1992 | www.yorvic.york.ac.uk/~oldfield/ squidmain.html |
| WHATCHECK | Hooft et al., 1996 | www.cmbi.kun.nl/gv/whatcheck |
| WHAT IF | Vriend, 1990 | www.cmbi.kun.nl/whatif |

A program specifically developed for checking the geometry of RNA structures, but that can also be used for DNA structures, is MC-Annotate (Gendron, Lemieux, and Berman, 2001). It computes a number of peculiarity factors, based on various metrics including torsion angles and root-mean-square deviations from standard conformations, that can highlight irregular regions in the structure that may be in error or merely under strain.

## Hetero Groups

The geometry of hetero compounds, as deposited in structures in the PDB, tends to be of widely varying quality. The HETZE program (Kleywegt and Jones, 1998) is one of the few validation methods that checks various geometrical parameters of the hetero compounds associated with PDB structures. These parameters include bond lengths, torsion angles, and some virtual torsion angles, the information principally coming from the small-molecule structures in the CSD (Allen et al., 1979).

## Software for Quality Checks

A large number of programs are freely available that can perform the sorts of quality checks described above on proteins, nucleic acids, and hetero compounds. Below are listed the most commonly used programs not requiring any specialist knowledge or additional specialist software. Details of how to obtain the programs are given in Table 14.2.

*PROCHECK.* PROCHECK (Laskowski et al., 1993) computes a number of stereo-chemical parameters for a given protein model and outputs the results in easy-to-understand colored plots in PostScript format. Significant deviation in the parameters from the standards that have been derived from a database of well-refined high-resolution proteins are highlighted as being unusual. The plots include: Ramachandran plots, both for the protein as a whole and for each type of amino acid; $\chi_1 - \chi_2$ plots for each amino acid type; main-chain bond lengths and bond angles; secondary structure plot; deviations from planarity of planar side chains; and so on.

*WHATCHECK and WHAT IF.* The WHATCHECK program (Hooft et al., 1996) is a subset of Gert Vriend's WHAT IF package (Vriend, 1990). It contains an enormous number of checks and produces a long and very detailed output of discrepancies of the given protein structure from the norms. The DACA method, mentioned above, for analyzing nonbonded contacts, is incorporated into the original WHAT IF program.

*PROVE.* PROVE compares atomic volumes against a set of precalculated standard values (Pontius, Richelle, and Wodak, 1996). Volumes are calculated using Voronoi polyhedra to define the space that each atom occupies by placing dividing planes between it and its neighbors.

*SQUID.* The SQUID program (Oldfield, 1992) displays two-dimensional and three-dimensional data derived from protein structures using many graph types. It can also be used for validation via ready-to-use scripts.

*ERRAT.* The ERRAT program has already been described. It analyzes nonbonded atom contacts in protein structures in terms of CC, CN, CO, and so forth contacts.

## QUALITY INFORMATION ON THE WEB

Rather than having to install and run one of the above packages, it is possible to obtain much of the information they provide from the Web. Several sites provide precomputed quality criteria for all existing structures in the PDB. Other sites allow you upload your own PDB file, via your Web browser, and will run their validation programs on it and provide you with the results of their checks.

### PDBsum—PROCHECK Summaries

The first site that provides precomputed quality criteria is the PDBsum Web site (Laskowski, 2001) at http://www.biochem.ucl.ac.uk/bsm/pdbsum. This Web site specializes in structural analyses and pictorial representations of all PDB structures. Each structure containing one or more protein chains has a PROCHECK and a WHAT CHECK button. The former gives a Ramachandran plot for all protein chains in the structure, together with summary statistics calculated by the PROCHECK program. These results can provide a quick guide to the likely quality of the structure, in addition to the structure's resolution, $R$-factor and, where available, $R_{\text{free}}$.

The WHATCHECK button links to the PDBREPORT for the structure, described below.

Occasionally the model of a protein structure is so bad that one can tell immediately from merely looking at the secondary structure plot on the PDBsum page. Most proteins have around 50–60% of their residues in regions of regular secondary structure, that is, in $\alpha$-helices and $\beta$–strands. However, if a model is really poor, the main-chain oxygen and nitrogen atoms responsible for the hydrogen-bonding that maintains the regular secondary structures can lie beyond normal hydrogen-bonding distances; so the algorithms that assign secondary structure (Chapter 17) may fail to detect some of the $\alpha$-helices and $\beta$–strands that the correct protein structure contains. Figure 14.11 gives an example of the secondary structure contents for a typical protein and for the protein that had the poor Ramachandran plot in Figure 14.9b.

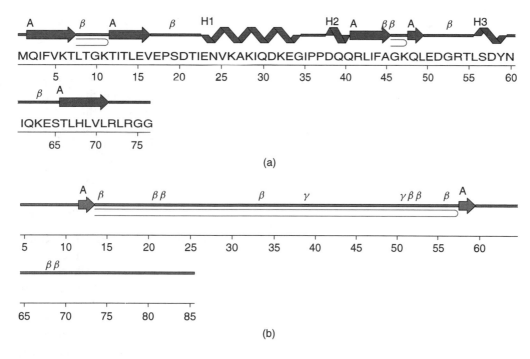

(a)

(b)

Figure 14.11. Schematic diagrams of two protein models in the PDB. (a) A typical protein showing an expected 50–60% of its residues in $\alpha$–helices (shown schematically by the sawtooth regions) and $\beta$–strands (shown by arrows). (b) A poorly defined model that has hardly any regions of secondary structure at all. The labels and symbols correspond to various secondary structure motifs. The $\beta$ and $\gamma$ symbols identify $\beta$- and $\gamma$–turns, while the hairpinlike symbols correspond to $\beta$–hairpins. The helices are labeled H1–H3 in a, and strands are labeled A for $\beta$-sheet A. The Ramachandran plots for both models are shown in Figure 14.7. The sequence of the protein in b has been removed to hinder identification. The above plots were obtained from the PDBsum database.

## PDBREPORT—WHATCHECK Results

The WHATCHECK button on the PDBsum page leads to the WHAT IF Check report on the given protein's coordinates. This report is a detailed listing (plus an even more detailed one, called the Full report) of the numerous analyses that have been precomputed using the WHATCHECK program. These analyses include space group and symmetry checks, geometrical checks on bond lengths, bond angles, torsion angles, proline puckers, bad contacts, planarity checks, checks on hydrogen-bonds, and more, including an overall summary report intended for users of the model. The PDBREPORT database can be accessed directly at http://www.cmbi.kun.nl/gv/pdbreport.

## PDB's Geometry Analyses

The PDB Web site (http://www.rcsb.org/pdb) also has geometrical analyses on each entry, consisting of tables of average, minimum, and maximum values for the protein's bond lengths, bond angles, and dihedral angles. Unusual values are highlighted. It is also possible to view a backbone representation of the structure in RasMol, colored according to the Fold Deviation Score—the redder the coloring the more unusual the residue's conformational parameters.

## Validation Servers on the Web

In addition to the sites mentioned above, there are a number of validation servers on the Web that allow you to submit a PDB file for analysis. Table 14.1 lists these servers. They are mostly for protein structures and most use programs that are freely available for in-house use (see Table 14.2). However, the servers can often be easier and more convenient to use, and of course save you having to download and install the programs, particularly the Biotech Validation server that runs the three most commonly used validation programs: PROCHECK, PROVE, and WHATCHECK.

## CONCLUSION

The main aim of this chapter is to impress on you that the macromolecular structures that form the very foundation of structural bioinformatics are not all of the same quality and can undermine that foundation if not carefully selected. All structures are just models devised to satisfy data obtained experimentally. As such, they will contain errors, both systematic and random. Some structures have been found to be seriously incorrect, that is, they are inaccurate models of the molecules they represent and in many cases have been replaced by more accurate models. Most structures are reasonably accurate but inevitably contain random errors, as is symptomatic of any experimental measurement. The quality of structures as a whole has improved over the past few years and this trend is expected to continue. However, determining which is a good structure and which is not is still not straightforward. Even traditional measures, such as the resolution and $R$-factor for X-ray structures, and number of restraints for NMR structures, do not always separate the good from the bad. Very often, other quality measures need to be taken into account when selecting a good data set.

The chapter has surveyed the information available, and some of the additional tests that can be performed to ensure that the reliability of any structures used is consistent with the conclusions to be drawn from them.

## ACKNOWLEDGMENTS

Thanks to Dr Mark Williams for his valuable comments on the text of this chapter.

## REFERENCES

Allen FH, Bellard S, Brice MD, Cartwright BA, Doubleday A, Higgs H, Hummelink T, Hummelink-Peters BG, Kennard O, Motherwell WDS, Rodgers JR, Watson DG (1979): The Cambridge Crystallographic Data Centre: computer-based search, retrieval, analysis and display of information. *Acta Crystallogr* B35:2331–9.

Bowie JU, Lüthy R, Eisenberg D (1991): A method to identify protein sequences that fold into a known three-dimensional structure. *Science* 253:164–70.

Brändén C-I, Jones TA (1990): Between objectivity and subjectivity. *Nature* 343:687–9. [One of the first papers to raise the issue of errors in protein structures, citing five examples of published protein structures that had been found to contain serious errors and had been replaced by more accurate models. The paper outlines the sources of errors that can arise during crystallographic structure determination and lists various means of limiting them.]

Brünger AT (1992): Free $R$ value: a novel statistical quantity for assessing the accuracy of crystal structures. *Nature* 355:472–5. [This paper represents an important milestone in accurate protein structure determination. It describes the application of cross-validation, a standard statistical technique, to the calculation of the $R$-factor, giving the $R_{free}$, which is unbiased and not vulnerable to artificial reduction. The $R_{free}$ is now a standard measure of a structure's goodness of fit to the data from which it was determined.]

Brünger AT (1997): Free $R$ value: cross-validation in crystallography. *Methods Enzymol* 277:366–96. [A more detailed exposition of the above.]

Clore GM, Garrett DS (1999): $R$-factor, free $R$, and complete cross-validation for dipolar coupling refinement of NMR structures. *J Am Chem Soc* 121:9008–12. [Description of measures equivalent to the crystallographic $R$-factor and $R_{free}$ for structures solved by solution NMR.]

Clore GM, Gronenborn AM (1998): New methods of structure refinement for macromolecular structure determination by NMR. *Proc Natl Acad Sci USA* 95:5891–8.

Colovos C, Yeates TO (1993): Verification of protein structures: patterns of nonbonded atomic interactions. *Protein Sci* 2:1511–9.

Esnouf RM (1997): An extensively modified version of MolScript that includes greatly enhanced coloring capabilities. *J Mol Graph Model* 15:132–4.

Esposito L, Vitagliano L, Zagari A, Mazzarella L (2000): Experimental evidence for the correlation of bond distances in peptide groups detected in ultrahigh-resolution protein structures. *Protein Eng* 13:825–8.

EU 3-D Validation Network (1998): Who checks the checkers? Four validation tools applied to eight atomic resolution structures. *J Mol Biol* 276:417–36. [An examination of the standard software tools used for validating protein structures by testing their performance on structures solved to atomic resolution that, by definition, are as good as can be obtained. For the most part, the validation parameters were found to hold for these high-resolution structures, although some modifications were called for, particularly in the tightening of the values. Surprisingly, the analysis also suggested certain modification to the refinement protocols used for such high-quality structures.]

Feng Z, Westbrook J, Berman HM (1998): *NUCheck. Computer Program*. New Brunswick, NJ: Rutgers University, NDB-407.

Gendron P, Lemieux S, Major F (2001): Quantitative analysis of nucleic acid three-dimensional structures. *J Mol Biol* 308:919–36.

Gronwald W, Kirchhöfer R, Görler A, Kremer W, Ganslmeier B, Neidig K-P, Kalbitzer HR (2000): RFAC, a program for automated NMR $R$-factor estimation. *J Biomol NMR* 17:137–51.

Hewat EA, Verdaguer N, Fita I, Blakemore W, Brookes S, King A, Newman J, Domingo E, Mateu MG, Stuart DI (1997): Structure of the complex of an Fab fragment of a neutralizing antibody with foot-and-mouth disease virus: positioning of a highly mobile antigenic loop. *EMBO J* 16:1492–500.

Hobohm U, Sander C (1994): Enlarged representative set of protein structures. *Protein Sci* 3:522–4.

Hooft RWW, Vriend G, Sander C, Abola EE (1996): Errors in protein structures. *Nature* 381:272. [This paper is the reference cited for the WHATCHECK program, although it hardly describes that program at all. Rather, it describes a detailed analysis, using the program, of "errors" in the 3442 entries in the PDB at the time; an analysis that stirred up not a little controversy among the crystallographic community at the time.]

Kelley LA, Gardner SA, Sutcliffe MJ (1996): An automated approach for clustering an ensemble of NMR-derived protein structures into conformationally-related subfamilies. *Protein Eng* 9:1063–5.

Kleywegt GJ (1997): Validation of protein models from C-alpha coordinates alone. *J Mol Biol* 273:371–6. [Derivation of the equivalent of the Ramachandran plot for protein structures for which only the coordinates of the $C^\alpha$ atoms have been determined.]

Kleywegt GJ (2000): Validation of protein crystal structures. *Acta Crystallogr* D56:249–65. [An excellent and detailed overview of the types and causes of errors in X-ray crystal structures and the measures that crystallographers need to take to reduce them as far as is possible.]

Kleywegt GJ, Jones TA (1995): Where freedom is given, liberties are taken. *Structure* 3:535–40.

Kleywegt GJ, Jones TA (1996): Phi/psi-chology: Ramachandran revisited. *Structure* 4:1395–400. [A detailed study of the Ramachandran plot—one of many, but a good one.]

Kleywegt GJ, Jones TA (1998): Databases in protein crystallography. *Acta Crystallogr* D54:1119–31.

Kraulis PJ (1991): MOLSCRIPT: a program to produce both detailed and schematic plots of protein structures. *J Appl Crystallogr* 24:946–50.

Laskowski RA (2001): PDBsum: summaries and analyses of PDB structures. *Nucleic Acids Res* 29:221–2.

Laskowski RA, MacArthur MW, Moss DS, Thornton JM (1993): PROCHECK—a program to check the stereochemical quality of protein structures. *J Applied Crystallogr* 26:283–91.

Lewis M, Chang G, Horton NC, Kercher MA, Pace HC, Schumacher MA, Brennan RG, and Lu P ((1996):): Crystal structure of the lactose operon repressor and its complexes with DNA and inducer. *Science* 271:1247–1254.

Luzzati PV (1952): Traitement statistique des erreurs dans la determination des structures cristallines. *Acta Crystallogr* 5:802–10. [A description, in French, of the Luzzati plot, which has come to be used for estimating the average positional errors in crystal structures, despite this not having been the paper's original purpose; its aim had been to estimate the positional changes required to reach an $R$-value of zero.]

Melo F, Feytmans E (1998): Assessing protein structures with a non-local atomic interaction energy. *J Mol Biol* 277:1141–52.

Merritt EA, Bacon DJ (1997): Raster3D: photorealistic molecular graphics. *Methods Enzymol* 277:505–24.

Morris AL, MacArthur MW, Hutchinson EG, Thornton JM (1992): Stereochemical quality of protein-structure coordinates. *Proteins Struct Func Genet* 12:345–64. [Derivation of a number of stereochemical parameters that appeared to be good indicators of protein structure quality in that they were well correlated with resolution. As most of the parameters are not among those restrained during structure refinement, they provide a useful independent measure of how well a structure agrees with what appears to be the norm for proteins. A few discrepancies from the norm are to be expected for any structure, but, if there are many, this suggests there may be something seriously wrong with it.]

Oldfield TJ (1992): SQUID: a program for the analysis and display of data from crystallography and molecular-dynamics. *J Molec Graphics* 10:247–52.

Pontius J, Richelle J, Wodak SJ (1996): Deviations from standard atomic volumes as a quality measure for protein crystal structures. *J Mol Biol* 264:121–36.

Ramachandran GN, Ramakrishnan C, Sasisekharan V (1963): Stereochemistry of polypeptide chain configurations. *J Mol Biol* 7:95–9. [The classic analysis of the distribution of $\phi$-$\psi$ torsion angles in protein main chains, which has given us the Ramachandran plot: one of the most powerful methods for checking whether a protein structure seems reasonable or contains severe errors or strained conformations.]

Read RJ (1986): Improved Fourier coefficients for maps using phases from partial structures with errors. *Acta Crystallogr* A42:140–9. [Derivation of the $\sigma_A$ plot that is used to estimate the average positional errors in crystal structures.]

Sheldrick GM, Schneider TR (1997): SHELXL: high resolution refinement. *Methods Enzymol* 277:319–43. [SHELX was a least-squares structure refinement method originally developed for small-molecule crystallography, but which has been adapted and developed over the years for handling macromolecular structures. In protein crystallography it is most commonly used for refining atomic resolution structures and can be used to calculate standard uncertainties in the atomic positions by full-matrix refinement in the final cycle.]

Tickle IJ, Laskowski RA, Moss DS (1998): Error estimates of protein structure coordinates and deviations from standard geometry by full-matrix refinement of $\gamma$B- and $\beta$B2-crystallin. *Acta Crystallogr* D54:243–52.

Vriend G (1990): WHAT IF: A molecular modeling and drug design program. *J Molec Graphics* 8:52–6.

Vriend G, Sander C (1993): Quality-control of protein models: directional atomic contact analysis. *J Appl Crystallogr* 26:47–60.

Weissig H, Bourne PE (1999): An analysis of the Protein Data Bank in search of temporal and global trends. *Bioinformatics* 15:807–31. [An interesting overview of the quality of structures in the PDB and how this has improved with time. It includes an analysis of obsolete entries, that is, entries that have since been superseded, and the various reasons for their replacement.]

Williamson MP, Kikuchi J, Asakura Y (1995): Application of 1H NMR chemical shifts to measure the quality of protein structures. *J Mol Biol* 247:541–6.

# 15

# ALL-ATOM CONTACTS: A NEW APPROACH TO STRUCTURE VALIDATION

Jane S. Richardson

The enormous wealth of macromolecular structure data already available and the even greater wealth soon to come—from structural genomics, from the push for atomic-resolution structures, and from the push to solve much larger biological complexes—provide a treasure trove of functional, interactional, and evolutionary data that will change how one can do biology. But, in order to make effective use of this great resource, it is important, among other things, to take into account the very large spread of accuracy in those data. Relatively low-resolution structures can be among the most valuable if they are of critical molecules or of large and complex cellular machinery. These structures show overall fold and relative positioning of their parts and they often illuminate function in surprising ways, but one should not expect to learn from them fine details in an active site or critical differences that determine substrate or inhibitor specificity. At the other extreme, increasing numbers of structures are being solved at better than 1Å resolution, where one can reliably detect minute changes and disentangle multiple conformations of side chains and waters. Within a given structure there can be even wider variability in quality. Regardless of resolution, most structures have some parts disordered enough that they are not visible in a crystallographic electron-density map (or have no observable NMR constraints). In some cases their coordinates will actually be missing, but more often disordered areas are indicated by a high crystallographic $B$-factor or highly divergent conformations in an NMR ensemble. If a particular part of a structure is important to the question being asked, such telltale signs should always be heeded.

*Structural Bioinformatics*
Edited by Philip E. Bourne and Helge Weissig
Copyright © 2003 by Wiley-Liss, Inc.

Many of the basic quality indicators such as resolution, $B$-factor, R and free R residuals (measures of how well the model accounts for the observed data), and model root-mean-square deviation (rmsd) are directly reported in the Protein Data Bank (PDB) coordinate file (Chapter 8; Berman et al., 2000). Beyond those indicators, the subject known as structure validation (Chapter 14) provides further tools for assessing both overall and local accuracy of structures. Standard validation programs such as ProCheck and WhatIf provide an excellent set of widely-used tools, centering especially on ideality of molecular geometry and on whether backbone $\phi$, $\psi$ angles occur outside the preferred *core* regions. Of special importance in validation are independent criteria not explicitly part of the target function optimized by the structure-refinement process, because their deviations are much more sensitive indicators of problems. The two classic such indicators are the $\phi$, $\psi$ or Ramachandran plot (Laskowski et al., 1993; Ramachandran et al., 1963; Lovell et al., 2002), since $\phi$, $\psi$ values are not in the target function, and the free R factor (Brunger 1992), the agreement with a designated 5–10% of the data that are deliberately kept out of refinement in order to provide an unbiased indicator of progress in model quality.

Recently we have discovered, in a surprisingly simple place, a new source of information for an unbiased and sensitive validation criterion: the hydrogen atoms. They constitute about half of the atoms, but they usually are ignored for technical or expediency reasons. H atoms are, of course, important and present in NMR structures (see Chapter 5), although often not treated with full radius. In macromolecular crystallography, polar H atoms are typically added to better define H bonds but with no van der Waals terms, while nonpolar H atoms are added and refined against the data only at ultra-high (near 1Å) resolution. The main reason for this omission is that hydrogens diffract X rays very poorly, so that they can be directly detected only under the best of conditions.[1] Another reason is that including hydrogens doubles the number of parameters if they are treated as fully independent, which is acceptable only when there is a large enough number of experimental observations. Finally, only recently has computer speed allowed the extra cost in time, either for structure refinements or for theoretical calculations. H atom volume is standardly accounted for by using larger united-atom radii for the other atoms, but the directionality and specificity of H interactions are not represented. The net result of all this is that the crystallographers have obligingly ignored half their atoms in refinement, managing to do quite well without them but now giving us the opportunity to use the correctness of the hydrogens' tight and specific packing interactions as both a global, and especially a local, validation criterion. This new method (Word et al., 1999a) is called all-atom contact analysis.

As applied to the structural database, all-atom contact analysis has two different goals. The first, long-term goal is to actually improve the accuracy of the data, by having structural biologists apply the criteria themselves and fix many errors before coordinates are deposited (a similar process occurred several years ago with routine application of free-R and Ramachandran-plot criteria). The second goal is to give users of the database an easy and effective way to assess local structural accuracy. The first goal would produce higher-grade ore for data mining, while the second improves the extraction process.

---

[1] The invisibility of hydrogens is actually very fortunate, because it produces the beautifully clear separation between aliphatic side chains in protein interiors at moderate resolution.

## THE METHOD OF ALL-ATOM CONTACT ANALYSIS

The all-atom method must start off with a reliable way to add H atoms and optimize their positions, which is done by the program Reduce (Word et al., 1999b). A
great many of the hydrogen positions are completely determined by the heavier atoms:
methylene, methine, backbone NH, aromatic H, and so forth. OH rotations and His pro-
tonation, however, must clearly be optimized relative to the surrounding structure. Less
obviously, the 180° flip orientation of Asn and Gln side-chain amides (and also flips of
His rings) need to be optimized; they are fairly often incorrect as reported, because the
N and O atoms of the amide are not easily distinguished by the experimental X-ray data.
However, the choice can reliably be made if both H bonding and potential clashes of the
$NH_2$ are considered (Word et al., 1999b). We have found, surprisingly, that most methyl
rotations do not actually need to be optimized because they are remarkably relaxed in
protein structures, with departures from staggered orientation seldom much above 10°.
$NH_3$ groups and Met side-chain methyls do however need rotational optimization. The
Reduce program handles nucleic acids and small-molecule ligands as well as proteins,
and interactions with individual bound waters are treated by a simplified model. The
reason hydrogen addition is a complex process is that the movable H atoms often occur
in interacting H-bond networks and must be optimized as a group rather than individually. In practice, such H-bonding cliques are small enough, given our simplified model
for water molecules, that exhaustive evaluation of all possible hydrogen positions is
computationally tractable. A single simple command runs Reduce rapidly and produces
a commented, properly formatted output PDB file with all H atoms present.

All-atom contacts are calculated by the program Probe (Word et al., 1999a) from
a Reduce-modified PDB file that now includes hydrogens. The usual output is contact
surfaces as color-coded dots in the "kinemage" format for display in the Mage graphics program (Richardson and Richardson 1992; Richardson and Richardson 2001) as
shown in the color figures for this chapter, but other display formats, numerical scores,
or lists of serious clashes can also be produced (see Current Facilities and Their Use
below). Typically, Probe is run on an entire PDB file, but it can also calculate the
internal contacts for a small region or just the contacts between two pieces (i.e., a
ligand and a protein), using a flexible syntax of atom selection.

Note: since color is the primary carrier of information in the displays produced by
all atom contact analysis, it is essential that the reader refer to the figure versions in the
Color Section. Figure 15.1 illustrates a simple example of all-atom contact surfaces for
a small region, to show the appearance of favorable van der Waals contacts, favorable
H-bond overlaps, and unfavorable atomic overlap, color-coded by the local gap distance between the two contacting atoms. The all-atom contact algorithm rolls a small
spherical probe (shown as a gray ball) on the surface of each atom, drawing a colored
dot only when the probe intersects another noncovalently-bonded atom. This method
is a bit like the inverse of solvent-exposed surface (Connolly 1983; Lee and Richards
1971), where here only occluded surface is shown; however, our much smaller probe
means that only atom pairs within 0.5Å of touching will count as contacts. These
contacts are extremely sensitive to fine details of how well the structure fits together.
If a local conformation is in the right energy well but not quite correct, it will usually produce just yellow and orange overlap dots. However, it is very difficult to fit
anything in a completely wrong conformation without producing red clash overlaps,
even after refinement has done its best at adjustments. Therefore, the primary way of
interpreting the all-atom contact results is simply that lots of soothing green (such as

**Figure 15.1.** Slice through a small section of protein structure (stick figure, backbone in white and side chains in cyan) showing the relation of all-atom contact surfaces (colored dots) to the atomic van der Waals surfaces (gray dots) and to the 0.25Å-radius probe sphere (gray ball) used in the calculation. The small probe sphere is rolled over the van der Waals surface of each atom, leaving a contact dot only when the probe touches another non-covalently-bonded atom. The dots are colored by the local gap width between the two atoms: blue when nearly maximum 0.5Å separation, shading to bright green near perfect van der Waals contact (0Å) gap. When suitable H-bond donor and acceptor atoms overlap, the dots are shown in pale green, forming lens or pillow shapes. When incompatible atoms interpenetrate, their overlap is emphasized with "spikes" instead of dots, and with colors ranging from yellow for negligible overlaps to bright reds and hot pinks for serious clash overlaps $\geq 0.4$Å. Kinemage-format contact dots also carry color information about their source atom (e.g., O's are red, S's are yellow, etc.); in Mage, one can toggle between the two color schemes. Figure also appears in Color Figure section.

seen in Fig. 15.2a and 15.2b) means the structure is correct, while an area of red spikes has some sort of problem. In fact, for an all-atom kinemage interactively displayed in Mage one can turn off everything but the bad clashes and quickly spot all problem areas even in a large structure, as shown for the 324-residue dimer in Figure. 15.2c.

In addition to graphic display, several scoring schemes suitable for different purposes produce numerical evaluations of the contact, H-bond, and clash terms (Word

**Figure 15.2.** All-atom contact examples from the dimer of 1MJH (Zarembinski et al., 1998), a well-determined structural-genomics protein at 1.7Å resolution. (a) All contacts for one of the typically well-packed and well-fit regions of aliphatic side chains, with the green of close van der Waals contacts predominant. (b) All contacts for an Arg side chain, with all 5 planar H-bonds (lens-shaped groups of pale green dots) of its guanadinium NH's formed either to protein O atoms or to waters (pink balls). (c) An overview of the dimer, with only the Cα backbone and the serious clashes ≥0.4Å (red spikes) shown. When interactively displayed in Mage, it is easy to locate and fix the small number of isolated problems, including two flipped-over His rings at the putative active site and a high-*B* Lys squeezed into insufficient space between two hydrophobic sidechains. Figure also appears in Color Figure section.

et al., 1999a). These scores are not energies, however, because the serious clash overlaps represent model errors, not real strains in the structure. When used to understand features of molecular architecture, such as side-chain packing, overlaps are treated simply as tight contacts, but for structure-validation and error-correction purposes, the clash overlaps are very much the dominant issue. We consider a serious clash (one that usually indicates some sort of misfitting) to occur where two incompatible atoms overlap by 0.4Å or more. The overall clash score of a structure is the number of serious clashes per 1000 atoms.

## RELATIONSHIP TO MORE TRADITIONAL CRITERIA

The well-ordered parts of the very best X-ray and NMR structures fit the all-atom contact criteria nearly perfectly, with extensive contacts throughout the interior, an absence of even modest clashes, and most atoms showing the green dot patches of ideal van der Waals contact as in Figure. 15.2a and 15.2b (and, at even higher resolution, in Fig. 15.5a below). Such agreement is strong confirmation that our algorithms and parameters have been chosen correctly. Clash score is strongly correlated with other indicators of structure quality: overall parameters such as resolution or number of NMR restraints correlate with overall score (Fig. 15.3a), and local crystallographic $B$-factor correlates especially strongly with locally measured clash score (Fig. 15.3b).

If structure factors are available, enabling examination of the electron-density map in the area of a serious clash, it usually turns out that the density is either weak or its shape is somewhat ambiguous, making a misfitting more likely than in clearer areas. For example, electron density for a side chain that branches at the $C\beta$ such as Thr or Val fairly often has a straight bar shape rather than a tetrahedral junction, making it possible to misfit the $\chi 1$ angle by 180°. When that happens (as for the Val in Fig. 15.4a), there are always clashes with the $H\beta$ or $H\gamma$ atoms, the side-chain rotamer will be poor, and the bond-angle geometry around the $C\alpha$ will almost always be badly distorted through forcing the $C\gamma$ atoms to fit into the bar-shaped electron density although connected to a $C\beta$ that has been fit on the wrong side of the bar. Figure. 15.4b shows both the original and the refit side chains, emphasizing the great difference in their geometry and conformations though occupying nearly the same space; Figure. 15.4c shows the excellent fit obtainable in a good rotamer with ideal geometry and no backbone movement. Traditionally, electron-density difference maps are used as indicators of this kind of problem (for instance, often but not always showing a pair of positive and negative peaks at the real and the misfit $C\beta$), but they are difficult for noncrystallographers to calculate and interpret. The Uppsala Electron Density Server (http://portray.bmc.uu.se/eds) is a valuable source of viewable electron density maps and related quality criteria, for those PDB files with available structure factors that could successfully be processed automatically.

In contrast to the technicalities of electron-density maps, user-friendly validation tools are available for assessing all-atom contacts, rotamers, geometric ideality, and whether backbone $\phi$, $\psi$ angles are unfavorable. These tools are best used in concert with one another, because a given problem usually shows up only in a subset of them.

---

**Figure 15.3.** Correlation of all-atom clash scores with other indicators of structure quality. (a) Overall clash score (number of serious overlaps $\geq 0.4\text{Å}$ per 1000 atoms, after correction of amide flips) as a function of resolution, for 328 protein structures between 0.8Å and 2.5Å resolution. The relationship is highly significant and is still improving down near 1Å. (b) Serious clashes per 1000 atoms, grouped into ranges of crystallographic $B$-factor values, for 100 proteins at 1.7Å resolution or better. Note that an atom with $B > 50$ is 10 times as likely to clash as one with $B$ between 10 and 20. Clashes fall off again at the very highest $B$ range, because those atoms are exposed at the surface with few neighbors. Note that for high-resolution structures only about 5% of the atoms have $B > 40$, so that the ones most prone to error can be omitted from empirical studies with little loss in sample size. Part b is reproduced from Word et al. (JMB, 1999a) by permission of Academic Press.

For instance, if the refinement terms for ideal geometry were heavily weighted relative to agreement with the experimental data, then bond angles will not be distorted but clashes will show; however, if clashes are between non-H atoms, then refinement may remove them at the expense of geometry. In our experience, the two most sensitive and

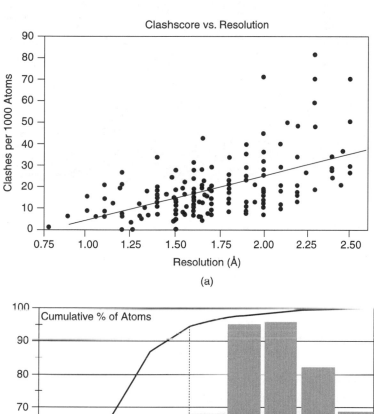

(a)

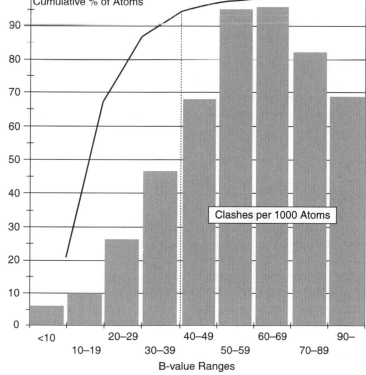

(b)

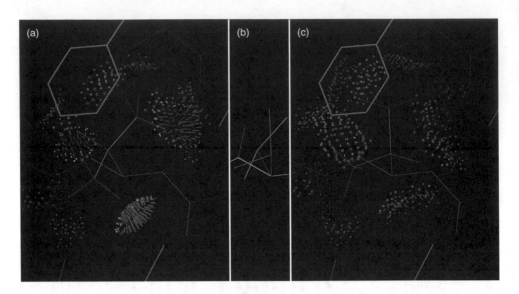

**Figure 15.4.** Diagnosis and correction of a backward-fit valine side chain. (a) All-atom contacts for the original side chain, with substantial clashes (hotpink) and an eclipsed $\chi 1$ angle. (b) Original and refit side chains, showing how both occupy the same space but in opposite orientations. Bond-angle distortions in the original put its C$\beta$ 0.48Å from the idealized position. (c) Good all-atom contacts for the refit Val, which has ideal geometry and staggered $\chi 1$ without backbone movement. Even without deposited structure-factor data, one can be fairly confident that the electron density must have been ambiguous and that the conformation shown in (c), not (a), is in the correct local energy well. From the 2SIM neuraminidase at 1.6Å resolution (Crennell et al., 1996). Figure also appears in Color Figure section.

reliable indicators of local problems in a structure are bond-angle distortions (Lovell et al., 2002) and all-atom clashes. Nonideal torsion angles or bond distances often tell more about how the refinement was set up than about the structural accuracy, whereas a bond angle off by 6–8° from ideal or a clash of 0.5Å (or both!) almost always means that something is seriously misfit.

In drawing conclusions from a structure or comparison, it seldom matters if one or two parameters are slightly off (e.g., a torsion angle by 15°), but it is often critical if the backbone or side chain are actually in the wrong conformation (e.g., a torsion off by 90–180°): that will change which atoms are in position to interact, say, with a ligand. Local problems in polypeptide chain-tracing, such as a sequence out of register by two within a $\beta$ strand, are often flanked at each end by clusters of all-atom clashes and bad bond angles. Neither all-atom contacts nor geometrical ideality are suitable in general, however, for identifying incorrect chain folds—both are too sensitive and too local. That task is probably best done by the sort of threading methods used in fold recognition (Chapter 26) and homology modeling (Chapter 25).

As an extra bonus, we have found that filtering experimental datasets by all-atom clashes and $B$-factors as well as by resolution can greatly improve the quality of Ramachandran-plot criteria (Lovell et al., 2002) and of side-chain rotamer libraries (Lovell et al., 2000), thus indirectly improving these more traditional validation tools. The new rotamers have no internal clashes and all occupy valid local energy minima. The Ramachandran plots are much cleaner, allowing defensible separation of disfavored but allowed regions from forbidden regions and the definition of core regions for Pro and Gly. Separate criteria for glycines are important to validation, because the lack of a C$\beta$ makes Gly $\phi$, $\psi$ the most error-prone for either X-ray or NMR structures. These new rotamers and Ramachandran plots are available from the Kinemage Web site (RichardsonLabWebSite 2001).

## CURRENT FACILITIES AND THEIR USE

Probe and Reduce for calculating all-atom contacts and Mage and Prekin for interactive display of molecules and contacts are available as free, open-source software from our Web site (RichardsonLabWebSite 2001). Probe and Reduce run on Unix, Linux, MacOSX, or (less conveniently) PC; the Mage/Prekin display runs on Unix, Linux, Mac, PC, or Java.

The most basic and general function of all-atom contact analysis for structure validation is to generate a clash report on a particular PDB file, either in graphic form or in list form. To set up for analysis of a PDB file called 1xyz, first add and optimize H atoms by running the command:

reduce -build  1xyz > 1xyzH

and make a kinemage file of the structure (including backbone, side chains or bases, H atoms, small-molecule ligands, and waters) with:

prekin -lots  1xyzH > 1xyzH.kin

Now calculate all atom contacts and append them to the kinemage file with:

probe 1xyzH ≫ 1xyzH.kin

This graphic contact report is viewed in Mage by typing:

mage 1xyzH.kin

For example, on a small, very high-resolution structure such as the 1BRF thermophilic rubredoxin in Figure 15.5 the contacts are excellent throughout (green, with some yellow and blue). In this case if everything is turned off except the bad overlaps, it is immediately obvious that there is a single serious clash between two surface side chains. If the crystallographer is looking at this clash report, he or she should investigate that region to see if it can be corrected; if a bioinformaticist is doing the evaluation, he or she now knows all of this structure is of extremely high quality except for the two clashing side chains, whose detailed conformation cannot be trusted. This example is very small for clarity of presentation in static two dimension (2D), but in the interactive display it is easy to locate the problem regions even on a large, lower-resolution structure and to zoom in and examine them.

The two areas in which such clash reports have had the greatest impact are both for crystal structures: detecting and fixing protein side chains fit in the wrong rotamer, and finding places where nucleic-acid backbone conformations are incorrect. The most common side-chain misfittings are for Asn/Gln flips (case below), Thr/Val/Leu tetrahedral branches (as in Fig. 15.4), and Met conformations. The reasons for problems

**Figure 15.5.** All-atom contacts for the entire structure of 1BRF rubredoxin (Bau et al., 1998), a highly accurate small protein structure at 0.95Å resolution. The dense green dot patches signifying well-packed contacts in the molecule and a well-fit model are seen consistently throughout the structure, except for a single red clash between two surface side chains. 1BRF thus illustrates both how precisely the all-atom contact criteria are satisfied in atomic-resolution protein structures and also how occasional local errors can be found even in such extremely high-quality structures. Figure also appears in Color Figure section.

with Thr/Val and with Leu are discussed in detail in Lovell et al. (2000); for validation purposes it suffices to know that these problems occur fairly often and that they almost always produce bad clashes and usually distort $C\alpha-C\beta$ geometry (Lovell et al., 2002). Met can be difficult because the heavy $S\delta$ atom produces diffraction ripples in the electron density that weaken the information for the nearby $C\gamma$ and $C\varepsilon$; all-atom clash and rotamer information can usually make the correct choice clear.

For nucleic-acid crystal structures, the bases are large, rigid, and well determined, and the P atom density is generally unambiguous, even at the moderate resolution (often around 2.5Å) typical of the most biologically interesting structures. In those same structures, however, the rest of the sugar-phosphate backbone has too many free parameters per observable atom (see Chapter 3 for description of the six backbone dihedrals per residue) and is quite prone to errors when in conformations less well understood than standard B-DNA or A-RNA. H atom clashes, however, mark the

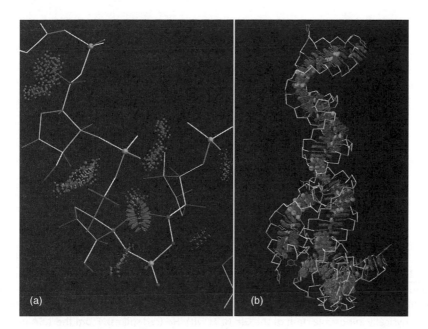

Figure 15.6. Base and backbone all-atom contacts in the 5S RNA from the 2.4Å ribosome structure of 1FFK (Ban et al., 2000). (a) A section of the backbone–backbone contacts, mostly very nicely packed but with one impossible overlap of C 3′ and C 5′ hydrogens (red spikes). (b) Base–base contacts, showing the long columns of well-fit base stacking. Figure also appears in Color Figure section.

incorrect conformations extremely clearly. Figure 15.6a shows all-atom contacts for just the backbone of part of a 5S RNA; most areas show excellent contacts, but one residue is in a physically impossible conformation. To calculate such a backbone contact display, use: probe-mc "mc" 1rnaH ≫ 1rnaH.kin. When analyzing nucleic acid structures, all-atom contacts also provide a quick and pleasing way to visualize base stacking (see Fig. 15.6b, done with: probe "base" 1rnaH ≫ 1rnaH.kin), and the NAContacts script (RichardsonLabWebSite 2001) gives a sensitive numerical measure of stacking quality.

As a clash report for those who prefer working with lists and scores rather than visual displays, the Clashlistcluster script produces a text file that lists all clashes ≥ 0.4Å, clustered into local groupings in space (often a single problem is responsible for several nearby clashes). After running Reduce, the command is:

clashlistcluster 1xyzH > 1xyzHclcl.txt

That file gives the total clash score, for an overall evaluation, or one can look to see whether the residues of interest have any bad clashes. Since all of these programs can be run and controlled from the command line, they can be combined into shell scripts that perform a desired sequence of operations on an entire list of structures. For more information on options, type reduce-help or probe-help.

As explained above, automatic correction of 180° flip alternatives for Asn, Gln, and His is done as part of the H-bond network optimization in Reduce (invoked by the "-build" flag). In order to see why Reduce made each change and to evaluate its

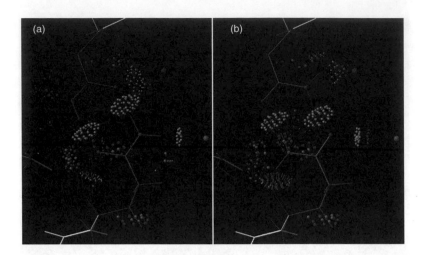

**Figure 15.7.** Resolving the ambiguity in a pair of doubly H-bonded side-chain amides, which have equivalent H bonds both to each other and to waters in the two possible flip states. (a) The correct flip orientation, with only a minor overlap. (b) The next-best, but incorrect, flip state with a large, physically impossible clash of the Gln N$\varepsilon$ H with H$\alpha$ (red spikes). From the 1.6Å peroxidase of 1ARU (Fukuyama et al., 1995). Figure also appears in Color Figure section.

level of certainty, a Perl script called Flipkin is used:

$$\text{flipkin 1xyzH} > \text{1xyzNQflip.kin}$$

This produces an output kinemage file with a preset view for each Asn/Gln in the structure (or His, if run with the "-h" flag added), with the ones Reduce chose to flip marked with "*" on the views menu. An animation is set up between the two possible flip states with display of the contacts, H bonds, and clashes in each state. Figure 15.7a and 15.7b show the two displays for a doubly interacting Asn-Gln pair whose H bonds are equally good in either flip state, but where the original choice has an impossibly bad clash with the Gln C$\alpha$H, whereas the flipped state fits well. The flip of a side-chain amide is a small change but can be crucial if it affects an H bond at an active, allosteric, or binding site.

In addition to assessing the reliability of particular pieces of three-dimensional structure, all-atom contact analysis can be used in the interactive Mage/Probe system (Word et al., 2000) to try out the plausibility of a proposed alternative conformation, or to see whether a modified sequence would be compatible with a known structure. This methodology was developed to tell whether or not a single-site mutation can be accommodated without movement of the original structure around it (essential information in determining whether observed mutant properties are actually due to the altered side chain itself); however, it would also be useful in structural bioinformatics for assessing the degree of structural change between closely related sequences, or deciding whether or not one protein could assume a different conformation seen for other members of the related family.

In the interactive Mage/Probe system, the programs are set up to communicate directly with each other, so that an all-atom contact display changes as you rotate bonds in the structure. When looking at a stick-figure kinemage of your structure (in

Unix or Linux), center on the residue of interest and choose "remote update" on the Tools menu. Ask for Prekin to mutate the residue (edit the three-letter code to what you want, in the command line proposed by Mage), and the new sidechain will appear in green, with idealized geometry, along with sliders to change its $\chi$ angles. (Note that the PDB file 1xyzH must be in the directory, for Prekin and Probe to use.) Go to "remote update" again, and ask for Probe to calculate contacts around the rotatable side chain. First you see contacts around the present position, and they will be updated automatically as you change the conformation.

Figure. 15.8a shows unsatisfactory Probe dots around a buried Tyr to Trp mutation that has been made rotatable, but not yet properly optimized. In the text window is a list of the rotamers for that amino acid (as defined in Lovell et al., 2000); if you click on one of the listed rotamers, the side chain will be put in that conformation. Usually most of the rotamers will have terrible clashes like the one in Figure. 15.8a, which is actually the second-best rotamer ($\chi_1$ *trans*, $\chi_2 - 105°$). Identify the rotamers that are most nearly acceptable, such as the good rotamer shown in Figure 15.8b ($\chi_1$ *trans*, $\chi_2 90°$). Then move the $\chi$ angles by modest amounts (up to 20–30°) to look for a position with green, blue, and yellow dots, and perhaps the pale green pillows of H-bond dots, without any appreciable amount of red spikes; in this case, there is a very well-fitting conformation only 3° away from the best rotamer. This Trp mutant was produced and found to have a stability and folding rate at least as good as, and an

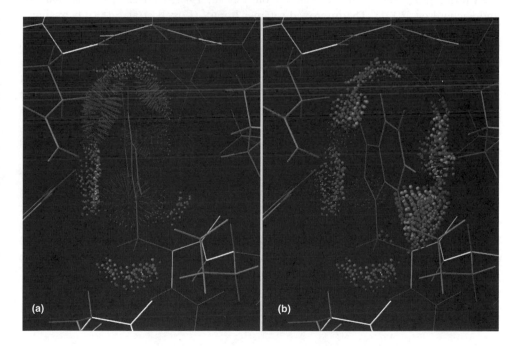

**Figure 15.8.** A test of alternative sequence possibilities substituting Trp for Tyr at a buried position in the N-terminal domain of $\lambda$ repressor, using the "remote update" function in the interactive Mage/Probe system. (a) One of the initial rotamer trials, with impossibly bad clashes on both sides of the Trp ring. (b) The best of the exact rotamers, with only two minor overlaps in orange, indicating that the Trp side chain can indeed fit without perturbing the structure significantly. Starting coordinates from 1LMB (Beamer and Pabo, 1992). Figure also appears in Color Figure section.

NMR spectrum very close to, that of the parent λ repressor domain (Ghaemmaghami et al., 1998). In general, if a satisfactory conformation can be found in Mage/Probe for the mutated side chain, that means that the new amino acid can be accommodated without changing anything else. If it looks as though moving another side chain would help, you can make it rotatable as well. If no acceptable conformation can be found, the mutation might still be stable and functional, but it could not be so without the structure rearranging. Predicting such rearrangements and their functional consequences is currently beyond the capabilities even of the most sophisticated modeling tools, and so this simple method has given you an answer nearly as good as can be done.

In a more systematic or formal context, the conformational search described above can be done by a function called Autobondrot built into the Probe program. It surveys all dihedral-angle values on a specified grid (e.g., for the χ angles of a mutated side chain) and outputs a contoured map of contact score (Word et al., 2000). If there is a sizable area in the map with score $> -1$, then it is considered that the mutant can be accommodated without significant structure change. Autobondrot is somewhat more complex to set up than the interactive Mage/Probe exploration, but it can then be run automatically.

For crystallographers solving new structures and wanting to improve database quality directly, our Web site has the tools and instructions for generating all-atom clash and H-bond displays interactively while rebuilding models in the commonly used fitting programs O and XtalView (Jones et al., 1991; McRee, 1999; Richardson and Richardson, 2001). The improved rotamer library (Lovell et al., 2000) is available as a drop-in replacement file for either program.

The all-atom contact tools are also valuable for NMR structures, but that use is less powerful and less straightforward. H atoms are explicitly included already in NMR refinement, and NMR structures are solved in terms of local distances not absolute Cartesian coordinates; injudicious application of contact criteria could just expand the structure undesirably. So far, the most general conclusion from all-atom contact analysis of NMR structures is that for the best-determined cases the interiors excellently fit all-atom criteria and the surface regions would then benefit from a final refinement step with all atoms at 100% radius (rather than the maximum of 75% radii currently standard). All-atom contact analysis can also be used for validation of theoretical model structures, but again its interpretation is much less robust than when applied to crystallographic structures. A serious clash still means that something must be wrong, but a lack of clashes does not mean the model is necessarily correct. One must also beware, when using software that may not output completely valid PDB format, of occasional apparent clashes produced by incorrectly resolved atom-name ambiguities such as hg_ for an Hγ hydrogen confused with hg__ for an $Hg^{++}$ ion.

## FUTURE DIRECTIONS

For the purposes of bioinformatics and structure validation, our most important plans are to develop service-provision on the kinemage Web site that will run all the functions described above on client-submitted files, now available in a preliminary form on the MolProbity subsite. It is always helpful to bypass the necessity for downloading, installing, and learning new programs, since even for very user-friendly software those steps are always more of a barrier than one feels they ever should be. It is now possible, from either specified PDB codes or uploaded files, to run clash reports with either visual or numerical output directly on the web, to receive a modified file with H atoms added and optimized, and to view directly on line in Java Mage an animated kinemage showing the Asn/Gln/His flips. Along with the all-atom contact functions, we will also provide visual and numerical evaluations, both in 3D on the structure and in one dimension on the sequence, of the ideality or favorability of sidechain rotamers, of $\phi$, $\psi$ values, and of bond-angle geometry, using our updated criteria.

Another important area for development will be further automation of the evaluation and correction functions for other side-chain types in the style now provided for Asn/Gln/His flips, and eventually for some kinds of backbone corrections. More complete automation is especially vital for use in the structural genomics effort, but it will help other users as well. We would appreciate feedback, both about patterns of use and effectiveness and also suggestions about needs and improvements.

## RELEVANT WEB SITES

http://kinemage.biochem.duke.edu. The Richardson laboratory Web site and kinemage home page is the primary source for up-to-date software, information, and other resources relevant to all-atom contact analysis; it includes documentation, datasets, validation examples interactively illustrated in JavaMage, and the MolProbity service that runs our software on a selected or uploaded file.

http://www.sdsc.edu/CCMS/Packages/XTALVIEW/xtalview. The source for XtalView version 4.0 and later, which supports real-time display of all-atom contacts during crystallographic model rebuilding.

http://origo.imsb.au.dk/~mok/o. The source for the O crystallographic rebuilding software, with links to the kinemage site for drop-in rotamer files and macros for updating an all-atom contact display.

## REFERENCES

Ban N, Nissen P, Hansen J, Moore PB, Steitz TA (2000): The complete atomic structure of the large ribosomal subunit at 2.4 Å resolution. *Science* 289:905–20. [1FFK].

Bau R, Rees DC, Kurtz DM, Scott RA, Huang HS, Adams MWW, Eidsness MK (1998): Crystal structure of rubredoxin from Pyrococcus furiosus at 0.95 angstrom resolution, and the structures of N-terminal methionine and formylmethionine variants of Pf Rd. Contributions of N-terminal interactions to thermostability. *J Biol Inorg Chem* 3:484–93. [1BRF]

Beamer LJ, Pabo CO (1992): Refined 1.8 Å crystal structure of the lambda repressor-operator complex. *J Mol Biol* 227:177–96. [1LMB]

Berman HM, Westbrook J, Feng Z, Gilliland G, Bhat TN, Weissig H, Shindyalov IN, Bourne PE (2000): The Protein Data Bank. *Nucleic Acids Res* 28:235–42.

Brünger AT (1992): Free *R* value: a novel statistical quantity for assessing the accuracy of crystal structures. *Nature* 355:472–5.

Connolly ML (1983): Solvent-accessible surfaces of proteins and nucleic acids. *Science* 221:709–13.

Crennell SJ, Garman EF, Philippon C, Vasella A, Laver WG, Vimr ER, Taylor GL (1996): The structures of Salmonella typhimurium LT2 neuraminidase and its complexes with three inhibitors at high resolution. *J Mol Biol* 259:264–80 [2SIM].

Fukuyama K, Kunishima N, Amada F, Kubota T, Matsubara H (1995): Crystal structures of cyanide- and triiodide-bound forms of Arthromyces ramosus peroxidase at different pH values. Perturbations of active site residues and their implication in enzyme catalysis. *J Biol Chem* 270:21884–92 [1ARU].

Ghaemmaghami S, Word JM, Burton RE, Richardson JS, Oas TG (1998): Folding kinetics of a fluorescent variant of monomerio λ repressor. *Biochemistry* 37:9179–85.

Jones TA, Zou J-Y, Cowan SW, Kjeldguard M (1991): Improved methods for building protein models in electron density maps and the location of erros in these models. *Acta Cryst, Sec A* 47:110–9.

Laskowski RA, Macarthur MW, Moss DS, Thornton JM (1993): PROCHECK—a program to check the stereochemical quality of protein structures. *J Applied Crystallogr* 26:283–91.

Lee BK, Richards FM (1971): The interpretation of protein structures: estimation of static accessibility. *J Mol Biol* 55:379–400.

Lovell SC, Word JM, Richardson JS, Richardson DC (2000): The penultimate rotamer library. *Proteins Struct Func Gen* 40:389–408. Development of an improved side-chain rotamer library using a large, high-resolution dataset filtered by all-atom clashes and *B*-factors; analyzes systematic side-chain misfittings prevalent in the structural database.

Lovell SC, Davis IW, Arendall WB III, de Bakker PIW, Word JM, Prisant MG, Richardson JS, Richardson DC (2002): Proteins struct Func Gen (in press).

McRee DE (1999): XtalView/Xfit: a versatile program for manipulating atomic coordinates and electron density. *J Struct Biol* 125:156–65.

Ramachandran GN, Ramakrishnan C, Sasisekharan V (1963): Stereochemistry of polypeptide chain configurations. *J Mol Biol* 7:95–9.

Richardson DC, Richardson JS (1992): The Kinemage: a tool for scientific illustration. *Protein Sci* 1:3–9.

Richardson JS, Richardson DC (2001): MAGE, PROBE, and Kinemages. In: Rossmann MG, Arnold E, editors. *International Tables for Crystallography, Volume F*. Dordrecht: Kluwer Academic Publishers, pp727–30. A brief survey of the crystallographic uses of all-atom contacts calculated by Reduce and Probe, and of the graphics system of Mage and kinemages.

Word JM, Lovell SC, LaBean TH, Taylor HC, Zalis ME, Presley BK, Richardson JS, Richardson DC (1999a): Visualizing and quantifying molecular goodness-of-fit: small-probe contact dots with explicit hydrogens. *J Mo Biol* 285:1711–33. Original description, test cases, and discussion of the all-atom contact method; explains parameter and algorithm choices in the Probe program and analyzes the contacts in a database of 100 high-resolution structures.

Word JM, Lovell SC, Richardson JS, Richardson DC (1999b): Asparagine and glutamine: using hydrogen atom contacts in the choice of side-chain amide orientation. *J Mol Biol* 285:1735–47. Primary reference for the program Reduce; explains H addition and optimization for protein, nucleic acid, and small-molecule "heterogens," including the analysis of H-bond networks and the correction of Asn/Gln/His flips.

Word JM, Bateman RC Jr, Presley BK, Lovell SC, Richardson DC (2000): Exploring steric constraints on protein mutations using Mage/Probe. *Protein Sci* 9:2251–9. Describes both interactive and batch-calculation methods for evaluating whether a specific sequence change is compatible with a given protein conformation.

Zarembinski TI, Hung LW, Mueller-Dieckmann HJ, Kim KK, Yokota H, Kim R, Kim, S-H (1998): Structure-based assignment of the biochemical function of a hypothetical protein: a test case of structural genomics. *Proc Natl Acad Sci USA* 95:15189–93 [1MJH].

# 16

# STRUCTURE COMPARISON AND ALIGNMENT

Philip E. Bourne and Ilya N. Shindyalov

Structure comparison refers to the analysis of two or more structures looking for similarities in their three-dimensional (3D) structures. Alignment refers to establishing equivalences between amino acid residues based on the 3D structure of two or more protein folds. All commonly used methods do a reasonably good job of recognizing the more obvious instances of similar 3D folds, but, as we shall see, structure alignments are much more variable. Most algorithms and resulting Web resources provide protein structure comparison and alignments at the level of domain or complete polypeptide chain.

It is important to clear up immediately any confusion between structure comparison and alignment versus structure superposition. These terms are sometimes used interchangeably in the literature; here, we make a stricter delineation. Structure superposition assumes that you already know of at least some residues that match between protein structures A and B. Typically, these C-alpha positions then become anchor points and the task becomes one of using a minimization technique or analytical procedure to find a transformation that minimizes the distance between aligned residues. The best solution is then the one that produces the lowest root mean square deviation (rmsd) between A and B. In other words, the alignment is already assumed and all that is required is tweaking to bring the two structures into register. Clearly, this is a much easier problem (and there is an exact solution to it) than having no a priori knowledge of what amino acids are equivalent. The latter is known as the structure alignment problem.

Structure superposition methods and codes have been around for some time, see for example, Diamond (1976); Kabsch (1976); Hendrickson (1979), and Kearlsey (1989). Finding proteins that exhibit 3D similarity and then finding the best alignment through

*Structural Bioinformatics*
Edited by Philip E. Bourne and Helge Weissig
Copyright © 2003 by Wiley-Liss, Inc.

gap insertions is much more difficult, and forms the basis of this chapter. Clearly structure superposition could play a role once the alignment is complete and the relationship between residues in the two structures has been established.

## WHY IS 3D STRUCTURE COMPARISON AND ALIGNMENT IMPORTANT?

The first question to ask is why is structure comparison and alignment important? This question has been answered several times in other chapters in this book; for example:

- Structure classification methods (Chapters 12 and 13) use structure alignments to help in the assignment of fold classes and can be used subsequently in establishing libraries of templates for use in proteome annotation.
- Structure alignments of a protein of known fold and function against a protein of unknown function can provide insight into the function of the unknown (Chapter 19). Structure alignments are of particular importance in an era of genomically driven structure determination where no a priori knowledge of the biological function may exist (Chapter 29).
- Structure prediction methods require that the predicted structure be evaluated against a variety of template structures (Chapters 25, 26, and 27).
- Structural alignments reveal distant sequence relationships not available from sequence alignments alone and can be used in protein engineering and protein modeling (this chapter).

As a general point, it is worth highlighting again here, even though it is introduced elsewhere in this book, that structure alignments provide us with information not available from current sequence alignment methods. The reason for this is the result of nature's ability to reduce complexity to manageable levels yet still maintain incredible diversity and adaptability. If you consider that an average protein consists of 300 amino acids, then there are $20^{300}$ possible proteins—more than the number of atoms in the universe. Nature has selected a very small subset of these—as few as 30,000 in humans for the functioning of a complex organism. Still greater reduction exists in three dimensions—all proteins from all species are believed to be represented by somewhere between 1000 and 5000 protein folds (Chothia, 1992). A brief discussion of the possible reasons for this limited number of folds is provided in Chapter 2. This remarkable reduction was first noted in the globins when only a small number of structures existed (Lesk and Chothia, 1980). It was later manifest in the hssp curve (Sander and Schneider, 1991), which was recently updated by Rost (1999).

Here, we take a slightly different look at the relationship using data from our own laboratory (Fig. 16.1). Each point on the graph in Figure 16.1 represents one of 1000 randomly selected polypeptide chains taken from the Protein Data Bank (PDB) that show a measure of structure similarity measured by CE (described below) with a z-score greater than 4.5, indicating structure similarity at the level of the protein superfamily. Plotted on the x-axis is the length of the polypeptide chain and on the y-axis the resulting sequence identity. Thus, while there are a number of matches with very high sequence identity (90% or greater) indicating post-translational modifications in the PDB, there are also many chains with low sequence identity. The region between

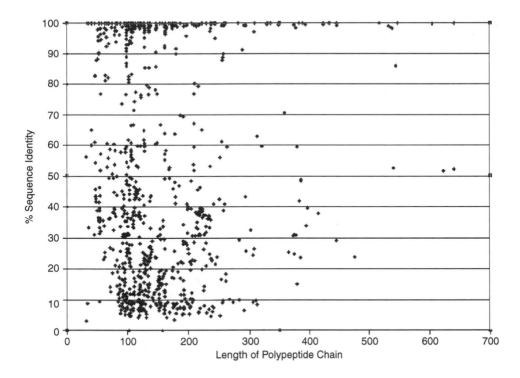

Figure 16.1. Structure similarity versus sequence similarity. Each data point represents one of 1000 randomly selected polypeptide chains from the PDB showing structure similarity as measured by CE with a z-score > 4.5.

30% and 20% sequence identity indicates a twilight zone where this relationship may be detectable by sequence methods alone. However, below 20% sequence identity is the so-called midnight zone where only structure comparison reveals the relationship between these two proteins at the level of the fold. Rost (1999) using an even larger data set showed that this relationship between sequence and structure, while having a length dependency, can be represented as a Gaussian distribution with a peak around the 9% sequence identity level. Certainly the relationship between these proteins, let alone an accurate sequence alignment, could not be achievable by sequence methods alone. Thus, structure alignments provide valuable insights not achievable from sequence alone.

As discussed elsewhere in this book, it is dangerous to consider these findings as absolute; they most certainly are not. The relationship between primary protein sequence, 3D structure, and biological function is more complex and still being interpreted. As George Bernard Shaw once said, "the golden rule is that there are no golden rules." For example, there are cases of structures that contain large regions of high sequence similarity yet share no structure similarity. The viral capsid protein (1PIV:1) shares an 80-residue stretch with glycosyltransferase (1HMP:A) where there is greater than 40% sequence identity, yet the structures within those regions are completely different (mostly beta versus mostly alpha, respectively). In short we have guidelines only; nevertheless, guidelines that prove very useful as we shall see subsequently.

## THE GENERAL APPROACH TO STRUCTURE COMPARISON AND ALIGNMENT

Structure comparison and alignment is an NP-hard problem that is solved heuristically by all methods. Although different heuristics employed by different methods tend to recognize similar folds, they will not provide exactly the same structure alignments. In fact two structure comparison methods may produce alignments that differ in every position (Godzik, 1996). We consider the impact of an NP problem and the fact that even if it were a tractable problem, it may provide the best analytical but not biological answer, after discussing the details of the most popular structure comparison and alignment methods.

There is a significant body of literature on methods of pairwise protein structure comparison and alignment. Orengo (1994) provided an overview of the field until 1994. Gibrat et al., (1996) highlighted some surprising results from structure alignment using different methods and Lemmen and Lengauer (2000) summarized protein structure alignment in the context of the general problem of structure alignment and superposition in drug design. In reading these reviews of various methods you will see that each methodology that has been tried can be boiled down to three or possibly four steps depending on how you count (steps 2 and 3 can be considered a single step):

1. Represent the proteins A and B (polypeptide chains, domains, or other amino acid fragments) in some coordinate independent space so that they can be readily compared
2. Compare A and B
3. Optimize the alignment between A and B
4. Measure the statistical significance of the alignment against some random set of structure comparisons

This methodology applies to pairwise structure comparison and alignment. Multiple structure alignment involves a somewhat different methodology and is considered separately later in this chapter.

Given this general approach there are two classes of problem that the defined algorithms try to solve. The first is to optimize the alignment between any given pair of proteins; the second is given a new target structure to determine, in some rank order, which structures in the PDB are most like the target. Pragmatically, even if a comparison between two proteins takes 30 seconds on a typical processor, today this still represents a significant computation with approximately 18,000 proteins in the PDB representing approximately 30,000 chains, some multidomain, resulting in over 428 years of computation. This compute bottleneck is solved in two ways by several of the resources listed in Table 16.1. First, as structures are released by the PDB each week, they are added to an all-by-all comparison database and so the computations are performed incrementally. Notwithstanding, even with 50 to 100 new structures appearing each week this is still a significant computation. Second, the known relationship between sequence and structure is employed to reduce the number of computations to be performed. A rough estimate (see the PDB site at http://www.rcsb.org/pdb/holdings.html) is that only 1 in 10 new structures represents a new fold. This ratio is dependant on the method used and how one defines a fold, but it serves to provide a rough estimate. To establish that ratio of 1:10, at least 5 of the 10 similarities can be inferred from sequence alone without the need

**TABLE 16.1. Web Resources Associated with Methods for Structure Comparison and Alignment**

| Name | Description | Citations[a] | URL and Web Resource Reference |
|------|-------------|----------|-------------------------------|
| CE | Combinatorial Extension of the Optimum Path (Shindyalov and Bourne, 1998) | 76 | http://cl.sdsc.edu/ce.html Shindyalov and Bourne (2001) |
| DALI | Distance Matrix Alignment (Holm and Sander, 1993a) | 890 | http://www.ebi.ac.uk/dali/ Deitmann et al., (2001) |
| HOMSTRAD | Homologous Structure Alignment Database (Mizuguchi et al., 1998) | 47 | http://www-cryst.bioc.cam.ac.uk/ ~homstrad/ Sowdhamini et al., (1998) |
| SARF2 | Spatial Arrangement of Backbone Fragments (Alexandrov, 1996) | 66 | http://123d.ncifcrf.gov/ sarf2.html |
| SSAP | Sequential Structure Alignment Program (Taylor and Orengo, 1989) | 248 | http://www.biochem.ucl.ac.uk/ ~orengo/-ssap.html |
| VAST | Vector Alignment Search Tool (Gibrat et al., 1996) | 122 | http://www.ncbi.nlm.nih.gov/ Structure/-VAST/vast.shtml Wang et al., (2001) |

[a]Citations of the original paper reporting the method as of May 20, 2002.

for structure comparison. Hence, this reduces the number of targets that need to be checked. Similarly, PDB structures can be grouped into a set of structural representatives so that the target is only compared against a subset of the complete PDB. To better understand the concept of a representative consider the CE algorithm, described subsequently, which uses the following criteria to define a single representative for a number of polypeptide chains being represented:

- The rmsd between two chains is less than 2Å
- The length difference between two chains is less than 10%
- The number of gap positions in alignment between two chains is less than 20% of aligned residue positions
- At least two-thirds of the residue positions in the represented chain are aligned with the representing chain

Other methods apply similar rules, but at the domain level. This reductionism somewhat reduces the accuracy of the comparison but provides the necessary gains in speed.

The discussion here considers steps 1–4 (steps 2 and 3 are discussed together) for several of the most popular methods of protein structure comparison and alignment (Table 16.1). By popular we refer to methods that have been cited a significant number of times and for which Web resources exist for the reader to immediately access these methods, and importantly, these resources are kept current. Web resources are of two types. First, there are those that allow you to compare two proteins using the specific method, and, second, there are those that provide a database of precalculated comparisons against all or a subset of the PDB. These two types are not mutually exclusive—you may be able to look up an existing database of comparisons and alignments and submit your own structure for comparison and alignment to the same resource.

It is beyond the scope of this chapter to deal with each method in detail. The intent is to give the reader a sense of the similarities and differences in approaches that have been employed. The reader is referred to the original papers outlined in Table 16.1 for a full treatment of the methods and the resulting Web resources.

## Protein Structure Representation

As stated, the first step is to suitably represent the two proteins to be compared. Certainly the methods presented here are not exhaustive for this step. For example, geometric hashing taken from computer vision is a technique applied by Nussinov and Wolfson (1991) and later refined by Fischer et al., (1994), but not part of any of the methodologies discussed.

Also at issue is what is being compared. Since domains are the functional units of currency for proteins, it makes sense to compare domains. Problems do arise here since, as described in Chapter 18, there is not always agreement on what constitutes a domain. However, a method such as CE, which focuses on polypeptide chains, can miss recognizing a domain if the domain is part of a long chain for which no similarity is found in other parts. Such a result reduces the significance of the match of those two domains.

**DALI.** Pragmatically, in terms of community availability and use (something these authors rate highly) while significant work had been done beforehand, notably from Taylor and Orengo (1989), protein structure comparison was popularized by the 1993 paper of Holm and Sander (1993a). This paper describes the use of distance matrices as embodied in the DALI method. Availability and use is defined by the relative number of citations presented in Table 16.1. Based on these numbers it is fitting to start with DALI.

DALI uses distance matrices to represent each structure to be compared. This idea was not new and dates back to the work of Phillips (1970). Each structure is represented as a two-dimensional (2D) array of distances between all C-alpha atoms. This representation has the advantage of placing all structures in a simplified common frame of reference. Conceptually, the problem is then straightforward, as if one is imagining each structure's contact map transparently overlaid. Overlap along the diagonal then represents similar backbone conformations (secondary structure) and off-diagonal similarity tertiary structure similarity. Moving one sheet of paper horizontally or vertically relative to the other to achieve overlap represents gap insertion into one or other of the structures. In a later version, a quick look-up was introduced that uses the alignment of secondary structure elements (SSEs) and is not dissimilar to VAST (see below).

**CE.** The combinatorial extension (CE) algorithm (Shindyalov and Bourne, 1998) also uses a distance approach to structure comparison, but at the level of octameric fragments; that is, a comparison of C-alpha distance matrices is made for every combination of eight residues in each protein chain. Each octameric fragment that aligns within the two structures being compared, based on a heuristic measure, is referred to as an aligned fragment pair (AFP).

**COMPARER.** COMPARER (Sali and Blundell, 1990) uses the comparison of residues' properties, segments (of residues), and relations between residues, and relations between segments. Examples of residue properties are identity and local conformation. Examples of segment properties are secondary structure type and orientation relative to the center of gravity. Examples of residues relationships are hydrogen bonds and hydrophobic clusters. Examples of segments relationships are distances to one or more nearest neighbors and the relative orientation of two or more segments.

**SARF2.** SARF2 (Alexandrov, Takahashi, and Go, 1992) operates at the level of SSEs represented by the C-alpha atoms of each residue. First, SSEs are detected by comparison with typical helix and strand templates to within 0.4Å and 0.8Å, respectively. Second, compatible pairs of SSEs and larger assembles of SSEs are constructed and analyzed.

**SSAP.** SSAP (Taylor and Orengo, 1989) uses the comparison of intraprotein $C^{\beta}$-$C^{\beta}$ vectors (calculated using dummy $C_{\beta}$ atoms in the case of glycine) to provide directionality.

**VAST.** The Vector Alignment Search Tool (VAST; Gibrat et al., 1996), as the name suggests, represents structures as a set of SSEs (vectors) whose type, directionality, and connectivity infer the topology of the structure.

## Comparison Algorithm and Optimization

**DALI.** The distance matrices are collapsed into regions of overlap (submatrices) of fixed size, which are then stitched together if there is overlap between adjacent fragments. A solution to the branch and bound algorithm for finding the overlap and optimal superposition is neatly described in Holm and Sander (1996), a later paper than that describing the original method.

**CE.** CE uses three thresholds in the alignment-building process. The first threshold detects AFPs (see above). The second threshold evaluates the suitability of a next candidate AFP relative to the current alignment (a single AFP in the beginning). The third threshold evaluates all alignments to find those that are optimal. The second and third thresholds define whether new AFPs are added to the alignment. Alignment extension is sought in a narrow area of the search space limited to a single gap of no longer than 30 residues in either of the two proteins being compared. This restriction permits computational tractability, but may miss nontopological alignments and also those with insertions of more than 30 residues. These thresholds are empirical, being based on observed comparison of intraresidue distances in structures known to align. If one or more significant alignments are found (up to 30 top-scoring alignments are retained), then further optimization is performed using dynamic programming and interprotein distances calculated based on superposition. The last step is repeated iteratively until the optimal alignment is found.

**COMPARER.** Fourteen properties and relationships are selected. For properties the dynamic programming algorithm is used to find the optimal alignment. For relationships the dynamic programming is not applicable since there is a dependence of scores for a given relationship on the assignment of other relationships. For this reason the so-called combinatorial simulated annealing technique is used for optimization (Sali and Blundell, 1990).

**SARF2.** Pairs of SSEs are evaluated for the angle between them, the shortest distance between their axes, the closest point on the axes, and the minimum and maximum distances from each SSE to their medium line. Then, searching for the largest ensembles of the mutually compatible pairs of SSEs is performed using an algorithm from graph theory used to solve the *maximum clique problem*. Further refinement and extension of the alignment by incorporation of additional residues is then performed.

**SSAP.** SSAP finds the optimal structure alignment by applying dynamic programming to the matrix of scores $S_{ik}$ for every pair of positions $i$ and $k$ from two proteins A and B, respectively. $S_{ik}$ is obtained from comparison of vectors between $C^\beta$ atoms at pairs of positions $i$ and $j$ to $C_\beta$ atoms in *selected matching positions*. The selected matching positions are in turn defined by applying dynamic programming to the matrix of differences of $C^\beta$-$C^\beta$ vectors from positions $i$ and $k$ to all other protein positions. Since dynamic programming is applied at two levels the whole procedure is called *double dynamic programming*.

**VAST.** Given the SSEs, alternative alignments are examined using a Gibbs sampling algorithm, beginning with a seed SSE-pair alignment. The optimal alignment is defined as that which is most surprising relative to the background distribution of alpha-carbon superposition residuals obtained by chance.

## Statistical Analysis of Results

**DALI.** The similarity score is derived from an all-against-all comparison of 225 representative structures with less than 30% sequence identity (Hobohm et al., 1992). The DALI score is then expressed as the number of standard deviations (z-score) from the average score derived from the database background distribution.

**CE.** Two distributions of rmsds and gaps are built and numerically tabulated for the 25% nonredundant set (Hobohm et al., 1992). The final z-score is calculated by combining z-scores from two tabulated distributions under the assumption of their normality.

**COMPARER.** Two scores $E$ and $A$ are introduced to measure residues' equivalence and gap penalties, respectively. Both scores are calculated based on scores seen in two unrelated proteins and two random sequences relationships.

**SARF2.** The similarity score is calculated as a function of rmsd and the number of matched C-alpha atoms. The significance of a particular comparison is evaluated by comparison of the score with the score distribution built from the comparison of the protein leghemoglobin with a set of 426 nonredundant structures (Fischer et al., 1996).

**SSAP.** So-called raw SSAP scores derived from the comparison are calibrated against known comparisons in the Classification, Architecture, Topology, Homology (CATH) database (Orengo et al., 1997). Thus, raw SSAP scores above 70–80 are indicative of topology level similarity if 60% of the residues of the larger protein are included in the alignment.

**VAST.** The significance of a VAST match is determined in a manner similar to its sequence counterpart, BLAST. VAST calculates a p-value for the best substructure superposition as the probability that this score would be seen by chance in drawing SSE pairs at random, multiplied by the number of alternative substructure alignments possible, given the SSEs in the protein pair under consideration.

## HOW WELL ARE WE DOING?

Using the preceding methods and others, our ability to recognize common folds, not anticipated from sequence alone, has led to interesting biological findings. A sample is presented in Sample Results from Structure Comparison and Alignment below. However, in using structure alignment as a tool in structural bioinformatics the answer is not so straightforward. Gaining new insights into such areas as structure prediction and biological function derived from remote homologs through structure alignment and not decipherable from sequence alignment alone is compelling. However, as Godzik (1996) has shown with respect to various structure alignment methods, we have shown (Scheeff, Bourne, and Shindyalov, 2002) in aligning the catalytic subunit of the protein kinases, and Jones has shown (Hadley and Jones, 1999) with respect to the comparison of structure classification methods, significant differences in structure alignments exist.

Differences in structure alignments are not surprising given the heuristics that each methodology applies to make an NP-hard problem computationally tractable. A simple view of these differences is as follows: Consider a spectrum that at one end maximizes the geometric relationship between two proteins and at the other provides the maximum amount of biological significance in the alignment. Depending on the task at hand, you may wish to be at one end of the spectrum or the other, or in the middle. Favoring the geometric end of the spectrum will likely lead to a better rmsd but more fragments, that is, a larger number of gaps and a loss of biological relevance. An example of this would be the breaking of hydrogen bonds in a beta sheet to better fit fragments of those sheets to each other in the two proteins being compared. Favoring the biological end of the spectrum will likely lead to a higher rmsd. Which methods favor which ends of the spectrum? An easy question, with a not so easy answer, since the answer is dependant on the parameters used when computing with each method and the particular proteins under study.

*An important consideration when using any structural alignment method is to consider the nature of the problem you are trying to solve and to experiment with a variety of methods.*

To illustrate the importance of the above statement, consider some results from our own work in comparing expert hand alignments of protein kinases against those produced by CE (Scheeff, Bourne, and Shindyalov, 2002) where the goal is to achieve the best functionally relevant alignment. The gold standard was the hand alignment of 18 protein kinases all with sequence identity below 40%—a substantial challenge to any sequence alignment method. In comparison to the gold standard, CE failed to make the best biological alignment in every case, partly because of significant spatial movements of secondary structural elements (as well as loops) relative to each other in the various structures.

However, these observations do not necessarily suggest that methods treating secondary structures as rigid bodies will yield superior results. In considering the alignment between cAMP-dependant protein kinase (1CDK:A) and an actin-fragmin kinase (1CJA:A) where the sequence identity is 13% and there is significant structure diversity, CE correctly inserted a gap in a beta strand to better preserve the orientation of side chains. Conversely, in the same structure pair, an aspartic acid residue, which is functionally critical as the catalytic base in the phosphotransfer reaction, is misaligned. This misalignment occurs as a result of the residue being adjacent to a loop region that presents a difficult challenge to CE. The reader is referred to the paper for further examples. What is clear is that better scoring functions are needed that can better incorporate what is known about the structures and function(s) of the respective molecules being fed into the alignment.

## SAMPLE RESULTS FROM STRUCTURE COMPARISON AND ALIGNMENT

As methods of structure comparison and alignment were published, they bought forth many previously unobserved structure comparisons (see, for example, Holm and Sander, 1993a), that were later captured in such resources as SCOP, CATH, and the DALI domain dictionary. We consider one such example here. Later, as structure comparison and alignment methods matured they became a standard experimental method integral to attempting to understand biological function. Here, we consider a second example taken from our own work where structure comparison was used in conjunction with other supporting evidence to provide a putative biological function subject to further experimental analysis. Together these two examples illustrate the importance of structure comparison and alignment to biology.

Holm and Sander (1993b) showed that the membrane insertion domain of the bacterial toxin colicin A (Parker et al., 1992) has the same topology of fold as the globins and phycocyanins with six helices sequentially aligned (Fig. 16.2). Both in terms of sequence and function there is no relationship and the original authors missed this structural similarity. The implication is that this similar fold represents structural convergence to a stable three-on-three helical sandwich.

An example from our own work illustrates the coming together of information from different methodologies, including structure comparison and alignment, to define a putative function. In this case the methodologies employed hidden Markov models (HMMs), site information from Prosite, and structure comparison found with CE. It was found that the alpha–beta hydrolase fold family that includes acetylcholinesterases contain putative $Ca2+$ binding sites, which in some family members may be critical for heterologous cell associations (Tsigelny et al., 2000). From a structure or sequence comparison perspective alone the evidence was not definitive. Structure alignments between acetylcholinesterases and classic calcium binding EF-hands, as found, for example, in calmodulin, are apparent but tentative—3Å to 4Å rmsd over approximately 80 residues with 10–20 residue gaps. Nevertheless, the Prosite signature for calcium binding is present and HMMs reveal potential calcium-binding sites for cholinesterases as well as neuroligins and gliotactins. It is postulated that with extracellular $Ca2+$ concentrations higher in the extracellular matrix than within the cell, binding associations could be weaker. While the outcome of this analysis is still in question, it does point to the need for experiments, for example mutagenensis at the site of calcium binding and monitoring of subsequent enzyme activity. Such synergy between in silico and in vivo and in vitro experiment is the future of structural bioinformatics.

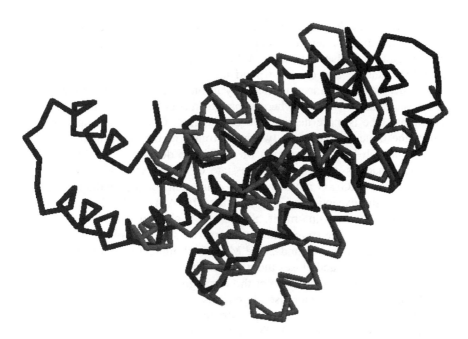

**Figure 16.2.** Structure alignment as computed by CE for colicin A (1COL:A) (grey) and c-phyco-cyanin (1CPC:A) (black). The alignment extends over 115 residues with 16 single residue gaps. The rmsd is 3.2Å with a z-score of 5.5. The sequence identity is 13.9%.

## MULTIPLE STRUCTURE ALIGNMENT

Our discussions thus far have involved only pairwise structure comparison and alignment, or at best, alignment of multiple structures to a single representative in a pairwise fashion—progressive structure alignment. True multiple structure alignment attempts to align all structures simultaneously to provide the best consensus alignment between all structures, which may not be the best alignment between any pair of structures. In principle, if it were accurate enough, multiple structure alignment could enhance the impact that profiles and HMMs have had from a purely sequence perspective by providing multiple alignments with weak yet definitive sequence relationships. A few approaches to multiple structure comparison and alignment have been undertaken (see, for example, Leibowitz, Nussinov, and Wolfson, 2001). Here, we outline one method to illustrate the principle of multiple structure alignment and compare the results to accurate hand alignments.

Our approach uses Monte Carlo optimization of an existing set of pairwise alignments derived using CE, and hence is referred to as MC-CE (Guda et al., 2001). It can be accessed at http://cl.sdsc.edu/mc/mc.html. From the starting alignment—a set of structures all aligned to a master structure—a set of moves were designed to address different alignment situations in a similar manner to that previously used by Mirny and Shakhnovich (1998) for sequence–structure alignment. Moves are then applied in a random manner to a constrained search space to seek the optimal alignment. Each step is tested against a scoring function and accepted with some level of probability. This procedure proceeds until convergence, that is, no random steps improve the optimal alignment as based on a distance score for each block of aligned residues across the multiple structures. Running MC-CE against 66 protein families produced a 12% increase in the number of aligned columns and a 22% decrease in total alignment length when compared to pairwise alignments. When compared to the hand alignments of the HOMSTRAD database and our own hand alignments for the protein kinase family there was widespread agreement, particularly in the more rigid C-terminal lobe of the protein kinase catalytic subunit where substrate recognition and binding take place. The N-terminal lobe is more flexible and challenging. Consider one specific example to illustrate the issues, namely, the two beta strands of the N-terminal lobe that contain the glycine-rich loop. This region of the kinase domain is flexible and often in different conformations depending on the state of the enzyme. However, it contains the well-conserved GxGxxG motif, which is important for the binding of ATP in the active site. With the exception of one structure (1CJA:A) it should be aligned without gaps to properly align this motif. Standard CE alignment splits off some of the sequence leading up to one strand and unnecessarily separates off a row of conserved glycines in the loop between strands. The MC-CE alignment compresses the sequence leading up to strand one, and closes the gap, which causes the glycine displacement in CE. However, MC-CE does not correct the misaligned glycine residues seen in some structures in the original CE pairwise alignments.

This example illustrates that multiple structure alignment techniques are in their infancy and that resources such as HOMSTRAD, with their human curation of specific protein families, are very valuable.

## MAPPING PROTEIN FOLD SPACE

Knowing that there are a finite number of folds and that at this time experimentally we have a reasonable percentage of those folds, estimated at 25–50%, and that structural genomics (Chapter 29) is aimed at giving us as many of the rest as possible, it is not surprising that we have attempted to establish "maps" of protein fold space, which we will continue to fill in. Structural Classification of Proteins (SCOP) and CATH are perhaps the best examples of those maps when other data, for example, sequence homology are included, but what about considering nothing but structure? The ultimate and almost certainly unanswerable question is, can we establish a structure-based phylogenetic tree that evolved from a single common ancestor—the original protein fold? Problems arise immediately since there are different views on what constitutes a protein fold. This problem is illustrated in Figure 16.3. The vertical axis is a count of the number of aligned residues broken into different cells of sequence identity and rmsd across the complete PDB as determine using the CE algorithm introduced previously. The rectangular slab of cells with greater than 20% sequence identity and in the 1–2Å rmsd range illustrate why comparative modeling works in defining a protein structure from sequence. What is surprising is the large number of residues in the 3–6Å rmsd range with very low (<20%) sequence identity. We have previously analyzed this information in one interpretation of protein fold space (Shindyalov and Bourne, 2000). Subsequently, we used the alignments of these regions that average 80 resides in length and are contained within domains (Reddy et al., 2001) to determine what specific conserved residue properties exist in these common substructures. Although some of these substructures are well known (the Rossman fold, the immunoglobulin fold, etc.), others have not been described explicitly in the past and yet occur frequently. Perhaps of most importance, these substructures

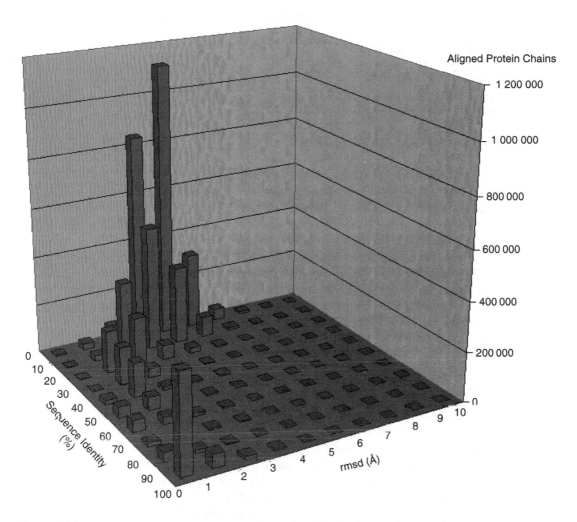

**Figure 16.3.** Structure versus sequence comparison. All residues in the PDB that partake in some measurable level of structure similarity are plotted on the vertical axis for various measures of sequence identity (y axis) and rmsd (x axis) as defined over the complete polypeptide chain.

overlap—part of one will be found in another, indicating the continuity of protein fold space. Thus, one answer to the question of what exactly constitutes a protein fold is a discrete and discernibly reused element of protein structure as determined systematically through comparative structure analysis. The history of characterizing protein folds has not been so systematic, but very valuable nevertheless. References such as Greek key motif, the jelly roll, the beta propeller, and so forth are well accepted yet were not defined systematically. Resources such as SCOP and CATH have undertaken a more systematic approach. Yet, if protein fold space is a continuum, then many systematic definitions are possible. The only point to make then to the student of protein structure space is to carefully define how you have systematically chopped up that space and be aware of the characterizations of the space that have been made before you. In short there are great opportunities for original thinking in defining and using protein fold space, particularly as that space gets more populated in the coming years.

For those who have already dabbled in the study of protein structure space, the challenge is to construct a map of the space at the right resolution to solve the problem at hand. For example, we continue to work with relatively high resolution maps, where many would call the discreet units substructures rather than folds, to address problems in protein engineering (see, for example, Reddy, Li, and Bourne, 2002). Others work with lower resolution maps—folds and larger structures.

Lower resolution maps often address the evolution of protein folds and must distinguish between convergent and divergent evolution. For protein engineering this distinction does not matter; it is the characteristics of the fold, notably its stability under mutagenesis that is of concern. Distinguishing between convergent and divergent evolution of protein folds must take into account sequence and/or functional considerations. We consider two characterizations of fold space that consider evolutionary constraints.

Partslist (Qian et al., 2001b; http://www.partslist.org) has many features, but includes mapping of folds to fully sequenced genomes by homology modeling, thereby providing a distribution of folds within a diverse set of species. Partslist reveals that both families, superfamilies and folds follow a power-law distribution (Qian, Lustcombe, and Gerstein, 2001a). In short, considering folds, a small number of folds are used many times and there is a steep fall-off where many folds are used only occasionally. Simpler organisms share these highly utilized folds.

Conversely, Holm and Sander (1996) have examined all of existing fold space—which they admit, and we emphasize here, is biased by the current contents of the PDB—and defined what they refer to as attractors in shape space. They hypothesize that attractors represent both dominant folding pathways and evolutionary sinks as a result of physical constraints. The result is five dominant attractors containing 40% of all known folds in 16 different fold classes.

The charting of protein fold space has strong analogies to the history of global maps. At present we would seem to be working at the level of Ptolemy's map (a map that existed prior to Columbus's voyage in 1492). This map was designed specifically for navigation, yet missed many of the vital elements. Structural genomics may well be the protein structure equivalent of Columbus's voyage opening a new world to be explored and utilized.

## THE FUTURE

The impetus for improved protein structure comparison, alignment, and characterization will be defined quite simply by quantity—the rate of increase in the number of experimentally determined new folds and the number of structures that conform to each fold. Hand comparisons and alignments will require automation while still retaining the level of quality provided by experts today, since no expert will be able to keep up with manual comparisons and alignments. Today we are in a mode where tools assist in the process with final decisions about comparisons and alignments made by human experts. This situation will not scale into the future. Quantity implies not just numbers but variety, complexity, and singularity. We can anticipate the structures of more membrane proteins, which because of issues of solubility are today underrepresented in the PDB. Complexity implies more structures of biological assemblies determined by conventional means. Finally, singularity will be defined by the structures of a large number of single-domain proteins determined by structural genomics from which protein–protein interactions will be ascertained and that will form pieces of a puzzle to be fitted to an outline defined by lower resolution techniques such as cryoelectron microscopy.

Several years ago one of us (Bourne, 1999) estimated that by the end of 2005 the PDB would contain 35,000 structures, almost double the number present at the time of writing (June 2002). While this number remains optimistic, structural genomics is moving from the engineering to the production phase so the number will increase rapidly.

In short the impetus to study and contribute to the field of structure comparison and alignment is already here. The result will be more structure alignment databases, either generic or comprised of specific folds, families, and superfamilies that will be tools for further research. Consider some examples of that research, some of which is already ongoing:

- Faster and more accurate protein fitting to electron density maps using consensus alignments.
- Improved functional characterization of proteins derived from structural genomics.
- Use of structure-based profiles and hidden Markov models (HMMs) to enhance sequence-only methods to find more distant sequence homologs.
- Better understanding of the relationship between sequence and fold in specific protein families.
- Better protein structure prediction through better fold libraries used in both homology modeling and fold recognition.

Exciting times indeed.

## ACKNOWLEDGMENTS

We thank Eric Scheeff and B.V.B. Reddy for a critical reading of this manuscript that includes elements of their published, and in some cases, unpublished work.

## REFERENCES

Alexandrov NN (1996): SARFing the PDB. *Protein Eng* 9(9):727–32.

Alexandrov NN, Takahashi K, Go N (1992): Common spatial arrangements of backbone fragments in homologous and non-homologous proteins. *J Mol Biol* 225:5–9.

Bourne PE, (1999): Editorial. *Bioinformatics* 15:715–6.

Chothia C (1992): Proteins. One thousand families for the molecular biologist. *Nature* 357:543–4.

Diamond R (1976): Comparison of conformations using linear quadratic transformations. *Acta Crystallogr* A32:1–10.

Dietmann S, Park J, Notredame C, Heger A, Lappe M, Holm L (2001): A fully automatic evolutionary classification of protein folds: Dali Domain Dictionary version 3. *Nucleic Acids Res* 29(1):55–7.

Fischer D, Wolfson H, Lin SL, Nussinov R (1994): Three-dimensional, sequence order-independent structural comparison of a serine protease against the crystallographic database reveals active site similarities: potential implications to evolution and to protein folding. *Protein Sci* 3(5):769–78.

Fischer D, Elofsson A, Rice D, Eisenberg D (1996): Assessing the performance of fold recognition methods by means of a comprehensive benchmark *Pac. Symp. Biocomp.* 300–18.

Gibrat JF, Madej T, Bryant SH (1996): Surprising similarities in structure comparison. *Curr Opin Struct Biol* 6(3):377–85.

Godzik A (1996): The structural alignment between two proteins: is there a unique answer? *Protein Sci* 5(7):1325–38.

Guda C, Scheeff ED, Bourne PE, Shindyalov IN (2001): A new algorithm for the alignment of multiple protein structures using Monte Carlo optimization. *Pacific Symp Biocomput* 6:275–86.

Hadley C, Jones DT (1999): A systematic comparison of protein structure classifications: SCOP, CATH and FSSP. *Structure Fold Des* 7(9):1099–112.

Hendrickson WA (1979): Transformations to optimize the superposition of similar structures. *Acta Crystallogr* A35:158–63.

Hobohm U, Scharf M, Schneider R, Sander C (1992): Selection of representative protein data sets. *Protein Sci* 1:409–17.

Holm L, Sander C (1993a): Protein structure comparison by alignment of distance matrices. *J Mol Biol* 233(1):123–38.

Holm L, Sander C (1993b): Structural alignment of globins, phycocyanins and colicin A. *FEBS Lett* 315:301–6.

Holm L, Sander C (1996): Mapping the protein universe. *Science* 273:595–602.

Kabsch W (1976): Solution for best rotation to relate two sets of vectors. *Acta Crystallogr* A32:922–3.

Kearsley SK (1989): Structural comparisons using restrained inhomogeneous transformations. *Acta Crystallogr* A45:628–35.

Leibowitz N, Nussinov R, Wolfson HJ (2001): MUSTA—a general, efficient, automated method for multiple structure alignment and detection of common motifs: application to proteins. *J Comput Biol* 8(2):93–121.

Lemmen C, Lengauer T (2000): Computational methods for the structural alignment of molecules. *J Comput-Aided Mol Des* 14:215–32.

Lesk AM, Chothia C (1980): How different amino acid sequences determine similar protein structures: the structure and evolutionary dynamics of the globins. *J Mol Biol* 136(3):225–70.

Mirny LA, Shakhnovich EI (1998): Protein structure prediction by threading. Why it works and why it does not. *J Mol Biol* 283(2):507–26.

Mizuguchi K, Deane CM, Blundell TL, Overington JP (1998): HOMSTRAD: a database of protein structure alignments for homologous families. *Protein Sci* 7:2469–71.

Nussinov R, Wolfson HJ (1991): Efficient detection of three-dimensional structural motifs in biological macromolecules by computer vision techniques. *Proc Natl Acad Sci USA* 88(23):10495–9.

Orengo CA (1994): Classification of protein folds. *Curr Biol* 4:429–40.

Orengo CA, Mitchie AD, Jones S, Jones DT, Swindells MB, Thornton JM (1997): CATH—A hierarchic classification of protein domain structures. *Structure* 5:1093–108.

Parker MW, Postma JP, Pattus F, Tucker AD, Tsernoglou D (1992): Refined structure of the pore-forming domain of colicin A at 2.4Å resolution. *J Mol Biol* 224(3):639–57.

Phillips DC (1970): The development of crystallographic enzymology. *Biochem Soc Symp* 30:11–28.

Qian J, Luscombe NM, Gerstein M (2001a): Protein family and fold occurrence in genomes: power-law behaviour and evolutionary model. *J Mol Biol* 313(4):673–81.

Qian J, Stenger B, Wilson CA, Lin J, Jansen R, Teichmann SA, Park J, Krebs WG, Yu H, Alexandrov V, Echols N, Gerstein M (2001b): PartsList: a web-based system for dynamically ranking protein folds based on disparate attributes, including whole-genome expression and interaction information. *Nucleic Acids Res* 29(8):1750–64.

Reddy BVB, Li W, Shindyalov IN, Bourne PE (2001): Conserved key amino acid positions (CKAAPs) derived from the analysis of common substructures in proteins. *Proteins* 42(2):148–63.

Reddy BVB, Li W, Bourne PE (2002): Conserved key amino acid positions used to morph protein folds. *Biopolymers* 64(3):139–45.

Rost B (1999): Twilight zone of protein sequence alignments. *Protein Eng* 12(2):85–94.

Sali A, Blundell TL (1990): Definition of general topological equivalence in protein structures. A procedure involving comparison of properties and relationships through simulated annealing and dynamic programming. *J Mol Biol* 212(2):403–28.

Sander C, Schneider R (1991): Database of homology-derived protein structures and the structural meaning of sequence alignment. *Proteins* 9:56–68.

Scheeff ED, Bourne PE, Shindyalov IN (2002): Comparative analysis of protein structure: automated vs. manual alignment of the protein kinase family. In: Tsigelny I, editor. *Protein Structure Prediction: Bioinformatic Approach.* La Jolla, CA: International University Line, pp. 463–75.

Shindyalov IN, Bourne PE (1998): Protein structure alignment by incremental combinatorial extension (CE) of the optimal path. *Protein Eng* 11(9):739–47.

Shindyalov IN, Bourne PE (2000): An alternative view of protein fold space. *Proteins* 38(3):247–60.

Shindyalov IN, Bourne PE (2001): CE: a resource to compute and review 3-D protein structure alignments. *Nucleic Acid Res* 29(1):228–9.

Sowdhamini R, Burke DF, Deane C, Huang JF, Mizuguchi K, Nagarajaram HA, Overington JP, Srinivasan N, Steward RE, Blundell TL (1998): Protein three-dimensional structural databases: domains, structurally aligned homologues and superfamilies. *Acta Crystallogr* D54:1168–77.

Taylor WR, Orengo CA (1989): Protein structure alignment. *J Mol Biol* 208(1):1–22.

Tsigelny I, Shindyalov IN, Bourne PE, Südhof TC, Taylor P (2000): Common EF-hand motifs in cholinesterases and neuroligins suggest a role for CA2+ binding in cell surface associations. *Protein Sci* 9(1):180–5.

Wang Y, Anderson JB, Chen J, Geer LY, He S, Hurwitz DI, Liebert CA, Madej T, Marchler GH, Marchler-Bauer A, Panchenko AR, Shoemaker BA, Song JS, Thiessen PA, Yamashita RA, Bryant SH (2002): MMDB: Entrez's 3D-structure database. *Nucleic Acids Res* 30(1):249–52.

# Section IV

## STRUCTURE AND FUNCTIONAL ASSIGNMENT

# 17

# SECONDARY STRUCTURE ASSIGNMENT

### Claus A. F. Andersen and Burkhard Rost

*The task.* When we look at a protein three-dimensional (3D) structure,[1] we notice regular macroelements that are repeated in all known structures: helices and strands. There is no unique physical definition to systematically assign secondary structure from 3D co-ordinates. Instead, there are many differing definitions, each capturing some aspects of reality. The relative spatial distances and orientations between two or more secondary structure segments are typically referred to as supersecondary structure. Here, we reviewed a number of the existing concepts to assign secondary structure from co-ordinates, that is, to label the secondary structure state for each residue. The terms *class*, *state*, and *regular secondary structure* are not used consistently in the literature. We used the following notation: (1) *states* are the types of secondary structure defined by a particular method, for example, G in DSSP; (2) *classes* are the groups of similar states, for example, the DSSP states H, G, and I all describing helices, and (3) *regular secondary structure* as positively defined state. Note that *nonregular* is usually defined as a negation, that is, by the absence of all the other criteria applied by a method to define the regular states.

   *The role of secondary structure assignment in structural genomics.* Typically, structural biologists assume the protein fold to be the basic unit for structure classification (see Chapter 4) (Lesk and Rose, 1981). The fold and other basic structural elements are classified by automatic systems, such as SCOP, CATH, FSSP, MMDB (Hogue and Bryant, 1998; Marchler-Bauer et al., 1999; Orengo et al., 1999; Lo Conte et al., 2000; Yang and Honig, 2000; Pearl et al., 2001). When classified by experts, the particular features of a given fold are often described by the overall secondary structure arrangements (Lesk and Rose, 1981; Lesk, 1991; Murzin, 1996), which therefore constitute a substantial step in protein classification. Functional aspects of proteins are

---

[1]Abbreviations and notations used in this chapter appear at the end of the chapter.

*Structural Bioinformatics*
Edited by Philip E. Bourne and Helge Weissig
Copyright © 2003 by Wiley-Liss, Inc.

also reflected in the secondary structure and occasionally function can be derived from secondary structure alone (see Chapter 19) (Przytycka, Aurora, and Rose, 1999; Young et al., 1999; Andersen et al., 2001). There are four main uses of secondary structure: (1) it is indicative of the fold, (2) it is an intuitive means of visualizing protein structures, (3) it influences the sequence alignment, and (4) it is related to function. In the context of structural genomics, these four features are important. One practical application is to use secondary structure segments to speed up large-scale all-against-all alignments of 3D structures. Another is the use of secondary structure segments for comparative modeling (Sternberg et al., 1999; Marti-Renom et al., 2000; Sauder, Arthur, and Dunbrack, 2000) and threading (Rost, 1995; Sippl, 1995; Fischer and Eisenberg, 1996; Russell, Copley, and Barton, 1996; Sippl and Floeckner, 1996; Rice and Eisenberg, 1997; Rost, Schneider, and Sander, 1997; Jaroszewski et al., 1998; de la Cruz and Thornton, 1999; Di Francesco, Munson, and Garnier, 1999; Jones, 1999a; Jones et al., 1999; Kolinski et al., 1999; Xu et al., 1999b). In turn, comparative modeling techniques and more sensitive sequence searches through threading are relevant for structural genomics. Firstly, these techniques assure that each experimental structure has the highest possible impact. Secondly, both methods are important to determine the areas of protein space that need to be explored as part of the target selection for structural genomics (Rost, 1998; Sali, 1998; Burley et al., 1999; Blundell and Mizuguchi, 2000; Shapiro and Harris, 2000; Liu and Rost, 2001a; Liu and Rost, 2001b).

*The role of secondary structure in sequence searches and structure prediction.* The relevance of secondary structure also explains why secondary structure prediction from sequence has become one of the most ardently pursued tasks in bioinformatics (Schulz, 1988; Barton, 1995; Lupas, 1996; Rost and Sander, 1996; Rost and O'Donoghue, 1997; Rost, 2001b). Various secondary structure assignment schemes exist, which differ considerably (as described below), so how do we evaluate and compare them? One idea is to use the particular secondary structure assignment that (1) agrees most between proteins of similar structure, and/or (2) is the most predictable from sequence. Obviously, the concepts *agree most* and most *predictable* have to be put into perspective: An assignment of X to all residues would be completely conserved and easy to predict although it would not carry any information. Hence, we would have to account for the information and relevance contained in an assignment. However, this simple concept has not been realized, yet. In fact, we have found that secondary structure prediction methods are reaching a level of accuracy at which the assignment problem becomes relevant (Andersen et al., 2001). Secondary structure prediction methods become increasingly important for prediction of general aspects of protein structure and function (Jones, Orengo, and Thornton, 1996; Rost and Sander, 1996; Finkelstein, 1997; Rost and O'Donoghue, 1997; Rost, 2001b) and for database searches (Fischer and Eisenberg, 1996; Rost, Schneider, and Sander, 1997; Xu et al., 1999a; Lindahl and Elofsson, 2000; Fain and Levitt, 2001; Jennings, Edge, and Sternberg, 2001). Thus, the assignment problem also influences these important fields of bioinformatics indirectly.

*History: from expert to automatic assignment of protein secondary structure.* Pauling and colleagues correctly predicted the idealized protein secondary structures of $\alpha$-helices (Pauling, Corey, and Branson, 1951), $\pi$-helices (Pauling, Corey, and Branson, 1951), and of $\beta$-sheets (Pauling and Corey, 1951) based on intrabackbone hydrogen bonds. Five decades later, we know that on average about half of the residues in proteins participate in helices or sheets (Berman et al., 2000). Pauling and colleagues incorrectly predicted that $3_{10}$-helices would not occur in proteins due to unfavourable bond angles; however, approximately 4% of the residues are observed in this conformation (Andersen, 2001). Initially, the crystallographers assigned secondary structure by eye from the 3D structures. At the time this was the only way to assign secondary structure. However, it lacked consistency, since experts occasionally disagree. This inconsistency was particularly problematic when comparing secondary structure predictions and was actually the primary objective for Kabsch and Sander (1983a, 1983b) to automate the assignment in their DSSP program. Originally developed to improve secondary structure prediction, DSSP has remained the standard in the field, most popular for its relatively reliable assignments. Curiously, the prediction method for which Kabsch and Sander (1983c) originally needed the automatic assignment was never published.

## HYDROGEN BOND MODELS

Since hydrogen bonds are used by many methods as the defining elements for secondary structure, we introduce both the concept of the hydrogen bond and the ways to define it. Pauling (1939) established the hydrogen bond as an important principle in chemistry. The rich network of hydrogen bonds in water creates a very particular environment in which polar molecules participate, while nonpolar molecules disrupt the network of hydrogen bonds. This disruption results in missing water–water hydrogen bonds and therefore is a relative energy cost compared to the hydrogen-bonded case (4 kcal/mol for Isoleucine and Leucine when compared to Glycine; Creighton, 1993). This energy cost is in the order of two hydrogen bonds (hydrogen bonds are in the range of $-2$ kcal/mol) and can be avoided/minimized by packing/agglomerating nonpolar molecules, thereby resulting in the hydrophobic effect.

For proteins the packing of nonpolar residues in the core is believed to be the main driving force in tertiary structure formation of proteins, while the specific secondary structures are governed by intraprotein hydrogen bonds (Hvidt and Westh, 1998). Packing the nonpolar residues in the core also means burying the polar backbone atoms and breaking the water–backbone hydrogen bonds. To avoid this heavy energy cost, the polarities are paired (forming hydrogen bonds) in the protein core, thereby fixing the protein conformation. If the protein backbone instead were non polar, the protein core elements would be free to move around changing the protein structure and thereby preventing the protein from functioning reliably and efficiently.

Approximately 90% of the backbone C=O and NH groups have hydrogen bonds (Baker and Hubbard, 1984). Using the Coulomb hydrogen bond definition (see below), we found that approximately 62% of the backbone C=O and NH groups have intrabackbone hydrogen bonds (Andersen, 2001). Pauling defined secondary structure by the intrabackbone hydrogen bonds, which has later become the prevalent means of assigning secondary structure. Thus, for simplicity we refer to intrabackbone hydrogen bonds when using the term *hydrogen bond*.

### Angle–Distance Hydrogen Bond Assignment

There are many different angles and distances that can be measured and used to identify the hydrogen bond. Baker and Hubbard (1984) assigned hydrogen bonds according to the angle NHO $= \theta$ and to the distance $r_{HO}$ in the hydrogen bond. A hydrogen bond is assigned when:

$$\theta > 120° \text{ and } r_{HO} < 2.5 \text{ Å} \tag{17.1}$$

This formula is similar to other rigid distance and angle constraints published (Bordo and Argos, 1994; Jeffrey and Saenger, 1994). Although a rather crude way of assigning hydrogen bonds, it has sufficed for several decades. In most applications, hydrogen bonds were only assigned visually for a few proteins, that is, explicit definitions of hydrogen bond energies were not necessary.

## Coulomb Hydrogen Bond Calculation

One way of finding hydrogen bonds is by calculating the Coulomb energy in the bond, as applied in DSSP (Kabsch and Sander, 1983a) focusing on the electrostatic attraction (Fig. 17.1). The Coulomb energy for the attraction and repulsion is given by:

$$E = f\delta^+\delta^- \left( \frac{1}{r_{NO}} + \frac{1}{r_{HC'}} + \frac{1}{r_{HO}} + \frac{1}{r_{NC'}} \right) \tag{17.2}$$

where $f = 332$ Å kcal/e² mol is the dimensional factor and $\delta^+$ and $\delta^-$ are the polar charges given in units of the elementary electron charges $e$. A cutoff level has been set for the weakest acceptable hydrogen bond so that the resulting energy is bound by: $E < -0.5$ kcal/mol in DSSP. The H-atom position is usually not given in PDB files requiring an extrapolation, in practice. The H-atom position that is needed to calculate the two distances $r_{OH}$ and $r_{HC'}$ in Eq. 17.2 is usually not given in PDB files. Hence, it must be extrapolated. DSSP uses an approximate position, assuming that the covalent bond between O=C′ is parallel to the covalent N-H bond adjacent to the same polypeptide bond. The direction of the O=C′ vector is kept while its length is set to 1 Å, that is, the length of the N-H bond (Creighton, 1993). The position of the H atom is extrapolated using the direction of the C′=O vector when starting out from the position of the N atom. These approximations made by DSSP simplify the calculation of the

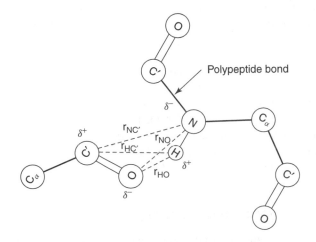

**Figure 17.1.** Distances used to calculate the Coulomb hydrogen bond.

H-atom position and appear to be rather accurate: Compared to the original bond angles and distances (Creighton, 1993), we found the DSSP approximation to yield an average error around 0.07 Å (Andersen, 2001). Both in the DSSP extrapolation and in our test the transpeptide bond, giving rise to the rigid peptide plane, was assumed. Partitioning *ab initio* energy calculations of the hydrogen bond into classical components showed that about 75% is electrostatic (Coulombic) and less than 5% comes from polarization and charge-transfer, for moderate strength bonds (Jeffrey and Saenger, 1994). Note that the Coulomb energy term does not incorporate atom–atom repulsion to penalize steric clashes and does not give rise to a characteristic hydrogen bond length.

## Empirical Hydrogen Bond Calculation

An empirical hydrogen bond energy calculation can be derived from the hydrogen bond geometry in crystal structures or from polypeptides, peptides, amino acids, and small organic compounds (Boobbyer et al., 1989; Wade, Clark, and Goodford, 1993) as applied in STRIDE (see below). The total energy $E_{hb}$ depends on the NO distance energy $E_r$, and on three bonding angles through the expressions $E_p$ and $E_t$:

$$E_{hb} = E_r \cdot E_t \cdot E_p \tag{17.3}$$

The distance dependency is similar to the Lennard-Jones potential for the van der Waals interaction, but uses powers of 8 and 6 instead of 12 and 6:

$$E_r = \left( \frac{4r_m^6}{r^6} - \frac{3r_m^8}{r^8} \right) E_m \tag{17.4}$$

where r is the NO distance, $r_m$ is the optimal distance, and $E_m$, the optimal energy. For intrabackbone hydrogen bonds $r_m = 3.0$ Å and $E_m, = -2.8$ kcal/mol is used. The two angular dependent terms are:

$$E_p = \cos^2(\theta)$$

$$E_t = \begin{cases} [0.9 + 0.1 \sin(2t_i)] \cos(t_o) & 0° < t_i \leq 90° \\ K_1[K_2 - \cos^2(t_i)] \cos(t_o) & 90° < t_i \leq 110° \\ 0 & 110° \leq t_i \end{cases} \tag{17.5}$$

where the angles $\theta$, $t_i$ and $t_0$ are specified in Figure 17.2.

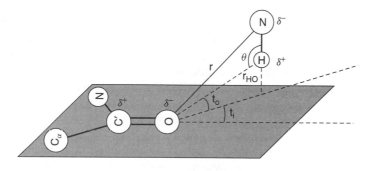

**Figure 17.2.** Angles and distances defining the empirical hydrogen bond. Note: figure similar to the one in Frishman and Argos (1995).

## ASSIGNMENT METHODS

### DSSP

The so-called Dictionary of Secondary Structure of Proteins (DSSP) by Kabsch and Sander (1983a) performs its sheet and helix assignments solely on the basis of backbone–backbone hydrogen bonds. The DSSP method defines a hydrogen bond when the bond energy is below $-0.5$ kcal/mol from a Coulomb approximation of the hydrogen bond energy (Eq. 17.2). The structure assignments are defined such that visually appealing and unbroken structures result. In case of overlaps, $\alpha$-helix is given first priority, followed by $\beta$-sheet. This procedure does not effect the Coulomb approximation, rather the realization of unbroken structures addresses the step from individual hydrogen bonds to assigning macrostructures to groups of such bonds.

An $\alpha$-helix assignment (DSSP state H) starts when two consecutive amino acids have $i \rightarrow i + 4$ hydrogen bonds, and ends likewise with two consecutive $i - 4 \leftarrow i$ hydrogen bonds. This definition is also used for $3_{10}$-helices (state G with $i \rightarrow i + 3$ hydrogen bonds) and for $\pi$-helices (state I with $i \rightarrow i + 5$ hydrogen bonds) as well. The helix definition does not assign the edge residue having the initial and final hydrogen bonds in the helix. A minimal size helix is set to have two consecutive hydrogen bonds in the helix, leaving out single helix hydrogen bonds, which are assigned as turns (state T).

$\beta$-sheet residues (state E) are defined as either having two hydrogen bonds in the sheet, or being surrounded by two hydrogen bonds in the sheet. These two alternatives imply three-sheet residue types: antiparallel and parallel with two hydrogen bonds or surrounded by hydrogen bonds. The minimal sheet consists of two residues at each partner segment. Isolated residues fulfilling this hydrogen bond criterion are labeled as $\beta$-bridge (state B). The recurring H-bonding patterns connecting the partnering strands in a $\beta$-sheet are occasionally interrupted by one or more so-called $\beta$-bulge residues. In DSSP these residues are also assigned as $\beta$-sheet E and may comprise up to four residues on one strand and one residue on the partnering strand. These interruptions in the $\beta$-sheet H-bonding pattern are only assigned as sheet if they are surrounded by H-bond-forming residues of the same type, that is, either parallel or antiparallel. The remaining two DSSP states S and (space) indicate a bend in the chain and unassigned/other, respectively.

### STRIDE

The secondary STRuctural IDEntification method (STRIDE) by Frishman and Argos (1995) uses an empirically derived hydrogen bond energy (Eq. 17.3) and phi-psi torsion angle criteria to assign secondary structure. Torsion angles are given $\alpha$-helix and $\beta$-sheet propensities according to how close they are to their regions in Ramachandran plots (Ramachandran and Sasisekharan, 1968). The method fixes five internal parameters for $\alpha$-helix and four for $\beta$-sheets. The parameters are optimized to mirror visual assignments made by crystallographers for a set of proteins. However, crystallographers often disagree in their assignment of secondary structure. This fact may challenge the concept of STRIDE. The annotations from crystallographers may be more similar to one another than all of them are to automatic assignments from, for example, DSSP. However, this remains to be shown. Since the secondary structure categories have different parameters, their assignment thresholds are independent for the hydrogen bond and phi-psi torsion angles. By construction, the STRIDE assignments agreed better with the expert assignments than DSSP, at least for the data set used to optimize the free parameters. In particular, the authors reported that every 11th $\beta$-sheet and every 32nd $\alpha$-helix were more in register with the expert assignments for the data set used.

Like DSSP, STRIDE assigns the shortest $\alpha$-helix (H$'$) if it contains at least two consecutive $i \rightarrow i + 4$ hydrogen bonds. In contrast to DSSP, helices are elongated to comprise one or both edge residues if they have acceptable phi-psi angles; similarly a short helix can be vetoed if the phi-psi angles are unfavorable. Therefore, hydrogen bond patterns may be ignored if the phi-psi angles are unfavorable. The sheet category does not distinguish between parallel and antiparallel sheets. The minimal sheet (E$'$) is composed of two residues each in one of five possible hydrogen bond conformations, that is, two more than for DSSP. The dihedral angles are incorporated into the final assignment criterion as was done for the $\alpha$-helix. Bulges are accepted applying the same criterion as DSSP. Single residue sheets, that is, $\beta$-bridges are labeled as B for the three DSSP hydrogen bond conformations and as b for the remaining two. Both $3_{10}$- (G$'$) and $\pi$-helices (I$'$) are implemented according to the DSSP scheme, but with the empirical hydrogen bond criterion. Turns are assigned according to the phi-psi angles of residue $i + 1$ and $i + 2$ as described in Wilmot and Thornton (1990). The C symbol is used whenever none of the above structure requirements are met.

## DEFINE

The algorithm DEFINE by Richards and Kundrot (1988) assigns secondary structures by matching $C_\alpha$-coordinates with a linear distance mask of the ideal secondary structures. First, strict matches are found, which subsequently are elongated and/or joined, allowing moderate irregularities or curvature. The algorithm locates the starts and ends of $\alpha$- and $3_{10}$-helices, $\beta$-sheets, sharp turns, and omega-loops. With these classifications the authors are able to assign 90–95% of all residues to at least one of the given secondary structure classes.

To assign $\alpha$-helices the linear mask is matched with each row in the distance matrix of the query protein (Fig. 17.3). If a segment longer than four residues matches the mask within the allowed cumulative discrepancy limit ($\varepsilon = 1$ Å) it is assigned. Assigned $\alpha$-helices are checked whether they start or end with a $3_{10}$-helix, but individual $3_{10}$-helices and $\pi$-helices are not investigated.

In order to assign $\beta$-sheets as a single category, the authors have applied a linear distance mask taken from ideal antiparallel sheets. The problems of the backbone bendability inside sheets and of the curvature for larger sheets has been solved by excluding nonrigid sheets from the definition. The minimum length of sheets is set to be four residues. According to Pauling's definition of a $\beta$-sheet, each strand must pair to another strand to form a sheet. In contrast, DEFINE may assign unpaired strands.

## P-Curve

Sklenar, Etchebest, and Lavery (1989) based their assignment scheme P-Curve on a mathematical analysis of protein curvature. Using differential geometry, they calculated a helicoidal axis on the basis of the fixed-axis systems of a series of peptide

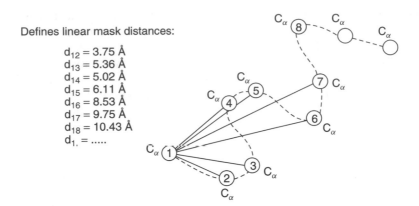

Defines linear mask distances:

$d_{12} = 3.75$ Å
$d_{13} = 5.36$ Å
$d_{14} = 5.02$ Å
$d_{15} = 6.11$ Å
$d_{16} = 8.53$ Å
$d_{17} = 9.75$ Å
$d_{18} = 10.43$ Å
$d_{1,} = .....$

**Figure 17.3.** DEFINE. The linear distance mask approach is visualized for an $\alpha$-helix. The mask is compared to the distances in the query protein. If the mask fits a certain segment, then this segment is assigned as $\alpha$-helix. The allowed root-mean-square difference between the distances in the mask and the ones observed in the query protein is determined by the cumulative discrepancy limit.

planes. The secondary structure assignments are performed by motif matching, where the parameters in the motif are the radius of the helicoidal system along with a series of tilting, rolling, and twisting measures describing geometrical differences between two peptide planes. This parameter analysis is achieved mainly by the use of the $C_\alpha$-coordinates. The P-Curve assignment differs significantly from those performed from phi/psi angles or H bonds, since different parameters are used (e.g., helicoidal radius, tilting, rolling, twisting). Furthermore, the degrees of freedom allowed when matching a P-Curve motif are quite different from those allowed when matching a DEFINE linear distance mask. For example, while the linear distance mask of DEFINE fits poorly to a curved $\beta$-strand, the local P-Curve parameters are likely to fit better.

The assigned secondary structures are recognized by matching known structural motifs. These motifs are based on standard values for the helicoidal parameters. The following motifs are used: right- and left-handed $\alpha$-helix, $3_{10}$- and $\pi$-helix, parallel and antiparallel $\beta$-sheets, and some other structures of little interest here. Note that like DEFINE, P-Curve may assign the category sheet to unpaired strands.

## DSSPcont

Continuous DSSP is a novel secondary structure assignment scheme described below (Emerging and Future Developments).

## PRACTICAL ASPECTS

### Programs and Databases

All methods described above have been coded. In some cases these programs are publicly available (DSSP, STRIDE, and DSSPcont, Table 17.1). For all these programs there are also available the assignments for all proteins deposited in PDB

T A B L E  17.1. Availability of Programs and Databases

| Program | World Wide Web | Platforms |
|---------|---------------|-----------|
| DSSP | www.cmbi.kun.nl/gv/dssp/ | IRIX (SGI) SOLARIS (SUN) LINUX |
| STRIDE | www.embl-heidelberg.de/argos/stride/stride_info.html | IRIX (SGI) |
| DSSPcont | www.cbs.dtu.dk cubic.bioc.columbia.edu/services/DSSPcont/ | IRIX (SGI) LINUX |

(Table 17.1). We explained the meaning of the output in Figure 17.4 (DSSP) and Figure 17.5 (STRIDE). The output of DSSPcont differs from DSSP on which it is based (see Emerging and Future Developments) only in the addition of eight extra columns, giving the continuous assignment to each of the eight DSSP states (G, H, I, T, E, B, S and ' ').

## Comparing Automatic Assignments

We used the simple structure of Crambin as an example to point out differences in three assignment schemes (Fig. 17.6, note that the P-Curve assignment was taken from the original publication; Sklenar, Etchebest, and Lavery, 1989). The secondary structure assignments of STRIDE and DSSP are identical except for one residue at the end of an $\alpha$-helix. P-Curve largely agrees with the positioning of the secondary structure elements, but not with their lengths. Looking at the sheet region in detail (Fig. 17.6b), we see that residues 39 and 40, assigned sheet by P-Curve only, are distant from any residue on the putatively pairing strand. According to Pauling, such an assignment would not be valid. The first sheet assignment by P-Curve covers residues 1–4, where residues 1 and 3 have one and two hydrogen bonds in the sheet, respectively. The extension of the strands/sheet by both DSSP and STRIDE appears reasonable.

Are the discrepancies observed for Crambin representative? Colloc'h and colleagues have compared DSSP, P-Curve, and DEFINE on a low homology data set consisting of 28,266 residues in 154 protein chains (Colloc'h, et al., 1993). The allowed cumulative discrepancy limit for DEFINE was set to $\varepsilon = 0.75$ Å for helix and to $\varepsilon = 0.5$ Å for sheet, in order to avoid an excess of secondary structure assignment. The authors found that all three algorithms agreed on the assignments of $\alpha$-helix, $\beta$-sheet, and nonregular structure for only 63% of all residues. Most disagreements were found between nonregular and regular (helix and sheet) structure (Fig. 17.7). In pairwise comparisons, DEFINE and P-curve and likewise DEFINE and DSSP agreed for 74% of the residues, while P-Curve and DSSP agreed in 79% of all residues. We have found that DSSP and STRIDE agree in 96% of all residues with 64% of the disagreements related to the helix assignment (unpublished results derived from a data set of 707 nonhomologous protein chains).

Not considering any assignment schemes superior, in principle, Colloc'h and colleagues suggested applying a consensus assignment: if two methods agree, use

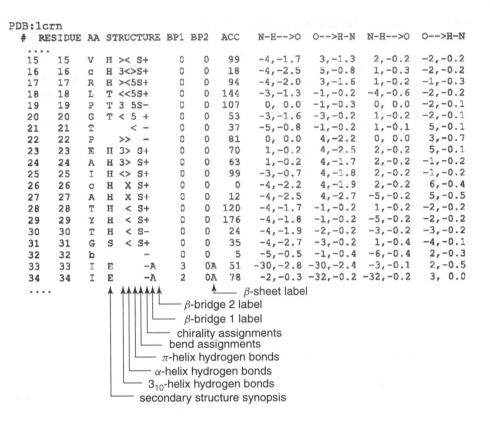

```
PDB:1crn
  #  RESIDUE AA STRUCTURE BP1 BP2 ACC  N-H-->O   O-->H-N   N-H-->O   O-->H-N
....
 15   15   V  H >< S+    0   0   99  -4,-1.7   3,-1.3    2,-0.2   -2,-0.2
 16   16   c  H 3<>S+    0   0   18  -4,-2.5   5,-0.8    1,-0.3   -2,-0.2
 17   17   R  H ><5S+    0   0   94  -4,-2.0   3,-1.6    1,-0.2   -1,-0.3
 18   18   L  T <<5S+    0   0  144  -3,-1.3  -1,-0.2   -4,-0.6   -2,-0.2
 19   19   P  T 3 5S-    0   0  107   0, 0.0  -1,-0.3    0, 0.0   -2,-0.1
 20   20   G  T < 5 +    0   0   53  -3,-1.6  -3,-0.2    1,-0.2   -2,-0.1
 21   21   T      < -    0   0   37  -5,-0.8  -1,-0.2    1,-0.1    5,-0.1
 22   22   P    >>   -   0   0   81   0, 0.0   4,-2.2    0, 0.0    3,-0.7
 23   23   E  H 3> S+    0   0   70   1,-0.2   4,-2.5    2,-0.2    5,-0.1
 24   24   A  H 3> S+    0   0   63   1,-0.2   4,-1.7    2,-0.2   -1,-0.2
 25   25   I  H <> S+    0   0   99  -3,-0.7   4,-1.8    2,-0.2   -1,-0.2
 26   26   c  H X S+     0   0    0  -4,-2.2   4,-1.9    2,-0.2    6,-0.4
 27   27   A  H X S+     0   0   12  -4,-2.5   4,-2.7   -5,-0.2    5,-0.5
 28   28   T  H < S+     0   0  120  -4,-1.7  -1,-0.2    1,-0.2   -2,-0.2
 29   29   Y  H < S+     0   0  176  -4,-1.8  -1,-0.2   -5,-0.2   -2,-0.2
 30   30   T  H < S-     0   0   24  -4,-1.9  -2,-0.2   -3,-0.2   -3,-0.2
 31   31   G  S < S+     0   0   35  -4,-2.7  -3,-0.2    1,-0.4   -4,-0.1
 32   32   b       -     0   0    5  -5,-0.5  -1,-0.4   -6,-0.4    2,-0.3
 33   33   I  E    -A    3  0A   51 -30,-2.8 -30,-2.4   -3,-0.1    2,-0.5
 34   34   I  E    -A    2  0A   78  -2,-0.3 -32,-0.2  -32,-0.2    3, 0.0
....
                    ↑ ↑↑↑↑↑↑↑        ↑_____ β-sheet label
                    | |||||| |_____ β-bridge 2 label
                    | ||||||_____ β-bridge 1 label
                    | |||||_____ chirality assignments
                    | ||||_____ bend assignments
                    | |||_____ π-helix hydrogen bonds
                    | ||_____ α-helix hydrogen bonds
                    | |_____ 3₁₀-helix hydrogen bonds
                    |_____ secondary structure synopsis
```

Figure 17.4. Explanation of DSSP output. Example: segment from Crambin. The two first columns contain the unique DSSP residue number and the corresponding PDB residue number. The third column (here empty) indicates the chain identifier if there are multiple chains. Then follows the amino acid "AA" in one-letter codes (note: lower case letters are all Cysteines, in order to mark Cysteine-bridges, e.g., residue 16 has a disulfide bond to residue 26). The "STRUCTURE" section starts with the secondary structure synopsis (HBEGITS listed in order of priority in case of overlaps) and is followed by helix hydrogen-bond indications for $3_{10}$-, $\alpha$-, and $\pi$-helix hydrogen bonds, where '>' indicates an acceptor, '<' a donor, and 'X' both. The bend and chirality are each given a column, followed by the $\beta$-bridge label columns (lower case labels are parallel $\beta$-bridges and upper case are antiparallel). The DSSP numbers of their partners are written in the "BP1" and "BP2" columns. Each $\beta$-sheet is also given a label (independent of the $\beta$-bridge labels) indicated in the adjacent column. The "ACC" column contains the solvent-accessible surface measured in $\text{Å}^2$ by estimating the number of water molecules in contact with the present residue. The two strongest backbone–backbone hydrogen bonds are then listed, where "$N - H \rightarrow O$" are donor hydrogen bonds and "$O \rightarrow N - H$" acceptor hydrogen bonds. The format indicates the relative position of the hydrogen bond partner followed by the energy in kcal/mol (e.g., "$-5, -0.8$" means that the partner residue DSSP number is 5 less than the present one and that the hydrogen bond energy is $-0.8$ kcal/mol). The remaining columns have been skipped in the figure, but are all labeled: "TCO" is cosine of the angle between the present C=O vector and that of the previous residue (close to 1 for helices and $-1$ for sheets), "KAPPA" is the bend angle $C_\alpha^{i-1} C_\alpha^i C_\alpha^{i+1}$ (when above 70° "S" is assigned), "ALPHA" is the dihedral angle $C_\alpha^{i-1} C_\alpha^i C_\alpha^{i+1} C_\alpha^{i+2}$ used to assign chirality ("+" when positive, "−" when negative), finally the "PHI", "PSI" angles are given followed by the (x, y, z) $C_\alpha$-coordinates.

```
PDB:1crn
REM   |---Residue---|      |--Structure--|   |-Phi-|   |-Psi-|   |-Area-|      1CRN
....
ASG   VAL -   15   15   H      AlphaHelix    -69.24    -41.22      93.8        1CRN
ASG   CYS -   16   16   H      AlphaHelix    -56.67    -36.00      18.4        1CRN
ASG   ARG -   17   17   H      AlphaHelix    -77.07    -16.13      94.1        1CRN
ASG   LEU -   18   18   H      AlphaHelix    -53.21    -46.17     143.0        1CRN
ASG   PRO -   19   19   C            Coil    -77.19     -7.60     108.9        1CRN
ASG   GLY -   20   20   C            Coil    106.26      7.31      52.1        1CRN
ASG   THR -   21   21   C            Coil    -52.67    136.34      38.4        1CRN
ASG   PRO -   22   22   C            Coil    -56.98    146.62      81.9        1CRN
ASG   GLU -   23   23   H      AlphaHelix    -56.41    -36.19      68.9        1CRN
ASG   ALA -   24   24   H      AlphaHelix    -63.43    -34.86      61.3        1CRN
ASG   ILE -   25   25   H      AlphaHelix    -74.77    -37.89      98.2        1CRN
ASG   CYS -   26   26   H      AlphaHelix    -64.95    -31.69       0.0        1CRN
ASG   ALA -   27   27   H      AlphaHelix    -62.04    -54.03      11.6        1CRN
ASG   THR -   28   28   H      AlphaHelix    -68.78    -25.49     121.1        1CRN
ASG   TYR -   29   29   H      AlphaHelix    -67.59    -36.30     174.0        1CRN
ASG   THR -   30   30   H      AlphaHelix   -108.96    -18.47      23.4        1CRN
ASG   GLY -   31   31   C            Coil     91.82     -3.07      36.1        1CRN
ASG   CYS -   32   32   C            Coil    -69.52    164.38       4.6        1CRN
ASG   ILE -   33   33   E          Strand   -129.76    157.03      51.0        1CRN
ASG   ILE -   34   34   E          Strand   -111.56    129.59      78.0        1CRN
....
```

Figure 17.5.  Explanation of STRIDE output. The STRIDE output for Crambin is shown to explain the format and for comparison to Figure 17.4. The format is simple and easily parsed, with "ASG" as the first word in the lines used for assignment. The residue columns comprise the three-letter amino acid code, the chain identifier ("−" for single chains), the PDB residue number, and the STRIDE residue number, which starts from one for every new chain. The two structure columns contain the one-letter structure assignments (HGIEBbTC) and its short description. The columns with phi psi are followed by the column with solvent accessibility (measured in $Å^2$).

that state, otherwise assign the nonregular state. They noticed several aspects of interest.

- The last residues of a sheet or a helix are often still in the same conformation, although they no longer have hydrogen bonds in the structure. This finding translates to the observation that ends (caps) of regular secondary structure segments are not well defined.
- It seems that $C_\alpha$-distance criteria (applied in DEFINE) alone can accommodate considerable distortion of the backbone, giving an excess of secondary structure assignments despite having reduced $\varepsilon$ considerably.
- DSSP is the only assignment scheme with a large peak for $\alpha$-helices of four residues, many of which constitute single helical turns.
- DEFINE assigns more than twice as many sheets of length four than the other methods.
- P-Curve has a tendency to assign overly long elements of regular secondary structure.

## Converting Secondary Structure States to Three Classes

Although no systematic analysis has attempted to compare secondary structure assignment methods in terms of their consistency, DSSP continues to be the most widely used method. In fact, most prediction methods are based on DSSP assignments. Typically,

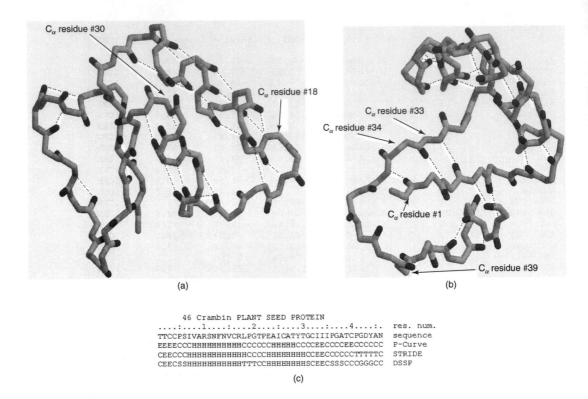

(a)                                              (b)

```
46 Crambin PLANT SEED PROTEIN
....:....1....:....2....:....3....:....4....:.  res. num.
TTCCPSIVARSNFNVCRLPGTPEAICATYTGCIIIPGATCPGDYAN  sequence
EEEECCCHHHHHHHHHHHHCCCCCCHHHHHCCCCEECCCCEECCCCC  P-Curve
CEECCCHHHHHHHHHHHHHCCCCHHHHHHHHCCEECCCCCCTTTTTC  STRIDE
CEECSSHHHHHHHHHHHHHTTTCCHHHHHHHHHSCEECSSSCCCGGGCC  DSSP
```

(c)

**Figure 17.6.** Protein secondary structure for crambin. The structure of the small protein Crambin (PDB identifier:1crn; Teeter, 1984) is shown from two angles: (a) the image of the two helices, and (b) the central short sheet. The automatic secondary structure assignment agrees well between the three methods shown (c).

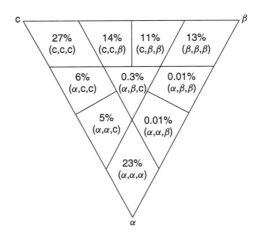

**Figure 17.7.** Comparison of three assignment schemes. The occurrences of three assignment classes ($\alpha$-helix, $\beta$-sheet, and nonregular) by three assignment methods: DSSP, P-Curve, and DEFINE give the 10 displayed categories, if the order is not regarded. When all schemes assign $\alpha$-helix, this is indicated by $(\alpha, \alpha, \alpha)$, when two assign $\alpha$-helix and one nonregular, this is indicated by $(\alpha, \alpha, c)$. The distinction between helix and sheet appears clear, since $(\alpha, \alpha, \beta)$ and $(\alpha, \beta, \beta)$ assignments are rare (<0.01%). Data are taken from (Colloc'h et al., 1993).

the 8 DSSP states are converted into three classes using the following convention: [GHI] → h, [EB] → e, [TS"] → c.

Usually, $3_{10}$-helices and $\beta$-bridges constitute short secondary structure segments that have some structural similarity to $\alpha$-helix and $\beta$-strand, respectively. However, they do have different sequence characteristics. Prediction methods, in general, are more precise in the core of regular secondary structure segments than at the termini (Rost and Sander, 1994; Cuff and Barton, 1999). Thus, $3_{10}$-helices and $\beta$-bridges are more difficult to predict than $\alpha$-helices and $\beta$-strands. Therefore, an alternative conversion that has been used more recently yields a seemingly higher level of prediction accuracy: [H] → h, [E] → e, [GITS"] → c.

## Assigning Secondary Structure for NMR Structures

Usually, NMR structures contain more than one model in a PDB file. By default, the available programs for DSSP and STRIDE read only the first model. Our recent work on extending secondary structure assignments to "continuous secondary structure" (DSSPcont, see below), suggested that this simplification throws away important information.

## Sequence Distributions for Secondary Structure

The amino acids typically found in $\alpha$-helices differ considerably from those found in $\beta$-sheets (Fig. 17.8). Alanine and Leucine often occur in $\alpha$-helices, whereas Proline and Glycine are rare. In $\beta$-sheets Valine and Isoleucine are over-represented, whereas Glycine, Aspartic acid, and Proline are under-represented. Shorter structures such as $3_{10}$-helices and $\beta$-bridges have distinct residue distributions. For $3_{10}$-helices, the Alanine and Leucine signal has disappeared; instead the sequences are dominated by Proline, which often is observed as a helix initiator and breaker. For $\beta$-bridges, we no longer find a preference for Valine and Isoleucine. This finding indicates the role of the side chain in defining secondary and tertiary structure, an observation that can be built into new assignment methods (see below). In general, these preferences have long been the basis of secondary structure prediction methods (Schulz, 1988; Fasman, 1989; Richardson and Richardson, 1989; Barton, 1995; Rost and Sander, 2000; Rost, 2001c).

## EMERGING AND FUTURE DEVELOPMENTS

### Concepts Involving Secondary Structure

Supersecondary structure, such as Greek-key and Zinc-finger motifs (Brändén and Tooze, 1991), describe the interaction and position of a few secondary structure elements. The recently developed I-sites library (Bystroff and Baker, 1998) is a collection of structure motifs for small segments with specific amino acid propensities. The main idea was to mine the structure database for recurring structural motifs and assign them as individual I-sites along with an amino acid propensity matrix covering the segment in question. The I-sites procedure can be viewed as a data-driven assignment process that classifies segments into structure motifs in the range of secondary and supersecondary structure. It has achieved considerable success in predicting protein structure (Lesk, Lo Conte, and Hubbard, 2001).

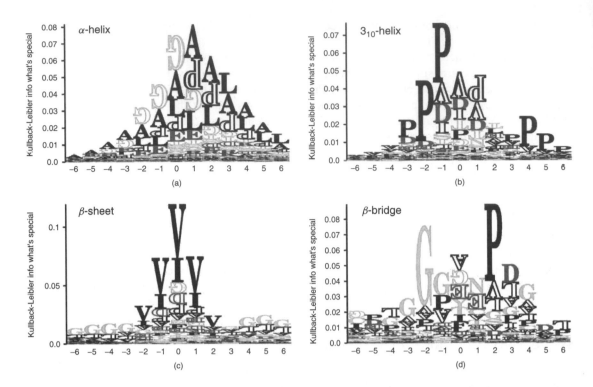

**Figure 17.8.** Sequence distributions for secondary structure. The four graphs show alignment statistics for (a) $\alpha$-helices, (b) $3_{10}$-helices, (c) $\beta$-sheets, and (d) $\beta$-bridges by the Kullback-Leibler information at positions surrounding the one assigned (position 0). The number of aligned segments are: (a) 41,803, (b) 4952, (c) 27,320, (d) 1851. These segments were retrieved from a data set of 707 nonhomologous protein chains using the DSSP assignment. At a given position, we therefore observed the 20 amino acids with a certain frequency; the Kullback-Leibler information calculates the information content of the observed frequencies with respect to the background frequencies (irrespectively of the structure). The more an observed set of frequencies differs from the background, the higher the respective letter. If an amino acid at a given position is observed less frequently than in the background, it is drawn upside-down and hollow.

If we could predict helices accurately, it also seems possible to predict their tertiary arrangement (Fain and Levitt, 2001), that is, the 3D structure. This thought gives rise to optimism, since secondary structure predictions are presently approaching a level of correctly predicting almost 80% of all residues in one of three classes: helix, strand, other (Eyrich et al., 2001; Rost and Eyrich, 2001). Most state-of-the-art prediction methods perform even better on a per-segment than on a per-residue basis. This fact is encouraging since the final assembly step on the way from secondary structure to 3D structure is more sensitive to missing a helix than to getting the ends slightly wrong.

The physical basis for secondary structure formation has not yet been fully described. The backbone–backbone hydrogen bonds used to assign secondary structure do not involve the side chains. Nevertheless, we observe strong preferences in the amino acids forming particular secondary structures. Simulating local interactions, Srinivasan and Rose (1999) found two competing forces that taken together explain this ostensible contradiction. These are local attractive interactions—mainly hydrogen bonds—versus side-chain conformational restrictions, constituting the enthalpic and entropic energy, respectively.

## STICK: Continuous Assignment Based on Geometry

The standard method used to define line segments is to fit an axis through each secondary structure element (DSSP, STRIDE, DEFINE). This approach has difficulties, both with inconsistent definitions of secondary structure and the problem of fitting a single straight line to a bent structure. STICK avoids these problems by finding a set of line segments independently of any external secondary structure definition (Taylor, 2001). This independence of explicit assumption allows the segments to be used as a novel basis for secondary structure definition by taking the average rise/residue along each axis to characterize the segment. This practice has the advantage that secondary structures are described by a single (continuous) value that is not restricted to the conventional classes of alpha–helix, $3_{10}$-helix, and beta-strand. This latter property allows structures without classic secondary structures to be encoded as line segments that can be used in comparison algorithms. When compared over a large number of pairs of homologous proteins, the current method was found to be slightly more consistent than a widely used method based on hydrogen bonds.

## DSSPcont: Continuous DSSP Assignment

Good secondary structure assignments are those that differ only between regions of protein structure not conserved between different NMR models for the same protein or between close homologues, and that distinguish between regions of thermal motion and less flexible regions (Andersen et al., 2001).

This concept led us to develop a continuous extension of DSSP (Andersen et al., 2001). This continuous assignment is based on multiple runs of DSSP with different hydrogen-bond thresholds. Then, we compiled a weighted average over the individual DSSP assignments to assign secondary structure to each residue. We determined the weights by applying the above criterion for *good* assignments starting with structural homologues from the FSSP (Holm and Sander, 1998) database. Inspecting the structural alignments in detail, we noted seven possible of reasons for observed structural differences:

1. Different solution composition, spatial grouping and/or environment of the proteins
2. Uncertainties/errors in the experimental setup
3. Minor thermal fluctuations (even though mostly averaged out)
4. Local amino acid substitutions causing the structural change
5. Insertions/deletions adjacent to the local stretch in question
6. Nonlocal changes forcing a new local conformation
7. Other less likely causes, for example, prionlike switching

Our objective was a secondary structure assignment method de-emphasizing the effects of 1 through 3 while capturing differences caused by sequence changes. However, for structural alignments of homologues, we cannot separate these effects as illustrated by a comparison between two related structures: periplasmic binding protein

(PDB: 4mbp; Quiocho, Spurlino, and Rodseth, 1997) and putrescine-binding protein (PDB: 1pot; Suugiyama et al., 1996). The structural alignment was obtained from FSSP with a Z-score of 23.2 and an root = mean = square deviation (rmsd) of 3.6 Å over 303 residues. We will focus on a small 10-residue segment (Fig. 17.9a) that has a spiraling structure ($\alpha$-helix, $3_{10}$-helix, or turn) and a $\beta$-bridge at the penultimate position

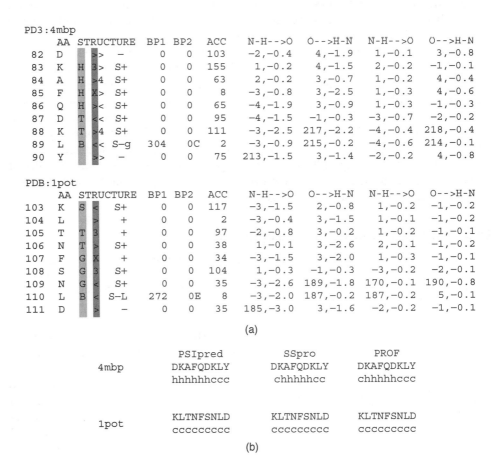

(a)

|  | PSIpred | SSpro | PROF |
|---|---|---|---|
| 4mbp | DKAFQDKLY | DKAFQDKLY | DKAFQDKLY |
|  | hhhhhhccc | chhhhhcc | chhhhhccc |
| 1pot | KLTNFSNLD | KLTNFSNLD | KLTNFSNLD |
|  | ccccccccc | ccccccccc | ccccccccc |

(b)

Figure 17.9. DSSP assignments for similar structures: 4mbp and 1pot. (a) The DSSP assignment for two segments taken from two structurally similar proteins (periplasmic-binding protein 4mbp (Quiocho, Spurlino, and Rodseth, 1997) and putrescine-binding protein 1pot (Suugiyama et al., 1996)) illustrates that the observed differences between these segments may originate from sequence differences. The boxed letters shown in the column next to the amino acid sequence give the final DSSP assignment: G = $3_{10}$-helix, H = $\alpha$-helix, T = turn, B = $\beta$-bridge, and S = bend. The next column shows the hydrogen bonds (>: hydrogen bond acceptor, <: hydrogen bond donor, and X: both), with indications of the hydrogen bond length, that is, $i \rightarrow i + (3, 4)$ for $3_{10}$ and $\alpha$-helices, respectively. (b) All the predictions from PSIPRED (Jones, 1999b), SSpro (Baldi et al., 1999) and PROFphd (Rost, 1996; Rost, 2001a) (See chapter 28) correctly spot the $\alpha$-helix signal in 4mbp, while missing this signal for 1pot. This result may indicate that the altered sequence changed the structure significantly in this region. Here, "h" refers to the DSSP class helix (H or G) and "c" to the DSSP nonregular class. Note: the predictions are cut out from those for the entire protein.

(Fig. 17.9a). Based on the assignment alone one might characterize the differences as problems in the assignment process, since both segments have $3_{10}$-helix hydrogen bonds over the entire stretch. However, 1pot has no $\alpha$-helix hydrogen bonds, resulting in the assignment of $3_{10}$-helix. The results from three high-quality prediction methods (Fig. 17.9b) suggest that the structural differences resulted from the sequence divergence, which means that the secondary structure assignments of the two segments should not necessarily be the same. This line of reasoning can be extended from short helices to short sheets and to the N- or C-terminal ends of helices and strands (caps). Therefore, we chose to optimize the weights for DSSPcont based on the comparisons between different NMR models for the same protein.

Our work on DSSPcont is still in progress. We briefly summarized below a few of the important results (for more details see Andersen et al., 2001). We found that the single residue rmsd between models of high-quality NMR structures correlated well with thermal fluctuations in water as independently measured by the order parameter. The resulting continuous DSSP assignments were constructed to reflect the differences between NMR models of the same protein, so that the assignments reflect segments with thermal fluctuations (Fig. 17.10). In other words the more a sequence segment fluctuates, the lower the probability for the assigned helix/sheet will become. Information of this type can also be obtained directly from crystal structures. Overall, we found that the continuous assignment of secondary structure reflected the average occupancy of secondary structure assignments. In particular, our continuous assignment for a single NMR structure is similar to the average obtained over all models. This result may indicate that for short intervals of time the concept of discrete secondary structure

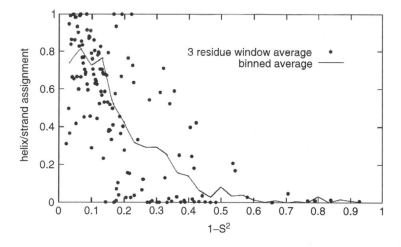

**Figure 17.10.** Protein motion and secondary structure. Using one set of coordinates from an ensemble of NMR models, the continuous DSSP assignment reproduces the segments in proteins that experimentally had a high degree of motion due to thermal fluctuations in water. Figure reproduced from (Andersen et al., 2001). Protein motion has been independently measured by the order parameter $1-S^2$, by the tumbling of the N–H backbone bond-vector. The order parameter is low when the amino acid is fixed as in the protein core and it is high when the residue fluctuates. The figure shows $1-S^2$ versus the continuous DSSP assignment grouping helices (GHI) and strands (EB). The points are averages over a window segment of three consecutive residues; the line gives an average of helix/strand assignments.

states is appropriate. However, thermal fluctuations change these states slightly. Hence, the average we actually observe is a continuous secondary structure. Our analysis also indicated that most secondary structure regions are crystal solid!

## CONCLUSION

Assigning secondary structure from 3D co-ordinates is an important problem. Many successful solutions have been proposed over since the 1980s. One of the oldest solutions is DSSP. There are many reasons why that program has become the standard in the field. In fact, secondary structure assignment may be one of the exceptional examples for tools in structural biology and bioinformatics that has not been revolutionized by the recent explosion of data. For most residues, most of the available methods agree in their assignment. Methods tend to differ mainly in locating the ends of regular secondary structure segments and in distinguishing between more subtle differences (e.g., alpha-, $3_{10}$-, or pi-helix). To oversimplify the data: The residues for which the assignment methods differ tend to be the residues for which structural homologues or different NMR models differ, too. This idea has recently led to a number of new concepts for the assignment task. Two of these new concepts introduce the idea of continuous secondary structure assignments (STICK and DSSPcont). This novel interpretation of secondary structure breaks with the early idea that there are secondary structure states. Since accurate secondary structure assignments are at the base of accurate comparisons between structures and predictions of protein structure, the story will continue.

**Abbreviations used: 3D**, three-dimensional; **DEFINE**, method assigning secondary structure from 3D co-ordinates based on linear distance masks to ideal secondary structure (Richards and Kundrot, 1988); **DSSP**, program and database assigning secondary structure and solvent accessibility for proteins of known 3D structure from hydrogen bonding patterns (Kabsch and Sander, 1983a); **DSSPcont**, continuous assignment of secondary structure for proteins of known 3D structure (Andersen et al., 2001); **NMR**, nuclear magnetic resonance; **P-Curve**, curvature based assignment of secondary structure from 3D (Sklenar et al., 1989); **PDB**, Protein Data Bank of experimentally determined 3D structures of proteins (Berman et al., 2000); **rmsd**, root-mean-square deviation; **STRIDE**, secondary STRuctural IDEntification method to assign secondary structure from 3D using hydrogen bonds and torsion angles (Frishman and Argos, 1995);

**Notations used:** *secondary structure assignment*, the step from 3D co-ordinates of residues to secondary structure states of residues (note: that; *state*, describes one of the eight DSSP assignments G, H, I, E, B, T, S, L (space in DSSP output); *class*, describes a group of states, a typical grouping of the eight DSSP states into three classes is: h = GHI, e = EB, and c = TSL.

## ACKNOWLEDGMENTS

Thanks to Jinfeng Liu (CUBIC, Columbia) for computer assistance and to Søren Brunak (CBS, Copenhagen) and Phil Borne (SCRIPPS) for helpful comments on the manuscript. The work of Burkhard Rost was supported by the grants 1-P50-GM62413-01 and RO1-GM63029-01 from the National Institutes of Health. Last, but not least, thanks to all those who deposit their experimental data in public databases, and to those who maintain those databases.

# REFERENCES

Andersen CAF (2001): Protein structure and the diversity of hydrogen bonds. The Technical University of Denmark Ph.D. Thesis.

Andersen CAF, Palmer AG, Brunak S, Rost B (2001): Continuous secondary structure assignment correlates with protein flexibility. *Structure*. [The evaluation criterion for structure assignments based on protein flexibility is described. The relationship between variations in NMR model coordinates and protein motion is demonstrated. The influence of motion on secondary structure is analyzed and an estimate for the upper limit of secondary structure prediction is presented.]

Baker EN, Hubbard RE (1984): Hydrogen bonding in globular proteins. *Prog Biophys Mol Biol* 44:97–179.

Baldi P, Brunak S, Frasconi P, Soda G, Pollastri G (1999): Exploiting the past and the future in protein secondary structure prediction. *Bioinformatics* 15:937–46.

Barton GJ (1995): Protein secondary structure prediction. *Curr Opin Struct Biol* 5:372–6.

Berman HM, Westbrook J, Feng Z, Gillliland G, Bhat TN, Weissig H, Shindyalov IN, Bourne PE (2000): The Protein Data Bank. *Nucleic Acids Res* 28:235–42.

Blundell TL, Mizuguchi K (2000): Structural genomics: an overview. *Prog Biophys Mol Biol* 73:289–95.

Boobbyer DN, Goodford PJ, McWhinnie PM, Wade RC (1989): New hydrogen-bond potentials for use in determining energetically favorable binding sites on molecules of known structure. *J Med Chem* 32:1083–94.

Bordo D, Argos P (1994): The role of side-chain hydrogen bonds in the formation and stabilization of secondary structure in soluble proteins. *J Mol Biol* 243:504–19.

Brändén C, Tooze J (1991): *Introduction to Protein Structure*. New York: Garland Publishing.

Burley SK, Almo SC, Bonanno JB, Capel M, Chance MR, Gaasterland T, Lin D, Sali A, Studier FW, Swaminathan S (1999): Structural genomics: beyond the human genome project. *Nat Genet* 23:151–7.

Bystroff C, Baker D (1998): Prediction of local structure in proteins using a library of sequence-structure motifs. *J Mol Biol* 281:565–77.

Colloc'h N, Etchebest C, Thoreau E, Henrissat B, Mornon J-P (1993): Comparison of three algorithms for the assignment of secondary structure in proteins: the advantages of a consensus assignment. *Protein Eng* 6:377–82. [A statistical comparison of secondary structure assignments between DSSP, P-Curve, and DEFINE is presented. The CF and GOR prediction algorithms are furthermore compared to the committee of the three assignment schemes analyzed.]

Creighton T (1993): *Proteins: Structures and Molecular Properties*. New York: W.H. Freeman.

Cuff JA, Barton GJ (1999): Evaluation and improvement of multiple sequence methods for protein secondary structure prediction. *Proteins* 34:508–19.

de la Cruz X, Thornton JM (1999): Factors limiting the performance of prediction-based fold recognition methods. *Protein Sci* 8:750–9.

Di Francesco V, Munson PJ, Garnier J (1999): FORESST: fold recognition from secondary structure predictions of proteins. *Bioinformatics* 15:131–40.

Eyrich V, Martí-Renom MA, Przybylski D, Fiser A, Pazos F, Valencia A, Sali A, Rost B (2002): EVA: continuous automatic evaluation of protein structure prediction servers. *Structure* 10:175–84.

Fain B, Levitt M (2001): A novel method for sampling alpha-helical protein backbones. *J Mol Biol* 305:191–201.

Fasman GD (1989): The development of the prediction of protein structure. In: Fasman GD, editor. *Prediction of Protein Structure and the Principles of Protein Conformation.* New York: Plenum Press, pp 193–303.

Finkelstein AV (1997): Protein structure: what is it possible to predict now? *Curr Opin Struct Biol* 7:60–71.

Fischer D, Eisenberg D (1996): Fold recognition using sequence-derived properties. *Protein Sci* 5:947–55.

Frishman D, Argos P (1995): Knowledge-based protein secondary structure assignment. *Proteins* 23:566–79. [The phi-psi statistics used for the $\alpha$-helix and $\beta$-sheet assignments are described along with a visualization of the five $\beta$-sheet hydrogen bond conformation used. The data set of 226 protein chains containing the crystallographers' assignments, used as the standard-of-truth, is also listed.]

Hogue CW, Bryant SH (1998): Structure databases. *Methods Biochem Anal* 39:46–73.

Holm L, Sander C (1998): Touring protein fold space with Dali/FSSP. *Nucleic Acids Res* 26:318–21.

Hvidt A, Westh P (1998): Different views on the stability of protein conformations, and hydrophobic effects. *J Solution Chem* 27:395–402.

Jaroszewski L, Rychlewski L, Zhang B, Godzik A (1998): Fold prediction by a hierarchy of sequence, threading, and modeling methods. *Protein Sci* 7:1431–40.

Jeffrey GA, Saenger W (1994): *Hydrogen Bonding in Biological Structures.* Berlin: Springer-Verlag.

Jennings AJ, Edge CM, Sternberg MJ (2001): An approach to improving multiple alignments of protein sequences using predicted secondary structure. *Protein Eng* 14:227–31.

Jones DT (1999a): GenTHREADER: an efficient and reliable protein fold recognition method for genomic sequences. *J Mol Biol* 287:797–815.

Jones DT (1999b): Protein secondary structure prediction based on position-specific scoring matrices. *J Mol Biol* 292:195–202.

Jones DT, Orengo CA, Thornton JM (1996): Protein folds and their recognition from sequence. In: Sternberg MJE, editor. *Protein Structure Prediction.* Oxford: Oxford University Press, pp 173–206.

Jones DT, Tress M, Bryson K, Hadley C (1999): Successful recognition of protein folds using threading methods biased by sequence similarity and predicted secondary structure. *Proteins* 37:104–11.

Kabsch W, Sander C (1983a): Dictionary of protein secondary structure: pattern recognition of hydrogen bonded and geometrical features. *Biopolymers* 22:2577–637. [The original description of DSSP describes the secondary structure definitions applied and the underlying rationale. The hydrogen bond model is analyzed along with the reasoning behind the chosen energy cutoff.]

Kabsch W, Sander C (1983b): How good are predictions of protein secondary structure? *FEBS Lett* 155:179–82.

Kabsch W, Sander C (1983c): Segment83. Unpublished.

Kolinski A, Rotkiewicz P, Ilkowski B, Skolnick J (1999): A method for the improvement of threading-based protein models. *Proteins* 37:592–610.

Lesk AM (1991): *Protein Architecture—A Practical Approach.* Oxford: Oxford University Press.

Lesk AM, Lo Conte L, Hubbard TJP (2001): Assessment of novel folds targets in CASP4: predictions of three-dimensional structures, secondary structures, and interresidue contacts. *Proteins.* 45(5):598–118.

Lesk AM, Rose GD (1981): Folding units in globular proteins. *Proc Natl Acad Sci USA* 78:4304–8.

Lindahl E, Elofsson A (2000): Identification of related proteins on family, superfamily and fold level. *J Mol Biol* 295:613–25.

Liu J, Rost B (2001a): Comparing function and structure between entire proteomes. *Protein Sci* 10:1970–9.

Liu J, Rost B (2001b): Target space for structural genomics revisited. *Bioinformatics*. 18:922–33.

Lo Conte L, Ailey B, Hubbard TJ, Brenner SE, Murzin AG, Chothia C (2000): SCOP: a structural classification of proteins database. *Nucleic Acids Res* 28:257–9.

Lupas A (1996): Coiled coils: new structures and new functions. *TIBS* 21:375–82.

Marchler-Bauer A, Addess KJ, Chappey C, Geer L, Madej T, Matsuo Y, Wang Y, Bryant SH (1999): MMDB: Entrez's 3D structure database. *Nucleic Acids Res* 27:240–3.

Marti-Renom MA, Stuart A, Fiser A, Sanchez R, Melo F, Sali A (2000): Comparative protein structure modeling of genes and genomes. *Ann Rev Biophys Biomol Struct* 29:291–325.

Murzin AG (1996): Structural classification of proteins: new superfamilies. *Curr Opin Struct Biol* 6:386–94.

Orengo CA, Pearl FM, Bray JE, Todd AE, Martin AC, Lo Conte L, Thornton JM (1999): The CATH Database provides insights into protein structure/function relationships. *Nucleic Acids Res* 27:275–9.

Pauling L (1939): *The Nature of the Chemical Bond*. New York: Cornell University Press.

Pauling L, Corey RB (1951): Configurations of polypeptide chains with favored orientations around single bonds: two new pleated sheets. *Proc Natl Acad Sci USA* 37:729–40.

Pauling L, Corey RB, Branson HR (1951): The structure of proteins: two hydrogen-bonded helical configurations of the polypeptide chain. *Proc Natl Acad Sci USA* 37:205–34.

Pearl FM, Martin N, Bray JE, Buchan DW, Harrison AP, Lee D, Reeves GA, Shepherd AJ, Sillitoe I, Todd AE, Thornton JM, Orengo CA (2001): A rapid classification protocol for the CATH domain database to support structural genomics. *Nucleic Acids Res* 29:223–7.

Przytycka T, Aurora R, Rose GD (1999): A protein taxonomy based on secondary structure. *Nat Struct Biol* 6:672–82.

Quiocho FA, Spurlino JC, Rodseth LE (1997): Extensive features of tight oligosaccharide binding revealed in high-resolution structures of the maltodextrin transport/chemosensory receptor. *Structure* 5:997.

Ramachandran GN, Sasisekharan V (1968): Conformation of polypeptides and proteins. *Adv Protein Chem* 23:284–438.

Rice DW, Eisenberg D (1997): A 3D–1D substitution matrix for protein fold recognition that includes predicted secondary structure of the sequence. *J Mol Biol* 267:1026–38.

Richards FM, Kundrot CE (1988): Identification of structural motifs from protein coordinate data: secondary structure and first-level supersecondary structure. *Proteins* 3:71–84. [The distance matrix masks used are listed for the different secondary structure elements. An overview of the methods for secondary structure assignment is presented.]

Richardson JS, Richardson DC (1989): Principles and patterns of protein conformation. In: Fasman GD, editor. *Prediction of Protein Structure and the Principles of Protein Conformation*. New York: Plenum Press, pp 1–98.

Rost B (1995): TOPITS: Threading one-dimensional predictions into three-dimensional structures. In: Rawlings C, Clark D, Altman R, Hunter L, Lengauer T, Wodak S, editors. *Third International Conference on Intelligent Systems for Molecular Biology*; Cambridge, England. Menlo Park, CA: AAAI Press, pp 314–21.

Rost B (1996): PHD: predicting one-dimensional protein structure by profile based neural networks. *Methods Enzymol* 266:525–39.

Rost B (1998): Marrying structure and genomics. *Structure* 6:259–63.

Rost B (2001b): Protein secondary structure prediction continues to rise. *J Struct Biol* 134:204–18.

Rost B (2001c): Rising accuracy of protein secondary structure prediction. In: Chasman D, editor. *Protein Structure Determination, Analysis, and Modeling for Drug Discovery.* New York: Dekker.

Rost B, Eyrich V (2001): EVA: large-scale analysis of secondary structure prediction. *Proteins.* 45(5):S192–S99.

Rost B, O'Donoghue SI (1997): Sisyphus and prediction of protein structure. *CABIOS* 13:345–56.

Rost B, Sander C (1994): 1D secondary structure prediction through evolutionary profiles. In: Bohr H, Brunak S, editors. *Protein Structure by Distance Analysis.* Amsterdam: IOS Press, pp 257–76.

Rost B, Sander C (1996): Bridging the protein sequence-structure gap by structure predictions. *Ann Rev Biophys Biomol Struct* 25:113–36.

Rost B, Sander C (2000): Third generation prediction of secondary structure. *Methods Mol Biol* 143:71–95.

Rost B, Schneider R, Sander C (1997): Protein fold recognition by prediction-based threading. *J Mol Biol* 270:471–80.

Russell RB, Copley RR, Barton GJ (1996): Protein fold recognition by mapping predicted secondary structures. *J Mol Biol* 259:349–65.

Sali A (1998): 100,000 protein structures for the biologist. *Nat Struct Biol* 5:1029–32.

Sauder JM, Arthur JW, Dunbrack Jr RL (2000): Large-scale comparison of protein sequence alignment algorithms with structure alignments. *Proteins* 40:6–22.

Schulz GE (1988): A critical evaluation of methods for prediction of protein secondary structures. *Ann Rev Biophys Biophys Chem* 17:1–21, PMID: 329:35–82.

Shapiro L, Harris T (2000): Finding function through structural genomics. *Curr Opin Biotech* 11:31–5.

Sippl MJ (1995): Knowledge-based potentials for proteins. *Curr Opin Struct Biol* 5:229–35.

Sippl MJ, Floeckner H (1996): Threading thrills and threats. *Structure* 4:15–9.

Sklenar H, Etchebest C, Lavery R (1989): Describing protein structure: a general algorithm yielding complete helicoidal parameters and a unique overall axis. *Proteins* 6:46–60. [The geometrical measures used for the assignment are visualized along with phi-psi plots of the helicoidal parameters. The differences to assignments based on phi-psi angle or hydrogen bond descriptions of secondary structure are also illustrated.]

Srinivasan R, Rose GD (1999): A physical basis for protein secondary structure. *Proc Natl Acad Sci USA* 96:14258–63.

Sternberg MJ, Bates PA, Kelley LA, MacCallum RM (1999): Progress in protein structure prediction: assessment of CASP3. *Curr Opin Struct Biol* 9:368–73.

Suugiyama S, Matsuo Y, Maenaka K, Vassylyev DG, Matsushima M, Kashiwagi K, Igarashi K, Morikawa K (1996): The 1.8-Å X-ray structure of the *Escherichia coli* potd protein complexed with spermidine and the mechanism of polyamine binding. *Protein Sci* 5:1984–90.

Taylor WR (2001): Defining linear segments in protein structure. *J Mol Biol* 310:1135–50. [STICK finds a set of line segments independently of any external secondary structure definition. This method allows the segments to be used as a novel basis for secondary structure definition by taking the average rise/residue along each axis to characterize the segment. This practice has the advantage that secondary structures are described by a single (continuous) value that is not restricted to the conventional classes of $\alpha$-helix and $\beta$-strand.]

Teeter MM (1984): Water structure of a hydrophobic protein at atomic resolution. Pentagon rings of water molecules in crystals of crambin. *Proc Natl Acad Sci USA* 81:6014.

Wade RC, Clark KJ, Goodford PJ (1993): Further development of hydrogen bond functions for use in determining energetically favorable binding sites on molecules of known structure. 1. Ligand probe groups with the ability to form two hydrogen bonds. *J Med Chem* 36:140–7.

Wilmot CM, Thornton JM (1990): Turns and their distortions: a proposed new nomenclature. *Protein Eng* 3:479–94.

Xu H, Aurora R, Rose GD, White RH (1999a): Identifying two ancient enzymes in Archaea using predicted secondary structure alignment. *Nat Struct Biol* 6:750–4.

Xu Y, Xu D, Crawford OH, Einstein J, Jr, Larimer F, Uberbacher E, Unseren MA, Zhang G (1999b): Protein threading by PROSPECT: a prediction experiment in CASP3. *Protein Eng* 12:899–907.

Yang AS, Honig B (2000): An integrated approach to the analysis and modeling of protein sequences and structures. I. Protein structural alignment and a quantitative measure for protein structural distance. *J Mol Biol* 301:665–78.

Young M, Kirshenbaum K, Dill KA, Highsmith S (1999): Predicting conformational switches in proteins. *Protein Sci* 8:1752–64. [A data set of 16 protein sequences having functions that involve substantial backbone rearrangements is analyzed with respect to the ambivalence of predicted secondary structure. The authors find all segments involved in conformational switches to have ambivalent predictions, measured by the similarity in prediction probabilities for helix, sheet, and loop, as reported by PHD (see Chapter 28).]

# 18

# IDENTIFYING STRUCTURAL DOMAINS IN PROTEINS

Lorenz Wernisch and Shoshana J. Wodak

The notion of domains in proteins plays a very important role in structural biology, genetics, biochemistry, and evolutionary biology. Often, however, this notion is defined differently in each of these subdisciplines. In structural biology, domains were initially defined as segments of the polypeptide chain that fold into globular units, which may also carry out specialized molecular functions (Janin and Wodak, 1983; Rose, 1979; Wetlaufer, 1973). In contrast, geneticists and biochemists define a domain as the minimal fragment of a gene, usually identified in a deletion experiment, that is still capable of performing a given function.

When the same or a closely similar domain is found in many different proteins, it is often called a module. Classic examples are the immunoglobulin domain (Go, 1983) or the SH2 domains (Sadowski, Stone, and Pawson, 1986). While the three-dimensional structure of the domain/module is conserved in the different contexts, the amino-acid sequence, and occasionally the function, may differ substantially.

Recent analyses of the fast-growing number of known genomes confirm that organisms as diverse as bacteria and human share many proteins and protein domains. This finding lends support to the view that the total number of genes/protein modules is small. The total number of different folds that protein modules can adopt was estimated to be about 1000, before any complete genome sequence was available (Chothia, 1992). More recently, these estimates were revised to be in the range of 1000–6000 (Brenner, Chothia, and Hubbard, 1997; Orengo, Jones, and Thornton, 1994). Many of the modules are thus highly recurrent. They may occur in isolation as small single-domain proteins, or they may be part of larger polypeptide chains, assembled by successive

*Structural Bioinformatics*
Edited by Philip E. Bourne and Helge Weissig
Copyright © 2003 by Wiley-Liss, Inc.

events of gene fusion. Combining a specific set of modules within a single polypeptide chain ensures that they are expressed together and localized in the same cells or cellular compartments (Tsoka and Ouzounis, 2000).

In some cases, the modules may participate in the same cellular process, and sometimes interact physically, forming specific protein–protein complexes, without being covalently linked. Thus, both the fused and separated arrangements exist in nature. When one arrangement is observed in some organism, the other is likely to be used in different ones. This observation was exploited in several methods recently proposed for detecting protein–protein interactions from the amino acid sequence (Enright et al., 1999; Marcotte et al., 1999a; Marcotte et al., 1999b). Although it is at present not clear what fraction of the interactions detected by these methods represent actual physical interactions between the modules, there is mounting evidence that these methods detect the module or protein involvement in common functional processes.

Whether the genes are fused or separate, there is thus a very close interplay between protein modules and their interactions. One major goal of the postgenomic era is therefore to systematically characterize the repertoire of protein modules and their interactions in terms of their biological function. This undertaking is a key component of the large scale proteomics efforts (Fields, 2001) and the related efforts in Structural Genomics (Montelione and Anderson, 1999, and Chapter 29). The latter aims at obtaining information about the three-dimensional (3D) structures of proteins and their complexes, one of the important prerequisites for understanding molecular function and cellular function. Since proteins can be composed of one or more modules, reliable methods for parsing protein 3D structures and sequences into their constituent modules are of great interest.

This chapter presents an overview of computational methods for parsing experimentally determined protein 3D structures, into substructures also termed structural domains. These domains may correspond to modules or to remnants of modules that existed in the past. The basic concepts underlying the domain-parsing methods were developed in the 1970s, and changed little since then. The actual algorithms, however, have undergone some new developments recently in terms of speed and generality. But probably more important still, we know much more today about the world of protein folds and their diversity. These different aspects will be described here, with illustrations taken from the authors' own work.


## HOW IT ALL STARTED

The first survey of structural domains in proteins was carried in the early 1970s (Wetlaufer, 1973), using visual inspection of the then available X-ray structures. Wertlaufer defined domains as regions of the polypeptide chain that form compact globular units, sometimes loosely connected to one another. At about the same time the first so-called $C\alpha$-$C\alpha$ distance plots were computed (Ooi and Nishikawa, 1973; Phillips, 1970) and shown to be useful for identifying structural domains. Domains were identified visually in these plots by looking for series of short $C\alpha$-$C\alpha$ distances in triangular regions near the diagonal, separated by regions outside the diagonal where few short distances occur, as illustrated in Figure 18.1. This approach was used by Rao and Rossmann (1973) to locate the nucleotide-binding domains of lactate dehydrogenase and flavodoxin. Ooi and Nishikawa (1973) applied this approach to $\alpha$-chymotrypsin and myoglobin and Rossman and Liljas (1974) to many other proteins.

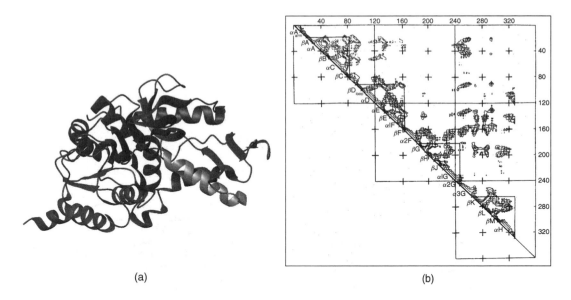

(a)                                        (b)

Figure 18.1. Domain structure of dogfish lactate dehydrogenase, determined using the Cα-Cα distance map. (a) Ribbon diagram of lactate dehydrogenase, showing the NAD binding (green) and catalytic domains (red). In gray is part of the helix, spanning residues 164–180, linking the two domains. (b) Distance map and structural domains in lactate dehydrogenase. Contours represent Cα-Cα distances of 4Å, 8Å, and 16Å within the subunit of dogfish lactate dehydrogenase. Elements of secondary structure are identified along the diagonal. Triangles enclose regions where short Cα-Cα distances are abundant. The NAD-binding domain comprises the first two triangles (counting from the N-terminus), which are subdomains. The catalytic domain comprises the last two triangles (the C-terminal domain). From Rossman and Liljas (1974) and reproduced by permission of Academic Press (London) Ltd. Figure also appears in Color Figure section.

## HOW ARE STRUCTURAL DOMAINS DEFINED?

The underlying concept of most if not all domain definition methods, in particular the earlier ones, has been that atomic interactions within domains are more extensive than between domains (Richardson, 1981; Wetlaufer, 1973). From this concept it follows that domains can be identified by looking for groups of residues with a maximum number of atomic contacts within a group, but a minimum number of contacts between the groups, as illustrated in Figure 18.2a.

This method was also considered as a way of predicting structural units that are likely to be stable on their own and possibly fold independently (Conejero-Lara et al., 1994), an important goal pursued by the initial domain surveys.

A problem, encountered rather frequently when using this definition to locate structural domain in protein structures, is that groups of residues satisfying the above criteria sometimes belong to noncontiguous segments of the polypeptide chain. This situation, which may arise as a result of gene insertion events or domain swapping (Bennett, Choe, and Eisenberg, 1994), entails that in partitioning the 3D structure, the polypeptide chain may be cut more than once, yielding noncontinuous domains, as shown in Figure 18.2b.

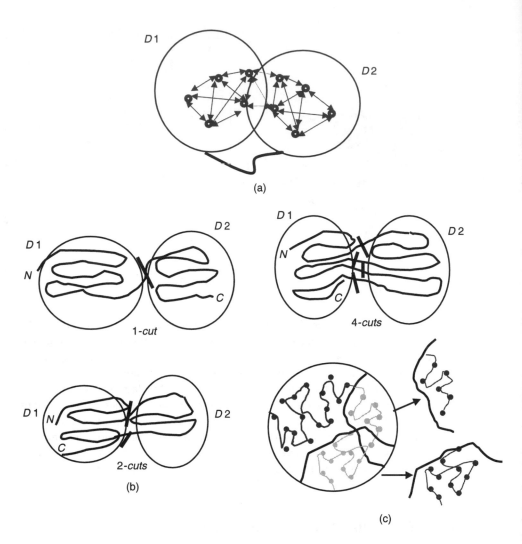

**Figure 18.2.** Illustration of the problem of parsing the protein 3D structure into structural domains. (a) The most common definition of structural domains, as groups of residues with a maximum number of contacts within each group and a minimum number of contacts between the groups. (b) Domains may be composed of one or more chain segments. Any domain-assignment procedure must therefore be able cut the polypeptide chains as many times as necessary. When both domains are composed of contiguous chain segments (continuous domains), only one chain cut (1-cut) is required. When one domain is continuous and the other discontinuous, a situation that may arise as a result of gene insertion, then the chain has to be cut in two places (2-cuts). When both domains are discontinuous, additional chain cuts may be required. In the example shown, the chain is cut in four places (4-cuts), and thus the domain on the left-hand side contains three chain segments, whereas that on the right contains two chain segments. (c) This drawing shows two solutions to the problem of partitioning the protein 3D structure into substructures. To partition the 3D structure into domains, many such solutions need to be examined in order to single out the one that satisfies the criterion given in (a). Figure also appears in Color Figure section.

A general method for identifying domains must therefore be able to handle this situation, and hence not use information about the covalent structure of the protein chain. The problem at hand then becomes an optimization problem, in which the optimal way of partitioning the 3D structure must be singled out from a large number of possibilities, as illustrated in Figure 18.2c. This is the challenge that all the methods for locating structural domains in proteins have to face.

Unfortunately, however, solving the optimization problem adequately and efficiently is often not sufficient, owing to the fact that the landscape of atomic interactions in proteins is inherently noisy. The optimization procedure therefore needs to be supplemented by additional criteria, which often include a description of some expected properties of previously characterized domains, as will be discussed below.

## FIRST GENERATION ALGORITHMS FOR DOMAIN ASSIGNMENTS

The first systematic survey of domains in a set of protein 3D structures, was performed by Rossman and Liljas (1974), by analyzing Cα-Cα distance maps. This work was followed a few years later, by three other studies by Crippen (1978), Rose (1979), and Janin and Wodak (1983); Wodak and Janin, (1981b). The methods described in these studies involved different algorithms and produced different results, but had one major aspect in common. They were used to partition the protein 3D structure in a hierarchic fashion, yielding not only domains but also smaller substructures, as illustrated in Figure 18.3. Surveys of these smaller substructures in a set of proteins revealed recurrent structural motifs comprising two or three secondary structure elements joined by loops (Wodak and Janin, 1981a). Interestingly, the systematic identification of similarly defined smaller substructures in proteins was recognized years later as a useful way to predict regions in proteins that would form first during folding (nucleation sites) (Moult and Unger, 1991).

Looking back at the early domain assignment studies of Wodak and Janin (1981b) one can find other valuable information, which could not be exploited at the time because the number of sequences and 3D structures was insufficient. Their algorithm, which was designed to detect essentially continuous domains, involved producing a surface area scan. This scan plotted the interface area B between an N-terminal segment of i residues and the complementary C-terminal segment, as a function of i. Domain boundaries were identified as minima of B in the scans.

Such surface area scan obtained for triose phosphate isomerase (TIM) (Janin and Wodak, 1983) is shown in Figure 18.4. A noteworthy feature of this scan is the minimum in B at residue 125, splitting the structure roughly in the middle (the other two minima, as 64 and 209, define further substructures). This feature was merely recorded at the time, but no interpretation was offered. More recently, however, the product of the HisA gene, also believed to adopt the TIM barrel fold, was found to display an internal sequence duplication (Fani et al., 1994). Subsequently, evidence of this duplication was also detected in other TIM barrel proteins of the histidine operon of *T. maritima* (Thoma et al., 1998). This finding, together with analysis of the larger number of available sequences and 3D structures for this ubiquitous fold (Copley and Bork, 2000), led to the suggestion that present-day barrels might have evolved from a common ancestor adopting a half-barrel structure, each featuring the glycine-rich phosphate binding site (Thoma et al., 1998).

This idea seems to suggest that a systematic search of "fault lines" in the protein 3D structure, such as the one enabled by the interface area scan in Figure 18.4, may

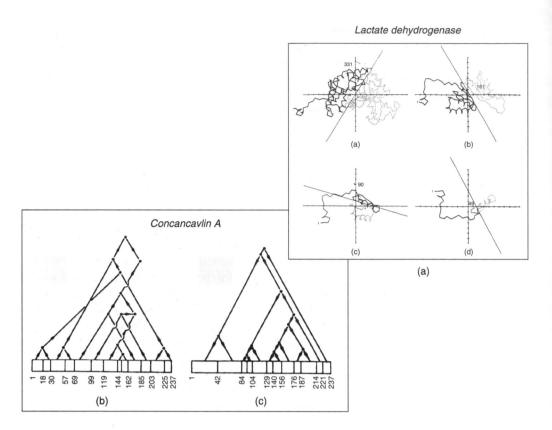

**Figure 18.3.** Illustration of the hierarchic partitioning of protein 3D structures into substructures performed by three of the early automatic domain-analysis methods. (a) The partitioning procedure of Rose (1979) applied to dogfish lactate dehydrogenase. The chain tracing of the subunit is projected into a 'disclosing plane' passing through its principal axes of inertia. A line is then found that divides the projection into two parts about equal size. The different projections represent (from left to right and from the bottom down) the whole subunit (residues 1–331) of the N-terminal domain (1–161), and the subdomain (residues 1–90) and (1–49). From Rose (1979), and reproduced with permission of Academic Press (London) Ltd. (b) Domain and subdomain hierarchies in concanavalin A adapted from Crippen (1978). It represents an ascending hierarchy of clusters, starting from 13 'reasonably straight segments' at the bottom of the hierarchy, and culminating with the entire concanavalin A subunit. (c) The domain and subdomain hierarchy in concanavalin A adapted from Wodak and Janin (1981b) describing the descending hierarchy derived from identifying the minimum interface area between substructures in successive interface area scans. Unlike the hierarchy in (b) this one generates continuous substructures. The agreement with the hierarchy in (b) is therefore poor.

reveal remnant features of the protein history that cannot be gleaned from the amino acid sequence alone. This is supported by the observation that many of the automatic domain assignment procedures, including some very recent ones, also tend to split in two some of the proteins adopting the TIM barrel structure (see below).

Several other methods were developed in the ten years, or so, following these initial works. Some of the earlier ones were aimed at identifying substructures, with reference to stability and folding, much along the original concept of Wetlaufer, whereas the later

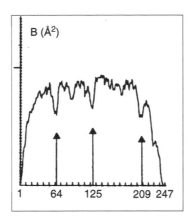

**Figure 18.4.** The interface area scan for Triose Phosphate Isomerase (TIM) computed as in Wodak and Janin (1981b). The area B in $\text{Å}^2$ of the interface between residues 1 to i and i to N, with N being the last residue in the chain, is plotted against the position I of the cleavage point. Residue numbering is as in the original file for the atomic coordinates. The scan has been produced using a model of the protein, where each residue is represented by a single interaction center. The arrows shown in the plot indicate the minima of B in the scan. These occur at residue 64, 125, and 209. The minimum at residue 125 splits the TIM barrel into two halves, which correspond well with the proposals for the common half barrel ancestor (see text).

ones, were aimed at parsing protein 3D structures into domains that can be used as a basis for structure classification.

We view them all as first-generation methods because they do not consider the problem of optimally partitioning the protein 3D structure in its full generality. Indeed, to identify optimal partitions, they invariably use the order of the residues along the sequence, and systematically (often recursively) split the protein into contiguous segments along the polypeptide chain, or assemble such segments. The main features of these methods are summarized next. A discussion of their performance is presented below in Domain Assignments: Is there a Unique Set of Criteria?

## Identifying Domains on the Basis of Physical Criteria

Among the earlier methods that identified domains on the basis of some physical criteria, that of Sander (1981) related domain identification to the motion of rigid bodies composed of continuous chain segments. Argos (1990) focused on analyzing the composition and conformation of linkers between domains, also composed of continuous chain segments, and used a graphic inspection to identify domain boundaries.

Rashin (1981) combined the method of Wodak and Janin (1981b) with a globularity measure for each substructure to identify domains that are likely to fold independently, whereas Zehfus and Rose (1986) used a related compactness measure to identify continuous domains, which Zehfus (1994) later extended to identify discontinuous ones also. Swindells (1995a, 1995b) with the program DETECTIVE, assigned domains by detecting distinct hydrophobic cores in a protein, using information on secondary structure, side-chain solvent accessibility and side-chain–side-chain contacts.

A somewhat ad hoc analysis of distance maps was used by Go (1983) to identify structural units in hen egg white lysozyme, which were correlated with the exon structure of the gene. The relation between exons and structural domains in proteins was actively debated during these years (see Janin and Wodak, 1983), and remains unconvincing to this day (Doolittle, 1995; Stoltzfus et al., 1994).

## Identifying Domains as Units for Protein Structure Classifications

With the significant increase in the number of known protein structures in recent years, the major incentive has been to devise automatic methods for identifying domains that can form the basis for a consistent protein structure classification. One of the first generation methods developed with this goal in mind was that of Islam, Luo, and Sternberg (1995).[1] This method starts with a hierarchy of recursive single cuts along the chain and subsequently assembles strongly interacting segments, thereby defining multisegment domains. The method of Siddiqui and Barton (1995), DOMAK,[2] involves partitioning the chain into three or four contiguous segments, but further chain cuts become prohibitively costly to compute, whereas Sowdhamini and Blundell (1995)[3] cluster secondary structure elements.

Several of these procedures are accessible on the Web in their original or updated versions.

## SECOND GENERATION METHODS FOR DOMAIN ASSIGNMENTS

The methods described in this section are those that consider the problem of partitioning the protein 3D structure in its full generality. The underlying algorithms usually find their inspiration from procedures developed in other disciplines (statistics, physics, graph theory), and are hence often computationally more efficient and general than those of the first generation methods.

The earliest method, in this category is the one by Holm and Sander (1994). This method uses a principal component analysis of a modified atomic contact matrix to find a partition with a low number of contacts, independently of the order of the residues along the sequence. This method, now called PUU,[4] is still being used today to assign domains for the FSSP fold classification (Holm and Sander, 1996).

## Domain Assignments Based on Graph Theoretical Methods

Of the more recent second-generation methods for domain assignment, several make elegant use of graph theoretical methods. The procedure implemented in STRUDL (STRUctural Domain Limits)[5] by Wernisch, Hunting, and Wodak (1999) provides a good example of such methods and their application to the problem at hand, and is therefore illustrated here with some detail.

This procedure views the protein as a 3D graph of interacting residues, with no reference to any covalent structure. The problem of identifying domains then becomes that of partitioning this graph into sets of residues such that the interactions between the sets

[1] Domain server: http://www.bmm.icnet.uk/~domains/

[2] DOMAK: http://www.compbio.dundee.ac.uk/

[3] http://www-cryst.bioc.cam.ac.uk/~mini/; http://www-cryst.bioc.cam.ac.uk/~ddbase/ (under construction since 1/3/99)

[4] http://www.es.embnet.org/Services/ftp/databases/puu/s/

[5] STRUDL: http://www.ucmb.ulb.ac.be/strudl/

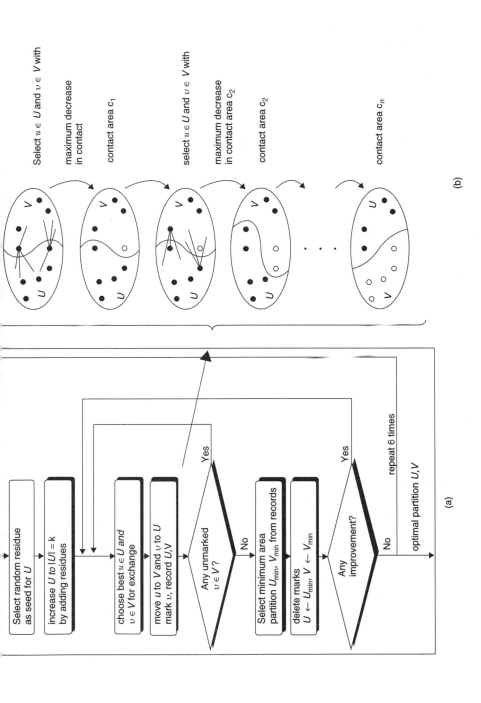

(a)

(b)

Figure 18.5. Domain identification using the graph heuristic procedure implemented in STRUDL (Wernisch, Hunting, and Wodak, 1999) (a) Overview of the major steps in the STRUDL algorithm. (b) The residue-exchange procedure in STRUDL. Residues $u \varepsilon U$ and $v \varepsilon V$ (filled circles) are selected so as to produce a maximal decrease or if that is not possible, a minimal increase in the contact area between $U$ and $V$ on exchange. Once moved to $V$, residue $u$ is flagged (empty circle) and can hence not be moved back to $U$. The exchange procedure stops when $V$ contains only flagged residues. Among all partitions with contact area $C_i$, with $i = 0, \ldots n$ the one with minimum contact area $C_{min}$ is selected.

373

is minimum. Since this problem is NP-hard, efficient heuristic procedures—procedures capable of approximating the exact solution with reasonable speed—are an attractive alternative. The algorithm used in this case was a slightly modified version of the Kernigan-Lin heuristic for graphs (Kernighan, 1970). The application of this heuristic to domain assignment is summarized in Figure 18.5.

A useful, though not essential aspect of this application is that the interactions between residue subsets were evaluated using contact areas between atoms. This area was defined as the area of intersection of the van der Waals sphere around each atom and the faces of its weighted Voronoi polyhedron. This contact measure is believed by the authors to be more robust than counting atomic contacts, due to its lower sensitivity to distance thresholds.

To identify domains for which the limits and size are not known in advance, the partitioning procedure described in Figure 18.5 is repeated k times, with k representing all the relevant values of the domain size, ranging from 1 to N/2, and N being the total number of residues in the protein. The partition with minimum contact area, identified for each value of k, is recorded. This information is then used to compute a *minimum contact density profile*. In this profile, the minimum contact area found for each k is normalized by the product of the sizes of the corresponding domains, in order to reduce noise (Holm and Sander, 1994; Islam, Lou, and Sternberg, 1995), and plotted against k. The domain definition algorithm then searches for the global minimum in this profile.

Figure 18.6a, illustrates the profile obtained for a variant of the p-hydroxy-benzoate hydroxylase mutant (PDB-RCSB code 1dob), a 394-residue protein composed of two discontinuous domains. The global minimum in the minimum contact density profile, although quite shallow, is clearly visible at k = 172, and yields the correct solution. The corresponding partition cuts the chain in five distinct locations yielding two domains, comprising six chain segments. The smallest of the two domains contains 172 residues; the largest contains 222 = 394 − 172, residues.

Once the global minimum is identified in the minimum contact density profile, a decision must be taken to either accept or reject the corresponding partition, with a

---

$\longrightarrow$

**Figure 18.6.** Minimum contact and minimum-contact density profiles computed by the procedure in STRUDL (Wernisch, Hunting, and Wodak, 1999). (a) The minimum contact density profile for the p-hydroxy-benzoate hydroxylase mutant (PDB-RCSB code 1dob). The $C_{dens}$ value in Å, is computed using the formula given in the figure. In this formula $c(U, V)$ is the contact area between the two residue groups $U$ and $V$, and $|U|$ and $|V|$ are the number of residues in $U$ and $V$, respectively. The plotted values represent the minimum of $C_{dens}$ computed for a given domain size k, where $k = 1, N/2$, with N being the total number of residues in the protein chain. Hence $|U|$ and $|V|$ equal k and $N - k$, respectively. The arrow at $k = 172$, indicates the global minimum of this profile. The dashed line delimits the value of $k = 20$ below which splits are not allowed, to avoid generating domains containing less that 20 residues. (b) The profiles of minimum contact area $c(U, V)$ in Å computed as a function of $k$ the number of residues in the smallest substructure $U$, for the same protein as in (a). Shown are the profile with no constraints on the number of chain cuts, as in the STRUDL procedure, and four other p-cut profiles, obtained by limiting the number of allowed chain cuts p (see Wernisch, Hunting, and Wodak, 1999 for details). Those cuts are 1-cut (- - - - ), 2-cuts (- - - - ) 3-cuts (-------) and 4-cuts ( · · · · ·). The global minimum of the contact area (arrow) can only be located in the unconstrained profile, illustrating the advantages of STRUDL over other procedures in which the number of allowed chain cuts is fixed.

rejection corresponding to classifying the structure as a one-domain protein. An obvious criterion on which to base such decision is the actual value of the contact area density in profiles such as that of Figure 18.6a: If this value is below a given threshold, the partition is accepted, otherwise it is rejected. But this simple criterion is unfortunately not reliable enough. Following other authors, additional criteria, representing expected properties of domains and of the interfaces between them (Rashin, 1981; Wodak and Janin, 1981b), were therefore used to guide the decision. The choice of these criteria

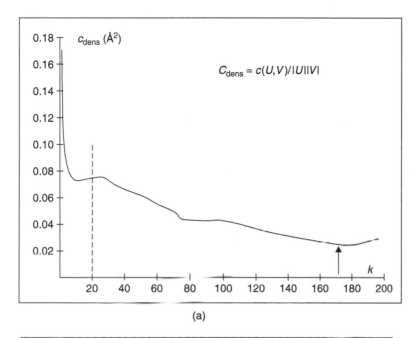

$$C_{dens} = c(U,V)/|U||V|$$

(a)

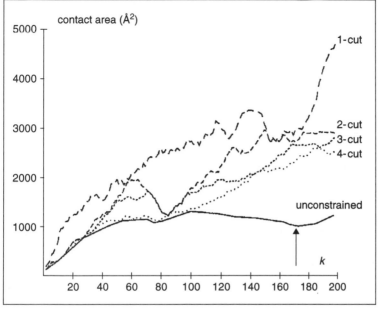

(b)

was carefully optimized on a training set of 192 proteins using a discriminant analysis, and tested on a different, much larger set of proteins, as summarized in Figure 18.7.

Finally, once a partition is accepted, the entire procedure is repeated recursively on each of the generated substructures until no further splits are authorized. This recursive approach was shown to successfully handle proteins composed of any number of continuous or discontinuous domains.

As always, assessing the performance of the method is a crucial requirement. Fortunately, the significant increase in the number of different proteins of known structure presently offers a more extensive testing ground. STRUDL was applied to a set of 787 representative protein chains from the Protein Data Bank (PDB) (Berman et al., 2000; Bernstein et al., 1977), and the results were compared with the domain definitions that were used as the basis for the CATH protein structure classification (Orengo et al., 1997). This definition was based on a consensus definition produced by Jones et al. (1998), using three automatic procedures, PUU (Holm and Sander, 1994), DETECTIVE[6] (Swindells, 1995a), and DOMAK (Siddiqui and Barton, 1995), as well as by manual assignments (see Domain Assignments: Is there a Unique Set of Criteria? below).

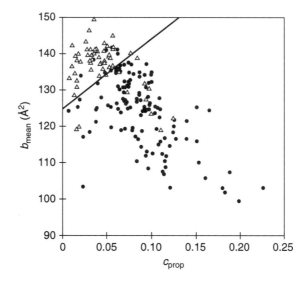

**Figure 18.7.** Threshold optimization for a pair of parameters used to evaluate proposed domain partitions with STRUDL (Wernisch, Hunting, and Wodak, 1999). This figure displays the plot of the mean burial $b_{mean}$ in Å, versus the contact area ratio $c_{prop}$. Both quantities are evaluated for domain partitions corresponding to the global minimum in the contact area density profiles computed by STRUDL. $b_{mean}$ is the average interresidue contact area of the two substructure. $C_{prop}$ Is the ratio of the contact area of the putative domains to the sum of the interresidue contact areas in the entire protein. The straight line optimally separates the single (filled circles) from the multidomain proteins (empty triangles). This optimal separation entails, however, 19 errors—proteins classified in the wrong category—out of the total of 192 protein considered in the set.

---

[6]http://www.biochem.ucl.ac.uk/bsm/cath_new

Results showed that domain limits computed by STRUDL coincide closely with the CATH domain definitions in 81% of the tested proteins, and hence that STRUDL performs as well as the best of the above-mentioned three methods. In contrast to these methods however, STRUDL uses no information on secondary structures in order to prevent splitting $\beta$-sheets, for instance. The 19% or so proteins for which the domain limits did not coincide with the CATH assignments represent interesting cases, which could for the most part be rationalized either on the basis of the intrinsic differences between the approaches (in this case, STRUDL versus CATH), or by the variability and complexity inherent in real proteins.

Several of these cases are illustrated in Figures 18.8–18.10. Figure 18.8 shows cases where STRUDL splits into two-domain proteins that CATH considers as being single-domain, single-architecture proteins. Among the shown examples are the DNA polymerase processivity factor PCNA (1plq), which clearly shows an internal duplication, and the lant seed protein narbonin (1nar), which adopts a TIM barrel fold, that many automatic domain-assignment procedures tend to split in two, as already mentioned. Differences of this type could be rationalized by the fact that CATH imposes criteria based on chain architecture and topology, whereas STRUDL does not.

Figure 18.9 illustrates two cases where the results of STRUDL and CATH differ by the assignment of a single, relatively short protruding chain segment. Such cases illustrate the inherent noisiness in the backbone chain trace, which invariably affects

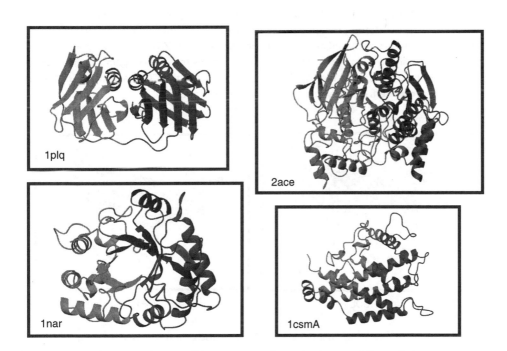

Figure 18.8. Examples of single-domain, single-architecture proteins in CATH (Jones et al., 1998), which STRUDL (Wernisch, Hunting, and Wodak, 1999) splits into two domains. Shown are the domain assignments produced by STRUDL. For the exact domain limits, the reader is referred to the STRUDL WEB site. The displayed protein ribbons belong to Torpedo californica acetylcholinesterase monomer (1ace), the plant seed protein narbonin (1nar), the eukaryotic DNA polymerase processivity factor PCNA (1plq), and Chorismate mutase chain A monomer (1csmA). Figure also appears in Color Figure section.

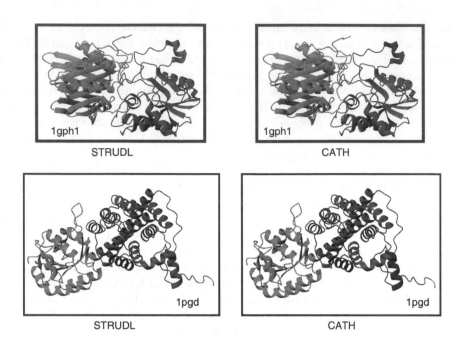

**Figure 18.9.** Different assignments by STRUDL (Wernisch, Hunting, and Wodak, 1999) and CATH (Jones et al., 1998), illustrating the effect of noise or decorations in the protein chain trace. The STRUDL assignments are displayed on the left-hand side, and the CATH assignments are displayed on the right. The short chain segments, which CATH assigns to separate domains are shown in blue. Some of the discrepancies may be due to simple 'slips' in the CATH assignments that have been or will be corrected. Figure also appears in Color Figure section.

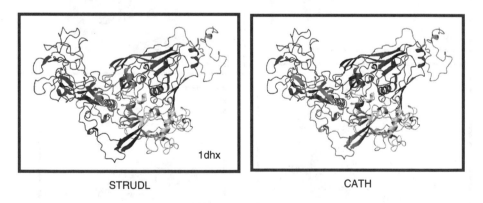

**Figure 18.10.** Different assignments by STRUDL (Wernisch, Hunting, and Wodak, 1999) (left) and CATH (Jones et al., 1998) (right), for the adenovirus hexon protein (1dhx), a protein with many domains of complex architecture. The different domains assigned by only method are displayed in different colors. Figure also appears in Color Figure section.

domain assignments. Figure 18.10 illustrates well the problem of treating proteins with many domains of different sizes and complex architecture. In such cases, often both the number of domains and their limits differ between CATH and STRUDL.

Another graph-theoretical procedure for domain assignment was recently proposed by Xu, Xu, and Grabow (2000) and implemented in the software DomainParser.[7] It also considers the protein 3D structure as a graph with residues as nodes and the contacts between the residues as the arcs. Parsing this network into domains is performed, as in Wernisch, Hunting, and Wodak (1999), by partitioning the 3D structure in two parts, in a recursive manner, but using a network-flow algorithm for finding the optimal cut. A set of rules for accepting or rejecting a cut, or for refining cuts, was also applied, but without performing a discriminant analysis. The procedure was tested on the 55 protein test set of Jones et al. (1998), and applied to all the chains in the FSSP database[8] (Holm and Sander, 1996). It was shown to yield very satisfactory results, which were similar to those obtained by Wernisch, Hunting, and Wodak (1999). The method is available on the Web.

## Other Methods

Two other methods, by Taylor (1999) and Xuan, Ling, and Chen (2000), have been reported. The approach proposed by Taylor is of particular interest. His method is akin to an Ising model. The basic algorithm is very simple. First, each residue in the chain is assigned numeric labels (the sequential residue number along the sequence). If a residue is surrounded by neighbors (defined using a distance threshold), with on average a higher label, its label increases, otherwise it decreases. The procedure is applied iteratively to all the residues in the chain. Several refinements were added to this simple scheme in order to circumvent a number of technical problems and to avoid unphysical situations such as frequent crossings of chain segments between domains, assigning domains that are too small, or cutting through $\beta$-sheets. The method was shown to yield results in good agreement with domain assignments extracted from the original literature, and compiled by Islam, Luo, and Sternberg (1995).

Interestingly, the author cites the trypsins, pepsins, lactate dehydrogenase, and a few TIM barrel structures as not yielding the accepted domain assignments with his procedure. His method split several of the TIM barrels, but failed to split the other proteins into their accepted domains. He then makes the appealing suggestion that the application of the procedure not only to one structure but to multiple structures from related proteins could be used to avoid such "errors".

## DOMAIN ASSIGNMENTS: IS THERE A UNIQUE SET OF CRITERIA?

To evaluate the performance of a new domain assignment method, its authors usually apply it to a set of proteins for which domain assignments have been produced manually, by experts (crystallographers or protein classifiers), or by other automatic domain-assignment methods, or both.

Today, the most common sources for such reference assignments are the domain-based protein classifications such as CATH (Orengo et al., 1997) and SCOP (Murzin et al., 1995), as well as resources such as FSSP (Holm and Sander, 1996). Although these sources feature very similar domain assignments for the majority of the proteins,

---

[7]DomainParser: http://compbio.ornl.gov/structure/domainparser/
[8]FSSP database: http://www2.embl-ebi.ac.uk/dali/fssp/

in a fraction of the proteins, which mostly comprises complex multidomain structures, the assignments do differ. Since the two major protein classifications use manual domain assignments, either alone or in addition to automatic ones, these discrepancies in domain assignments cannot be readily interpreted, adding to the difficulty of assessing new domain assignment procedures.

The difficulty in producing consistent assignments with different methods are well illustrated in two studies. One by Jones et al. (1998), which compares and combines the results of different domain-assignment methods to produce the domain definitions used in CATH. The other is by Wernisch, Hunting, and Wodak (1999), in which the differences between the results of one algorithm (STRUDL) is systematically compared to the CATH assignments.

Jones et al. (1998) applied four methods, PUU (Holm and Sander, 1994), DETECTIVE (Swindells, 1995b), DOMAK (Siddiqui and Barton, 1995), and the method of Islam, Luo, and Sternberg (1995) to generate domain assignments for a subset of 55 proteins with published assignments, taken from the repertoire of Islam, Luo, and Sternberg (1995). For the evaluation, any two assignments were considered as similar if they had the same number of domains, and at least 85% of the residues were assigned to the same domain.

Comparison of the domain assignments made by the individual methods with the published assignments showed a good overlap for 72% of the structures, on average, with scores of individual methods ranging between 67–76%. Single-domain proteins were predicted with the highest accuracy (mean 85%). Assignment for two-domain proteins was also quite accurate with a mean of 75%. The accuracy was seen to drop sharply as the number of domains increased.

Combining the assignments from several methods led to a clear improvement of the assignment accuracy. Consensus assignments, defined as those that were the same by all four methods, yielded the highest level of accuracy (100%), but these assignments could be made for only 52% of the protein chains. Using consensus assignments from three methods (DETECTIVE, PUU, and DOMAK) raised this percentage to 67%, with only a slight drop in accuracy level to 97%. The CATH domain assignments were then generated using consensus assignments from the three methods for a set of 787 representative proteins from the PDB. For the still sizable fraction of these proteins for which consensus assignments could not be obtained, mainly complex multidomain proteins, manual assignments were made.

Analysis of the assignments produced by this composite approach showed that there was little difference in the performance of the three algorithms, as in general all three produced good assignments of some structures and poor ones for others. When PUU and DOMAK made the wrong assignment, they tended to "over cut," that is, to divide structure into too many domains, rather than too few. DETECTIVE displayed an opposite trend. One general conclusion reached from this analysis, was that automatic procedures could probably be improved through the use of information on known protein folds, as already suggested (Holm and Sander, 1994). It was also cautioned (Islam et al. 1995) that the performance of automatic procedures is likely to remain limited due to the inherent variability in the details of natural protein structures, which often leads to exceptions.

The analysis of Wernisch, Hunting, and Wodak (1999), in which the assignments produced by STRUDL for the same set of 787 representative proteins were compared to those in CATH, reached similar conclusions but presented them in a somewhat different light. The analysis emphasized the fact that most automatic procedures incorporate a "postprocessing" step, which applies criteria based on our knowledge about the physical or topological parameters of domains. However, the most appropriate set of criteria may very well depend on the purpose for which the domain assignments are made. If the goal is to use the assigned domains as the basis for structural classification, as, for example, in CATH, then our knowledge about domain topologies, architecture, and function is useful. But if the goal is to identify compact structural domains that might be stable on their own, or might be remnants of ancient domains, then physical criteria must be considered in priority.

Note that using criteria based on domain topology or architecture requires procedures for identifying and defining acceptable folding topologies, which are not available today. In the meantime, a practical alternative, suggested and used by other authors (Jones et al. 1998; Holm and Sander, 1994) is to verify that a newly assigned domain corresponds to an already known folding motif. This suggestion may well take care of the majority of the cases, especially once the structural databases become populated with enough examples of all the possible folding motifs. Until then, however, it harbors the potential danger of biasing the domain assignments toward what is already known.

## CONCLUSION AND PERSPECTIVES

In this chapter we presented an overview of the principles underlying the detection of structural domain in proteins, and of the computational procedures that implement these principles in order to assign domains from the atomic coordinates of complete proteins. This overview showed that significant progress has been achieved over the years in the generality and reliability of the algorithms for domain detection. We should add that progress has also been achieved in calculation speed. The more recent second-generation methods, which cut the polypeptide chain in many places simultaneously, are orders of magnitude faster than the older methods that produce these cuts sequentially, so much so that calculation speed of domain-assignment methods has ceased to be an issue with present-day computer speeds.

However, we showed that some important limitations remain. All so-called second-generation algorithms elegantly solve the problem of partitioning the structures into domains composed of several chain segments, and can detect any number of domains. But a postprocessing step, or additional criteria, are always needed in order to deal with the inherent variability of natural protein structures, as well as with the inherent fuzziness of how we define domains. The latter was illustrated by the conflicts that may arise between defining domains as compact units with minimum interactions, versus units displaying a distinct chain topology.

Clearly, the rapid increase in the number of known protein structures expected in the coming years, should enable the refinement of our criteria for defining domains and for dealing with their inherent variability. The proposal of performing domain assignments on multiple structures belonging to related proteins and then averaging the results (Taylor, 1999), is a promising avenue in this regard. After all, a similar strategy of averaging the results from related proteins has become the rule for secondary structure predictions, or for identifying relationships between proteins using sequence data.

Another interesting direction, not sufficiently explored yet, would be to analyze domain–domain interfaces. If we believe that a protein containing more than one structural domain arose by gene duplication and/or gene fusion events followed by mutations, then we may expect the properties of interfaces between domains (ratio of hydrophobic/polar residues, H-bonds, packing, etc.), to resemble more those of interfaces between subunits than to the protein core. A recent study of Jones, Marin, and Thornton (2000) provides interesting insights in this regard.

Last but not least, it would be very useful if our knowledge and understanding of structural domains could be applied to identify domains from the amino acid sequence. This identification has many potential applications for structural genomics efforts (see Chapter 29) that could, for example, be directed in priority toward novel protein domains or toward compact domains, which should in principle be easier to crystallize. Another application would be the analysis of the evolution of protein structure and function.

Methods for identifying protein domains from the amino acid sequence presently rely mainly on sequence patterns, which are more directly associated with the domain function than with their structural properties. Several very valuable resources on the Web offer a compilation of such patterns and of sequence profiles for known domains (Bateman et al., 2000; Corpet et al., 2000; Hofmann et al., 1999).

But extremely few methods are presently available for the prediction of structural domain from the amino acid sequence without reference to function, clearly reflecting the difficulty of the task. The earliest attempt in this direction was by Kikuchi, Nemethy, and Scheraga (1988), who used information on average C$\alpha$-C$\alpha$ distances in proteins to define domain limits from the amino acid sequence. Of the handful of other more recent studies, those of Gracy and Argos (1998) and Murvai, Vlahovicek, and Ponger (2000) used information on sequence similarity to define domains from sequence, whereas that of Wheelan, Marchler-Bauer, and Bryant (2000) shows that information on the size distribution of structural domain in proteins can be a useful guide in identifying, or guessing domain boundaries from the amino acid sequence. The predictive capacity of these methods is still quite low, but will undoubtedly improve, as better ways of combining structural and sequence information for families of homologous proteins are devised.

## ACKNOWLEDGMENTS

The authors acknowledge support for this work from the Action de Recherches concertées de la Communauté Française de Belgique, project N° 97/01-211 (S. J. Wodak), and are grateful to Jean Richelle in Brussels and to Ricardo Valente at the European Bioinformatics Institute (EBI) for help with the computer systems.

## REFERENCES

Argos P (1990): An investigation of oligopeptides linking domains in protein tertiary structures and possible candidates for general gene fusion. *J Mol Biol* 211:943–58.

Bateman A, Birney E, Durbin R, Eddy SR, Howe KL, Sonnhammer EL (2000): The Pfam protein families database. *Nucleic Acids Res* 28:263–6.

Bennett MJ, Choe S, Eisenberg D (1994): Domain swapping: entangling alliances between proteins. *Proc Natl Acad Sci USA* 91:3127–31. [The original paper describing the domain swapping principle and suggesting it as a general principle for the evolution of multimeric proteins.]

Berman HM, Westbrook J, Feng Z, Gilliland G, Bhat TN, Weissig H, Shindyalov IN, Bourne PE (2000): The Protein Data Bank. *Nucleic Acids Res* 28:235–42.

Bernstein FC, Koetzle TF, Williams GJ, Meyer EE, Jr., Brice MD, Rodgers JR, Kennard O, Shimanouchi T, Tasumi M (1977): The Protein Data Bank: a computer-based archival file for macromolecular structures. *J Mol Biol* 112:535–42.

Brenner SE, Chothia C, Hubbard TJ (1997): Population statistics of protein structures: lessons from structural classifications. *Curr Opin Struct Biol* 7:369–76.

Chothia C (1992): Proteins. One thousand families for the molecular biologist. *Nature* 357:543–4. [Before any complete genome sequences was available, this short note presents arguments in support of the idea that the number of protein folds shared by all organisms is limited to about 1000.]

Conejero-Lara F, De Filippis V, Fontana A, Mateo PL (1994): The thermodynamics of the unfolding of an isolated protein subdomain. The 255-316 C-terminal fragment of thermolysin. *FEBS Lett* 344:154–6.

Copley RR, Bork P (2000): Homology among ($\beta\alpha/8$) barrels: implications for the evolution of metabolic pathways. *J Mol Biol* 303:627–41.

Corpet F, Servant F, Gouzy J, Kahn D (2000): ProDom and ProDom-CG: tools for protein domain analysis and whole genome comparisons. *Nucleic Acids Res* 28:267–9.

Crippen GM (1978): The tree structural organization of proteins. *J Mol Biol* 126:315–32. [One of the very first systematic procedure for identifying structural domains in proteins.]

Doolittle RF (1995): The multiplicity of domains in proteins. *Ann Rev Biochem* 64:287–314.

Enright AJ, Iliopoulos I, Kyrpides NC, Ouzounis CA (1999): Protein interaction maps for complete genomes based on gene fusion events. *Nature* 402:86–90.

Fani R, Lio P, Chiarelli I, Bazzicalupo M (1994): The evolution of the histidine biosynthetic genes in prokaryotes: a common ancestor for the hisA and hisF genes. *J Mol Evol* 38:489–95.

Fields S (2001): Proteomics. Proteomics in genomeland. *Science* 291:1221–4.

Go M (1983): Modular structural units, exons, and function in chicken lysozyme. *Proc Natl Acad Sci USA* 80:1964–8.

Gracy J, Argos P (1998): Automated protein sequence database classification. II. Delineation of domain boundaries from sequence similarities. *Bioinformatics* 14:174–87.

Hofmann K, Bucher P, Falquet L, Bairoch A (1999): The PROSITE database, its status in 1999. *Nucleic Acids Res* 27:215–9.

Holm L, Sander C (1994): Parser for protein folding units. *Proteins* 19:256–68. [The first truly general algorithm for defining structural domains that can identify domains comprising multiple chain segments.]

Holm L, Sander C (1996): Mapping the protein universe. *Science* 273:595–603.

Islam SA, Luo J, Sternberg MJ (1995): Identification and analysis of domains in proteins. *Protein Eng* 8:513–25.

Janin J, Wodak SJ (1983): Structural domains in proteins and their role in the dynamics of protein function. *Prog Biophys Mol Biol* 42:21–78.

Jones S, Stewart M, Michie A, Swindells MB, Orengo C, Thornton JM (1998): Domain assignment for protein structures using a consensus approach: characterization and analysis. *Protein Sci* 7:233–42. [A useful systematic comparison between different methods for assigning structural domains in proteins.]

Jones S, Marin A, Thornton JM (2000): Protein domain interfaces: characterization and comparison with oligomeric protein interfaces. *Protein Eng* 13:77–82.

Kernighan BW (1970): An efficient heuristic procedure for partitioning graphs. *Bell Systems Tech J* 49:291–307.

Kikuchi T, Nemethy G, Scheraga HA (1988): Prediction of the location of structural domains in globular proteins. *J Protein Chem* 7:427–71.

Marcotte EM, Pellegrini M, Ng HL, Rice DW, Yeates TO, Eisenberg D (1999a): Detecting protein function and protein–protein interactions from genome sequences. *Science* 285:751–3.

Marcotte EM, Pellegrini M, Thompson MJ, Yeates TO, Eisenberg D (1999b): A combined algorithm for genome-wide prediction of protein function. *Nature* 402:83–6.

Montelione GT, Anderson S (1999): Structural genomics: keystone for a Human Proteome Project. *Nat Struct Biol* 6:11–2.

Moult J, Unger R (1991): An analysis of protein folding pathways. *Biochemistry* 30:3816–24.

Murvai J, Vlahovicek K, Pongor S (2000): A simple probabilistic scoring method for protein domain identification. *Bioinformatics* 16:1155–6.

Murzin AG, Brenner SE, Hubbard T, Chothia C (1995): SCOP: a structural classification of proteins database for the investigation of sequences and structures. *J Mol Biol* 247:536–40.

Ooi T, Nishikawa K (1973): *Conformation of biological molecules and polymers*. In: Bergmann A, Pullmann B, editors. New York: Academic Press, pp 173–87.

Orengo CA, Jones DT, Thornton JM (1994): Protein superfamilies and domain superfolds. *Nature* 372:631–4.

Orengo CA, Michie AD, Jones S, Jones DT, Swindells MB, Thornton JM (1997): CATH—a hierarchic classification of protein domain structures. *Structure* 5:1093–108.

Phillips DC (1970): Past and present. In: Goodwin TW, editors. *British Biochemistry* London: Academic Press, pp 11–28. [The pioneering study in which the idea of structural domains was first described.]

Rao ST, Rossmann MG (1973): Comparison of super-secondary structures in proteins. *J Mol Biol* 76:241–56.

Rashin AA (1981): Location of domains in globular proteins. *Nature* 291:85–7.

Richardson JS (1981): The anatomy and taxonomy of protein structure. *Adv Protein Chem* 34:167–339.

Rose GD (1979): Hierarchic organization of domains in globular proteins. *J Mol Biol* 134:447–70. [One of the first systematic procedures for defining structural domains in proteins from the atomic coordinates.]

Rossman MG, Liljas A (1974): Letter: Recognition of structural domains in globular proteins. *J Mol Biol* 85:177–81. [One of the pioneering studies in which the idea of structural domains was first described.]

Sadowski I, Stone JC, Pawson T (1986): A noncatalytic domain conserved among cytoplasmic protein–tyrosine kinases modifies the kinase function and transforming activity of Fujinami sarcoma virus P130gag-fps. *Mol Cell Biol* 6:4396–408.

Sander C (1981): Physical criteria for folding units of globular proteins. In Balaban M, editor. *Structural Aspects of Recognition and Assembly in Biological Macromolecules. Vol I: Proteins and Protein Complexes, Fibrous Proteins*. Jerusalem: Alpha Press, pp 183–95.

Siddiqui AS, Barton GJ (1995): Continuous and discontinuous domains: an algorithm for the automatic generation of reliable protein domain definitions. *Protein Sci* 4:872–84.

Sowdhamini R, Blundell TL (1995): An automatic method involving cluster analysis of secondary structures for the identification of domains in proteins. *Protein Sci* 4:506–20.

Stoltzfus A, Spencer DF, Zuker M, Logsdon JM, Jr., Doolittle WF (1994): Testing the exon theory of genes: the evidence from protein structure. *Science* 265:202–7.

Swindells MB (1995a): A procedure for detecting structural domains in proteins. *Protein Sci* 4:103–12.

Swindells MB (1995b): A procedure for the automatic determination of hydrophobic cores in protein structures. *Protein Sci* 4:93–102.

Taylor WR (1999): Protein structural domain identification. *Protein Eng* 12:203–16. [An elegant heuristic procedure, inspired by the Ising model of solid-state physics, for assigning structural domains in proteins.]

Thoma R, Schwander M, Liebl W, Kirschner K, Sterner R (1998): A histidine gene cluster of the hyperthermophile Thermotoga maritima: sequence analysis and evolutionary significance. *Extremophiles* 2:379–89.

Tsoka S, Ouzounis CA (2000): Prediction of protein interactions: metabolic enzymes are frequently involved in gene fusion. *Nat Genet* 26:141–2.

Wernisch L, Hunting M, Wodak SJ (1999): Identification of structural domains in proteins by a graph heuristic. *Proteins* 35:338–52. [This paper describes a novel graph theoretical procedure for assigning structural domains in proteins. Handles any number of noncontiguous chain segments, uses no information on secondary structure. A discriminant analysis is used to derive a set of criteria that define physically meaningful domains.]

Wetlaufer DB (1973): Nucleation, rapid folding, and globular intrachain regions in proteins. *Proc Natl Acad Sci USA* 70:697–701. The first definition and systematic analysis of structural domains in proteins.]

Wheelan SJ, Marchler-Bauer A, Bryant SH (2000): Domain size distributions can predict domain boundaries. *Bioinformatics* 16:613–8. [One of the few attempts to define structural domains from sequence information.]

Wodak SJ, Janin J (1981a): Defining compact domains in globular proteins. In Balaban M. editors. *Structural Aspects of Recognition and Assembly in Biological Macromolecules.* Rehovot: Israel International Science Services, pp 149–67. [One of the first approaches for identifying small, compact substructures in proteins that are likely to be stable.]

Wodak SJ, Janin J (1981b): Location of structural domains in protein. *Biochemistry* 20:6544–52. [This paper describes the first approach for defining structural domains in proteins that relies on the evaluation of some physical property (size of the domain interface) as opposed to purely geometric criteria.]

Xu Y, Xu D, Gabow HN (2000): Protein domain decomposition using a graph-theoretic approach. *Bioinformatics* 16:1091–104.

Xuan ZY, Ling LJ, Chen RS (2000): A new method for protein domain recognition. *Eur Biophys J* 29:7–16.

Zehfus MH (1994): Binary discontinuous compact protein domains. *Protein Eng* 7:335–40.

Zehfus MH, Rose GD (1986): Compact units in proteins. *Biochemistry* 25:5759–65.

# 19

# INFERRING PROTEIN FUNCTION FROM STRUCTURE

Gail J. Bartlett, Annabel E. Todd, and Janet M. Thornton

## THE IMPORTANCE OF PREDICTING FUNCTION FROM STRUCTURE

With the advent of structural genomics, prediction of biological function from structure has become one of the major goals of structural biology and bioinformatics (Shapiro and Harris, 2000). Assignment of biological function provides a valuable first step toward experimental characterization of cellular and physiological roles of gene products. Ultimately, this assignment would improve genome analysis and annotation, and aid in the design of proteins with novel or modified functions.

Large-scale genome sequencing projects have provided us with details of all the genes an organism needs to survive. From this information we can translate the amino acid sequences of all the proteins that the genome could produce. Structural genomics projects aim to solve the structures of all these proteins, but their functions will be unknown. This process is a reversal of the usual experimental investigation of proteins, which involves taking a protein of interest, carrying out biochemical experiments to determine functional information about it, and then using the structure to rationalize this functional information (Thornton, Todd, and Milburn, 1999). For example, the tyrosine kinases were known to be signaling molecules long before the crystal structure revealed molecular mechanisms of their function (Hubbard et al., 1991). If structural genomics projects are to achieve their full scientific potential, it is vital to develop methods for predicting function from structure, in order that we can annotate the genomes with functional information.

Traditionally, identification of similar amino acid sequences is used to infer both the structure and the function of a protein. It is believed that structure and function

*Structural Bioinformatics*
Edited by Philip E. Bourne and Helge Weissig
Copyright © 2003 by Wiley-Liss, Inc.

can be transferred between similar sequences because they have been conserved over long periods of time. This belief has been confirmed for structure (Chothia and Lesk, 1986), but is more difficult to justify for function. It has been shown (Todd, Orego, and Thornton, 2001) that above 40% sequence identity, homologous proteins tend to have the same function, but below this threshold, conservation of function falls rapidly. Even at high levels one must be cautious in inferring function, as some sequence relatives with 35% or more sequence identity can have differing catalytic activities. Moreover, structure and function start to diverge at the same percentage sequence identity.

Protein functions (e.g., an enzyme active site) are often conferred by a few conserved residues, which sequenced-based methods often fail to detect. These residues will, however, be related in three dimensions, so a comparison of structural similarities between proteins has the potential to identify functional similarities in nonhomologous proteins.

## PROTEIN FUNCTIONS

### What Is the Function of a Protein?

The function of a protein is not always well defined (Skolnick and Fetrow, 2000). The term *function* covers a multitude of features a protein may exhibit (Table 19.1). The functional definitions given in Table 19.1 are different and distinct. Different experimental techniques can elucidate different aspects of function.

### Enzyme/Nonenzyme Classification

In this chapter we focus mainly on enzymes. They have a useful and well-established hierarchical classification (see later in this section) that has provided a useful starting point in exploring protein structure and function relationships. Additionally, enzymes are over-represented in the Protein Data Bank (PDB); therefore, several structure–function analyses have focused on this group of proteins.

It is easier to classify enzymes than to classify nonenzymes, partially because enzymes catalyze chemical reactions, and it is the reactions that can be classified.

TABLE 19.1. Different Aspects of Protein Function

| | |
|---|---|
| Biochemical | The chemical interactions occurring in a protein. For example, in an enzyme, the biochemical function would be the chemical reaction catalyzed by the enzyme, the substrates it binds, which ligands and/or cofactors are required to complete the reaction, and which regulators might affect its action. |
| Biological | The role within the cell of the protein, including cellular and physiological aspects. This includes the localization of a protein to a particular cell organelle or cell type. It also tells us which biological pathways the protein might be involved in, and under which conditions (e.g., heat shock) the protein may become active. |
| Phenotypic | The role played by the protein in the organism as a whole. This role can be investigated by deleting or mutating the gene encoding the protein and observing the effect on the organism. |

It is difficult to classify the nonenzymes as they may take part in many different protein–protein interactions, in different signaling pathways that may operate independently of each other. The Enzyme Commission (EC) scheme (NC-IUBMB, 1992) is the best developed and most widely used of all the protein functional classification schemes. It is a four-level hierarchy that classifies different aspects of chemical reactions catalyzed by enzymes (Table 19.2). The first digit denotes the class of the reaction, and subsequent levels classify the substrate, the type of bond involved, cofactors, and other specificities. The EC number of glyceraldehyde-3-phosphate dehydrogenase (GAPDH) is 1.2.1.12, which has the following detailed classification:

| | |
|---|---|
| EC 1.-.-.- | Oxidoreductases |
| EC 1.2.-.- | Acting on the aldehyde or oxo group of donors |
| CE 1.2.1.- | With NAD(+)or NADP(+) as acceptor |
| EC 1.2.1.12 | Glyceraldehyde 3-phosphate dehydrogenase (phosphorylating) |

TABLE 19.2. Description of the Different Levels in the EC Classification

| First figure | Second figure | Third Figure |
|---|---|---|
| A. OXIDOREDUCTASES<br><br>Substrate is oxidized–regarded as the hydrogen or electron donor | Describes substrate acted on by enzyme | Type of acceptor |
| B. TRANSFERASES<br><br>Transfer of a group from one substrate to another | Describes group transferred | Further information on the group transferred |
| C. HYDROLASES<br><br>Hydrolytic cleavage of a bond | Describes type of bond | Nature of substrate |
| D. LYASES<br><br>Cleavage of bonds by elimination | Type of bond | Further information on the group eliminated |
| E. ISOMERASES | Type of reorganization | Type of substrate |
| F. LIGASES<br><br>Enzyme catalyzing the joining of two molecules in concert with hydrolysis of ATP | Type of bond formed | Type of compound formed |

*Note*: An enzyme reaction is assigned a four-digit EC number, where the first digit denotes the class of reaction. Note that the meaning of subsequent levels depends on the primary number, for example, the substrate acted on by the enzyme is described at the second level for oxidoreductases, whereas it is described at the third level for hydrolases. Different enzymes clustered together at the third level are given a unique fourth number, and these enzymes may differ in substrate/product specificity or cofactor-dependency, for example. Note that the EC is a classification of overall enzyme reactions and not enzymes. Adapted from Todd et al., 2001.

Each of the protein databases discussed in the next section lists an enzyme's EC number.

## Database Classification of Protein Function

Information about protein function is contained in several different databases. Some of them mention function in passing, whereas others attempt to classify function. Some (e.g., SWISS-PROT) cover all organisms, while others (e.g., GenProtEC) are concerned with just one organism. Many of the functional schemes combine different aspects of function (Table 19.3). Consequently, functional information is only partially captured by these databases, which makes it difficult to transfer functional information from one homologous protein to another. Many functions have only been inferred from sequence similarities. These inferences may be incorrect and cannot be relied upon until confirmed by experiment. Indeed, some annotations in the databases are known to be wrong. These in accuracies and errors reduce the reliability of functional annotations. Table 19.3 shows the functional information obtained for glyceraldehyde-3-phosphate dehydrogenase (GAPDH) from four different databases.

## Multifunctional Proteins

Multifunctional proteins are even harder to classify in the functional schemes. For example, methylenetetrahydrofolate dehydrogenase/cyclohydrolase catalyzes the conversion of methylenetetrahydrofolate to formylfolate in two separate reactions, thought to proceed using the same or overlapping active sites (Allaire et al., 1998). This enzyme has two EC numbers associated with it—1.5.1.5 and 3.5.4.9. There are many multifunctional proteins where the function varies as a consequence of changes in expression and environment (Jeffery, 1999). Oligomerization and cellular localization are examples of such changes. Phosphoglucose isomerase acts as a neuroleukin, a cytokine, and a differentiation and maturation mediator in its monomeric, extracellular form, but as a dimer inside the cell, it has a role in glucose metabolism, catalyzing the interconversion of glucose-6-phosphate and fructose-6-phosphate. The function of multifunctional proteins also can vary according to cell type and cellular concentrations of ligand, substrate, cofactor, or product.

## Gene Ontologies

While this chapter is chiefly concerned with protein function, it is relevant to mention some of the functional classification schemes concerned with genes, as it is the gene product that will form the protein and carry out the function.

Functional ontology schemes try to organize genes according to the biological processes they perform—a necessary part of genome annotation. The Gene Ontology scheme (http://www.geneontology.org; Ashburner et al., 2000) consists of three independent functional ontology schemes for genes from different organisms: biological process, molecular function, and cellular component (see Table 19.4). The scheme uses a controlled vocabulary for describing the roles of genes and gene products in any organism. The relationship between gene product and biological process, molecular function and cellular component is often one-to-many, and by separating and independently assigning these attributes, relationships between gene product and function can be clarified more easily.

In a comparison of six different functional classification schemes (Rison, Hodgman, and Thornton 2000), each was mapped onto an iteratively generated functional

TABLE 19.3. Functional Classification of Glyceraldehyde-3-Phosphate in Protein Databases

| Database | GAPDH classification | | Comments |
|---|---|---|---|
| SWISS-PROT | Catalytic activity | D-glyceraldehyde-3-phosphate + orthophosphate + NAD+ → diphosphateglycerate + NADH | SWISS-PROT covers multiple species. |
| http://ca.expasy.org/sprot | | | The comments section contains information about catalytic activity, pathways, and subcellular localization. |
| | Pathway | First step in the second phase of glycolysis | |
| | Subcellular localization | Cytoplasmic | |
| | Keywords | Glycolysis; oxidoreductase; NAD; Multigene family | No attempt is made to classify functions, although each entry is accompanied by manually annotated keywords, chosen from a carefully controlled vocabulary. |
| ENZYME | Reaction catalyzed | D-glyceraldehyde-3-phosphate +phosphate + NAD+ ⇔ 3-phospho-D-glyceroyl phosphate +NADH | ENZYME covers multiple species. |
| http://ca.expasy.org/enzyme | | | Functional information is limited to the EC number, the reaction catalyzed, and notes on variable specifity. |
| | Comments | Acts very slowly on D-glyceraldehyde and some other aldehydes. Thiols can replace phosphate. | |
| Yeast Proteome Database (YPD) | Reaction | Catalyzes the reversible oxidation and phosphorylation of D-glyceraldehyde-3-phosphate to 1,3-diphosphoglycerate in glycolysis. | YPD covers *Sacchromyces cerevisiae* only. |
| http://www.proteome.com | Localization | | |
| | Mutant phenotype | | YPD is manually curated with a functional classification that contains more information pertaining to biological and phenotypic function than biochemical function. |

(*Continued overleaf*)

391

**T A B L E  19.3.** (*Continued*)

| Database | GAPDH classification | | Comments |
|---|---|---|---|
| | Localization | Cytoplasmic | The reaction catalyzed is annotated in the form of a sentence, unlike ENZYME and SWISS-PROT, which use the equation format. |
| | Mutant phenotype | Null: viable | |
| | Cellular role | Carbohydrate metabolism; Cell stress | Each gene product is annotated with cellular role, biochemical function, cellular localization, and mutant phenotype. |
| | Biochemical function | Oxidoreductase | |
| | Function | Function may be required in stressed cells | An additional function section indicates physiological function. |
| GenProtEC http://genprotec.mbl.edu | Metabolism | Energy metabolism, carbon: Glycolysis | GenProtEC contains information about *E. coli* only. |
| | | Cofactor, small molecule carrier: Building block biosynthesis: Pyridoxine (vitamin B6) | Uses three-level hierarchical classification, which starts very simply and becomes more detailed moving down through the levels. |
| | Location of gene products | Central intermediary metabolism: gluconeogenesis | One gene product can have multiple entries in each level. |
| | | Cytoplasm | Emphasizes biological functional information, but does provide links to other databases for other functional aspects. |

**TABLE 19.4. Gene Ontology Functional Classifications**

| Category | Description |
| --- | --- |
| Biological process | A biological objective to which the gene product contributes. A process (which often involves a chemical or physical transformation) is accomplished via one or more assemblies of molecular functions. A biological process can be high level (or less specific), for example, cell growth and maintenance, or low level (or more specific), for example, glycolysis. |
| Molecular function | The biochemical activity of a gene product, describing what it actually does without alluding to where or when. A molecular function can be broad (or less specific), for example, enzyme, or narrow (or more specific), for example, hexokinase. |
| Cellular component | Refers to the place in the cell where a gene product is active. |

ontology scheme in order to compare them. Each scheme fared differently, with some schemes showing broader functional coverage than others. It is important, therefore, when using information from the databases, that the features of each database are taken into consideration. One database may give you more accurate information than another, depending on how that information has been obtained and classified.

## WHAT INFORMATION CAN BE OBTAINED FROM THREE-DIMENSIONAL PROTEIN STRUCTURES?

### Basic Structure

The structure comes in the form of a PDB file (see Chapter 8), which is a list of three-dimensional (3D) coordinates of all the atoms in the protein. The PDB file itself contains little, if any, functional data. There is sometimes a "site" record, but this is used for various purposes, such as ligand-binding sites, metal-binding sites, or active sites and is not consistent. Some PDB files contain no functional information at all, except for the name of the protein. However, from the structure we can derive information relating to biological function; this information is summarized in Figure 19.1.

Looking at a visual representation of a protein structure tells us the overall organization of the protein chain in three dimensions. We can identify buried residues that make up the core of the protein, and residues on the surface exposed to the solvent. We can see the shape and molecular composition of the surface, as well as the juxtaposition of individual groups. We can also see the quaternary structure present in the crystal, which can tell us the oligomeric state of the protein, and may throw light on protein—protein interactions.

### Protein-Ligand Complexes

Protein-ligand complexes generally yield more functional information than the structure of the protein alone. The ligand, which is usually constructed with some knowledge of the function (e.g., a transition-state analogue), can give us clues as to the identity of

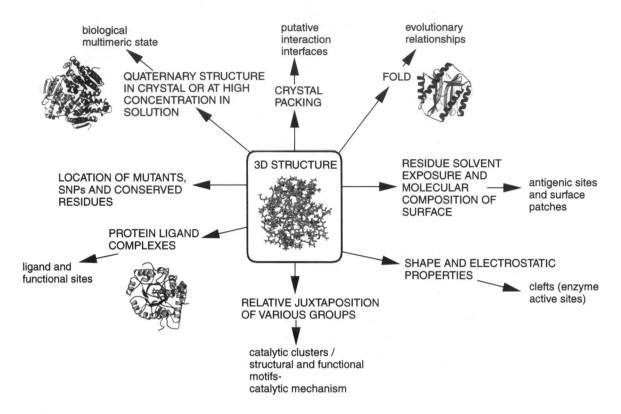

**Figure 19.1.** From structure to function: A summary of information that can be derived from 3D structure, relating to biological function.

groups binding the substrate, or residues that may be involved in other aspects of protein function such as catalysis or regulation. From this information, it may be possible to postulate a catalytic mechanism for an enzyme. In structures from structural genomics projects, however, the only ligands will tend to be co-enzymes and occasionally natural ligands picked up from the cell of the organism in which the protein has been cloned. (see Structural Genomics at work—Prediction of Function in Earnest below).

Essentially, structures provide information about the biochemical function of a protein, but not other aspects of function. However, if the protein in question is of unknown function, knowledge of the structure may help guide biochemical experiments to establish other functional aspects.

## RELATIONSHIP BETWEEN STRUCTURE AND FUNCTION

Analysis of the PDB suggests that a limited number of protein folds/families occur in nature (Chothia, 1992); (Orengo, Jones, and Thornton, 1994). This limitation is probably due to physicochemical constraints on protein folding that favor particular packing arrangements, combined with evolutionary selection. We can examine the relationship between structure and function on different levels—class, fold, homology, and analogy.

## Protein Structural Class and Enzyme Function

All structural classes of proteins form enzymes (Martin et al., 1998; Hegyi and Gerstein, 1999). $\alpha/\beta$ folds are over-represented in enzymes compared with the normal distribution. This over-representation is primarily due to the large number of nucleotide-binding domains in enzymes. The all-$\alpha$ and small folds are generally associated with nonenzymes. It is thought that the mainly-$\alpha$ class of proteins is under-represented in enzymes due to the inability of main-chain polar groups to be involved in catalysis, since their hydrogen-bonding potential is satisfied by other residues in the helix. Edges of $\beta$-sheets are thought to be more accessible.

Martin et al., (1998) found no correlation between protein class/architecture and enzyme function (according to the EC classification). This is presumably because enzymic activity is defined by a few amino acids in a precise location and orientation. However, Hegyi and Gerstein (1999), using additional sequence data, found statistically significant correlations between class and enzyme type. They found transferases and hydrolases to be particularly common among the $\alpha/\beta$ folds, although the origin of this bias is unclear and probably reflects evolutionary selection. Therefore, when trying to elucidate enzyme function from structure, the gross structural classification is unlikely to be of much help.

## Protein Fold and Function

As the number of folds in nature is limited, and there are multiple superfamilies within each fold group, proteins with similar structure can have totally different functions. The most promiscuous fold (or superfold) is the $(\beta/\alpha)_8$ barrel, which appears to have recurred multiple times in evolution and has many diverse functions. It is associated with 61 different EC numbers, with representatives in EC top-level classifications 1–5 (Nagano et al., 2001), showing a clear lack of correlation between EC number and topology. The top five most versatile folds in all proteins are summarized in Figure 19.2. All are $\alpha/\beta$ or $\alpha + \beta$ folds, and have many different functions associated with each.

As the wealth of structural and functional information grows, the number of functions per fold will undoubtedly increase.

## Homologous Families and Function

To date, most folds have one homologous family associated with them (Todd, Orengo, and Thornton, 1999). Within homologous protein families, it is expected that family members will have related functions. However, this is not always the case, and considerable diversity has been seen within homologous superfamilies.

The classic example of divergence of function within a homologous family is that of lysozyme and $\alpha$-lactalbumin (Acharya et al., 1991). These two proteins have high sequence identity between them, but vary in their function (Fig. 19.3). At the other end of the scale, during evolution, the globin family of proteins has been subjected to multiple amino acid changes, but their function has remained unchanged (Fig. 19.4).

In a study of conservation of function in 31 diverse enzyme superfamilies (Todd, Orengo, and Thornton, 2001), it was found that these superfamilies were associated with over 200 protein functions. On identification of the sequence relatives of all

## TIM barrel fold

The structure consists of an eightfold repeat of beta/alpha units. Eight parallel beta strands on the inside are covered by eight alpha helices on the outside.  The fold was first seen in triose phosphate isomerase.  All known TIM barrel structures are enzymes, except for the narbonin family. Many of these enzymes are glycosyl hydrolases (EC 3.2.x.x).  The fold is higly versatile, being found in single-domain monomeric enzymes and as the catalytic domain of larger enyzmes.  The active site is found at the C-terminal end of the barrel in a series of loops, hence it is very easy to alter the function and/or specificity without altering the core structure.

Number of EC numbers associated with this fold (to the third level): 29

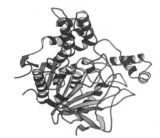

## Alpha/beta hydrolase fold

The structure is an eight stranded, mostly parallel alpha/beta structure. The  fold is tolerant to large insertions and is a very plastic. All proteins known so far containing this fold are enzymes.  The enzymatic properties of this fold are formed by a catalytic triad of a nucleophile, acid and a histidine residue.  The nucleophile is found in a "nucleophilic elbow" turn located just after the fifth beta strand.

Number of EC numbers associated with this fold (to the third level): 17

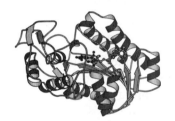

## NAD binding domain

This is a double beta-alpha-beta-alpha-beta motif, and is a common structural motif of enzymes binding NAD, NADP and other related cofactors, for example, NAD is found in dehydrogenases as the hydrogen acceptor. The domain is found as a common core unit in many structures, with other structural units at the periphery.

Number of EC numbers associated with this fold (to the third level): 5

## P-loop NTP hydrolase fold

This fold consists of alpha/beta/alpha, parallel or mixed beta sheets of variable size. The fold binds the phosphate of ATP or GTP and is found in ATP and GTP binding proteins such as adenylate kinase.  The P-loop is a phosphate binding loop that binds the phosphate groups of ATP and GTP,and is a glycine-rich sequence with the consensus sequence (A,G)xxxxGK(T,S).  The P-loop residues are shown in detail (left) in guanylate kinase.

Number of EC numbers associated with this fold (to the third level): 5

## Ferredoxinlike fold

This fold consists of an alpha/beta sandwich with an antiparallel beta sheet. The ferredoxinlike fold is associated  predominantly with nonenzymatic ferredoxins, like the example shown (Ferredoxin ii from *D. gigas*, left). Ferredoxins are iron-sulphur clusters invovled in electron transport, and often form part of multisubunit assemblies.  An example of an enzyme with this fold is muconolactone isomerase (EC 5.3.3.4).

Number of EC numbers associated with this fold (to the third level): 5

**Figure 19.2.**  A summary of the top five most versatile folds in all proteins (according to Hegyi and Gerstein, 1999). EC numbers quoted are from Nagano et al., 2001 (TIM barrels), Todd, 2001 ($\alpha/\beta$ hydrolases), and Hegyi and Gerstein, 1999 (P-loop NTP hydrolases, NAD binding domains, ferredoxins).

(a) Lysozyme EC 3.2.1.17      (b) Alpha-lactalbumin (nonenzyme)

**Figure 19.3.** Lysozyme (a) and $\alpha$-lactalbumin (b) have 40% sequence identity between them, and similar structures, but they have different functions. Lysozyme is an O-glycosyl hydrolase, but $\alpha$-lactalbumin does not have this catalytic activity. Instead, it regulates the substrate specificity of galactosyl transferase through its sugar-binding site, which is common to both $\alpha$-lactalbumin and lysozyme. Both the sugar-binding site and catalytic residues have been retained by lysozyme during evolution, but in $\alpha$-lactalbumin the catalytic residues have changed and it is no longer an enzyme.

(a) *V. stercoraria* hemoglobin      (b) *P. marinus* hemoglobin

**Figure 19.4.** The globin fold is resilient to amino acid changes. *V. stercoraria* (bacterial) hemoglobin (a) and *P. marinus* (eukaryotic) hemoglobin (b) share just 8% sequence identity, but their overall fold and function is identical.

the members of these protein families, the number of associated functions more than tripled. Some families contained proteins with differing numbers of associated enzyme functions, and some contained proteins that did not function as enzymes at all. This suggests divergence of function during enzyme evolution, with re-use of the same molecular architecture again and again. Such economy must have simplified the process of metabolic evolution.

## Analogues

Some functions have different structural solutions; these proteins are analogous to one another, sharing structural similarity but no sequence similarity. Such analogues are examples of convergent evolution toward the same function. The classic example is that of trypsin and subtilisin (Wallace, Laskowski, and Thornton, 1996) (Fig. 19.5).

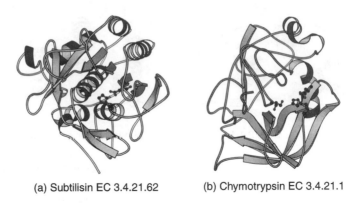

(a) Subtilisin EC 3.4.21.62          (b) Chymotrypsin EC 3.4.21.1

**Figure 19.5.** Subtilisin (a) and chymotrypsin (b) are both serine endopeptidases. They share no sequence identity, and their folds are unrelated. However, they have an identical, three-dimensionally conserved Ser-His-Asp catalytic triad, which catalyzes peptide bond hydrolysis. These two enzymes are a classic example of convergent evolution.

The most promiscuous functions are the glycosidase and carboxylase enzyme families, which are associated with seven different structural folds each (Hegyi and Gerstein, 1999).

It must be noted that these studies are biased by the content of the databases, for example, the PDB is biased toward smaller proteins that are easier to crystallize. They also depend on the accuracy of functional annotation in databases, which, as we have already shown, has variations. In addition, there is not always a 1:1 relationship between gene, protein, and reaction. What is more, the EC classification system has several limitations. It classifies chemical reactions, not the underlying biological mechanism, and the reaction direction is chosen arbitrarily, for example, 4.1.1.31 PEP carboxylase is classified as a lyase, when it catalyzes irreversible carbon–carbon bond formation. EC numbers also give no details of the reaction chemistry, so two enzymes sharing a common reaction chemistry and mechanistic strategy may have different EC numbers.

## ASSIGNING FUNCTION FROM STRUCTURE

There are several methods for assigning function from structure. One can make comparisons with proteins of known structure from the database, or use local structural motifs that capture the essence of biochemical function to search for a function in a new structure. Alternatively, the *ab initio* method relies on information gained solely from the solved structure and not on other sources. Each of these methods is discussed below.

### *Ab initio* Prediction

One of the main factors determining how a protein interacts with other molecules is the size of clefts on the protein surface. Clefts provide an increased surface area from which the solvent may be excluded, and therefore an increased opportunity for the protein to form complementary hydrogen bonds and hydrophobic contacts with small ligands.

A protein-ligand binding site (active site) is often found to be the largest cleft in the protein, and this cleft is often significantly larger than other clefts in the protein (Laskowski et al., 1996). There are two advantages to having an active site in a cleft. Firstly, it enables precise positioning of the substrate in order to facilitate catalysis. Secondly, burial of the substrate in such a cleft will seal it off from the bulk solvent, which effectively decreases the dielectric constant and allows the enzyme to generate electrostatic forces necessary for catalysis.

## Structural Comparisons

We can compare the structure of a protein of unknown function to structures of proteins of known function in structural databases such as CATH (Orengo et al., 1999) or SCOP (Lo Conte et al., 2000), and within defined bounds inherit the functional information from the closest match. This method of assigning function from structure is by far the most powerful, because proteins with a similar structure and sequence identity are evolutionarily related and are likely to share a similar function. However, caution must be exercised in transferring function from one homologous protein to another. If two proteins share structural similarity, but do not share sequence identity at all, their structural similarities might be the result of convergent evolution. Although each protein performs a similar function and has the same structure they may not be evolutionarily related. Moreover, two structurally homologous, evolutionarily related proteins might have different functions, (as in the case of the crystallins (Wistow, Mulders, and de Jong, 1987; Cooper, Isola, Stevenson and Baptist 1993) (Fig. 19.6).

## Structural Motifs

In order to use local structural motifs to search a new protein structure, in the case of enzymes, detailed knowledge of the active site is required. Enzyme active site residues are often more conserved than the overall fold, (e.g., subtilisin and chymotrypsin, Fig. 19.5). Structural motifs can be used to identify ancestors with the same global fold

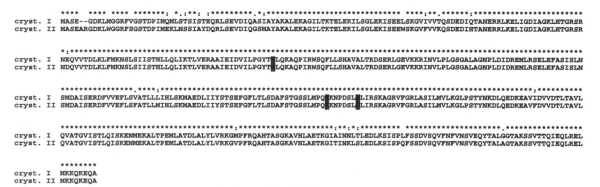

Duck crystallin delta-I *nonenzyme*
Duck crystallin delta-II/arginosuccinate lyase *enzyme*

**Figure 19.6.** Sequence alignment of duck crystallins δ-I and δ-II, proteins found in the eye lens that contribute to its refractive properties. The δ-II crystallin has arginosuccinate lyase enzyme activity, but the δ-I crystallin has lost this activity, even though it shares 94% sequence identity with the δ-II crystallin, and the active site residues are conserved.

and the same activity, as well as proteins with similar functions but different folds. Six different methods using this general strategy for assigning function are discussed below.

### 1. SITE and Site-Match.

SITE (Zhang et al., 2000), is a database containing information from the SITE entries of PDB files, as well as SWISS-PROT functional annotations, and ligand-interacting residues, defined by using a cutoff a certain distance from the ligand in the PDB file. The program Site-Match correlates this information with that produced from a sequence alignment, and verifies conservation of functional site residues. Conservation of these residues can be taken as an indicator of functional conservation. For clearly homologous proteins with significant sequence identity, approximately 10% did not contain conserved functional site residues. This percentage rises to 50% in weakly homologous proteins. These results show that care must be taken when transferring function between seemingly homologous proteins.

### 2. TESS.

TESS stands for Template Search and Superposition, and is a geometric-hashing algorithm for deriving 3D template coordinates from structures deposited in the PDB (Wallace, Borkakoti, and Thornton, 1997). The templates contain all the atoms considered essential for the enzyme to perform its catalytic function. The information for a template is acquired by mining the primary literature and assessing which residues form the active site. Given a set of 3D coordinates as a template and a protein structure, TESS looks for a match between them.

A TESS template for the serine protease active site, containing the vital atoms from the Ser-His-Asp catalytic triad, was able to describe the active site of all the serine proteases, acetylcholinesterase and haloalkane dehalogenase. The template could distinguish between catalytic triads found in enzymes and noncatalytic triads (i.e., atoms found in that configuration purely by chance) on the basis of the root-mean-square deviation (rmsd) of the atoms from the initial template. This result suggests that convergent evolution draws functional atoms into their optimal catalytic positions.

### 3. Fuzzy Functional Forms (FFFs).

The FFF method uses 3D structural information to identify biologically relevant sites in protein structures (Fetrow et al., 2001). The resulting active site descriptors are called fuzzy functional forms. An FFF describing disulphide oxidoreductase function identified 27 sequences in S. cerevisiae as potential disulphide oxidoreductases. The FFF was based on the common active site of the glutaredoxin, thioredoxin, and disulphide isomerase protein family, which consists of two cysteine residues essential for redox activity, and a structurally conserved cis-proline. The FFF method is similar to that of TESS, except that where TESS uses 3D atom coordinates of functional protein side chains, the FFF method uses the distances between alpha carbons with a small variance. The FFF method can therefore be used with inexact models as well as with high resolution structures, unlike TESS, but is less specific. All previously known thioredoxins, glutaredoxins, and disulphide isomerases were correctly identified, with just three false positives. Three of the novel predictions made were subsequently validated:

YERH4C —glutaredoxin 4
YDR098C—glutaredoxin 3
YPL059W—disulphide oxidoreductase

The FFF also postulated a disulphide oxidoreductase regulatory mechanism for two subunits of the yeast oligosaccharyltransferase complex. Via homology, this prediction could be extended to a potential tumor-suppressor gene N33 in humans, whose biochemical function was previously unknown.

**4. SPASM, RIGOR.** SPASM and RIGOR are tools for studying constellations of small numbers of residues (Kleywegt, 1999). SPASM stands for Spatial Arrangements of Side-chains and Main-chain, and can be used to find matches in the structural database for any user-defined motif. SPASM is similar to the FFF and TESS methods, using C-alpha and side-chain pseudoatoms as its template, but it has the advantage of being very easy to use. RIGOR compares a database of predetermined motifs against a newly determined structure, which could have an unknown function.

A SPASM template of catalytic residues from cellulobiohydrolase I from *Trichoderma reesei* hit four PDBs in the database that were expected to contain a similar set of residues.

**5. Molecular Recognition.** This method searches for similar spatial arrangements of atoms around a particular chemical moiety in proteins by superposing them (Kobayashi and Go, 1997). Arrangements in a pair of proteins are said to be similar when there are many corresponding overlapping atoms. This method can detect similar binding sites in proteins unrelated by sequence or overall fold.

A comparison of atoms surrounding adenine moieties in proteins highlighted structural similarity between protein kinases, cAMP-dependent protein kinase, casein kinase-1, and D-Ala-D-Ala ligase at their adenine-binding sites, in spite of the fact that these enzymes showed a lack of similarity in overall fold and sequence.

The same method was applied to phosphate-binding sites (Kinoshita et al., 1999), and found four frequently occurring structural motifs of protein atoms interacting with phosphate groups. Each motif appeared in different protein superfamilies with different folds. The most common motif is the P-loop GXXX, which interacts with the phosphate group via the backbone atoms, and is shared by 13 superfamilies (including the P-loop NTP hydrolases discussed previously (see Fig. 19.2)).

**6. Protein Side Chain Patterns.** Another method similar to the TESS, FFF, and SPASM templates detects active sites in proteins via recurring amino acid side-chain patterns (Russell, 1998). However, this method requires only protein structural data and associated multiple sequence alignments. The search is constrained by distance constraints and amino acid conservation, and amino acids unlikely to be involved in protein active sites (i.e., hydrophobic residues) are ignored. Matches are scored by rmsd, which is itself assessed by statistical significance, unlike the other methods listed above.

An all-against-all comparison of representatives of the PDB revealed previously unknown, convergently evolved (i.e., sequence independent) similarities, which point to possible functional similarities. These include a di-zinc binding pattern (Asp/Asp/His/His/Ser) common to alkaline phosphatase and bacterial aminopeptidase, and an Asp/Glu/His/His/Asn/Asn pattern common to the active sites of DNAse I and endocellulase E1. These functional similarities can now be investigated by experimental means.

## STRUCTURAL GENOMICS AT WORK—PREDICTION OF FUNCTION IN EARNEST

Large scale genome sequencing projects have led to the concept of structural genomics, which is the idea that one can determine 3D protein structures on a genomewide scale. In July 2001, there were 49 complete genomes sequenced (according to the Institute for Genomic Research, http://www.tigr.org/). Additionally, technological advances in PCR-based recombinant DNA technology, high-level protein expression systems, and structural characterization methods have increased the rapidity with which protein structures can be solved. Taken together, this suggests that high-throughput expression, crystallization, and subsequent structure determination should be possible on a genomewide scale (Burley et al., 1999). It is possible to envisage structure becoming an early part of biological analysis, to determine the function of a protein encoded by a particular gene, or at least guide biochemical experiments in elucidating its function.

A study of 424 nonmembrane proteins (excluding proteins with a clear sequence homologue in the PDB) cloned from *Methanobacterium thermoautotrophicum* (*M.th*), was carried out to test the feasibility of structural genomics projects (Christendat et al., 2000). These 424 proteins represent approximately one-third of all the proteins produced by *M.th*. Of these proteins, 20% were found to be suitable for either X-ray crystallographic or NMR spectroscopic analysis. Of the first 10 structures determined, several provided a model for interpretation of existing functional data, while some of the structures, including those containing protein-ligand complexes, provided enough functional hints to generate hypotheses for biochemical function that could be tested in the laboratory. This study showed that high-throughput structure determination is indeed feasible. However, the percentage of proteins suitable for structural analysis was low, indicating that improvements still need to be made in expression and crystallization techniques if projects such as these are to achieve their full potential.

A review of 15 hypothetical proteins of known structure and their functional assignment (Teichmann, Murzin, and chothia, 2001) gives some idea of the quality of functional assignments that can be made from structure. For each of the proteins, functional information was inferred by researchers from structural similarity to proteins of known structure and function. The extent of functional similarity was assessed by the extent of conservation of functional site residues. The structure and sequences of homologues of known function was used to find surface cavities/grooves in which conserved residues indicated an active site. Bound cofactors in the structure also provided functional information. This information, combined with experimental work, was assessed according to the depth of functional information that could be obtained. For the 15 proteins, detailed functional information was obtained for a quarter of them. For another half, some functional information was obtained, and for another quarter, no functional information could be obtained. This suggests that analysis of proteins of known structure but unknown function is often able to yield basic functional information, which can then be verified and built on using experimental techniques. It is much more difficult to assign detailed function to a hypothetical protein of known structure.

### Specific Examples of Functional Assignment from Structure

*Mj0577—Putative Atp Molecular Switch.* Mj0577 is an open reading frame (ORF) of previously unknown function from *Methanococcus jannaschii*. Its structure was determined at 1.7 Å (Fig. 19.7a) (Zarembinski et al., 1998). The structure contains

(a) *M. jannaschii* Mj0577 ATPase

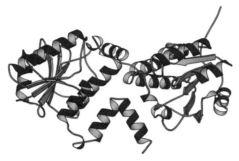

(b) *E. coli* YcaC gene product-putative hydrolase

(c) Archaeal inositol-monophosphatase with
fructose-1,6-bisphosphatase activity

**Figure 19.7.** Prediction of function in earnest: Three structures solved in the absence of functional information. (a) Putative archaeal ATPase molecular switch; (b) putative bacterial hydrolase; (c) bifunctional archaeal inositol monophosphatase/fructose-1,6-bisphosphatase.

a bound ATP molecule, picked up from the *E. coli* host. The presence of bound ATP led to the proposition that Mj0577 is either an ATPase, or an ATP-binding molecular switch. Further experimental work showed that Mj0577 cannot hydrolyze ATP by itself, and can only do so in the presence of *M. jannaschii* crude cell extract. Therefore, it is more likely to act as a molecular switch, in a process analogous to ras-GTP hydrolysis in the presence of GTPase activating protein.

***YcaC—A Bacterial Hydrolase.*** YcaC is a 621bp ORF found in *E. coli*. Its gene product was previously uncharacterized and had no assigned function. The structure of the YcaC gene product (YcaCgp) was determined at 1.8 Å (Fig. 19.7b) and was shown to form an octameric complex (Colovos, Casico, and Yeates, 1998). Structural comparisons showed it to be closely homologous to carbamoylsarcosine aminohydro-lase (CSHase), a bacterial hydrolase. However, the sequence identity between the two proteins was only 20%. The catalytic residues of YcaCgp were predicted by homology to those of CSHase. However, other aspects of its function are unknown and remain to be elucidated.

**IMPase—A Bifunctional Protein.** Fructose-1,6-bisphosphatase (FBPase) was not found in the gene sequence of *M. jannaschii*, in spite of the fact that experiments show FBPase activity in crude cell extracts of this organism. However, inositol monophosphotase (IMPase) is present. IMPase is a distant relative of FBPase. Structural determination and analysis of this enzyme (Fig. 19.7c), showed that one of its loops has the same conformation as the catalytic metal-binding loop of FBPase (Johnson et al., 2001), and it has been shown that IMPase exhibits both IMPase and FBPase activity.

## CONCLUSION

We have seen in this chapter that structure–function relationships are key to under-standing in molecular terms how a protein works. Structural data can complement experimental work, for example, if it is known from biochemical experiments that a particular protein of interest binds ATP, the structure of the protein complexed with an analogue of ATP will reveal exactly where ATP binds. It will also identify the residues on the protein that might stabilize the interaction between ligand and pro-tein, and the potential structural consequences of ligand binding. Structural data can also guide experimental work in eliciting the function of a protein. For example, if one can infer from structural homology that a particular protein is a hydrolase with a nucleotide-binding domain, one can carry out experiments to confirm this and identify possible substrates with some idea of in which direction to proceed.

As structural genomics projects progress, determining protein function from struc-ture with no prior knowledge of the function will become increasingly important, as will the development of new methods to implement this determination. At the moment, one can readily obtain basic functional information using the methods described in this chapter. It is rare to obtain detailed functional information, but this may become more common as the wealth of information in the databases is increased.

Care must be taken in functionally annotating proteins that are distantly related, in order to maintain accuracy in the databases. Ultimately, experiments will be needed to knock out or inhibit the protein function to be sure of its biological role.

## REFERENCES

Acharya K, Ren J, Stuart D, Phillips D, Fenna R (1991): Crystal structure of human alpha-lactalbumin at 1.7 A resolution. *J Mol Biol* 221(2):571–81.

Allaire M, Li Y, MacKenzie R, Cygler M (1998): The 3-D structure of a folate-dependent dehydrogenase/cyclohydrolase bifunctional enzyme at 1.5 A resolution. *Structure* 6(2):173–82.

Ashburner M, Ball CA, Blake JA, Botstein D, Butler H, Cherry JM, Davis AP, Dolinski K, Dwight SS, Eppig JT, Harris MA, Hill DP, Issel-Tarver L, Kasarskis A, Lewis S, Matese JC,

Richardson JE, Ringwald M, Rubin GM, Sherlock G (2000): Gene ontology: tool for the unification of biology. The Gene Ontology Consortium. *Nat Genet* 25(1):25–9. [This paper describes the role of GO in producing a vocabulary of functional information that can be applied to all eukaryotes, and its application to genome annotation and transfer of functional information between genomes.]

Burley S, Almo S, Bonanno J, Capel M, Chance M, Gaasterland T, Lin D, Sali A, Studier F, Swaminathan S (1999): Structural genomics: beyond the human genome project. *Nat Genet* 23(2):151–7. [A good review of structural genomics, describing the rationale behind it, the goals, technical challenges, and potential pitfalls ahead, and the state of ongoing pilot projects.]

Chothia C (1992): Proteins: one thousand families for the molecular biologist. *Nature* 357(6379):543–4. [This paper estimates that only 1000 or so different protein folds will occur in nature. This number was estimated by taking the proportion of genome sequences with a relative in the sequence database, and the proportion of these database sequences belonging (at the time) to one of 120 protein families, assuming that sequence comparisons find 80% of related proteins.]

Chothia C, Lesk A (1986): The relation between the divergence of sequence and structure in proteins. *EMBO J* 5(4):823–6. [The original demonstration that structural similarity is directly related to sequence similarity.]

Christendat D, Yee A, Dharamsi A, Kluger Y, Savchenko A, Cort J, Booth V, Mackereth C, Saridakis V, Ekiel I, Kozlov G, Maxwell K, Wu N, McIntosh L, Gehring K, Kennedy M, Davidson A, Pai E, Gerstein M, Edwards A, Arrowsmith C (2000): Structural proteomics of an archaeon. *Nat Struct Biol* 7(10):903–9. [This work assesses a set of proteins purified from *M. thermoautotrophicum* in terms of their suitability for structural determination by X-ray crystallography and/or NMR spectroscopy, and demonstrates both the feasibility of structural proteomics and its potential in aiding identification of biochemical function of novel proteins.]

Colovos C, Cascio D, Yeates T (1998): The 1.8 A crystal structure of the ycaC gene product from *Escherichia coli* reveals an octameric hydrolase of unknown specificity. *Structure* 6(10):1329–37.

Cooper DL, Isola N, Stevenson K, Baptist EW (1993): Members of the ALDH gene family are lens and corneal crystallins. *Adv Exp Med Bio* 328:169–79. [This paper provides a discussion of the recruitment of proteins from the aldehyde dehydrogenase gene family to structural proteins in the eye, and the evolutionary implications of this phenomenon.]

Fetrow JS, Siew N, Di Gennaro JA, Martinez-Yamout M, Dyson HJ, Skolnick J (2001): Genomic-scale comparison of sequence- and structure-based methods of function prediction: does structure provide additional insight? *Protein Sci* 10:1005–14. [This work uses functional motifs based on distance and angle separations derived from a family of structures to identify disulphide oxidoreductases in *S. cerevisiae*. The work shows that such motifs, combined with structural information, provide a better prediction of function than sequence comparison alone.]

Hegyi H, Gerstein M (1999): The relationship between protein structure and function: a comprehensive survey with application to the yeast genome. *J Mol Biol* 288(1):147–64. [Extensive analysis of the relationship between protein structure and function based on single domain proteins in the SCOP database (see Lo Conte L et al., 2000).]

Hubbard S, Wei L, Ellis L, Hendrickson W (1994): Crystal structure of the tyrosine kinase domain of the human insulin receptor. *Nature* 372(6508):746–54.

Jeffery C (1999): Moonlighting proteins. *Trends Biochem Sci* 24(1):8–11. [This review discusses multifunctional proteins and mechanisms by which cells switch between these different functions.]

Johnson K, Chen L, Yang H, Roberts M, Stec B (2001): Crystal structure and catalytic mechanism of the MJ0109 gene product: a bifunctional enzyme with inositol monophosphatase and fructose 1,6-bisphosphatase activities. *Biochemistry* 40(3):618–30.

Kobayashi N, Go N (1997): A method to search for similar protein local structures at ligand binding sites and its application to adenine recognition. *Eur Biophys J* 26(2):135–44. [This paper describes a molecular recognition method for comparing similar sites in a pair of proteins. Chemical moieties were superimposed and the geometric properties of the elements in the space around them were compared. The method was used to identify modes of adenine recognition by proteins.]

Kinoshita K, Sadanami K, Kidera A, Go N (1999): Structural motif of phosphate-binding site common to various protein superfamilies: all-against-all structural comparison of protein-mononucleotide complexes. *Protein Eng* 12(1):11–4. [This paper describes an attempt to search for a common structural motif in phosphate-binding sites in protein mononucleotide complexes. All such binding sites in the PDB were compared, and four frequently occurring motifs were found, each of which occurs in different superfamilies with different folds.]

Kleywegt G (1999): Recognition of spatial motifs in protein structures. *J Mol Biol* 285(4):1887–97. [This work describes the programs SPASM and RIGOR and their application to function prediction, comparative structural analysis, and design of novel functional sites.]

Laskowski R, Luscombe N, Swindells M, Thornton J (1996): Protein clefts in molecular recognition and function. *Protein Sci* 5(12):2438–52. [This analysis shows how cleft volume relates to molecular interaction and function. The authors found that both ligand binding and protein–protein interactions usually involved the largest cleft in the protein.]

Lo Conte L, Ailey B, Hubbard T, Brenner S, Murzin A, Chothia C (2000): SCOP: a structural classification of proteins database. *Nucleic Acids Res* 28(1):257–9.

Martin A, Orengo C, Hutchinson E, Jones S, Karmirantzou M, Laskowski R, Mitchell J, Taroni C, Thornton J (1998): Protein folds and functions. *Structure* 6(7):875–84. [A statistical analysis of the correlation between protein function and class, architecture and topology of protein structure, based on proteins classified in the CATH domain database (see Orengo et al. 1999).]

Nagano N, Porter CT, Thornton JM (2001): The $(\beta/\alpha)_8$ glycosidases: sequence and structure analyses suggest distant evolutionary relationships. *Protein Eng* 14(11):845–55.

Orengo C, Jones D, Thornton J (1994): Protein superfamilies and domain superfolds. *Nature* 372(6507):631–4. [An estimate of the number of naturally occurring protein folds, based on the number of known sequence families and statistical analysis of the recurrence of known folds.]

Orengo C, Pearl F, Bray J, Todd A, Martin A, Lo Conte L, Thornton J (1999): The CATH database provides insights into protein structure/function relationships. *Nucleic Acids Res* 27(1):275–9.

Rison SCG, Hodgman TC, Thornton JM (2000): Comparison of functional annotation schemes for genomes. *Func Integr Genomics* 1:56–69. A survey of a variety of functional annotation schemes, including GenProtEC, GO, EcoCyc, Kegg and WIT. A combination of schemes is analyzed in terms of the coverage of functional space.]

Russell R (1998): Detection of protein three-dimensional side-chain patterns: new examples of convergent evolution. *J Mol Biol* 279(5):1211–27. [This paper presents an automated method for detecting recurring 3D side-chain patterns in protein structures without any prior knowledge of functional residues.]

Shapiro L, Harris T (2000): Finding function through structural genomics. *Curr Opin Biotech* 11(1):31–5. [A review of methods of elucidating the function of proteins identified through structural genomics projects. Also discussed are the medical and agricultural benefits of focusing on certain targets such as disease genes, and future perspectives.]

Skolnick J, Fetrow J (2000): From genes to protein structure and function: novel applications of computational approaches in the genomic era. *Trends Biotech* 18(1):34–9. [A good review of computational approaches to structure and function prediction, which emphasizes the need for structural descriptors of functional sites in order to make use of sequence and structural information.]

Teichmann S, Murzin A, Chothia C (2001): Determination of protein function, evolution and interactions by structural genomics. *Curr Opin Struct Biol* 11(3):354–63. [A review of the quality of functional assignments that can be made from structure.]

Thornton J, Todd A, Milburn D, Borkakoti N, Orengo C (2000): From structure to function: approaches and limitations. *Nat Struct Biol* 7 Suppl:991–4. [A short review presenting the sort of functional information obtainable from a protein structure. It also describes some examples of prediction of function in earnest, and discusses implications for rational drug design.]

Todd A (2001): University of London. Evolution of function in protein superfamilies (Dissertation), p. 437.

Todd A, Orengo C, Thornton J (1999): Evolution of protein function, from a structural perspective. *Curr Opin Chem Biol* 3:548–556. [A review of the evolution of protein function that focuses mainly on enzymes. The relationship of protein class, fold, and homologous families to function is discussed, as are mechanisms of functional evolution, with respect to substrate specificity and catalytic mechanism.]

Todd A, Orengo C, Thornton J (2001): Evolution of function in protein superfamilies, from a structural perspective. *J Mol Biol* 307(4):1113–43. [This work combines sequence and structural information to identify superfamily relatives, and assesses variation in enzyme function at different levels of sequence identity. Additionally, with reference to 31 diverse enzyme superfamilies, the paper presents a detailed review of how functional variation is implemented in terms of sequence and structural changes.]

Wallace A, Laskowski R, Thornton J (1996): Derivation of 3D coordinate templates for searching structural databases: application to Ser-His-Asp catalytic triads in the serine proteinases and lipases. *Protein Sci* 5(6):1001–13. [Describes the use of a 3D coordinate template of atoms from functional residues to search for proteins of a similar function in the PDB.

Wallace A, Borkakoti N, Thornton J (1997): TESS: a geometric hashing algorithm for deriving 3D coordinate templates for searching structural databases. Application to enzyme active sites. *Protein Sci* 6(11):2308–23. [An extension of previous work (Wallace, Laskowski, and Thornton, 1996), this paper presents a geometric-hashing approach to searching for a predefined cluster of residues, and its application to the Ser-His-Asp catalytic triad. Other functional sites are also explored.]

Nomenclature Committee of the International Union of Biochemistry and Molecular Biology (NC-IUBMB) (1992): *Enzyme Nomenclature*. New York: Academic Press.

Wistow G, Mulders J, de Jong WW (1987): The enzyme lactate dehydrogenase as a structural protein in avian and crocodilian lenses. *Nature* 326:622–4.

Zarembinski T, Hung L, Mueller-Dieckmann H, Kim K, Yokota H, Kim R, Kim S (1998): Structure-based assignment of the biochemical function of a hypothetical protein: a test case of structural genomics. *Proc Natl Acad Sci USA* 95(26):15189–93.

Zhang B, Rychlewski L, Pawlowski K, Fetrow J, Skolnick J, Godzik A (1999): From fold predictions to function predictions: automation of functional site conservation analysis for functional genome predictions. *Protein Sci* 8(5):1104–15. [Describes SITE and SITEMATCH, methods for automating functional site conservation analysis. The method is used to analyze the relationship between fold and function similarity for a large number of fold predictions.]

# Section V

## PROTEIN INTERACTIONS

# 20

# PREDICTION OF PROTEIN–PROTEIN INTERACTIONS FROM EVOLUTIONARY INFORMATION

Alfonso Valencia and Florencio Pazos

The more we know about the molecular biology of the cell, the more we see genes and proteins as part of networks or pathways instead of as isolated entities, and their function as a variable dependent of the cellular context and not only of the individual properties.

Genomic information can be seen as the first catalog of building blocks for the challenging task of understanding the functions of the genes and proteins within their situation in networks and pathways. Protein interactions are a first key step in this direction, even if other complex genetic regulatory mechanisms and issues related with the genetic specificity (cell-type specificity, individual differences, etc.) also will have to be addressed in the future.

The theoretical study of protein interactions has two aspects: the prediction of the residues or regions implicated in the interaction and the prediction of interaction partners (which protein interacts with which one). These two problems typically have been addressed by biophysical and biochemical techniques, such as binding studies (chromatographic isolation of complexes, co-immunoprecipitation, protection, cross-linking studies, etc.) and indirect genetic methods (gene suppression studies, systematic mutagenesis and interspecies exchanges). The development of genomic and postgenomic technologies has changed the panorama considerably, with the possibility of obtaining a massive amount of data about protein interactions faster and more systematically. Progress has been done in the automation of experimental approaches such as yeast-two-hybrid based methods, and mass spectrometry determination of components of

*Structural Bioinformatics*
Edited by Philip E. Bourne and Helge Weissig
Copyright © 2003 by Wiley-Liss, Inc.

macromolecular complexes. At the same time a number of new bioinformatics techniques have been developed based on the considerable amount of information about sequences and genomes that is being accumulated in databases. In this chapter we review the status of the bioinformatics approaches to the study of protein interactions.

## EVOLUTIONARY FEATURES RELATED WITH STRUCTURE AND FUNCTION

Multiple sequence alignments are rich sources of evolutionary information. Looking at the mutational behavior of the positions, a lot of information about protein structure and function can be extracted (Fig. 20.1).

### Conservation

The information most widely extracted from multiple sequence alignments are the conserved positions (Zuckerandl and Pauling, 1963). These invariable positions are interpreted as important residues for the structure or function of the protein since no changes where allowed on them during evolution. Conserved positions usually are located in structural cores (structural importance) and active sites (functional importance). Some authors have studied the relation between binding sites, conserved positions, and type of amino acid in those positions (Ouzounis et al., 1998; Villar and Kauvar, 1994).

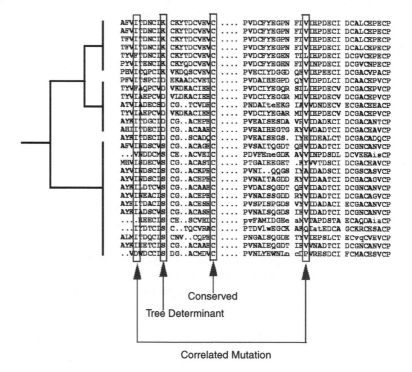

**Figure 20.1.** Sequence features related with structure and function extracted from multiple sequence alignments.

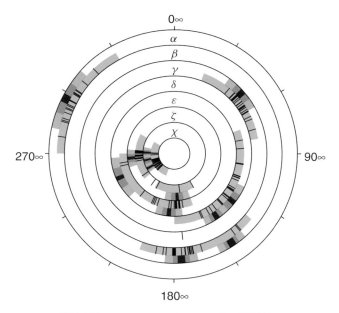

BDL001 over expected ranges for B-DNA.

Figure 3.10. Conformational wheel (Schneider, Neidle, and Berman, 1997) showing the torsion angles for BDL001 (Drew et al, 1981). Black lines show actual values of torsion angles, cyan background their allowed range in the B-type DNA conformation (Schneider, Neidle, and Berman, 1997). The grey shades in the outer rings show the average value(s) of the torsions in dark grey flanked by values of one and two estimated standard deviations in lighter grey.

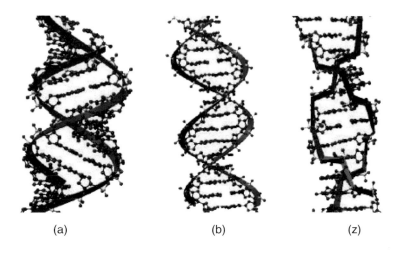

(a)                                    (b)                                    (z)

Figure 3.11. Canonical helical types of A-, B-, and Z-DNA from (Berman, Gelbin, and Westbrook, 1996) with permission from Elsevier Science.

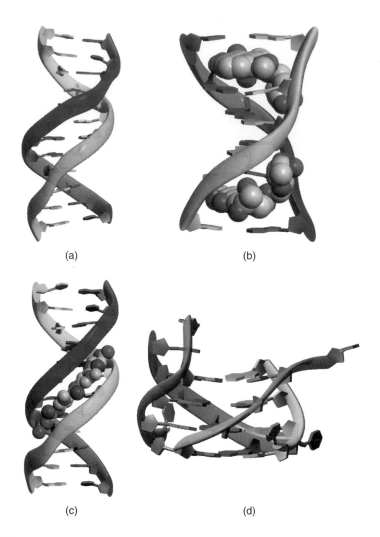

(a)　　　　　　　　　　(b)

(c)　　　　　　　　　　(d)

**Figure 3.12.** Examples of B-DNA. (a) The Dickerson dodecamer (Drew et al., 1981); (b) B-DNA daunomycin (Frederick et al., 1990). The drug is intercalated in between the CG base pairs; (c) Netropsin-DNA complex (Goodsell, Kopka, and Dickerson, 1995). The drug is bound in the minor groove; (d) DNA tetraplex (Phillips et al., 1997). Images are colored by strand.

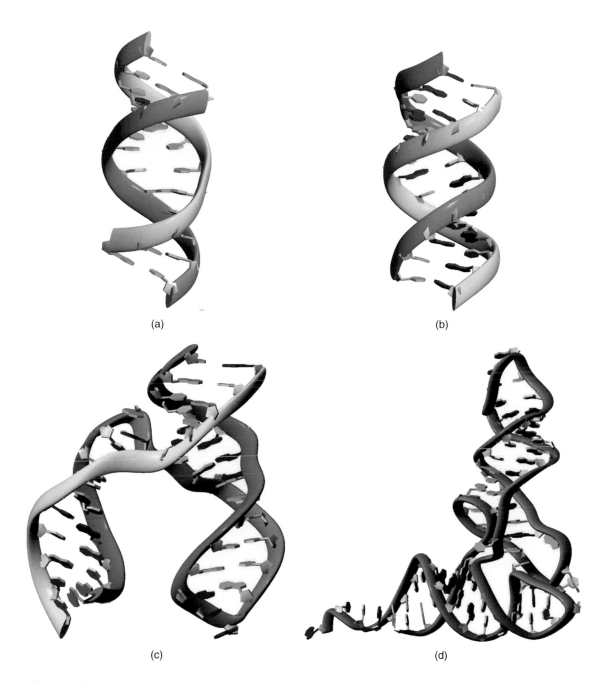

**Figure 3.13.** Examples of RNA. (a) RNA duplex with mismatches r(GGACUUCGGUCC) (Holbrook et al., 1991); (b) A-RNA duplex (Dock-Bregeon et al., 1989); (c) Hammerhead ribozyme (Pley, Flaherty, and McKay, 1994); (d) tRNA (Sussman et al., 1978). Backbone is colored blue (for structures with one strand) and blue and gold (for structures with two strands). Bases are colored green for guanine, yellow for cytosine, red for adenine, and cyan for uracil. Modified bases follow the same color scheme.

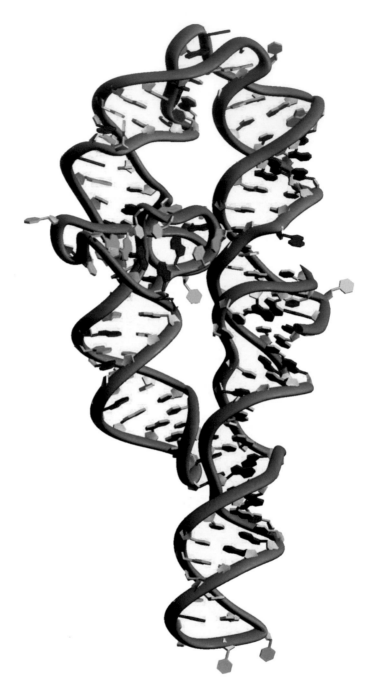

**Figure 3.14.** Group I intron ribozyme (Cate et al., 1996). Backbone is colored blue; bases are colored green for guanine, yellow for cytosine, red for adenine, and cyan for uracil.

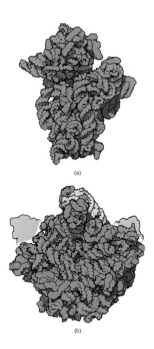

(a)

(b)

Figure 3.15. The (a) 30s and (b) 50s ribosome structure (Ban et al., 2000; Schluenzen et al., 2000; Wimberly et al., 2000). Image created by David Goodsell for the Protein Data Bank's Molecule of the Month series at http://www.pdb.org/.

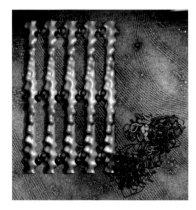

Figure 6.4. Example for a hybrid study that combines elements of electron crystallography and helical reconstruction with homology modeling and molecular docking approaches to elucidate the structure of an actin-fimbrin crosslink (Volkmann et al., 2001b). Fimbrin is a member of a large superfamily of actin-binding proteins and is responsible for cross-linking of actin filaments into ordered, tightly packed networks such as actin bundles in microvilli or stereocilia of the inner ear. The diffraction patterns of ordered paracrystalline actin-fimbrin arrays (background) were used to deduce the spatial relationship between the actin filaments (white surface representation) and the various domains of the crosslinker (the two actin-binding domains of fimbrin are pink and blue, the regulatory domain cyan). Combination of this data with homology modeling and data from docking the crystal structure of fimbrin's N-terminal actin-binding domain into helical reconstructions (Hanein et al., 1998), allowed us to build a complete atomic model of the cross-linking molecule (foreground, color scheme as in surface representation of the array).

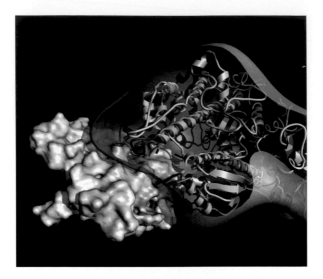

Figure 6.5. Example of a combination of high-resolution structural information from X-ray crystallography and medium-resolution information from electron cryomicroscopy (here 2.1 nm). Actin and myosin were docked into helical reconstructions of actin decorated with smooth-muscle myosin (Volkmann et al., 2000). Interaction of myosin with filamentous actin has been implicated in a variety of biological activities, including muscle contraction, cytokinesis, cell movement, membrane transport, and certain signal transduction pathways. Attempts to crystallize actomyosin failed due to the tendency of actin to polymerize. Docking was performed using a global search with a density correlation measure (Volkmann and Hanein, 1999). The estimated accuracy of the fit is 0.22 nm in the myosin portion and 0.18 nm in the actin portion. One actin molecule is shown on the left as a molecular surface representation. The yellow area denotes the largest hydrophobic patch on the exposed surface of the filament, a region expected to participate in actomyosin interactions. The fitted atomic model of myosin is shown on the right. The transparent envelope represents the density corresponding to myosin in the 3D reconstruction. The solution set concept (see text) was used to evaluate the results and to assign probabilities for residues to take part in the interaction. The tone of red on the myosin model is proportional to this statistically evaluated probability (the more red, the higher the probability).

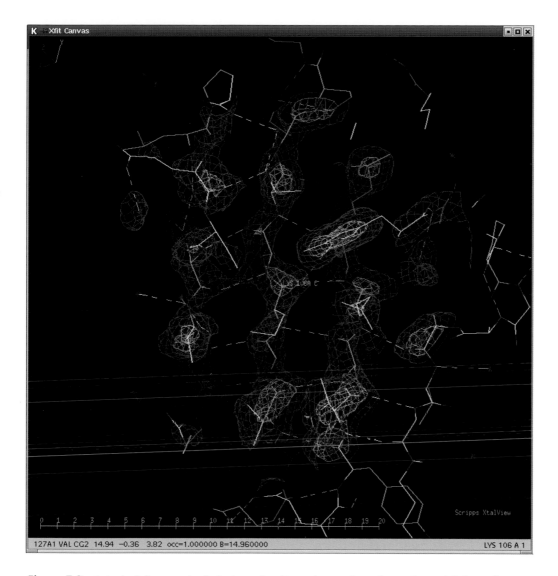

**Figure 7.2.** A typical fragment of electron density and a section of atomic model from the structure of the CuA domain from cytochrome BC3 (Williams et al., 1999) (PDB ID code 2CUA), displayed using *XFit* from the *XtalView* package. Bonds are colored according to the atoms that they join, in this case with carbon atoms colored yellow, oxygen atoms red, and nitrogen atoms blue. Also represented are putative hydrogen bonds, which are drawn as dashed white lines. The electron density map is shown here with two active contour levels, in salmon and purple.

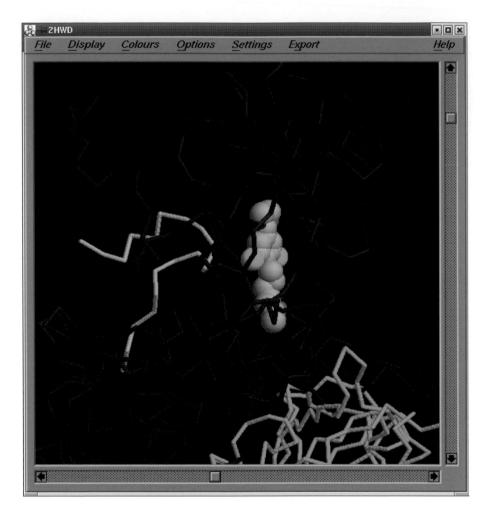

Figure 7.3. A region of human rhinovirus 1A (HRV-1A), including a bound drug molecule (Kim et al., 1993) (PDB ID code 2HWD). The virus proteins are shown as a simple backbone trace, with the drug represented as space-filling spheres and colored according to atom type.

Figure 7.4. The structure of the reduced form of human thioredoxin (Weichsel et al., 1996) (PDB ID code 1ERT), drawn in the Richardson-style schematic secondary structure representation. The protein chain is colored smoothly from blue at the N-terminus to red at the C-terminus, with $\beta$-strands represented by arrows pointing from the N- to the C-terminus, and $\alpha$-helices are drawn as spiral ribbons. Regions without defined secondary structure are shown as a simple, smooth tube. The four $\beta$-strands form a $\beta$-sheet at the center of the structure, which is easily visible in this kind of schematic representation. The image was generated using MolScript and render.

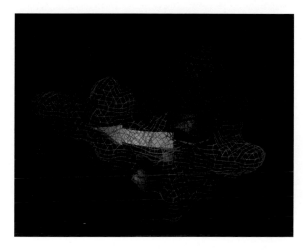

Figure 7.5. A molecular surface drawn as a mesh, overlaid on a secondary structure representation of the toxin LQ2 from *Leiurus Quinquestriatus* (Renisio et al., 1999) (PDB ID code 1LIR). The image was prepared entirely within PyMol.

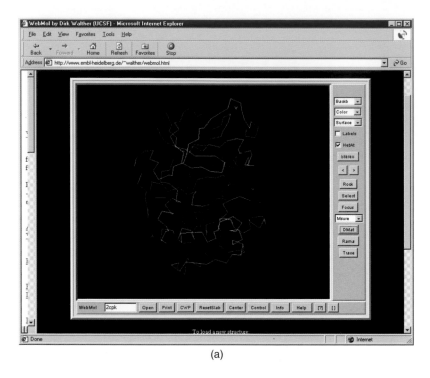

(a)

**Figure 7.6.** (a) The structure of c-AMP-Dependent protein kinase (Knighton et al., 1991) (PDB ID code 2CPK) displayed using *WebMol*.

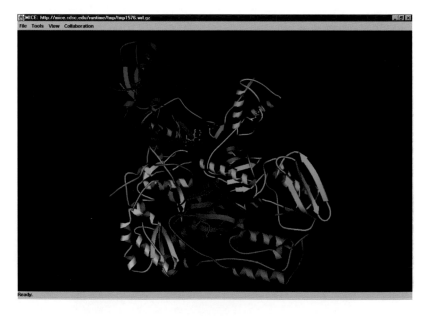

**Figure 7.7.** The structure of reverse transcriptase (RT) from the human immunodeficiency virus (HIV) (Hopkins et al., 1996) (PDB ID code 1RT1), displayed using *MICE*. This particular RT structure includes a drug molecule, just visible at the base of the cleft between the "finger" and "thumb" domains.

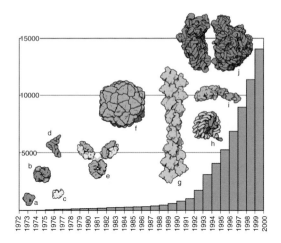

Figure 9.2. Growth chart of the PDB showing the total number of structures available in the PDB archive per year and highlighting example structures from different time periods: a) myoglobin, b) hemoglobin, c) lysozyme, d) transfer RNA, e) antibodies, f) entire viruses, g) actin, h) the nucleosome, i) myosin, and j) 30s ribosomal subunits. Images were created by Dr. David Goodsell, who authors the PDB's Molecule of the Month series. Figure originally appeared in the International Union of Crystallography Newsletter (2001). Images, descriptions and the molecules, and links to related information can be found at http://www.rcsb.org/pdb/molecules/molecule_list.html.

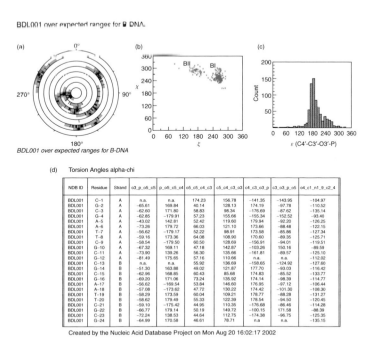

Figure 10.2. Examples of torsion angle reports generated from the NDB: (a) conformation wheel showing the torsion angles for structure BDL001 (Drew et al., 1981) over the average values for all B-DNA; (b) scattergram graph showing the relationship of $\chi$ vs. $\zeta$ for all B-DNA. Two clusters, BI and BII, are labeled; (c) histogram for $\varepsilon$ (C4'-C3'-O3'-P) for all B-DNA; (d) a torsion angle report for BDL001.

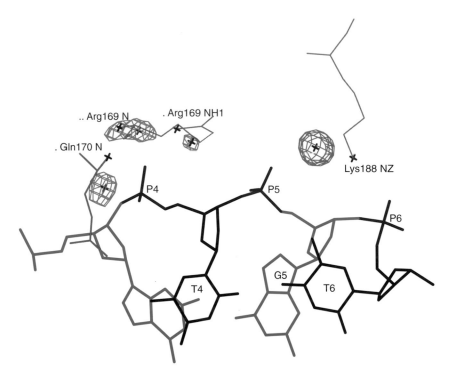

**Figure 10.6.** A view of the three residues in the consensus region for the high resolution CAP-DNA$_{GCE}$ complex (Parkinson et al., 1996b). The predicted phosphate hydration is drawn as pseudoelectron density in cyan, the interacting protein residues are shown in dark brown, and the phosphate groups are red. The protein atoms that contact the DNA shown as blue crosses. The predicted sites are the red crosses. Reprinted from (Woda et al., 1998) with permission from the Biophysical Society.

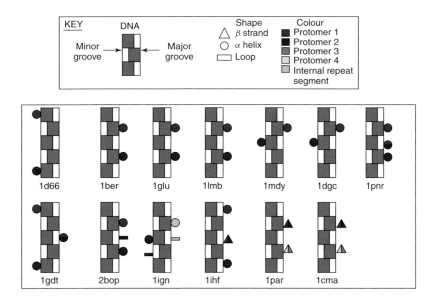

**Figure 10.8.** Simple model diagrams of protein-DNA complexes for double-headed binding proteins. The diagrams give an indication of the predominant secondary structure of the binding motif, protein symmetry and the type and relative position of the DNA groove bound. The secondary structure of the predominant binding motifs are indicated using different symbols analogous to those used in TOPS diagrams (Westhead & Thornton, 1998). Only one symbol of each type is indicated in any one groove, hence both a single sheet and two sheets are indicated by a single colored triangle. The symmetry of each protein is indicated by using a different color for each symmetry (or pseudo symmetry) related element. A single symbol shaded in two colors indicates that there are secondary structures of this type contributed by more than one symmetry-related element. Reprinted with permission from (Jones et al., 1999).

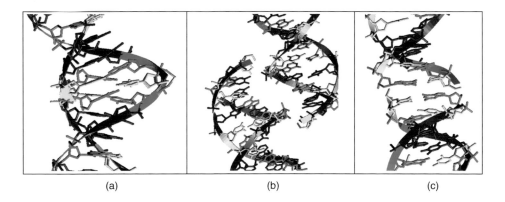

(a)                              (b)                              (c)

**Figure 10.9.** Examples of packing motifs in DNA duplexes in a B- and A-DNA. From left to right: (a) minor groove-minor groove interactions in BDL042 (Leonard & Hunter, 1993); (b) major groove-backbone interactions in BDJ060 (Goodsell et al., 1995); (c) stacking interactions in BDJ025 (Grzeskowiak et al., 1991); The bases are colored green for guanine, yellow for cytosine, red for adenine, and blue for thymine. Reprinted from (Berman et al., 1996) with permission from Elsevier Science.

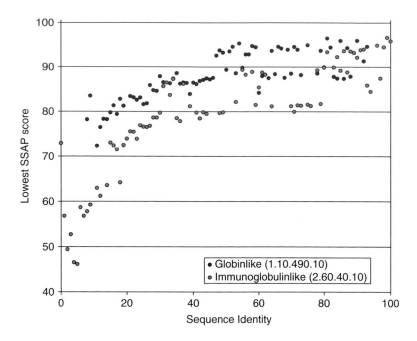

**Figure 13.4.** Structural plasticity plots showing the structural similarity for pairs of relatives in selected CATH superfamilies, as measured by the SSAP algorithm (Taylor and Orengo, 1989), as a function of sequence identity measured after structural alignment. To highlight the maximum deviation for each pairwise sequence identity the minimum SSAP value recorded is shown. Red dots are mainly-$\alpha$ globin superfamily and blue dots are the mainly-$\beta$ Immunoglobulin.

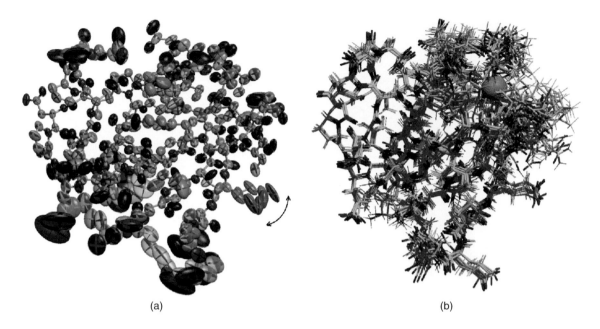

(a)                                                                              (b)

Figure 14.1. The different types of model generated by X-ray crystallography and NMR spectroscopy. Both are representations of the same protein: rubredoxin. (a) In X-ray crystallography
the model of a protein structure is given in terms of atomic coordinates, occupancies, and
B-factors. The side chain of Glu50 has two alternative conformations, shown in the paler colors,
with the change from one conformation to the other identified by the double-headed arrow.
The B-factors on all the atoms are illustrated by "thermal ellipsoids," which give an idea of
each atom's anisotropic displacement about its mean position. The larger the ellipsoid, the more
disordered the atom. Note that the main chain atoms tend to be better defined than the side
chain atoms, some of which exhibit particularly large uncertainty of position. Carbons are shown
in green, oxygens in red, nitrogens blue, sulfur yellow, and the bound zinc ion in deep pink.
The region around the bound zinc ion appears well ordered. This is in stark contrast with the
NMR case in (b). The coordinates and B-factors come from PDB entry 1irn, which was solved at
1.2Å and refined with anisotropic B-factors. (b) The result of an NMR structure determination is
a whole ensemble of model structures, each of which is consistent with the experimental data.
The ensemble shown here corresponds to 10 of the 20 structures deposited for as PDB code
1bfy. In this case the metal ion, shown in pink, is iron. Hydrogens are colored white. The more
disordered regions represent either regions that are more mobile, or regions with a paucity of
experimental data, or a combination of both. The region around the iron-binding site appears
particularly disordered. Both diagrams were generated with the help of the Raster3D program
(Merritt and Bacon,1997).

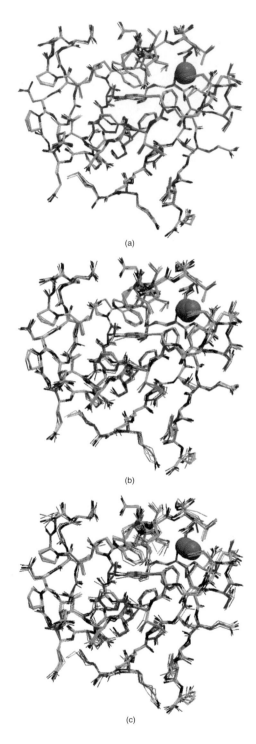

Figure 14.7. Examples of typical uncertainties in atomic positions for (a) an *s.u.* of 0.2Å, (b) 0.3Å, and (c) 0.39Å. The protein is the same rubredoxin from Figure 14.1a. Of course, as shown in Figure 14.1a, the distribution of uncertainties would not normally be so uniform, with higher variability in the surface side-chain atoms than, say, the buried main-chain atoms.

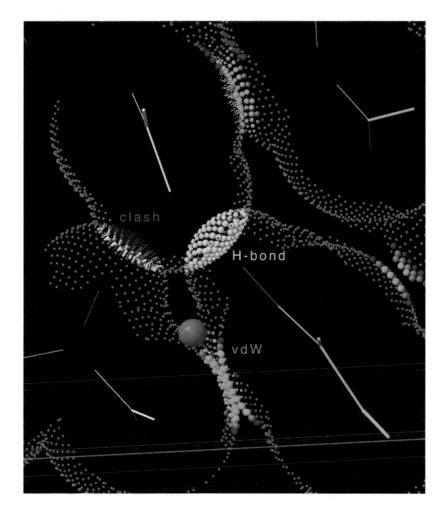

Figure 15.1. Slice through a small section of protein structure (stick figure, backbone in white, and side chains in cyan) showing the relation of all-atom contact surfaces (colored dots) to the atomic van der Waals surfaces (gray dots) and to the 0.25Å-radius probe sphere (gray ball) used in the calculation. The small probe sphere is rolled over the van der Waals surface of each atom, leaving a contact dot only when the probe touches another noncovalently-bonded atom. The dots are colored by the local gap width between the two atoms: blue when nearly maximum 0.5Å separation, shading to bright green near perfect van der Waals contact (0Å) gap. When suitable H-bond donor and acceptor atoms overlap, the dots are shown in pale green, forming lens or pillow shapes. When incompatible atoms interpenetrate, their overlap is emphasized with spikes instead of dots, and with colors ranging from yellow for negligible overlaps to bright reds and hot pinks for serious clash overlaps $\geq$0.4Å. Kinemage-format contact dots also carry color information about their source atom (e.g., Os are red, Ss are yellow, etc.); in Mage, one can toggle between the two color schemes.

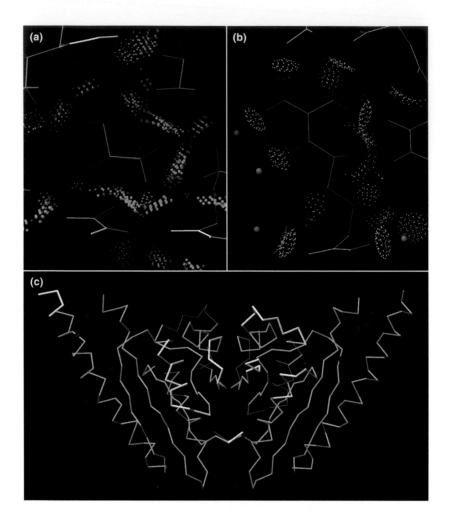

Figure 15.2. All-atom contact examples from the dimer of 1MJH (Zarembinski et al., 1998), a well-determined structural-genomics protein at 1.7Å resolution. (a) All contacts for one of the typically well-packed and well-fit regions of aliphatic side chains, with the green of close van der Waals contacts predominant. (b) All contacts for an Arg side chain, with all 5 planar H-bonds (lens-shaped groups of pale green dots) of its guanadinium NHs formed either to protein O atoms or to waters (pink balls). (c) An overview of the dimer, with only the Cα backbone and the serious clashes ≥0.4Å (red spikes) shown. When interactively displayed in Mage, it is easy to locate and fix the small number of isolated problems, including two flipped-over His rings at the putative active site and a high-B Lys squeezed into insufficient space between two hydrophobic side chains.

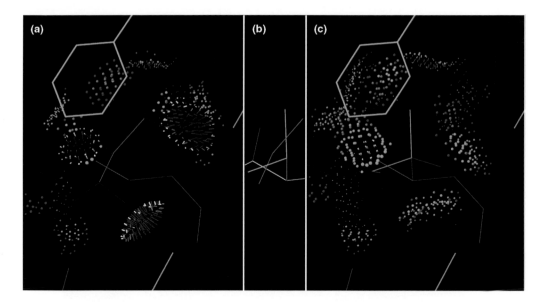

Figure 15.4. Diagnosis and correction of a backward-fit valine side chain. (a) All-atom contacts for the original side chain, with substantial clashes and an eclipsed $\chi 1$ angle. (b) Original and refit side chains, showing how both occupy the same space but in opposite orientations. Bond-angle distortions in the original put its $C\beta$ 0.48Å from the idealized position. (c) Good all-atom contacts for the refit Val, which has ideal geometry and staggered $\chi 1$ without backbone movement. Even without deposited structure-factor data, one can be fairly confident that the electron density must have been ambiguous and that the conformation shown in (c), not (a), is in the correct local energy well. From the 2SIM neuraminidase at 1.6Å resolution Crennell, 1996 #672.

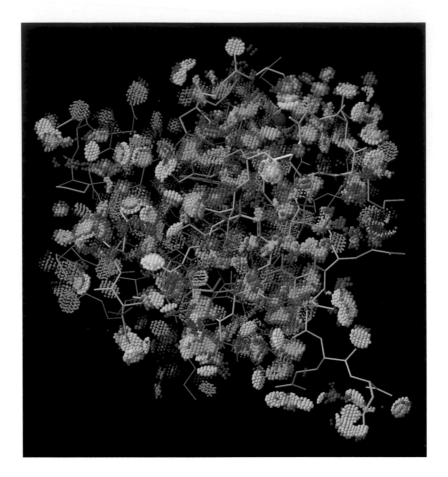

**Figure 15.5.** All-atom contacts for the entire structure of 1BRF rubredoxin (Bau et al., 1998), a highly accurate small protein structure at 0.95Å resolution. The dense green dot patches signifying well-packed contacts in the molecule and a well-fit model are seen consistently throughout the structure, except for a single red clash between two surface side chains. 1BRF thus illustrates both how precisely the all-atom contact criteria are satisfied in atomic-resolution protein structures and also how occasional local errors can be found even in such extremely high-quality structures.

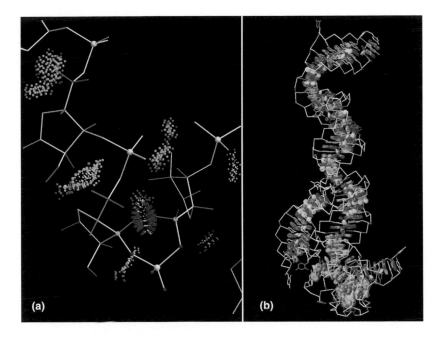

Figure 15.6. Base and backbone all-atom contacts in the 5S RNA from the 2.4Å ribosome structure of 1FFK (Ban et al., 2000). (a) A section of the backbone–backbone contacts, mostly very nicely packed but with one impossible overlap of C3′ and C5′ hydrogens (red spikes). (b) Base–base contacts, showing the long columns of well-fit base stacking.

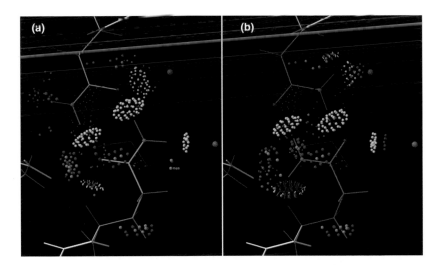

Figure 15.7. Resolving the ambiguity in a pair of doubly H-bonded side-chain amides, which have equivalent H-bonds both to each other and to waters in the two possible flip states. (a) The correct flip orientation, with only a minor overlap. (b) The next-best, but incorrect, flip state with a large, physically impossible clash of the Gln N$\varepsilon$ H with H$\alpha$ (red spikes). From the 1.6Å peroxidase of 1ARU (Fukuyama et al., 1995).

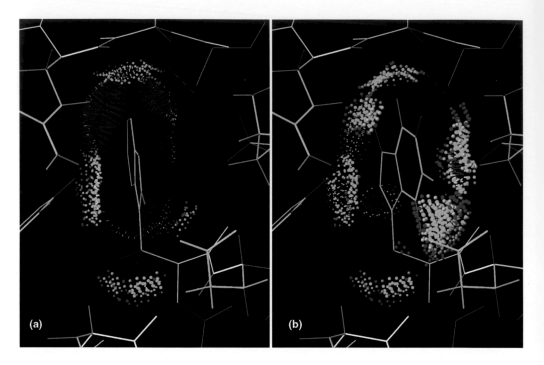

**Figure 15.8.** A test of alternative sequence possibilities substituting Trp for Tyr at a buried position in the N-terminal domain of λ repressor, using the "remote update" function in the interactive Mage/Probe system. (a) One of the initial rotamer trials, with impossibly bad clashes on both sides of the Trp ring. (b) The best of the exact rotamers, with only two minor overlaps in orange, indicating that the Trp side chain can indeed fit without perturbing the structure significantly. Starting coordinates from 1LMB (Beamer and Pabo, 1992).

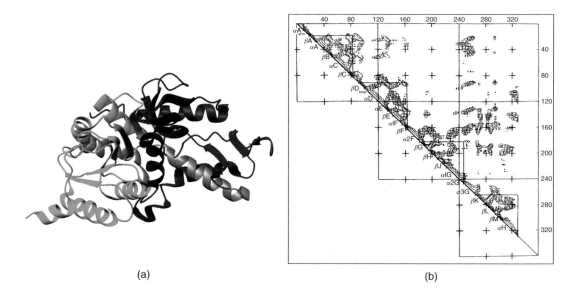

(a)  (b)

**Figure 18.1.** Domain structure of dogfish lactate dehydrogenase, determined using the $C\alpha$-$C\alpha$ distance map. (a) Ribbon diagram of lactate dehydrogenase, showing the NAD binding (green) and catalytic domains (red). In gray is part of the helix, spanning residues 164–180, linking the two domains. (b) Distance map and structural domains in lactate dehydrogenase. Contours represent $C\alpha$-$C\alpha$ distances of 4Å, 8Å, and 16Å within the subunit of dogfish lactate dehydrogenase. Elements of secondary structure are identified along the diagonal. Triangles enclose regions where short $C\alpha$-$C\alpha$ distances are abundant. The NAD binding domain comprises the first two triangles (counting from the N-terminus), which are subdomains. The catalytic domain comprises the last two triangles (the C-terminal domain). From Rossman and Liljas (1974) and reproduced by permission of Academic Press (London) Ltd.

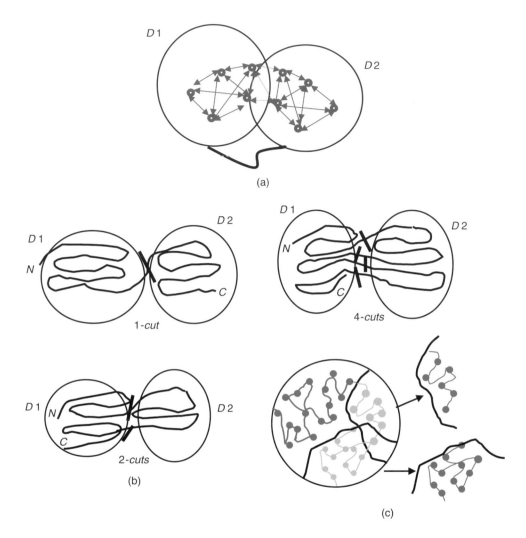

Figure 18.2. Illustration of the problem of parsing the protein 3D structure into structural domains. (a) The most common definition of structural domains, as groups of residues with a maximum number of contacts within each group and a minimum number of contacts between the groups. (b) Domains may be composed of one or more chain segments. Any domain-assignment procedure must therefore be able cut the polypeptide chains as many times as necessary. When both domains are composed of contiguous chain segments (continuous domains), only one chain cut (1-cut) is required. When one domain is continuous and the other discontinuous, a situation that may arise as a result of gene insertion, then the chain has to be cut in two places (2-cuts). When both domains are discontinuous, additional chain cuts may be required. In the example shown, the chain is cut in four places (4-cuts), and thus the domain on the left-hand side contains three chain segments, whereas that on the right contains two chain segments. (c) This drawing shows two solutions to the problem of partitioning the protein 3D structure into substructures. To partition the 3D structure into domains, many such solutions need to be examined in order to single out the one that satisfies the criterion given in (a).

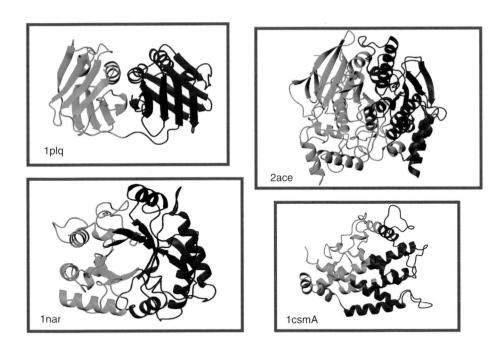

**Figure 18.8.** Examples of single-domain, single-architecture proteins in CATH (Jones et al., 1998), which STRUDL (Wernisch, Hunting, and Wodak, 1999) splits into two domains. Shown are the domain assignments produced by STRUDL. For the exact domain limits, the reader is referred to the STRUDL WEB site. The displayed protein ribbons belong to Torpedo californica acetylcholinesterase monomer (1ace), the plant seed protein narbonin (1nar), the eukaryotic DNA polymerase processivity factor PCNA (1plq), and Chorismate mutase chain A monomer (1csmA).

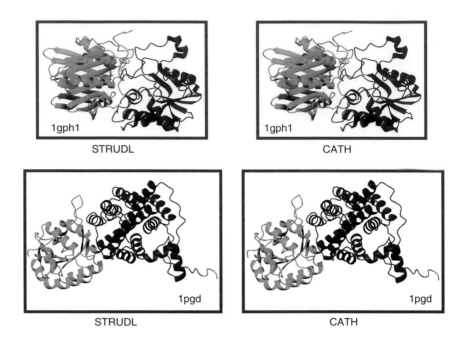

**Figure 18.9.** Different assignments by STRUDL (Wernisch, Hunting, and Wodak, 1999) and CATH (Jones et al., 1998), illustrating the effect of noise or decorations in the protein chain trace. The STRUDL assignments are displayed on the left-hand side, and the CATH assignments are displayed on the right. The short chain segments, which CATH assigns to separate domains are shown in blue. Some of the discrepancies may be due to simple 'slips' in the CATH assignments that have been or will be corrected.

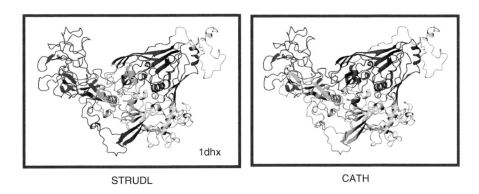

STRUDL                                          CATH

**Figure 18.10.** Different assignments by STRUDL (Wernisch, Hunting, and Wodak, 1999) (left) and CATH (Jones et al., 1998) (right), for the adenovirus hexon protein (1dhx), a protein with many domains of complex architecture.

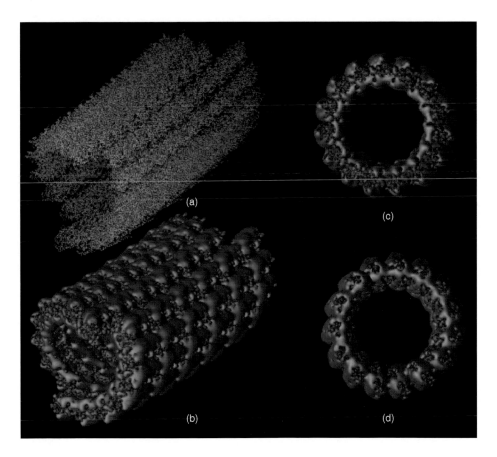

**Figure 21.4.** Electrostatic properties of a 1.2-million atom, 400 × 300 × 300 Å microtubule fragment illustrating the current state of the art for continuum electrostatics calculations. The potential was calculated using APBS to solve the PBE at 150 mM ionic strength. (a) The backbone atoms of the microtubule. (b) Electrostatic potential isocontours for microtubule shown at +1.0 (blue) and −1.0 (red) kT/e. (c) Potential isocontours (as in B) for so-called "−" end of microtubule. (d) Potential isocontours (as in B) for so-called "+" end of microtubule.

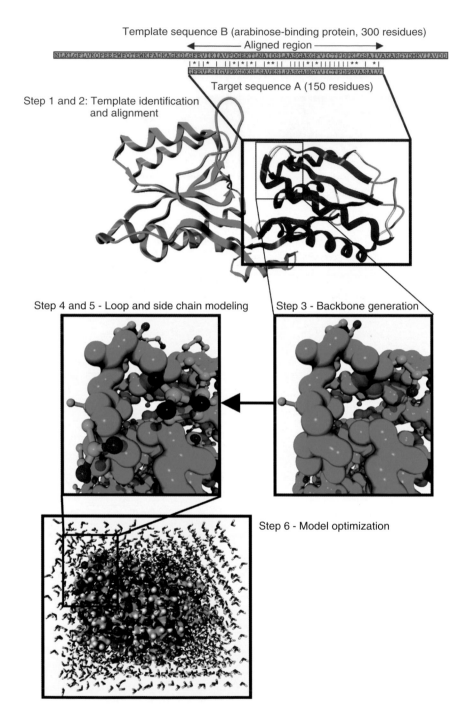

**Figure 25.2.** The steps to homology modeling. The fragment of the template (arabinose-binding protein) corresponding to the region aligned with the target sequence forms the basis of the model (including conserved side chains). Loops and missing side chains are predicted, then the model is optimized (in this case together with surrounding water molecules). Images created with Yasara (www.yasara.com).

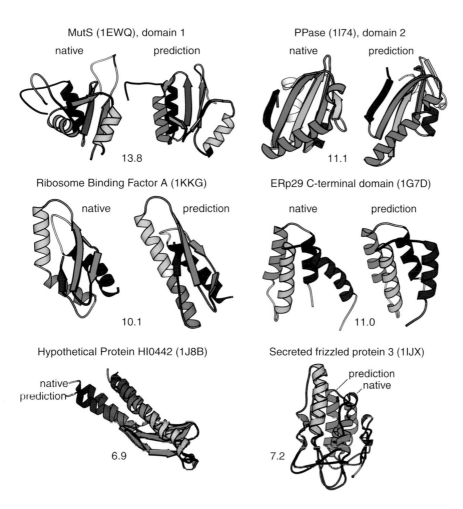

MutS (1EWQ), domain 1

native    prediction

13.8

PPase (1I74), domain 2

native    prediction

11.1

Ribosome Binding Factor A (1KKG)

native    prediction

10.1

ERp29 C-terminal domain (1G7D)

native    prediction

11.0

Hypothetical Protein HI0442 (1J8B)

native
prediction

6.9

Secreted frizzled protein 3 (1IJX)

prediction
native

7.2

Figure 27.1. Examples of ROSETTA structure predictions from CASP4 (see Chapter 24). Native/prediction pairs are shown left-to-right, except for 1J8B and 1IJX, which are displayed as a superposition of native and predicted structures. Values indicate Calpha root-mean-square (rms) deviations between native and predicted structures, in angstroms. Colors represent position along the chain from blue (N terminus) to red (C terminus).

native          prediction          homolog (1NKL)

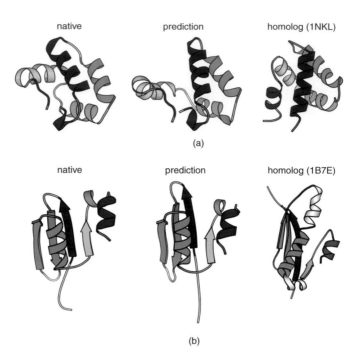

(a)

native          prediction          homolog (1B7E)

(b)

**Figure 27.2.** Potential of *ab initio* predcitions to detect distant protein homologies. (a) The native structure of bacterial-lysis protein Bacteriocin AS-48 (left, PDB id 1E68) is compared to the best ROSETTA prediction for the structure (center), and the native structure of NK-Lysin (right, PDB id 1NKL), a functionally similar protein. (b) The native structure of domain 2 of the DNA mismatch repair protein MutS (left, PDB id 1EWQ), is compared to the best ROSETTA prediction for the domain (center), and a domain from the native structure of the Tn5 transposase inhibitor (right, PDB id 1B7E). In both (a) and (b) the *ab initio* models of the proteins were of sufficient quality to detect these functional homologs by the similarity of the folds in the absence of significant sequence similarity.

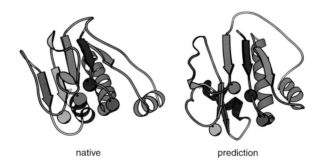

native                    prediction

**Figure 27.3.** An example of active-site conservation in *ab initio* models. The ROSETTA predicted structure of domain 1 from an inorganic pyrophosphatase from Streptococcus mutans is compared to the corresponding domain in the native structure (PDB id 1I74). Strongly conserved active site residues are rendered as spheres along the backbone. Note the similar relative orientation of these residues in the native and predicted structures, implying that *ab initio* models may be sufficient to detect functional homologies using methods that search for functionally significant residue arrangements.

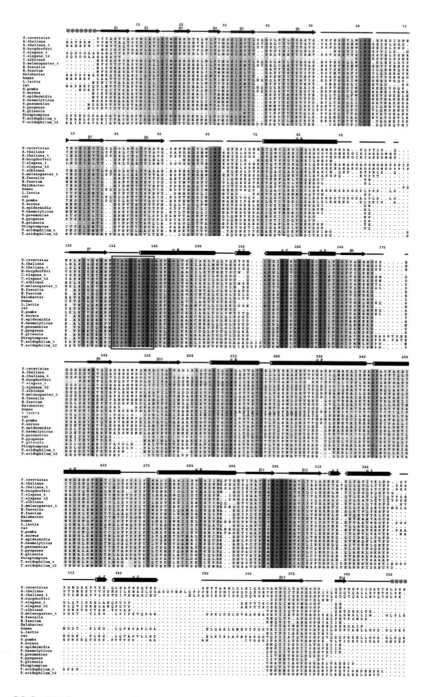

Figure 29.2. MDD sequence alignments. Proteins similar to *S. cerevisiae* MDD (E-value $<10^{-4}$) were identified using PSI-BLAST (Altschul et al., 1997), and aligned with CLUSTAL (Higgins, Bleasby, and Fuchs, 1992). Secondary structural elements of *S. cerevisiae* MDD are shown with cylinders (α-helices) and arrows (β-strands). Grey dots denote poorly resolved residues in the final electron density map. Color coding denotes sequence conservation among MDDs (white → green ramp, 30 → 100% identity). Red box denotes the ATP-binding P-loop. Adapted from (Bonanno et al., 2001).

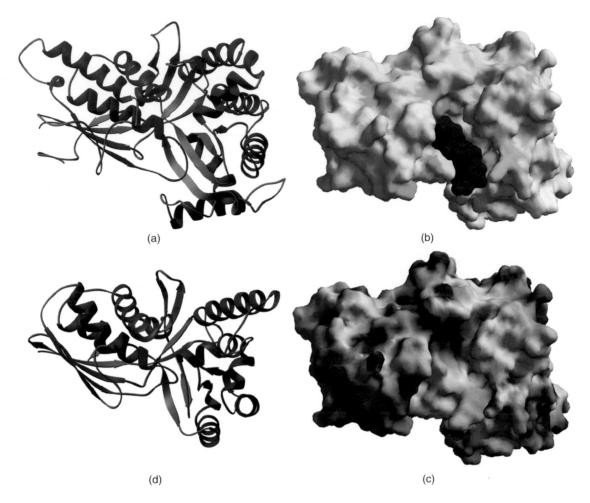

(a)

(b)

(d)

(c)

Figure 29.4. *S. cerevisiae* MDD and *M. jannaschii* HSK. Ribbon drawings of MDD (a) and HSK (d) in the same orientation. P-loops are colored red. Surface representations of MDD color coded for (b) sequence conservation using the white → green color ramp from Figure 29.2 and (c) calculated surface electrostatic potential (red ← 10 to blue > +10kBT, where kB is the Boltzmann constant and T is the temperature (Gilson, Sharp, and Honig, 1988).

## Family-Dependent Conservation (Tree Determinants)

There is a more subtle kind of conservation, the family-dependent conservation. These type of residues are called *tree determinants*. These positions are conserved in the subfamilies that form well-defined branches of the phylogenetic tree and are different in the chemical type of the amino acids that characterize each subfamilies, hence, they would contain key information for determining the structure of the phylogenetic tree of the family, and would be most likely be related with the specific features of each subfamily, that is, differential binding to other proteins and substrates.

One of the first approaches to the prediction of these kinds of functional residues, implemented in the *sequencespace* program was developed by Casari, Sander, and Valencia (1995) and was later followed by other similar methods (Lichtarge, Bourne, and Cohen, 1996; Andrade et al., 1997). The underlying principle in all these methods is the detection of positions in multiple sequence alignments characteristic of the different groups of sequences that form part of a larger protein family. In those works, the relation between the tree-determinant residues and functional residues is discussed for some protein families. Only recently, with the availability of more protein sequences and structures, has it been possible to test systemically the implication of tree-determinant residues in the formation of functional sites in a large enough collection of known protein structures (del Sol, Pazos, and Valencia, 2002).

## Coevolution (Correlated Mutations)

Another sequence-based approach for the prediction of protein structure and molecular complexes is based on the detection of correlated mutations in multiple sequence alignments and their use as distance constraints between residues belonging to the same or different proteins. Correlated mutations correspond to pairs of positions with a clear pattern of covariation. The underlying evolutionary model for explaining their relation with space neighboring is related with the covarion model, and assumes that part of the detected correlated pairs correspond to compensatory mutations, where in particular sequences of the multiple sequence alignments the mutation of one residue was compensated along the evolution by a mutation of a neighbor residue (Fig. 20.2), most likely to keep proteins (or protein complexes) in permissible limits of protein stability.

The method proposed in 1994 (Göbel et al., 1994) was a weak predictor of proximity between residues in protein structures. Later this accuracy of correlated mutations in predicting residue contacts was improved, combining them with other sequence-based features, such as conservation or hydrophobicity (Olmea and Valencia, 1997; Pazos, Olmea, and Valencia, 1997b). In spite of the low accuracy, these predicted contacts have been demonstrated to be very useful, for example, in filtering structural models (Olmea, Rost, and Valencia, 1999) or driving *ab initio* simulations (Ortiz et al., 1999).

## PREDICTION OF INTERACTING REGIONS

### Structure-Based Methods (Physical Docking)

The problem of determining the physical structure of protein complexes when the structure of the members is known is part of the problem of docking molecules, such as proteins with small molecules (Chapter 22). Despite the considerable efforts that have gone into solving this problem, directed at the design of new drugs, the solutions in the

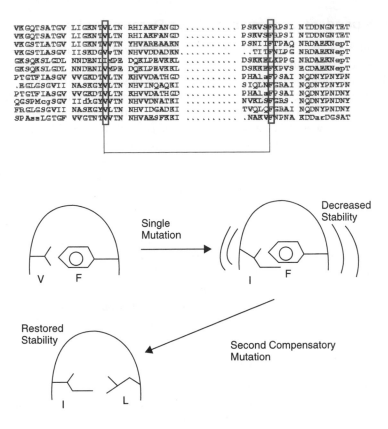

```
VKGQTSATGV LIGKNIVLTN RHIAKFANGD .......... PSKVSFRPSI NTDDNGNTET
VKGQTSATGV LIGKNIVLTN RHIAKFANGD .......... PSKVSFRPSI NTDDNGNTET
VKGSTIATGV LIGKNIVVTN YHVAREAARN .......... PSNIIFTPAQ NRDAEKNapT
VKGSTIASGV IISKDGVVTN NHVVDDADKN........... ..TITWNLPG NRDAEKNapT
GKSQKSLGDL NNDENIIMPE DQKLPEVKKL .......... DSKKKLKPPG NRDAEKNapT
GKSQKSLGDL NNDENIVMPE DQKLPEVKKL .......... DSKKFFKPVS ECDAEKNapT
PTGTFIASGV VVGKDIVLTN KHVVDATHGD .......... PHAlaFPSAI NQDNYPNYPN
.EGLGSGVII NASKGYVLTN NHVINQAQKI .......... SIQLNFGRAI NQDNYPNYPN
PTGTFIASGV VVGKDIVLTN KHVVDATHGD .......... PHAlaFPSAI NQDNYPNDNY
QGSPMcgSGV IIdkGYVVTN NHVVDNATKI .......... NVKLSFGRS. NQDNYPNDNY
FRGLGSGVII NASKGYVLTN NHVIDGADKI .......... TVQLQFGRAI NQDNYPNDNY
SPAssLGTGF VVGTNIVVTN NHVAESFKKI .......... .NAKVFNPNA KDDarDGSAT
```

**Figure 20.2.** Hypothesis for explaining the relation found between correlated mutations and space neighboring.

case of protein–protein interaction are still far from optimal. The basic problem is that the interaction surfaces have few differential characteristics that can be captured with statistical methods, and are almost statistically indistinguishable from other surfaces (see Lo Conte, Chothia, and Janin, 1999; Jones and Thornton, 1997).

Despite these difficulties various groups have achieved considerable progress in the development of physical docking programs. Most of the current approaches consider proteins as rigid bodies and have the physical matching of the surfaces as their main guide. Only a few packages integrate conformational flexibility, allowing the interacting surfaces to adapt one to the other, normally at the expenses of reducing the search space for possible interacting surfaces. Some programs take into account other features for predicting regions of interaction, such as hydrophobicity or electrostatics (Chapter 21).

For reviews about docking see Lengaguer and Rarey (1996), Halperin, Ma, Wolfson, Nussinov (2002), Smith, Sternberg (2002), where different programs are compared, and Chapters 21 and 22. An interesting effort to compare the various protein docking approaches in a blind test is organized by J. Janin and colleagues (CAPRI: http://capri.ebi.ac.uk/). The results of this experiment, if enough structures of protein complexes become available for the comparison, will be important for updating our view of the capability of current docking methods, a pressing question now that the structural genomics efforts are on the way to solve a substantially larger number of isolated proteins and protein domains (Chapter 29).

## Sequence-Based Methods

***Tree-Determinant Residues.*** The relation between tree-determinant residues (see Family-Dependent Conservation above) and interacting surfaces has been analyzed in detail in a few well-characterized systems (Casari, Sander, and Valencia, 1995; Lichtarge, Bourne, and Cohen, 1996; Pazos et al., 1997c; Atrian et al., 1997; Caffrey, O'Neill, and Shields, 2000; Pereira-Leal and Seabra, 2001). Of more importance, it has been demonstrated by direct experiments how exchanging tree-determinant residues between protein subfamilies switch the specificity of interaction of the corresponding proteins. This has been the case of two separate studies on different proteins of the *ras* family of small GTPases (Stenmark et al., 1994; Azuma et al., 1999). In Figure 20.3, the tree-determinant residues extracted from the multiple sequence alignments of *Ran* and *Rcc1* are marked in the structure of the complex formed by these two proteins. It can be seem that many of these residues, predicted from sequence information alone and before the structure of the complex was solved, map in the interaction surface, especially in *Rcc1*.

***Correlated Mutations.*** Interprotein correlated mutations can point to the residues and regions implicated in the interaction between the two proteins. In spite of the weak power of correlated mutations in predicting residue neighboring, it was demonstrated (Pazos et al., 1997a) that it is enough to predict the tendency of pairs of residues to be part of the interacting surface in interacting proteins.

The advantage of correlated mutations is their independence of the structural information, which opens the possibility of predicting interprotein neighbor residues when only the protein sequences are known. A related advantage is their independence of structural features such as conformational changes between the free and bound forms, a problem for physical docking programs. Figure 20.4 shows how correlated mutations were used to predict the interaction between the two structural domains of the chaperon *DnaK* in the absence of structural information. The structure of the two domains of the protein *DnaK* has now been solved independently. In this case correlated mutations were used to predict pairs of residues of both domains with a tendency to be closer in the structure of the complex (continuous lines). Most of the predicted pairs correspond to possible contacts between the upper subdomain of the Nt domain and two loops of the Ct domain. These two loops, far apart in sequence, are close in the three-dimensional (3D) structure of the Ct domain (Fig. 20.4, right). Even if this way of interaction is still a model which has not been confirmed yet, a number of experimental details fit well with it, including the close interaction between the ATP-binding sites on the Nt domain and the peptide-binding cleft on the Ct domain.

## Hybrid Methods

Some methods use both structure and sequence information to predict interaction surfaces. Correlated mutations can be used to predict interaction regions in combination with structural information, although they can work without that information as discussed previously (Correlated Mutations) and that is their real advantage. In the case of protein complexes where the structure of the two members is known, the information about correlated mutations can be interpreted as distance restraints used for filtering a set of complexes looking for the real one. This set of complexes can be obtained from a docking program, or generated by moving and rotating the two molecules to get all

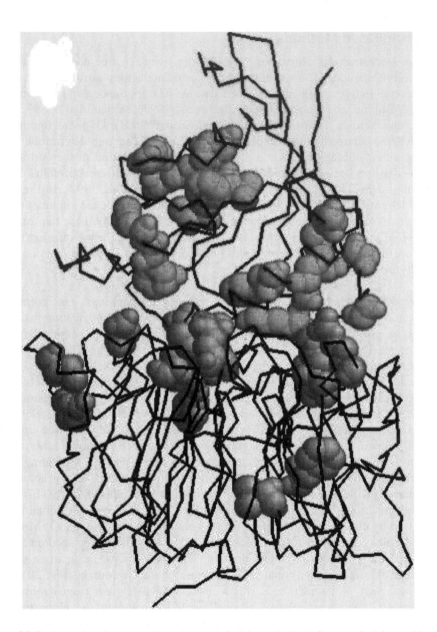

**Figure 20.3.** Complex between *Ran* (upper chain) and *Rcc1* (lower chain) marking the tree-determinant residues (in spacefill) for the two proteins. Protein Data Bank code of the complex: 1i2m. Tree-determinant residues were calculated with the *sequencespace* algorithm (Casari, Sander, and Valencia, 1995). Figure courtesy of J. A. García-Ranea.

possible arrangements between them. In Figure 20.5 a number of artificial complexes between the two chains of hemoglobin are constructed and plotted according with their structural similarity with the real complex (rms: root mean square), and the closeness of interprotein correlated mutations ($Xd$) in that conformation. It can be seen that most of the artificial complexes have $Xd$ values lower than the real one, that is, the correlated pairs tend to be closer in the real complex than in any other one.

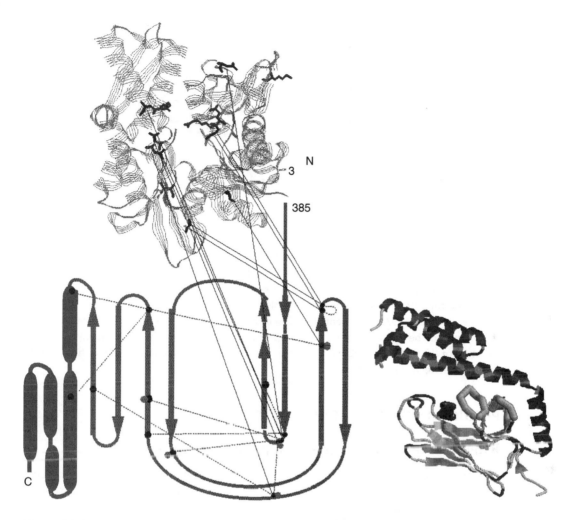

**Figure 20.4.** Prediction of a the interaction between two domains of *DnaK* with information about correlated mutations without using structural information. Left panel: 3D structure of the Nt domain, supersecondary structure of the Ct domain, and interdomain correlated mutations predicting the interaction regions. Right panel: 3D structure of the Ct domain, solved later, marking the two loops predicted to interact with the Nt domain.

Recent publications (Zhou and Shan, 2001; Fariselli et al., 2002) have shown that it is possible to train neural networks with structural and sequence information to predict protein–protein binding surfaces. These two methods use the structure of a protein to define surface patches of neighbor residues and the multiple sequence alignment to obtain the sequence profile for the members of the patch. A neural network is trained with that information coming from proteins with known interaction surfaces. After the training process, surface patches (plus sequence profiles) of proteins with known structure but unknown interaction surface are presented to the network, and it predicts whether or not that patch is implicated in the interaction. The accuracy of these methods is higher than 70% (two-state, interaction surface or not).

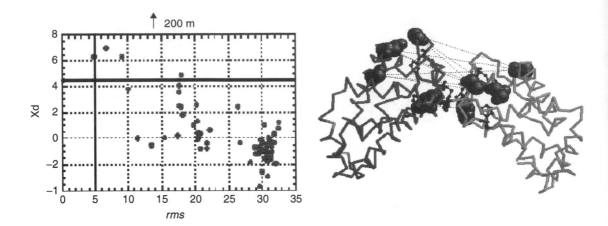

**Figure 20.5.** Selection among docking models using interprotein correlated mutations as distance constraints. Each dot represents a different structural complex between the two chains of hemoglobin ($\alpha$ and $\beta$) produced by a docking program. The rms value (X axis) represents the structural difference between that complex and the real structure (right). The *Xd* value (Y axis; Pazos et al., 1997a) represents the closeness between the interprotein correlated pairs in that complex. The *Xd* value of the real complex (rms = 0) is marked with a horizontal line. The structure of the real hemoglobin $\alpha/\beta$ complex is represented on the right showing the correlated residues between the two chains.

## PREDICTION OF INTERACTION PARTNERS

### Experimental Approaches and High-Throughput Applications

The experimental approaches for the determination of interaction partners has undergone dramatic improvement since the late 1980s, particularly through the systematic application of different strategies based on the yeast-two-hybrid protocol (Fields and Song, 1989) scanning for all the interacting pairs of proteins in the complete proteome of an organism, like in yeast (Ito et al., 2000; Uetz et al., 2000) or *H. pylori* (Rain et al., 2001). Other proteomic techniques, such as mass spectrometry (Gavin et al., 2002) or surface plasmon resonance, are also being used for obtaining the full "interactome" (see Supplement to *Trends in Biotechnology*, Vol. 19, 2001 for a review). Even if the accuracy of the different experimental approaches is still controversial, and there is a surprising lack of overlap of some of the interacting maps produced (Legrain, Wojcik, and Gauthier, 2001), it is also obvious that in a very short time we will have available large collections of pair-wise interactions with attached confidence values.

### Databases and Collections of Interacting Proteins

To reduce the problem posed by the lack of standard large collections of interacting proteins, a number of initiatives are underway for the construction of interaction databases.

- SPIN-PP collects complexes of proteins of known structure, and the analysis of the physicochemical characteristics of their interfaces (http://trantor.bioc. columbia.edu./cgi-bin/SPIN/).

- MIPS contains a large collection of loosely annotated interactions between yeast proteins (http://www.mips.biochem.mpg.de/proj/yeast/tables/interaction/).
- ProNet stores interactions between human proteins (http://pronet.doubletwist.com).
- DIP contains different types of interacting proteins annotated and linked to original references (http://www.ampere.doe-mbi.ucla.edu:8801/dip.html).
- BIND contains interactions involving not only proteins but nucleic acids and small molecules as well. A lot of information about the interaction is included, such as conformational changes, pathway where the interaction is included, and so forth (http://www.bind.ca).

Complementary to these efforts some groups are developing data-mining approaches directed at the automatic extraction of known interactions from literature sources (Blaschke et al., 1999; Frieman et al., 2001; Thomas et al., 2000).

For example, the *Suiseki* system (Blaschke et al., 1999) is designed to first automatically extract protein and gene names and use them to search for typical grammatical constructions indicating interactions (e.g., "proteinA *binds* proteinB"). Even if problems such as the detection of protein names, the implicit information used in the construction of sentences, and the directionality of the interactions are still far from solved, the current implementation of *Suiseki* is able to extract highly scoring interactions (well represented in the text corpus and phrased in grammatical construction well characterized by the system) with less than 20% errors. An error rate that makes the results of practical utility for researchers in the field, but not yet for the completely automatic construction of interaction databases.

## Computational Methods Based on Genomic Information

In parallel to the exciting experimental approaches for the detection of interacting pairs of proteins, a number of bioinformatic techniques are also being developed. This first generation of methods has focused on the more general problem of predicting sets of functionally related proteins, instead of predicting physically interacting proteins; for example, predicting proteins that form part of a signaling pathway but not the order of the interactions in the pathway. The three main genomic approaches in the direction of predicting functional interactions are shown in the upper panel of Figure 20.6.

***Phylogenetic Profiles.*** This method is based on the detection of genes that have a similar specie distribution, that is, they are present/absent in the same species (Fig. 20.6a; Pellegrini et al., 1999; Gaasterland and Ragan, 1998). The hypothesis behind this approach is that the corresponding proteins are functionally related since their distribution seems to indicate that one protein cannot work without the other. Even if there are a number of complex cases that would not fit with this schema, for example, multidomain proteins, the idea is powerful enough for detecting an interesting number of potential functional relations.

***Conservation of Gene Neighboring.*** These methods are based on the conservation of the proximity of genes along the genome between distantly related species to predict interaction (Fig. 20.6b; Dandekar et al., 1998; Overbeek et al., 1999). A typical example would be proteins coded by genes of a conserved operon that, if present

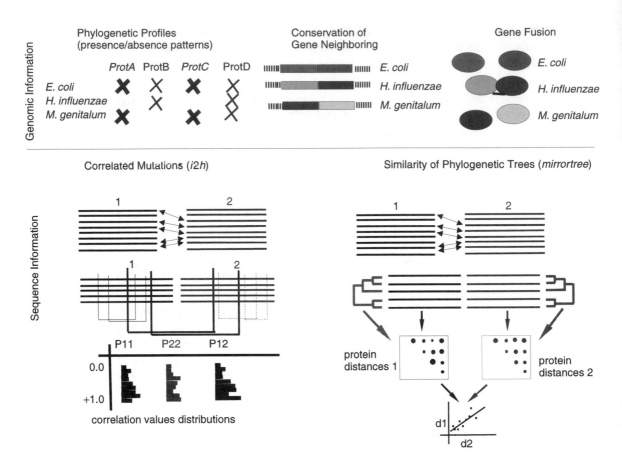

**Figure 20.6.** Methods for predicting protein interaction partners. Upper panel: methods based on genomic information. Lower panel: methods based on sequence information.

together in different bacteria, would be proposed as components of a functional mechanism related with the activity of that operon. In a previous publication, Tamames et al. (1997) demonstrated that pairs of genes belonging to the same functional class tend to be neighbors in genomes.

***Gene Fusion.*** A third group of methods is related to the presence of fused genes in various genomes (Marcotte et al., 1999; Enright et al., 1999). In this case, if two proteins are coded by two independent genes in some organisms, and by a single gene in other organisms (Fig. 20.6c), it is logical to conclude that the two proteins, as independent entities or as domains of the same protein, would be functionally related. Tsoka and Ouzounis (2000) have extended this observation to demonstrate that this is the case for many metabolic proteins.

## Computational Methods Based on Sequence Information

The methods discussed above do not use the information contained in the sequences of the proteins to predict interaction partners; they use just genomic features (gene presence/absence, gene position). Recently, two independent bioinformatics methods have emerged that address the problem of predicting interactions based on evolutionary information derived from sequence (Fig. 20.6, lower panel).

***Correlated Mutations (i2h).***  The first method is based on the idea of correlated mutations, which were discussed earlier. In this case, the prediction of interacting regions is extended to the detection of interacting partners in large collections of alternatives (Pazos and Valencia, 2002).

The multiple sequence alignments of two proteins are reduced, leaving only common species (Fig. 20.6d) and correlated mutations are calculated for three types of pairs: internal to one of the proteins, internal to the other, and interprotein pairs. Based on the distribution of correlation values for these three sets of pairs, an interaction index is calculated for these two proteins (Fig. 20.6d). A clear relation was found between high values of that interaction index and real interacting pairs of proteins (Pazos and Valencia, 2002).

***Similarity of Phylogenetic Trees (mirrortree).***  The second method is based on the similarity of phylogenetic trees of interacting or functionally related proteins. That similarity was qualitatively observed in sporadic cases such as insulin and insulin-receptors (Fryxell, 1996) or dockerins/cohesins (Pages et al., 1997) and first quantified for two proteins by Goh et al. (2000). Pazos and Valencia (2001) statistically demonstrated the relation between the similarity of phylogenetic trees and interaction in large sets of interacting proteins. The hypothesis behind this relationship is that interacting proteins would be subject to a process of coevolution that would be translated into a stronger than expected similarity between their phylogenetic trees.

The fist step in the *mirrortree* method is the same as in *i2h*, that is, the reduction of the alignments of the two proteins, leaving only common species. Then, a matrix containing the distances between all the proteins in the alignment is constructed for both proteins (Fig. 20.6e). These matrices can be considered as representations of the phylogenetic trees of these two proteins. The similarity between the phylogenetic trees is indirectly evaluated as the similarity between the two data sets of these matrices, using a correlation formulation (Pazos and Valencia, 2001).

This method goes one step beyond the phylogenetic profiles method described above, although both are based on coevolution of interacting proteins. In this case, the length of the branches and the structure of the trees are taken into account, whereas in the phylogenetic profiles method, only the pattern of presence/absence (leaves of the trees) is considered.

These two sequence-based approaches were tested in different systems including a large set of two domain proteins, in which the domains are treated as independent proteins, various collections of proteins previously published to interact, and in a genome-wise collections of alignments. In all the cases both methods were able to predict a relevant number of interactions with a clear concentration of known interactions among the top scoring pairs.

As an example, Figure 20.7 shows the results obtained applying the *mirrortree* system to a set of proteins reported to interact by Dandekar et al. (1998). It was possible to calculate 244 pairs where 8 pairs of known interaction and 8 pairs of possible interaction (pairs of ribosomal proteins) were included. True and possible interactions are clustered at high *mirrortree* scores, whereas the bulk of noninteracting

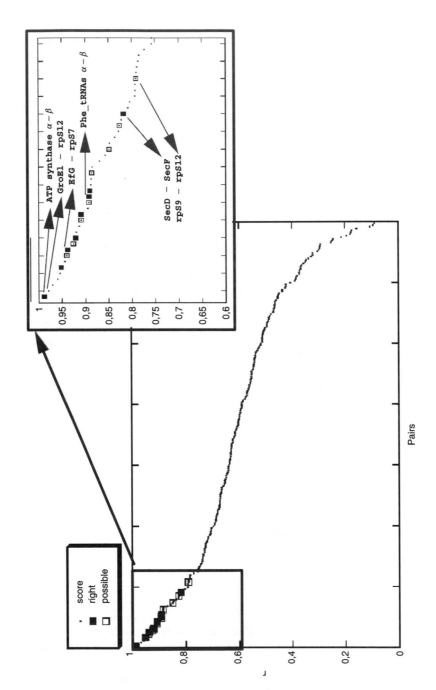

Figure 20.7. Results obtained with the *mirrortree* system for a set of 56 proteins that includes a number of interacting pairs, taken from Dandekar et al., 1998. The *mirrortree* score for the 244 pairs is plotted, known interactions are marked with black boxes, and possible interactions with open boxes. Some of the pairs are labeled with the names of the proteins.

pairs falls at low scores. There are some false positives and negatives. The pair of noninteracting proteins with highest score is the one formed by the chaperonin GroEl and the ribosomal protein S12.

The disadvantage of both methods is the need of big multiple sequence alignments for the two proteins to evaluate, since only sequences from common species can be used for the calculations (Fig. 20.6d and 20.6e). This problem can be alleviated in the future with the continuous stream of complete genome sequences.

## THE FUTURE

In the near future we will see a combination of results from the new powerful experimental techniques (yeast two hybrid, mass spectrometry and others), the new bioinformatics approaches for the prediction of interacting partners, the systematic docking studies applied to the structures obtained from structural genomics projects, perhaps with the help of the sequence-based approaches (correlated mutations and tree determinants), the relations established between genes with similar expression patterns in DNA array experiments, and the systematic mining of available information about protein interactions in databases and literature repositories (Fig. 20.8). With all this information in hand we will quickly approach a situation in which, for simple model systems (yeast and bacteria), it will be possible for the first time to glimpse the structure of the complex network of protein interactions that govern cell life.

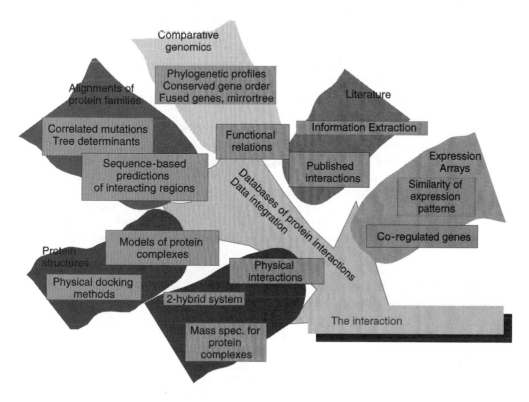

Figure 20.8. Getting the whole map of protein interactions by combining different techniques.

We have already seen how many new interactions will point to unknown functional and structural relations far beyond what we know from current biochemistry and molecular biology. The promise is that this information will lead not only to a better understanding of cell function, but also to better ways of manipulating cell function with new and more specialized drugs designed specifically for that purpose.

## REFERENCES

Andrade MA, Casari G, Sander C, Valencia A (1997): Classification of protein families and detection of the determinant residues with an improved self-organizing map. *Biol Cybern* 76:441–50.

Atrian S, Sanchez-Pulido L, González-Duarte R, Valencia A (1997): Shaping of *Drosophila* alcohol dehydrogenase through evolution. Relationship with enzyme functionality. *J Mol Evol* 47:211–21.

Azuma Y, Renault L, Garcia-Ranea JA, Valencia A, Nishimoto T, Wittinghofer A (1999): Model of the ran-RCC1 interaction using biochemical and docking experiments. *J Mol Biol* 289:1119–30.

Blaschke C, Andrade MA, Ouzounis C, Valencia A (1999): Automatic extraction of biological information from scientific text: protein–protein interactions. *ISMB* 99:60–7.

Caffrey DR, O'Neill LA, Shields DC (2000): A method to predict residues conferring functional differences between related proteins: application to MAP kinase pathways. *Protein Sci* 9:655–70.

Casari G, Sander C, Valencia A (1995): A method to predict functional residues in proteins. *Nat Struct Biol* 2:171–8. [An elegant method for detecting tree-determinant residues. Functional positions predicted with this program successfully served as guide in many experimental studies.]

Dandekar T, Snel B, Huynen M, Bork P (1998): Conservation of gene order: a fingerprint of proteins that physically interact. *Trends Biochem Sci* 23:324–8.

Del Sol A, Pazos F, Valencia A (2002): Automatic methods for the prediction of functionally important residues. Submitted to *J Mol Biol*.

Enright AJ, Iliopoulos I, Kyrpides NC, Ouzounis CA (1999): Protein interaction maps for complete genomes based on gene fusion events. *Nature* 402:86–90.

Fields S, Song O (1989): A novel genetic system to detect protein–protein interactions. *Nature* 340:245–6. [The original work describing the Yeast Two-Hybrid system for the detection of interacting pairs of proteins. The automatization of this system is the basis of the high-throughput detection of interacting pairs of proteins (i.e., for full proteomes).]

Frieman C, Kra P, Krauthammer M, Yu H, Rzhetsky A (2001): GENIS: A natural-language processing system for the extraction of molecular pathways from journal articles. *ISMB* 01:74–82.

Fariselli P, Pazos F, Valencia A, Casadio R (2002): Prediction of protein–protein interaction sites in heterocomplexes with neural networks. *Eur J BioChem* 269:1356–61.

Fryxell KJ (1996): The coevolution of gene family trees. *Trends Genet* 12:364–9.

Gaasterland T, Ragan MA (1998): Microbial genescapes: phyletic and functional patterns of ORF distribution among prokaryotes. *Microbiol Comp Genomics* 3:199–217.

Gavin AC, Bosche M, Krause R, Grandi P, Marzioch M, Bauer A, Schultz J, Rick JM, Michon AM, Cruciat CM, Remor M, Hofert C, Schelder M, Brajenovic M, Ruffner H, Merino A, Klein K, Hudak M, Dickson D, Rudi T, Gnau V, Bauch A, Bastuck S, Huhse B, Leutwein C, Heurtier MA, Copley RR, Edelmann A, Querfurth E, Rybin V, Drewes G, Raida M, Bouwmeester T, Bork P, Seraphin B, Kuster B, Neubauer G, Superti–Furga G (2002):

Functional orgranization of the yeast proteome by systematic analysis of protein complexes. *Nature* 415:141–7.

Gob CS, Bogan AA, Joachimiak M, Walther D, Cohen FE (2000): Co-evolution of proteins with their interaction partners. *J Mol Biol* 299:283–93.

Göbel U, Sander C, Schneider R, Valencia A (1994): Correlated mutations and residue contacts in proteins. *Proteins* 18:309–17.

Halperin I, Ma B, Wolfson H, Nussinor R (2002): Principles of docking: An overview of search algorithms and a guide to scoring functions. *Proteins* 47:409–43.

Ito T, Tashiro K, Muta S, Ozawa R, Chiba T, Nishizawa M, Yamamoto K, Kuhara S, Sakaki Y (2000): Toward a protein–protein interaction map of the budding yeast: a comprehensive system to examine two-hybrid interactions in all possible combinations between the yeast proteins. *Proc Natl Acad Sci USA* 97:1143–7. [One of the first attempts trying to establish the full protein interaction map of an organism, in this case, yeast.]

Jones S, Thornton JM (1997): Analysis of protein–protein interaction sites using surface patches. *J Mol Biol* 272:121–32.

Legrain P, Wojcik J, Gauthier JM (2001): Protein–protein interaction maps: a lead towards cellular functions. *Trends Genet* 17:346–52.

Lengauer T, Rarey M (1996): Methods for predicting molecular complexes involving proteins. *Curr Opin Struct Biol* 5:402–6.

Lichtarge O, Bourne HR, Cohen FE (1996): An "Evolutionary Trace" method defines binding surfaces common to protein families. *J Mol Biol* 257:342–58.

Lo Conte L, Chothia C, Janin J (1999): The atomic structure of protein–protein recognition sites. *J Mol Biol* 285:2177–98.

Marcotte EM, Pellegrini M, Ho-Leung N, Rice DW, Yeates TO, Eisenberg D (1999): Detecting protein function and protein–protein interactions from genome sequences. *Science* 285:751–3.

Olmea O, Valencia A (1997): Improving contact predictions by the combination of correlated mutations and other sources of sequence information. *Folding & Design* 2:S25–S32.

Olmea O, Rost B, Valencia A (1999): Effective use of sequence correlation and conservation in fold recognition. *J Mol Biol* 293:1221–39.

Ortiz A, Kolinski A, Rotkiewicz P, Ilkowski B, Skolnick J (1999): Ab initio folding of proteins using restraints derived from evolutionary information. *Proteins* S3:177–85.

Ouzounis C, Perez-Irratxeta C, Sander C, Valencia A (1998): Are binding residues conserved? *Pacific Symp Biocomput* 3:399–410.

Overbeek R, Fonstein M, D'Souza M, Pusch GD, Maltsev N (1999): Use of contiguity on the chromosome to predict functional coupling. *In Silico Biol* 1:93–108.

Pages S, Belaich A, Belaich JP, Morag E, Lamed R, Shoham Y, Bayer EA (1997): Species-specificity of the cohesin-dockerin interaction between *Clostridium thermocellum* and *Clostridium cellulolyticum*: prediction of specificity determinants of the dockerin domain. *Proteins* 29:517–27.

Pazos F, Helmer-Citterich M, Ausiello G, Valencia A (1997a): Correlated mutations contain information about protein–protein interaction. *J Mol Biol* 271:511–23. [Evolutionary information is used to predict specific protein interaction sites.]

Pazos F, Olmea O, Valencia A (1997b): A graphical interface for correlated mutations and other structure prediction methods. *Comp Applied Biol Sci* 13:319–21.

Pazos F, Sanchez-Pulido L, García-Ranea JA, Andrade MA, Atrian S, Valencia A (1997c): Comparative analysis of different methods for the detection of specificity regions in protein families. In: Lundh D, Olsson B, and Narayanan A, editors. *Biocomputing and Emergent Computation*. River Edge, NJ: World Scientific, pp 132–45.

Pazos F, Valencia A (2001): Similarity of phylogenetic trees as indicator of protein–protein interaction. *Protein Eng* 14:609–14.

Pazos F, Valencia A (2002): In silico two-hybrid system for the selection of physically interacting protein pairs. *Proteins* 47:219–27.

Pellegrini M, Marcotte EM, Thompson MJ, Eisenberg D, Yeates TO (1999): Assigning protein functions by comparative genome analysis: protein phylogenetic profiles. *Proc Natl Acad Sci USA* 96:4285–8.

Pereira-Leal JB, Seabra MC (2001): Evolution of the rab family of small GTP-binding proteins. *J Mol Biol* 313:889–901.

Rain JC, Selig L, De Reuse H, Battaglia V, Reverdy C, Simon S, Lenzen G, Petel F, Wojcik J, Schächter V, Chemana Y, Labigne A, Legrain P (2001): The protein–protein interaction map of Helicobacter pylori. *Nature* 409:211–5.

Smith GR, Sternberg MJ (2002): Prediction of protein–protein interactions by docking methods. *Curr Opin Struct Biol* 12:28–35. [Good review on protein–protein interactions and docking methods, including descriptions of the available programs and servers].

Stenmark H, Valencia A, Martinez O, Ullrich O, Goud B, Zerial M (1994): Distinct structural elements of rab5 define its functional specificity. *EMBO J* 13:575–83. [First rational replacement of specificities based on a sequencespace analysis.]

Tamames J, Casari G, Ouzounis C, Valencia A (1997): Conserved clusters of functionally related genes in two bacterial genomes. *J Mol Evol* 44:66–73.

Thomas J, Milward D, Ouzounis C, Pulman S, Carrol M (2000): Automatic extraction of protein interactions from scientific abstracts. *Pacific Symp Biocomput* 5:541–52.

Tsoka S, Ouzounis CA (2000): Prediction of protein interactions: metabolic enzymes are frequently involved in gene fusion. *Nat Genet* 26:141–2.

Uetz P, Giot L, Cagney G, Mansfield TA, Judson RS, Knight JR, Lockshon D, Narayan V, Srinivasan M, Pochart P, Qureshi-Emili A, Li Y, Godwin A, Conover D, Kalbfleisch T, Vijayadamovar G, Yang M, Johnston M, Fields S, Rothberg J (2000): A comprehensive analysis of protein–protein interactions in *Saccharomyces cerevisiae*. *Nature* 403:623–31.

Villar HO, Kauvar LM (1994): Amino acid preferences at protein binding sites. *FEBS Lett* 349:125–30.

Zhou HX, Shan Y (2001): Prediction of protein interaction sites from sequence profile and residue neighbor list. *Proteins* 44:336–43.

Zuckerkandl E, Pauling L (1965): Evolutionary divergence and convergence in proteins. In: Bryson V, and Vogel HJ, editors. *Evolving Genes And Proteins*. New York: Academic Press, pp 97–166.

# 21

# ELECTROSTATIC INTERACTIONS

## Nathan A. Baker and J. Andrew McCammon

An understanding of electrostatic interactions is essential for the full development of structural bioinformatics. The structures of proteins and other biopolymers are being determined at an increasing rate through structural genomics and other efforts. Specific linkages of these biopolymers in cellular pathways or supramolecular assemblages are being detected by genetic and other experimental efforts. To integrate this information in physical models for drug discovery or other applications requires the ability to evaluate the energetic interactions within and among biopolymers. Among the various components of molecular energetics, the electrostatic interactions are of special importance due to the long range of these interactions and the substantial charges of typical components of biopolymers. Indeed, electrostatics can be used to help assign biopolymers such as proteins to functional families, since particular kinds of ligand-binding sites may be indicated by the spatial distribution of the charges in the proteins.

We provide a brief overview of the role of electrostatics in biopolymers and supramolecular assemblages, and then outline some of the methods that have been developed for analyzing electrostatic interactions.

## OVERVIEW OF FUNCTIONAL ROLES OF ELECTROSTATICS

Electrostatic interactions help to determine the structure and flexibility of biopolymers, and the strength and kinetics of their associations with small molecules, other biopolymers, and biological membranes. Such interactions are of key importance for nucleic acids, since each nucleotide subunit carries a negative charge on its phosphate group. But proteins are also rich in charged groups, and the cumulative contributions to the electrostatic potential of a protein from its dipolar groups (such as the peptide linkages) can be substantial.

*Structural Bioinformatics*
Edited by Philip E. Bourne and Helge Weissig
Copyright © 2003 by Wiley-Liss, Inc.

In physiological settings, biopolymers are typically immersed in a solution comprising water and small, diffusible ions. The high dielectric coefficient of water, together with the tendency of diffusible ions to move toward biopolymer charges of opposite sign, reduces the effective interactions among the biopolymer charges. Nevertheless, these "solvent-screened" interactions strongly influence biopolymer behavior, especially within the physiological Debye length of about 1 nm. For biopolymers such as DNA that have high charge densities, counterions condense near the surface of the biopolymer. The resulting effective charge of DNA, for example, is about 25% of what it would be in the absence of condensation.

The general tendencies of charges to prefer a high dielectric environment (due to favorable free energy of solvation), and of opposite charges to attract, are reflected in the structures of most globular proteins: charged side chains are typically at the surface of the protein, and the relatively few buried charges often are salt-bridged with opposite charges. Similar principles influence the structure and thermodynamics of protein–protein complex formation. Although the advantage of ion-pairing in the formation of protein folds or complexes is substantially offset by the disadvantage of loss of aqueous solvation, the thermodynamic penalty of charge desolvation dictates that ion-pairing and other favorable electrostatic interactions within or between proteins are common features of protein structure.

As for kinetics, it has been firmly established that the rates of association of many biopolymers with one another or with small ligands are greatly increased by electrostatic steering of the diffusional encounters. This phenomenon is commonly observed in situations where an evolutionary advantage has likely been conferred by great speed. Even with the combined dielectric and ionic screening expected in a typical physiologic (150 mM ionic strength) solution, electrostatic-steering effects can lead to increases in the rate constant of association by two orders of magnitude.

## BRIEF HISTORY

The importance of electrostatic interactions in protein behavior was recognized early in the twentieth century by Linderstrom-Lang, who introduced a simple spherical model for protein titration in 1924. In this model, the protein was regarded as impenetrable, and the charges of the acidic and basic groups were treated as being uniformly distributed on the surface of the protein sphere. Thus, substantial cancellation of charge occurred, and the work of charging the protein sphere was approximated as the self-interaction energy of the net charge on the spherical surface. During subsequent decades, more detailed models that retained the approximation of spherical symmetry were developed. The first model that included discrete locations for the interacting charges, still located within a spherical body, but now including such features as dielectric heterogeneity and a nonzero ionic strength, was presented by Tanford and Kirkwood in 1957. Such models were used to account for the titration properties of proteins, the effects of pH and ionic strength on the activity of enzymes, and, as late as 1981, in work by Flanagan et al., the electrostatic contributions to the energetics of dimer-tetramer assembly in hemoglobin.

A new era of electrostatic models was ushered in by a 1982 paper in the *Journal of Molecular Biology* by Warwicker and Watson. Drawing on the increased knowledge of the three-dimensional structure of proteins, and especially on increased computer power, Warwicker and Watson introduced a grid-based, finite-difference approach for calculating the electrostatic potential of a nonspherical protein. The interior of the protein had a dielectric coefficient of 2, and the surrounding solvent had a dielectric coefficient of 80. This work provided the first hints concerning the possible functional importance of the shaping of the electrostatic potential and its gradient by the topography of the protein. Zauhar and Morgan introduced a boundary element approach for the analysis of this model in 1985. An important advance was described by Klapper et al. in 1986, who allowed for the inclusion of ionic strength effects by finite-difference solution of the linearized Poisson–Boltzmann equation.

A much simpler approximate model for describing electrostatic contributions to solvation energies and forces was introduced by Still et al. in 1990. This method is based on the Born ion, a canonical electrostatics model problem describing the electrostatic potential and solvation energy of a spherical ion (Born, 1920). The generalized Born method of Still et al. (1990) uses an analytical expression based on the Born ion model to approximate the electrostatic potential and solvation energy of small molecules. Although it fails to capture all the details of molecular structure and ion distributions provided by more rigorous models, such as the Poisson–Boltzmann equation, it has gained popularity as a very rapid method for evaluating approximate forces and energies for solvated molecules and continues to be vigorously developed.

The kinetic effects of electrostatics in steering biomolecular encounters are usually studied in the context of the diffusion equation, since the motions of the solutes are overdamped. The most detailed such studies make use of the Brownian dynamics simulation method of Ermak and McCammon (1978), which allows for structure and flexibility of the biomolecules, and hydrodynamic as well as potential-derived interactions. Rate constants for diffusion-controlled encounters of a protein with other small or large molecules can be determined by simulating their Brownian motion and analyzing their trajectories using a procedure introduced by Northrup et al. in 1984.

## NEED FOR FASTER METHODS, FOR HIGH-THROUGHPUT, AND LARGER STRUCTURES

The era of structural bioinformatics has created an urgent need for faster methods to solve problems in biomolecular electrostatics. As the structures of more proteins and other biopolymers become available through structural genomics and other initiatives, there will be a corresponding need to calculate the physical properties of these molecules to help assign them to families and functions. The need is even greater when one considers that any given biopolymer typically acts in concert with many others. Thus, there are combinatorial factors that increase the number of calculations that must be done, either to assess the thermodynamics of association of biopolymers, or—even more dramatically—to model the dynamics of association of such molecules, for example, with frequent updates of the electrostatic forces in the course of Brownian dynamics simulations.

An excellent example of the functional analysis of proteins aided by electrostatic calculations is provided in recent work by Murray and Honig (2002). This work focused on C2 domains, a large group of sequentially varied but structurally conserved modules that target the binding of proteins involved in signal transduction, membrane trafficking and fusion, and other cellular activities. Murray and Honig demonstrated that the targeting of C2 modules is determined in large part by electrostatics. For example, the membrane-binding face of C2 modules from protein kinase C become positively charged when the modules coordinate calcium ions, causing a calcium-triggered binding to negatively charged patches of membrane. By contrast, the corresponding face of C2 modules from cytosolic phospholipase A2 switch from a negative to neutral character on coordination of calcium ions; the binding of these ions triggers binding to neutral membranes, by reducing the unfavorable free energy of dehydration of the charged face on contact with the membrane. Murry and Honig (2002) show how these principles can be used to rationalize or predict the binding properties of other C2 modules, including ones whose structures are based on homology modeling.

The thermodynamics and kinetics of protein–protein association and larger-scale supramolecular assembly can be analyzed and predicted in many cases with the aid of electrostatic calculations, supplemented in the kinetics area by simulations of the diffusional motion of the proteins. Recent reviews of work in this area have been provided by Elcock, Sept, and McCammon (2001) and by Gabdoulline and Wade (2001). It has been possible with such calculations to replicate the experimentally observed rates of association of such protein pairs as barnase–barstar and fasciculin 2-acetylcholinesterase, including the variations in the rates as functions of ionic strength and protein mutagenesis. Similar calculations, by Sept and McCammon (2001), have provided a basis for understanding the nucleation and growth of polar actin filaments.

To handle very large numbers of binding partners or large supramolecular systems and to improve on the current diffusional encounter simulations by frequent updating of the electrostatic forces will require faster methods for solving the electrostatic equations. The remainder of this chapter outlines the corresponding theory and methods used, and illustrates recent progress in this area.

## POISSON–BOLTZMANN THEORY

Although methods such as generalized Born have found uses in several aspects of structural bioinformatics, we will confine the remainder of this discussion to Poisson–Boltzmann types of methods because of their relatively rigorous framework for inclusion of biomolecular topology and ionic strength effects.

### Introduction to the Equation

The canonical expression for the electrostatic potential in a continuum setting is the Poisson equation

$$-\nabla \cdot \epsilon(\mathbf{x})\nabla\phi(\mathbf{x}) = \varrho(\mathbf{x}), \qquad (21.1)$$

where $\epsilon(\mathbf{x})$ is a spatially varying dielectric coefficient, $\phi(\mathbf{x})$ is the electrostatic potential, and $\varrho(\mathbf{x})$ is the charge distribution that generates $\phi(\mathbf{x})$. The dielectric coefficient $\epsilon(\mathbf{x})$ typically assumes different values inside the solute and in the bulk solvent to reflect the relative polarizabilities of the two media. For biomolecules in an aqueous environment, $\epsilon$ generally is given a value of 2–20 inside the solute and a value of 80 in the solvent. Figure 21.1a shows the traditional definition of $\epsilon(\mathbf{x})$, which includes a jump discontinuity across the molecular surface while $\epsilon(\mathbf{x})$ changes between the protein and solvent dielectric values. However, more recent work (see, for example, Im, Beglov, and Roux, 1988) has proposed smoother definitions for $\epsilon(\mathbf{x})$ to reduce artifacts arising from the rapidly changing coefficient.

Likewise, the charge distribution $\varrho(\mathbf{x})$ has typically been given a very discontinuous definition, which can pose numerical difficulties for solution of the Poisson equation. In the absence of mobile counterions, $\varrho(\mathbf{x})$ is often treated as a collection of Dirac delta functions that model the $N_f$ atomic partial charges of the solute:

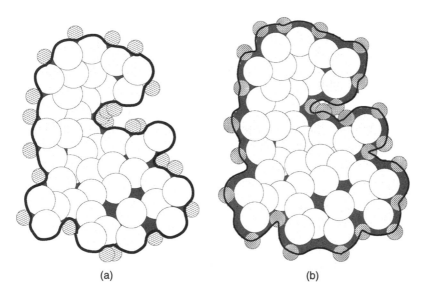

Figure 21.1. Popular definitions of the PBE coefficients. (a) The molecular surface (solid black line) often used to define the dielectric coefficient $\epsilon(\mathbf{x})$. This surface can be constructed by rolling solvent probes (small hatched circles) over the macromolecule (large white spheres). Gray regions show areas outside the atomic volume that are treated as inside the molecular surface. (b) The ion-accessible volume is the region outside the solid black line. This volume is defined as the region of space accessible to ion probe spheres (hashed circles).

$$\varrho(\mathbf{x}) = \varrho_f(\mathbf{x}) = 4\pi e_c^2 \beta \sum_{i}^{N_f} z_i \delta(\mathbf{x} - \mathbf{x}_i), \qquad (21.2)$$

where $e_c$ is the electron charge, $\beta = 1/(k_B T)$ is the inverse thermal energy, $k_B$ is the Boltzmann constant, $T$ is the temperature, $z_i$ are the magnitudes of the atomic partial charges (in units of $e_c$), and $\mathbf{x}_i$ are the partial charge positions. The Dirac delta function is a point distribution function with the property $\int f(\mathbf{x})\delta(\mathbf{y} - \mathbf{x}) \, d\mathbf{x} = f(\mathbf{y})$. The collection of constants scaling the delta functions implicitly assumes a dimensionless $\phi(\mathbf{x}) = e_c \beta \Phi(\mathbf{x})$, where $\Phi(\mathbf{x})$ is the electrostatic potential with the desired units.

The Poisson–Boltzmann equation (PBE) is a variant of the Poisson equation where mobile counterion charges are introduced to the charge distribution function in a mean field fashion, $\varrho(\mathbf{x}) = \varrho_f(\mathbf{x}) + \varrho_m(\mathbf{x})$, where $\varrho_m(\mathbf{x})$ denotes the mobile charge distribution. Mean field, or Debye–Hückel, electrolyte theory describes the distribution of each counterion species $i$ as $\rho_i(\mathbf{x}) = \overline{\rho}_i e^{-z_i\phi(\mathbf{x}) - V_i(\mathbf{x})}$, where $\overline{\rho}_i$ is the bulk concentration of species $i$, $z_i\phi(\mathbf{x})$ is the (dimensionless) energy of placing a counterion with partial charge $z_i$ at position $\mathbf{x}$ in the potential $\phi(\mathbf{x})$, and $V_i(\mathbf{x})$ is a (dimensionless) steric energy function that prevents mobile charges from entering the interior of the solute. This representation allows the mobile charge distribution function for $N_m$ counterion species to be written as

$$\varrho_m(\mathbf{x}) = 4\pi e_c^2 \beta \sum_{i}^{N_m} z_i \rho_i(\mathbf{x}) = 4\pi e_c^2 \beta \sum_{i}^{N_m} z_i \overline{\rho}_i e^{-z_i\phi(\mathbf{x}) - V_i(\mathbf{x})}. \qquad (21.3)$$

In the case of a $1:1$ monovalent ion distribution where $V_1 = V_2$, equation (21.3) can be simplified to $\varrho_m(\mathbf{x}) = -\bar{\kappa}^2(\mathbf{x})\sinh\phi(\mathbf{x})$, where $\sinh x = (e^x - e^{-x})/2$ was used and the coefficient is defined as $\bar{\kappa}^2(\mathbf{x}) = \epsilon_s e^{-V(\mathbf{x})}\kappa^2$. Here $\kappa$ is the Debye–Hückel parameter, defined for a general $N_m$-component electrolyte solution as $\kappa = \left(4\pi e_c^2\beta/\epsilon_s \sum_i^{N_m} \bar{\rho}_i z_i^2\right)^{1/2}$, where $\epsilon_s$ is the dielectric constant of the bulk solvent. As illustrated in Figure 21.1b, the function $e^{-V_i(\mathbf{x})}$ is usually treated as a discontinuous characteristic function that is unity within a ion-accessible volume (typically slightly larger than the protein volume) and zero otherwise. The PBE for a $1:1$ monovalent electrolyte is therefore

$$-\nabla \cdot \epsilon(\mathbf{x})\nabla\phi(\mathbf{x}) + \bar{\kappa}^2 \sinh\phi(\mathbf{x}) = 4\pi e_c^2 \beta \sum_i^{N_f} z_i \delta(\mathbf{x} - \mathbf{x}_i). \qquad (21.4)$$

For sufficiently small values of $\phi(\mathbf{x})$, the approximation $\sinh\phi(\mathbf{x}) \sim \phi(\mathbf{x})$ is often applied to this equation to give the linearized PBE:

$$-\nabla \cdot \epsilon(\mathbf{x})\nabla\phi(\mathbf{x}) + \bar{\kappa}^2(\mathbf{x})\phi(\mathbf{x}) = 4\pi e_c^2 \beta \sum_i^{N_f} z_i \delta(\mathbf{x} - \mathbf{x}_i). \qquad (21.5)$$

All of these equations are solved in conjunction with a Dirichlet condition, which specifies the value of potential at the boundary of some domain. For a sufficiently large domain, this condition is typically zero or some asymptotic form of the solution, such as the Debye–Hückel potential.

The PBE can also be derived from statistical mechanics using a continuum representation of the solvent dielectric properties (see, for example, Netz and Orland, 2000). While such treatments are too complicated to present here, one important aspect of these derivations is the development of a free energy expression for electrostatic interactions and the construction of the PBE as the "saddle-point" equation for the potential that minimizes this free energy.

## Energies

As discussed above, the PBE defines an electrostatic energy that can be derived from physical chemistry arguments (Sharp and Honig, 1990) or field theory saddle-point approximations. The free energy is a function of the electrostatic potential as well as the atomic positions, charges, and radii. For a $1:1$ monovalent electrolyte, this function has the form

$$G = \int \left[\varrho_f \phi - \frac{\epsilon}{2}(\nabla\phi)^2 - \bar{\kappa}^2(\cosh\phi - 1)\right] d\mathbf{x}. \qquad (21.6)$$

The first term $\int \varrho_f \phi \, d\mathbf{x}$ is the energy of inserting the protein charges into the electrostatic potential and can be interpreted as the energy of interaction for the fixed charges. The second term $-\int \epsilon(\nabla\phi)^2/2 \, d\mathbf{x}$ represents electrostatic stresses in the dielectric medium. Finally, the third term includes the effects of the mobile charge configuration and can be interpreted in terms of the excess osmotic pressure of the system. The subtraction of unity from the exponential in this term makes this an excess osmotic pressure and is necessary to cause the energy to vanish in the absence of a potential. Like the PBE, this energy expression can be linearized (see 21.5) for sufficiently small $\phi$ by assuming $\cosh\phi \sim 1 + \phi^2/2$. This linearized form of the energy leads to an additional simplification; Gauss' Law allows the second term to be rewritten $-\int \epsilon (\nabla\phi)^2/2 \, d\mathbf{x} = \int \phi/2 \nabla \cdot \epsilon \nabla\phi \, d\mathbf{x}$ and gives two equivalent free energy expressions

$$G = \frac{1}{2} \int \varrho_f \phi \, d\mathbf{x} = \frac{1}{2} \int [\epsilon (\nabla\phi)^2 + \bar{\kappa}^2 \phi^2] \, d\mathbf{x}. \qquad (21.7)$$

These free energy expressions can be used for a variety of static calculations on biomolecules, including the determination of binding constants, $pK_a$s, and solvation energies. These calculations are typically performed from a series of Poisson–Boltzmann energy evaluations that are then analyzed by free energy cycles. Figure 21.2 shows the specific case of $pK_a$ calculations, where the energy of protonating a functional group in a biomolecule is calculated in a stepwise fashion by determining the energies of the isolated biomolecule without the functional group, the isolated functional group in its protonated and unprotonated state, the biomolecule with the protonated functional group, and the biomolecules with the unprotonated functional group. These energies are then combined (as shown in Fig. 21.2) to give the free energy of protonating the functional group in the biomolecular environment, which can be converted to a $pK_a$ value. Similar cycles are used to calculate binding and solvation energies.

## Forces

Poisson–Boltzmann calculations have also found an increasingly important role in force evaluation for implicit solvent dynamics simulations. In such simulations, the dynamic

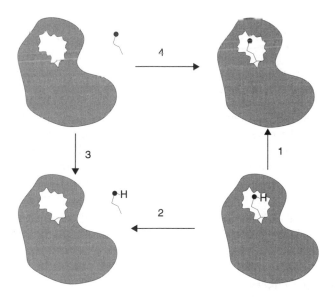

**Figure 21.2.** Titration calculation for a biomolecular functional group. The free energy of protonating a functional group (ball and stick moiety) in the presence of a biomolecule (gray object) is calculated by a thermodynamic cycle. Specifically the free energy of interest $\Delta G_1$ is calculated in terms of the other steps in the cycle: $\Delta G_1 = \Delta G_4 + \Delta G_2 - \Delta G_3$, where $\Delta G_4$ and $-\Delta G_2$ are the energies of inserting the unprotonated and protonated groups into the biomolecule, respectively, and $\Delta G_3$ is the energy of protonating the isolated functional group.

trajectory of a solute is calculated without the inclusion of the numerous explicit solvent molecules required for traditional molecular dynamics simulations. Instead, the solvent effects are modeled by stochastic forces applied to the biomolecule to mimic solvent buffeting, damping forces to provide the effect of solvent viscosity, continuum approximations of apolar interactions, and continuum electrostatics calculations (such as PBE) to include the effects of implicit solvent and salt on electrostatic forces in the solute.

To derive forces from the PBE, we simply differentiate the free energy with respect to atomic positions. As mentioned previously, a potential $\phi$ that solves the PBE is a saddle point of $G$, that is, $\partial G / \partial \phi = 0$. Therefore, the force $F_i$ on atom $i$ can be written solely in terms of variations in the coefficients with respect to atomic displacements $\partial \mathbf{y}_i$

$$F_i = -\int \left[ \phi \left( \frac{\partial \varrho_f}{\partial \mathbf{y}_i} \right) - \frac{1}{2} (\nabla \phi)^2 \left( \frac{\partial \epsilon}{\partial \mathbf{y}_i} \right) - (\cosh \phi - 1) \left( \frac{\partial \bar{\kappa}^2}{\partial \mathbf{y}_i} \right) \right] d\mathbf{x}. \qquad (21.8)$$

The terms of the integrand in this force expression have the same interpretation as for the free energy. The first term is the force density for atomic displacements in the potential $\phi$, the second is the dielectric boundary pressure on atom $i$, and the third is osmotic pressure on atom $i$.

The mechanics of evaluating atomic forces from equation (21.8) have been discussed in detail by Gilson et al. (1993) and Im, Beglov, and Roux (1998). These authors present excellent reviews of this topic, including the effects of discontinuities in the PBE coefficients $\epsilon$ and $\bar{\kappa}^2$ on the methods for force evaluation and the accuracy of the numerical results.

## NUMERICAL SOLUTION OF THE POISSON–BOLTZMANN EQUATION

Very few analytical solutions of the PBE exist for realistic biomolecular geometries and charge distributions. Therefore, this equation is usually solved numerically by a variety of computational methods (Table 21.1). These methods typically rely on a discretization to project the continuous solution down onto a finite-dimensional set of basis functions. In the case of the linearized PBE (21.5), the resulting equations are the usual linear matrix-vector equation, which can be solved directly. However, the nonlinear equations obtained from the full PBE require more specialized techniques, such as Newton methods, to determine the solution to the discretized algebraic equation. Specifically, Newton methods start with an initial solution guess and iteratively improve this guess by solving related linear equations for corrections to the current solution. Newton methods, as well as other popular methods for solution of nonlinear equations, have been reviewed by Holst and Saied (1995).

### Finite Difference Discretization

Some of the most popular discretization techniques employ Cartesian meshes to subdivide the domain in which the PBE is to be solved. Of these, the finite difference method has been at the forefront of PBE solvers. In its most general form, the finite difference method solves the PBE on a nonuniform Cartesian mesh, as shown in Figure 21.3a for a two-dimensional domain. While Cartesian meshes offer relatively simple problem setup, they provide little control over how unknowns are placed in the solution

**T A B L E 21.1.** Some of the Software That Implements the Concepts Described in This Chapter

| Program | Description | URL |
|---------|-------------|-----|
| DelPhi | Solves the PBE using highly optimized finite difference methods. | http://trantor.bioc.columbia.edu/delphi/ |
| APBS | Solves the PBE using parallel multigrid and parallel adaptive finite element methods. | http://agave.wustl.edu/apbs |
| MEAD | Solves the PBE using finite difference methods and determines $pK_a$ values while incorporating conformational flexibility of the macromolecule. | http://www.scripps.edu/bashford/ |
| UHBD | Solves the PBE using finite difference methods, calculates binding and solvation energies, determines $pK_a$'s, and performs Brownian dynamics simulations. | http://mccammon.ucsd.edu/uhbd.html |
| MacroDox | Solves the PBE using finite difference methods, determines $pK_a$'s, and performs Brownian dynamics simulations. | http://pirn.chem.tntech.edu/macrodox.html |
| AMBER | In addition to providing explicit solvent simulation tools, this package implements generalized Born and PBE-based implicit solvent methods in both dynamics and free energy evaluation simulations. | http://www.amber.ucsf.edu/amber/amber.html |
| CHARMM | In addition to providing explicit solvent simulation tools, this package implements generalized Born and PBE-based implicit solvent methods in both dynamics and free energy evaluation simulations. | http://yuri.harvard.edu/ |

domain. Specifically, as shown by Figure 21.3a, the Cartesian nature of the mesh makes it impossible to locally increase the accuracy of the solution in a specific region without increasing the number of unknowns across the entire grid.

Differential operators for problems discretized by finite difference methods are typically approximated using Taylor expansions. For example, a discretized one-dimensional Laplacian operator has the form

$$-\nabla^2 u(x_i) \approx \frac{-u(x_{i+1}) + 2u(x_i) - u(x_{i-1})}{h^2}, \qquad (21.9)$$

where $x_j$ denotes the grid point coordinates and $h$ is the mesh spacing. Given vectors u and f representing the values of the solution and source terms at the grid points, it is straightforward to develop a matrix form of the problem Au = f, where A is a sparse symmetric matrix with 2 on the main diagonal, $-1$ on the first off-diagonal elements, and 0 elsewhere in the matrix. Discretization of the differential operator for the PBE

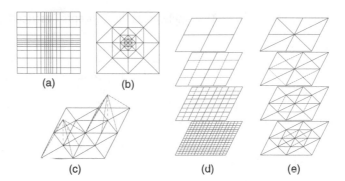

**Figure 21.3.** Meshes and hierarchies used in Poisson–Boltzmann solvers. (a) Cartesian mesh suitable for finite difference calculations; nonuniform mesh spacing can be used to provide a limited degree of adaptivity. (b) Finite element mesh exhibiting adaptive refinement. (c) Examples of typical piecewise linear basis functions used to construct the solution in finite element methods. (d) The multilevel hierarchy used to solve the PBE for a finite difference discretization; red lines denote the additional unknowns added at each level of the hierarchy. (e) The multilevel hierarchy used to solve the PBE for a finite element discretization; red lines denote simplex the subdivisions used to introduce additional unknowns at each level of the hierarchy.

yields a matrix with a similar sparse symmetric structure, but with more nonzeros per row.

## Adaptive Finite Element Discretization

Unlike finite difference methods, adaptive finite element discretizations offer the ability to place computational effort in specific regions of the problem domain. Finite element meshes (see Fig. 21.3b) are composed of simplices that are joined at edges and vertices. The solution is constructed from piecewise polynomial basis functions (see Fig. 21.3c) that are associated with mesh vertices and typically are nonzero only over a small set of neighboring simplices. Solution accuracy can be increased in specific areas by locally increasing the number of vertices through simplex refinement (subdivision). As shown in Figure 21.3b, the number of unknowns (vertices) is generally increased only in the immediate vicinity of the simplex refinement and not throughout the entire problem domain, as in finite difference methods. This ability to locally increase the solution resolution is called *adaptivity* and is the major strength of finite element methods applied to the PBE (see Holst, Baker, and Wang, 2000).

Typically, the algebraic system is assembled using a Galerkin discretization, where a weak form of the PBE is imposed for each basis function in the system. Specifically, the original differential form of the PBE is transformed by integration with a basis function $v$ to give an integral equation

$$\int (\epsilon \nabla \phi \cdot \nabla v + \bar{\kappa}^2 v \sinh \phi - \varrho_f v) \, dx = 0. \tag{21.10}$$

The algebraic system is implicitly assembled by representing the solution as a linear combination of finite element basis functions $\phi = \sum_i u_i \psi_i$ and imposing the weak PBE (21.10) using every basis function $\psi_j$ as test functions. As with the finite difference method, this discretization scheme leads to sparse symmetric matrices with a small number of nonzero entries in each row.

## Multilevel Solvers

Multilevel solvers, in conjunction with the Newton methods described above, have been shown to provide most efficient solution of the algebraic equations obtained by discretization of the PBE with either finite difference or finite element techniques. Most sizable algebraic equations are solved by iterative methods that repeatedly apply a set of operations to improve an initial guess until a solution of the desired accuracy is reached. However, the speed of traditional iterative methods has been limited by their inability to quickly reduce low-frequency (long-range) error in the solution. Multilevel methods overcome this problem by projecting the discretized system onto meshes (or grids) at multiple resolutions (see Fig. 21.3). The advantage of this multiscale representation is that the slowly converging low-frequency components of the solution on the finest mesh are quickly resolved on coarser levels of the system. This gives rise to a multilevel solver algorithm, where the algebraic system is solved directly on the coarsest level then used to accelerate solutions on finer levels of the mesh.

As shown in Figure 21.3, the assembly of the multiscale representation, or multi-·level hierarchy, depends on the method used to discretize the PBE. For finite difference types of methods, the nature of the grid lends itself to the assembly of a hierarchy with little additional work. In the case of adaptive finite element discretizations, the most natural multiscale representation is constructed by refinement of an initial mesh that typically constitutes the coarsest level of the hierarchy.

## FUTURE DIRECTIONS

Recent developments in Poisson–Boltzmann solver technology (Baker et al., 2000, 2001; Holst et al., 2000 and Wang, 2000) have extended the applicability of continuum electrostatics methods to biomolecular systems consisting of hundreds of thousands of atoms by facilitating solution of the PBE on massively parallel computers. For example, Figure 21.4 shows the electrostatic potential of a 1.2-million atom microtubule fragment roughly 400 Å in length. Such large-scale calculations on microtubules and other cellular components are the starting point for computational investigation into the molecular aspects of cellular function.

In the future, it should be possible to apply these techniques to help determine and investigate macromolecular interactions at the cellular scale. Additional research is likely to focus on the computational evaluation of protein–protein interactions on a genomewide scale. As increasingly larger numbers of biomolecular structures are being determined, high-throughput continuum electrostatics methods will facilitate the development of computational proteomics to determine the network of biomolecular reactions in living organisms.

## ACKNOWLEDGMENTS

Support for this work was provided, in part, by the Howard Hughes Medical Institute and grants to J. Andrew McCammon from NIH, NSF, and NPACI/SDSC. Additional

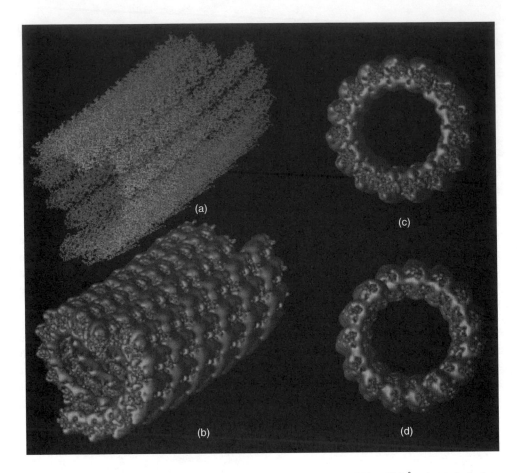

**Figure 21.4.** Electrostatic properties of a 1.2-million atom, $400 \times 300 \times 300$ Å microtubule fragment illustrating the current state of the art for continuum electrostatics calculations. The potential was calculated using APBS to solve the PBE at 150 mM ionic strength. (a) The backbone atoms of the microtubule. (b) Electrostatic potential isocontours for microtubule shown at $+1.0$ and $-1.0$ kT/e. (c) Potential isocontours (as in B) for so-called "$-$" end of microtubule. (d) Potential isocontours (as in B) for so-called "$+$" end of microtubule. Figure also appears in Color Figure section.

support has been provided by the W. M. Keck Foundation and the National Biomedical Computation Resource.

## FURTHER READING

Bokris JO'M, Reddy AKN (1998): *Modern Electrochemistry I. Ionics.* New York: Plenum Press. [Review of electrostatics in aqueous ionic solutions.]

Braess D (1997): *Finite Elements. Theory, Fast Solvers, and Applications in Solid Mechanics.* New York: Cambridge University Press. [Provides descriptions of finite difference and finite element methods as well as several solution methods for partial differential equations.]

Gilson M (2002): Introduction to continuum electrostatics, with molecular applications. In: Beard DA, editor. *Computational and Theoretical Biophysics.* Biophysics textbook online at http://www.biophysics.org/btol/compute.html#3C. [Review of continuum electrostatics concepts and methods.]

Jackson JD (1975): *Classical Electrostatics*. New York: John Wiley & Sons. [Standard reference text for continuum electrostatics.]

## REFERENCES

Baker NA, Holst MJ, Wang F (2000): Adaptive multilevel finite element solution of the Poisson–Boltzmann equation II: refinement at solvent accessible surfaces in biomolecular systems. *J Comput Chem* 21:1343–52. [This article describes the application of finite element methods to the solution of the PBE.]

Baker NA, Sept D, Joseph S, Holst MJ, McCammon JA (2001a): Electrostatics of cellular components: application to microtubules and the ribosome. *Proc Natl Acad Sci* USA 89:10037–41. [This article describes the efficient parallel solution of the Poisson-Boltzmann equation for large biomolecules.]

Baker NA, Sept D, Holst MJ, McCammon JA (2001b): The adaptive multilevel finite element solution of the Poisson–Boltzmann equation on massively parallel computers. *IBM J Res Dev* 45:427–38. [Both Baker et al. 2001 papers describe efficient parallel algorithms for both the finite difference and finite element solution of the PBE for biomolecular systems at the near-cellular scale.]

Born M (1920): Volumen und hydratationswärme der ionen. *Z Phys* 1:45–8. [Introduced an electrostatic model for ion solvation.]

Elcock AH, Sept D, McCammon JA (2001): Computer simulation of protein–protein interactions. *J Phys Chem* B 105:1504–18. [A review of modern methods for simulating interactions between proteins.]

Ermak DL, McCammon JA (1978): Brownian dynamics with hydrodynamic interactions. *J Chem Phys* 69:1352–60. [Introduced the method for simulating the diffusional motion of biopolymers or other molecules, including the effects of electrostatic interactions.]

Flanagan MA, Ackers GK, Matthew JB, Hanania GIH, Gurd FRN (1981): Electrostatic contributions to the energetics of dimer-tetramer assembly in human hemoglobin: pH dependence and effect of specifically bound chloride ions. *Biochemistry* 20:7439–49. [Uses the Tanford–Kirkwood model for protein titration to determine pH effects on hemoglobin dimer-tetramer assembly.]

Gabdoulline RR, Wade RC (2001): Protein–protein association: investigation of factors influencing association rates by Brownian dynamics simulations. *J Mol Biol* 306:1139–55. [Provides an analysis of the diffusional encounter simulations of a variety of different pairs of proteins.]

Gilson MK, Davis ME, Luty B, McCammon JA (1993): Computation of electrostatic forces on solvated molecules using the Poisson–Boltzmann equation. *J Phys Chem* 97:3591–600. [Description of the calculation of forces from solutions to PBE.]

Holst MJ, Baker NA, Wang F (2000a): Adaptive multilevel finite element solution of the Poisson–Boltzmann equation I: algorithms and examples. *J Comput Chem* 21:1319–42.

Holst MJ, Saied F (1995): Numerical solution of the nonlinear Poisson–Boltzmann equation: developing more robust and efficient methods. *J Comput Chem* 16:337–64. [Provides a description of an efficient multigrid algorithm for solving the PBE on a Cartesian mesh and reviews several methods for solution of nonlinear partial differential equations.]

Im W, Beglov D, Roux B (1998): Continuum solvation models: computation of electrostatic forces from numerical solutions to the Poisson–Boltzmann equation. *Comput Phys Chem* 111:59–75. [Introduces alternative definitions for the dielectric and ionic accessibility coefficients in the PBE and describes the calculation of electrostatic forces.]

Klapper I, Hagstrom R, Fine R, Sharp K, Honig B (1986): Focusing of electric fields of the active site of Cu-Zn superoxide dismutase: effects of ionic strength and amino-acid modification. *Proteins Struct Func Genet* 1:47–59. [Introduced the finite-difference method for solving the linearized PBE for a protein with a realistic surface topography.]

Linderstrom-Lang K (1924): *C. R. Travl. Lab. Carlsberg* 15(7): [Describes protein titration in terms of a simple spherical model.]

Murray D, Honig B (2002): Electrostatic control of the membrane targeting of C2 domains. *Molec Cell* (Forthcoming) Poisson–Boltzmann electrostatics are used to predict the binding properties of C2 domains from a variety of proteins. *Molec Cell* 9:145–54.

Netz RR, Orland H (2000): Beyond Poisson–Boltzmann: fluctuation effects and correlation functions. *Eur Phys J* 1:203–14. [Field theory derivation of the PBE and higher-order electrostatics formulations.]

Northrup SH, Allison SA, McCammon JA (1984): Brownian dynamics of diffusion-influenced bimolecular reactions. *J Chem Phys* 80:1517–24. [Presents the basic method for analyzing Brownian dynamics trajectories to calculate the rate constants of diffusional encounter of biopolymers or other molecules.]

Sept D, McCammon JA (2001): Thermodynamics and kinetics of actin filament nucleation. *Biophys J* 81:667–74. [Poisson–Boltzmann and Brownian dynamics calculations are used to analyze the initial steps in the formation of actin filaments.]

Sharp KA, Honig B (1990): Calculating total electrostatic energies with the nonlinear Poisson–Boltzmann equation. *J Phys Chem* 94:7684–92. [Describes the calculation of electrostatic energies from the PBE.]

Still WC, Tempczyk A, Hawley RC, Hendrickson T (1990): Semianalytical treatment of solvation for molecular mechanics and dynamics. *J Am Chem Soc* 112:6127–9. [Describes the development of the generalized Born method for biomolecular electrostatics.]

Tanford C, Kirkwood JG (1957): Theory of protein titration curves. I. General equations for impenetrable spheres. *J Am Chem Soc* 79:5333–9. [Describes the electrostatics of a spherical protein with low dielectric coefficient and discrete, buried charges, immersed in a dilute electrolyte solution.]

Warwicker J, Watson HC (1982): Calculation of the electric potential in the active site cleft due to alpha-helix dipoles. *J Mol Biol* 157:671–9. [First application of a finite-difference method to solve the Poisson equation for a protein with realistic surface topography.]

Zauhar R, Morgan RJ (1985): A new method for computing the macromolecular electric potential. *J Mol Biol* 186:815–20. [Solution of the PBE by boundary element methods.]

# Section VI

## PROTEINS AS DRUG TARGETS

# 22

# PRINCIPLES AND METHODS OF DOCKING AND LIGAND DESIGN

J. Krumrine, F. Raubacher, N. Brooijmans, and I. Kuntz

Structural bioinformatics can facilitate the discovery, design, and optimization of new chemical entities. These new chemicals can range from drugs and biological probes to biomaterials, catalysts, and new macromolecules. Molecular design is important in fields as diverse as organic chemistry, physical chemistry, chemical engineering, chemical physics, bioengineering, and molecular biology. No single strategy or method has come forward that provides an optimum solution to the many different challenges involved in designing materials with new properties. Our goal in this chapter is to set forth the general principles that are likely to be of greatest influence in the next several years and then to focus on structure-based drug design as an example where there has been enough experience to make a critical evaluation.

What are the essential concepts needed to embark on ligand design? "Ligand" comes from "ligare" meaning a "band" or "tie." It is currently used to mean a molecule (of any size) that binds or interacts with another molecule through noncovalent forces—that is, the interaction (usually) does not involve chemical bond formation. The second molecule—the "target" or "receptor"—is typically the larger species. The resulting molecular complex may contain multiple copies of the ligand and/or the receptor. There are many physical, chemical, and biological properties of the complex that will be influenced by changes in the ligand. The nature of the interaction between ligand and receptor depends on a balance in the chemical/physical forces between them and the forces between each of these molecules and the solvent or environment. These forces basically arise from the interaction of electrons and are studied at the most fundamental level using quantum mechanics (QM). However, the direct application of quantum theory to molecules of biological interest remains limited by computational resources for systems larger than a few amino acids. Thus,

*Structural Bioinformatics*
Edited by Philip E. Bourne and Helge Weissig
Copyright © 2003 by Wiley-Liss, Inc.

most computational approaches involve significant empirical adjustments and may lack generality. Nevertheless, a firm comparison between theory and experiment is possible. Much of the focus of ligand design deals with properties directly connected to thermodynamics—free energy of binding, solubility in aqueous and nonaqueous environments, and so forth. Kinetic issues are less often considered, being more complicated to measure and to calculate, but kinetics clearly plays a major role in many systems of pharmacological interest such as enzyme catalysis, signaling cascades, and molecular rearrangements.

Of particular interest to us in this chapter is the free energy of binding, $\Delta G_{bind}$, with its associated components the enthalpy and entropy of binding. The free energy of binding is defined as:

$$\Delta G_{bind} = \Delta G_{complex} - (\Delta G_{ligand} + \Delta G_{receptor})  \qquad (22.1)$$

$\Delta G_{bind}$, defined in this way (for further discussion, see Atkins, 1997), is a function of the temperature, pressure, ionic strength, pH, solvent, and concentrations of all the chemical species present. The $\Delta G$ terms on the right hand side of Eq. 22.1 remind us that absolute free energies are not available experimentally. Instead, the free energy of a substance under a particular set of conditions is defined as its free energy difference from a reference state. Ideally, one should carry out both experiments and calculations under the same reference conditions so that work in different laboratories can be readily compared. This is rarely done, with consequences that we return to later. For a review of thermodynamics for chemical and biophysical applications see Atkins (1997). For direct application of thermodynamic measurements to complex systems see Plum and Breslauer (1995). The most common measurement for $\Delta G_{bind}$ is through the equilibrium constant for the complex:

$$\Delta G_{bind} = -RT \ln K_{eq} = RT \ln K_d  \qquad (22.2)$$

where R is the gas constant and T is the absolute temperature in Kelvin. Of course, this is of great interest in drug design because prediction of $K_{eq}$ is a direct prediction of ligand affinity. The pharmacological literature frequently reports the dissociation constant, $K_d$, which is simply the reciprocal of the equilibrium constant, or the $IC_{50}$, the concentration of ligand that achieves a 50% change in the normal activity. The relationship between $IC_{50}$ and $K_i$ must be worked out for each system. Simple formulas are available for enzyme inhibition (Winzor and Sawyer, 1995).

Our goal for this chapter is to summarize the current computational approaches in the field of structure-based drug design. The object of such efforts is to identify/design a molecule that will bind with high affinity and specificity to a biological target of known (or predictable) three-dimensional (3D) structure. Such molecules may need to be further modified to have other desirable properties such as the appropriate solubility, the appropriate molecular weight, and proper metabolic characteristics, and so forth (see ADME below) before one would consider them true "clinical candidates." The primary reason for interest in structure-based strategies is the appreciation that knowledge of the receptor (and ligand) tertiary structures can be tapped either to speed the discovery process or to enhance the qualities of the ligands. However, it should be clear at the onset that such knowledge is neither necessary (many drugs have been discovered with no information about the receptor) nor sufficient (structures of HIV-integrase, for example, have not yet led to useful drug candidates). Nevertheless, we anticipate that the role of structural information and the use of computational tools will become ever more important in the molecular design process.

## METHODS AND TOOLS IN COMPUTER-AIDED MOLECULAR DESIGN

The first challenge for computer-aided design is to identify one or more lead compounds—compounds that show activity in an appropriate assay. Until recently, most drugs on the market came from lead compounds discovered from the screening of natural products, insight into fundamental biochemistry, or exploring analogs of known substrates or ligands. Only in the last few years has it been feasible to expect structures of the drug target to be available during a drug discovery project. Structures of many targets, especially membrane-bound proteins, are still in very short supply. Once discovered, leads must be modified in an iterative cycle (see Figure 22.1) in order to enhance their potency and selectivity. Moreover, the compound and its metabolites must be non-toxic and be available at the site of action for a sufficient amount of time.

Computational methods are beginning to play a major role in the process of drug design involving areas such as virtual screening of combinatorial libraries, 3D structure determination of target macromolecules and computational design of new or improved lead compounds.

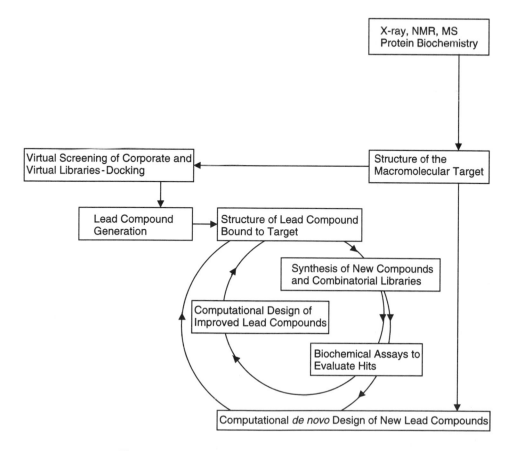

Figure 22.1. Docking and *de novo* drug design cycles.

## Computer-Aided Drug Design

Computational ligand design can be divided into two different strategies: ligand-based (analog-based) or target-based (structure-based) design that can be used together or independently. Analog-based design relies on a set of known ligands and is particularly valuable if no structural information about the receptor is available. Hence, it is generally applicable to all classes of drugs. Target-based design usually starts with the structure of a receptor site, such as the active site in a protein. This structure can be generated from direct experimentation or can be deduced from experimental structures through homology modeling (Al-Lazikani et al., 2001).

Computer-aided drug design (CADD) can call on a broad range of molecular computational techniques (Leach, 2001). One important approach uses the laws of classical mechanics to calculate molecular geometries, energies, and motion. Typically, interactions are described on an atomic level using force fields. Molecular mechanics is often used to optimize the energy of a molecule by finding the molecular conformation that gives a minimum energy on the potential energy hypersurface. Molecular dynamics allows the evolution of the simulated system in time. Newton's first law is solved numerically, yielding a trajectory—that is, a collection of closely related "snap shots" of the simulated molecules. Solvent and counter ions can be considered explicitly. In a Monte Carlo simulation, the system evolves from one state to another in a stochastic process satisfying the Boltzmann distribution.

***Analog-Based Design.*** The analog-based approach mainly uses pharmacophore maps and quantitative structure-activity relationships (QSAR) to identify or modify a lead in the absence of a known three-dimensional structure of the receptor. It is necessary to have experimental affinities and molecular properties of a set of active compounds, for which the chemical structures are known.

PHARMACOPHORES. A pharmacophore is an explicit geometric *hypothesis* of the critical features of a ligand (Guener, 2000). Standard features include hydrogen-bond donors and acceptors, charged groups, and hydrophobic patterns. The hypothesis can be used to screen databases for compounds and to refine existing leads. For a geometric alignment of the functional groups of the leads, it is necessary to specify the conformations that individual compounds adopt in their bound state. To construct this consensus arrangement of pharmacophoric points, a constrained systematic search is performed with the most rigid molecule first.

Since the simple presence of a pharmacophoric fingerprint is not sufficient for predicting activity, inactive compounds possessing the required pharmacophoric features must also be considered. By comparing the volume of the active and the inactive compounds, a common volume can be constructed in order to approximate the shape of the (unknown) receptor site to further refine the pharmacophore model and to screen out additional compounds. Each binding mode might require a different pharmacophore model.

QSAR. The goal of QSAR studies is to predict the activity of new compounds based solely on their chemical structure. The underlying assumption is that the biological activity can be attributed to incremental contributions of the molecular fragments determining the biological activity. This assumption is called the linear free energy principle. Information about the strength of interactions is captured for each compound by, for example, steric, electronic, and hydrophobic descriptors. One of the earliest QSAR linear free energy approaches defined the effect of substituent properties on the ionization of benzoic acids leading to the well-known Hammet equation (Hammet, 1970). These ideas were applied to drug activities by Hansch, Leo, and Hoekman (1995). An extension of QSAR is the use of additional conformational information for a 3D-QSAR study Oprea and Waller (1991). The predictive nature of a QSAR approach is limited to new compounds that are similar to the compounds from the training set. There is also a risk of chance correlations.

***Structure-Based Design.*** The first step in site-directed drug design (Fig. 22.2) is the determination of the 3D structure of the target macromolecule, primarily by X-ray crystallography and NMR spectroscopy or computational methods such as homology modeling or *ab initio* methods (see Chapters 25 and 27 in this book).

The negative image of the receptor defines the space available for ligand binding. There may be many potential binding sites. The actual binding site can be located by comparison with known protein–ligand complexes or through homology to related complexes. Mutational data are evaluated, as is simple geometry: the binding site is often the largest cavity in the protein.

A critical issue is conformational analysis. All except the simplest ligands explore different geometries as their atoms move under thermal forces. Of interest are the lowest energy conformations for a particular ligand when it is free in solution and when it is bound to the receptor. Differences in internal energy enter directly into the computation of net binding energies. Further, the selection of a small number of bound conformations from a larger set of available conformations in solution causes an entropy loss on binding. Finally, conformational analysis is important because the geometric differences and physical properties can be large for different conformations of a specific ligand. To obtain 3D structures of multiple low-energy conformations, a variety of conformational search strategies have been developed (Leach, 1997).

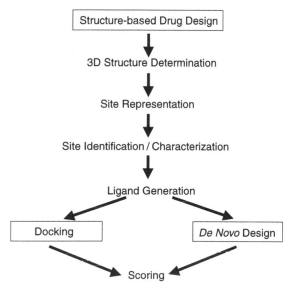

Figure 22.2. Flow chart of the docking procedure.

Site-directed ligand generation branches into two main approaches: docking and building (*de novo* design). Docking methods search available databases for matches to an active site, whereas *de novo* design seeks to generate new ligands by connecting atoms or molecular fragments uniquely chosen for a particular receptor. Docking is the computational equivalent of high-throughput screening. Given an appropriate database, compounds are retrieved and are often immediately available for testing. *De novo* design can suggest chemically novel ligand classes that are not limited to previously synthesized compounds. However, one must deal with the question of synthetic feasibility of the generated compounds.

DOCKING. The aim of molecular docking is to evaluate the feasible binding geometries of a putative ligand with a target whose 3D structure is known. The binding geometries, often called binding modes or poses include, in principle, both the positioning of the ligand relative to the receptor (ligand *configuration*) and the conformational state(s) of the ligand and the receptor. The exploration of the configurational and conformational space (the sampling) *and* the energetic evaluation of each discrete geometry (the scoring) are separable tasks. Docking methods can therefore be evaluated by their ability to rapidly and accurately dock large numbers of small molecules into the binding site of a receptor, allowing for a rank ordering in terms of strength of interaction with a particular receptor. Another concern is the prediction of selectivity toward various receptors. Therefore, the essential feature of any treatment of ligand-receptor interaction is the correct estimation of free energy of binding.

There are three basic tasks any docking procedure must accomplish: (1) characterization of the binding site; (2) positioning of the ligand into the binding site (orienting); and (3) evaluating the strength of interaction for a specific ligand-receptor complex ("scoring") (see Fig. 22.2). The first two tasks will be discussed in this section; scoring will be discussed in Flow Charts and Fundamental Problems below. In order to screen large databases, automated docking is required. Automated searching methods can be classified into two general approaches: geometric search methods, which include systematic search grids as well as descriptor matching, and energy search methods. Matching methods apply heuristic rules for pruning the combinatorial search tree of matches between ligand and receptor descriptors. An energy search accomplishes the alignment of the ligands by minimizing the ligand-receptor interaction energy using Monte Carlo or molecular dynamics simulations or genetic algorithms.

*Site Characterization.* To characterize the binding site, the DOCK suite of programs (Ewing et al., 2001; Meng, Shoichet, and Kuntz, 1992) derives a negative image of the binding site defined by a set of overlapping spheres. Using the sphere centers as matching points, the ligands are oriented in the binding site. Steric complementarity is then improved by minimizing the ligand-receptor interaction energy.

FlexX also uses a matching procedure for the placement of an initial ligand fragment. Triangles of interaction centers in the base fragment are mapped onto triangles of interaction points lying on the surface of the receptor (Rarey, Wefing, and Lengauer, 1996a).

CLIX (Lawrence and Davis, 1992) utilizes chemical descriptors for receptor site features. The matching points are energetically favorable sites for certain functional groups, resulting in a good ligand score when a matching ligand is placed in this region.

For protein–protein docking, improved computational efficiency is obtained by concentrating on the molecular surfaces for the description of the interacting sites between the two proteins. The computational requirements can be further reduced by using a low-resolution presentation of the surfaces (Vakser, 1995). The atomic details are then evaluated in a later step. Like chemical descriptors, it is also possible to map physicochemical properties on these surfaces, such as hydrophobicity or electrostatic potentials.

*Orientation Procedures.* If the ligand (or part of the ligand) is assumed to be rigid, a systematic grid search of the six-dimensional space of the mutual ligand-receptor orientations is possible. Accuracy is limited by the step size of the search grid and the time available for the search. To speed up the search process the Fast Fourier Transform (FFT) can be used (Katchalski-Katzir et al., 1992). The FFT replaces the convolution of the grids by a much faster multiplication in Fourier space to rapidly search the orientation space. This method (implemented for instance in FTDock (Gabb, Jackson, and Sternberg, 1997) is mostly used to dock large structures with significant geometric complementarity such as two proteins. The protein–protein docking program DOT additionally uses convolution products for evaluating the electrostatic and van der Waals interaction energies of a complex (Mandell et al., 2001). Other methods for macromolecular docking rely on methods derived from Computer Vision for discovering known objects in a scene (Sandak, Nussiniov, and Wolfson, 1995).

For efficiency, most methods for docking small molecules adopt a descriptor-matching approach. Descriptors (points in space that may have properties assigned to them) of the ligand are superimposed with some tolerance to descriptors (points) of the binding pocket. Initial matches are then used to orient the whole molecule in the pocket to generate a reasonable ligand configuration, which may be subsequently refined by optimization of a suitable scoring function. The ligand and receptor descriptors not only have to match geometrically but also chemically. Due to the combinatorial character of matching the search is rarely exhaustive, but is usually sufficient for recovering the binding geometry.

Energy search methods employ complete molecular force fields for exploring the configurational energy surface of a ligand interacting with a receptor. By locating minima on the surface, possible binding configurations are evaluated. The multiple minima problem often requires long simulation times, limiting energy-based methods to a small number of ligands per search. Often, ligand flexibility is explored directly in the search procedure.

Genetic algorithms as implemented in AUTODOCK (Morris et al., 1998) or HAMMERHEAD (Welch, Ruppert, and Jain, 1996) allow for a very thorough conformational search. Limited flexibility can also be included. During each iteration, selective pressure is applied to encourage high-scoring conformational features to be carried over from the current to the next generation. Random mutations introduce new microconformations while crossover steps allow for an exchange of conformational subsets.

In Monte Carlo searches (implemented in FLO98 (Bohacek and McMartin, 1994) or MCDOCK (Liu and Wang, 1999) either the internal conformation of the ligand is changed by a random rotation about a bond or the entire molecule is randomly translated and rotated. The energy of this new configuration is then either accepted or rejected using the standard Metropolis criterion for accepting or rejecting a move. In a Tabu search, such as used by PRO_LEADS (Baxter et al., 1998), records are kept of the visited regions in search space, which focus the next moves to a less explored region in search space. The Mining Minima (David, Luo, and Gilson, 2001) approach combines a Tabu search with a genetic algorithm. Low-energy conformations of the ligand-receptor system are located; subsequently, the minima are locally explored by gradually narrowing down a random search that is centered around the lowest minima found so far.

DOCKING FLEXIBLE LIGANDS. In addition to the implicit treatment of ligand flexibility by energy search methods, four basic schemes have been developed to explore ligand flexibility: pregeneration of rigid conformers, docking rigid fragments, either independently or interconnected, and incremental growth from a rigid anchor.

The first approach carries out a conformational search of the ligand and then rigidly docks this pregenerated ensemble of conformers into the receptor. In the second method, fragments are docked and later joined to reassemble the ligand. Thirdly, joined fragments are docked allowing hinge-bending structural movement (Sandak, Wolfson, and Nussinov, 1998). The last method, using incremental construction (Ewing and Kuntz, 1997; Kramer et al., 1999), starts with a rigidly docked anchor fragment followed by a depth-first or breadth-first conformational search inside the binding site (anchor-and-grow). Such a search can be heavily constrained by the geometry of the receptor. A concern, however, is that the ligand conformations during the search may have relatively high internal energies compared with the lowest energy conformers. Neither approach guarantees complete sampling of the ligand conformational space. Some implementations of the last two methods have significant overlap with *de novo* design strategies. They are also slower than the conformational pregeneration methods.

DE NOVO DESIGN. The central concept of *de novo* design is the construction of molecules that have not, necessarily, been synthesized previously. There are three basic classes of *de novo* design methods: fragment-positioning methods, fragment-connecting methods, and sequential-grow methods.

*Fragment Placement.* Instead of completely building up a new ligand, these methods determine favorable binding positions for single atoms or small fragments (see GRID [Goodford, 1985] or MCSS [Miranker and Karplus, 1991] for implementation). The underlying assumption is that a small number of well-placed fragments will account for significant binding interaction, while the rest of the molecule serves as a scaffold that links active fragments together. A clustering of the generated fragment orientations is usually performed.

The fragments are chosen to capture the basic molecular interactions such as hydrogen bonding (donor/acceptor) and hydrophobicity, and to optimally represent the functional groups and structural subunits present in a larger diverse library. The placement procedure uses either a molecular mechanics force field or a rule-based approach derived from an analysis of structural databases. Both the fragment connection method and the anchor-and-grow approach rely on a set of previously placed fragments as starting points.

*Connection Methods.* Site point connection methods (as implemented in CLIX [Lawrence and Davis, 1992] or LUDI [Bohm, 1994]) attempt to place small molecules in the binding pocket to match site points that provide favorable interactions. The site points are either derived directly by rules or by previous fragment placement, as described above.

Fragment connection methods (such as used by CAVEAT [Lauri and Bartlett, 1994], HOOK [Eisen, Wiley, and Karplus, 1994], or PRO_LEADS [Baxter et al., 1998]) retrieve scaffolds from a database in order to connect isolated fragments by overlaying corresponding bond vectors. A suitable linker provides a compatible geometry for connecting the critical fragments. The linker itself may be either rigid or flexible. In a final step, the linker has to be tested for overlap with the receptor.

The large number of available programs using connection strategies reflects the fact that molecular fragments are a standard tool of chemists.

*Sequential Grow.* The step-by-step construction of a putative ligand within a binding pocket is another useful approach for generating new potential leads or optimizing the functionality of a known inhibitor. First, a seed atom or fragment is placed in the binding site and then the new ligand is successively built up by bonding additional structural elements (see LEAPFROG (2001) or GROW [Moon and Howe, 1991]). Flexibility is introduced by conformational searching and minimization or by random orientations accepted by Monte Carlo criteria. The building procedure is guided by scoring the growing ligand at each step. The final results often depend on the selection of the initial position. Since the selection of each added unit is based on its binding score, smaller binding ligands are generated compared to fragment joining methods. A problem is the inherent liability of the growing procedure to combinatorial explosion. Another, less obvious, difficulty is the vastness of chemical space compared with the (relatively) small number of compounds that are feasible from the standpoint of synthetic chemistry (Clark, Murray, and Li, 1997).

An extension to the sequential-grow procedures is the connecting/disconnecting approach (used by DLD [Miranker and Karplus, 1995]). By accounting for bond breaking and bond formation, it allows a fine-tuning of the generated ligands beyond the supplied building units. Chemical mutations can create a more diverse ligand set that is optimized to fit the particular binding site.

All the *de novo* methods face a common set of problems. Since the overall shape of the generated compounds is imposed by the binding site, it is not guaranteed that the generated conformations of the ligands are energetically optimal. Point charges (used in force fields) are constantly changing during the building process. Also, as noted, the synthetic accessibility has to be addressed. Often, the first-generation molecules tend to have extensive interlinked ring systems, multiple chiral centers, and chemical instability due to the complexity of incorporating organic chemistry principles into the generation process. Linking methods have not yet been thoroughly explored.

At present, docking approaches are the methods of choice for new lead discovery due to the ability to rapidly screen readily available molecules from a database. *De novo* methods are suitable to suggest ligand modifications for lead optimization. The two strategies can be combined in various library design protocols (Murray et al., 1997; Waszkowycz, Perkin, and Sykes, 2001).

**Virtual Library Design.** The advent of combinatorial chemistry (Gallop et al., 1994; Gordon et al., 1994) has stimulated the development of computational screening of libraries of compounds that, themselves, might either be real or assembled on the computer. It is possible to make many more compounds computationally than can be synthesized or screened experimentally. Virtual screening and the use of library design principles are thus being used to prioritize experimental efforts to make the best use of chemical and screening resources.

Library design incorporates different strategies depending on the project at hand. One goal is to provide a *diverse* set of compounds. Various measures of chemical diversity are available (Blaney and Martin, 1997). Alternatively, one can design or identify molecules that test one or more pharmacophore hypotheses. These methods often involve an extensive conformational search for each ligand. A third approach is to use geometrical and chemical features of a known target site to generate putative scaffolds and substituents. Programs such as CombiDock and CAVEAT are used for these purposes (Lauri and Bartlett, 1994; Sun et al., 1998). These so-called virtual combinatorial libraries (Bohm and Stahl, 2000; Leach and Hann, 2000) achieve their tremendous diversity by attaching all possible combinations from a selected basis set of substituents to connecting sites on a central scaffold. The substituents are selected from readily available starting molecules subject to their ability of undergoing a molecular reaction connecting them to the scaffold.

The advantage of virtual screening over random high-throughput screening is the generation of directed libraries considering molecular properties that meet criteria required for drug-likeness (see ADME section below) and exhibit specificity for the selected target. The limiting aspect in designing virtual libraries is the synthetic accessibility of the products by combinatorial library synthesis techniques.

## FLOW CHARTS AND FUNDAMENTAL ISSUES

### Complex Basis for Free Energies of Binding

*Free Energies and "Energies".* The key to successful evaluation of ligand placement is an accurate (free) energy calculation. The calculation must be compared to an accurate experiment, ideally under the same conditions.

Experimentally the strength of the interactions between two species can be quantified by measuring the equilibrium association constant, $K_a$ or the dissociation constant, $K_d$ (Eq. 22.2). These thermodynamic parameters can be connected to chemical kinetics through Eq. 22.3,

$$K_a = k_{on}/k_{off} \tag{22.3}$$

where $k_{on}$ is the forward rate of the reaction between R and L, at equilibrium, and $k_{off}$ is the off rate or dissociation rate, at equilibrium.

Computationally, the goal is to estimate the binding free energy, $\Delta G_{bind}$, which is directly related to the experimentally measured $K_a$ as shown in Eqs. 22.2 and 22.4 (below). It is important to note that $\Delta G_{bind}$ is the *difference* in the free energy of the complex and the free energy of its components, the receptor and the ligand. Thus, we will need to do more than just compute the free energy of the complex in solution as is illustrated in Figure 22.3.

*Physical Principles of Complex Formation.* A useful way to consider the binding free energy is in terms of the changes in enthalpy and entropy on formation of the complex, as expressed in Eq. 22.4,

$$\Delta G_{bind} = \Delta H - T\Delta S \tag{22.4}$$

Changes in enthalpy arise from alterations in Van der Waals interactions and Coulombic interactions as the atoms of the complex replace atoms from the solvent on complex formation. The effects of changes in the internal energy of the receptor and ligand as their conformational preferences alter on complexation must also be considered. The

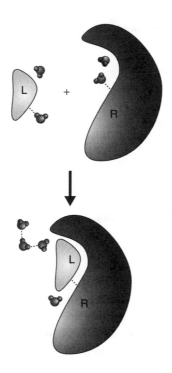

**Figure 22.3.** Receptor-ligand binding process.

change in entropy on binding reflects differences in translational and rotational degrees of freedom for the ligand, receptor, and solvent molecules (configurational entropy) and the loss of conformational and vibrational entropy on binding for the receptor and ligand.

We now describe the thermodynamics of the binding process in more detail. As is illustrated in Figure 22.3, before complex formation, the ligand and the active site of the receptor both make interactions with the solute molecules. The polar and charged groups on the surface of the unbound receptor and the unbound ligand form hydrogen bonds with the water molecules in the solvent, and all groups interact through Van der Waals interactions (see Table 22.1). On complex formation, the receptor binding site and the ligand become at least partially desolvated and the hydrogen bonds with the solvent are replaced with hydrogen bonds between the receptor and the ligand, as illustrated in Figure 22.3. It has been shown that burying a hydrogen bond donor or acceptor group (either neutral or charged) in the protein or complex interior without formation of a hydrogen bond can be detrimental to stability (Bogan and Thorn, 1998; Hendsch et al., 1996). Polar and nonpolar groups of the receptor and the ligand form Van der Waals interactions on complex formation, and charged groups interact strongly through Coulomb interactions. The release of the ordered water molecules around the ligand and in the receptor active site on complex formation and the resulting increase in entropy of these water molecules favors binding and is what underlies the hydrophobic effect. Entropy losses that occur on complex formation are partially due to the reduction in translational and rotational degrees of freedom of the ligand. The translational and rotational degrees of freedom of the complex are also slightly different from those of the receptor in isolation. Another source of entropy loss is due to the freezing out of the

T A B L E  22.1. Noncovalent Interactions and Their Distance Dependence

| Force | Distance |
|---|---|
| Short-range repulsion | $\sim 1/r^{12} - 1/r^{10}$ |
| Electrostatic interactions | |
|   Coulombic interactions: | |
|     charge-charge | $\sim 1/r$ |
|   Charge-dipole | $\sim 1/r^2$ |
|   Charge-induced dipole | $\sim 1/r^4$ |
| Nonelectrostatic interactions | |
|   Van der Waals interactions: | |
|     dipole-induced dipole | $\sim 1/r^6$ |
|     induced dipole-induced dipole   (London or dispersion force) | $\sim 1/r^6$ |
|     dipole-dipole | $\sim 1/r^6$ |
| Hydrogen bonds | |
|   $-D-H...A-$ | <2.0E-10  m(H-A) |

side chains on the surface of the ligand and the receptor on interaction with each other. Overall, there is a delicate balance between exchange of hydrogen bonds, establishment of Van der Waals and Coulomb interactions, entropy losses of the receptor and ligand, and gain in entropy of the solvent.

There are a number of ways to calculate directly the binding free energy, $\Delta G_{bind}$. A classical approach is through the partition function. Promising efforts are underway, although the computer time required to reach convergence is quite large (Head, Given, and Gilson, 1997). For special systems, the use of thermodynamic integration and/or free energy perturbation can be employed (Kollman, 1993). Alternatively, one can use Eq. 22.4 and calculate the enthalpic and the entropic changes on complex formation. The enthalpy can be calculated using a molecular mechanics force field (Table 22.2), and the entropy can be obtained using Boltzmann's law

$$S = -k \sum_{j} P_j \ln P_j \qquad (22.5)$$

where k is the Boltzmann constant and $P_j$ is defined as

$$P_j = \frac{e^{-E_j/kT}}{\sum_{j} e^{-E_j/kT}} \qquad (22.6)$$

These equations show that in order to estimate the entropy, all the states accessible to the system and the energy E of each state need to be known. Unfortunately, molecules of biological interest have a large number of degrees of freedom making it impossible to sample all the conformations using molecular dynamics (MD) or Monte Carlo (MC) simulations. Because of the large number of accessible states, the entropy of the solvent is also hard to obtain, which is one reason why a continuum solvent model that treats the degrees of freedom of the solvent implicitly is often used (see below).

TABLE 22.2. Functional Form of Different Scoring Functions

| Name of function | Function |
| --- | --- |
| Molecular Mechanics FF | |

$$E_{MM} = \sum_{bonds} K_r (r - r_{eq})^2 + \sum_{angles} K_\vartheta (\vartheta - \vartheta_{eq})^2$$

$$+ \sum_{dihedrals} \frac{V_n}{2} \left[ 1 + \cos(n\phi - \gamma) \right]$$

$$+ \sum_{i<j} \left[ \frac{A_{ij}}{R_{ij}^{12}} - \frac{B_{ij}}{R_{ij}^6} + \frac{q_i q_j}{\varepsilon R_{ij}} \right]$$

| | |
| --- | --- |
| LIE | |

$$\Delta G_{bind} = \beta \langle \Delta V_{Coul} \rangle + \alpha \langle \Delta V_{vdW} \rangle$$

| | |
| --- | --- |
| Extended LIE | |

$$\Delta G_{bind} = \beta \langle \Delta V_{Coul} \rangle + \alpha \langle \Delta V_{vdW} \rangle + \gamma \langle \Delta SASA \rangle$$

| | |
| --- | --- |
| LUDI | |

$$\Delta G_{bind} = \Delta G_0 + \Delta G_{rot} \times N_{rot}$$

$$+ \Delta G_{hb} \sum_{neutr.H\text{-}bonds} f(\Delta R, \Delta \alpha)$$

$$+ \Delta G_{io} \sum_{ionic\ int.} f(\Delta R, \Delta \alpha)$$

$$+ \Delta G_{aro} \sum_{aro\ int.} f(\Delta R, \Delta \alpha) + \Delta G_{lipo} \sum_{lipo} f^*(\Delta R)$$

| | |
| --- | --- |
| MM/PBSA or IS/ES | |

$$\overline{G} = \overline{E}_{MM} + \overline{G}_{PBSA} - T S_{solute}$$

The total entropy change can be split into four terms:

$$\Delta S_{total} = \Delta S_{trans} + \Delta S_{rot} + \Delta S_{conf} + \Delta S_{vibr} \tag{22.7}$$

where $\Delta S_{trans}$ and $\Delta S_{rot}$ are the translational and rotational entropy changes on complex formation, $\Delta S_{conf}$ is the change in conformational entropy, and $\Delta S_{vibr}$ is the change in vibrational entropy. The translational and rotational entropy changes can be calculated from statistical thermodynamics (Hill, 1986). The conformational entropy change can be estimated with an empirical scale developed by Pickett and Sternberg (1993), where the entropy loss of a particular amino acid side chain is related to the loss of accessible surface area (ASA) on complex formation. The most time-consuming computation is the change in vibrational entropy calculation (Case, 1994).

A major omission in the above analysis is direct consideration of the entropy changes in the solvent. This important issue is only roughly approximated through the use of the empirical buried surface area term. It is quite possible that the uncertainty in the entropy terms is currently limiting our ability to predict equilibrium constants.

***Experimental Conditions.*** We now turn to a discussion of the experimental conditions that can influence the measured binding constants. Both the entropy contribution and the enthalpy change on complex formation are strongly temperature dependent (Murphy, 1999). Other conditions that can influence the measured $K_a$ are pH, ionic strength, and water activity. Table 22.3 shows binding data from a well-studied protein–ligand interaction, namely, methotrexate binding to dihydrofolate reductase (DHFR), from different references in the literature. The biggest difference in measured $K_{eq}$ is the difference between the bacterial and eukaryotic enzymes. The factor of $10^4$-fold difference in $K_{eq}$ corresponds to ca. 5 kcal/mol in $\Delta G_{bind}$. The pH differences also influence the measured $K_{eq}$ significantly, which is not surprising since methotrexate contains several polar groups that will change ionization state. The presence or absence of NADPH can change the measured K about 100-fold.

The problem for the modeler is how to take these experimental variables into account. Rarely is a crystal structure available that has been obtained under similar conditions to those used to measure binding. The pH can be taken into account by determining the ionization state of the ionizable side chains of the amino acids and ligand groups. Modeling the temperature dependence of the enthalpic contributions to the binding free energy is impossible with most methods used in drug design, although molecular dynamics simulations can be performed at different temperatures.

Generally, the precise experimental conditions under which the binding constant was measured are ignored when binding free energies are being calculated. This neglect comes about for two reasons. First, through well-known compensation effects, the binding free energy is relatively less affected by changing some experimental conditions (Tame, 1999). Second, to carry out simulations as a function of temperature is laborious, and to model pH effects, except at the simplest level of fixing the charge states of ionizable groups, is difficult to do accurately (Borjesson and Hunenberger, 2001; Mertz and Pettitt, 1994).

***Solvation/Hydration Effects.*** Biomolecular interactions generally take place in aqueous environments; and, in order to calculate binding free energies, the influence of the solvent on the binding process needs to be included (Honig and Nicholls, 1995). There are two ways in which solvent influences biomolecular interaction processes. The first way is a short-range effect and arises through local solute–solvent interactions. The nonpolar solvation free energies, which account for transferring a nonpolar solute from the gas phase to the solvent, are often assumed to be proportional to the solvent accessible surface area (SASA). The second part of solvent effects is due to long-range electrostatic interactions. Water molecules are highly polarizable, that is, the orientation and electronic distribution of a water molecule readily fluctuates in the presence of the electrostatic field generated by the distributed charges on the protein and ligand. As a result, the Coulomb interactions between solvated charges is attenuated, or "screened," dramatically. This screening, represented most simply by a macroscopic

**TABLE 22.3. Experimental Binding Data of Methotrexate and DHFR**

| Enzyme source | pH | T(°C) | Ionic strength I | NADPH present | MTX conc | K μM | ΔG (calc. at T = 300 K) |
|---|---|---|---|---|---|---|---|
| L. Casei | 6.5 | 25 | 0.18 | yes | 0.6 | 0.2 | 9.20 |
| | | | | | 0.6 | 0.22 | 9.14 |
| | | | | | 0.3 | 0.28 | 9.00 |
| | | | | | 0.1 | 0.27 | 9.00 |
| | | | | | 0.1 | 0.27 | 9.00 |

*ref: Biochem. Pharmacol., 37, 541- ('88)*
*method: affinity chromatography*

| Enzyme source | pH | T(°C) | Ionic strength I | NADPH present | MTX conc | K μM | ΔG (calc. at T = 300 K) |
|---|---|---|---|---|---|---|---|
| E. coli | 6.0 | 25 | ? | yes | ? | 0.000023 | 14.60 |
| | 8.0 | 25 | ? | yes | ? | 0.00096 | 12.40 |
| | 9.5 | 25 | ? | yes | ? | 0.031 | 10.30 |

*ref: J. Med. Chem. 31, 129- ('88)*
*method: fluorescence titration*

| Enzyme source | pH | T(°C) | Ionic strength I | NADPH present | MTX conc | K $M^{-1}$ | ΔG (calc. at T = 300 K) |
|---|---|---|---|---|---|---|---|
| Bovine liver | 6.8 | 10 | ? | no | ? | 3.00E+08 | 11.60 |
| | | 10 | ? | yes | ? | 3.00E+10 | 14.40 |
| | | 37 | ? | no | ? | 1.60E+07 | 9.90 |
| | | 37 | ? | yes | ? | 3.00E+10 | 14.40 |

*ref: BBA 1040, 245- ('90)*
*method: calorimetry*

| Enzyme source | pH | T(°C) | Ionic strength I | NADPH present | MTX conc | K $M^{-1}$ | ΔG (calc. at T = 300 K) |
|---|---|---|---|---|---|---|---|
| Murine | 6.8 | 37 | ? | no | ? | 6.00E+07 | 10.70 |
| | 6.8 | 37 | ? | yes | ? | 3.50E+10 | 14.50 |
| E. coli | 6.8 | 37 | ? | no | ? | 3.70E+10 | 14.50 |
| | 6.8 | 37 | ? | yes | ? | 4.00E+10 | 14.60 |

*ref: BBA 1207, 74- ('94)*
*method: microcalorimetry*

dielectric constant, reduces electrostatic interactions in water by 80-fold from their vacuum values (see Solvent Representation below).

## Scoring Functions

Earlier, we considered how the binding constant, $K_a$, can be related to a theoretically obtainable binding free energy and what the fundamental forces are underlying the binding process. The only practical and rigorous way to calculate the binding free energy is the free energy perturbation (FEP) method (see Kollman, 1993 for a review), which calculates relative binding free energies by slowly mutating a ligand from one state to another. Although it is theoretically possible to calculate absolute binding free energies (Helms and Wade, 1998), the methodology is generally only used to calculate the relative binding free energy of two very similar ligands. The major hurdle is the amount of computer time required for the system to adjust to even minor mutations.

The need for methods that can be used in high-throughput settings dealing with hundreds of thousands of diverse compounds has led to the development of a wide variety of methods, which can be subdivided in four major approaches.

***First Principles Methods.*** These methods generally use a molecular mechanics force field (Table 22.2), which contains intramolecular forces between the atoms that are bonded to each other (bond, angle, and dihedral terms), and intermolecular forces, which describe the forces between nonbonded atoms (Van der Waals and Coulomb terms). Application of a molecular mechanics force field to the calculation of complex stability will only result in an energy prediction, since no entropic contributions are included in the force field. Further, most applications of a molecular mechanics force field do not explicitly consider the interaction of the unbound ligand and receptor with the solvent. Despite the omission of entropic terms, the evaluation of molecular mechanics force fields is time consuming. The original DOCK force field (Gschwend, Good, and Kuntz, 1996a) for example, only evaluates the intermolecular Van der Waals and Coulomb interaction energies.

Different groups have implemented methods to estimate the contribution of the solvent to the binding process to obtain a more complete energy evaluation. We discuss these implementations below in the Solvent Representation and Better Scoring Functions.

***Semiempirical Methods.*** The linear interaction energy (LIE) method (Aqvist, Medina, and Samuelsson, 1994) was developed to calculate absolute binding free energies without the need for sampling nonphysical transitional states such as those generated in FEP (Table 22.2). The basis of this method lies in the linear response approximation for electrostatic forces, which was shown to give $\beta = 1/2$ (Table 22.2). The coefficient $\alpha$ is then empirically derived from known binding data, making the method semiempirical (Aqvist, Medina, and Samuelsson, 1994). Later it was found that $\beta = 1/2$ is only valid for ligands containing charged groups, while for dipolar molecules $\beta$ is dependent on the system under consideration (Aqvist and Hansson, 1996). Though less time-consuming than FEP, it still requires either a MD or a MC simulation for both the inhibitor free in solution and in the complex to obtain the average electrostatic and Coulombic responses. Jorgenson's group has extended the LIE method and made it more empirical by adding a surface area (SA) term (Pierce and Jorgensen, 2001; Rizzo, Tirado-Rives, and Jorgensen, 2001); $\beta$ is derived empirically as well.

***Empirical Methods.*** There are numerous other empirically derived scoring functions, of which the LUDI scoring function (Bohm, 1994) is probably the most well known. Empirical scoring functions have been developed to be able to score ligands very rapidly. First, a number of structural descriptors, which represent the physical principles underlying complex formation, are selected (see Table 22.2). Next, weights are derived for each of the descriptors by regression methods, using standard statistical methods, with a training set. For the receptor-ligand complexes present in the training set, both structural data and experimental binding data are available. Although these scoring functions can still be interpreted in terms of the physical principles underlying complex formation, much of the atomic detail is lost.

While empirical scoring functions with sufficient parameters can fit training data as precisely as desired, it has proven difficult to derive functions that are general enough to describe the full range of organic diversity and to yield transferable parameters as new data become available. The central assumption that each occurrence of a certain interaction always gives the same contribution has been recently challenged (Tame, 1999). Another problem is the relatively sparse experimental binding data that is available for protein–ligand complexes of known structure.

***Knowledge-Based Potentials.*** To avoid deriving weights from experimental binding data, several groups have used so-called knowledge-based potentials based on interatomic contact preferences between atoms (Gohlke and Klebe, 2001; Muegge et al., 1999; Verkhivker et al., 1995; Wallqvist, Jernigan, and Covell, 1995). The potentials are obtained by statistical analysis of atom-pairing frequencies observed in crystal structures of protein–ligand complexes. This approach is closely related to a classical statistical physics method, which uses potentials of mean force (PMF) to account for all the physical forces in radially averaged representations.

## Parametrization of Molecular Mechanics Scoring Functions

First principles scoring functions require parameterization of the atomic charges and van der Waals radii. Also some choice must be made for solvent representation.

***Charge Representation.*** In general, atoms and molecules can be characterized by their charge distributions, which are most accurately calculated using *ab initio* techniques. The electrostatic force between atoms or molecules is given by the integration of Coulomb's law over the total charge distribution. Electrostatic forces are necessarily present even for interactions involving neutral atoms.

For convenience, the nuclear and electronic charge distributions are often approximated as point charges; that is, the charge is assumed to be located at the nuclear center and to occupy no volume. In this way, atoms, molecules, molecular fragments, even amino and nucleic acids are described by a configuration of point charges that, ideally, reproduce the electrostatic properties of the atom or molecule. The point charges are usually placed at atom centers and are called partial atomic charges. The electrostatic interaction between molecules then reduces to a simple pairwise sum of Coulomb interactions over all atomic charges. Point charge models are also convenient for simulations that require the calculation of forces, such as MD, because in this approximation electrostatic forces act directly on nuclei.

A number of methods have been developed to construct partial atomic charges or charge models to represent small organic molecules, amino acids and nucleic acids.

Two well-known charge models are those of Mulliken (1955) and Bader (1994). Bader developed the theory of atoms in molecules (AIM), which can be applied to partitioning electron density. Mulliken assumes that each basis function in an *ab initio* calculation is centered on a nucleus, then assigns the electrons in each orbital to the appropriate center. Electron density associated with an overlap integral is assigned half to each nuclear center. Mulliken charges are sometimes problematic in that the derived charges can depend on the choice of the basis set in the *ab initio* calculation. Bader partitions electron density based on the topology of the charge distribution. For example, the points of minimum charge density along a bond are defined as critical points. By using critical points and other topological features to define atomic regions and then numerically integrating the charge density within a region, a population is assigned to each atom. AIM charges are more reliable than Mulliken charges in the sense that they have been found to be invariant to the basis set used to derive them. Further, AIM charges reproduce electronegativity trends and give appropriate dipole moments.

Another approach is based on the idea that the electrostatic potential (ESP) produced by an atom or molecule determines how it will interact with the particles around it. The ESP-fitting scheme, in which charges are subjected to a least-squares fit to reproduce the ESP at a number of grid points outside the van der Waals surface of the molecule, has produced two widely used charge models for amino acids and nucleic acids, the restrained electrostatic potential (RESP) (Bayly et al., 1993) and CHELPG (Breneman and Wiberg, 1990) models. An inherent problem with ESP-fitting models is that charges on buried atoms can vary widely without significantly affecting the quality of the ESP fit. This numerical instability is partially alleviated by restraining the magnitude of unstable charges in the RESP model. RESP and CHELPG charges can be thought of as a set of empirical quantities designed to represent an electrostatic potential rather than as charges in the usual sense. The RESP charge model is used to parametrize the AMBER force field (Cornell et al., 1995).

The methods described above have focused on amino acids, nucleic acids, and a few ligands of particular interest. A general quantum mechanical treatment of the widely diversified molecules found, for example, in the Available Chemical Directory (ACD), would be very time consuming. Instead, empirical methods for rapidly assigning charges to molecules have been developed. One example is the Gasteiger and Marsili (1980) method, which derives charges based on the atom types and atom connectivities. Gasteiger–Marsili and other quick methods are often used to assign charges for ligands in docking studies because they are rapidly calculated, but they perform poorly in condensed-phase systems. Recently, a more accurate alternative to the quickest methods has been proposed. The AM1–BCC model (Jakalian et al., 2000) is reported to provide atomic charges of comparable quality to HF/6-31G* RESP charges. The AM1–BCC method captures the underlying features of the electron distribution and formal charge by taking Mulliken charges as a starting point. Bond charge corrections are then calculated, which are parameterized against the HF/6-31G* electrostatic potential of a training set of compounds containing the functional groups of interest.

An inherent source of error in using charge models is that they do not consider that charges are dependent on the conformation of the molecule. This issue is related to the more general problem of incorporating polarizability, the tendency of electrons to redistribute in response to surrounding electric fields. Polarizability has generally been ignored in the past, but considerable efforts are under way to develop nonadditive force fields that treat polarizability (Banks et al., 1999; Cieplak, Caldwell, and Kollman,

2001). Because polarizability is a fundamental concern in solvated systems, we return to a discussion of polarizability below in Solvent Representation.

***Van der Waals Radii.*** In addition to the charge representations, van der Waals parameters representing the effective size of atoms must also be determined. The Pauli exclusion principle mandates strong repulsion as two atoms approach each other. The interaction energy between a pair of atoms often passes through a minimum due to attractive dispersion forces and becomes zero at infinite distance (Fig. 22.4).

The attractive and repulsive interactions are combined to give the van der Waals potential function that appears in Table 22.1. The function, called the Lennard–Jones potential or 6–12 potential, contains adjustable parameters representing the separation of minimum energy, $r_m$, and the well depth, $\varepsilon$, shown in Figure 22.4. While powers other than 6–12 have been used in other force fields for the attractive and repulsive terms (Veith et al., 1998), the repulsive exponent is often taken to be twice the attractive exponent so that the repulsive term can be rapidly calculated as the square of attractive term.

Van der Waals interactions depend on atom type. For a polyatomic system containing N different atom types, a total of N(N-1)/2 parameters would be required. To eliminate the need for extensive parameterization when calculating Lennard–Jones functions, mixing rules have been devised. In the Lorentz–Berthelot scheme, for example, the separation of minimum energy is taken to be the arithmetic mean of the two species and the well depth is taken to be the geometric mean (Leach, 2001).

***Solvent Representation.*** The most accurate and physically realistic simulations of solvated ligand-receptor complexes (e.g., FEP or MD simulations) represent the solvent by an explicit collection of individual water molecules where each molecule is treated as a configuration of point charges. The simplest water models comprise three, four, or five charge sites with a rigid geometry. Some examples include TIP3P and SPC (Berendsen et al., 1981; Jorgensen et al., 1983). More sophisticated models include polarization effects.

Generally, explicit water models work well for charge densities found in protein–ligand complexes, although there are occasionally difficulties in simulations of highly charged nucleic acid–ligand complexes, most likely due to the neglect of polarization effects. Another limitation occurs in systems where quantum mechanical

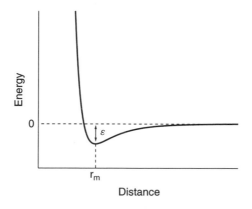

**Figure 22.4.** The Lennard-Jones (6–12) potential describing van der Waals interactions.

interactions of the water molecules and, say, a multivalent metal ion, would be poorly represented by point charge descriptions. The most profound drawback of explicit water models is the computational expense, due to the large number of pairwise interactions.

Continuum water models are based on the observation that, in most cases, water molecules do not play a direct role in ligand–complex binding. Thus, in general, the intramolecular solvent potential, the solute–solvent potential, and the solvent–solvent potential can be described by potentials of mean force that are independent of the solvent degrees of freedom (Roux and Simonson, 1999; Roux, Yu, and Karplus, 1990). Obvious exceptions include metalloproteases that contain a zinc atom in the active site, which sometimes coordinates an essential water molecule that plays a structural role in ligand binding. In this case, a reasonably accurate solvation model might incorporate one explicit water molecule, while using a continuum solvation model to treat the effects of all other water molecules.

A simple approximation to take into account the screening effects of the solvent is by using a distance-dependent dielectric constant, for which there is no real physical basis (Leach, 2001). A more sophisticated approach begins with scaling the permittivity constant. A solvated receptor-ligand system can be characterized by two regions with permittivity constants $\varepsilon_{water}$ and $\varepsilon_{complex}$. An appropriate value for $\varepsilon_{water}$ is simply the bulk dielectric constant for water, which varies from 88 at $0\,°C$ to 55 at $100\,°C$. The usual choice is 80. Amino acids and organic molecules are less polarizable than water because the motion of atoms in a protein, particularly in the core, are largely constrained, and the charge distributions of biological and organic molecules are far less susceptible to perturbation than water. Therefore, $\varepsilon_{complex}$ is taken to be between 1, the dielectric constant of vacuum, and 4, the dielectric constant of many nonpolar organic liquids.

Using this construct, the screening effects of solvated systems can be calculated. The most widely used approaches are the Generalized–Born model (Still et al., 1990) and the Poisson–Boltzmann method (Davis et al., 1991: Sitkoff, Sharp, and Honig, 1994; Warwicker and Watson, 1982; CAPRI, 2001; Grant, Pickup, and Nicholls, 2001), which reduces to the Poisson equation when there is no salt present in the solvent (see Chapter 21). Of importance, these methods include reaction field energies, which are the electrostatic contributions of induced surface charge at the solute–solvent boundary due to the high polarizability of water.

Methods have been developed to take into account the short-range, nonpolar, solvent effects. The nonpolar solvation free energy term is often assumed to be proportional to the SASA, and is intended to account for the free energy of cavity formation in the solvent and for the solute–solvent dispersion interactions. One of these methods uses atomic solvation parameters (ASP) for either amino acid side chains or chemical groups, which have been derived from experimental free energies of transfer from water to octanol. This free energy of transfer is related to the SASA of each amino acid side chain or chemical group (see Juffer et al., 1995 for a review). By using the SASA as a measure of the nonpolar solvation term, the assumption is made that the interior of the molecule does not contribute to the favorable dispersion interactions with the solvent. Recently, Pitera and Van Gunsteren (2001) showed that this assumption is not necessarily correct for large biomolecules.

A potential of mean force approach, based on residue–residue contacts, has also been used to account for nonpolar solvation effects (Miyazawa and Jernigan, 1985; Miyazawa and Jernigan, 1996). The described methods for estimating solvation effects are often used in combination with first-principles methods to get a more reliable energy prediction (Kuhn and Kollman, 2000; Wang et al., 2001).

## Combinatorial Nature of Calculation

*The Sampling Problem.* An important issue in all structure-based methods is the compromise between speed and exploration of the search space, either the configurational/conformational space in docking or the combinatorial chemical space in *de novo* design. At a minimum, for rigid docking, three degrees of translational freedom and three degrees of rotational freedom must be sampled. In addition, there are many conformational degrees of freedom of the ligand and the receptor that may be explored in flexible docking. A priori, bond lengths, bond angles, and torsional angles are internal degrees of freedom of a molecule. While bond lengths and bond angles are generally held constant, torsional angles can vary significantly. Currently, most methods can handle molecular flexibility of the ligand and limited, if any, flexibility of the receptor.

*Receptor Flexibility.* Ideally, the conformational degrees of freedom for both the ligand and the receptor are explored, since many complexes show induced fit. To incorporate the flexibility of the binding site or even large scale conformational rearrangements of the protein by domain motions, several approaches have been reported. The simplest models use soft scoring functions (Gschwend and Kuntz, 1996b), allowing for some overlap between the ligand and the receptor. This possible overlap permitted accounts for structural uncertainties or small side-chain movements of the receptor. Receptor flexibility can also be addressed by searching a tree of possible side-chain rotamer libraries for complementing a docked ligand. The conformational sampling of side-chain rotamers can by improved by incorporating minimization of the receptor side chains in the ligand-receptor complex (Schaffer and Verkhivker, 1998). Switching techniques (Apostolakis, Pluckthun, and Caflisch, 1998), first apply soft interaction potentials to smooth the energy surface. Then, the full interaction potential is gradually introduced. This soft potential facilitates the initial matching and the overcoming of conformational barriers. Alternatively, a composite receptor structure can be generated by the overlay of multiple crystal structures from different complexes of the same receptor (Knegtel, Kuntz, and Oshiro, 1997). This structure is used to capture the different configurations of an ensemble of receptor structures accessible on ligand binding. The most sophisticated methods generate an expanded ensemble of substates using MD or MC simulations of a docked complex (Lamb and Jorgensen, 1997). Additionally, by extracting different conformations from a MD trajectory, multiple docking targets or a composite structure can be generated even in the lack of appropriate crystal structures.

Loop flexibility and domain motions are accounted for by hinge-bending (Sandak, Nussiniov, and Wolfson, 1995) of larger rigid protein regions. Other models incorporate low frequency harmonic modes (Bahar et al., 1999) to describe large-scale protein movements.

While we often picture ligand-receptor complexes as single geometry snapshots, the real situation is much more complex. The ligand may adopt dozens or even hundreds of conformations; the receptor has many more thermally accessible geometries available. Thus, even a tightly bound complex would be more accurately thought of as an ensemble of microstates (Kumar et al., 2000). The more completely this ensemble can be described and incorporated into searching and scoring procedures, the more accurate the final results are likely to be.

## TESTS OF METHODS

### What Are the Proper Tests of These Theories?

Docking consists of two parts: searching and scoring. In this section we discuss how each has been evaluated for most programs developed, explain how to assess whether they are working correctly, give some docking results obtained using different programs, and outline what tests for these methods are most appropriate for applications such as high-throughput screening.

*Reassembling Complexes.* One traditional test of docking programs is the retrospective generation of the experimental structure of the complex as the best scoring geometry. Such a test examines both the sampling procedures and the scoring functions.

GEOMETRY ISSUES. To assess how close the predicted binding mode is to the experimental binding mode, most laboratories use the root-mean-square distance (Rmsd) between the calculated and observed complexes.

$$Rmsd = \sqrt{\frac{\sum_{i=1}^{N_{atoms}} d_i^2}{N_{atoms}}} \qquad (22.8)$$

where $d_i$ is the distance between the coordinates of atoms $i$ in the two structures when overlaid. An Rmsd for the top-scoring binding mode of $<2.0\text{Å}$ is generally considered to be acceptable for small ligands (Ewing et al., 2001; Vieth et al., 1998; Rarey, Kramer, and Lengauer, 1999). This measure can be readily generalized to Rmsd for the predicted and observed geometry of the receptor as well if the receptor is treated as flexible.

The reassembly of a specific ligand-receptor complex is much easier than the general problem of how conformations of ligand and receptor adapt to each other. Several algorithms have been proposed for incorporating flexible receptor information (Apostolakis, Pluckthun and Caflisch, 1998; Jones, Willett, and Glen, 1995; Leach, 1994).

The Rmsd may not be the best metric for systems that exhibit conformational changes. Reporting the differences between observed and calculated structures in terms of a translation of center-of-mass, a rigid rotation, and internal dihedral angle changes has the advantage of providing much more information about the nature of the differences, especially the generation of higher energy conformational states of ligand and receptor (Lee and Levitt, 1997).

A deeper question is what the real purpose of the calculation is. That is, the scientific issue in drug design is rarely, if ever, the reproduction of a known structure. Rather, one is normally asked how to modify a ligand or receptor, or how to generate a totally new motif. Thus, one often wants to know not only the best scoring mode, but how robust the geometry is to changes in ligand or changes (mutations) in the receptor. It is often helpful to have a list of alternative binding modes for these purposes. One of the encouraging aspects of the earliest applications of DOCK to heme binding in globins was the identification of alternate heme geometries that were, in fact, experimentally accessible (Kuntz et al., 1982). An interesting start to examining the robustness of the predicted geometry is the selectivity measure introduced by the Brooks group at Scripps (Vieth et al., 1998) in which Z-scores of the lowest-energy docked and misdocked structures are compared. The bigger the energy gap between the docked and misdocked structure, the more selective the energy function (Vieth et al., 1998) and the more robust the prediction of the docking mode.

DOCKING SUCCESSES AND FAILURES. Several docking programs have been tested on numerous complexes; the Rmsd compared to the crystal structure is often reported both for the best-scoring binding mode, as well as for the closest docked configuration. The top-scoring configuration is found to be within 2Å in 45% (Ewing et al., 2001; Diller and Mertz, 2001), 65% (Jones et al., 1997), and 73% (Rarey, Wefing, and Lengauer, 1996) of the test cases for the different programs. Both FlexX and DOCK find configurations within 2Å for all test cases (19 and 15, respectively) and Diller and Merz achieve this for 85% of their test cases (103 in total). These results show that retrieving the correct binding mode can be accomplished in most cases, but assigning the lowest energy score to the correct binding is more difficult.

RANK ORDERING OF ENERGIES. An important capability of a docking program is to rank order different ligands that (potentially) bind to the same receptor. Rank ordering is used in both library screening and lead optimization. While any scoring protocol should limit the number of false positives and minimize the false negatives, these applications place different requirements on the scoring function. In library screening, one expects a wide diversity of compounds. The scoring function must deal with large differences in net charge, hydrogen bonding potential, and hydrophobic interactions. Small errors in any component can weight an interaction disproportionately, swapping out the true positives. The challenge in lead optimization is to rank correctly ligands that are very similar to each other. Muegge's group at Bayer has performed an interesting study in which they compared three different scoring functions in ranking 61 inhibitors against stromelysin (Ha et al., 2000). They found that the in-house developed PMF scoring function performed not only better in ranking the compounds than both the DOCK score and the LUDI score, but also that the PMF scoring function did not give rise to any false negatives for this target (Ha et al., 2000). Bissantz, Folkers, and Rognan (2000) show that the success of a docking program and a scoring function is highly dependent on the target. Further, a number of scoring functions have been recently compared with a new method that attempts to bridge the gulf between high-accuracy, low-throughput free energy simulations and empirical approaches. This method, OWFEG (one-window free energy grid), generates from a single MD simulation a grid surrounding a molecule of interest that represents the free energy for insertion of a probe group at any point on that grid, and it was shown to be superior to each of the scoring functions to which it was compared (Pearlman, 1999; Pearlman, and Charifson, 2001).

A relatively straightforward way to limit the number of false positives is by using consensus scoring. In consensus scoring several different scoring functions are applied and the intersection of all low-scoring hits is taken, which improves the hit rates significantly (Charifson et al., 1999).

At present, there does not appear to be one scoring function that is significantly better over a wide range of targets (Bissantz, Folkers, and Rognan, 2000). There is a true dichotomy in opinions on the merits of scoring formulas based on molecular theory versus those developed on empirical, statistical grounds. While this difference will not be easily resolved in the near future, what is clear is that to be successful in library screening and most optimization tasks, a free energy-based approach is required.

FREE ENERGIES OF BINDING. As discussed earlier, to calculate or predict $\Delta G_{bind}$ both the enthalpic and entropic contributions need to be taken into account.

The DOCK scoring function (Gschwend and Kuntz, 1996), which is a simplification of the AMBER force field (Cornell et al., 1995) calculates only enthalpic contributions to binding, which means that the DOCK score would not be expected to correlate strongly with experimental binding data. The same can be said about knowledge-based potentials, which also do not contain entropy information directly. The weights in empirical scoring functions can be optimized to reproduce experimental binding constants. Depending on the generality of the training set and the validity of the underlying assumptions, these functions should be able to predict binding free energies.

The LUDI (Bohm, 1994) scoring function has been implemented in FlexX Rarey et al., 1996b and predicted binding free energies have been compared to experimental values. It was found that for some of the complexes, the predicted binding free energy deviated substantially from the experimental value, indicating that certain energetic contributions are not modeled (correctly) by the scoring function (Rarey et al., 1996b).

***Virtual Screening.*** To evaluate virtual screening, two metrics are used: the hit rate—that is, the recovery of true positives, and the enrichment factor—the number of true positives divided by the number of true positives plus false positives. A major difficulty in most situations is to set up a proper control experiment to measure the amount of productive information used in the library design process. In an experiment from our laboratory, in collaboration with Jonathan Ellman, structure-based and diversity-based designed libraries against an aspartyl protease, Cathepsin D, were synthesized and assayed. The experiments showed that the structure-based method yields higher hit rates (three- to sevenfold) and more potent affinities (three- to fourfold) as well (Kick et al., 1997).

The enrichment factor ($EF$) is defined as

$$EF = \frac{a/n}{A/N} \tag{22.9}$$

where $a$ is the number of active compounds in the top $n$ compounds, $N$ is the number of compounds in the library, and $A$ is the number of active compounds in the library. The $EF$ can be used only if the activities of all the compounds in the library are known.

Knegtel and Wagener (1999) used this metric to assess the success of virtual screening with two different targets. They screened more than 1000 compounds for each target using two scoring functions and different amounts of conformational sampling during docking. Their results suggest that less conformational sampling yields higher enrichment, because scoring for both actives and inactives improves with more sampling. This is an indication that the scoring function cannot distinguish between true positives and chemically similar false positives. Diller and Merz (2001) found maximal enrichments of threefold using different scoring functions and statistical analysis in examining a set of known activities against a particular target.

***Common Docking Failures.*** Because most of the docking protocols are deterministic, it is possible to ask why the programs do not return the experimental test result. If we exclude experimental errors, the two major computational failure modes are incomplete searching and inaccurate scoring functions. A search-algorithm failure, sometimes called a "soft failure" (Verkhivker et al., 2000), occurs when the search algorithm is unable to find the native binding mode. A search-algorithm failure can be identified by comparing the energy of the minimized experimental structure with the energy of the lowest-energy binding mode that was found. If the energy of the minimized crystal structure is lower than the energy of the most favorable binding mode, the search algorithm did not explore the binding energy landscape thoroughly enough (Ewing et al., 2001; Verkhivker et al., 2000). The most likely cause of a search failure is the rugged, multiminima character of the binding energy landscapes, arising from the geometric complexity of docking and the short-range and (in some cases) discontinuous nature of the scoring functions.

A scoring-function failure, or hard failure (Verkhivker et al., 2000), is indicated when the minimized crystal structure has a higher energy than the predicted lowest-energy binding mode (Ewing et al., 2001; Verkhivker et al., 2000). It has generally proven difficult to identify which term(s) of the scoring function are responsible for hard failures. One approach to understand better the nature of hard failures is to use a simple energy function to generate potential binding modes, followed by rescoring of the binding modes with a more elaborate energy function. The simple energy function makes the binding energy landscape less rugged, allowing for a more exhaustive search, while the second scoring function is a better representation of the physical principles of molecular recognition (Verkhivker et al., 2000).

*Ligand Design Successes.* It is no longer possible to provide a complete list of successful efforts to design ligands. A reasonably comprehensive review through 1996 is available (Charifson and Kuntz, 1997). Suffice it to say that dozens of clinical candidates (Charifson and Kuntz, 1997), a variety of mutant proteins (Marshall and Mayo, 2001), and several libraries of ligands (Kick et al., 1997) have been reported. While none of the methods used in these studies can be viewed as universally applicable, the number of structure-based results now in the literature can certainly be taken as an indication of the practical utility of molecular design.

## FUTURE

In this section we give examples of how the current force field-based scoring functions can be improved and illustrate recent developments in library design. Also, we address some further applications of theoretical methods, which will become more important in the future due to our rapidly increasing knowledge of protein structures and the human genome.

### Better Scoring Functions

The most obvious approximation made in most force field-based scoring functions is the neglect of solvent polarization and screening effects and the hydrophobic effect. Physics-based methods that model solvent effects are the Generalized-Born (GB) method and the Poisson-Boltzmann (PB) equation, both of which calculate the screened Coulombic interaction directly (by different methods). Further, both use a surface area term to estimate the hydrophobic effect. The GB model has recently been implemented in DOCK and has been used during the postprocessing of the top-ranked compounds (Zou, Sun, and Kuntz, 1999). ICM (Totrov and Abagyan, 1997) uses both a modified image electrostatic approximation (MIMEL) and a boundary element solution of the Poisson equation in a two-step energy calculation. PB calculations have been found to be sensitive to small changes in atomic positions (Schapira, Totrov, and Abagyan, 1999), which suggests a need for sampling and averaging the calculated energies.

Both the Hermans group and the Kollman group have recently published papers describing free energy calculations using molecular mechanics in combination with continuum solvent models (ES/IS or MM/PBSA) (Kollman et al., 2000; Vorobjev, Almagro, and Hermans, 1998). These methods both use an explicit water molecular dynamics trajectory of the protein or complex of interest. This trajectory is subsequently postprocessed with a continuum solvent model. The free energy is then calculated by estimating the terms contributing to the free energy (Table 22.2). This method thus takes into account enthalpic contributions through the MM force field terms, screening of the Coulomb interactions due to the solvent by solving the PB equation, and the hydrophobic effect using an SA term. Finally, the entropy of the solute is taken into account by performing normal mode or quasi-harmonic analysis. Averaging over the trajectory is necessary because the fluctuations in particularly the Coulomb term and the solvation term from PB can be quite large. This methodology has been successfully used in a wide variety of interesting applications, underscoring the potential and general applicability of first-principles approaches. Unfortunately, the amount of computer time required to obtain an MD trajectory prohibits use of this method in early-stage, structure-based drug design, when hundreds of thousands of compounds are screened. It is certainly feasible, however, to use this methodology when optimizing lead compounds or to investigate different binding modes found by docking more thoroughly.

David, Lou, and Gilson (2000) compared GB, PB, and the distance-dependent dielectric approach. They suggest that the PB equation is the method of choice, time permitting, but under time constraints GB is certainly preferable over a distance-dependent dielectric constant approach.

## Better Database Organization

The structural genomics and functional genomics projects will significantly increase the availability of structures, the knowledge of their functions, and the knowledge of functions of genes for which the structure might not yet have been solved. Homology models can be used to generate structures for proteins for which a crystal or NMR structure is not yet available. These developments will lead to a rapid increase in available drug targets with which computational drug design methods will be used to screen databases and design targeted libraries. This overflow of information will require the intelligent use of resources.

***Organization of Receptors/Targets.*** The availability of structures of both the drug target and its closest homologues will allow screening against several structures in order to increase the selectivity of the hits. Only one study has been published so far showing the promise of screening against multiple proteins at a time (Lamb et al., 2001), which was discussed in the Library Design section.

***Organization of Ligands.*** The number of available compounds in both real libraries and virtual libraries and the number of available libraries is increasing rapidly. To be able to screen as many compounds as possible in a short period of time, we are forced to screen the databases differently.

Shoichet's group has developed an approach to organize ligand databases into families (Su et al., 2001). They identify the largest rigid fragment in each compound and group compounds sharing common fragments into families, followed by overlaying the rigid fragments. Their docking strategy starts with generating poses for the rigid fragment of each family, followed by scoring each molecule and orientation separately. The best scoring compound of each family is the representative molecule and is included in the hit list. Their results show that docking in families increases the number of known ligands and analogs between 45% and 500%, depending on the target (three different targets were used), compared to docking and ranking each molecule independently. The diversity of the hits was also significantly higher (between 20–300%) (Su et al., 2001). They also suggest that family-based docking can reduce dependence on the accuracy of the scoring function since related hits are presented together. One can subsequently choose to just screen the representative molecule of each family or multiple ligands in the same family (Su et al., 2001).

## Macromolecular Docking

Reproducing protein–protein complexes by docking methods has been shown to be difficult when separately crystallized proteins are used, due to the flexibility of surface side-chains (Betts and Sternberg, 1999) and the difficulty of taking flexibility into account. But the expected impact of the current structural and functional genomic efforts will lead to an increased importance of macromolecular docking for evaluating protein–protein interactions to investigate cellular pathways.

In order to test prediction methods and to assess their validity, the Critical Assessment of Predicted Interactions (CAPRI) experiment was initiated (CAPRI, 2001). It aims for a comparative evaluation of protein–protein docking algorithms in the field of structure prediction. Organized as a blind test, macromolecular docking is used to predict the binding mode of two proteins based on their 3D structure in the unbound state.

## Landscape Models

Molecular recognition and protein folding share several common aspects, such as the existence of a thermodynamically stable native structure, a large number of accessible conformational states and the complex nature of interactions. It has been found that a critical factor in determining the success of predicting the structure of a binding complex is the shape of the binding energy landscape, which can be compared to the folding funnel of the protein folding problem (Betts and Sternberg, 1999).

Using simplified, short-range interaction functions in docking experiments, energy funnels near the conformation of the native binding site could be revealed (Zhang, Chen, and DeLisi, 1999). The energy decreases as the degree of similarity between the native and the docked near native structures increases. Funnels dominated by short-range interactions can fine tune long-range electrostatic steering forces to determine protein association rates. Landscape models exploring the whole binding landscape will allow for a better understanding of the mechanism of molecular recognition in ligand-receptor interactions.

## ADME

An optimized lead structure binding to a target with high affinity still has a long way to undergo in order to become an effective drug. The compound has to pass animal and clinical trials during which factors such as toxicity, bioavailability, and resistance are considered. In order to preclude a later failure of a possible drug candidate in this time- and cost-intensive process, recent efforts try to incorporate some of these factors in the initial drug-design process.

In addition to pharmacodynamics issues (biological effects of the drug), the pharmacokinetic profile constituted by the so-called ADME (absorption, distribution, metabolism and excretion) properties of the compound have to be considered (van de Waterbeemd et al., 2001). This incorporates substructural filters or the evaluation on molecular properties accounting for drug-likeness. The employed methods include rather simple models such as the Lipinski's rule of five for oral bioavailability (Lipinski et al., 1997) or, alternatively, rely on sophisticated filters intended to capture a more detailed ADME behavior by calculating relevant properties from the two- or three-dimensional molecular structure using, for instance, neural networks Sadowski and Kubinyi (1998).

## Pharmacogenetics

Pharmacogenetics can be defined as the study of differences in drug response of individuals as a result of differences in their genetic makeup. One of the most well-studied examples of drug response related to genetic variations is due to polymorphisms occurring in the drug-metabolizing enzymes cytochrome P450s (CYPs).

The individual genotype can influence the daily dose required for certain drugs and the kind and severity of adverse drug reactions, or determine whether there will be a drug response at all. Knowledge of how different genotypes influence drug response will change the future of drug development at all stages and the practice of medicine as well. More knowledge of genotype-related drug responses can lead to more effective drugs with fewer side effects.

## REFERENCES

Al-Lazikani B, Jung J, Xiang Z, Honig B (2001): Protein structure prediction. *Curr Opin Chem Biol* 5(1):51–6.

Apostolakis J, Pluckthun A, Caflisch A (1998): Docking small ligands in flexible binding sites. *J Comp Chem* 19(1):21–37.

Aqvist J, Medina C, Samuelsson J-E (1994): A new method for predicting binding affinity in computer-aided drug design. *Protein Eng* 7(3):385–91.

Aqvist J, T, Hansson (1996): On the validity of electrostatic linear response in polar solvents. *J Phys Chem A* 100:9512–21.

Atkins PW (1995): *Physical Chemistry*, 6th ed. New York: W H Freeman.

Bader RFW (1994): *Atoms in Molecules: A Quantum Theory*. Oxford: Clarendon Press.

Bahar I, Erman B, Jernigan RL, Atilgan AR, Covell DG (1999): Collective motions in HIV-1 reverse transcriptase: examination of flexibility and enzyme function. *J Mol Biol* 285:1023–37.

Banks JL, Kaminski GA, Zhou RH, Mainz DT, Berne BJ, Friesner RA (1999): Parametrizing a polarizable force field from ab initio data. I. The fluctuating point charge model. *J Chem Phys* 110(2):741–54.

Baxter CA, Murray CW, Waszkowycz B, Young SS (1998): Flexible docking using tabu search and an empirical estimate of binding affinity. *Proteins* 33:367–82.

Bayly CI, Cieplak P, Cornell WD, Kollman PA (1993): A well-behaved electrostatic potential based method using charge restraints for deriving atomic charges: the RESPmodel. *J Phys Chem* 97:10269–80.

Berendsen HJC, Postma JPM, van Gunsteren WF, Hermans J (1981): Interaction models for water in relation to protein hydration. In: Pullman B, editor. *Intermolecular Forces*, Reidel: Dordrecht: pp 331–42.

Betts MJ, Sternberg MJE (1999): An analysis of conformational changes on protein–protein association: implications for predictive docking. *Protein Eng* 12(4):271–83.

Bissantz C, Folkers G, Rognan D (2000): Protein-based virtual screening of chemical databases. 1. Evaluation of different docking/scoring combinations. *J Med Chem* 43:4759–67.

Blaney JM, Martin EJ (1997): Computational approaches for combinatorial library design and molecular diversity analysis. *Curr Opin Chem Biol* 1:54–9.

Bogan AA, Thorn KS (1998): Anatomy of hot spots in protein interfaces. *J Mol Biol* 280(1):1–9.

Bohacek RS, McMartin C (1994): Multiple highly diverse structures complementary to enzyme binding sites: results of extensive application of de novo design method incorporating combinatorial grow. *J Am Chem Soc* 116:5560–71.

Bohm H-J (1994): The development of a simple empirical scoring function to estimate the binding constant for a protein-ligand complex of known three-dimensional structure. *J Comput-Aided Mol Des* 8:243–56.

Bohm H-J, Stahl M (2000): Structure-based library design: molecular modelling merges with combinatorial chemistry. *Curr Opin Chem Biol* 4:283–6.

Borjesson U, Hunenberger PH (2001): Explicit-solvent molecular dynamics simulation at constant pH: methodology and application to small amines. *J Chem Phys* 114(22):9706–19.

Breneman CM, Wiberg KB (1990): CHELPG. *J Comp Chem* 11:361–73.

CAPRI: Critical Assessment of Prediction of Interactions, http://capri.ebi.ac.uk.2001.

Case DA (1994): Normal mode analysis of protein dynamics. *Curr Opin Struc Biol* 4:285–90.

Charifson PS, Kuntz ID (1997): Recent successes and continuing limitations in computer-aided drug design. In: Charifson PS, editor. *Practical Application of Computer-Aided Drug Design.* New York: Marcel Dekker, pp 1–37.

Charifson PS, Corkery JJ, Murcko MA, Walters WP (1999): Consensus scoring: a method for obtaining improved hit rates from docking databases of three-dimensional structures into proteins. *J Med Chem* 42:5100–9. [Comparison of different scoring functions.]

Cieplak P, Caldwell JW, Kollman PA (2001): Molecular mechanics models for organic and biological systems going beyond the atom centered two body additive approximation: aqueous solution free energies of methanol and N-methyl acetamide, nucleic acid base, and amide hydrogen bonding and chloroform/water partition coefficients of the nucleic acid bases. *J Comp Chem* 22(10):1048–57.

Clark DE, Murray CW, Li J (1997): Current issues in de novo molecular design. In: Lipkowitz KB, Boyd DB, editor. *Reviews in Computational Chemistry*. New York: Wiley-VCH, pp 66–125.

Cornell WD, Cieplak P, Bayly CI, Gould IR, Merz KM, Jr, Ferguson DM, Spellmeyer DC, Fox T, Caldwell JW, Kollman PA (1995): A second generation force field for the simulation of proteins, nucleic acids, and organic molecules. *J Am Chem Soc* 117:5179–97.

David L, Luo R, Gilson MK (2000): Comparison of generalized Born and Poisson models: energetics and dynamics of HIV protease. *J Comp Chem* 21:295–309.

David L, Luo R, Gilson MK (2001): Ligand-receptor docking with the Mining Minima optimizer. *J Comput-Aided Mol Des* 15:157–71.

Davis ME, Madura JD, Luty BA, McCammon JA (1991): Electrostatics and diffusion of molecules in solution: simulations with the University of Houston Brownian Dynamics program. *Comput Physics Commun* 62:187–97.

Diller DJ, Merz KM, Jr, (2001): High throughput docking for library design and library prioritization. *Proteins* 43:113–24.

Eisen MB, Wiley DC, Karplus M (1994): HOOK: a program for finding novel molecular architectures that satisfy the chemical and steric requirements of a macromolecule binding site. *Proteins* 19:199–221.

Ewing TJA, Kuntz ID (1997): Critical evaluation of search algorithms for automated molecular docking and database screening. *J Comp Chem* 18:1175–89.

Ewing TJA, Makino S, Skillman AG, Kuntz ID (2001): DOCK 4.0: search strategies for automated molecular docking of flexible molecule databases. *J Comput-Aided Mol Des* 15:411–28.

Gabb HA, Jackson RM, Sternberg MJE (1997): Modelling protein docking using shape complementarity, electrostatics, and biochemical information. *J Mol Biol* 272:106–20.

Gallop MA, Barrett RW, Dower WJ, Fodor SPA, Gordon EM (1994): Applications of combinatorial technologies to drug discovery. 1. Background and peptide combinatorial libraries. *J Med Chem* 39(9):1233–51.

Gasteiger J, Marsili M (1980): Iterative partial equalization of orbital electronegativity—rapid access to atomic charges. *Tetrahedron* 36:3219–88.

Gohlke H, Klebe G (2001): Statistical potentials and scoring functions applied to protein-ligand binding. *Curr Opin Struc Biol* 11:231–5.

Goodford PJ (1985): A computational procedure for determining energetically favorable binding sites on biologically important macromolecules. *J Med Chem* 28:849–57.

Gordon EM, Barrett RW, Dower WJ, Fodor SPA, Gallop MA (1994): Applications of combinatorial technologies to drug discovery. 2. Combinatorial organic synthesis, library screening strategies, and future directions. *J Med Chem* 37(10):1385–1401.

Grant AJ, Pickup BT, Nicholls A (2001): A smooth permittivity function for Poisson-Boltzmann solvation models. *J Comp Chem* 22(6):608–40.

Gschwend DA, Good AC, Kuntz ID (1996a): Molecular docking towards drug discovery. *J Mol Recognit* 9:175–86.

Gschwend DA, Kuntz ID (1996b): Orientational sampling and rigid-body minimization in molecular docking revisited: on-the-fly optimization and degeneracy removal. *J Comput-Aided Mol Des* 10:123–32.

Guener O, editor. (2000): *Pharmacophore Perception, Development, and Use in Drug Design*. La Jolla, CA: International University Line USA.

Ha S, Andreani R, Robbins A, Muegge I (2000): Evaluation of docking/scoring approaches: a comparative study based on MMP3 inhibitors. *J Comput-Aided Mol Des* 14:435–48.

Hammet LP (1970): *Physical Organic Chemistry*, 2nd ed. New York: McGraw-Hill.

Hansch C, Leo A, Hoekman DH (1995): *Exploring QSAR*. ACS Professional Reference Book, Vol. I and II. New York: Oxford University Press USA.

Head MS, Given JA, Gilson MK (1997): "Mining minima": direct computation of conformational free energies. *J Phys Chem A* 101:1609–18.

Helms V, Wade RC (1998): Computational alchemy to calculate absolute protein-ligand binding free energy. *J Am Chem Soc* 120:2710–3.

Hendsch ZS, Jonsson T, Sauer RT, Tidor B (1996): Protein stabilization by removal of unsatisfied polar groups: computational approaches and experimental tests. *Biochemistry* 35(24):7621–5.

Hill TL (1986): *An Introduction to Statistical Thermodynamics*. Mineola, NY: Dover Publications.

Honig B, Nicholls A (1995): Classical electrostatics in biology and chemistry. *Science* 268:1144–9. [Review of electrostatic contributions to solvation and binding free energies.]

Jakalian A, Bush BL, Jack DB, Bayly CI (2000): Fast, efficient generation of high-quality atomic charges. AM1-BCC model: I. Method. *J Comp Chem* 21(2):132–46.

Jones G, Willett P, Glen RC (1995): Molecular recognition of receptor sites using a genetic algorithm with a description of desolvation. *J Mol Biol* 245:43–53.

Jones G, Willett P, Glen RC, Leach AR, Taylor R (1997): Development and validation of a genetic algorithm for flexible docking. *J Mol Biol* 267:727–48.

Jorgensen WL, Chandrasekhar J, Madura JD, Impey RW (1983): Comparison of simple potential functions for simulating liquid water. *J Chem Phys* 79:926–35.

Juffer AH, Eisenhaber F, Hubbard SJ, Walther D, Argos P (1995): Comparison of atomic solvation parametric sets: applicability and limitations in protein folding and binding. *Protein Sci* 4:2499–509.

Katchalski-Katzir E, Shariv I, Eisenstein M, Friesem AA, Aflalo C, Vakser IA (1992): Molecular surface recognition: determination of geometric fit between proteins and their ligands by correlation techniques. *Proc Natl Acad Sci USA* 89:2195–9.

Kick EK, Roe DC, Skillman AG Jr, Liu G, TJA, Ewing Sun Y, Kuntz ID, Ellman JA (1997): Structure-based design and combinatorial chemistry yield low nanomolar inhibitors of cathepsin D. *Chem Biol* 4:297–307.

Knegtel RMA, Kuntz ID, Oshiro CM (1997): Molecular docking to ensembles of protein structures. *J Mol Biol* 266(2):424–40.

Knegtel RMA, Wagener M (1999): Efficacy and selectivity in flexible database docking. *Proteins* 37:334–45.

Kollman PA (1993): Free energy calculations: applications to chemical and biochemical phenomena. *Chem Rev* 93:2395–417.

Kollman PA, Massova I, Reyes C, Kuhn B, Huo S, Chong L, Lee M, Lee T, Duan Y, Wang W, Donini O, Cieplak P, Srinivasan J, Case DA, Cheatham TE, III (2000): Calculating structures and free energies of complex molecules: combining molecular mechanics and continuum models. *Acc Chem Res* 33(12):889–97.

Kramer B, Metz G, Rarey M, Lengauer T (1999): Ligand docking and screening with FlexX. *Med Chem Res* 9(7–8):463–78.

Kuhn B, Kollman PA (2000): A ligand that is predicted to bind better to avidin than biotin: insights from computational fluorine scanning. *J Am Chem Soc* 122(16):3909–16.

Kumar S, Ma B, Tsai C, Sinha N, Nussinov R (2000): Folding and binding cascades: dynamic landscapes and population shifts. *Protein Sci* 9:10–9.

Kuntz ID, Blaney JM, Oatley SJ, Langridge R, Ferrin TE (1982): A geometric approach to macromolecular-ligand interactions. *J Mol Biol* 161:269–88.

Lamb MD, Jorgensen WL (1997): Computational approaches to molecular recognition. *Curr Opin Chem Biol* 1:449–57.

Lamb ML, Burdick KW, Toba S, Young MM, Skillman AG Jr, Zou X, Arnold JR, Kuntz ID (2001): Design, docking, and evaluation of multiple libraries against multiple targets. *Proteins* 42:296–318.

Lauri G, Bartlett PA (1994): CAVEAT—a program to facilitate the design of organic molecules. *J Comput-Aided Mol Des* 8(1):51–66.

Lawrence MC, Davis PC (1992): CLIX: A search algorithm for finding novel ligands capable of binding proteins of known three-dimensional structure. *Proteins* 12:31–41.

Leach AR (1994): Ligand docking to proteins with discrete side-chain flexibility. *J Mol Biol* 235:345–56.

Leach AR (1997): A survey of methods for searching the conformational space of small and medium-sized molecules. In: Lipkowitz KB, Boyd DB, editors. *Reviews in Computational Chemistry*. New York: Wiley-VCH, pp 1–55.

Leach AR (2001): *Molecular Modelling: Principles and Applications*, 2nd ed. Englewood Cliffs, NJ: Prentice Hall. [Comprehensive text on major techniques of molecular modeling and computational chemistry.]

Leach AR, Hann MM (2000): The in silico world of virtual libraries. *Drug Discovery Today* 5(8):326–36.

Lee CE, Levitt M (1997): Packing as a structural basis of protein stability: understanding mutant properties from wildtype structure. *Pacific Symp Biocomput* pp 245–255.

LEAPFROG, Sybyl Receptor-based Design v6.8, pp 1–159, Tripos Inc., 1699 South Hanley Road, St. Louis, MO 63144.

Lipinski CA, Lombardo F, Dominy BW, Feeney PJ (1997): Experimental and computational approaches to estimate solubility and permeability in drug discovery and development settings. *Adv Drug Del Rev* 23:3–25.

Liu M, Wang SM (1999): MCDOCK: A Monte Carlo simulation approach to the molecular docking problem. *J Comput-Aided Mol Des* 13(5):435–51.

Mandell JG, Roberts VA, Pique ME, Kotlovyi V, Mitchell JC, Nelson E, Tsigelny I, Ten Eyck LF (2001): Protein docking using continuum electrostatics and geometric fit. *Protein Eng* 14(2):105–13.

Marshall SA, Mayo SL (2001): Achieving stability and conformational specificity in designed proteins via binary patterning. *J Mol Biol* 305(3):619–31.

Meng EC, Shoichet BK, Kuntz ID (1992): Automated docking with grid-based energy evaluation. *J Comput Chem* 13:505–24.

Mertz JE, Pettitt BM (1994): Molecular dynamics at constant pH. *Int J Supercomput Ap* 8(1):47–53.

Miranker A, Karplus M (1991): Functionality maps of binding sites: a multiple copy simultaneous search method. *Proteins* 11:29–34.

Miranker A, Karplus M (1995): An automated method for dynamic ligand design. *Proteins* 23:472–90.

Miyazawa S, Jernigan RL (1985): Estimation of effective interresidue contact energies from protein crystal structures: quasi-chemical approximation. *Macromolecules* 18:534–52.

Miyazawa S, Jernigan RL (1996): Residue-residue potentials with a favorable contact pair term and an unfavorable high packing density term, for simulation and threading. *J Mol Biol* 256:623–44.

Moon JB, Howe JW (1991): Computer design of bioactive molecules: a method for receptor-based de novo ligand design. *Proteins* 11:314–28.

Morris GM, Goodsell DS, Halliday RS, Huey R, Hart WE, Belew RK, Olson AJ (1998): Automated docking using a lamarckian genetic algorithm and an impirical binding free energy function. *J Comput Chem* 19:1639–62.

Muegge I, Martin YC, Hajduk PJ, Fesik SW (1999): Evaluation of PMF scoring in docking weak ligands to the FK506 binding protein. *J Med Chem* 42:2498–503.

Mulliken RS (1955): Electronic population analysis on LCAO-MO molecular wave functions I. *J Chem Phys* 23:1833–46.

Murphy KP (1999): Predicting binding energetics from structure: looking beyond $\Delta G^o$. *Med Res Rev* 19(4):333–9.

Murray CW, Clark DE, Auton TR, Firth MA, Li J, Sykes RA, Waszkowycz B, Westhead DR, Young SC (1997): PRO_SELECT: combining structure-based drug design and combinatorial chemistry for rapid lead discovery. 1. Technology. *J Comput-Aided Mol Des* 11(2):193–207.

Oprea TI, Waller CL (1991): Theoretical aspects of three-dimensional quantitative structure-activity relationships. In: Lipkowitz KB, Boyd DB, editors. *Reviews in Computational Chemistry*. New York: Wiley-VCH, pp 127–82.

Pearlman DA (1999): Free energy grids: A practical qualitative application of free energy perturbation to ligand design using the OWFEG method. *J Med Chem* 42:4313–24.

Pearlman DA, Charifson PS (2001): Improved scoring of ligand-protein interactions using OWFEG free energy grids. *J Med Chem* 44:502–11.

Pickett SD, Sternberg MJE (1993): Empirical scale of side-chain conformational entropy in protein folding. *J Mol Biol* 231:825–39.

Pierce AC, Jorgensen WL (2001): Estimation of binding affinities for selective thrombin inhibitors via Monte Carlo simulations. *J Med Chem* 44:1043–50.

Pitera JW, van Gunsteren WF (2001): The importance of solute-solvent van der Waals interactions with interior atoms of biopolymers. *J Am Chem Soc* 123:3163–4.

Plum GE, Breslauer KJ (1995): Calorimetry of proteins and nucleic acids. *Curr Opin Struct Biol* 5:682–90.

Rarey M, Wefing S, Lengauer T (1996a): Placement of medium-sized molecular fragments into active sites of proteins. *J Comput-Aided Mol Des* 10:41–54.

Rarey M, Kramer B, Lengauer T, Klebe G (1996b): A fast flexible docking method using an incremental construction algorithm. *J Mol Biol* 261:470–89.

Rarey M, Kramer B, Lengauer T (1999): Docking of hydrophobic ligands with interaction-based matching algorithms. *Bioinformatics* 15(3):243–50.

Rizzo RC, Tirado-Rives J, Jorgensen WL (2001): Estimation of binding affinities for HEPT and Nevirapine analogues with HIV-1 reverse transcriptase via Monte Carlo simulations. *J Med Chem* 44:145–54.

Roux B, Yu H-A, Karplus M (1990): Molecular basis for the Born model of ion solvation. *J Phys Chem* 94:4683–8.

Roux B, Simonson T (1999): Implicit solvent models. *Biophys Chem* 78:1–20. [Review of continuum solvent models and the underlying statistical mechanical basis.]

Sadowski J, Kubinyi H (1998): A scoring scheme for discriminating between drugs and nondrugs. *J Med Chem* 41:3325 9.

Sandak B, Nussiniov R, Wolfson HJ (1995): An automated computer vision and robotics-based technique for 3-D flexible biomolecular docking and matching. *Comput App Biosci.* 11:87–99.

Sandak B, Wolfson HJ, Nussinov R (1998): Flexible docking allowing induced fit in proteins: Insights from an open to closed conformational isomers. *Proteins* 32:159–74.

Schaffer L, Verkhivker GM (1998): Predicting structural effects in HIV-1 protease mutant complexes with flexible ligand docking and protein side-chain optimization. *Proteins* 33:295–310.

Schapira M, Totrov M, Abagyan R (1999): Prediction of the binding energy for small molecules, peptides and proteins. *J Mol Recognit* 12:177–90.

Sitkoff D, Sharp KA, Honig B (1994): Accurate calculation of hydration free energies using macroscopic solvent models. *J Phys Chem* 98:1978–88.

Still WC, Tempczyk A, Hawley RC, Hendrickson T (1990): Semianalytical treatment of solvation for molecular mechanics and dynamics. *J Am Chem Soc* 112:6127–9.

Su AI, Lorber DM, Weston GS, Baase WA, Matthews BW, Shoichet BK (2001): Docking molecules by families to increase the diversity of hits in database screens: computational strategy and experimental evaluation. *Proteins* 42:279–93.

Sun Y, Ewing TJA, Skillman AG Jr, Kuntz ID (1998): CombiDOCK: structure-based combinatorial docking and library design. *J Comput-Aided Mol Des* 12:597–604.

Tame JRH (1999): Scoring functions: a view from the bench. *J Comput-Aided Mol Des* 13:99–108.

Totrov M, Abagyan R (1997): Flexible protein-ligand docking by global energy optimization in internal coordinates. *Proteins* 29(Suppl. 1):215–20.

Vakser IA (1995): Protein docking for low-resolution structures. *Protein Eng* 8:371–7.

van de Waterbeemd H, Smith DA, Beaumont K, Walker DK (2001): Property-based design: optimization of drug absorption and pharmacokinetics. *J Med Chem* 44(9):1313–33.

Verkhivker GM, Appelt K, Freer ST, Villafranca JE (1995): Empirical free energy calculations of ligand-protein crystallographic complexes. I. Knowledge-based ligand-protein interaction potentials applied to the prediction of human immunodeficiency virus 1 protease binding affinity. *Protein Eng* 8(7):677–91.

Verkhivker GM, Bouzida D, Gehlaar DK, Rejto PA, Arthurs S, Colson AB, Freer ST, Larson V, Luty BA, Marrone T, Rose PW (2000): Deciphering common failures in molecular docking of ligand-protein complexes. *J Comput-Aided Mol Des* 14:731–51.

Vieth M, Hirst JD, Kolinski A, Brooks CL, III (1998): Assessing energy functions for flexible docking. *J Comp Chem* 19(14):1612–22.

Vorobjev YN, Almagro JC, Hermans J (1998): Discrimination between native and intentionally misfolded conformations of proteins: ES/IS, a new method for calculating conformational free energy that uses both dynamics simulations with an explicit solvent and an implicit solvent continuum model. *Proteins* 32:399–413.

Wallqvist A, Jernigan RL, Covell DG (1995): A preference-based free-energy parameterization of enzyme-inhibitor binding. Applications to HIV-1-protease inhibitor design. *Protein Sci* 4:1881–903.

Wang J, Wang W, Huo S, Lee M, Kollman PA (2001): Solvation model based on weighted solvent accessible surface area. *J Phys Chem B* 105:5055–67.

Warwicker J, Watson HC (1982): Calculation of the electric potential in the active site cleft due to $\alpha$-helix dipoles. *J Mol Biol* 157:671–9.

Waszkowycz B, Perkin TDJ, Sykes RA, Li J (2001): Large-scale virtual screening for discovering leads in the postgenomic era. *IBM Sys J* 40(2):360–76.

Welch W, Ruppert J, Jain AN (1996): HAMMERHEAD: fast, fully automated docking of flexible ligands to protein binding sites. *Chem Biol* 3:449–63.

Winzor DJ, Sawyer WH (1995): *Quantitative Characterization of Ligand Binding*. New York: Wiley-Liss.

Zhang C, Chen J, DeLisi C (1999): Protein–protein recognition: exploring the energy funnels near the binding sites. *Proteins* 34:255–67.

Zou X, Sun Y, Kuntz ID (1999): Inclusion of solvation in ligand binding free energy calculations using the generalized-Born model. *J Am Chem Soc* 121:8033–43.

# 23

# STRUCTURAL BIOINFORMATICS IN DRUG DISCOVERY

Eric B. Fauman, Andrew L. Hopkins, and Colin R. Groom

Modern pharmaceutical discovery has benefited from both the rigor of scientific discovery and the acceleration of technological advancements. The pharmaceutical industry had its origins at the beginning of the twentieth century. Scientific advancement over the past 100 years has seen the discovery of DNA, the understanding of proteins as specific molecular entities, and the harnessing of X-rays to understand proteins at the atomic level. Structural bioinformatics now is poised to do its part to accelerate the drug discovery process. This chapter follows the historic development of the current paradigm for pharmaceutical drug discovery, and highlights how structural bioinformatics is influencing this process.

## HISTORICAL DEVELOPMENT OF DRUG DISCOVERY

The current dominant paradigm in pharmaceutical drug discovery seeks to find a particular small molecule inhibitor to bind to a specific receptor, a macromolecular target. Our ability to pursue this paradigm rests on scientific and technological achievements in the twentieth century, particularly with regard to our ability to manipulate organic small molecules on the one hand, and to study the biological targets on the other (Sneader, 1985; Drews, 2000).

Humanity has, of course, been looking for remedies for its ailments long before there was a drug discovery industry. The use of willow bark as a treatment for pain relief, for example, can be traced back to Hippocrates and earlier. Such use is entirely empiric—a certain recipe gave relief to certain symptoms. Many such folk remedies

*Structural Bioinformatics*
Edited by Philip E. Bourne and Helge Weissig
Copyright © 2003 by Wiley-Liss, Inc.

were known, the progeny of some of which, such as willow bark, have found their place in our modern medicine chests.

The first step toward our modern approach to drug discovery was the suggestion that these remedies generally contained an active ingredient that could be isolated and purified. This idea can be traced back to 1530 to Paracelsus, a Swiss physician. It would be nearly another 300 years, in 1829, however, before the active ingredient in willow bark, salicin, was purified.

The synthesis the year before of urea by Fredrich Wöhler ushered in organic synthesis, which gave chemists the ability to manipulate these small organic compounds. Salicylic acid itself was first synthesized in 1852.

Along with the power to create a specific molecule came the ability to create many closely related compounds. These techniques were first put to profitable use in the dye industry in the mid 1800s, creating for the first time numerous low-cost dye compounds. Using such dyes for histological staining, Paul Ehrlich recognized that related molecules often exhibit related biological effects, a concept referred to today as SAR or structure-activity relationships. This concept was applied to derivatives of salicylic acid to try to discover forms of the drug that were less unpleasant for the patient. This research eventually led to the development of acetylsalicylic acid in 1897 by Felix Hoffmann at Bayer. Bayer named this compound Aspirin: "a" for acetyl, "spir" from Spiraea ulmaria, the meadowsweet plant, and "in," a common suffix for medicines at the time.

In noticing how some compounds more readily stained bacterial cells than human cells, Paul Ehrlich eventually developed another major cornerstone of modern drug discovery, the concept of a *therapeutic index*. All drugs have a minimal dose at which they demonstrate beneficial effects, and a minimal dose at which they demonstrate harmful effects. The therapeutic index is simply the ratio of these two doses. The interplay between trying to make compounds more effective and trying to make them safer is one that continues be at the center of pharmaceutical discovery to this day.

The recognition of activity being associated with specific molecular entities was paralleled (much later) by John Langley's suggestion in 1878 that there must be specific "receptors" for such compounds in the host, which bind to these entities. Knowing that there is a host receptor however is not the same as knowing what that receptor is. It would be another 100 years, for example, before John Vane and his colleagues discovered the link between aspirin and prostaglandin synthesis, establishing cyclooxygenase (COX) as aspirin's site of action. This work earned John Vane the 1982 Nobel Prize in Physiology or Medicine.

Cyclooxygenase was first given a structural face by Michael Garavito and colleagues in 1994 (Picot Loll Garavito et al., 1994). This structure, and that of the inducible COX-2 (Luong et al., 1996; Kurumbail et al., 1996), has made possible the first forays into structure-based design against this venerated target (Marnett and Kalgutkar, 1999).

## MODERN DRUG DISCOVERY

Presently, most pharmaceutical drug discovery programs begin with a known macromolecular target, and seek to identify a suitable small molecule modulator (Ratti and Trist, 2001; Dean, Zanders, and Bailey, 2001). The advent of the postgenomic era has started to point the way to novel targets, about which more will be said later. Typically, however, the target (usually a protein) has already been identified through biological or genetic investigations to be important in the disease of interest. The approach of modern drug discovery is rational and reductionist with a defined hypothesis of how the chosen mechanism of action could be beneficial against disease.

Following the identification of the target of interest, confidence in the approach is built with a variety of genetic and chemical target validation experiments. The process to discover a lead molecule begins with the development of an assay to look for modulators (either inhibitors, antagonists, or agonists) of the target's activity, followed by a high-throughput screen (HTS) of a large number of small molecules, in some cases up to a million or more (Landro et al., 2000). In the best cases, this method identifies one or more small molecule "hits" in the micromolar range, that is, having binding constants from 10 micromolar to the low nanomolar range.

Elaboration of the initial small molecule hit through medicinal chemistry is next used to try to improve the potency, ideally lowering the $K_i$ to the low nanomolar range to produce a potent lead molecule (Foye, 1989). There are an estimated $10^{62}$ possible small molecules, of which obviously only a tiny fraction can be created and tested. Recently, the techniques of combinatorial chemistry have been developed to rapidly generate hundreds and thousands of derivative compounds from a common scaffold in the hit-to-lead optimization stage. Recent years have seen the successful use of a variety of computational techniques, from quantitative structure activity relationships (QSAR) to computer-aided drug design (CADD) and structure-based drug design.

The process of optimizing the lead molecule into a "candidate" drug is usually the longest and most expensive stage in the drug discovery process (although this is still a fraction of the costs of drug development). Although the candidate is usually an analog of the original lead, it is still considered an art to successfully synthesize and select the exact compound that fulfills all the required properties of potency, absorption, bioavailability, metabolism, and safety. In many ways the lead-to-candidate stage of drug discovery is a multidimensional optimization problem searching within the relatively limited chemical space of analogs of the lead compound.

Following the selection of the candidate molecule, the drug development scientists develop large-scale production methods, and conduct the preclinical animal safety studies. Investigational new drugs (INDs) must pass through a set of three clinical trials: Phase I, a small study on healthy subjects to confirm safety; Phase II, a slightly larger study on a patient population to confirm efficacy; and Phase III, a large study of patients to gather additional information about safety and efficacy (CDER Handbook, 1998). However, even after the medicinal chemist has carefully crafted and balanced the properties of potency, bioavailability, and metabolism, over 90% of the compounds entering clinical trials fail to make it to market, most often due to poor biopharmaceutical properties, toxicity, or lack of efficacy (Venkatesh and Lipper, 1999). Due to the attrition of so many potential pharmaceuticals and the rising costs of drug discovery, the average cost to bring each new chemical entity (NCE) to market is estimated to be $770 million (Kettler, 1999).

## THE IMPACT OF STRUCTURAL BIOINFORMATICS ON DRUG DISCOVERY

### Structural Bioinformatics in a Pharmaceutical Context

Informatics and knowledge-based methods play an important role in the framework of the postgenomic drug discovery paradigm, in support of the traditional roles of screening and medicinal chemistry (Fig. 23.1). Genomics and bioinformatics support genetic methods of target identification and validation (Cunningham, 2000). The ability of chemoinformatics to process the properties of millions of virtual compounds for selection for synthesis and screening is an enabling technology for combinatorial chemistry and HTS. Biological structural information can be usefully exploited from the identification of the target protein all the way to the design of a bioavailable drug via structure-based drug design with suitable drug metabolism properties aided by ADMET (absorption, distribution, metabolism, excretion, and toxicology) modeling.

The techniques of structural bioinformatics are particularly valuable in the area from target identification to lead discovery. As depicted in Figure 23.2, a structural bioinformatics group in a pharmaceutical company can serve to link resources and results among bioinformatics, structural biology, and structure-based drug design (SBDD) groups.

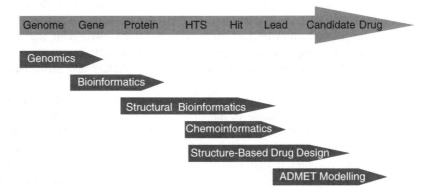

Figure 23.1. The roles informatics plays in the postgenomic drug discovery.

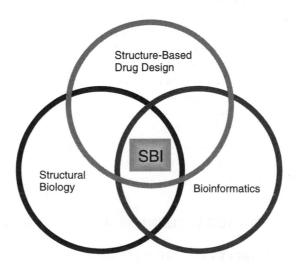

Figure 23.2. The relationship between structural bioinformatics (SBI) and other disciplines in drug discovery.

As will be seen, the most powerful effective strategies involve a tight integration of the computational and experimental techniques at each stage of drug discovery.

## Target Assessment

As structural inferences become available at the very earliest stages of a drug discovery program, structural bioinformatics can provide an a priori assessment of the ability of a target to be inhibited by a druglike molecule.

Designing compounds with appropriate biopharmaceutical properties that are still able to bind to their targets with an appropriate affinity is the challenge for the medicinal chemist. As detailed below, the challenge for structural bioinformaticians is to determine the magnitude of the medicinal chemists' task.

*Assessing Target Druggability.* Not all small molecules can be drugs, and not all proteins can be drug targets. A small molecule must have certain properties, and a protein must contain a binding site that is complementary or compatible with these properties.

Binding sites on proteins usually exist out of functional necessity. Due to hydrophobic forces, the energetically optimal protein would be spherical, with all its hydrophobic residues pointing inward (see Chapter 2). The majority of successful drugs achieve their activity by competing for a binding site on a protein with an endogenous small molecule. Drugs exploiting allosteric binding sites, with no known natural endogenous ligand, are relatively rare (e.g., the nonnucleoside binding site on HIV-1 reverse transcriptase), and these binding sites are usually not exposed since this is energetically expensive.

EXAMPLE: KINASES AND OTHER ATPASES. Examination of the natural ligands of a protein can be valuable in assessing the capacity of a binding site to bind a druglike molecule. The numerous types of ATPases present an interesting example. ATP is a common cofactor for many enzymes. It is recognized by a number of protein folds in a variety of ways. The adenosine portion of ATP (the adenine and ribose rings) has properties one would expect to be able to mimic in a drug. In contrast, it would be difficult to mimic the three charged phosphate groups in a druglike molecule, because charged compounds typically cannot penetrate cell membranes. Thus, when one is considering an ATPase as a potential target it is helpful to determine the way in which the ATP is recognized.

In protein kinases, the adenine ring of ATP fits into a well-defined, relatively hydrophobic pocket, forming a number of important hydrogen bonds. The phosphate groups play a relatively minor role in this recognition (Johnson et al., 1998). It has proven to be relatively straightforward to generate potent, druglike inhibitors of protein kinases that are competitive for ATP, making use of this attractive binding pocket (Bridges, 2001; Dumas, 2001). In ATPases that rely on coordination of the phosphate group, for example, those containing a so-called Walker A or B motif (Walker et al., 1982), inhibition by druglike molecules has proven difficult (except in exceptional cases where, for example, inhibitors bind to two such motifs) (Wigley et al., 1991). In addition, it may also be considerably more difficult to achieve selectivity if the drug predominately exploits main-chain interactions.

In many cases, the exact structure of the ATPase may not be known. Even when the structural motif used to recognize ATP is not known one can gain clues as to the attractiveness of the ATP site by reference to the binding affinities for ATP, ADP, AMP, adenosine, and adenine as gleaned from biochemical studies or from the literature. Differences in dissociation constants in this series may allow one to predict which regions of ATP are important recognition features.

EXAMPLE: PROTEASES. The proteases present additional subtleties involved in assessing the tractability of individual molecular targets. The substrates of proteases, that is peptides, represent reasonable chemical leads for a drug discovery process. Crucial to success, however, is the ability to "depeptidize" these leads, that is, remove the peptide bonds in order to avoid absorption and metabolic stability issues. For some proteases, this is relatively straightforward. One example is the serine protease thrombin, where many druglike inhibitors, barely resembling peptides, have been developed (Steinmetzer, Hauptmann, and Sturzebecher, 2001). In the case of the aspartyl protease renin, it has proven much more challenging to develop nonpeptidic inhibitors.

This difference in druggability can be explained by analyzing the way in which substrate is recognized by these two types of proteases. Serine proteases typically recognize both main-chain and side-chain features in their substrates, forming hydrogen bonds to relatively few main-chain groups on only one side of the scissile bond. Aspartyl proteases are typically tolerant of side-chain substitutions in their substrates and rely on main-chain hydrogen bonds to bind their substrates. In this case, it has proven difficult to retain potency in an inhibitor while reducing the peptidic character of leads. This has also been the case for many viral proteases, where substrate peptides are often weakly bound in shallow surface depressions on the enzyme surface (Chen et al., 1996).

***Quantitative Assessment of Target Druggability.*** The foregoing discussion assumes the opportunity to perform an in depth individual analysis of the relevant protein structures. For a quick assessment of a large number of potential targets, a more quantitative approach may be more appropriate.

Such a quantitative approach is already well established for assessing the druglike properties of a small molecule. The rule-of-five (Table 23.1) is a set of properties to suggest which compounds are likely to show poor absorption or permeation, since such compounds are unlikely to show good oral bioavailability (Lipinski et al., 1997). More recent work has further refined what distinguishes drugs from other compounds (Walters and Murcko, 2002; Lipinski, 2000; Sadowski and Kubinyi, 1998; Gillet et al., 1999; Blake, 2000, Clark and Pickett, 2000).

As a receptor binding site must be complementary to a drug, it is reasonable to assume that equivalent rules could be developed to describe physicochemical properties

TABLE 23.1. The Rule-of-Five

---

A compound is likely to show poor absorption or permeation if:
1. It has more than five hydrogen bond donors
2. The molecular weight is over 500
3. The Clog P (calculated octanol/water partition coefficient) is over five
4. The sum of nitrogens and oxygens is over 10
5. Weak inhibition (<100 nM) is observed

---

of binding sites with the potential to bind rule-of-five compliant inhibitors with a potent binding constant (e.g., $K_i < 100$ nM). A number of properties complementary to the rule of five can be calculated; for example, the surface area and volume of the pocket, hydrophobic and hydrophilic character, and the curvature and shape of the pocket. Programs such as SURFACE (Lee and Richards, 1971; CCP4), CAST (Liang, Edelsbrunner, and Woodward, 1998), ms (Connolly, 1993), and GRASP (Nicholls, Sharp, and Honig, 1991) can be used to calculate these and other parameters. Following the assumption that properties of the drug are complementary to those of the binding site, analysis of the calculated physicochemical properties of the putative drug-binding pocket on the target protein can provide an important guide to the medicinal chemist in predicting the likelihood of discovering a drug against the particular target site.

The logarithmic relationship between the free energy of binding ($\Delta G$) and the binding constant ($K_i$) (23.1) means every 10-fold increase in potency is due to a $-1.363$ kcal/mol change in binding energy

$$\Delta G = -RT \ln(K_i) \qquad (23.1)$$

Thus a drug with a typical dissociation constant of 10 nM binds with a free energy of $-11$ kcal/mol and a 1 μM HTS hit binds with $-8.4$ kcal/mol.

The strength of binding is predominately driven by burying of hydrophobic surfaces (van der Waals and entropy). The free energy gained from burying hydrophobic surfaces is estimated at around 0.03 kcal/mol/$\text{Å}^2$, with buried polar surfaces giving up about 0.1 kcal/mol/$\text{Å}^2$. A drug with a 10 nM dissociation constant needs to bury 370 $\text{Å}^2$ of hydrophobic surface area. Therefore, every 46 $\text{Å}^2$ of buried hydrophobic surface (the surface area of a methyl group) buys a 10-fold increase in potency, approximately equivalent to the maximal affinity per nonhydrogen atom defined by Kuntz et al. (1999). Encapsulated cavities are capable of binding low molecular weight compounds with high affinities since they maximize the ratio of the surface area to the volume.

In addition to the predominantly hydrophobic contribution to the binding of many drugs, ionic interactions, such as those found in zinc proteases (for example, ACE inhibitors) allow low molecule weight molecules to bind strongly.

***The Druggable Genome.*** Biological systems contain only four types of macromolecules with which we can interfere using small molecule therapeutic agents: proteins, polysaccharides, lipids, and nucleic acids. Toxicity, specificity, and the inability to obtain potent compounds against the latter three types means that the majority of successful drugs achieve their activity by modifying the activity of a protein by competing for a binding site on a protein with an endogenous small molecule. Thus, there is a limited number of molecular targets for which commercially viable compounds can currently be developed, leading to the concept of the "druggable genome." The druggable genome is the subset of the genes in the human genome that express proteins that are capable of binding small druglike (i.e., rule-of-five compatible) molecules.

In a comprehensive review of the accumulated portfolio of targets in the pharmaceutical industry, Drews identified 483 proteins that have been exploited to date (Drews, 1996; Drews and Ryser, 1997). In a critical review of Drews's estimates we have analyzed the sequences of all targets of marketed and investigational drugs or leads (Investigational drugs database (IDdb), Current Drugs Ltd.) that are rule-of-five compatible. These 400 targets fall into a few gene families, as shown in Figure 23.3. Interestingly, only about 120 InterPro domains define all the ligand-binding domains for all proteins for which rule-of-five compliant inhibitors are available (Hopkins and Groom, 2002). This distribution of exploited targets actually changes rather slowly. On average, new drugs are launched against only about four novel targets each year (see Fig. 23.4).

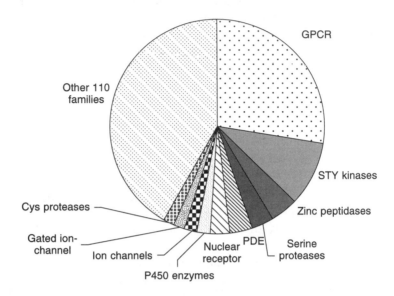

**Figure 23.3.** Gene family distribution of the molecular targets of current investigational and marketed drugs (data derived from a combination of sources including the literature and Investigational drugs database (IDdb), Current Drugs Ltd. (Hopkins and Groom).).

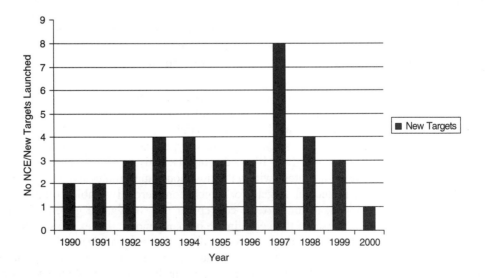

**Figure 23.4.** The number of new targets, for which small molecule drugs have been developed and taken to market, launched in the last decade (data derived from *Drug News Prospect.*, Prous Science).

Existing targets are at the intersection of two necessary attributes: an ability to bind compounds with acceptable properties (druggability), and a link to disease (relevance). The total number of targets that possess both of these attributes can be estimated in a number of ways.

The druggable genome is a subset of the total human genome. The completion of the draft human genome sets the total number of human genes at about 30,000 (Venter et al., 2001; Genome International Sequencing Consortium, 2001). Estimates of the number of druggable targets in the genome based on the idea of assessing the total number of ligand-binding domains have produced figures in excess of 10,000 (Bailey, Zanders, and Dean, 2001). Using a more conservative gene family approach, focused on proteins that share greater than 30% sequence identity to a known target in the 120 druggable gene family domains, suggests around 3000 presumed druggable targets in the human genome. Gene families are not equally populated, with genes distributed among a few very large gene families, and many sparsely populated gene families. This distribution suggests that there may be very few large druggable gene families left to discover.

Separately, one can assess how many genes are likely to affect some disease process. Considering about 100 major human diseases, and assuming there are 10 genes directly involved in any given disease process, with another 5 to 10 genes influencing the activity of those genes, yields an estimated 5000–10,000 disease-related genes (Drews, 2000).

The universe of exploitable small molecule targets for drugs is the intersection between the druggable genome and those genes related to diseases (see Fig. 23.5). Structural bioinformatics has a great role to play in identifying all the druggable proteins coded for in genomes of interest.

## Target Triage

The availability of the sequences from complete genomes has revealed many more potential targets than could possibly be prosecuted using current experimental technologies. As such it is sometimes necessary to prioritize targets from a large potential subset. Such a subset may arise, for example, from a gene expression study (Sallinen et al., 2000) or by analyzing the genomes of disease-causing organisms (McDevitt and

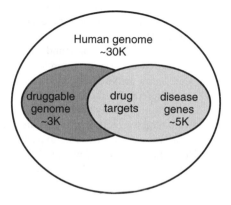

**Figure 23.5.** The effective number of exploitable drug targets can be determined by the intersection of the number of genes linked to disease and the druggable subset of the human genome.

Rosenberg, 2001). Properties of interest when considering large numbers of targets go beyond the assessment of the active site described above and include those summarized in Table 23.2.

***Computational Approaches to Target Triage.*** A number of these factors can be analyzed computationally. One of the first examples of this is the work of Bruccoleri and others at Bristol-Myers Squibb (Bruccoleri, Dougherty, and Davison, 1998). In this work, described as congenomics, those genes common to a number of bacterial species but absent from higher organisms were identified as potential targets for antibacterial agents.

This work was taken further at Bayer (Spaltmann, Blunck, Ziegelbauer, 1999), where an automated procedure for target prioritization was applied to the recently completed genome of the yeast *S. cerevisiae*, which served as a model for related pathogenic species of fungi. The system developed, CATS (computer-aided target selection), scored targets based on the importance of a gene for the organism, the occurrence of the gene in multiple target species, whether specificity of inhibition could be achieved by reference to sequence similarity with vertebrates, and whether assay development was facile. Of note is that a number of proteins that are the target of antifungal agents scored very highly in this approach, thereby validating this computational approach.

**TABLE 23.2. Considerations When Selecting Molecular Targets**

| Property | Consideration |
| --- | --- |
| Confidence in rational | How strong is the evidence to indicate that modulating the activity of the target will produce the desired response? |
| Sense of modulation | Is inhibition or agonism necessary to correctly modulate disease process? Note that while GPCRs can effectively be turned on by agonists this is difficult with most enzymes. |
| Ability to bind druglike molecules | Does the target have the potential to recognize compounds with appropriate properties to a reasonable affinity? |
| Ease of screening | Can the target protein be obtained in quantities sufficient to facilitate high-throughput screening? Are reagents such as substrates and cofactors known and available? |
| Availability of functional assays | Can the activity of compounds against the protein targets be followed up in a disease-relevant functional assay? |
| Availability of protein structure | Is an X-ray or NMR structure of the protein available to allow structure-assisted drug design? Is the protein amenable to structural biology? |
| Pathway | Is the target protein in a redundant signaling or metabolic pathway where it can be bypassed? |
| Potential for resistance | Do pathogens contain mutant isoforms of the target protein? What level of modulation is required for therapeutic activity? |
| Availability of chemical leads | Are there chemical leads available with suitable properties? |
| Selectivity | Are there related proteins (in the host) that might be affected by inhibitors against the target? |

SELECTIVITY. Most potential target have related host proteins whose function must not be affected by a successful therapeutic. This is true, for example, for proteins from the identified gene families, discussed above.

In considering selectivity issues, certainly one must first look at related sequences in the same protein family. However, there is not necessarily a direct correlation between the similarities one would infer from homology and those one would infer from compound action. This discrepancy exists because only a small fraction of the residues in a protein interact directly with any one ligand. Although long-range inter-actions and conformational changes can play a role in inhibitor binding, in general it is those residues lining a ligand-binding site that are of most importance.

The ATP cleft of the protein kinases serves as a good example of this situation. Although all protein kinases bind ATP in the same region and conformation, only a few of the residues facing the cleft are highly conserved (see Fig. 23.6). The remaining residues are subject to random drift, and thus distantly related kinases can actually end up having more similar clefts than more closely related kinases. This is depicted graph-ically in Figure 23.7. The tree on the left was constructed using sequence identities calculated across the entire kinase domain for a diverse set of kinase structures. The tree on the right used identities across only the 16 residues lining the ATP-binding site (as deduced from the CDK2 structure, 1FIN; Jeffrey et al., 1995). Note for example that the *Zea mays* CK2 kinase, 1A6O, which is grouped with the cyclin-dependent and MAP kinases on the basis of overall sequence identity, is actually seen to be relatively isolated on the basis of its active site alone. This distinction reflects the remarkable ability of this enzyme to hydrolyze either ATP or GTP, due to a number of unique features at the ATP-binding cleft (Niefind et al., 1998).

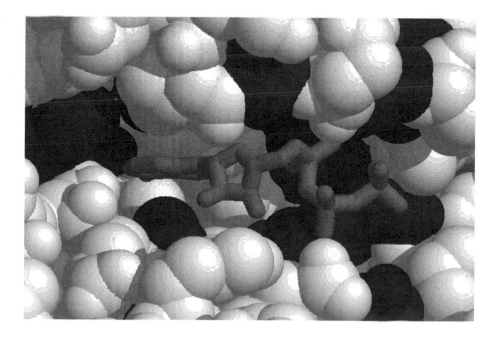

**Figure 23.6.** ATP (light gray sticks) bound to the active site of CDK2 (PDB code 1FIN). Residue positions highly conserved across all protein kinase domains are shown in dark gray; variable positions shown in white.

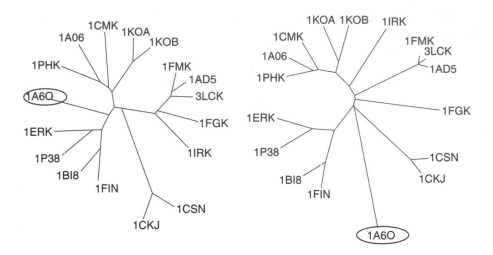

**Figure 23.7.** Comparison of phylogenetic trees based either on the entire sequence of the kinase domain (left) or of just the 16 side chains that interact with the ATP (right).

This use of noncontiguous regions of sequences to establish relationships was first introduced by Sandberg et al. (1998) and has recently been described as structure-activity relationship homology (SARAH; Frye, 1999). In SARAH, protein targets are grouped both by their sequence similarities and by their ability to bind an array of compounds, such as may be determined from pharmaceutical HTS. SARAH builds on earlier ideas around affinity fingerprinting (Kuvar et al., 1995). As Frye points out, protein targets were often grouped according to their response to ligands before we had sequence information at our disposal (Lefkowitz, Hoffman, and Taylor, 1990).

## Target Validation

Once a target has been selected for a drug discovery program, it must be experimentally validated. A validated target is one where the premise that modulating the activity of the target has been proven to affect the disease process.

A genetic approach to target validation can use a knock-out of the gene of interest, or use RNA antisense technology to inactivate the gene (McClurg and Keenan, 1999).

The most rigorous way to validate a target is to use a compound, for example, a known small-molecule inhibitor. Clearly at the start of a drug discovery program such compounds may not be available. For some targets the techniques of chemical genetics give us another route. Here, site-directed mutants of the target protein are made in order to make that target sensitive to an existing compound. Essentially this technique uses protein engineering to do what a medicinal chemist would normally be tasked with. Structural bioinformatics clearly plays a central part in this approach, which relies on successfully predicting a match between an existing compound and engineered binding site.

The origins of chemical genetics lie in work by Cohen and others (Eyers et al., 1998), in which members of the MAP kinase family were mutated in order to make them sensitive to the well-understood inhibitor SB-203580. Others have since used these techniques on other members of the kinase family (Bishop et al., 2000). The most attractive aspect of this approach is that the modified gene can be put back into the organism and the activity of compound compared between wild-type and mutant organisms or cell lines. Thus, it allows us to answer questions such as, "If I generated a selective inhibitor of protein X, would it have the desired effect?"

## Lead Identification

After a target has been identified and validated, chemical leads (of low or medium potency) must be identified to supply a starting point for the medicinal chemistry efforts. Such leads can come from knowledge of the natural ligands or from *de novo* design, but most often are found through high-throughput screening.

*Assay Design.* In order to design an assay for a particular target, one obviously needs to know what function that target performs and what ligands it recognizes. In the case of novel targets, structural bioinformatics can be used for function and ligand prediction. As described in Chapter 19, a variety of techniques exist to help deduce the function of a protein given knowledge of its structure. Although such methods seldom pinpoint the exact biochemical activity, they can help define what sorts of experiments should be tried (Bugl et al., 2000).

As discussed earlier, the function of a protein can be dictated by just a handful of residues in a binding pocket. For example, if we were to try to determine the endogenous ligand that might be recognized by a G-protein coupled receptor we may wish to focus on a selection of residues based on mutagenesis data or structure prediction.

Counter-screens can also be used at this point in the drug discovery program as an experimental means to address selectivity concerns. A structural bioinformatics approach can be used to help select those host proteins that are most likely to bind to inhibitors of the primary target. For example, if the primary target were a kinase, hits identified through HTS might be subsequently tested against a panel of related kinases (Davies et al., 2000). As described earlier, the proteins most likely to bind the identified inhibitors may not be those most closely related by homology since ligand binding can be dominated by just those residues in the binding cleft. A structural bioinformatics approach, identifying first the obvious and remote homologs and then comparing the presumed ligand-binding residues, can help guide the selection of appropriate counter screen targets.

*Using Structural Similarity to Find Chemical Leads.* In general, if two gene products have similar structures, they may bind similar ligands. This relationship enables one to use chemical leads for one protein target against another Even a drug discovery program against a relatively well-understood target can be "jump-started" by recognizing a relationship between the current target and other targets studied in the company (see Chapter 16, Structure-Structure Comparison and Alignment and Chapter 26, Fold Recognition Methods). Discovering such a connection can facilitate the transfer of institutional knowledge across the traditional therapeutic area divisions. For example the early work on the antiviral target HIV protease, benefited from the body of knowledge derived from the search for inhibitors of renin, a distantly related aspartyl protease important in cardiovascular disease (Appelt, 1997).

Many other examples of such situations can be found among the targets of known drugs, such as angiotensin-converting enzyme, neutral endopeptidase, and thermolysin. In these cases there is no detectable similarity at the sequence level.

However all three proteases bind similar substrates and have been shown to bind similar inhibitors (Roques, 1985). The structural similarity between two of these enzymes has recently been confirmed by the structure determination of neutral endopeptidase (Oefner et al., 2000).

Using structural similarity to find chemical leads typically causes concern regarding the potential selectivity resultant compounds will have. In practice, however, where the proteins share less than 30% sequence identity the active sites are usually nonidentical and leads can be optimized toward one particular target through computational and medicinal chemistry approaches.

***Virtual Screening.*** Due to the high costs of HTS in terms of staffing and compound stock depletion it is highly desirable to attempt to find leads for a protein target computationally. Protein structure can be used in two distinct ways to search databases of compound structures.

The first is to derive a pharmacophore describing the functionally important sites in a ligand-binding site. These sites can be determined by reference to the protein structures, using software such as GRID (Goodford, 1985), X-site (Laskowki et al., 1996), and hotspots (Mills, Perkins, and Dean, 1997). Structures of compounds in chemical libraries (whether existing or not) can then be computationally assessed as to whether they have the ability to adopt a conformation that matches the pharmacophore.

The second method is docking and scoring (Chapter 22; Gohlke and Klebe, 2001). Here problems are still encountered in correctly docking a small molecule structure to a protein and calculating the binding energy. Current methods can often enrich a set of compounds by identifying a subset of compounds that are more likely to bind. Although showing much promise, the most reliable methods are slow and challenges still exist in being able to distinguish between compounds that do and do not bind to the target protein.

***Chemical Library Design.*** Chemical libraries, consisting of hundreds to thousands of related compounds, can be generated through combinatorial chemistry by starting with a scaffold or template and decorating it with a variety of functional groups or monomers. A chemical library designed around an early chemical lead can help establish the SAR for that series, pointing the way to more potent compounds. Protein structure can play a great role in the design of such chemical libraries (Bone and Salemme, 1997; Caflisch and Ehrhardt, 1997).

Chemical libraries can be designed by reference simply to binding data for small molecule hits. However, if a known or predicted three-dimensional structure for the target is available, or, even better, the structure of a representative compound correctly bound to the protein, such information can be used to design more efficient chemical libraries. In general far more compounds can be designed or enumerated than can actually be produced and tested. Knowledge of the structure of the active site can focus the design and synthesis toward a collection of compounds more likely to bind. In practice, a pharmacophore is usually first defined from the ligand-binding site as is done in virtual screening. Such a pharmacophore can then be applied either to the selection of monomers from which to build the chemical library or to determine which compounds to actually synthesize from an enumerated list of potential compounds.

## Lead Optimization

The longest phase in preclinical drug discovery is the process of deriving a high-potency inhibitor from the chemical lead, while optimizing its physicochemical properties to maximize its chances for success as a drug. By bridging and building on resources in bioinformatics, structural biology, and structure-based drug design, structural bioinformatics can accelerate the quest for a potential drug.

***Structural Biology for Structure-Based Drug Design.*** The most relevant structure-based drug design programs involve repeated cycles of determining the structure of the target in complex with a number of lead compounds and their analogs. One hurdle in establishing a rapid cycle of crystallography and structure-based design is in actually obtaining crystals of the target protein of sufficient quality.

A genetic construct encoding the exact full-length sequence of a protein is not necessarily the ideal one to use in order to obtain material for screening or structural studies. Firstly, the full-length protein may be large and contain protein domains that are not of relevance to the studies being performed. The protein may express poorly, be insoluble, or fail to crystallize, at least in a time frame compatible with the discovery process.

In order to avoid such problems, structural bioinformatics should be used to design suitable constructs at the outset of the project. As described in Chapter 18 (domain assignments), proteins are frequently composed of discrete domains. The computational techniques of domain detection can be combined with experimental techniques of domain assignment (for example, limited proteolysis followed by mass spectroscopy analysis). The designed constructs can be evaluated on the basis of their expression levels, solubility, activity, and, of course, crystallizability.

The selection of an appropriate domain for structural studies often begins by aligning the sequence of the target protein with that of a protein of known structure. One can then determine where it may be possible to truncate the protein at the amino- and carboxy-termini. Where structures are not available, secondary structure prediction can be used. Combining secondary structure prediction based on multiple sequence alignments with analysis of sequence conservation can be particularly successful. One can often express a protein from a construct designed to start and finish where sequence conservation is high and at the end of predicted elements of secondary structure

It should be noted that the same process can be used to guide construct design for the production of the target protein for HTS assays. In fact, in the best cases the same construct can be designed for both screening and structural biology.

As described in Chapter 4 (crystallography), a second hurdle in structure determination by crystallography is the so-called "phase problem." One common solution to this problem is molecular replacement. In molecular replacement, a model representing some or all of the new protein is rotated and translated into the new unit cell in an attempt to find a solution. Originally molecular replacement was only used for cases of a specific protein in different space groups, or in cases of high sequence identity. Greater computational power, as well as the greater wealth of known folds, has resulted in molecular replacement being successfully applied even in cases where the starting model exhibited only 20% sequence identity or less (Storici et al., 1999; Hong et al., 2000). The extent to which the core of a protein is distorted with greater sequence divergence (Chothia and Lesk, 1986) presumably will place a limit on when molecular replacement can be used. However, even when phases are determined experimentally (for example, through heavy atoms or selenomethionine techniques), identification of a suitable structural homolog at even 10% sequence identity can greatly accelerate the interpretation of the electron density maps and of the protein structure (Bugl et al., 2000).

*Use of Protein Surrogates.* Structural bioinformatics can play a key role in structure-based drug design approaches even when the structure of the target protein is not available; for example through the generation of a homology model or the use of a surrogate protein. Clearly the more similar the sequences of the target protein the better the homology model is likely to be (see Chapter 25). However, these models are of more value if there is experimental evidence to validate their use.

When a particular target is inaccessible to structural biology, a project may rely on the use of a related protein for structure determinations. In many cases, a surrogate may simply be an orthologous protein from another species or a similar member of the same gene family. Sometimes a surrogate may exhibit similarity at the structural and functional level that does not extend to sequence, for example in the case of thermolysin and NEP (Barclay et al., 1994).

## ADMET Modeling

Potency and simple property filters such as the rule-of-five are often the main criteria in the lead discovery stage of a drug. To design a candidate medicine from the initial lead, however, one needs to consider a host of additional parameters that can affect the biopharmaceutical and safety properties of the drug such as the *in vivo* absorption, distribution, metabolism, excretion, and toxicology (ADMET). The tools of structural bioinformatics, namely sequence-structure relationships and protein homology modeling, can be employed in the field of ADMET modeling.

The most developed work in this field is in the area of cytochrome P450 modeling to predict drug metabolism (Ekins, de Groot, and Jones, 2001). The metabolism of a drug by various cytochrome P450 enzymes is an important factor in the development of a drug. The route of metabolism can affect the drug's half-life, dose, and even safety, since P450 polymorphisms result in differential metabolism. In the absence of a structure of a human P450 enzyme, homology models can be employed to model the P450 active sites. Combining structure-activity relationships with P450 protein homology modeling enables the production of pharmacophores that are capable of predicting compound metabolism with success (de Groot et al., 1999a; de Groot et al., 1999b).

Although the structure of no human cytochrome P450 is known, the crystal structure of the rabbit CYP2C5 was recently determined. (Williams et al., 2000). The availability of a mammalian P450 has considerably improved many of the models and sequence alignments of the human P450s and thus should enable the constructions of more predictive pharmacophores for a range of metabolizing enzymes (de Groot, Alex, and Jones).

## CONCLUSION AND FUTURE DIRECTIONS

With the advent of the genome era and the expected increase in the availability of protein structures from the introduction of structural genomics (see Chapter 29), it is likely that in the next few years most drug discovery programs against a soluble protein will begin with a structure or model of the target protein.

The early use of structural information in a drug discovery project can provide a great deal of insight for the lead discovery program. Quantitative target assessment can play a great role in the investment decision of whether or not to pursue a project. Many medicinal chemistry projects have floundered for the simple reason that the binding pocket of the target did not have the required physicochemical properties complementary to binding a potent small molecule. Thus, initial analysis of binding site can provide a significant guide to the ultimate success of the target. A robust equivalent to the rule-of-five for molecular targets will enable automatic target assessment for the large number of novel structures expected from the structural genomics initiatives.

In additional to assessing targets, comparative structural bioinformatics holds promise in identifying new drug targets from combined study of pathogen and human genomes. In comparing active sites, computational target triaging methods can simultaneously assess target druggability and species selectivity. Identifying interspecies differences in the binding sites of drug targets can also be exploited in either the choice of *in vivo* models or by specifically engineering animal receptors by site directed mutagenesis to mimic the human receptor binding pocket.

Knowledge-based approaches, combined with the current explosion in sequence and structure data, may move us to a new *prospective* paradigm in which it may be possible to discover a suitable drug against a given target long before any application is known. Combined with advances in single-nucleotide polymorphism detection, this may make possible individualized medicines in which each patient gets a drug designed against his or her particular form of the target (Pfost, Boyce-Jacino, and Grant, 2000).

As we move toward a situation where drug discovery projects are bathed in structural and sequence information, it is the role of the structural bioinformatician to integrate this wealth of data accelerating drug discovery.

## ACKNOWLEDGMENTS

We should like to thank Dr. John Overington of Inpharmatica and Dr. Alexander Alex for extensive input into the concepts of druggable targets and beautiful binding sites. Dr. Huifen Chen developed much of the methodology for kinase active site analysis.

## FURTHER READING

de Groot MJ, Alex AA, Jones BC Development of a combined protein and pharmacophore model for CYP2C9, *J Med Chem* 45(10):1983–93.

Dean PM, Zanders ED, Bailey DS, (2001): Industrial-scale, genomics-based drug design and discovery. *Trends Biotech* 19(8):288–92.

Picot D, Loll PJ, Garavito RM, (1994): The X-ray crystal structure of the membrane protein prostaglandin H2 synthease-1. *Nature* 367(6460):243–9.

## REFERENCES

Appelt K (1997): Inhibitors of HIV-1 protease. in: Veerapandian P, editor. *Structure-Based Drug Design*. New York: Marcel Dekker, pp 1–39.

Bailey D, Zanders E, Dean P (2001): The end of the beginning for genomic medicine. *Nat Biotech* 19:207–9.

Barclay PL, Danilewicz JC, Matthews BW, James K (1994): Inhibition of thermolysin and neutral endopeptidase 24.11 by a novel glutaride derivative. *Biochemistry* 33:51–6.

Bishop C, Ubersax JA, Petsch DT, Matheos DP, Gray NS, Blethrow J, Shimizu E, Tsien JZ, Schultz PG, Rose MD, Wood JL, Morgan DO, Shokat KM (2000): A chemical switch for inhibitor-sensitive alleles of any protein kinase. *Nature* 407(6802):395–401.

Blake JF (2000): Chemoinformatics—predicting the physicochemical properties of "drug-like" molecules. *Curr Opin Biotech* 11(1):104–7.

Bone R, Salemme FR (1997): The integration of structure-based design and directed combinatorial chemistry for new pharmaceutical discovery. In: Veerapandian P, editor. *Structure-Based Drug Design.* New York: Marcel Dekker, pp 525–39.

Bridges AJ (2001): Chemical inhibitors of protein kinases. *Chem Rev* 101(8):2541–71.

Bruccoleri RE, Doughtery TJ, Davison DB (1998): Concordance analysis of microbial genomes. *Nucleic Acids Res* 26(19):4482–6.

Bugl H, Fauman EB, Staker BL, Zheng F, Kushner SR, Saper MA, Bardwell JC, Jakob U (2000): RNA methylation under heat shock control. *Molec Cell* 6(2):349–60.

Caflisch A, Ehrhardt C (1997): Structure-based combinatorial ligand design. In: Veerapandian P, editor. *Structure-Based Drug Design.* New York: Marcel Dekker, pp 541–58.

CDER Handbook (1998): http://www.fda.gov/cder/handbook/index.html.

Chen P, Tsuge H, Almassy RJ, Gribskov CL, Katoh S, Vanderpool DL, Margosiak SA, Pinko C, Matthews DA, Kan C-C (1996): Structure of the human cytomegalovirus protease catalytic domain reveals a novel serine protease fold and catalytic triad. *Cell* 86(5):835–43.

Chothia C, Lesk AM (1986): The relation between the divergence of sequence and structure in proteins. *EMBO J* 5(4):823–6.

Clark DE, Pickett SD (2000): Computational methods for the prediction of "drug-likeness." *Drug Discovery Today* 5:49–58.

Collaborative Computational Project, Number 4. (1994): The CCP4 suite: programs for protein crystallography. *Acta Cryst* D50:760–3.

Connolly ML (1993): The molecular surface package. *J Mol Graph* 11(2):139–41.

Cunningham MH (2000): Genomics and proteomics: the new millennium of drug discovery and development. *J Pharmacol Toxicol Methods* 44(1):291–300.

Davies SP, Reddy H, Caivano M, Cohen P (2000): Specificity and mechanism of action of some commonly used protein kinase inhibitors. *Biochem J* 351:95–105.

de Groot MJ, Ackland MJ, Horne VA, Alex AA, Jones BC (1999a): A novel approach to predicting P450 mediated drug metabolism. CYP2D6 catalyzed N-dealkylation reactions and qualitative metabolite predictions using a combined protein and pharmacophore model for CYP2D6. *J Med Chem* 42(20):4062–70.

de Groot MJ, Ackland MJ, Horne VA, Alex AA, Jones BC (1999b): Novel approach to predicting P450-mediated drug metabolism: development of a combined protein and pharmacophore model for CYP2D6. *J Med Chem* 42(9):1515–24.

Drews J (1996): Genomic sciences and the medicine of tomorrow. *Nat Biotech* 14:1516–8.

Drews J (2000): Drug discovery: a historical perspective. *Science* 287:1960–4.

Drews J, Ryser ST (1997): Classic drug targets. Special Pullout, *Nat Biotech* 15.

Dumas J (2001): Protein kinase inhibitors: emerging pharmacophores. *Expert Opin Ther Pat* 11(3):405–29.

Ekins S, de Groot MJ, Jones JP (2001): Pharmacophore and three-dimensional quantitative structure activity relationship methods for modeling cytochrome P450 active sites. *Drug Met Disp* 29(7):936–44.

Eyers PA, Craxton M, Morrice N, Cohen P, Goedert M (1998): Conversion of SB 203580-insensitive MAP kinase family members to drug-sensitive forms by a single amino-acid substitution *Chem Biol* 5(6):321–8.

Foye WO (1989): *Principles of Medicinal Chemistry*, 3rd ed. Philadelphia: Lea & Febiger.

Frye SV (1999): Structure-activity relationship homology (SARAH): a conceptual framework for drug discovery in the genomic era. *Chem Biol* 6:R3–R7.

Genome International Sequencing Consortium (2001): Initial sequencing and analysis of the human genome. *Nature* 409:860–921.

Gillet VJ, Willet P, Bradshaw J, Green DVS (1999): Selecting combinatorial libraries to optimize diversity and physical properties. *J Chem Inf Comput Sci* 39(1):169–77.

Gohlke H, Klebe G (2001): Statistical potentials and scoring functions applied to protein-ligand binding. *Curr Opin Struct Biol* 11(2):231–5.

Goodford PJ (1985): A computational procedure for determining energetically favorable binding sites on biologically important molecules. *J Med Chem* 28:849–57.

Hong L, Koelsch G, Lin X, Wu S, Terzyan S, Ghosh AK, Zhang XC, Tanj J (2000): Structure of the protease domain of memapsin 2 (beta-secretase) complexed with inhibitor *Science* 290(5489):150–3.

Hopkins AL, Groom CR (2002): The Druggable Genome. *Nat Rev Drug Discov* 1(9):727–30.

Jeffrey RD, Russo AA, Polyak K, Gibbs E, Hurwitz J, Massague J, Pavletich NP (1995): Mechanism of CDK activation revealed by the structure of a cyclinA-CDK2 complex. *Nature* 376(6538):313–20.

Johnson LN, Lowe ED, Noble MEM, Owen DJ (1998): The structural basis for substrate recognition and control by protein kinases. *FEBS Lett* 430(1,2):1–1.

Kettler HE (1999): Updating the cost of a new chemical entity. London: Office of Health Economics.

Kuntz ID, Chen K, Sharp KA, Kollman PA (1999): The maximal affinity of ligands. *Proc Natl Acad Sci USA* 96:997.

Kurumbail RG, Stevens AM, Gierse JK, McDonald JJ, Stegeman RA, Pak JY, Gildehaus D, Miyashiro JM, Penning TD, Seibert K, Isakson PC, Stallings WC (1996): Structural basis for selective inhibition of cyclooxygenase-2 by anti-inflammatory agents. *Nature* 384:644–8.

Kuvar LM, Higgins DI, Villar HO, Sportsman JR, Engqvist-Goldstein A, Bukar R, Bauer KE, Dilley H, Rocke DM (1995): Predicting ligand binding to proteins by affinity fingerprinting. *Chem Biol* 2:107–18.

Landro JA, Taylor ICA, Stirtan WG, Osterman DG, Kirstie J, Hunnicutt EJ, Rae PMM, Sweetnam PM (2000): HTS in the new millennium—the role of pharmacology and flexibility. *J Pharmacol Toxicol Methods* 44:273–89.

Laskowski RA, Thornton JM, Humblet C, Singh J (1996): X-site: use of empirically derived atomic packing preferences to identify favorable interaction regions in the binding sites of proteins. *J Mol Biol* 259(1):175–201.

Lee B, Richards FM (1971): The interpretation of protein structures: estimation of static accessibility. *J Mol Biol* 55(3):379–400.

Lefkowitz RJ, Hoffman BB, Taylor P (1990): Neurohumoral transmission: the autonomic and somatic motor nervous systems. In: Gilman AG, Rall TW, Nies AS, Taylor P, editors. *Goodman and Gilman's The Pharmacological Basis of Therapeutics*. New York: Pergamon Press.

Liang J, Edelsbrunner H, Woodward C (1998): Anatomy of protein pockets and cavities: measurement of binding site geometry and implications for ligand design. *Protein Sci* 7:1884–97.

Lipinski CA, Lombardo F, Dominy BW, Feeney PJ (1997): Experimental and computational approaches to estimate solubility and permeability in drug discovery and development settings. *Adv Drug Del Rev* 23:3–25.

Lipinski CA (2000): Drug-like properties and the causes of poor solubility and poor permeability. *J Pharmacol Toxicol Methods* 44(1):235–49.

Luong C, Miller A, Barnett J, Chow J, Ramesha C, Browner MF (1996): Flexibility of the NSAID binding site in the structure of human cyclcooxygenase-2. *Nat Struct Biol* 3(11):927–33.

Marnett LJ, Kalgutkar AS (1999): Cyclooxygenase 2 inhibitors: discovery, selectivity and the future. *Trends Pharmacol Sci* 20(11):465–9.

McClurg M, Keenan C (1999): Antisense as a drug discovery tool. *Innovations Pharm Tech* 99(3):14–9.

McDevitt D, Rosenberg M (2001): Exploiting genomics to discover new antibiotics. *Trends Microbiol* 9(12):611–7.

Mills JEJ, Perkins TDJ, Dean PM (1997): An automated method for predicting the positions of hydrogen-bonding atoms in binding sites. *J Comput-Aided Mol Des* 11(3):229–42.

Nicholls A, Sharp KA, Honig B (1991): Protein folding and association: insights from the interfacial and thermodynamic properties of hydrocarbons. *Proteins Struct Func Gen* 11(4):281–96.

Niefind K, Guerra B, Pinna LA, Issinger O-G, Schomburg D (1998): Crystal structure of the catalytic subunit of protein kinase CK2 from Zea mays at 2.1 A resolution. *EMBO J* 17:2451–62.

Oefner C, D'Arcy A, Hennig M, Winkler FK, Dale GE (2000): Structure of human neutral endopeptidase (neprilysin) complexed with phosphoramidon. *J Mol Biol* 296(2):341–9.

Pfost DR, Boyce-Jacino MT, Grant DM (2000): A SNPshot: pharmacogenetics and the future of drug therapy. *Trends Biotech* 18(8):334–8.

Ratti E, Trist D (2001): The continuing evolution of the drug discovery process in the pharmaceutical industry. *Il Farmaco* 56:13–19.

Roques BP (1985): Enkephalinase inhibitors and molecular exploration of the differences between active sites of enkephalinase and the angiotensin converting enzyme. *J Pharmacol* 16(Suppl. 1):5–31.

Sadowski J, Kubinyi H (1998): A scoring scheme for discriminating between drugs and nondrugs. *J Med Chem* 41(18):3325–9.

Sallinen S-L, Sallinen PK, Haapasalo HK, Helin HJ, Helén PT, Schraml P, Kallioniemi O-P, Kononen J (2000): Identification of differentially expressed genes in human gliomas by DNA microarray and tissue chip techniques. *Cancer Res* 60:6617–22.

Sandberg M, Erikkson L, Jonsson J, Sjostrom M, Wold S (1998): New chemical descriptors relevant for the design of biologically active peptides. A multivariate characterization of 87 amino acids. *J Med Chem* 41(14):2481–91.

Shapiro L, Harris T (2000): Finding function through structural genomics. *Curr Opin Biotech* 11(1):31–5.

Smith WL, DeWitt DL, Garavito RM (2000): Cyclooxygenases: structural, cellular and molecular biology. *Ann Rev Biochem* 69:145–82.

Sneader W (1985): *The Evolution of Modern Medicines*. New York: John Wiley & Sons.

Spaltmann F, Blunck M, Ziegelbauer. K (1999): Computer-aided target selection—prioritizing targets for antifungal drug discovery. *Drug Discovery Today* 4(1.):

Steinmetzer T, Hauptmann J, Sturzebecher J (2001): Advances in the development of thrombin inhibitors. *Expert Opin Invest Drugs* 10(5):845–64.

Storici P, Capitani G, De Biase D, Moser M, John RA, Jansonius JN, Schirmer T (1999): Crystal structure of GABA-Aminotransferase, a target for antiepileptic drug therapy. *Biochemistry* 38:8628–34.

Venkatesh S, Lipper RA (1999): Role of the development scientist in compound lead selection and optimization. *J Pharmac Sci* 89(2):145–54.

Venter et al. (2001): The sequence of the human genome. *Science* 291:1304–51.

Walker JE, Saraste M, Runswick MJ, Gay NJ (1982): Distantly related sequences in the $\alpha$-
and $\beta$-subunits of ATP synthase, myosin, kinases and other ATP-requiring enzymes and a
common nucleotide binding fold. *EMBO J* 1:945–51.

Walters WP, Murcko MA (2002): Prediction of 'drug-likeness'. *Adv Drug Deliv Rev*
54(3):255–71.

Wigley DB, Davies GJ, Dodson EJ, Maxwell A, Dodson G (1991): Crystal structure of an N-
terminal fragment of the DNA gyrase B protein. *Nature* 351:624–9.

Williams PA, Cosme J, Sridhar V, Johnson EF, McRee DE (2000): Mammalian microsomal
cytochrome P450 monoxygenase: structural adaptations for membrane binding and functional
diversity. *Molec Cell* 5:121–31.

# Section VII

## STRUCTURE PREDICTION

# 24

# CASP AND CAFASP EXPERIMENTS AND THEIR FINDINGS

Philip E. Bourne

The prediction of the three-dimensional (3D) structure of a protein from its one-dimensional (1D) protein sequence is a much published and debated area of structural bioinformatics. This prediction involves the kind of fold that the given amino acid sequence may adopt; in other words, whether it takes a *new fold* or one of the existing folds. If the sequence takes one of the existing folds, which is the most suitable fold among the known folds (*fold recognition*)? When fold recognition is apparent because of good sequence similarity to one of the known structures, then the question is how best can one model the structure of the given sequence, taking the relevant information from existing homologous structures in the Protein Data Bank (PDB) (*comparative modeling*). Alternatively, what if none of this information is available and the structure is modeled from first principles (*ab initio*)?

What makes structure prediction (at least to this author's knowledge) unique among scientific endeavors is the manner in which progress in the field is measured. The Critical Assessment of Structure Prediction (CASP) and the Critical Assessment of Fully Automated Structure Prediction (CAFASP) experiments provide a measure of this progress by attempting to measure in a quantitative way the success of many research groups on a predefined set of structures. Beyond a measure of progress, as is true of all good experiments, these experiments suggest new ways of addressing the problem, and influence the results presented as subsequent CASPs and CAFASPs.

The approach is for an independent group to solicit protein targets for use in CASP and CAFASP months in advance on their availability. The targets are NMR and X-ray protein structures comprising one or more domains either determined and not published or anticipated to be determined in time for review and to provide the sequence of those targets to the groups competing in CASP and CAFASP. These groups then make a

*Structural Bioinformatics*
Edited by Philip E. Bourne and Helge Weissig
Copyright © 2003 by Wiley-Liss, Inc.

to the groups competing in CASP and CAFASP. These groups then make a series of blind predictions of the 3D structure based on the protein sequence and submit those results to a server for independent and comparative review. CAFASP predictions are collected from prediction servers registered with CAFASP in an automated manner within a very short (48 hours) period following the release of the target and may also be used for subsequent CASP experiments by some groups (all CAFASP predictions are available from the CAFASP site). CASP predictions take longer and the results of both are ranked by a number of criteria that depend on the rules decided by the official CASP organizers and then members of the groups gather at the biannual Asilomar meeting to review their results and hear from members of the groups who did best or came up with a significantly new methodology and those who assessed the different classes of predictions.

I do not know another scientific endeavor that is so open and, from a predictor's perspective, so blatantly competitive. I see it as a testament to the character of those scientists in the field of structural bioinformatics who are willing to share their formative ideas and, data and have them all openly reviewed. Having a laboratory that has competed in a minor way in two past CASPs I can tell you that it becomes a compelling exercise and, when arriving at Asilomar and getting the proceedings containing the results, many different emotions—surprise, joy, disappointment—all bubble to the surface. My laboratory is not alone in participating in this endeavor. Figure 24.1 indicates the number of groups participating in the four CASPs held since 1994. Clearly structure prediction is considered a compelling area of study by many, who are affectionally known as "CASPers." It is human nature to at least in some measure regard this as an individual or at least a team effort against stiff competition. However, to the organizers, assessors, and the U.S. federal agencies (National Institutes of Health, National Library of Medicine and Department of Energy) that fund these efforts, it provides a measure of global improvement and an indication of what specific areas of protein structure prediction need to be improved. Consider an example. CASP4 reports that the protein's structure prediction community has been unable to significantly improve

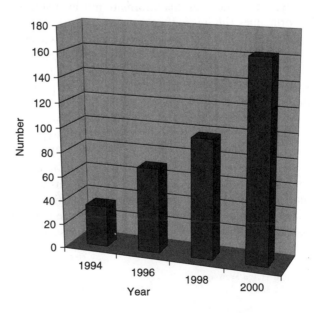

Figure 24.1. Number of groups participating in CASP in 1994, 1996, 1998, and 2000.

the ability to derive a model structure from a target recognized by comparative modeling that is significantly closer to the real structure than the target from which it was derived. This conclusion is an example of a bottleneck in our progress and will likely receive significant attention in CASP5. Other bottlenecks are outlined below.

Much has been written about CASP and CAFASP results. There is no intent here to repeat or even review what has already been written. Here I provide a synopsis of what CASP and CAFASP mean to the field of structural bioinformatics and as such serves as a forward to the three chapters that follow; chapters that review the three key areas of protein structure prediction: comparative (also known as homology) modeling, fold recognition (also known as threading) and ab initio, the latter having been reclassified as "new fold methods" in CASP4 in recognition of the fact that existing structure information is used in some way even by these methods.

The CASP experiments were run in 1994, 1996, 1998, and 2000, and CASP5 will be run in the summer of 2002. CASPs 2–4 have been extensively documented in supplements to the journal *Proteins: Structure Function and Genetics* (CASP2 29(S1); CASP3 37(S3); CASP4 45(S5)). CAFASP arose in part through recognition of the emergence of high-throughput structure prediction techniques. Although a better prediction of a single structure might be made by an expert with reference to the literature, this manual approach takes time and does not scale to, for example, predicting structure of all genes or open reading frames in the genome of a higher organism. Fully automated procedures are practical for large-scale predictions and they too should be assessed, hence the emergence of CAFASP run independently for the same set of targets as available as CASP4. In 2002 CAFASP will be more fully integrated with CASP.

## WHAT PROGRESS HAS BEEN MADE?

Progress over all CASPs is summarized by Venclovas et al. (2001). In summary, some areas have advanced and some have remained almost static. We explore this further below for each major methodology. At CASP3 the question was posed, "When will we be able to reliably predict a protein structure?" The answer returned at that time was when we have determined all structures experimentally or at least have a representative of each fold. A statement that in part led to the emergence of the structure genomics initiative (Chapter 29), which has as one objective the filling in of protein fold space so that comparative modeling can be more useful. As we see below, the situation is not quite that simple. Then, as if in some form of retort, the quality of new fold predictions improved in CASP4 with some of the contact predictions approaching a useful level of accuracy. Detection of homologous folds at lower levels of sequence identity has also improved. Nevertheless, the best models are still not good enough for defining function. In part this relates to the follow-on steps of homology modeling, which are discussed in detail in Chapter 25. For example, both comparative modeling and fold recognition alignment of the proposed structure with the template remain a problem, as well as subsequent improvement of the model with the template as a starting point.

Asking what progress has been made is a problem in itself, since some would say it depends on how you measure success or failure. The CASP organizers have been innovative in their evolving approach to this problem. Nevertheless, a quantitative answer is hard to come by. Consider a couple of crude examples. First, for a multidomain protein, a group might do very well on predicting a single domain, but when measured over the whole protein the rmsd is poor. The assessors have attempted to address this situation with a variety of measures relating to, for example, the percentage of correctly aligned residues, measurement by domain, correct positioning of biologically important residues, positions of side chains in the core and overall, and so forth. Second, a complete genome analysis done in batch mode identifies the structure of a protein by matching it to a template using comparative modeling. A group spends all summer on a single prediction of the same protein using manual techniques and all available biological knowledge. When superimposed the two models differ by approximately 1 Å rmsd overall. However, the "summer" group were able to detect features of the active site that led to some functional predictions of value to biologists. Meanwhile, the "high-throughput" group missed the functional prediction, but they made useful discoveries by modeling easier proteins located elsewhere in the genome. How do you measure which approach is better in such a context? The simple answer is that both approaches have merit and eventually each will contribute to progress in the other.

## COMPARATIVE MODELING

As described in Chapter 25 comparative modeling can be used when there is a clear relationship between the sequence of a protein of unknown structure to that of a sequence of a known structure, most likely found in the Protein Data Bank (PDB; Chapter 9). The most recent discussion on the results of comparative modeling comes from the CASP4 experiment by Tramontano, Leplae, and Morea (2001) who undertook a detailed analysis of those predictions in this category. While they rightly took great pains to emphasize the difficulties in making assessments, they concluded:

- Overall little progress was made since CASP3.
- Alignment of target to template remains a problem and more importantly the quality of the alignment does not correlate well with the level of sequence identity between template and target even at levels of sequence identity approaching 50%. The best methods rarely achieve over an 80% correct alignment with sequence identities below 50%.
- On average biologically important regions are predicted better than the protein as a whole. However, this finding has more to do with the spatial conservation of key residues important to function than a testament of the methods applied. Better prediction of biologically important regions assumes of course that the best template is chosen on which to model.
- Loop modeling remains a significant problem.
- Improvements could occur as the database of available targets continues to grow. The plus side is that a better template may be devised from multiple experimental structures; the negative side is that there is more opportunity to select a completely incorrect template. Thus, correct template(s) selection becomes a greater challenge as the databases of experimental structures increase.
- Some automated servers perform as well as individual efforts.
- Prediction of the relative orientation of domains relative to those seen in the templates remains elusive.

In summary, we have a way to go before comparative models prove consistently useful surrogates for experimental structures. At this time it would appear impossible to consistently use such models in rational drug design experiments or mutagenesis experiments other than in active site regions, especially when the sequence identity to the template is low.

## FOLD RECOGNITION

As described in Chapter 26, fold recognition techniques deal with finding relationships between sequence and structure that do exist, but are not immediately obvious, that is, a successful model will be proven to have structural similarity to a known fold, but no immediately obvious sequence similarity. Targets in this category generally fall within the twilight and midnight zones of sequence–structure relationships (Rost, 1999). Thus, fold recognition depends on advanced sequence comparison methods, comparisons of secondary structure, and the threading of sequences onto a variety of templates looking for a favorable hit. One measure of the popularity of the approach is that from CASP3 to CASP4 the number of predictions in this category rose from 3807 to 11,136. Conclusions from CASP4 in the fold recognition category are:

- Several groups submitted models that were much closer to the true structure than any of the existing templates within the PDB. But at the same time some of the same groups made completely incorrect predictions. Nevertheless, there was a qualitative assessment that the top scoring groups had made significant progress since CASP3.
- As is true for comparative modeling, prediction of multidomain proteins is more difficult than that for single-domain proteins.
- Predictions varied widely, with a large number of poor predictions. Several of the public servers performed better than more than one-half of the predicting groups. As the assessors pointed out, this result would seem to indicate some groups are less interested in their relative performance than the low probability that they will achieve a valuable prediction. Taking this inference further, there would seem to be a relatively small number of predictors with significant experience in both the process and the techniques for good prediction. Or even further, the best predictors use or have been able to provide their own methodologies, at least to some extent, in automated servers for the benefit of the whole community.

## NOVEL FOLD RECOGNITION

This class of prediction was known in earlier CASPs as *ab initio* fold prediction, but was renamed in CASP4 to better define the current methodologies that are being applied, particularly, to separate methodologies that are using sequence homology from those that are not, by simply testing the latter ones on targets where no sequence homology is actually known. Now *ab initio* is reserved for those methods that rely only on physical principles and not on any existing structure on sequence data. Clearly this is a fine line since those physical principles are themselves derived from known structure and sequence, but it is meant to imply that they are used to define general principles, rather than used directly. Chapter 28 discusses this further. The results from CASP4 for novel fold recognition can be found in Lesk, Lo Conte, and Hubbard (2001). Success in this category was measured in terms of tertiary structure prediction, secondary structure prediction, and residue-residue contacts. The conclusions in this category from CASP4 are:

- Progress has been made in the areas of tertiary structure prediction and in contact prediction, both using *ab initio* methods and knowledge-based methods.
- Secondary structure prediction still breaks down with the appearance of unusual secondary structures, for example, very long helices that were broken into fragments by all participating groups.

- Assessment should be performed with some consideration for the difficulty of the target even though this difficulty is hard to measure. Specifically, with the continuity in fold space there is ambiguity in what can be considered a new fold, but if a fold is believed to be truly new that should be weighted higher than a fold that at least has partial similarity to a known fold.

## CAFASP

In recognition of the value of automated prediction servers—which in part reflect progress in structure prediction influenced by previous CASPs—the results of CAFASP2 were published along with the CASP4 results (Fischer et al., 2001). Overall, according to the CAFASP assessors, only 11 groups in CASP performed better that the automated servers and a number of those groups clearly used the automated servers as part of their prediction strategy. However, the best human predictions do much better than the best automated predictions. Moreover, perhaps stating the obvious, difficult targets for humans are also difficult targets for automated methods, and there is much room for improvement in both categories. This comparison between CASP and CAFASP results is useful in a number of ways; most notably, it indicates to structure predictors what elements of the expert contribution need still to be added to automated approaches—clearly a nontrivial exercise—and for a biologist how valuable are the predictions compared to the best expert opinion and which Internet-accessible servers perform best.

CAFASP2 characterized five classes of server: fold recognition (19), secondary structure prediction (8), contacts prediction (2), *ab initio* (2), and homology modeling (3). The numbers in parentheses indicate how many servers were in each category. Some general observations are:

- Targets were divided into two classes: homology modeling (15) and fold recognition (26). The top ranking servers produced correct models for all homology modeling targets, but for only 5 of the 26 fold recognition targets (of the 21 not well predicted, 4 had new folds).
- In the fold recognition server group, the servers combined found approximately twice the number of targets than did any server alone, speaking to the value of a well-evaluated consensus approach.
- Secondary structure prediction accuracy was measured at 76% overall, but there was insufficient data to provide a detailed comparative analysis.

## SUMMARY

This short introductory chapter is intended simply to introduce a sense of the progress, limitations, challenges, and likely future developments in the field of protein structure prediction through what seems to be a unique scientific process. CASP and CAFASP represent a direct challenge and careful assessment of a field of study that has captured the interest of many scientists. Three of the best scientists in the field and their colleagues provide a more detailed description of the field and how it is developing in Chapters 25, 26, and 27.

As prediction methods have advanced the distinction between comparative modeling, fold recognition, and novel fold recognition have blurred somewhat. It is a testament to the community that as the knowledge of the algorithms evolved, World Wide Web servers providing access to these algorithms appeared. Thus, making it relatively straightforward for any investigator to apply a melting pot of methods to the prediction process. What all approaches need are more targets and a continued refinement to the evaluation process. The first need is being met in part by the PDB, which is, with depositors' approval, releasing sequences ahead of structure release (see http://www.rcsb.org/pdb/status.html). Further, the structural genomics projects are reporting their progress for all targets on a weekly basis (see http://targetdb.pdb.org/). While there is no indication that the sequences of the latter will lead to a structure, it is a rich source of targets (17,000 in October 2002).

Not only do CASP and CAFASP measure progress, they help define where efforts should be directed to move the field forward. It is a testament to how far the field has come that investigators are now turning to the unknown. Although attempting to predict a structure that will appear experimentally helps improve the methods applied to structure prediction, it does not further our understanding of living systems directly. Attempts at defining the "The Most Wanted" (Abbott, 2001)—the structures most in need of prediction to help further our understanding of the biology, and the efforts to make those predictions, speak to a healthy future for the field of protein structure prediction. To the many individuals who help define the CASP and CAFASP processes, serve the community as assesors and compete in the experiments this is a tribute.

## REFERENCES

Abbott A (2001): Computer modelers seek out 'Ten Most Wanted' proteins. *Nature* 409(6816):4.

Fischer D, Elofsson A, Rychlewski L, Pazos F, Valencia A, Rost B, Ortiz AR, Dunbrack RL Jr (2001): CAFASP2: the second critical assessment of fully automated structure prediction methods. *Proteins* 45 Suppl 5:171–83.

Lesk AM, Lo Conte L, Hubbard TJ (2001): Assessment of novel fold targets in CASP4: predictions of three-dimensional structures, secondary structures, and interresidue contacts. *Proteins* 45 Suppl 5:98–118.

Rost B (1999): Twilight zone of protein sequence alignments. *Protein Eng* 12(2):85–94.

Tramontano A, Lcplae R, Morea V (2001): Analysis and assessment of comparative modeling predictions in CASP4. *Proteins* 45 Suppl 5:22–38.

Venclovas ZA, Fidelis K, Moult J (2001): Comparison of performance in successive CASP experiments. *Proteins* 45 Suppl 5:163–70.

# 25

# HOMOLOGY MODELING

Elmar Krieger, Sander B. Nabuurs, and Gert Vriend

The ultimate goal of protein modeling is to predict a structure from its sequence with an accuracy that is comparable to the best results achieved experimentally. This would allow users to safely use rapidly generated *in silico* protein models in all the contexts where today only experimental structures provide a solid basis: structure-based drug design, analysis of protein function, interactions, antigenic behavior, and rational design of proteins with increased stability or novel functions. In addition, protein modeling is the only way to obtain structural information if experimental techniques fail. Many proteins are simply too large for NMR analysis and cannot be crystallized for X-ray diffraction.

Among the three major approaches to three-dimensional (3D) structure prediction described in this and the following two chapters, homology modeling is the easiest one. It is based on two major observations:

1. The structure of a protein is uniquely determined by its amino acid sequence (Epstein, Goldberger, and Anfinsen, 1963). Knowing the sequence should, at least in theory, suffice to obtain the structure.

2. During evolution, the structure is more stable and changes much slower than the associated sequence, so that similar sequences adopt practically identical structures, and distantly related sequences still fold into similar structures. This relationship was first identified by Chothia and Lesk (1986) and later quantified by Sander and Schneider (1991). Thanks to the exponential growth of the Protein Data Bank (PDB), Rost (1999) could recently derive an accurate limit for this rule, shown in Figure 25.1. As long as the length of two sequences and the percentage of identical residues fall in the region marked as "safe," the two sequences are practically guaranteed to adopt a similar structure.

*Structural Bioinformatics*
Edited by Philip E. Bourne and Helge Weissig
Copyright © 2003 by Wiley-Liss, Inc.

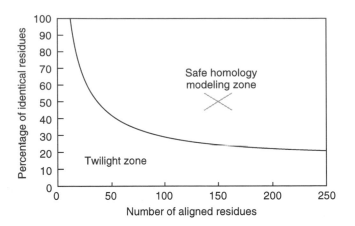

**Figure 25.1.** The two zones of sequence alignments. Two sequences are practically guaranteed to fold into the same structure if their length and percentage sequence identity fall into the region marked as "safe." An example of two sequences with 150 amino acids, 50% of which are identical, is shown (gray cross).

Imagine that we want to know the structure of sequence A (150 amino acids long, Figure 25.2, steps 1 and 2). We compare sequence A to all the sequences of known structures stored in the PDB (using, for example, BLAST), and luckily find a sequence B (300 amino acids long) containing a region of 150 amino acids that match sequence A with 50% identical residues. As this match (alignment) clearly falls in the safe zone (Fig. 25.1), we can simply take the known structure of sequence B (the template), cut out the fragment corresponding to the aligned region, mutate those amino acids that differ between sequences A and B, and finally arrive at our model for structure A. Structure A is called the *target* and is of course not known at the time of modeling. In practice, homology modeling is a multistep process that can be summarized in seven steps:

1. Template recognition and initial alignment
2. Alignment correction
3. Backbone generation
4. Loop modeling
5. Side-chain modeling
6. Model optimization
7. Model validation

At almost all the steps choices have to be made. The modeler can never be sure to make the best ones, and thus a large part of the modeling process consists of serious thought about how to gamble between multiple seemingly similar choices. A lot of research has been spent on teaching the computer how to make these decisions, so that homology models can be built fully automatically. Currently, this allows modelers to construct models for about 25% of the amino acids in a genome, thereby supplementing the efforts of structural genomics projects (Sanchez and Šali, 1999, Peitsch, Schwede, and Guex, 2000). This average value of 25% differs significantly

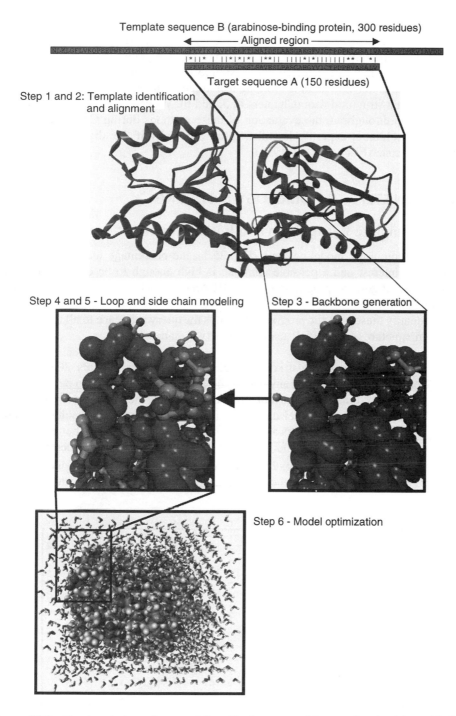

Figure 25.2. The steps to homology modeling. The fragment of the template (arabinose-binding protein) corresponding to the region aligned with the target sequence forms the basis of the model (including conserved side chains). Loops and missing side chains are predicted, then the model is optimized (in this case together with surrounding water molecules). Images created with Yasara (www.yasara.com). Figure also appears in Color Figure section.

between individual genomes, ranging from 16% (*Mycoplasma pneumoniae*) to 30% (*Haemophilus influenzae*) and increasing steadily thanks to the continuous growth of the PDB. For the remaining ~75% of a genome, no template with a known structure is available (or cannot be detected with a simple BLAST run), and one must use fold recognition (Chapter 26), *ab initio* folding techniques (Chapter 27), or simply an experiment to obtain structural data (Chapters 4, 5, and 6). While automated model building provides high throughput, the evaluation of these methods during CASP (Chapter 24) indicated that human expertise is still helpful, especially if the alignment is close to the twilight zone (Fischer et al., 1999).

## THE SEVEN STEPS TO HOMOLOGY MODELING

### Step 1: Template Recognition and Initial Alignment

In the safe homology modeling zone (Fig. 25.1), the percentage identity between the sequence of interest and a possible template is high enough to be detected with simple sequence alignment programs such as BLAST (Altschul et al., 1990) or FASTA (Pearson, 1990).

To identify these hits, the program compares the query sequence to all the sequences of known structures in the PDB using mainly two matrices:

1. A residue exchange matrix (Fig. 25.3). The elements of this $20 * 20$ matrix define the likelihood that any two of the 20 amino acids ought to be aligned. It is clearly seen that the values along the diagonal (representing conserved residues) are highest, but one can also observe that exchanges between residue types with similar physicochemical properties (for example $F \rightarrow Y$) get a better score than exchanges between residue types that widely differ in their properties.

2. An alignment matrix (Fig. 25.4). The axes of this matrix correspond to the two sequences to align, and the matrix elements are simply the values from the

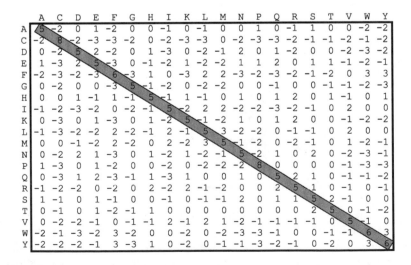

**Figure 25.3.** A typical residue exchange or scoring matrix used by alignment algorithms. Because the score for aligning residues A and B is normally the same as for B and A, this matrix is symmetric.

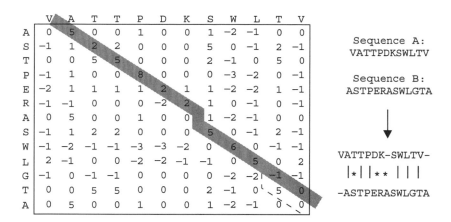

|   | V | A | T | T | P | D | K | S | W | L | T | V |
|---|---|---|---|---|---|---|---|---|---|---|---|---|
| A | 0 | 5 | 0 | 0 | 1 | 0 | 0 | 1 | -2 | -1 | 0 | 0 |
| S | -1 | 1 | 2 | 2 | 0 | 0 | 0 | 5 | 0 | -1 | 2 | -1 |
| T | 0 | 0 | 5 | 5 | 0 | 0 | 0 | 2 | -1 | 0 | 5 | 0 |
| P | -1 | 1 | 0 | 0 | 8 | 0 | 0 | 0 | -3 | -2 | 0 | -1 |
| E | -2 | 1 | 1 | 1 | 1 | 2 | 1 | 1 | -2 | -2 | 1 | -1 |
| R | -1 | -1 | 0 | 0 | 0 | -2 | 2 | 1 | 0 | -1 | 0 | -1 |
| A | 0 | 5 | 0 | 0 | 1 | 0 | 0 | 1 | -2 | -1 | 0 | 0 |
| S | -1 | 1 | 2 | 2 | 0 | 0 | 0 | 5 | 0 | -1 | 2 | -1 |
| W | -1 | -2 | -1 | -1 | -3 | -3 | -2 | 0 | 6 | 0 | -1 | -1 |
| L | 2 | -1 | 0 | 0 | -2 | -2 | -1 | -1 | 0 | 5 | 0 | 2 |
| G | -1 | 0 | -1 | -1 | 0 | 0 | 0 | 0 | -2 | -2 | -1 | -1 |
| T | 0 | 0 | 5 | 5 | 0 | 0 | 0 | 2 | -1 | 0 | 5 | 0 |
| A | 0 | 5 | 0 | 0 | 1 | 0 | 0 | 1 | -2 | -1 | 0 | 0 |

Sequence A:
VATTPDKSWLTV

Sequence B:
ASTPERASWLGTA

↓

VATTPDK-SWLTV-
|*||**  |||
-ASTPERASWLGTA

**Figure 25.4.** The alignment matrix for the sequences VATTPDKSWLTV and ASTPERASWLGTA, using the scores from Figure 25.3. The optimum path corresponding to the alignment on the right side is shown in gray. Residues with similar properties are marked with a star (*). The dashed line marks an alternative alignment that scores more points but requires opening a second gap.

residue exchange matrix (Fig. 25.3) for a given pair of residues. During the alignment process, one tries to find the best path through this matrix, starting from a point near the top left, and going down to the bottom right. To make sure that no residue is used twice, one must always take at least one step to the right and one step down. A typical alignment path is shown in Figure 25.4. At first sight, the dashed path in the bottom right corner would have led to a higher score. However, it requires the opening of an additional gap in sequence A (Gly of sequence B is skipped). By comparing thousands of sequences and sequence families, it became clear that the opening of gaps is about as unlikely as at least a couple of nonidentical residues in a row. The jump roughly in the middle of the matrix, however, is justified, because after the jump we earn lots of points (5,6,5), which would have been (1,0,0) without the jump. The alignment algorithm therefore subtracts an "opening penalty" for every new gap and a much smaller "gap extension penalty" for every residue that is skipped in the alignment. The gap extension penalty is smaller simply because one gap of three residues is much more likely than three gaps of one residue each.

In practice, one just feeds the query sequence to one of the countless BLAST servers on the web, selects a search of the PDB, and obtains a list of hits—the modeling templates and corresponding alignments (Fig. 25.2).

## Step 2: Alignment Correction

Having identified one or more possible modeling templates using the fast methods described above, it is time to consider more sophisticated methods to arrive at a better alignment.

Sometimes it may be difficult to align two sequences in a region where the percentage sequence identity is very low. One can then use other sequences from homologous

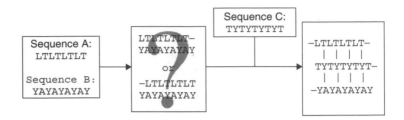

Figure 25.5. A pathological alignment problem. Sequences A and B are impossible to align, unless one considers a third sequence C from a homologous protein.

proteins to find a solution. A pathological example is shown in Figure 25.5: Suppose you want to align the sequence LTLTLTLT with YAYAYAYAY. There are two equally poor possibilities, and only a third sequence, TYTYTYTYT, that aligns easily to both of them can solve the issue.

The example above introduced a very powerful concept called "multiple sequence alignment." Many programs are available to align a number of related sequences, for example CLUSTALW (Thompson, Higgins, and Gibson, 1994), and the resulting alignment contains a lot of additional information. Think about an Ala → Glu mutation. Relying on the matrix in Figure 25.3, this exchange always gets a score of 1. In the 3D structure of the protein, it is however very unlikely to see such an Ala → Glu exchange in the hydrophobic core, but on the surface this mutation is perfectly normal. The multiple sequence alignment implicitly contains information about this structural context. If at a certain position only exchanges between hydrophobic residues are observed, it is highly likely that this residue is buried. To consider this knowledge during the alignment, one uses the multiple sequence alignment to derive position-specific scoring matrices, also called *profiles* (Taylor, 1986, Dodge, Schneider, and Sander, 1998).

When building a homology model, we are in the fortunate situation of having an almost perfect profile—the known structure of the template. We simply know that a certain alanine sits in the protein core and must therefore not be aligned with a gluta-mate. Multiple sequence alignments are nevertheless useful in homology modeling, for example, to place deletions (missing residues in the model) or insertions (additional residues in the model) only in areas where the sequences are strongly divergent. A typical example for correcting an alignment with the help of the template is shown in Figures 25.6 and 25.7. Although a simple sequence alignment gives the highest score for the wrong answer (alignment 1 in Fig. 25.6), a simple look at the structure of the template reveals that alignment 2 is correct, because it leads to a small gap, compared to a huge hole associated with alignment 1.

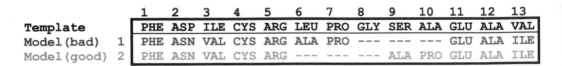

| | | 1 | 2 | 3 | 4 | 5 | 6 | 7 | 8 | 9 | 10 | 11 | 12 | 13 |
|---|---|---|---|---|---|---|---|---|---|---|---|---|---|---|
| **Template** | | PHE | ASP | ILE | CYS | ARG | LEU | PRO | GLY | SER | ALA | GLU | ALA | VAL |
| **Model (bad)** | 1 | PHE | ASN | VAL | CYS | ARG | ALA | PRO | --- | --- | --- | GLU | ALA | ILE |
| Model (good) | 2 | PHE | ASN | VAL | CYS | ARG | --- | --- | --- | ALA | PRO | GLU | ALA | ILE |

Figure 25.6. Example of a sequence alignment where a three-residue deletion must be modeled. While the first alignment appears better when considering just the sequences (a matching proline at position 7), a look at the structure of the template leads to a different conclusion (Figure 25.7).

**Figure 25.7.** Correcting an alignment based on the structure of the modeling template (Cα-trace shown in black). While the alignment with the highest score (dark gray, also in Figure 25.6) leads to a gap of 7.5 Å between residues 7 and 11, the second option (white) creates only a tiny hole of 1.3 Å between residues 5 and 9. This can easily be accommodated by small backbone shifts. (The normal Cα–Cα distance of 3.8 Å has been subtracted).

## Step 3: Backbone Generation

When the alignment is ready, the actual model building can start. Creating the backbone is trivial for most of the model: One simply copies the coordinates of those template residues that show up in the alignment with the model sequence (Fig. 25.2). If two aligned residues differ, only the backbone coordinates (N,Cα,C and O) can be copied. If they are the same, one can also include the side chain (at least the more rigid side chains, since rotamers tend to be conserved).

Experimentally determined protein structures are not perfect (but still better than models in most cases). There are countless sources of errors, ranging from poor electron density in the X-ray diffraction map to simple human errors when preparing the PDB file for submission. A lot of work has been spent on writing software to detect these errors (correcting them is even more difficult), and the current count is at more than 10,000,000 problems in the 17,000 structures deposited in the PDB by the end of 2001. It is obvious that a straightforward way to build a good model is to choose the template with the fewest errors (the PDBREPORT database [Hooft et al., 1996] at www.cmbi.nl/gv/pdbreport can be very helpful). But what if two templates are available, and each has a poorly determined region, but these regions are not the same? One should clearly combine the good parts of both templates in one model—an approach known as multiple template modeling. (The same applies if the alignments between the model sequence and possible templates show good matches in different regions). Although in principle multiple template modeling is a nice idea (and done by automated modeling servers such as Swiss-Model [Peitsch, Schwede, and Guex, 2000]), it is difficult in practice to achieve results that are really closer to the true structure than all the templates. Nevertheless, it is possible, as has been shown by AndrejŠalis' group in CASP4.

## Step 4: Loop Modeling

In the majority of cases, the alignment between model and template sequence contains gaps. Either gaps in the model sequence (deletions as shown in Figs. 25.6 and 25.7) or in the template sequence (insertions). In the first case, one simply omits residues from the template, creating a hole in the model that must be closed. In the second case, one takes the continuous backbone from the template, cuts it, and inserts the missing residues. Both cases imply a conformational change of the backbone. The good news is that conformational changes cannot happen within regular secondary structure elements. It is therefore safe to shift all insertions or deletions in the alignment out of helices and strands, placing them in loops and turns. The bad news is that these changes in loop conformation are notoriously difficult to predict (one big unsolved problem in homology modeling). To make things worse, even without insertions or deletions we often find quite different loop conformations in template and target. Three main reasons can be identified (Rodriguez, http://www.cmbi.kun.nl/gv/articles/text/gambling.html):

1. Surface loops tend to be involved in crystal contacts, leading to a significant conformational change between template and target.
2. The exchange of small to bulky side chains underneath the loop pushes it aside.
3. The mutation of a loop residue to proline or from glycine to any other residue. In both cases, the new residue must fit into a more restricted area in the Ramachandran plot, which most of the time requires conformational changes of the loop.

There are two main approaches to loop modeling:

1. Knowledge based: one searches the PDB for known loops with endpoints that match the residues between which the loop has to be inserted, and simply copies the loop conformation. All major molecular modeling programs and servers support this approach (e.g., 3D-Jigsaw [Bates and Sternberg, 1999], Insight [Dayringer, Tramontano, and Fletterick, 1986], Modeller [Šali and Blundell, 1993], Swiss-Model [Peitsch, Schwede, and Guex, 2000], or WHAT IF [Vriend, 1990]).
2. Energy based: as in true *ab initio* fold prediction, an energy function is used to judge the quality of a loop. Then this function is minimized, using Monte Carlo (Simons et al., 1999) or molecular dynamics techniques (Fiser, Do, and Šali, 2000) to arrive at the best loop conformation. Often the energy function is modified (e.g., smoothed) to facilitate the search (Tappura, 2001).

At least for short loops (up to 5–8 residues), the various methods have a reasonable chance of predicting a loop conformation that superimposes well on the true structure. As mentioned above, surface loops tend to change their conformation due to crystal contacts. So if the prediction is made for an isolated protein and then found to differ from the crystal structure, it might still be correct.

## Step 5: Side-Chain Modeling

When we compare the side-chain conformations (rotamers) of residues that are conserved in structurally similar proteins, we find that they often have similar $\chi_1$-angles (i.e., the torsion angle about the $C_\alpha - C_\beta$ bond). It is therefore possible to simply copy conserved residues entirely from the template to the model (see also Step 3) and achieve a higher accuracy than by copying just the backbone and repredicting the side chains. In practice, this rule of thumb holds only at high levels of sequence identity, when the conserved residues form networks of contacts. When they get isolated (<35%

sequence identity), the rotamers of conserved residues may differ in up to 45% of the cases (Sanchez and Šali, 1997).

Practically all successful approaches to side-chain placement are at least partly knowledge based. They use libraries of common rotamers extracted from high-resolution X-ray structures. The various rotamers are tried successively and scored with a variety of energy functions. Intuitively, one might expect rotamer prediction to be computationally demanding due to the combinatorial explosion—the choice of a certain rotamer automatically affects the rotamers of all neighboring residues, which in turn affect their neighbors and so on. With 100 residues and on average ~5 rotamers per residue, one would already end up at $5^{100}$ different combinations to score. A lot of research has been spent on the development of methods to make this enormous search space tractable (Desmet et al., 1992). The number of combinations is in fact so large, that even nature could not try all of them during the folding process, which indicates that there must exist mechanisms to shrink down the search space.

Beside the trivial fact that copying conserved rotamers from the template often splits up the protein into distinct regions where rotamers can be predicted independently, the key to handling the combinatorial explosion lies in the protein backbone. Certain backbone conformations strongly favor certain rotamers (allowing, for example, a hydrogen bond between side chain and backbone) and thus greatly reduce the search space. For a given backbone conformation, there may be only one strongly populated rotamer that can be modeled right away, thereby providing an anchor for surrounding, more flexible side chains. An example for a backbone conformation that favors two different tyrosine rotamers is shown in Figure 25.8. These position specific rotamer libraries are widely used today (de Filippis, Sander, and Vriend, 1994, Stites, Meeker, and Shortle, 1994, Dunbrack and Karplus, 1994). To build such a library, one takes high-resolution structures and collects all stretches of three to seven residues (depending on the method) with a given amino acid at the center. To predict a rotamer, the corresponding backbone stretch in the template is superposed on all the collected examples,

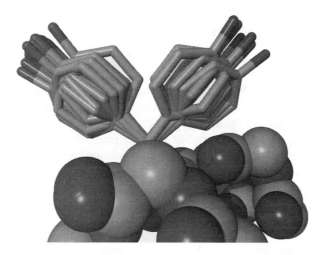

**Figure 25.8.** Example of a backbone-dependent rotamer library. The current backbone conformation (space-filling display) favors two different rotamers for Tyrosine (sticks), which appear about equally often in the database.

and the possible side-chain conformations are selected from the best backbone matches (Chinea et al., 1995).

Further evidence that the combinatorial problem of rotamer prediction is far smaller than originally believed was found recently. Xiang and Honig (2001) first removed one single side chain from known structures and repredicted it. In a second step, they removed all the side chains and added them again using the same simple search strategy. Surprisingly, it turned out that the accuracy was only marginally higher in the much easier first case.

The prediction accuracy is usually quite high for residues in the hydrophobic core where more than 90% of all $\chi_1$-angles fall within $\pm20°$ from the experimental values, but much lower for residues on the surface where the percentage is often even below 50%. There are two reasons for this:

1. Experimental reasons: flexible side chains on the surface tend to adopt multiple conformations, which are additionally influenced by crystal contacts. So even experiment cannot provide one single correct answer.
2. Theoretical reasons: the energy functions used to score rotamers can easily handle the hydrophobic packing in the core (mainly Van der Waals interactions), but are not accurate enough to get the complicated electrostatic interactions on the surface right, including hydrogen bonds with water molecules and associated entropic effects.

It is important to note that the prediction accuracies given in most publications cannot be reached in real-life applications. The reason is that the methods are evaluated by taking a known structure, removing the side chains and repredicting them. The algorithms thus rely on the correct backbone, which is not available in homology modeling. The backbone of the template often differs significantly from the target. The rotamers must thus be predicted based on an incorrect backbone and prediction accuracies tend to be lower in this case.

## Step 6: Model Optimization

The problem just mentioned above leads to a classical chicken-and-egg situation. To predict the side-chain rotamers with high accuracy, we need the correct backbone, which in turn depends on the rotamers and their packing. The common approach to such a problem is an iterative one: predict the rotamers, then the resulting shifts in the backbone, then the rotamers for the new backbone, and so on, until the procedure converges. This boils down to a sequence of rotamer prediction and energy minimization steps. The latter use the methods from the loop-modeling step above, but this time they must be applied to the entire protein structure, not just an isolated loop. Optimizing a complete protein requires an enormous accuracy in the energy function, because there are many more paths leading away from the answer (the target structure) than toward it, which is why energy minimization must be used carefully. At every minimization step, a few big errors (like bumps, i.e., too short atomic distances) are removed while many small errors are introduced. When the big errors are gone, the small ones start accumulating and the model moves away from the target (Fig. 25.9). As a rule of thumb, today's modeling programs therefore either restrain the atom positions and/or apply only a few hundred steps of energy minimization. In short, model optimization

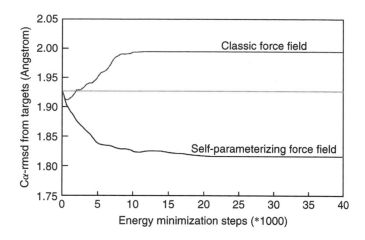

Figure 25.9. The average rmsd between models and targets during an extensive energy min-imization of 14 homology models with two different force fields. Both force fields improve the models during the first ~500 energy minimization steps but then the small errors sum up in the classic force field and guide the minimization in the wrong direction, away from the target while the self-parameterizing force field goes in the right direction. To reach experimental accuracy, the minimization would have to proceed all the way down to ~0.5 Å, which is the uncertainty in experimentally determined coordinates.

does not work until energy functions (force fields) get more accurate. Two ways to achieve that accuracy are currently being pursued:

1. Quantum force fields: protein force fields must be fast to handle these large molecules efficiently, energies are therefore normally expressed as a function of the positions of the atomic nuclei only. The continuous increase of computer power has now finally made it possible to apply methods of quantum chemistry to entire proteins, arriving at more accurate descriptions of the charge distribution (Liu et al., 2001). It is however still difficult to overcome the inherent approximations of today's quantum chemical calculations. Attractive Van der Waals forces are, for example, so difficult to treat, that they must often be completely omitted. While providing more accurate electrostatics, the overall accuracy achieved is still about the same as in the classical force fields.

2. Self-parameterizing force fields: the accuracy of a force field depends to a large extent on its parameters (e.g., Van der Waals radii, atomic charges). These parameters are usually obtained from quantum chemical calculations on small molecules and fitting to experimental data, following elaborate rules (Wang, Cieplak, and Kollman, 2000). By applying the force field to proteins, one implicitly assumes that a peptide chain is just the sum of its individual small molecule building blocks—the amino acids. Alternatively, one can just state a goal, for example, improve the models during an energy minimization, and then let the force field parameterize itself while trying to optimally fulfill this goal (Krieger, Koraimann, and Vriend, 2002). This method

leads to a computationally rather expensive procedure. Take initial parameters (for example, from an existing force field), change a parameter randomly, energy minimize models, see if the result improved, keep the new force field if yes, otherwise go back to the previous force field. With this procedure, the force field accuracy increases enough to go in the right direction during an energy minimization (Fig. 25.9), but experimental accuracy is still far out of reach.

The most straightforward approach to model optimization is simply to run a molecular dynamics simulation of the model. Such a simulation follows the motions of the protein on a femtosecond ($10^{-15}$ s) timescale and mimics the true folding process. One thus hopes that the model will complete its folding and "home in" to the true structure during the simulation. The advantage is that a molecular dynamics simulation implicitly contains entropic effects that are otherwise difficult to treat; the disadvantage is that the force fields are again not accurate enough to make it work. (One must in fact be happy if the model is not messed up during the simulation). Nevertheless, one of the main tasks of *Blue Gene*, the forthcoming fastest computer in the world, will be to run exactly this type of molecular dynamics simulations (IBM Blue Gene team, 2001). More accurate force fields will have to be available when *Blue Gene* goes online in 2005.

## Step 7: Model Validation

Every homology model contains errors. The number of errors (for a given method) mainly depends on two values:

1. The percentage sequence identity between template and target. If it is greater than 90%, the accuracy of the model can be compared to crystallographically determined structures, except for a few individual side chains (Chothia and Lesk, 1986; Sippl, 1993). From 50% to 90% identity, the rms error in the modeled coordinates can be as large as 1.5 Å, with considerably larger local errors. If the sequence identity drops to 25%, the alignment turns out to be the main bottleneck for homology modeling, often leading to very large errors.

2. The number of errors in the template.

Errors in a model become less of a problem if they can be localized. It is, for example, hardly important that a loop far away from an enzyme's active site is placed incorrectly. An essential step in the homology modeling process is therefore the verification of the model. There are two principally different ways to estimate errors in a structure:

1. Calculating the model's energy based on a force field: This method checks if the bond lengths and bond angles are within normal ranges, and if there are lots of bumps in the model (corresponding to a high Van der Waals energy). Essential questions such as "Is the model folded correctly?" cannot yet be answered this way, because completely misfolded but well-minimized models often reach the same force field energy as the target structure (Novotny, Rashin, and Bruccoleri, 1988). This surprising finding is mainly due to the fact that molecular dynamics force fields do not explicitly contain entropic terms (such as

the hydrophobic effect), but rely on the simulation to generate them. Although this problem can be addressed by extending the force field and adding, for example, a solvation term the major drawback is that one always obtains a single number for the entire protein and cannot easily trace problems down to individual residues.

2. Determination of normality indices that describe how well a given characteristic of the model resembles the same characteristic in real structures. Many features of protein structures are well suited for normality analysis. Most of them are directly or indirectly based on the analysis of interatomic distances and contacts. Some published examples are:

- General checks for the normality of bond lengths, bond and torsion angles (Morris et al., 1992; Czaplewski et al., 2000) are good checks for the quality of experimentally determined structures, but are less suitable for the evaluation of models because the better model-building programs simply do not make this kind of error.

- Inside/outside distributions of polar and apolar residues can be used to detect completely misfolded models (Baumann, Frommel, and Sander, 1989).

- The radial distribution function for a given type of atom (i.e., the probability to find certain other atoms at a given distance) can be extracted from the library of known structures and converted into an energylike quantity, called a "potential of mean force" (Sippl, 1990). Such a potential can easily distinguish good contacts (e.g., between a $C\gamma$ of valine and a $C\delta$ of isoleucine) from bad ones (e.g., between the same $C\gamma$ of valine and the positively charged amino group of lysine).

- If not only the distance, but also the direction of atomic contacts is taken into account, one arrives at 3D distribution functions that can also easily identify misfolded proteins and are good indicators of local model building problems (Vriend and Sander, 1993).

Most methods used for the verification of models can also be applied to experimental structures (and hence to the templates used for model building). A detailed verification is essential when trying to derive new information from the model, either to interpret or predict experimental results or plan new experiments.

In summary, it is safe to say that homology modeling is unfortunately not as easy as stated in the beginning. Ideally, homology modeling uses threading (Chapter 26) to improve the alignment, and *ab initio* folding (Chapter 27) to predict the loops and molecular dynamics simulations with a perfect force field to home in to the true structure. Taking these steps correctly will keep researchers busy for a long time, leaving lots of fascinating discoveries to good old experiment.

## ACKNOWLEDGMENTS

We thank Rolando Rodriguez, Chris Spronk, and Rob Hooft for stimulating discussions and practical help. We apologize to the numerous crystallographers who made all this work possible by depositing structures in the PDB for not referring to each of the 17,000 very important articles describing these structures.

## FURTHER READING

Gregoret LM, Cohen FE (1990): Novel method for the rapid evaluation of packing in protein structures. *J Mol Biol* 211:959–74.

Holm L, Sander C (1992): Evaluation of protein models by atomic solvation preference. *J Mol Biol* 225:93–105.

## REFERENCES

Altschul SF, Gish W, Miller W, Myers EW, Lipman DJ (1990): Basic local alignment search tool. *J Mol Biol* 215:403–10.

Bates PA, Sternberg MJE (1999): Model building by comparison at CASP3: using expert knowledge and computer automation. *Proteins* (Suppl. 3):47–54.

Baumann G, Frommel C, Sander C (1989): Polarity as a criterion in protein design. *Protein Eng* 2:329–34.

Chinea G, Padron G, Hooft RWW, Sander C, Vriend G (1995): The use of position specific rotamers in model building by homology. *Proteins* 23:415–21.

Chothia C, Lesk AM (1986): The relation between the divergence of sequence and structure in proteins. *EMBO J* 5:823–36.

Czaplewski C, Rodziewicz-Motowidlo S, Liwo A, Ripoll DR, Wawak RJ, Scheraga HA (2000): Molecular simulation study of cooperativity in hydrophobic association. *Protein Sci* 9:1235–45.

Dayringer HE, Tramontano A, Fletterick RJ (1986): Interactive program for visualization and modelling of proteins, nucleic acids and small molecules. *J Mol Graph* 4:82–7.

de Filippis V, Sander C, Vriend G (1994): Predicting local structural changes that result from point mutations. *Protein Eng* 7:1203–8.

Desmet J, De Maeyer M, Hazes B, Lasters I (1992): The dead-end elimination theorem and its use in protein side-chain positioning. *Nature* 356:539–42.

Dodge C, Schneider R, Sander C (1998): The HSSP database of protein structure–sequence alignments and family profiles. *Nucleic Acids Res* 26:313–5.

Dunbrack RL Jr, Karplus M (1994): Conformational analysis of the backbone dependent rotamer preferences of protein side chains. *Nat Struct Biol* 5:334–40.

Epstein CJ, Goldberger RF, Anfinsen CB (1963): The genetic control of tertiary protein structure: studies with model systems. *Cold Spring Harb Symp Quant Biol* 28:439.

Fischer D, Barret C, Bryson K, Elofsson A, Godzik A, Jones D, Karplus KJ, Kelley LA, Mac-Callum RM, Pawowski K, Rost B, Rychlewski L, Sternberg MJE (1999): CAFASP1: Critical assessment of fully automated structure prediction methods. *Proteins* (Suppl. 3):209–17.

Fiser A, Do RK, Šali A (2000): Modeling of loops in protein structures. *Protein Sci* 9:1753–73.

Hooft RWW, Vriend G, Sander C, Abola EE (1996): Errors in protein structures. *Nature* 381:272.

IBM Blue Gene team (2001): Blue Gene: a vision for protein science using a petaflop supercomputer. *IBM Sys J* 40:310–27.

Krieger E, Koraimann G, Vriend G (2002): Increasing the precision of comparative models with YASARA NOVA—a self-parameterizing force field. *Proteins* 47:393–402.

Liu H, Elstner M, Kaxiras E, Frauenheim T, Hermans J, Yang W (2001): Quantum mechanics simulation of protein dynamics on long timescale. *Proteins* 44:484–9.

Morris AL, MacArthur MW, Hutchinson EG, Thorton JM (1992): Stereochemical quality of protein structure coordinates. *Proteins* 12:345–64.

Novotny J, Rashin AA, Bruccoleri RE (1988): Criteria that discriminate between native proteins and incorrectly folded models. *Proteins* 4:19–30.

Pearson WR (1990): Rapid and sensitive sequence comparison with FASTP and FASTA. *Methods Enzymol* 183:63–98.

Peitsch MC, Schwede T, Guex N (2000): Automated protein modelling—the proteome in 3D. *Pharmacogenomics* 1:257–66.

Rost B (1999): Twilight zone of protein sequence alignments. *Protein Eng* 12:85–94.

Šali A, Blundell TL (1993): Comparative protein modelling by satisfaction of spatial restraints. *J Mol Biol* 234:779–815.

Sanchez R, Šali A (1997): Evaluation of comparative protein structure modeling by MODELLER-3. *Proteins* (Suppl. 1):50–8.

Sanchez R, Šali A (1999): ModBase: a database of comparative protein structure models. *Bioinformatics* 15:1060–1.

Sander C, Schneider R (1991): Database of homology-derived protein structures and the structural meaning of sequence alignment. *Proteins* 9:56–68.

Simons KT, Bonneau R, Ruczinski I, Baker D (1999): Ab initio structure prediction of CASP III targets using ROSETTA. *Proteins* (Suppl. 3):171–6.

Sippl MJ (1990): Calculation of conformational ensembles from potentials of mean force. *J Mol Biol* 213:859–83.

Sippl MJ (1993): Recognition of errors in three dimensional structures of proteins. *Proteins* 17:355–62.

Stites WE, Meeker AK, Shortle D (1994): Evidence for strained interactions between side-chains and the polypeptide backbone. *J Mol Biol* 235:27–32.

Tappura K (2001): Influence of rotational energy barriers to the conformational search of protein loops in molecular dynamics and ranking the conformations. *Proteins* 44:167–79.

Taylor WR (1986): Identification of protein sequence homology by consensus template alignment. *J Mol Biol* 188:233–58.

Thompson JD, Higgins DG, Gibson TJ (1994): ClustalW: improving the sensitivity of progressive multiple sequence alignments through sequence weighting, position-specific gap penalties and weight matrix choice. *Nucleic Acids Res* 22:4673–80.

Vriend G (1990): WHAT IF—A molecular modeling and drug design program. *J Molec Graphics* 8:52–6.

Vriend G, Sander C (1993): Quality control of protein models: directional atomic contact analysis. *J Applied Crystallogr* 26:47–60.

Wang J, Cieplak P, Kollman PA (2000): How well does a restrained electrostatic potential (RESP) model perform in calculating conformational energies of organic and biological molecules? *J Comput Chem* 21:1049–74.

Xiang Z, Honig B (2001): Extending the accuracy limits of prediction for side-chain conformations. *J Mol Biol* 311:421–30.

# 26

# FOLD RECOGNITION METHODS

Adam Godzik

Despite a good qualitative understanding of the forces that shape the folding process, present knowledge is not enough to be used for direct prediction of protein structure from first principles, such as the fundamental equations of physics. A related but easier problem is to recognize which of the known protein folds is likely to be similar to the (unknown) fold of a new protein when only its amino acid sequence is known. This has been variously called an inverse folding problem (find a sequence fitting a structure), threading (since a sequence is being threaded through a known structure), and finally a fold recognition problem. Solution of the fold recognition problem is a necessary prerequisite to the solution of the general folding problem. If we are unable to recognize a structure similar to the correct one, how could we possibly arrive at the correct structure starting for a random one? At the same time, even the partial solution to the fold recognition problem offers the immediate advantage of an efficient and fast structure prediction tool.

Such considerations lead to the development of methods that attempt to recognize possible structural similarities even in the absence of recognizable sequence similarity. Of course, advances in sequence analysis are constantly changing the threshold between fold recognition and sequence-based homology recognition. Therefore, in this chapter, (example, see Figure 26.2) both sequence and structure/energy-based fold recognition methods are discussed together.

The practical importance of the fold recognition approach to protein structure prediction stems from the fact that very often apparently unrelated proteins adopt similar folds. As discussed elsewhere in this volume, more than half of the newly solved proteins thought to be unrelated to any of the known proteins turn out to have a well-known fold. Only a few years ago, such cases were viewed as a curiosity, but now they become almost a rule. Some folds such as a beta barrel triose phosphate isomerase (TIM) fold, were discovered in over 20 protein superfamilies thought to be

*Structural Bioinformatics*
Edited by Philip E. Bourne and Helge Weissig
Copyright © 2003 by Wiley-Liss, Inc.

unrelated. Other popular folds often found in apparently unrelated protein families are greek-key beta barrels, which are found in immunoglobulins, copper binding proteins, several families of receptors, adhesion molecules, and so forth. Similarly, the ferrodoxin fold is present in 36 superfamilies (SCOP, 1995). Overall, only 100 folds account for about half of all protein superfamilies in one of the more popular protein structure classifications, the structural classification of proteins (SCOP) database (discussed in detail in Chapter 12 in this volume). There are several possible explanations for this phenomenon, and it is almost certain that some examples may be found for each:

- Divergent evolution. Proteins with similar folds are actually related, but our current sequence analysis tools are not sensitive enough to recognize very distant homologies. Several new algorithms, such as PSI-BLAST (Altschul et al., 1997), Hidden Markov Models (Bateman et al., 2000), and profile–profile alignment tools (Rychlewski et al., 2000) redefined sequence similarity. Many protein families that were thought to be unrelated a few years ago are now firmly in the distant homology class. With the continuous improvement of such algorithms, we can expect that a majority of cases of "unexplained structural similarity" will eventually be included in this class.
- Convergent evolution. Common functional requirements, such as binding to the same classes of substrates, lead to similar structural solutions. There are very few undisputed examples of convergent evolution and most involve similarities of small subfragments, such as active site residues (serine proteases) or binding patterns (DNA-binding proteins).
- Limited number of folds. Unrelated proteins end up having similar folds because the space of possible folds is small and nature is simply running out of solutions. Despite strong theoretical arguments (Ptitsyn and Finkelstein, 1980), there are not many examples for such "accidentally" similar structures except for very small proteins, such as three and four helical bundles. In fact, many theoretically predicted arrangements of secondary structure elements (Chothia and Finkelstein, 1990) are still not seen in nature despite rapid growth of known protein structures.
- Misguided analysis. Apparent structural similarity may result from deficiencies in our analysis tools and not from any actual similarity between protein structures. For instance, SCOP (Structured Classifications of Proteins), CATH (Class, Architecture, Topology in Homologous super family), and FSSP (Fold classification based on Structure–Structure alignments of Proteins) structure classifications agree in only about 60% of cases (Hadley and Jones, 1999). Detection of structural similarity is usually based on empirical criteria, most often on fitting similarity score to an empirical distribution of random similarity scores. Therefore, a false positive is possible when two structures are very different from all other structures, but not really similar. Of course, predicting such similarities is not really feasible, and also not particularly useful.

A correct choice between these possibilities is fundamental because it not only influences our thinking about the protein sequence/structure/function relationship, but also indicates the most efficient structure prediction strategies. Even more important, the first two possibilities suggest that there could be functional similarities between proteins with similar structures, either due to the evolutionary relationship between proteins or due to convergent evolution. This, in turn, increases the practical importance of fold recognition that can be now viewed not only as a structure prediction, but also a function prediction tool.

Closer analysis suggests that the first mechanism is responsible for most of the known examples of "unexpected" structural similarity. Often the homology was postulated only after a structural similarity was discovered (Taylor, 1986; Russell and Barton, 1992), but later accepted based on other similarities, including similarities in function. The case of the "enolase superfamily" (Babbitt et al., 1995) is particularly interesting because it illustrates the practical importance of establishing a homology relationship between protein families. In this case, distant homology between enolase and mandelate racemase was postulated based on extensive structural similarities between both enzymes, and despite the lack of (then) recognizable sequence similarity and significant differences between their biochemical function. This hypothesis led to the reevaluation of the enzymatic mechanisms of both proteins and the discovery that a crucial step in two different reactions catalyzed by these two proteins is highly similar, involving the abstraction of the $\alpha$-proton of a carboxylic acid to form an enolic intermediate. This discovery firmly established the homology of both enzymes and added to our understanding of two different enzymatic reactions, opening a new venue in designing inhibitors for both enzymes. At the same time, it made possible the structure and function prediction for several newly discovered enzymes. At the time of this discovery, fold recognition was in its infancy and none of the then available algorithms was able to recognize a homology that distant; now this would be considered a medium difficulty problem.

Our current understanding of the evolution at the molecular level is good enough to describe the process of small changes and adaptations in proteins. This understanding allows us to reliably assess relationships between proteins when their sequences are similar. It also means that in proteins for which reliable evolutionary relationships can be established, functions and structures had no time to diverge dramatically. At the same time, there is no general consensus about how new protein folds and completely new functions have emerged. Consequently, it is not clear how to study relations between proteins when there is no clear similarity between their sequences and only other arguments suggest that they might be related. The confusion extends to the nomenclature used to describe specific cases. Terms such as superfamily, fold family, and so on are used by different groups in different contexts (Doolittle, 1994). Authors variously claim distant evolutionary relationships (Farber and Petsko, 1990), convergent structural evolution (Lesk, 1995), or random similarities (Ptitsyn and Finkelstein, 1980) in seemingly similar cases.

Fold recognition can be successful in each of the first three scenarios discussed above, but it is the first one that makes it particularly useful. Protein structure prediction is rarely a goal in itself and the real questions usually concern possible functions of new proteins. If the predicted fold comes from a homologous (even distant) protein, there is a good chance that some aspects of function could also be conserved. More recent analyses of fold families suggest that usually the arrangement of active site residues, identity of cofactors, and general features of the reaction being catalyzed are often conserved for enzymes sharing the same fold (Todd, Orengo, and Thornton, 2001). The possibility of predicting even some aspects of function for new proteins adds practical importance to fold recognition, despite its humble origins as a poor man version of the protein-folding problem.

## THEORETICAL BACKGROUND FOR FOLD RECOGNITION

Two different perspectives dominate theoretical studies of proteins and as a result, there are two different classes of fold recognition algorithms. Roughly, these algorithms could be called biological and physical because to some extent they embody the research philosophies of these two disciplines. A biologist strives to explain the natural world in terms of patterns of evolution. In a tradition that reaches to Aristotle and Carolus Linnaeus, biologists identify, describe and classify the diversity of life to reveal the patterns of evolution. Following in this spirit, molecular biologists study proteins very much like eighteenth and nineteenth century naturalists studied plant and animal species. Proteins are described and classified into families and analyzed for patterns of mutations at various positions along the sequence. The sequence similarity between proteins from different species forms the basis of molecular phylogenetics, which now rivals traditional morphological phylogenetics in terms of analysis of the relationships between species. It is often extended to study relations between entire processes, such as regulatory networks and metabolic pathways or between subpopulations within species. In contrast, a physicist seeks to explain nature in terms of fundamental laws, with similarities between systems being just manifestations of the same laws at work. In this spirit, protein structure is seen as a complex shape defined by specific interactions between amino acids along the chain. Different sequences adopting similar folds are viewed as multiple solutions to the same minimization problem and fold recognition problem can be formulated as a constrained minimization where only some points in conformational space are being considered.

There are many problems that are easily understood and studied within one approach. For closely related proteins, it is possible to build reliable evolutionary trees and analyze the relationships between the organisms from which they came without any reference to the fact that these proteins fold to the same structure adopting a free energy minimum in solution. For studying the short-time dynamics of side-chain movements in proteins or for predicting the results of a point mutation we may not care about how this enzyme evolved. But many problems require insights from both perspectives. Why do some proteins have very similar structures and yet perform very different functions? Why do others have similar functions but their structures are different? How can we design a good drug that would bind to the target even if the target would undergo a mutation? How can we predict when and how function would diverge between distantly related proteins?

In this chapter, we show how these two views of the protein led to the development of two distinct classes of fold recognition algorithms. We discuss how the further progress of this and other areas of protein structure prediction relies on successfully merging these two approaches. The first part discusses the molecular evolution of proteins and how understanding of this process led to the development of more sensitive algorithms for comparison of protein sequences. The second part presents a physicist's view of the protein world, concentrating on ideas of energy, potentials of mean force, and free energy and describes prediction methods that use this language to recognize a possible fold of a new sequence.

In most of this chapter, we study questions of similarities and differences between proteins and the question of relations between them. Therefore, the term *homology* will be used only to denote an implied evolutionary relationship between proteins. In contrast, the terms *analogy* or *similarity* will be used to describe the similarity between two sequences or structures without implying any relationship between them.

## PROTEINS AS SEEN BY A BIOLOGIST

### Molecular Evolution, Sequence Similarity, and Protein Homology

Since protein sequences first became available, researchers realized that sequences of homologous proteins from related organisms are very similar. The difference grows with an increasing evolutionary distance between species, which corresponds very well to the known mechanisms of DNA replication and repair. This observation led to the development of sequence alignment methods (Doolittle, 1996), which attempt to find an optimal alignment between the sequences of two (or more) proteins being compared. Since homologous proteins by definition evolved from a common ancestor, we expect that a series of mutations, deletions, and insertions led from the common ancestor to both modern sequences. A list of such elementary steps is equivalent to a unique alignment. When possible homology between two proteins is considered, the null hypothesis is that the two sequences do not come from the common ancestor. The comparison of their sequences should be similar to that of a comparison between two random strings of letters. Therefore, if a larger than random similarity is found, it is generally assumed that such proteins are homologous. This connection is so strong that sequence similarity is de facto used as a synonym of homology, despite the fact that homology is a much stronger concept and the two are not equivalent. At the same time, two proteins may be homologous despite lack of an easily recognizable sequence similarity.

Once the homology is established, we can predict the structure and function of new proteins reasoning by analogy, assuming that in evolution at least some aspects of function are conserved. The rule that strong sequence similarity is equivalent to strong structure and function similarity is the only reliable prediction rule discovered so far. Homology modeling discussed in Chapter 25 in this volume turned this general rule into a powerful prediction method. Since the 1960s, the efforts of X-ray crystallographers and, more recently, NMR spectroscopists yielded thousands of protein structures, and biochemists have characterized tens of thousands of proteins. These proteins, with their sequences and/or structures available in public databases, form a rich source of knowledge that can be used to identify newly sequenced proteins. To apply analogy reasoning, two problems must be solved. First, the similarity must be recognized, which for distant homologs may not be trivial. Second, the detailed alignment, that is, residue-by-residue equivalence table between the two proteins must be constructed. Interestingly, the former problem turns out to be much more difficult that the first (Jaroszewski, Rychlewski, and Godzik, 2000).

### Protein Sequence Analysis

Sequence comparison is a well-developed scientific field (Doolittle, 1996; Waterman, 1995; Gribskov and Dereveux, 1991). With rigorous mathematical techniques similar to those in telecommunication signal analysis, two protein sequences are treated as two strings of characters and the similarity between them is compared to that expected by chance between random strings. If the similarity is larger than that expected by chance, common ancestry is assumed and both proteins are identified as homologous, with subsequent assertions concerning their structure and function.

The similarity between two sequence strings is usually defined as the sum of similarities between residues in both proteins at equivalent positions. In the example illustrated in Figure 26.1, identical residues are denoted by vertical bars. A scoring

Figure 26.1. An example of a sequence alignment between two proteins.

matrix, giving a numerical score for aligning any two amino acids, defines a similarity. The similarity score ranges from large and positive (for the same residue in both positions), to smaller and positive (for residues with similar features, such as valine and leucine) to large and negative (for very different residues). Scoring based on minimizing the difference between two protein sequences is also possible.

$$S = \sum_{i}^{n} \text{MM} (A_i, B_{AB(i)}), \tag{26.1}$$

where MM is the mutation matrix and AB(i) is the residue equivalent to i in sequence B. Many different similarity matrices are used in literature with several of the best quite similar to each other despite different assumptions used in their derivation (Tomii and Kanehisa, 1996; Frishman and Argos, 1996). An important feature of the scoring function such as in Eq. (26.1) is that it is additive or local; in other words, score for one position does not depend on the alignment (or on residues) in another position.

Gaps and insertions in both sequences, such as seen in Figure 26.1, are necessary for the optimal alignment. This is in full agreement with our knowledge of evolution at the molecular level, where mutations as well as deletions and insertions in DNA sequences are possible. The optimal alignment with gaps can be found by dynamic programming (Needelman and Wunsch, 1970; Smith and Waterman, 1981). Alternatively, similar sequences can be identified by searching for high scoring fragments (HSF) (Altschul et al., 1990), the uninterrupted alignment fragments, presented as highlighted boxes in Figure 26.1. Software tools, such as BLAST (Altschul et al., 1990 (now updated to PSI-BLAST [Altschul et al., 1997]) or FASTA (Pearson and Miller, 1992) became standards in searching for similar sequences in protein databases and are easily available as software packages (GCG, 1991) or WEB servers. Other sets of tools, such as CLUSTALW (Higgins and Gibson, 1995) or PileUp (GCG, 1991), address questions of organizing a family of homologous proteins into a family tree.

Unfortunately, despite their solid theoretical foundations, all methods and algorithms used in sequence analysis face the same problem. With increasing evolutionary distance, sequence similarity between homologous proteins fades. Using simple alignment tools and mutation matrix scoring, it is increasingly difficult to distinguish the homology from the null hypothesis of random similarity. This is referred to as the "twilight zone" of sequence similarities and corresponds to about 25% of identical amino acids in an optimal alignment between protein pairs. In other words, at the level of about 25% sequence identity it is equally likely to find a spurious homology as it is to find a true homology. This value strongly depends on the length of the alignment, as illustrated by well-known examples of identical pentapeptides with different structures (Kabsch and Sander, 1985; Argos, 1987) and analyzed in detail for pairs of similar structures (Sander and Schneider, 1991). Even for whole proteins, there are spurious sequence similarities at such levels. For instance, hypoxantine guanine phosphosibosyltransferase (Protein Data Bank [PDB] code 1 hmp) and the coat protein of a poliovirus (1 piv) share an 80-amino acid fragment with over 40% sequence identity, despite the lack of any structure or function similarity. This and many other examples of spurious similarities around 25% sequence identity illustrate how dangerous it is to assume homology using sequence similarity as the only argument.

In general, homologous proteins that can be reliably identified using simple sequence similarity searches are usually closely related, with little or no variation in function and generally very similar structures.

## Protein Families and Multiple Alignments

The diversity of sequences in a family of homologous proteins captures successful biological experiments in mutating a protein coding sequence without destroying its function, and what follows, its structure. We can assume that with very few exceptions of pseudogenes or dramatic changes of function between paralogous proteins, the mutations destroying a structure of a protein would not be represented among proteins existing in nature. Therefore, the analysis of a pattern of mutations in homologous families can provide us with information about the importance of various positions along the sequence and, indirectly, of types of restrictions placed on a given position by the protein function and structure. For instance, we can expect that positions that are easily mutated are not important either for function or for structure and are most likely located in exposed loops or turns. In contrast, a position in the hydrophobic core of a protein would easily accommodate only some types of mutations (hydrophob to hydrophob) but not others (hydrophob to hydrophil). In the same way, residues in active sites, on the surface, on the interface between protein domains—all have their own rules, stemming from the fact that similar mutations at different positions would lead to different effects for the entire protein and some would be easily accepted, while others will not. For this reason, a uniform mutation matrix, the same at every position along the sequence, does not provide a good description of the evolutionary process under strong pressure of preserving the structure and function of a protein. A set of position-specific mutation rules can be derived from the analysis of a multiple align-

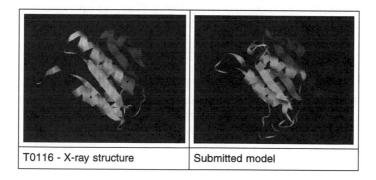

| T0116 - X-ray structure | Submitted model |

**Figure 26.2.** A successful example of a fold prediction using a profile–profile alignment program FFAS. A comparison of the predicted and experimental structure of CASP4 target 116 (see the text for the discussion of the CASP experiment). The score of the alignment was statistically significant with the e.value of e-2, despite the very low sequence similarity between the target and the template of 10% identical residues.

```
LRRLLPDDTHIMAVV ANAYGHGDVQVARTALEAGASRLAVAFLDEALALREKGIEAP        pdb|1SFT|A
FRQYVGPKTNLMAVV ADAYGHGAVRVAQTALQAGADWLAIATLGEGIELREAGITAP       ALR_SYNY3
MKKHIGEHVHLMAVE ANAYGHGDAETAKAALDAGASCLAMAILDEAISLRKKGLKAP       ALR_BACSU
LRE-LAPASKLVAVV ANAYGHGLLETART-LPD-ADAFGVARLEEALRLRAGGITQP       ALR1_SALTY
LRE-LAPASKMVAVV ANAYGHGLLETART-LPD-ADAFGVARLEEALRLRAGGITKP       ALR1_ECOLI
```

**Figure 26.3.** An example of a multiple alignment: the small part of the family of alanine racemase.

ment of a homologous family and subsequently used to align new sequences in this family. This idea, in various forms and under different names introduced by several groups (profile, position-specific mutation matrix, or Hidden Markov Models of protein families), allowed sequence analysis methods to break through the twilight zone and reliably recognize distant homologs even when their sequence identity was much below the 25% identity and comparison of single sequences appeared random.

An entire class of distant homology recognition methods evolved from the analysis of mutation patterns in homologous families (see Figure 26.3 for an example of a multiple alignment). A pattern of sequence variation along the sequence can be used to identify positions where some specific structural and/or functional requirements restrict variation, even without a full understanding of these restrictions. It is important to note that techniques used to recognize protein folds by comparing sequences (or sequence profiles), while often treated as part of fold recognition field, also can be used for more general distant homology recognition problem whether or not the distant homologs have known structure. When applied to fold recognition, these methods explicitly search for proteins from the first of the groups discussed above, the diverging homologous proteins. Distant homology recognition methods closely compete with threading, that is, energy-based fold recognition and in recent years seems to be gaining the upper hand (see later in the chapter).

From the time this idea was introduced in 1987 (Gribskov, McLachlan, and Eisenberg, 1987), it remained on the forefront of the sequence analysis field. For instance, several top algorithms in the last Critical Assessment of Structure Prediction (CASP4) meeting belong to this category. In recent years, it gained even more popularity as it was implemented in PSI-BLAST, the newest variant of the most popular sequence alignment program BLAST (Altschul et al., 1997). There are many variants and specific implementations of this basic idea (see Table 26.2) with most differences occurring in the following areas:

- Multiple alignment construction. Simultaneous alignment of several sequences is an NP-hard computational problem (Just, 2001), so most algorithms use heuristic approaches, ranging from hierarchical build-up procedures (PSI-BLAST, Pile-Up) through constructing an approximate phylogenetic tree and using it as a guide in alignment calculation (clustalw Higgins and Gibson, 1995; Jeanmougin et al., 1998) to stochastic minimization techniques, such as simulated annealing (Godzik and Skolnick, 1994) or Hidden Markov Models (Karplus et al., 1997).
- How to analyze the multiple alignment? How to extract the most relevant information from the multiple alignment? For instance, there are large groups of closely related proteins that do not add much information. Some algorithms simply average the composition at the aligned positions (GCG-Profile) or try to maximize the information content at each position (PSI-BLAST), whereas others calculate sequence weights from the matrix of interfamily similarities (FFAS).

- How is the similarity between a representation of a family and a sequence (or a second family) calculated? Some methods compare a representation of a family (profile, position-specific mutation matrix, Hidden Markov Model) to a sequence (GCG-Profile or PSI-BLAST), others compare two families (BLOCKS, FFAS). Also specifics of the scoring methods vary between methods.

Table 26.1 summarizes differences between several leading profile alignment algorithms. It is interesting to note that despite very different mathematical formulation (profile methods, position-specific mutation matrix methods, or Hidden Markov Model based methods), methods are essentially equivalent and use very similar concepts despite very different mathematical notation.

## PROTEINS AS SEEN BY A PHYSICIST

All the fold recognition methods discussed so far are based on homology recognition, that is, they assume that structural similarity results from the distant relation between the two proteins. Thus the hypothesis being tested was whether or not a new protein sequence belongs to a given family of proteins with a specific set of mutation rules. The structure was not used directly and entered the picture only by restricting accepted mutations in different ways at different positions. At the same time, most proteins fold on their own (sometimes with the help of chaperones acting as catalysts of folding), without checking what the structure of their homologs is in databases but following physical laws governing their behavior.

According to the widely accepted "thermodynamic hypothesis," the native conformation of a protein corresponds to a global free energy minimum of the protein/solvent system (Anfinsen, 1973; Privalov and Gill, 1988). Therefore, having a correct energy function, one could use the tools of computational physics to search for the native structure in conformational space. Despite many important advances (Bonneau et al., 2001), this approach is still unable to reliably predict a previously unknown structure of a protein for which only a sequence is known. Two principal problems facing the *ab initio* prediction of protein structure are the lack of adequate molecular potentials and the enormous size of the conformational space of even the smallest protein. Comparing the energy of the same system in two (or more) conformations, as done in fold recognition methods, avoids the latter problem, but unfortunately, as will be discussed later, introduces many new complications.

Energy-based fold recognition methods can be compared to minimization by a grid search, where the grid points where the energy is being calculated are based on known protein structures. Because of the visual analogy of energy calculations using a sequence of one protein forced (threaded through) to adopt a structure of another, energy-based fold recognition was called threading (Bryant and Lawrence, 1993; Godzik and Skolnick, 1992a). Since large structural databases must be scanned, energy calculations in threading algorithms by necessity must be optimized for speed. Many different threading algorithms have been developed (Bryant and Lawrence, 1993; Finkelstein and Reva, 1990; Bowie, Luethy, and Eisenberg, 1991; Sippl and Weitckus, 1992; Jones, Taylor, and Thornton, 1992; Godzik, Skolnick, and Kolinski, 1992b; Maiorov and Crippen, 1992; Ouzounis et al., 1993; Matsuo and Nishikawa, 1994; Yi and Lander, 1994; Wilmanns and Eisenberg, 1995; Thiele, Zimmer, and Lengauer, 1995; Selbig, 1995; Lathrop and Smith, 1996; Alexandrov, Nussinov, and Zimmer,

TABLE 26.1. A Short Overview of Major Sequence-Only Fold Recognition/Distant Homology Recognition Algorithms

| | Profiles[a] | PSI-BLAST[b] | Hidden Markov Models[c] | Intermediate Sequence Search[d] | Prof-sim[f] | FFAS[e] |
|---|---|---|---|---|---|---|
| Multiple alignment | Hierarchical, user-controlled iterations | Hierarchical, user-controlled iterations and e.value threshold | Clustalw, edited by hand | No multiple alignment built | PSI-BLAST on specially prepared database[f] | PSI-BLAST 5 iterations with $10^{-3}$ e-value threshold |
| Profile | Simple averaging | —preclustering with 98% identity cutoff<br>—pseudocount-based variability estimation<br>—background amino acid frequencies | Stochastic search for an optimal model describing position specific a.a. distributions | Iterative search for homologs of already identified homologs of the prediction target<br>—run until convergence | Profiles treated as distributions | —preclustering with 97% identity cutoff<br>—amino acid composition filter<br>—sequence diversity base weight |
| Template database | Database of nonredundant sequences (nr) | Database of nonredundant sequences (nr) | Database of HMMs for PDB proteins | Database of nonredundant sequences | Profiles of PDB proteins | Profiles of proteins from PDB, PFAM, COG and several genomes |

[a] See Gribskov, McLachlan, and Eisenberg (1987) for description of profiles.
[b] See Altschul et al. (1997) for description of PSI-BLAST.
[c] See Karplus et al. (1997) for description of Hidden Markov Models.
[d] See Park et al. (1997) for description of Intermediate Sequence Search.
[e] See Rychlewski, Jaroszewski, and Godzik (2000) for description of FFAS.
[f] See Yona and Levitt (2002) for description of prof-sim.

1996; Tropsha et al., 1996; Koretke, Luthey-Shulten, and Wolynes, 1996; Russell, Copley, and Barton, 1996; Jaroszewski et al., 1998). In all cases, threading algorithms followed the paradigm of sequence alignment with its basic steps of identifying the possible template and building the alignment. As a result, the threading approach to structure prediction has limitations similar to sequence-based fold recognition. First and foremost, an example of the correct structure must exist in the structural database that is being screened. If not, the method will fail. Then, the quality of the model is limited by the extent of actual structural similarity between the template and the probe structure.

## Force Fields for Simulations and Threading

To speed up energy calculations, the full three-dimensional structure of a protein is usually simplified. Each level of simplification effects the way the energy of the system is calculated. There are less possible interaction centers and more degrees of freedom are averaged. The interaction energy between generalized centers becomes a potential of mean force (Hill, 1960). By averaging over fast changing degrees of freedom, such as bond vibrations and positions of solvent molecules, potentials of mean force are more adequate to describe long time processes such as folding, despite some loss of accuracy because of the loss of many details. For instance, it can be shown that potentials of mean force can easily distinguish grossly misfolded proteins from their correctly folded counterparts, something that atom–atom molecular potentials are unable to do (Novotny, Brucolleri, and Karplus, 1984).

In principle, it is possible to derive potentials of mean force from simulation by explicitly averaging fast degrees of freedom. This averaging is done routinely in simulations for simple molecular liquids where accurate potentials of mean force can be obtained by averaging vibrational degrees of freedom. However, for complicated systems such as proteins, averaging is not possible and parameters are usually obtained from the analysis of regularities in experimentally determined protein structures. There have been many derivations of empirical interaction parameter sets, starting from 1976 (Tanaka and Scheraga, 1976) and continuing until today. Several detailed reviews were published recently (Rooman and Wodak, 1995; Godzik, Kolinski, and Skolnick, 1995; Tobi et al., 2000) and a compilation of existing parameter sets is available through the author's Web page at bioinformatics.burnham.org. There are still many unanswered questions lingering over the theoretical foundations of derivations of knowledge-based interaction parameters and we can expect significant progress in this area.

Despite the lack of complete success as measured by the ability to predict protein structures from their amino acid sequence alone, existing energy parameters adequately capture many features of interactions within proteins. Potentials of mean force derived from the statistical analysis of interaction regularities in proteins can reliably recognize grossly misfolded structures or wrong crystallographic models (Luethy, Bowie, and Eisenberg, 1992), assess the quality of models prepared in homology modeling (Jaroszewski, Pawlowski, and Godzik, 1998), and capture subtle changes to protein structure models introduced during refinement of crystallographic structures (Szczesny et al., 2002). And of course, the same potentials can be used in fold recognition.

## Threading Approximations

Using energy to recognize similarity between distant homologs leads to several unique challenges. One of the most important ones is that the energy stabilizing a protein structure comes from interactions between side chains distant in sequence. Scoring of alignments in a sequence-based comparison is based on Eq. (26.1) or its variants, where all contributions to the total score come from comparing residues (or PSMMs or single steps in HMM) at single positions and do not depend on gaps or deletions introduced elsewhere in the alignment. In other words, the score is local, allowing the fast and powerful dynamic programming algorithm to be used for alignments. Energy-based scores are not local and alignment with nonlocal functions is an NP-complete problem, that is, it has the same level of computational complexity as the traveling salesman problem and other famous minimization problems (Lathrop, 1994). From the early days of threading the non-lower nature of energy based scores forced the use of many approximations.

The most obvious approach is to use an alignment technique that could work with nonlocal scoring functions. This solution was used by a few groups (Bryant and Lawrence, 1993) because of the enormous computational cost and slow convergence. Even then it was necessary to limit the space of possible alignments by eliminating deletions and insertions inside secondary structure elements and restricting lengths. By making these approximations a little stronger, it was possible to use combinatorial brand-and-bound minimization algorithms to find a global alignment minimum (Lathrop and Smith, 1996).

Another solution was to use two-level dynamic programming to optimize interaction partners for each possible pair of aligned residues (Jones, Taylor, and Thornton, 1992). By explicitly considering only the most important interactions between strongly interacting residues, the computational overhead was manageable and the Threader algorithm, which used this approach, was one of the most successful early threading algorithms.

Most other groups used approximations to energy calculations that allowed them to use it in dynamic programming. The most common approximation was a "frozen approximation" (Godzik, Skolnick, and Kolinski, 1992b) where interaction partners for energy calculations were "frozen" to be the same as in the template and were updated only after the alignment was made. Several other groups adopted this approach, which could be iterated (interaction partners updated after alignment is calculated used to calculate the alignment, etc. [Godzik, Skolnick, and Kolinski, 1992b; Wilmanns and Eisenberg, 1995] or relaxed for some interactions [Thiele, Zimmer, and Lengauer, 1995]). This allowed fast alignment calculations but for a price of introducing yet another simplification to the energy calculations. A detailed analysis of various approximations and errors made in a specific threading algorithm is discussed below.

Differences between various threading algorithms are usually found in three areas:

1. Protein model and interaction description. To speed up energy calculations, the full three-dimensional structure of a protein is usually simplified, which profoundly effects the way the energy of the system is calculated. Side chains are described by interaction points, which could be located at $C\alpha$ or $C\beta$ positions, special interaction points, or can encompass the entire side chain. The interaction energy can be distant dependent or not, and only some parts of the protein molecule can be included in the energy calculations.

2. Energy parameterization. There are many variants of the empirical energy parameters derivation, that mostly differ in the assumptions about the reference state (Godzik, 1996).

3. Alignment algorithms. Threading energy is a nonlocal function of the alignment between the prediction target sequence and the template structure. Dynamic programming with frozen approximation (Godzik, Skolnick, and Kolinski, 1992b), two-dimensional dynamic programming (Jones, Taylor, and Thornton, 1992). Monte Carlo minimization (Bryant and Lawrence, 1993), branch-and-bound algorithm (Lathrop and Smith, 1996), and various hybrid approaches can be used for the alignment.

Table 26.2 brings together a short summary and comparison of various threading algorithms, with emphasis on "pure threading" algorithms. However, in practice many of these algorithms still rely heavily on sequence information mixing elements of classical threading and homology recognition algorithms. Most of the recently developed algorithms or most recent updates of old algorithms, such as 3D-PSSM (Kelley, MacCallum, and Sternberg, 2000), GenThreader (Jones, 1998), Bioinbgu (Fischer, 2000), and others can be characterized a hybrid threading/homology recognition algorithms. Also many other technical choices influence relative performance of different algorithms, which are compared as "package deals" and it is difficult to establish relative importance of various specific choices. Therefore, despite significant success of many of these algorithms in fold prediction competitions (CASP meetings) and in providing structural insights in many specific biological problems, they have not contributed significantly to our understanding of folding and forces that influence protein structures.

## What Are the Major Sources of Errors in Threading?

In an attempt to study the limits of the topology fingerprint threading we studied in detail the effects of various approximations on the threading results for the small benchmark of 68 pairs (Zhang et al., 1997). Structural alignments of all pairs were prepared using the combinatorial extension algorithm (Shindyalov and Bourne, 1998); an example of a structural alignment is presented on Figure 26.4, with some interactions identified by lines above and below the sequence.

The correct energy of the target protein within its own structure is the reference point used to compare all other values. Figure 26.4 is used to illustrate the discussion consisting of five approximations with the target protein residues identified by underlined bold:

1. The correct self-threading energy, but calculated only for structure fragments that do have corresponding fragments in the template protein with their entire interaction environment. Only interactions that are entirely within nonaligned fragments will be omitted.

2. The same as number 1 but only interactions with structure fragments that were aligned to the template protein were used. Therefore, the interaction **IA** from Figure 26.4 is now omitted. To contrast it with the following approximations, we call it the "correct partners–correct interactions" (CC) approximation.

3. Interactions from the target protein and the interaction partners from the template protein according to the structural alignment are used to calculate the energy. Therefore, pair interactions contributing to the energy would be **F**L + **V**L, **T**L + **D**T and **I**E + **F**R. This approximation is the "wrong partners–correct interactions" (WC) approximation.

T A B L E  26.2. A Short Overview of Several Threading Algorithms

| | Model | Alignment | Energy | S | PSS | Other Features | Use of Homologues |
|---|---|---|---|---|---|---|---|
| Bryant | $C\alpha$ | MC | Distance dependent | Y | N | Library of conserved cores of families | In "position-specific scoring matrices" |
| Sippl | Interaction centers | Dynamic programming | Distance dependent | N | N | | Yes—independent runs |
| Jones (Threader) | $C\alpha$ | 2D dynamic programming | Distance dependent | Y | Y | | N |
| Jones (GenThreader) | $C\alpha$, N, C, O | Dynamic programming | None in alignment | Y | N | Neural network measuring alignment and model quality | Yes—as a profile |
| Eisenberg | Residue surface contacts | Dynamic programming | Environment classes | Y | Y | | N |
| Godzik (I) | | Dynamic programming and thawing | Contact based | N | N | | N |
| Honig | $C\alpha$ | Mean-field dynamic programming | Distance dependent | N | N | Hydrogen bonding term | N |
| Torda | N, $C\alpha$, C, O, C | Dynamic programming | Neighbor nonspecific score | N | N | Two scoring functions, for alignment and ranking | N |

*Note*: Columns S and PSS denote use of sequence and predicted secondary structure, respectively.

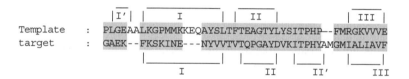

**Figure 26.4.** An example of a structural alignment between two proteins. Some specific interactions, discussed in the text, are identified in each structure.

4. Interactions from the template protein and the interaction partners from the target protein according to the structural alignment are used to calculate the energy. In this calculation, interactions contributing to the energy would be **FV**, **VD** and **IF**. This approximation is the "correct partners–wrong interactions" (CW) approximation.

5. Finally, both interactions and interaction partners from the template protein were used, that is, interactions **FL** + **VL**, **VL** + **DF** and **IE** + **FR**. This approximation is the "wrong partners–wrong interactions" (WW) approximation.

The correct energy, as well as approximations 1 through 3, could only be calculated if the experimental structure of the target is known. Approximations 4 and 5 require the correct structural alignment, so indirectly they also rely on knowing the target structure. Therefore, all these energies are unknown for genuine predictions and again can only be estimated using models of the target structure obtained by comparative modeling. In practice, both the comparative modeling and the alignment procedure based only on the target sequence are likely to introduce errors of their own.

Approximations 1 and 2 are introduced to analyze the extent of target–template structural similarity. Approximation 3 is not particularly interesting (once we know the correct structure, what is the point in using wrong partners). All three approximations are included here only for the sake of completeness, and would not be used much in the subsequent analysis.

At first the energy of one of the proteins in the pair was calculated using its own structure and, in several steps, interaction information as supplemented by information derived from the structure of the other protein with the same fold. The last step was equivalent to the energy as calculated in threading. The goal of this experiment was to identify the source of errors made in threading energy calculations. The important point is that approximations 1–4 can be calculated only in the context of a benchmark. Specific residue names are used in examples below to identify and differentiate different approximations.

Approx 1: All interactions and amino acid partners from the target protein sequence and structure FV, TD, IA, and IF were used. This is the correct self-threading energy.

Approx 2: Interactions and interaction partners from the prediction target protein were used. However, now only interactions with structural fragments aligned to the template protein are allowed. Therefore, the interaction IA from Figure 26.2 is now omitted. This is the "correct amino acid partners–correct interactions" approximation.

Approx 3: Interactions from the target protein are used, but the amino acid partners are taken from the template protein according to the structural alignment. Therefore, interactions contributing to the energy would be F**L** + V**L**, T**L** + D**T**, and **I**E + F**R**. This is the "wrong amino acid partners–correct interactions" approximation.

Approx 4: Interactions present in the target protein are used, but the amino acid partners from the probe protein were used according to the structural alignment. In this calculation, interactions contributing to the energy would be FV, VD, and IF. This is the "correct amino acid partners–wrong interactions" approximation. Note that the "wrong" interaction II from the template is used.

Approx 5: Finally, interactions and interaction partners from the target protein were used, that is, interactions F$\underline{\text{L}}$ + V$\underline{\text{L}}$, V$\underline{\text{L}}$ + D$\underline{\text{F}}$ and $\underline{\text{I}}$E + F$\underline{\text{R}}$. Note that the "wrong" interaction II from the template is used. This is the "wrong amino acid partners–wrong interactions" approximation.

Approximation 5 is equivalent to the "frozen approximation" introduced to eliminate the nonlocal character of the scoring function in threading (Godzik, Skolnick, and Kolinski, 1992b). "Thawing" the interactions (updating the interaction environment) can bring the energy calculation resulting from 5 into 4, but only if the alignment is correct. Approximation 2 could be used if the correct alignment was known and the structure correctly repacked to allow for changes in the interaction patterns along with changes in sequence. Finally, the self-threading energy (approximation 1) corresponds to the stability test of a complete, correct structure of the probe sequence. Our current generation of the topology fingerprint threading algorithm calculates the energy according to approximation 5. It is possible to iteratively converge to 4.

The results for the 68 pairs are presented in Figure 26.5. Of crucial importance is the observation that the use of the correct partners–wrong interactions (approximation 4), gives a very good approximation of the correct energy. It differs, on the average, by only 1.2 energy units per alignment fragment and 0.25 energy units per residue. Clearly, the interaction patterns in the conserved structural fragments are close

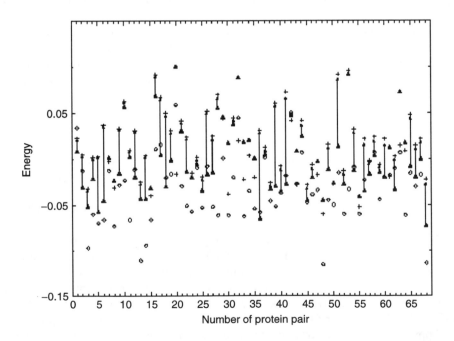

**Figure 26.5.** Differences between various approximations for energy calculations. Open circles correspond to the real energy, triangles to approximation 4, and crosses to approximation 5 (see text for details).

enough to be used for energy calculations. Finally, the approximation used as a first step in the topology fingerprint threading algorithm (approximation 5) is clearly the worst, resulting in errors that are about 6 energy units per aligned fragment on average. In other words, the frozen approximation introduced for computational speed in the context of the current interaction definition is terrible. Basically, the pair contribution to the interaction energy using the frozen approximation is wrong, and the error is of same the magnitude as the pair interaction itself.

These results suggest at least two possible ways of improving the sensitivity of threading. One is to move beyond the "frozen approximation" to at least the "correct partners–wrong interactions" approximation. Unfortunately, the scoring function becomes nonlocal, prohibiting the use of dynamic programming alignment (Lathrop, 1994). Another possibility is to change the interaction definition so that the energy differences resulting from approximations 2–5 are smaller. The set of aligned proteins can be used to select the protein representation and associated interaction scheme that minimizes this difference.

## Comparing and Assessing Various Fold Recognition Algorithms

The ultimate test of fold recognition methods is the prediction of the folds of new proteins when only their sequences are known and before any structural information is available. Dedicated meetings, such as the Critical Assessment Structure Prediction (CASP) meeting in Asilomar, California, bring together almost all groups actively developing fold recognition algorithms. In these meetings, structure prediction groups are provided with sequences of proteins, which structures are about to be solved, but are not yet publicly available. Therefore, all structure predictions are done blind, without any knowledge of the actual structure. This scenario provides a perfect opportunity to compare the performance of various structure prediction algorithms. The last CASP meeting took place on December 2000, and the most interesting result from that meeting was that methods based only on the sequence information, such as Hidden Markov Model methods (Karplus et al., 1997) or profile–profile alignment methods (Rychlewski et al., 2000) compete with energy-only threading methods (Bryant and Lawrence, 1993; Sippl and Weitkus, 1992) as well as with hybrid methods combining contributions both from sequence and structure (Koretke et al., 1999; Jones et al., 1999). In the regular CASP meeting the predictions are submitted by research groups that are free to combine results from fold prediction algorithms with other approaches, their own intuition, biochemical knowledge, and so forth. To focus on direct comparison of algorithms, an automated comparison of fold prediction servers (Critical Assessment of Fully Automated Structure Prediction [CAFASP] experiment) was initiated at CASP3 (Kelley et al., 1999; Fischer et al., 2001). To avoid comparison based on a small number of examples, an ongoing test and comparison of fold prediction algorithms LiveBench (Bujnicki et al., 2001) was initiated and is now in its fourth year.

However, for development of new methods, another choice is to use benchmarks, or sets of proteins, whose structure is predicted and is known. We call them prediction *targets*. For each target, its sequence is matched against a large number of proteins with known structures (*templates*). The goal is to identify the most appropriate template protein. In a benchmark, the quality of a given prediction method can be measured by the number of targets for which the template chosen by the algorithm was indeed similar to its real structure. One of the early benchmarks was based on 68 proteins identified by Fisher et al. (1995) and was used to evaluate several variants of three-dimensional profile methods developed at UCLA (1986), the RFSRV method (Fischer,

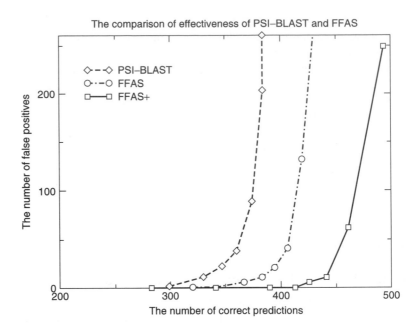

**Figure 26.6.** Sensitivity plot for several sequence only fold recognition algorithms on a benchmark of 929 proteins identified by SCOP and DALI as being structurally similar. Number of correct predictions for every method is shown as a function false prediction with the same level of significance. PDB BLAST is a specific strategy of using PSI-BLAST for fold recognition, FFAS is a profile–profile alignment described previously (Rychlewski et al., 2000), FFAS+ is the newest algorithm developed in our group.

2000), and the GeneFold algorithm (Jaroszewski et al., 1998). Progress in automated structure comparison and easy availability of fold classification databases make it possible to develop larger benchmarks—most popular benchmarks are based on existing classifications of protein structures, such as SCOP.

An example of such a comprehensive benchmark of over 900 protein pairs was built using structure clustering from DALI (1995) and SCOP (1995) databases (see Figure 26.6). The DALI database was used for selection of protein pairs of significant structural similarity but low sequence similarity and SCOP was used to verify the structural similarity of the pair and to assess the level of similarity (fold, superfamily, family). The full benchmark list (as well as a full list of results for all methods discussed here) is available from our Web server at bioinformatics.burnham-inst.org/benchmarks on the Fold and Function Assignment System (FFAS) page.

## SUMMARY

We are still missing a basic understanding of sequence/structure/function relationships in proteins. Analogy-based prediction algorithms remain the only reliable fold prediction tools. New methods, such as threading and hybrid threading/sequence fold recognition, can often recognize even the most distant homologues and, in some cases, even unrelated proteins with similar overall structures. This knowledge pushed the envelope of analogy-based function analysis to the point that the majority of newly sequenced genomes can be tentatively assigned to already characterized protein super-families. However, at this evolutionary distance, fold prediction is no longer equivalent to function prediction. Instead of having the same exact function, distantly related proteins might share some functional analogy that is not obvious to the casual observer. The main challenge facing the fold recognition field is to develop tools to follow the structure prediction with function prediction and analysis.

## REFERENCES

Alexandrov NN, Nussinov R, Zimmer RM (1996): Fast fold recognition via sequence to structure alignment and contact capacity potentials. *Pacific Symp Biocomput.* 1996:53–72

Altschul SF, Gish W, Miller W, Myers EW, Lipman DJ (1990): Basic local alignment search tool. *J Mol Biol* 215:403–10.

Altschul SF, Madden TL, Schaeffer AA, Zhang J, Zhang Z, Miller W, Lipman DJ (1997): Gapped BLAST and PSI-BLAST: a new generation of protein database search programs. *Nucleic Acid Res* 25:3389–402.

Anfinsen CB (1973): Principles that govern the folding of protein chains. *Science* 181:223–30.

Argos P (1987): Analysis of sequence-similar pentapeptides in unrelated protein tertiary structures. *J Mol Biol* 197:331–48.

Babbitt PC, Mrachko GT, Hasson MS, Hiusman GW, Kolter R, Ringe D, Petsko GA, Kenyon GL, Gerlt JA (1995): Functionally diverse enzyme superfamily that abstracts the $\alpha$ proton of carboxylic acids. *Science* 267:1159–161.

Bateman A, Birney E, Durbin R, Eddy SR, Howe KL, Sonnhammer EL (2000): The Pfam protein families database. *Nucleic Acids Res* 28(1):263–6.

Bonneau R, Tsai J, Ruczinski I, Chivian D, Rohl C, Strauss CE, Baker D (2001): Rosetta in CASP4: progress in *ab initio* protein structure prediction. *Proteins* 45(Suppl 5):119–26.

Bowie JU, Luethy R, Eisenberg D (1991): A method to identify protein sequences that fold into a known three dimensional structure. *Science* 253:164–70.

Bryant SH, Lawrence CE (1993): An empirical energy function for threading protein sequence through folding motif. *Proteins* 16:92–112.

Bujnicki JM, Elofsson A, Fischer D, Rychlewski L (2001): LiveBench-1: continuous bench-marking of protein structure prediction servers. *Protein Sci* 10(2):352–61.

Chothia C, Finkelstein A (1990): The classification and origins of protein folding patterns. *Ann Rev Biochem* 59:1007–39.

DALI (1995): *Protein structure comparison by alignment of distance matrices.* Heidleberg: EMBL.

Doolittle RF (1994): Convergent evolution: the need to be explicit. *TIBS* 19:15–18.

Doolittle RF (1996): Molecular evolution: Computer analysis of protein and nucleic acids sequences. Abelson JN, Simon MI, editors. *Methods in Enzymology*, Vol. 183. San Diego: Academic Press.

Farber GK, Petsko GA (1990): The evolution of $\alpha/\beta$ enzymes. *TIBS* 15:228–34.

Finkelstein AV, Reva BA (1990): Determination of globular protein chain fold by the method of self-consistent field (in Russian). *Biofizika* 35:402–6.

Fischer D (2000): Hybrid fold recognition: combining sequence derived properties with evolutionary information. *Pacific Symp Biocomput.* 2002:119–130

Fischer D, Tsai C-J, Nussinov R, Wolfson H (1995): A 3D sequence independent representation of the protein data bank. *Protein Eng* 8:981–97.

Fischer D, Elofsson A, Rychlewski L, Pazos F, Valencia A, Rost B, Ortiz AR, Dunbrack RL, Jr. (2001): CAFASP2: The second critical assessment of fully automated structure prediction methods. *Proteins* 45(Suppl 5):171–83.

Frishman D, Argos P (1996): Incorporation of non-local interactions in protein secondary structure prediction from the amino acid sequence. *Protein Eng* 9:133–42.

[GC Group] (1991): *Program Manual for the GCG Package.*

Godzik A (1996): Knowledge-based potentials for protein folding: what can we learn from protein structures? *Structure Curr Biol* 4:363–6.

Godzik A, Skolnick J (1992): Sequence structure matching in globular proteins: application to supersecondary and tertiary structure prediction. *Proc Natl Acad Sci USA* 89:12098–102.

Godzik A, Skolnick J, Kolinski A (1992): A topology fingerprint approach to the inverse folding problem. *J Mol Biol* 227:227–38.

Godzik A, Skolnick J (1994): Flexible algorithm for direct multiple alignment of protein structures and sequences. *CABIOS* 10:587–96.

Godzik A, Kolinski A Skolnick J (1995): Are proteins ideal mixtures of amino acids? Analysis of energy parameter sets. *Protein Sci* 4:2107–17.

Gribskov M, McLachlan M, Eisenberg D (1987): Profile analysis: detection of distantly related proteins. *Proc Natl Acad Sci USA* 84:4355–8.

Gribskov M, Dereveux J (1991): *Sequence Analysis Primer.* UWBC Biotechnical Resource Series, Burgess RR, editor. New York: Stocton Press.

Hadley C, Jones DT (1999): A systematic comparison of protein structure classifications: SCOP, CATH and FSSP [In Process Citation]. *Structure Fold Des* 7(9):1099–12.

Higgins JD, Gibson TJ (1995): CLUSTAL W: improving the sensitivity of progressive multiple sequence alignment through sequence weighting, position specific gap penalties and weight matrix choice. *Nucleic Acid Res* 22:4673–80.

Hill TL (1960): *An Introduction to Statistical Thermodynamics.* New York: Dover Publications.

Jaroszewski L, Rychlewski L, Zhang B, Godzik A (1998): Fold prediction by a hierarchy of sequence and threading methods. *Protein Sci* 7:1431–40.

Jaroszewski L, Pawlowski K, Godzik A (1998): Multiple model approach: exploring the limits of comparative modeling. *J Mol Model* 4:294–309.

Jaroszewski L, Rychlewski L, Godzik A (2000): Improving the quality of twilight-zone alignments. *Protein Sci* 9(8):1487–96.

Jeanmougin F, Thompson JD, Gouy M, Higgins DG, Gibson TJ (1998): Multiple sequence alignment with Clustal X. *Trends Biochem Sci* 23:403–5.

Jones D (1998): *GenTHREADER.* http://globin.bio.warwick.ac.uk/genome: Warwick.

Jones DT, Taylor WR, Thornton JM (1992): A new approach to protein fold recognition. *Nature* 358:86–9.

Jones DT, Tress M, Bryson K, Hadley C (1999): Successful recognition of protein folds using threading methods biased by sequence similarity and predicted secondary structure [In Process Citation]. *Proteins* (**Suppl** 3): 104–11.

Just W (2001): Computational complexity of multiple sequence alignment with SP-score. *J Comput Biol* 8(6):615–23.

Kabsch W, Sander C (1985): Identical pentapeptides with different backbones. *Nature* 317:207.

Karplus K, Sjolander K, Barret C, Cline M, Haussler D, Hughey R, Holm L, Sander C (1997): Predicting protein structure using Hidden Markov Models. *Proteins* (**Suppl** 1): 134–9.

Kelley LA, MacCallum RM, Sternberg M, Karplus K, Fischer D, Elofsson A, Godzik A, Rychlewski L, Pawlowski K, Jones D, Bryson K (1999): CAFASP-1: critical assessment of fully automated structure prediction. *Proteins* (**Suppl** 3): 209–17

Kelley LA, MacCallum RM, Sternberg MJ (2000): Enhanced genome annotation using structural profiles in the program 3D- PSSM. *J Mol Biol* 299(2):499–520.

Koretke KK, Luthey-Shulten Z, Wolynes PG (1996): Self-consistently optimized statistical mechanical energy functions for sequence structure alignment. *Protein Sci* 5:1043–59.

Koretke KK, Russell RB, Copley RR, Lupas AN (1999): Fold recognition using sequence and secondary structure information [In Process Citation]. *Proteins* (**Suppl** 3): 141–8.

Lathrop RH (1994): The protein threading problem with sequence amino acid interaction preferences is NP-complete. *Protein Eng* 7:1059–68.

Lathrop R, Smith TF (1996): Global optimum protein threading with gapped alignment and empirical pair scoring function. *J Mol Biol* 255:641–65.

Lesk AM (1995): Systematic representation of protein folding patterns. *J Molec Graphics* 13:159–64.

Luethy R, Bowie JU, Eisenberg D (1992): Assessment of protein models with three dimensional profiles. *Nature* 356:83–5.

Maiorov VN, Crippen GM (1992): Contact potential that recognizes the correct folding of globular proteins. *J Mol Biol* 277:876–88.

Matsuo Y, Nishikawa K (1994): Protein structural similarities predicted by a sequence-structure compatibility method. *Protein Sci* 3:2055–63.

Needelman SB, Wunsch CD (1970): A general method applicable to the search for similarities in the amino acid sequence of two proteins. *J Mol Biol* 48:443–53.

Novotny J, Brucolleri R, Karplus M (1984): An analysis of incorrectly folded protein models. Implications for structure prediction. *J Mol Biol* 177:787–818.

Ouzounis C, Sander C, Scharf M, Schneider R (1993): Prediction of protein structure by evaluation of sequence-structure fitness. aligning sequences to contact profiles derived from 3D structures. *J Mol Biol* 232:805–25.

Park J, Teichmann SA, Hubbard T, Chothia C (1997): Intermediate sequences increase the detection of homology between sequences. *J Mol Biol* 273(1):349–54.

Pearson WR, Miller W (1992): Dynamic programming algorithms for biological sequence comparison. *Methods Enzymol* 210:575–601.

Privalov PL, Gill SJ (1988): Stability of Protein Structure and Hydrophobic Interaction. *Adv Protein Chem* 39:191–235.

Ptitsyn OB, Finkelstein AV (1980): Similarities of protein topologies: evolutionary divergence, functional convergence or principles of folding? *Quart Rev Biophys* 13:339–86.

Rooman MJ, Wodak SJ (1995): Are database derived potentials valid for scoring both forward and inverted protein folding? *Protein Eng* 3:849–58.

Russell R, Barton G (1992): Multiple protein sequence alignment from tertiary structure comparison: assignment of global and residue confidence levels. *Proteins* 14:309–23.

Russell RB, Copley RR, Barton GJ (1996): Protein fold recognition by mapping predicted secondary structures. *J Mol Biol* 259:349–65.

Rychlewski L, Jaroszewski L, Li W, Godzik A (2000): Comparison of sequence profiles: strategies for structural predictions using sequence information. *Protein Sci* 9:232–41.

Sander C, Schneider R (1991): Database of homology-derived protein structures and the structural meaning of sequence alignment. *Proteins Struct Func Genet* 9:56–68.

[SCOP] *Structural classification of proteins* (1995): MRC Cambridge.

Selbig J (1995): Contact pattern induced pair potentials for protein fold recognition. *Protein Eng* 8:339–51.

Shindyalov IN, Bourne PE (1998): Protein structure alignment by incremental combinatorial extension (CE) of the optimal path. *Protein Eng* 11:739–47.

Sippl MJ, Weitckus S (1992): Detection of native-like models for amino acid sequences of unknown three-dimensional structure in a database of known protein conformations. *Proteins* 13:258–71.

Smith TF, Waterman MS (1981): Comparison of biosequences. *Adv Applied Math* 46:473–500.

Szczesny P, Jaroszewski L, Grzechnik S, Godzik A (2002): PSQS—protein structure quality score—a flexible tool for protein model quality evaluation. In preparation.

Tanaka S, Scheraga HA (1976): Medium and long range interaction parameters between amino acids for predicting three dimensional structures of proteins. *Macromolecules* 9:945–50.

Taylor WR (1986): Identification of protein sequence topology by consensus template alignment. *J Mol Biol* 188:233–58.

Thiele R, Zimmer R, Lengauer T (1995): Recursive dynamic programming for adaptive sequence and structure alignment. *ISMB* 3:384–92.

Tobi D, Shafran G, Linial N, Elber R (2000): On the design and analysis of protein folding potentials. *Proteins* 40(1):71–85.

Todd AE, Orengo CA, Thornton JM (2001): Evolution of function in protein superfamilies, from a structural perspective. *J Mol Biol* 307(4):1113–43.

Tomii K, Kanehisa M (1996): Analysis of amino acid indices and mutation matrices for sequence comparison and structure prediction of proteins. *Protein Eng* 9:27–36.

Tropsha A, Singh RK, Vaisman II, Zheng W (1996): An algorithm for prediction of structural elements in small proteins. *Pacific Symp Biocomput.* 1996:614–623

[UCLA] (1996): *The UCLA-DOE benchmark to assess the performance of fold recognition methods*. University of California, Los Angeles.

Waterman MS (1995): *Introduction to Computational Biology: Maps, Sequences and Genomes (Interdisciplinary Statistics)*. New York: Chapman & Hall.

Wilmanns M, Eisenberg D (1995): Inverse protein folding by the residue pair preference profile method. *Protein Engi* 8:626–39.

Yi TM, Lander ES (1994): Recognition of related proteins by iterative template refinements. *Protein Sci* 3:1315–28.

Yona G, Levitt M (2002): Within the twilight zone: a sensitive profile–profile comparison tool based on information theory. *J Mol Biol* 315(5):1257–75.

Zhang B, Jaroszewski L, Rychlewski L, Godzik A (1997): Similarities and differences between non-homologous proteins with similar folds. Evaluation of threading strategies. *Folding & Design* 12:307–17.

# 27

# *AB INITIO* METHODS

Dylan Chivian, Timothy Robertson, Richard Bonneau, and David Baker

*Ab initio* structure prediction seeks to predict the native conformation of a protein from the amino acid sequence alone. Such attempts are both a fundamental test of our understanding of protein folding, and an important practical challenge in this era of large scale genome sequencing projects, which are producing large numbers of protein sequences for which no three-dimensional structural information is available.

Anfinsen showed forty years ago that all of the information necessary for a protein to fold to the native state resides in the protein's amino acid sequence (Anfinsen et al., 1961; Anfinsen, 1973). In the absence of large kinetic barriers in the free energy landscape, Anfinsen's results and those of large numbers of researchers in the intervening years suggest that the native conformations of most proteins are the lowest free energy conformations for their sequences (for a description of some notable exceptions, see Baker and Agard, 1994).

Successful structure prediction requires a free energy function sufficiently close to the true potential for the native state to be at one of the lowest free energy minima, as well as a method for searching conformational space for low energy minima. *Ab initio* structure prediction is challenging because current potential functions have limited accuracy, and the conformational space to be searched is vast. Many methods use reduced representations, simplified potentials, and coarse search strategies in recognition of this resolution limit (Simons et al., 1997; Samudrala et al., 1999; Ortiz et al., 1999; Pillardy et al., 2001). Encouragingly, these simplified methods are starting to show some success in protein structure prediction (Murzin, 2001; Lesk, Lo Conte, and Hubbard, 2001) and have advanced to the point where genome scale modeling may become useful.

*Structural Bioinformatics*
Edited by Philip E. Bourne and Helge Weissig
Copyright © 2003 by Wiley-Liss, Inc.

## REPRESENTATIONS OF THE POLYPEPTIDE CHAIN

The most detailed representations include all atoms of the protein and the surrounding solvent molecules. However, representing this large number of atoms and the interactions between them is quite computationally expensive, and it is not clear that this level of detail is necessary during the phase of the search far from the native conformation.

To streamline the calculations, representations can be simplified in a variety of ways. The use of explicit solvent molecules is usually replaced by employing implicit solvent models. United atom representations are frequently used in which hydrogens are drawn into their base carbon, oxygen, and nitrogen atoms. Side chains can be represented using a limited set of conformations (Dunbrack and Karplus, 1994) that are found to be prevalent in structures from the Protein Data Bank (PDB; see Chapter 9), without any great loss in predictive ability. Alternatively, side-chain atoms can be replaced entirely by locating the side-chain properties at either the centroid of the side chain or at the beta carbon (Simons et al., 1997), which amounts to averaging over the side-chain degrees of freedom and permits a significant performance enhancement at the loss of some degree of specificity.

The size of the conformational space to be searched can be further reduced by restricting the conformations available to the polypeptide backbone. Certain torsion angle pairs are preferred by amino acids in particular local structures (Marqusee, Robbins, and Baldwin, 1989; Blanco, Rivas, and Serrano, 1994; Callihan and Logan, 1999). One may restrict the torsion angles to discrete values commonly seen in known structures, either by utilization of a small set of phi–psi pairs (Park and Levitt, 1995), by selecting pairs from an ideal set based on predicted regular secondary structure, or by the use of fragments from known protein structures (Sippl, Hendlich, and Lackner, 1992; Bowie and Eisenberg, 1994; Jones, 1997; Simons et al., 1997).

A method developed by our group that builds structures from protein fragments, called Rosetta (examples of Rosetta predictions in Critical Assessment of Structure Prediction 4 (CASP4) are shown in Figure 27.1), is based on a model of folding in which short segments of the protein chain flicker between different local structures, consistent with their local sequence, and folding to the native state occurs when these local segments are oriented such that low free energy interactions are made throughout the protein (Simons et al., 1997). In simulating this process, it is assumed that the ensemble of local structures sampled by a given sequence segment during folding is roughly approximated by the distribution of local structures sampled by that sequence segment in native protein structures. A list of possible conformations is extracted from experimental structures for each nine residue segments of the chain, and protein tertiary structures are assembled by searching through the combinations of these short fragments for conformations that have buried hydrophobic residues, paired beta strands, and other low free energy features of native proteins. This strategy resolves some of the typical problems with both the conformational search and the free energy function: The search is greatly accelerated as switching between different possible local structures can occur in a single Monte Carlo step, and less demands are placed on the free energy function since local interactions are accounted for in the fragment libraries.

In the most simplified models, entire segments of contiguous regular secondary structure are represented as rigid bodies, allowing only freedom at the junctions (Eyrich, Standley, and Friesner, 1999). Such methods perform searches of probable arrangements of the elements, thus significantly decreasing the conformational search. However, such representations lack enough detail to allow for more subtle features such as strand twist and do not accommodate packing issues well.

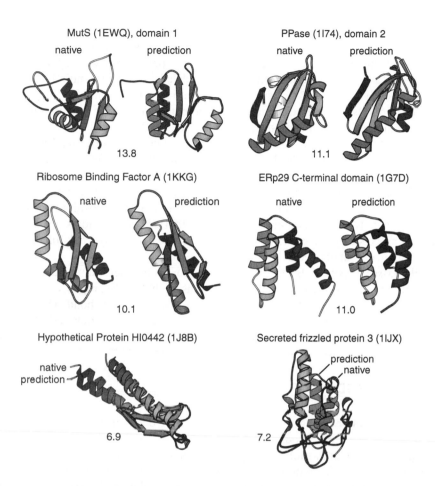

MutS (1EWQ), domain 1

native      prediction

13.8

PPase (1I74), domain 2

native      prediction

11.1

Ribosome Binding Factor A (1KKG)

native      prediction

10.1

ERp29 C-terminal domain (1G7D)

native      prediction

11.0

Hypothetical Protein HI0442 (1J8B)

native
prediction

6.9

Secreted frizzled protein 3 (1IJX)

prediction
native

7.2

**Figure 27.1.** Examples of ROSETTA structure predictions from CASP4 (see Chapter 24). Native/prediction pairs are shown left-to-right, except for 1J8B and 1IJX, which are displayed as a superposition of native and predicted structures. Values indicate Calpha root-mean-square (rms) deviations between native and predicted structures, in angstroms. Colors represent position along the chain from blue (N terminus) to red (C terminus). Figure also appears in Color Figure section.

An alternative model with a long history is that of the lattice representation, in which residues are restricted to points on a regular three-dimensional lattice, with residues proximal in sequence occupying adjacent lattice points (Skolnick and Kolinski, 1991; Hinds and Levitt, 1994; Dill et al. 1995; Ishikawa, Yue, and Dill, 1999). Such methods allow for very fast sampling of conformational space, but are limited in their ability to represent some of the finer details of backbone conformations (Reva et al., 1996).

## POTENTIAL FUNCTIONS

There are two categories of potentials that may be employed in evaluating the free energy of the peptide chain and the surrounding solvent. Molecular mechanics potentials seek to model the forces that determine protein conformation using physically based functional forms parameterized from small molecule data or *in vacuo* quantum mechanical (QM) calculations. For example, van der Waals interactions are usually represented using a standard 6–12 potential with parameters derived from simple liquids, whereas electrostatic interactions are modeled using Coulomb's law with partial charges derived from QM calculations on peptide substructures or from chemical intuition. In contrast, protein structure-derived potentials or scoring functions are empirically derived from experimental structures from the PDB (Sippl, 1995; Koppensteiner and Sippl, 1998). Usually a functional form is not specified and instead pseudoenergies are obtained by taking the logarithm of probability distribution functions. Such structure-derived potentials are particularly useful in conjunction with reduced complexity models, where they may be viewed as representing the interactions between, for example, side-chain centroids after averaging over all plausible positions of the atoms not represented (Kocher, Rooman, and Wodak, 1994). Such potentials are also useful in treating aspects of protein thermodynamics, particularly the hydrophobic effect, that are not completely understood.

Both classes of potentials must represent the forces that determine macromolecular conformation: solvation, electrostatic interactions including hydrogen bonds and ion pairs, Van der Waals interactions, and, in certain cases, covalent bonds (Park, Huang, and Levitt, 1997). Additionally, they must be applicable at a granularity that is in keeping with that of the representation selected and the target resolution of the method.

## SEARCH METHODS

In searching, as in selecting the appropriate level of detail in the representation and in the potential, one must choose the granularity of the search based on the resolution desired from the method. Molecular dynamics directly integrates Newton's equations of motion to derive the motion of a molecule in a given potential. However, the very small step size required for numerical stability makes molecular dynamics with full atom representation of protein and solvent impractical for *de novo* generation of low-resolution models.

To accelerate conformational searching, one must employ techniques that permit coarse sampling of the energy landscape. A variety of methods may be used in conjunction with reduced complexity models and simplified potentials to perform broad searches through low-resolution structures, including Metropolis Monte Carlo simulated annealing (Simons et al., 1997), simulated tempering (Hansmann and Okamoto, 1997), evolutionary algorithms (Bowie and Eisenberg, 1994), and genetic algorithms (Pedersen and Moult, 1997). Individual moves in these procedures can involve quite large perturbations, and allow much more rapid (and more coarse) sampling of conformational space in a relatively short time. For example, simple torsion space Monte Carlo procedures involve changing the backbone torsion angles of one or a small number of residues by several degrees, which can produce quite large changes in the Cartesian coordinates of the protein. Fragment insertion-based procedures (see above) can speed sampling by allowing jumps between different local structures in a single step.

A single search is unlikely to find the global minimum of the free energy landscape, and may instead yield a structure that has become trapped in a local minimum. In an effort to correct for this possibility, many current methods perform numerous conformational searches, generating an ensemble of candidate structures. Numerous techniques have been used to select those structures most likely to be close to the native from the ensemble (Park and Levitt, 1996; Huang et al., 1996; Samudrala and Moult, 1998), and future insights into features of native protein structures and properties of near-native ensembles will undoubtedly add to the arsenal of methods of selecting the most nativelike structures. Ultimately, improvements in potential functions may make identification of the most accurate models a straightforward procedure of selecting those conformations possessing the lowest free energy (Vorobjev, Almagro, and Hermans, 1998; Lazaridis and Karplus, 1999; Rapp and Friesner, 1999; Petrey and Honig, 2000; Lee et al., 2001). It is possible that improved energy functions for discrimination will ultimately involve a fusion of molecular mechanics-based and protein database-derived potentials.

## APPLICATIONS

Genome functional annotation and structural genomics initiatives are two areas of research where *ab initio* protein structure prediction could make important contributions.

### Genome Annotation

While traditionally genome annotation has been accomplished using sequence-similarity search tools, many factors reduce the ability of sequence homology to identify distant homologs (Russell and Pontig, 1998). Domain insertions, circular permutations, exchange of secondary structure elements, and genetic drift all contribute to the divergence of functionally related proteins over time. Thus, the annotation of open reading frames lacking detectable sequence homology to proteins of known function represents a promising application for *ab initio* models. Low-resolution *ab initio* predicted structures may be able to reveal structural and functional relationships between proteins not apparent from sequence similarity alone. This concept is well illustrated by some examples of predictions from CASP4. In the first examples (Figs. 27.2a and 27.2b), the predicted structures were each found to be structurally related to a protein with a similar function, but no significant sequence similarity. In the second example (Fig. 27.3), functionally important residues were found clustered in the predicted structures. In both cases, some of the most important insights into these proteins' function could have been obtained from the predicted structures alone.

Structural similarities like these may be detected using several different methods. First, predicted structures may be compared against the PDB, using a general structure–structure comparison tool (Chapter 16). Recent experiments have found significant matches of *ab initio* predictions to structural homologs of the native structures for a variety of sequences, suggesting that current techniques may be sufficient to detect evolutionarily distant functional homologies in this manner (Simons, Strauss, and Baker, 2001; Bonneau et al., 2002, see also Chapter 20).

Second, *ab initio* structures could be probed for the presence of residues adopting conserved geometric motifs (e.g., serine protease catalytic triads). While this approach has been applied to *ab initio* models with some success (Fetrow and Skolnick, 1998a, Fetrow, et al., 1998b), it remains unclear how to best apply the technique to low-resolution structures. In particular, some question remains as to how ambiguous structural motifs must be in order to detect homologies in low-resolution models.

native                         prediction                    homolog (1NKL)

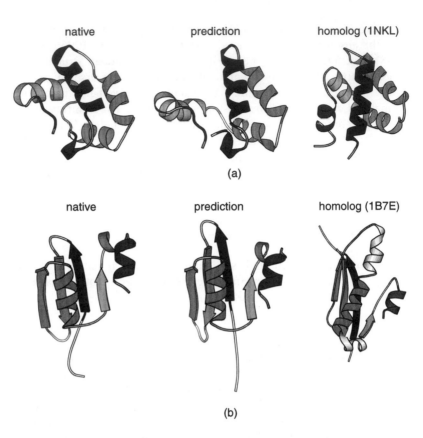

(a)

native                         prediction                    homolog (1B7E)

(b)

**Figure 27.2.** Potential of *ab initio* predcitions to detect distant protein homologies. (a) The native structure of bacterial-lysis protein Bacteriocin AS-48 (left, PDB id 1E68) is compared to the best ROSETTA prediction for the structure (center), and the native structure of NK-Lysin (right, PDB id 1NKL), a functionally similar protein. (b) The native structure of domain 2 of the DNA mismatch repair protein MutS (left, PDB id 1EWQ), is compared to the best ROSETTA prediction for the domain (center), and a domain from the native structure of the Tn5 transposase inhibitor (right, PDB id 1B7E). In both (a) and (b) the *ab initio* models of the proteins were of sufficient quality to detect these functional homologs by the similarity of the folds in the absence of significant sequence similarity. Figure also appears in Color Figure section.

Third, predicted structures could be used to improve the sensitivity and reliablity of matches to sequence-based motif libraries, such as the PROSITE database (Bucher and Bairoch, 1994). Previous work has shown that weak matches to functional motif patterns may be filtered effectively by requiring similarity between the structures of pattern matches and the known structural environments of particular motifs (Jonassen et al., 2000). Therefore, it seems possible that *ab initio* models could provide this structural information when high-resolution structures are unavailable.

## Structural Genomics Initiatives

Structural genomics initiatives present a second opportunity for the application of *ab initio* methods in several ways. First, *ab initio* structure prediction can help guide target

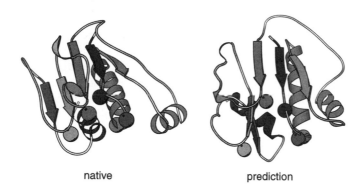

native                          prediction

**Figure 27.3.** An example of active-site conservation in *ab initio* models. The ROSETTA predicted structure of domain 1 from an inorganic pyrophosphatase from Streptococcus mutans is compared to the corresponding domain in the native structure (PDB id 1I74). Strongly conserved active site residues are rendered as spheres along the backbone. Note the similar relative orientation of these residues in the native and predicted structures, implying that *ab initio* models may be sufficient to detect functional homologies using methods that search for functionally significant residue arrangements. Figure also appears in Color Figure section.

selection by focusing experimental structure determination on those proteins likely to adopt novel folds or to be of particular biological importance.

Second, although homology modeling methods have been applied on a genomic scale (Sanchez and Sali, 1998, Sanchez and Sali, 1999), these approaches are inherently limited by their need for at least one homolog of known structure with good coverage and sufficient sequence similarity to be structurally equivalent (Marti-Renom et al.; see also Chapter 25). Homologs of this quality are not always available, and therefore homology methods tend to leave significant fractions of both sequences and genomes improperly modeled. *Ab initio* techniques do not face this limitation, and thus may be a valuable adjunct to homology methods, filling in structural gaps and producing much more complete sets of models than could be obtained by either technique alone.

Third, even small amounts of experimental data can dramatically improve the quality and reliability of *ab initio* structure prediction with the application of spatial constraints. For example, the Rosetta method can produce moderate- to high-resolution structures when combined with limited NMR constraints (Standley et al., 1999; Bowers, Strauss, and Baker, 2000; Rohl and Baker, 2002). In addition, other sources of experimental data such as chemical cross-linking experiments could be used, allowing rapid structure determination for proteins not readily amenable to X-ray or NMR analysis (e.g., membrane-bound proteins). *Ab initio* structure prediction may therefore be useful for increasing the speed of structure determination, which is particularly important for structural genomics.

## FUTURE WORK

What are the prospects for improvement in *ab initio* protein structure prediction methods? Improvement in potential functions should permit the generation of more precise and accurate structures. All atom potentials in particular seem promising for the refinement of low-resolution models. Additionally, more detailed structures may require better fine search strategies. Even for coarse models, the sampling rate of protein conformational space has been a limitation, as demonstrated by the tendency of *ab initio* models to adopt low contact order conformations (Plaxco, Simons, and Baker, 1998). Correcting for this contact order bias through focused sampling of higher-order conformations will require significantly more computational resources, but is likely to improve the prediction of larger, more complicated proteins. Ideally, the development of search strategies that do not face this local-contact bias would provide a boost to *ab initio* methods.

*Ab initio* protein structure prediction has traditionally been an area of primarily academic interest, attaining only slow progress. Recently, however, there have been significant advancements in the field. There is hope that *ab initio* methods will continue to improve, and that this improvement will provide both fundamental insights into the physics underlying protein folding and a valuable, practical resource for genome analysis.

## FURTHER READING

CASP3 (1999): Results from the Comparative Assessment of Techniques for Protein Structure Prediction. *Proteins* 37(S3):149–208.

CASP4 (Forthcoming) Results from the Comparative Assessment of Techniques for Protein Structure Prediction. *Proteins* 45(S5):98–162.

Chothia C (1984): Principles that determine the structure of proteins. *Ann Rev Biochem* 53:537–72.

Kabsch W, Sander C (1984): On the use of sequence homologies to predict protein structure: identical pentapeptides can have completely different conformations. *Proc Natl Acad Sci USA* 81:1075–8.

Lazaridis T, Karplus M (2000): Effective energy functions for protein structure prediction. *Curr Opin Struct Biol* 10:139–45.

Simons KT, Strauss C, Baker D, (2001): Prospects for ab initio protein structural genomics. *J Mol Biol* 306:1191–9.

Sippl MJ (1995): Knowledge-based potentials for proteins. *Curr Opin Struct Biol* 5:229–35.

Wallace AC, Borkakoti N, Thornton JM (1997): TESS: a geometric hashing algorithm for deriving 3D coordinate templates for searching structural databases. Application to enzyme active sites. *Protein Sci* 6:2308–23.

## REFERENCES

Anfinsen CB (1973): Principles that govern the folding of protein chains. *Science* 181:223–30.

Anfinsen CB, Haber E, Sela M, White FW Jr (1961): The kinetics of the formation of native ribonuclease during oxidation of the reduced polypeptide domain. *Proc Natl Acad Sci USA* 47:1309–14.

Baker D, Agard DA (1994): Kinetics versus thermodynamics in protein folding. *Biochemistry* 33:7505–9.

Blanco FJ, Rivas G, Serrano L (1994): A short linear peptide that folds into a native stable beta-hairpin in aqueous solution. *Nat Struct Biol* 1:584–90.

Bonneau R, Strauss CE, Rohl CA, Chivian D, Bradley P, Malonstrom L, Robertson T, Baker D (2002): De novo prediction of three-dimensional structures for major protein families. *J Mol Biol* 322:65–78.

Bowers PM, Strauss CE, Baker D (2000): De novo protein structure determination using sparse NMR data. *J Biomol NMR* 18:311–8.

Bowie JU, Eisenberg D (1994): An evolutionary approach to folding small alpha-helical proteins that uses sequence information and an empirical guiding fitness function. *Proc Natl Acad Sci USA* 91:4436–40.

Bucher P, Bairoch A (1994): A generalized profile syntax for biomolecular sequence motifs and its function in automatic sequence interpretation. *Proc Int Conf Intell Syst Mol Biol* 2:53–61.

Callihan DE, Logan TM (1999): Conformations of peptide fragments from the FK506 binding protein: comparison with the native and urea-unfolded states. *J Mol Biol* 285:2161–75.

Dann CE, Hsieh JC, Rattner A, Sharma D, Nathans J, Leahy DJ (2001): Insights into Wnt binding and signaling from the structures of two frizzled cysteine-rich domains. *Nature* 12:86–90.

Davies DR, Braem LM, Reznikoff WS, Rayment I (1999): The three-dimensional structure of a Tn5 transposase-related protein determined to 2.9-Å resolution. *J Biol Chem* 274:11904–13.

Dill KA, Bromberg S, Yue K, Fiebig KM, Yee DP, Thomas PD, Chan HS (1995): Principles of protein folding—a perspective from simple exact models. *Protein Sci* 4:561–602.

Dunbrack RL Jr, Karplus M (1994): Conformational analysis of the backbone-dependent rotamer preferences of protein sidechains. *Nat Struct Biol* 1:334–40.

Eyrich VA, Standley DM, Friesner RA (1999): Prediction of protein tertiary structure to low resolution: performance for a large and structurally diverse test set. *J Mol Biol* 288:725–42.

Fetrow JS, Skolnick J (1998a): Method for prediction of protein function from sequence using the sequence-to-structure-to-function paradigm with application to glutaredoxins/thioredoxins and T1 ribonucleases. *J Mol Biol* 281:949–68.

Fetrow JS, Godzik A, Skolnick J (1998b): Functional analysis of the Escherichia coli genome using the sequence-to-structure-to-function paradigm: identification of proteins exhibiting the glutaredoxin/thioredoxin disulfide oxidoreductase activity. *J Mol Biol* 282:703–11.

Gonzalez C, Langdon G, Bruix M, Galvez A, Valdivia E, Maqueda M, Rico M (2000): Bacteriocin AS-48, a microbial cyclic polypeptide structurally and functionally related to mammalian NK-lysin. *Proc Nat Acad Sci* 97:11221–6.

Hansmann UH, Okamoto Y (1997): Numerical comparisons of three recently proposed algorithms in the protein folding problem. *J Comput Chem* 18:920–33.

Hinds DA, Levitt M (1994): Exploring conformational space with a simple lattice model for protein structure. *J Mol Biol* 243:668–82.

Huang ES, Subbiah S, Tsai J, Levitt M (1996): Using a hydrophobic contact potential to evaluate native and near-native folds generated by molecular dynamics simulations. *J Mol Biol* 257:716–25.

Huang YJ, Swapna GV, Shukla K, Ke H, Xia B, Inovye M, Montalione GT (Forthcoming).

Ishikawa K, Yue K, Dill KA (1999): Predicting the structures of 18 peptides using Geocore. *Protein Sci* 8:716–21.

Jonassen I, Eidhammer I, Grindhaug SH, Taylor WR (2000): Searching the protein structure databank with weak sequence patterns and structural constraints. *J Mol Biol* 304:599–619.

Jones DT (1997): Successful ab initio prediction of the tertiary structure of NK-lysin using multiple sequences and recognized supersecondary structural motifs. *Proteins* 29(S1):185–91.

Kocher JP, Rooman MJ, Wodak SJ (1994): Factors influencing the ability of knowledge-based potentials to identify native sequence-structure matches. *J Mol Biol* 235:1598–613.

Koppensteiner WA, Sippl MJ (1998): Knowledge-based potentials—back to the roots. *Biochemistry* (Mosc) 63:247–52.

Lazaridis T, Karplus M (1999): Discrimination of the native from misfolded protein models with an energy function including implicit solvation. *J Mol Biol* 288:477–87.

Lee MR, Tsai J, Baker D, Kollman PA (2001): Molecular dynamics in the endgame of protein structure prediction. *J Mol Biol* 313:417–30.

Lesk AM, Lo Conte L, Hubbard T (2001): Assessment of novel fold targets in CASP4: predictions of three-dimensional structures, secondary structures, and interresidue contacts. *Proteins* 45(S5):98–118.

Liepinsh E, Andersson M, Roysschaert JM, otting G (1997): Saposin fold revealed by the NMR structure of NK-lysin. *Nat Struct Biol* 4:793–5.

Liepinsh E, Barishev M, Shapiro A, Ingelman-Sundberg M, Otting G, Mkrtchian S (2001): Thioredoxin fold as a homodimerization module in the potative chaperone Erp29: NMR structures of the domains and experimental model of the 51 kDa dimer. *Structure* 9:457–71.

Lim K, Tempcyzk A, Toedt J, Parsons J, Howard A, Eisenstein E, Herzberg O (Forthcoming).

Marqusee S, Robbins VH, Baldwin RL (1989): Unusually stable helix formation in short alanine-based peptides. *Proc Natl Acad Sci USA* 86:5286–90.

Marti-Renom MA, Stuart AC, Fiser A, Sanchez R, Melo F, Sali A (2000): Comparative protein structure modeling of genes and genomes. *Ann Rev Biophys Biomol Struct* 29:291–325.

Merckel MC, Fabrichniy IP, Salminen A, Kalkkinen N, Baykov AA, Lahti R, Goldman A (2001): Crystal structure of Streptococcus mutans pyrophosphatase: a new fold for an old mechanism. *Structure* 9:289–97.

Murzin AG (2001): Progress in protein structure prediction. *Nat Struct Biol* 8:110–2.

Obmolova G, Ban C, Hsieh P, Yang W (2000): Crystal structures of mismatch repair protein MutS and its complex with a substrate DNA. *Nature* 407:703–10.

Ortiz AR, Kolinski A, Rotkiewicz P, Ilkowski B, Skolnick J (1999): Ab initio folding of proteins using restraints derived from evolutionary information. *Proteins* 37(S3):177–85.

Park BH, Levitt M (1995): The complexity and accuracy of discrete state models of protein structure. *J Mol Biol* 249:493–507.

Park B, Levitt M (1996): Energy functions that discriminate X-ray and near native folds from well-constructed decoys. *J Mol Biol* 258:367–92.

Park BH, Huang ES, Levitt M (1997): Factors affecting the ability of energy functions to discriminate correct from incorrect folds. *J Mol Biol* 266:831–46.

Pedersen JT, Moult J (1997): Protein folding simulations with genetic algorithms and a detailed molecular description. *J Mol Biol* 269:240–59.

Petrey D, Honig B (2000): Free energy determinants of tertiary structure and the evaluation of protein models. *Protein Sci* 9:2181–91.

Pillardy J, Czaplewski C, Liwo A, Lee J, Ripoll DR, Kazmierkiewicz R, Oldziej S, Wede-meyer WJ, Gibson KD, Arnautova YA, Saunders J, Ye YJ, Sheraga HA (2001): Recent improvements in prediction of protein structure by global optimization of a potential energy function. *Proc Natl Acad Sci USA* 98:2329–33.

Plaxco KW, Simons KT, Baker D (1998): Contact order, transition state placement and the refolding rates of single domain proteins. *J Mol Biol* 277:985–94.

Rapp CS, Friesner RA (1999): Prediction of loop geometries using a generalized born model of solvation effects. *Proteins* 35:173–83.

Reva BA, Finkelstein AV, Sanner MF, Olson AJ (1996): Adjusting potential energy functions for lattice models of chain molecules. *Proteins* 25:379–88.

Rohl CA, Baker D (2002): De novo determination of protein backbone structure from residual dipolar couplings using Rosetta. *J Am Chem Soc* 124:2723–9.

Russell RB, Ponting CP (1998): Protein fold irregularities that hinder sequence analysis. *Curr Opin Struct Biol* 8:364–71.

Samudrala R, Moult J (1998): An all-atom distance-dependent conditional probability discriminatory function for protein structure prediction. *J Mol Biol* 275:895–916.

Samudrala R, Xia Y, Huang E, Levitt M (1999): Ab initio protein structure prediction using a combined hierarchical approach. *Proteins* 37(S3):194–8.

Sanchez R, Sali A (1998): Large-scale protein structure modeling of the Saccharomyces cerevisiae genome. *Proc Natl Acad Sci USA* 95:13597–602.

Sanchez R, Sali A (1999): Comparative protein structure modeling in genomics. *J Comp Phys* 151:388–401.

Simons KT, Kooperberg C, Huang E, Baker D (1997): Assembly of protein tertiary structures frcm fragments with similar local sequences using simulated annealing and Bayesian scoring functions. *J Mol Biol* 268:209–25.

Simons KT, Strauss C, Baker D (2001): Prospects for ab initio protein structural genomics. *J Mol Biol* 306:1191–9.

Sippl MJ (1995): Knowledge-based potentials for proteins. *Curr Opin Struct Biol* 5:229–35.

Sippl MJ, Hendlich M, Lackner P (1992): Assembly of polypeptide and protein backbone conformations from low energy ensembles of short fragments: development of strategies and construction of models for myoglobin, lysozyme, and thymosin beta 4. *Protein Sci* 1:625–40.

Skolnick J, Kolinski A (1991): Dynamic Monte Carlo simulations of a new lattice model of globular protein folding, structure and dynamics. *J Mol Biol* 221:499–531.

Standley DM, Eyrich VA, Felts AK, Friesner RA, McDermott AE (1999): A branch and bound algorithm for protein structure refinement from sparse NMR data sets. *J Mol Biol* 285:1691–710.

Vorobjev YN, Almagro JC, Hermans J (1998): Discrimination between native and intentionally misfolded conformations of proteins: ES/IS, a new method for calculating conformational free energy that uses both dynamics simulations with an explicit solvent and an implicit solvent continuum model. *Proteins* 32:399–413.

# 28

# PREDICTION IN 1D: SECONDARY STRUCTURE, MEMBRANE HELICES, AND ACCESSIBILITY

Burkhard Rost

*No general prediction of three-dimensional (3D) structure from sequence yet.* The hypothesis that the 3D structure[1] of a protein (the fold) is uniquely determined by the specificity of the sequence has been verified for many proteins (Anfinsen, 1973). While it is now known that particular proteins (chaperones) often play an important role in folding (Corrales and Fersht, 1996; Martin and Hartl, 1997; Ellis, Dobson, and Hartl, 1998), it is still generally assumed that the final structure is at the free-energy minimum (Dobson and Karplus, 1999). Thus, all information about the native structure of a protein is coded in the amino acid sequence, plus its native solution environment. Can we decipher the code? Hence, can we predict 3D structure from sequence? In principle, the code could by deciphered from physicochemical principles (Levitt and Warshel, 1975; Hagler and Honig, 1978). In practice, the inaccuracy in experimentally determining the basic parameters and the limited computing resources prevent prediction of protein structure from first principles (van Gunsteren, 1993). Therefore, the only successful structure prediction tools are knowledge-based, using a combination of statistical theory and empirical rules. The field of protein structure prediction advanced significantly during the 1990s (see Chapter 27). However, we can still not predict structure from sequence. Rather, the best methods now get the basic characteristics about a fold right some of the time (CASP4, 2000; Lesk, Lo Conte, and Hubbard, 2001).

---

[1] Abbreviations and symbols used in this chapter appear at the end of the chapter.

*Structural Bioinformatics*
Edited by Philip E. Bourne and Helge Weissig
Copyright © 2003 by Wiley-Liss, Inc.

*Structure prediction in 1D becomes increasingly accurate and important.* An extreme simplification of the prediction problem is to project 3D structure onto strings of structural assignments. For example, we can assign a secondary structure state—marked by one symbol—for each residue, or we can assign a number for the accessibility of that residue. Such strings of per-residue assignments are essentially one-dimensional (1D). In fact, arguably the most surprising improvements in bioinformatics since the early 1990s may have been achieved by methods predicting protein structure in 1D. The key to this breakthrough came through the wealth of information about evolution contained in ever-growing databases. Moreover, prediction accuracy continues to rise (Rost, 2001b)! This success is crucial for target selection in structural genomics, for using structure prediction to get clues about function, and for using simplified predictions for more sensitive database searches and predictions of higher-dimensional aspects of protein structure (see below).

*Apologies to developers!* This brief synopsis of methods predicting protein structure in 1D has no chance of being fair to all developing methods for 1D protein structure prediction. Even a restricted MEDLINE search revealed over 200 publications in the last 12 months. Consequently, the review will be somewhat unfair to the majority of developers. Instead, the focus lies on the small subset of most accurate or most widely used methods.

## METHODS

### Secondary Structure Prediction Methods

*Basic concept.* The principal idea underlying most secondary structure prediction methods is the fact that segments of consecutive residues have preferences for certain secondary structure states (Bränden and Tooze, 1991; Rost, 1996). Thus, the prediction problem becomes a pattern classification problem tractable by pattern recognition algorithms. The goal is to predict whether a residue is in a helix, strand, or in neither of the two (no regular secondary structure, often referred to as the *coil* or *loop* state). The first generation prediction methods in the 1960s and 1970s were all based on single amino acid propensities (Chou and Fasman, 1974; Robson, 1976; Garnier, Osguthorpe, and Robson, 1978; Schulz and Schirmer, 1979; Fasman, 1989). Basically, these methods compiled the probability of a particular amino acid for a particular secondary structure state. The second-generation methods dominating the scene until the early 1990s extended the principle concept to compiling propensities for segments of adjacent residues, that is, taking the local environment of the residues into consideration. Typically methods used segments of 3–51 adjacent residues (Nishikawa and Ooi, 1982; Nishikawa and Ooi, 1986; Deleage and Roux, 1987; Biou et al., 1988; Bohr et al., 1988; Gascuel and Golmard, 1988; Levin and Garnier, 1988; Qian and Sejnowski, 1988; Garnier and Robson, 1989). Basically any imaginable theoretical algorithm had been applied to the problem of predicting secondary structure from sequence: physicochemical principles, rule-based devices, expert systems, graph theory, linear and multilinear statistics, nearest-neighbor algorithms, molecular dynamics, and neural networks (Schulz and Schirmer, 1979; Fasman, 1989; Rost and Sander, 1996b; Rost and Sander, 2000). However, it seemed that prediction accuracy stalled at levels around 60% of all residues correctly predicted in either of the three states: helix, strand, or other. It was argued that the limited accuracy resulted from the fact that all methods used only information local in sequence (input: about 3–51 consecutive residues). Local information was estimated to account for roughly 65% of the secondary structure formation. Two additional problems were common to most methods developed from 1957 to 1993a. First, predicted secondary structure segments were, on average, only half as long as observed segments. Historically, this problem was solved for the first time through a particular combination of neural networks (Rost and Sander, 1992; Rost and Sander, 1993a). Second, strands were predicted at levels of accuracy only slightly superior to random predictions. Again, the argument for this deficiency was that the hydrogen bonds determining the formation of sheets (note: paired strands form a sheet) are less local in sequence than the bonds responsible for helices (Chapter 17). Again, this problem was first solved through neural networks (Rost and Sander, 1992; Rost and Sander, 1993a). The solution was rather simple: we realized that about 20% of the correctly predicted residues were in strands, about 30% in helices, and about 50% in nonregular secondary structure. These values are similar to the percentage of the respective classes in proteins. This observation prompted us to simply bias the database used for training neural networks by presenting each class equally often. The result was a prediction well balanced between the three classes, that is, about 60% of the strand residues were predicted correctly. In practice, this was an important advance. However, it also cast an important spotlight onto the explanation that secondary structure formation is partially determined by nonlocal interactions. Clearly, sheets are nonlocal structures. Nevertheless, the preferences for a segment to form a strand or a helix appear similarly strong because both can be predicted at similar levels of accuracy designing the appropriate prediction method (Rost and Sander, 1993a; Rost and Sander, 1994a; Rost, 1996).

*Evolutionary information key to significantly improved predictions.* On the one hand, about 67 out of 100 residues can be exchanged in a protein without changing structure (Rost, 1999b). On the other hand, exchanges of very few residues often destabilize a protein structure. The explanation for this ostensible contradiction is simple: evolution has realized the unlikely by exploring all "neutral" mutations that do not prevent structure formation.[2] Thus, the residue exchange patterns extracted from an aligned protein family are highly indicative of specific structural details. Furthermore, also implies that a profile of N consecutive residues taken from alignments implicitly contains nonlocal information since the evolutionary selection on the level of proteins work on a 3D object, rather than on sequence. Early on it was realized that this information can improve predictions (Dickerson, Timkovich, and Almassy, 1976; Maxfield and Scheraga, 1979; Zvelebil et al., 1987). However, the breakthrough of the third-generation methods to levels above 70% accuracy required a combination of larger databases with more advanced algorithms (Rost and Sander, 1993a; Rost and Sander, 2000). It was also recognized very early on that information from the position-specific evolutionary exchange profile of a particular protein family facilitates discovering more distant members of that family (Dickerson, Timkovich, and Almassy, 1976). Automatic database search methods successfully used position-specific profiles

---

[2]Russell Doolittle coined the term "twilight zone" for the region in which sequence similarity ceases to imply similarity in 3D structure (Doolittle, 1986). Typically, this region begins around 33% pairwise sequence identity for proteins that align over 100 residues (Rost, 1999b). However, the vast majority of all proteins of similar structure have levels of sequence identity far below this mark; they populate the "midnight zone" in which sequences diverged to random levels of similarity (Rost, 1997; Yang and Honig, 2000). This observation may indicate that evolution had enough time to reach an equilibrium at which we can in fact not distinguish between two different events, namely, the convergence of two different sequences to the same structure and the divergence of sequences while maintaining structure (Rost, 1997; Rost, 1999b).

for searching (Barton, 1996). However, the breakthrough to large-scale routine searches has been achieved by the development of PSI-BLAST (Altschul et al., 1997) and Hidden Markov models (Eddy, 1998; Karplus, Barrett, and Hughey, 1998). Since the improvement of secondary structure prediction relies significantly on the information content of the family profile used, today's larger databases and better search techniques resulted in pushing prediction accuracy even higher. The current top-of-the-line secondary structure prediction methods are all based on extended profiles (Rost, 2001b; Przybylski and Rost, 2002).

*The key players.* PHD was the program that surpassed the level of 70% accuracy first (Rost and Sander, 1993a; Rost and Sander, 1994a). It uses a system of neural networks to achieve a performance well balanced between all secondary structure classes (Fig. 28.1). Although still widely used, PHD is no longer the most accurate

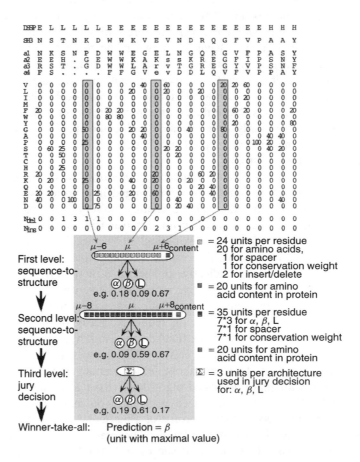

**Figure 28.1.** Neural network system for secondary structure prediction (PHDsec). From the multiple alignment (here guide sequence SH3 plus four other proteins a1-a4, note that lower case letters indicate deletions in the aligned sequence) a profile of amino acid occurrences is compiled. To the resulting 20 values at one particular position $\mu$ in the protein (one column) three values are added: the number of deletions and insertions, as well as the conservation weight (CW). Thirteen adjacent columns are used as input. The whole network system for secondary structure prediction consists of three layers: two network layers and one layer averaging over independently trained networks.

method (Rost and Eyrich, 2001; Przybylski and Rost, 2002). Similar in performance is JPred2 (Cuff and Barton, 2000); it combines the results from various prediction methods, in particular from JNet (Cuff and Barton, 2000), NSSP (Salamov and Solovyev, 1997), PREDATOR (Frishman and Argos, 1996) and PHD (Rost, 1996). David Jones pioneered using automated, iterative PSI-BLAST searches (Jones, 1999b). The most important step climbed by the resulting method PSIPRED has been the detailed strategy to avoid polluting the profile through unrelated proteins. To avoid this trap, the database searched has to be filtered first (Jones, 1999b). Other than the advanced use of PSI-BLAST, PSIPRED achieves its success through a neural network system similar to that implemented in PHD. At the Critical Assessment for Structural Prediction (CASP) meeting at which David Jones introduced PSIPRED, Kevin Karplus and colleagues presented their prediction method (SAM-T99sec) finding more diverged profiles through Hidden Markov models (Karplus et al., 1999). The most important prediction method used by SAM-T99sec is a simple neural network with two layers of hidden units. However, the major strength of the method appears to be the quality of the alignment used. The only method published recently that improves prediction accuracy significantly not through more divergent profiles but through the particular algorithm is SSpro. Instead, SSpro1 is successful through the particular algorithmic improvement implemented (Baldi et al., 1999). The principle idea of the method is to overcome the limitations of feed-forward neural networks with an input window of relatively small and fixed size with bidirectional recurrent neural networks (BRNN) capable of taking the entire protein chain as input (Baldi et al., 1999; Baldi and Brunak, 2001). The most recent improvement realized in SSpro2 resulted from combining the advanced network architectures with PSI-BLAST profiles (Pollastri et al., 2001). Quite a different route toward secondary structure prediction is taken by the HMMSTR/I-sites programs (Bystroff, Thorsson, and Baker, 2000, described in more detail in Chapter 27).

*Specialized method: coiled-coil predictions.* A coiled coil is a bundle of several helices assuming a side-chain packing geometry often referred to as "knobs-into-holes" (Crick, 1953). The knobs are the side chains of one helix that pack into the hole created by four side chains surrounding the facing helix. This supercoil slightly alters the helix periodicity from 3.6 to 3.5 and results in the coiled-coil specific symmetry in which every seventh residue occupies a similar position on the helix surface. The first and fourth of the seven residues are typically hydrophobic, the other four hydrophilic, frequently exposing the helix to solvent. These specific sequence features are at the base of accurate predictions for coiled-coil helices (Lupas, 1996; Lupas, 1997). The most widely used program is COILS that bases on amino acid preferences compiled for the few coiled-coil proteins that were known at high resolution a decade ago (Lupas, Van Dyke, and Stock, 1991). The program detects coiled-coil preferences in windows of 14, 21, and 28 residues. The longer the window the better the distinction between proteins that have coiled-coil regions and those that do not (Lupas, 1996). If we know the precise location of the coiled-coil regions and the multimeric state, we can predict 3D structure for coiled-coil regions at levels of accuracy that resemble experimentally determined structures (below 2.5Å; Nilges and Brünger, 1993; O'Donoghue and Nilges, 1997). O'Donoghue and Nilges used the experimentally known boundaries of the coiled-coil regions and their known multimeric state for prediction. It remains to be tested how sensitive that 3D prediction is with respect to errors in predicting the coiled-coil regions. Recently, Wolf, Kim, and Berger (1997) developed a method to predict the multimeric state of a coiled-coil region. When labeling all likely coiled-coil proteins in entire proteomes, we found that about 8–10% of all eukaryotic proteins and 2–10% of all proteins in archae and prokaryotes contain at least one coiled-coil region (Liu and Rost, B. 2001b).

## Solvent Accessibility Prediction Methods

*Basic concept.* It has long been argued that if the segments of secondary structure could be accurately predicted, the 3D structure could be predicted by simply trying different arrangements of the segments in space (Cohen, Sternberg, and Taylor, 1981; Monge, Friesner, and Honig, 1994; Mumenthaler and Braun, 1995; Cohen and Presnell, 1996). One criterion for assessing each arrangement could be to use predictions of residue solvent accessibility (Lee and Richards, 1971; Chothia, 1976; Connolly, 1983). The principal goal is to predict the extent to which a residue embedded in a protein structure is accessible to solvent. Solvent accessibility can be described in several ways (Lee and Richards, 1971; Chothia, 1976; Connolly, 1983). The most detailed fast method compiles solvent accessibility by estimating the volume of a residue embedded in a structure that is exposed to solvent (Fig. 28.2; note: this method was developed by [Connolly, 1983] and later implemented in DSSP [Kabsch and Sander, 1983]). Different residues have a different possible accessible area. The most extreme simplification for accessibility accounts for this difference by normalizing (dividing observed value by maximally possible value) to a two-state description, distinguishing between residues that are buried (relative solvent accessibility <16%) and exposed (relative solvent

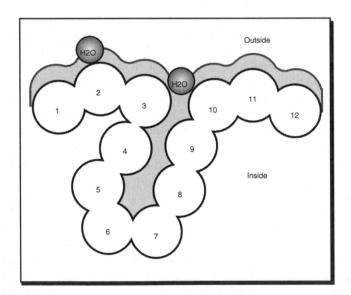

**Figure 28.2.** Measure accessibility. Residue solvent accessibility is usually measured by rolling a spherical water molecule over a protein surface and summing the area that can be accessed by this molecule on each residue (typical values range from 0–300 $\text{Å}^2$). To allow comparisons between the accessibility of long extended and spherical amino acids, typically relative values are compiled (actual area as percentage of maximally accessible area). A more simplified descriptions distinguishes two states: buried (here residues numbered 1–3 and 10–12) and exposed (here residues 4–9) residues. Since the packing density of native proteins resembles that of crystals, values for solvent accessibility provide upper and lower limits to the number of possible inter-residue contacts.

accessibility 16%). The precise choice of the threshold is not well defined (Hubbard and Blundell, 1987; Rost and Sander, 1994b). The classical method to predict accessibility is to assign either of the two states, buried or exposed, according to residue hydrophobicity, that is, very hydrophobic stretches are predicted to be buried (Richards, 1977; Tanford, 1978; Kyte and Doolittle, 1982; Sweet and Eisenberg, 1983). However, more advanced methods have been shown to be superior to simple hydrophobicity analyses (Holbrook, Muskal, and Kim, 1990; Mucchielli-Giorgi, Hazout, and Tuffery, 1999; Carugo, 2000; Li and Pan, 2001; Naderi-Manesh et al., 2001). Typically, these methods use similar ways of compiling propensities of single residues or segments of residues to be solvent accessible, as secondary structure prediction methods. For particular applications, such as using predicted solvent accessibility to predict glycosylation sites, it seems beneficial to train neural networks on different definitions of accessibility (Hansen et al., 1998; Gupta et al., 1999). In particular, Hansen et al. (1998) realized alternative compilations by changing the size of the water molecule used in DSSP (Fig. 28.2). In contrast to the situation for secondary structure, most of the information needed to predict accessibility is contained in the preference of single residues (Rost and Sander, 1994b). Nevertheless, using windows of adjacent residues also improves solvent accessibility prediction significantly (Rost and Sander, 1994b; Thompson and Goldstein, 1996).

*Evolutionary information improves accessibility prediction.* Solvent accessibility at each position of the protein structure is evolutionarily conserved within sequence families. This fact has been used to develop methods for predicting accessibility using multiple alignment information (Rost and Sander, 1994b; Wako and Blundell, 1994; Rost, 1996; Thompson and Goldstein, 1996; Cuff and Barton, 2000; Rost, 2001a). The two-state (buried, exposed) prediction accuracy is above 75%, that is, more than four percentage points higher than for methods not using alignment information. Predictions of solvent accessibility have also been used successfully for prediction-based threading, as a second criterion toward 3D prediction by packing secondary structure segments according to upper and lower bounds provided by accessibility predictions, and as basis for predicting functional sites (Rost and O'Donoghue, 1997).

*Available key players.* It is possible, that the exclusion of methods predicting solvent accessibility from the CASP meetings (see Practical Aspects) slowed down the progress of the field. In particular, few of the methods developed are readily available through public servers. Prominent exceptions are the solvent accessibility predictions by PHD (Rost and Sander, 1994b) and PROFphd (Rost, 2001a) available through the PredictProtein server (Rost, 1996; Rost, 2000). Both use systems of neural networks with alignment information. The improvement of PROFphd over PHD was achieved by (1) training the neural networks only on high-resolution structures and by (2) using predicted secondary structure as additional input (Rost, 2001a). Technically, both PROFphd and PHD are the only available methods predicting real values for relative solvent accessibility on a grid of 0, 1, 4, 9, 16, 25, 36, 49, 64, 81 (percentage relative accessibility). Another method that improved prediction accuracy considerably over older programs is embedded in the JPred2 server (Cuff and Barton, 2000). It uses PSI-BLAST profiles as input to neural networks predicting accessibility in two states (buried/exposed).

## Transmembrane Helix Prediction Methods

*The task.* Even in the optimistic scenario that in the near future most protein structures will be experimentally determined, one class of proteins will still represent a challenge for experimental determination of 3D structure: transmembrane proteins. The major obstacle with these proteins is that they do not crystallize, and are hardly tractable by NMR spectroscopy. Consequently, for this class of proteins structure prediction methods are even more needed than for globular water-soluble proteins. Fortunately, the prediction task is simplified by strong environmental constraints on transmembrane proteins: the lipid bilayer of the membrane reduces the degrees of freedom making the prediction almost a 2D problem (Taylor, Jones, and Green, 1994). Two major classes of membrane proteins are known: proteins that insert helices into the lipid bilayer (Fig. 28.3), and proteins that form pores by $\beta$-strand barrels (von Heijne, 1996; Seshadri et al., 1998; Buchanan, 1999). Since there is not much experimental information available on different porinlike (beta-strand barrel) membrane proteins, we can hardly estimate prediction accuracy for this class. The situation is quite different for helical membrane proteins. Knowing the precise location of transmembrane helices, we can predict 3D structure by simply exploring all possible conformations (Taylor, Jones, and Green, 1994). Although predicting transmembrane helices is simpler than predicting globular helices, there is ample evidence that prediction accuracy has been significantly overestimated (Möller, Croning, and Apweiler, 2001; Chen, Kerngtsky, and Rost, 2002a).

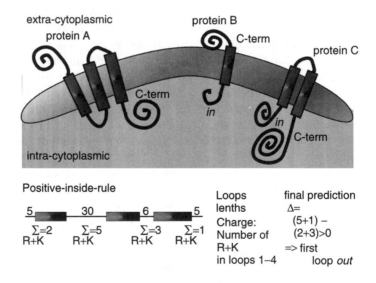

**Figure 28.3.** Topology of helical membrane proteins. In one class of membrane proteins, typically apolar helical segments are embedded in the lipid bilayer oriented perpendicular to the surface of the membrane. The helices can be regarded as more or less rigid cylinders. The orientation of the helical axes, that is, the topology of the transmembrane protein, can be defined by the orientation of the first N-terminal residues with respect to the cell. Topology is defined as *out* when the protein N-term (first residue) starts on the extracytoplasmic region (protein A), and as *in* if the N-term starts on the intracytoplasmic side (proteins B and C). The lower part explains the inside-out-rule. The differences between the positive charges are compiled for all even and odd nonmembrane regions. If the even loops have more positive charges, the N-term of the protein is predicted outside. This rule holds for most proteins of known topology.

*Basic concept.* We can use a number of observations that constrain the problem of predicting membrane helices: (1) TM helices are predominantly apolar and between 12 and 35 residues long (Chen and Rost, 2002b). (2) Globular regions between membrane helices are typically shorter than 60 residues (Wallin and von Heijne, 1998; Liu and Rost, 2001). (3) Most TMH proteins have a specific distribution of the positively charged amino acids Arginine and Lysine coined the "positive-inside-rule" by Gunnar von Heijne (von Heijne, 1986; von Heijne, 1989). Connecting loop regions at the inside of the membrane have more positive charges than loop regions at the outside (Fig. 28.3). (4) Long globular regions (>60 residues) differ in their composition from those globular regions subject to the "inside-out-rule". Most methods simply compile the hydrophobicity along the sequence and predict a segment to be a transmembrane helix if the respective hydrophobicity exceeds some given threshold (Tanford, 1980; Eisenberg et al., 1984; Klein, Kanehisa, and De Lisi, 1985; Engelman, Steitz, and Goldman, 1986; Jones, Taylor, and Thornton, 1994; von Heijne, 1994; Hirokawa, Boon-Chieng, and Mitaku, 1998; Phoenix, Stanworth, and Harris, 1998; Tusnady and Simon, 1998; Harris, Wallace, and Phoenix, 2000; Lio and Vannucci, 2000). Additionally, some methods also explore the hydrophobic moment (Eisenberg et al., 1984; von Heijne, 1996; Liu and Deber, 1999), or other membrane-specific amino acid preferences (Ben-Tal et al., 1997; Monne, Hermannson, and von Heijne, 1999; Pasquier et al., 1999; Pilpel, Ben-Tal, and Lancet, 1999). The most important step is to adequately average hydrophobicity values over windows of adjacent residues (von Heijne, 1992; von Heijne, 1994). One of the major problems of hydrophobicity-based methods appears to be the poor distinction between membrane and globular proteins (Rost et al., 1995; Möller, Croning, and Apweiler, 2001). A number of methods use the positive-inside-rule to also predict the orientation of membrane helices (Sipos and von Heijne, 1993; Jones, Taylor, and Thornton, 1994; Persson and Argos, 1997; Sonnhammer, von Heijne, and Krogh, 1998; Harris, Wallace, and Phoenix, 2000; Tusnady and Simon, 2001).

*Evolutionary information improves prediction accuracy.* Using evolutionary information also improves TMH predictions significantly (Neuwald, Liu, and Lawrences, 1995; Rost et al., 1995; Rost, Casadio, and Fariselli, 1996a; Persson and Argos, 1997). However, the growth of the sequence databases seems to have reversed the advantage of using evolutionary information (Chen, Kernytsky, and Rost, 2002a). Until around 1997, most membrane helices were conserved in the following sense. Assume protein A has a TMH at positions N1–N2. Since the number of membrane helices is important for the function of the protein, we expect that all proteins A' that are found to be similar to A in a database search will also have a membrane helix at the corresponding positions N1–N2. However, precisely this assumption no longer proves correct (Chen and Rost, 2002b). The practical result is that alignment-based predictions are much less accurate when based on the large merger of SWISS-PROT and TrEMBL (Bairoch and Apweiler, 2000) than when based on the smaller SWISS-PROT, only (Chen and Rost, 2002b). Interestingly, we can explore the power of using evolutionary information by carefully filtering the results from PSI-BLAST searches (Chen, Kernytsky, and Rost, 2002a).

*Available key players.* TopPred2 is one of the classics in the field. It averages the GES-scale of hydrophobicity (Engelman, Steitz, and Goldman, 1986) using a trapezoid window (von Heijne, 1992; Sipos and von Heijne, 1993). MEMSAT (Jones, Taylor, and Thornton, 1994) introduced a dynamic programming optimization to find the most likely prediction based on statistical preferences. TMAP (Persson and Argos, 1996) uses statistical preferences averaged over aligned profiles. PHD combines a neural network using evolutionary information with a dynamic programming optimization of the final prediction (Rost et al., 1995; Rost, Casadio, and Fariselli, 1996a). DAS optimizes the use of hydrophobicity plots (Cserzö et al., 1997). SOSUI (Hirokawa, Boon-Chieng, and Mitaku, 1998) uses a combination of hydrophobicity and amphiphilicity preferences to predict membrane helices. TMHMM is the most advanced—and seemingly most accurate—current method to predict membrane helices (Sonnhammer, von Heijne, and Krogh, 1998). It embeds a number of statistical preferences and rules into a Hidden Markov model to optimize the prediction of the localization of membrane helices and their orientation (note: similar concepts are used for HMMTOP [Tusnady and Simon, 1998]).

## PROGRAMS AND PUBLIC SERVERS

All methods described are available through public servers. A list of URLs and the contact addresses are summarized in Table 28.1. Most programs listed in Table 28.1 (except HMMSTR and PSIPRE) are also available by single-click: META-PP allows you to fill out a form with the sequence and your e-mail address once and to simultaneously submit your protein to a number of high-quality servers (Eyrich and Rost, 2000). This concept of accessing many servers through one has been pioneered by the BCM-Launcher (Smith et al., 1996), supposedly accessing the largest number of different methods. Other combinations are given by NPSA (Combet et al., 2000), META-Poland (Rychlewski, 2000), and ProSAL (Kleywegt, 2001). In contrast to all others, META-PP attempts (1) to return as few results as possible by filtering out technical messages and (2) to combine only high-quality methods. Note that both the BCM launcher and the current GCG package (Devereux, Haeberli, and Smithies, 1984) return predictions of secondary structure from methods that are neither state-of-the-art nor competitive with the best method from a decade ago without indicating this to the user. A generalization of the common interface idea is implemented in the sequence retrieval system SRS (Etzold and Argos, 1993; Etzold, Ulyanov, and Argos, 1996), enabling simultaneous access of most existing databases. Successively SRS starts to also incorporate the direct access to prediction methods.

## PRACTICAL ASPECTS

### Evaluation of Prediction Methods

*Correctly evaluating protein structure prediction is difficult.* Developers of prediction methods in bioinformatics may significantly overestimate their performance because of the following reasons. First, it is difficult and time-consuming to correctly separate data sets used for developing and testing. Second, estimates of performance of the different methods are often based on different data sets. This problem frequently originates from the rapid growth of the sequence and structure databases. Third, single scores are usually not sufficient to describe the performance of a method. The lack of clarity is particularly unfortunate at a time when an increasing number of tools are made easily available through the Internet and many of the users are not experts in the field of protein structure prediction. Two prominent examples illustrate this problem: (1) Transmembrane helix predictions have been estimated to yield

TABLE 28.1. Availability of Prediction Methods

| Method | Type | Server | Program |
|---|---|---|---|
| JPred2 | | jura.ebi.ac.uk:8888 | James Cuff james@ebi.ac.uk |
| PHD | acc | cubic.bioc.columbia.edu/predictprotein | Burkhard Rost rost@columbia.edu |
| PROFphd | acc | cubic.bioc.columbia.edu/predictprotein | Burkhard Rost rost@columbia.edu |
| ASP | sec+ | cubic.bioc.columbia.edu/predictprotein | Malin Young mmyoung@sandia.gov |
| COILS | sec | cubic.bioc.columbia.edu/predictprotein | Andrei Lupas andrei.lupas@tuebingen.mpg.de |
| HMMSTR | sec+ | | Chris Bystroff bystrc@rpi.edu |
| JPred2 | sec | jura.ebi.ac.uk:8888 | James Cuff james@ebi.ac.uk |
| PHDpsi | sec | cubic.bioc.columbia.edu/predictprotein | Burkhard Rost rost@columbia.edu |
| PHD | sec | cubic.bioc.columbia.edu/predictprotein | Burkhard Rost rost@columbia.edu |
| PROFking | sec | www.aber.ac.uk/~phiwww/prof | Ross King rdk@aber.ac.uk |
| PROFphd | sec | cubic.bioc.columbia.edu/predictprotein | Burkhard Rost rost@columbia.edu |
| PSIPRED | sec | insulin.brunel.ac.uk/psiform.html | David Jones d.jones@cs.ucl.ac.uk |
| SAM-T99sec | sec | www.cse.ucsc.edu/research/compbio/HMM-apps/T99-query.html | Kevin Karplus karplus@cse.ucsc.edu |
| SSpro2 | sec | promoter.ics.uci.edu/BRNN-PRED | Pierre Baldi pfbaldi@ics.uci.edu |
| DAS | tmh | www.sbc.su.se/~miklos/DAS | |
| HMMTOP | tmh | www.enzim.hu/hmmtop | Gábor E. Tusnády tusi@enzim.hu |
| MEMSAT | tmh | insulin.brunel.ac.uk/psipred | David Jones d.jones@cs.ucl.ac.uk |
| PHD | tmh | cubic.bioc.columbia.edu/predictprotein | Burkhard Rost rost@columbia.edu |
| SOSUI | tmh | sosui.proteome.bio.tuat.ac.jp/sosuiframe0.html | Takatsugu Hirokawa sosui@biophys.bio.tuat.ac.jp |
| SPLIT | tmh | www.mbb.ki.se/tmap/index.html | Davor Juretic juretic@mapmf.pmfst.hr |
| TMAP | tmh | www.mbb.ki.se/tmap/index.html | Bengt Persson Bengt.Persson@ibp.vxu.se |
| TMHMM | tmh | www.cbs.dtu.dk/services/TMHMM-1.0 | Anders Krogh krogh@cbs.dtu.dk |
| Tmpred | tmh | www.ch.embnet.org/software/TMPRED_form.html | Philipp Bucher pbucher@isrecsun1.unil.ch |
| TopPred2 | tmh | www.sbc.su.se/~erikw/TopPred22 | Gunnar von Heijne gunnar@dbb.su.se |

levels above 95% per-residue accuracy more than 18 percentage points more than seems to hold up (Chen and Rost, 2002b). (2) Many publications on predicting the secondary structural class from amino acid composition allowed correlations between training and testing sets. Consequently, levels of prediction accuracy published—close to 100%—exceeded by far the theoretical possible margins—around 60% (Wang and Yuan, 2000).

*CASP: how well do experts predict protein structure?* The CASP experiments attempt to address the problem of overestimated performance (Moult et al., 1995; Moult et al., 1997; Moult et al., 1999; Zemla, Venclovas, and Fidelis, 2001). The procedure used by CASP is: (1) Experimentalists who are about to determine the structure of a protein send the sequence to the CASP organizers (Zemla, Venclovas, and Fidelis, 2001). (2) Sequences are distributed to the predictors. The deadline for returning results is given by the date that the structure will be published. (3) All predictions are evaluated in a meeting at Asilomar. CASP resolves the bias resulting from using known protein structures as targets. However, it often cannot provide statistically significant evaluations since the number of proteins tested is too small (Marti-Renom et al., 2001; Rost and Eyrich, 2001a). Nevertheless, CASP provides valuable insights into the performance of prediction methods, and has become the major source of development in the field of protein structure prediction. Due to the fact that "failing at CASP is bad for the CV," most predictions are submitted only after experts have studied the data in detail. Thus, CASP intrinsically evaluates how well the best experts in the field can predict structure.

*CAFASP: how well do computers predict structure?* Critical Assessment of Fully Automated Structure Prediction (CAFASP) has recently extended CASP by testing automatic prediction servers on the CASP proteins (Fischer et al., 1999). Although CAFASP aims at evaluating programs rather than experts, it is still limited to a small number of test proteins (Fischer et al., 2001; Zemla, Vanclovas, and Fidelis, 2001). In fact, in most categories (comparative modeling, fold recognition, novel folds) did not suffice to distinguish between the top 10 methods in a statistically significant way (see also Chapter 27). Furthermore, for most categories, we could not even conclude whether the field had improved over a period of two years.

*EVA and LiveBench: automatic, large-scale evaluation of performance.* The limitations of CASP and CAFASP prompted two efforts at creating large-scale and continuously running tools that automatically assesses protein structure prediction servers: EVA (Eyrich et al., 2001a; Eyrich et al., 2001b) and LiveBench (Bujnicki et al., 2001). LiveBench specializes in the evaluation of fold recognition, whereas EVA analyzes comparative modeling, contact prediction, fold recognition, and secondary structure prediction. The EVA results for secondary structure prediction methods were essential to conclude that these methods have improved significantly and to isolate the particular reasons for the improvements (mostly due to growing databases) (Rost, 2001b; Rost and Eyrich, 2001a; Przybylski and Rost, 2002).

## Secondary Structure Prediction in Practice

*77% right means 23% wrong!* The best current methods (PSIPRED, PROFphd, SSpro) reach levels of around 77% accuracy (percentage of residues predicted correctly in one of the three states helix, strand, or other) (Eyrich et al., 2001b; Rost, 2001b; Rost and Eyrich, 2001a).[3] Five observations are important for using prediction methods: (1) Levels of accuracy are averages over many proteins (Fig. 28.4a). Hence, the accuracy for the prediction of your protein may be much lower—or much higher—than 77%. (2) Stronger predictions are usually more accurate (Fig. 28.4b). This allows you—to some extent—to find out whether or not the prediction for your protein is more likely to be above or below average. (3) Often predictions go badly wrong, that is, helices are incorrectly predicted as strands and vice versa. In fact, the best current methods confuse helices and strands for about 3% on average of all residues (Eyrich et al., 2001b). Encouragingly, some of these "bad errors" are in fact not so severe after

[3]By the time you read this methods may have already been improved. Thus, consult the EVA Web pages at cubic.bioc.columbia.edu for the latest statistics.

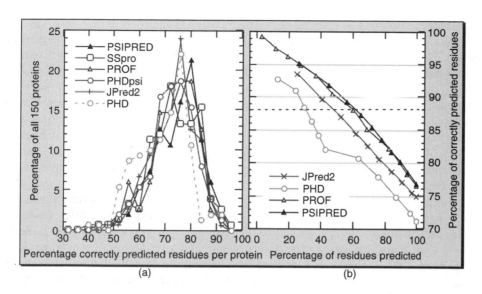

**Figure 28.4.** Prediction accuracy varies but stronger predictions are better! All results are based on 150 novel protein structures not used to develop any of the method shown (Eyrich et al., 2001b; Rost, 2001b). For all methods shown, the three-state per-residue accuracy varies significantly between these proteins, with one standard deviation on the order of 10% (A). This implies that it is difficult for users to estimate the *actual* accuracy for their protein. However, most methods now provide an index measuring the reliability of the prediction for each residue. Shown is the accuracy versus the cumulative percentages of residues predicted at a given level of reliability (coverage versus accuracy). For example, PSIPRED and PROFphd reach a level above 88% for about 60% of all residues (dashed line). This particular line is chosen since secondary structure assignments by DSSP agree to about 88% for proteins of similar structure. Although JPred2 is only marginally less accurate than PSIPRED and PROFphd, it reaches this level of accuracy for less than half of all residues.

all, since some of them are due to regions that can switch structural conformations in response to environmental changes (see below). (4) Prediction accuracy is rather sensitive to the information contained in the alignment used for the prediction: differences between single-sequence-based predictions and optimal alignment-based predictions can exceed more than 25 percentage points (Przybylski and Rost, 2002). (5) If on average 77% of the residues are correctly predicted, this trivially implies that 33% are wrong. Often it is extremely instructive to form an expert opinion about where these wrong predictions are (Hubbard et al., 1996).

*Sources of latest improvement: four parts database growth, three extended search, two other.* Jones solicited two causes for the improved accuracy of PSIPRED: (1) training and (2) testing the method on PSI-BLAST profiles. Cuff and Barton (2000) examined in detail how different alignment methods improve. However, which fraction of the improvement results from the mere growth of the database, which from using more diverged profiles, and which from training on larger profiles? Using the PHD version from 1994 to separate the effects (Przybylski and Rost, 2002), we first compared a noniterative standard BLAST (Altschul and Gish, 1996) search against SWISS-PROT (Bairoch and Apweiler, 2000) with one against SWISS-PROT + TrEMBL (Bairoch and Apweiler, 2000) + PDB (Berman et al., 2000). The larger database improves performance by about two percentage points (Przybylski and Rost, 2002). Secondly, we compared the standard BLAST against the big database with an iterative PSI-BLAST search. This search yielded less than two percentage points additional improvement (Przybylski and Rost, 2002). Thus, overall, the more divergent profile search against today's databases supposedly improves any method using alignment information by almost four percentage points. The improvement through using PSI-BLAST profiles to develop the method are relatively small: PHDpsi was trained on a small database of not very divergent profiles in 1994; for example, PROFphd was trained on PSI-BLAST profiles of a 20 times larger database in 2000. The two differ by only one percentage point, and part of this difference resulted from implementing new concepts into PROF (Rost, B. 2001b).

*Averaging over many methods may help.* All methods predict some proteins at lower levels of accuracy than others (Rost, Sander, and Schneider, 1993b; Lesk, Lo Conte, and Hubbard, 2001; Rost and Eyrich, 2001a). Nevertheless, for most proteins there is a method that predicts secondary structure at a level higher than average (Rost and Eyrich, 2001a). The latter is applied when averaging over prediction methods. In fact, such averages are helpful as long as compiled over *good* methods (Rost et al., 2001b). Thus, using ALL available programs is a rather bad idea!

## Solvent Accessibility Prediction in Practice

Very few of the seemingly more accurate methods predicting solvent accessibility are publicly available. Furthermore, there is no EVA-like evaluation of methods based on large sets and identical conditions. The first case scenario in which methods are compared based on different data sets and different ways to define accessibility is the rule rather than the exception. Thus, it is important to view values published with caution. Most methods predict accessibility in two states (exposed and buried). Levels of prediction accuracy vary significantly according to choice of the thresholds to distinguish between the two states (Cuff and Barton, 2000). If we define all residues that are less than 16% solvent accessible as exposed, the best current methods reach levels around 75% ± 10% accuracy (Rost and Sander, 1994a; Lesk, 1997; Cuff and Barton, 2000; Przybylski and Rost, 2002). Using alignment information improves prediction accuracy significantly. However, accessibility predictions are more sensitive to alignment errors than are secondary structure predictions (Rost and Sander, 1994a; Przybylski and Rost, 2002). A reason for this may be that accessibility is evolutionarily less well conserved than is secondary structure (Przybylski and Rost, 2002).

## Transmembrane Helix Prediction in Practice

*Caution: no appropriate estimate of performance available!* The appropriate evaluation of methods predicting membrane helices is even more difficult than the evaluation of other categories of structure prediction. Three major problems prevent adequate analyses: (1) We do not have enough high-resolution structures to allow a statistically significant analysis (Chen and Rost, 2002b). (2) Low-resolution experiments (gene fusion) differ from high-resolution experiments (crystallography) almost as much as prediction methods do (Chen and Rost, 2002b). Thus, low-resolution experiments do not suffice to evaluate prediction accuracy. (3) All methods optimize some parameters. Since there are so few high-resolution structures, all methods use as many of the known ones as possible. However, that methods perform much better on proteins for which they were developed than on new proteins was impressively demonstrated and overlooked in a recent analysis of prediction methods (Möller, Croning, and Apweiler, 2001).

*Crude estimates for where we are at in the field.* The best current methods (HMM-TOP2, PHDhtm, and TMHMM2) predict all helices correct for about 70% of all proteins (Chen and Rost, 2002b). For more than 60% of the proteins, the topology is also predicted correctly (Möller, Croning, and Apweiler, 2001; Chen and Rost, 2002b). The most accurate per-residue prediction is achieved by PHDhtm, getting about 70% of the observed TMH residues right (Chen and Rost, 2002b). All methods based on advanced algorithms tend to underestimate transmembrane helices; thus, about 86% of the TMH residues predicted by the best methods in this category PHD and DAS are correct (Chen and Rost, 2002b). Most method tend to confuse signal peptides with membrane helices; the best separation is achieved by a system predicting subcellular localization ALOM2 (Nakai and Kanehisa, 1992). Almost as accurate are PHD and TopPred2 (followed by TMHMM) (Möller, Croning, and Apweiler, 2001). Surprisingly, most methods have also been overestimated in their ability to distinguish between globular and helical membrane proteins; particularly, most methods based only on hydrophobicity scales incorrectly predict membrane helices in over 90% of a representative set of globular proteins (Chen, Kernytsky, and Rost, 2002a). TMHMM, SOSUI, and PHDhtm appear to yield the most accurate distinction between membrane and globular proteins (<2% false positives) (Chen and Rost, 2002b). The most accurate hydrophobicity index appears to be the one recently developed in the Ben-Tal group (Kessel and Ben-Tal, 2002). All methods fail to distinguish membrane helices from signal peptides to the extent that the best methods still falsely predict membrane helices for 25% (PHDhtm) to 34% (TMHMM2) of all signal peptides tested. The good news for the practical application is that we have an accurate method to detect signal peptides (Nielsen et al., 1997), and that most incorrectly predicted membrane helices start closer than 10 residues to N-terminal Methionine residues, that is, they could be corrected by experts.

*Genome analysis: many proteins contain membrane helices.* Despite the overestimated performance, predictions of transmembrane helices are valuable tools to quickly scan entire genomes for membrane proteins. A few groups base their results only on hydrophobicity scales, known to have extremely high error rates in distinguishing globular and membrane proteins. Nevertheless, the averages published for entire genomes are surprisingly similar between different authors (Goffeau et al., 1993; Rost, Casadio, and Fariselli, 1996a; Arkin, Brünger, and Engelman, 1997; Frishman and Mewes, 1997; Jones, 1998; Wallin and von Heijne, 1998; Liu and Rost, 2001). About 10–30% of all proteins appear to contain membrane helices. One crucial difference, however, is that more cautious estimates do not perceive a statistically significant difference in the percentage of TMH proteins across the three kingdoms: eukaryotes, prokaryotes, and archae (Liu, Tan, and Rost, 2002b). However, the preferences between particular types of membrane proteins differs; in particular, eukaryotes have more 7TM proteins (receptors), whereas prokaryotes have more 6- and 12TM proteins (ABC transporters) (Wallin and von Heijne, 1998; Liu and Rost, 2001).

## EMERGING AND FUTURE DEVELOPMENTS

*Regions likely to undergo structural change predicted successfully.* Young, Kirshenbaum, Dill, and Highsmith (Young et al., 1999) have unravelled an impressive correlation between local secondary structure predictions and global conditions. The authors monitor regions for which secondary structure prediction methods give equally strong preferences for two different states. Such regions are processed combining simple statistics and expert rules. The final method is tested on 16 proteins known to undergo structural rearrangements, and on a number of other proteins. The authors report no false positives, and identify most known structural switches. Subsequently, the group applied the method to the myosin family identifying putative switching regions that were not known before, but that appeared to be reasonable candidates (Kirshenbaum, Young, and Highsmith, 1999). I find this method most remarkable in two ways: (1) it is the most general method using predictions of protein structure to predict some aspects of function, and (2) it illustrates that predictions may be useful even when structures are known (as in the case of the myosin family).

*Classifying proteins based on secondary structure predictions in the context of genome analysis.* Proteins can be classified into families based on predicted and observed secondary structure (Gerstein and Levitt, 1997; Przytycka, Aurora, and Rose, 1999). However, such procedures have been limited to a very coarse-grained grouping only exceptionally useful to infer function (Rost B, 2002). Nevertheless, in particular, predictions of membrane helices and coiled-coil regions are crucial for genome analysis. Recently, we came across an observation that may have important implications for structural genomics in particular: More than one fifth of all eukaryotic proteins appeared to have regions longer than 60 residues, apparently lacking any regular secondary structure (Liu, Tan, and Rost, 2002b). Most of these regions were not of low complexity, that is, not composition biased (Dunker and Obradovic, 2001b; Romero et al., 2001; Liu, Tan, and Rost, 2002b). Surprisingly, these regions appeared evolutionarily as conserved as all other regions in the respective proteins. This application of secondary structure prediction may aid in classifying proteins, and in separating domains, possibly even in identifying particular functional motifs.

*Aspects of protein function predicted based on expert-analysis of secondary structure.* The typical scenario in which secondary structure predictions help us to learn more about function come from experts combining predictions and their intuition, most often to find similarities to proteins of known function but insignificant sequence similarity (Brautigam et al., 1999; Davies et al., 1999; de Fays et al., 1999; Di Stasio et al., 1999; Gerloff et al., 1999; Juan et al., 1999; Laval et al., 1999; Seto et al., 1999; Xu et al., 1999; Jackson and Russell, 2000; Paquet et al., 2000; Shah et al., 2000; Stawiski et al., 2000). Usually, such applications are based on very specific details about predicted secondary structure (Rost, 2002). Thus, successful correlations of secondary structure and function appear difficult to incorporate into automatic methods.

*Exploring secondary structure predictions to improve database searches.* Initially, three groups independently applied secondary structure predictions to fold recognition, that is, the detection of structural similarities between proteins of unrelated sequences (Rost, 1995; Fischer and Eisenberg, 1996; Russell, Copley, and Barton, 1996). A few years later, almost every other fold recognition/threading method has adopted this concept (Ayers, et al., 1999; de la Cruz and Thornton, 1999; Di Francesco, Munson, and Garnier, 1999; Hargbo and Elofsson, 1999; Jones, 1999a; Jones et al., 1999; Koretke et al., 1999; Ota et al., 1999; Panchenko, Marchler-Bauer, and Bryant, 1999; Kelley, MacCallum, and Sternberg, 2000). Two recent methods extended the concept by not only refining the database search, but by actually refining the quality of the alignment through an iterative procedure (Heringa, 1999; Jennings, Edge, and Sternberg, 2001). A related strategy has been implored by Ng and the Henikoffs to improve predictions and alignments for membrane proteins (Ng, Henikoff, and Henikoff, 2000).

*From 1D predictions to 2D and 3D structure.* Are secondary structure predictions accurate enough to help in predicting higher order aspects of protein structure automatically? Two-dimensional (interresidue contacts) predictions: Baldi, Pollastri, Andersen, and Brunak (Baldi et al., 2000) have recently improved the level of accuracy in predicting beta-strand pairings over earlier work (Hubbard and Park, 1995) through using another elaborate neural network system. Three-dimensional predictions: the following list of five groups exemplifies that secondary structure predictions have now a popular first step toward predicting 3D structure. (1) Ortiz et al. (1999) successfully use secondary structure predictions as one component of their 3D structure prediction method. (2) Eyrich et al. 1999a; Eyrich, Standley, and Friesner, 1999b) minimizes the energy of arranging predicted rigid secondary structure segments. (3) Lomize et al. (Lomize, Pogozheva, and Mosberg, 1999) also start from secondary structure segments. (4) Chen et al. (Chen, Singh, and Altman, 1999) suggest using secondary structure predictions to reduce the complexity of molecular dynamics simulations. (5) Levitt et al. (Samudrala et al., 1999; Samudrala, et al., 2000) combine secondary structure-based simplified presentations with a particular lattice simulation attempting to enumerate all possible folds.

*Using accessibility to predict aspects of function.* Features of multiple alignments can reveal aspects of protein function (Casari, Sander, and Valencia, 1995; Lichtarge, Bourne, and Cohen, 1996; Lichtarge, Yamamoto, and Cohen, 1997; Pazos et al., 1997; Marcotte et al., 1999; Irving et al., 2000). To simplify the story: Residues can be conserved because of structural and functional reasons. If we could distinguish between the two, we could predict the functional residues. Obviously, residues that are exposed *and* conserved are likely to reveal functional constraints. This result suggests using predicted accessibility and combination with alignments to predict functional residues (Rost, 2002 unpublished). Another possible application of predicted accessibility is the prediction of subcellular localization: the surface compositions differ significantly between extracellular, cytoplasmic, and nuclear proteins (Andrade, O'Donoghue, and Rost, 1998). Currently, we use predicted accessibility to improve the prediction of subcellular localization (Nair and Rost, 2002 unpublished).

*Using 1D predictions for target selection in structural genomics.* Structural genomics proposes to experimentally determine one high-resolution structure for every known protein (Gaasterland, 1998; Rost, 1998; Sali, 1998; Burley et al., 1999; Shapiro and Harris, 2000; Thornton, 2001). Obviously, this goal could be reached faster if we could avoid all proteins of known structure, which is relatively straightforward (Sali, 1998; Liu and Rost, 2002; Liu, Tan, and Rost, 2002b). More difficult is the task of avoiding proteins that do not express, do not purify, or are longer than 200 residues and do not crystallize (or do not diffract well enough). One way toward this goal is to exclude all proteins with membrane helices (Liu and Rost, 2001; Liu and Rost, 2002a). Can bioinformatics do more than that? In a preliminary analysis, we used predicted accessibility to predict the globularity of a protein (Rost, 1999a). Although prediction accuracy is rather low, we observe some correlation between the percentage of surface residues and the globularity of a protein. One important task to choose structural genomics targets most effectively is the prediction of structural domains from sequence. Currently, we are exploring ways of using predicted accessibility and secondary structure toward this end (Liu and Rost, 2002 unpublished).

*Eukaryotes full of floppy proteins?* Recently it has been shown that regions of low complexity—as predicted by the program SEG (Wootton and Federhen, 1996)—are the rule rather than the exception in the protein universe (Saqi, 1995; Garner et al., 1998; Romero et al., 1998; Wright and Dyson, 1999; Dunker et al., 2001a; Dunker and Obradovic, 2001b; Romero et al., 2001). Using predictions of secondary structure, we found that there are many proteins that do not have low-complexity regions but nevertheless appear to have long (>70 residues) regions without regular secondary structure (helix/strand, dubbed NORS). Such NORS proteins appear to be significantly more abundant in the eukaryotes than in all other kingdoms, reaching levels around 25% of the entire genome (Liu, Tan, and Rost, 2002b). We found many of the NORS to be evolutionarily conserved, suggesting that these may in fact be proteins with induced structure rather than without structure.

## NOTATIONS

**Abbreviations used: 1D structure**, one-dimensional, for example, sequence, or strings of secondary structure or solvent accessibility; **2D structure**, two-dimensional (e.g., interresidue distances); **3D structure**, three-dimensional co-ordinates of protein structure; **ASP**, method identifying regions of structure ambivalent in response to global changes (Kirshenbaum, Young, and Highsmith, 1999; Young et al., 1999); **BLAST**, fast sequence alignment method (Altschul and Gish, 1996); **CASP**, Critical Assessment of Protein Structure Prediction (Zemla, Venclovas, and Fidelis, 2001); **COILS**, coiled-coil prediction (Lupas, Van Dyck, and Stock, 1991); **DSSP**, program and database assigning secondary structure and solvent accessibility for proteins of known 3D structure (Kabsch and Sander, 1983); **EVA**, server automatically evaluating structure prediction methods (Eyrich et al., 2001a; Eyrich et al., 2001b); **HMM**, Hidden Markov Model; **HMMSTR**, Hidden Markov model-based prediction of secondary structure (Bystroff, Thorsson, and Baker, 2000); **HMMTOP**, Hidden Markov model predicting transmembrane helices (Tusnady and Simon, 1998); **JPred2**, divergent profile (PSI-BLAST) based neural network prediction of secondary structure and solvent accessibility (Cuff and Barton, 2000); **MEMSAT**, dynamic-programming based prediction of transmembrane helices (Jones, Taylor, and Thornton, 1994); **META-PP**, internet service allowing access to a variety of bioinformatics tools through one single interface (Eyrich and Rost, 2000); **PHD**, Profile-based neural network prediction of secondary structure, solvent accessibility, and transmembrane helices (Rost, 1996); **PHDpsi**, divergent profile-(PSI-BLAST) based neural network prediction (Przybylski and Rost, 2002); **PSI-BLAST**, position specific iterated database search (Altschul et al., 1997); **PROFphd**, Advanced profile-based neural network prediction of secondary structure (Rost, 2001a); **PSIPRED**, divergent profile-(PSI-Blast) based neural network prediction (Jones, 1999b); **SAM-T99sec**, neural network prediction, using hidden Markov models as input (Karplus et al., 1999); **SOSUI**, hydrophobicity and amphiphilicity based transmembrane helix prediction (Hirokawa, Boon-Chieng, and Mitaku, 1998); **SPLIT**, transmembrane helix prediction (Juretic et al., 1998); **SSpro**, profile-based advanced neural network prediction method (Baldi et al., 1999); **SSpro2**, divergent profile-based advanced neural network prediction method (Pollastri et al., 2001); **TM**, transmembrane; **TMAP**, alignment-based prediction of transmembrane helices (Persson and Argos, 1996); **TMH**, transmembrane helix; **TMHMM**, Transmembrane prediction using cyclic hidden Markov models (Sonnhammer, von Heijne, and Krogh, 1998); **TMpred**, prediction of transmembrane helices (Hofmann and

Stoffel, 1993); **TopPred2**, hydrophobicity-based membrane helix prediction (von Heijne, 1992; Cserzö et al., 1997);

**Symbols used: secondary structure: H** = helix, **E** = strand, **L** = other; **transmembrane helices: T** = transmembrane, **N** = globular; **solvent accessibility: e** = exposed (16% relative accessible surface), **b** = buried (<16%);

## ACKNOWLEDGMENTS

Thanks to Jinfeng Liu and Volker Eyrich (Columbia) for computer assistance. Thanks also to the EVA team who enabled me to quote some of the numbers given: Volker Eyrich (Columbia), Marc Marti-Renom and Andrej Sali (both Rockefeller), Florencio Pazos and Alfonso Valencia (Madrid). The work of BR was supported by the grants 1-P50-GM62413-01 and RO1-GM63029-01 from the National Institute of Health. Last, but not least, thanks to all those who deposit their experimental data in public databases, and to those who maintain those databases.

## REFERENCES

Altschul S, Madden T, Shaffer A, Zhang J, Zhang Z, Miller W, Lipman D (1997): Gapped Blast and PSI-Blast: a new generation of protein database search programs. *Nucleic Acids Res* 25:3389–402.

Altschul SF, Gish W (1996): Local alignment statistics. *Methods Enzymol* 266:460–80.

Andrade MA, O'Donoghue SI, Rost B (1998): Adaptation of protein surfaces to subcellular location. *J Mol Biol* 276:517–25.

Anfinsen CB (1973): Principles that govern the folding of protein chains. *Science* 181:223–30.

Arkin IT, Brünger AT, Engelman DM (1997): Are there dominant membrane protein families with a given number of helices? *Proteins* 28:465–6.

Ayers DJ, Gooley PR, Widmer-Cooper A, Torda AE (1999): Enhanced protein fold recognition using secondary structure information from NMR. *Protein Sci* 8:1127–33.

Bairoch A, Apweiler R (2000): The SWISS-PROT protein sequence database and its supplement TrEMBL in 2000. *Nucleic Acids Res* 28:45–8.

Baldi P, Brunak S, Frasconi P, Soda G, Pollastri G (1999): Exploiting the past and the future in protein secondary structure prediction. *Bioinformatics* 15:937–46. [The most complicated and seemingly most successful architecture for using neural networks predicting secondary structure is presented.]

Baldi P, Pollastri G, Andersen CA, Brunak S (2000): Matching protein beta-sheet partners by feedforward and recurrent neural networks. *ISMB* 8:25–36.

Baldi P, Brunak S (2001): *Bioinformatics: The Machine Learning Approach.* Cambridge: MIT Press.

Barton GJ (1996): Protein sequence alignment and database scanning. In: Sternberg MJE, editor. *Protein Structure Prediction.* Oxford: Oxford University Press, pp 31–64.

Ben-Tal N, Honig B, Miller C, McLaughlin S (1997): Electrostatic binding of proteins to membranes. Theoretical predictions and experimental results with charybdotoxin and phospholipid vesicles. *Biophys J* 73:1717–27.

Berman HM, Westbrook J, Feng Z, Gillliland G, Bhat TN, Weissig H, Shindyalov IN, Bourne PE (2000): The Protein Data Bank. *Nucleic Acids Res* 28:235–42.

Biou V, Gibrat JF, Levin JM, Robson B, Garnier J (1988): Secondary structure prediction: combination of three different methods. *Protein Eng* 2:185–91.

Bohr H, Bohr J, Brunak S, Cotterill RMJ, Lautrup B, Nørskov L, Olsen OH, Petersen SB (1988): Protein secondary structure and homology by neural networks. *FEBS Lett* 241:223–8.

Brändén C, Tooze J (1991): *Introduction to Protein Structure*. New York: Garland Publishing.

Brautigam C, Steenbergen-Spanjers GC, Hoffmann GF, Dionisi-Vici C, van den Heuvel LP, Smeitink JA, Wevers RA (1999): Biochemical and molecular genetic characteristics of the severe form of tyrosine hydroxylase deficiency. *Clin Chem* 45:2073–8.

Buchanan SK (1999): β-Barrel proteins from bacterial outer membranes: structure, function and refolding. *Curr Opin Struct Biol* 9:455–61.

Bujnicki JM, Elofsson A, Fischer D, Rychlewski L (2001): LiveBench-1: continuous benchmarking of protein structure prediction servers. *Protein Sci* 10:352–361.

Burley SK, Almo SC, Bonanno JB, Capel M, Chance MR, Gaasterland T, Lin D, Sali A, Studier FW, Swaminathan S (1999): Structural genomics: beyond the human genome project. *Nat Genet* 23:151–7.

Bystroff C, Thorsson V, Baker D (2000): HMMSTR: a hidden Markov model for local sequence-structure correlations in proteins. *J Mol Biol* 301:173–90.

Carugo O (2000): Predicting residue solvent accessibility from protein sequence by considering the sequence environment. *Protein Eng* 13:607–9.

Casari G, Sander C, Valencia A (1995): A method to predict functional residues in proteins. *Nat Struct Biol* 2:171–8.

CASP4 (2000): Fourth meeting on the critical assessment of techniques for protein structure prediction. Prediction Center, Lawrence Livermore National Lab WWW document: http://PredictionCenter.llnl.gov/casp4/Casp4.html.

Chen CC, Singh JP, Altman RB (1999): Using imperfect secondary structure predictions to improve molecular structure computations. *Bioinformatics* 15:53–65.

Chen CP, Kernytsky A, Rost B (2002): Transmembrane helix predictions revisited. *Prot Sci*, in press.

Chen CP, Rost B (2002b): Long membrane helices and short loops predicted less accurately. *Prot Sci*, in press.

Chen CP, Rost B (2002a): State-of-the-art in membrane prediction. *Appl Bioinf* 1:21–35.

Chothia C (1976): The nature of the accessible and buried surfaces in proteins. *J Mol Biol* 105:1–12.

Chou PY, Fasman GD (1974): Prediction of protein conformation. *Biochemistry* 13:211–5.

Cohen FE, Sternberg MJE, Taylor WR (1981): Analysis of the tertiary structure of protein β-sheet sandwiches. *J Mol Biol* 148:253–72.

Cohen FE, Presnell SR (1996): The combinatorial approach. In: Sternberg MJE, editor. *Protein Structure Prediction*. Oxford: Oxford University Press, pp 207–28.

Combet C, Blanchet C, Geourjon C, Deléage G (2000): NPS@: Network Protein Sequence Analysis. *TIBS* 25:147–50.

Connolly ML (1983): Solvent-accessible surfaces of proteins and nucleic acids. *Science* 221:709–13.

Corrales FJ, Fersht AR (1996): Kinetic significance of GroEL₁₄. (GroES₇)₂ complexes in molecular chaperone activity. *Folding & Design* 1:265–73.

Crick FHC (1953): The packing of a-helices: simple coiled-coils. *Acta Crystallogr* A6:689–97.

Cserzö M, Wallin E, Simon I, von Heijne G, Elofsson A (1997): Prediction of transmembrane α-helices in prokaryotic membrane proteins: the dense alignment surface method. *Protein Eng* 10:673–6.

Cuff JA, Barton GJ (2000): Application of multiple sequence alignment profiles to improve protein secondary structure prediction. *Proteins* 40:502–11.

Davies GP, Martin I, Sturrock SS, Cronshaw A, Murray NE, Dryden DT (1999): On the structure and operation of type I DNA restriction enzymes. *J Mol Biol* 290:565–79.

de Fays K, Tibor A, Lambert C, Vinals C, Denoel P, De Bolle X, Wouters J, Letesson JJ, Depiereux E (1999): Structure and function prediction of the Brucella abortus P39 protein by comparative modeling with marginal sequence similarities. *Protein Eng* 12:217–23.

de la Cruz X, Thornton JM (1999): Factors limiting the performance of prediction-based fold recognition methods. *Protein Sci* 8:750–9.

Deleage G, Roux B (1987): An algorithm for protein secondary structure prediction based on class prediction. *Protein Eng* 1:289–94.

Devereux J, Haeberli P, Smithies O (1984): GCG package. *Nucleic Acids Res* 12:387–95.

Dickerson RE, Timkovich R, Almassy RJ (1976): The cytochrome fold and the evolution of bacterial energy metabolism. *J Mol Biol* 100:473–91.

Di Francesco V, Munson PJ, Garnier J (1999): FORESST: fold recognition from secondary structure predictions of proteins. *Bioinformatics* 15:131–140.

Di Stasio E, Sciandra F, Maras B, Di Tommaso F, Petrucci TC, Giardina B, Brancaccio A (1999): Structural and functional analysis of the N-terminal extracellular region of beta-dystroglycan. *Biochem Biophys Res Commun* 266:274–8.

Dobson CM, Karplus M (1999): The fundamentals of protein folding: bringing together theory and experiment. *Curr Opin Struct Biol* 9:92–101.

Doolittle RF (1986): *Of URFs and ORFs: A Primer on How to Analyze Derived Amino Acid Sequences*. Mill Valley, CA: University Science Books.

Dunker AK, Lawson JD, Brown CJ, Williams RM, Romero P, Oh JS, Oldfield CJ, Campen AM, Ratliff CM, Hipps KW, Ausio J, Nissen MS, Reeves R, Kang C, Kissinger CR, Bailey RW, Griswold MD, Chiu W, Garner EC, Obradovic Z (2001): Intrinsically disordered protein. *J Mol Graph Model* 19:26–59.

Dunker AK, Obradovic Z (2001): The protein trinity-linking function and disorder. *Nat Biotech* 19:805–6.

Eddy SR (1998): Profile hidden Markov models. *Bioinformatics* 14:755–63.

Eisenberg D, Schwartz E, Komaromy M, Wall R (1984): Analysis of membrane and surface protein sequences with the hydrophobic moment plot. *J Mol Biol* 179:125–42.

Ellis RJ, Dobson C, Hartl U (1998): Sequence does specify protein conformation. *TIBS* 23:468.

Engelman DM, Steitz TA, Goldman A (1986): Identifying nonpolar transbilayer helices in amino acid sequences of membrane proteins. *Ann Rev Biophys Biophys Chem* 15:321–53.

Etzold T, Argos P (1993): SRS—an indexing and retrieval tool for flat file data libraries. *Comp Applied Biol Sci* 9:49–57.

Etzold T, Ulyanov A, Argos P (1996): SRS: information retrieval system for molecular biology data banks. *Methods Enzymol* 266:114–28.

Eyrich VA, Standley DM, Felts AK, Friesner RA (1999a): Protein tertiary structure prediction using a branch and bound algorithm. *Proteins* 35:41–57.

Eyrich VA, Standley DM, Friesner RA (1999b): Prediction of protein tertiary structure to low resolution: performance for a large and structurally diverse test set. *J Mol Biol* 288:725–42.

Eyrich V, Rost B (2000): The META-PredictProtein server. CUBIC, Columbia University, Dept. of Biochemistry & Molecular Biophysics WWW document (http://cubic.bioc.columbia.edu/predictprotein/submit_meta.html).

Eyrich V, Martí-Renom MA, Przybylski D, Fiser A, Pazos F, Valencia A, Sali A, Rost B (2001a): EVA: continuous automatic evaluation of protein structure prediction servers. *Bioinformatics* 17:1242–3.

Eyrich V, Martí-Renom MA, Przybylski D, Fiser A, Pazos F, Valencia A, Sali A, Rost B (2001b): EVA: continuous automatic evaluation of protein structure prediction servers. Columbia University WWW document (http://cubic.bioc.columbia.edu/eva).

Fasman GD (1989): The development of the prediction of protein structure. In: Fasman GD, editor. *Prediction of Protein Structure and the Principles of Protein Conformation.* New York: Plenum Press, pp 193–303. [Gold mine for finding citations for very early prediction methods.]

Fischer D, Eisenberg D (1996): Fold recognition using sequence-derived properties. *Protein Sci* 5:947–55.

Fischer D, Barret C, Bryson K, Elofsson A, Godzik A, Jones D, Karplus KJ, Kelley LA, MacCallum RM, Pawowski K, Rost B, Rychlewski L, Sternberg M (1999): CAFASP-1: critical assessment of fully automated structure prediction methods. *Proteins* IS Suppl 3:209–17.

Fischer D, Elofsson A, Rychlewski L, Pazos F, Valencia A, Rost B, Ortiz AR, Dunbrack RL (2001): CAFASP2: the second critical assessment of fully automated structure prediction methods. *Proteins* Suppl 5:171–83.

Frishman D, Argos P (1996): Incorporation of non-local interactions in protein secondary structure prediction from the amino acid sequence. *Protein Eng* 9:133–42.

Frishman D, Mewes HW (1997): Protein structural classes in five complete genomes. *Nat Struct Biol* 4:626–8.

Gaasterland T (1998): Structural genomics taking shape. *TIBS* 14:135.

Garner E, Cannon P, Romero P, Obradovic Z, Dunker AK (1998): Predicting disordered regions from amino acid sequence: common themes despite differing structural characterization. *Genome Inform* 9:201–14.

Garnier J, Osguthorpe DJ, Robson B (1978): Analysis of the accuracy and implications of simple methods for predicting the secondary structure of globular proteins. *J Mol Biol* 120:97–120.

Garnier J, Robson B (1989): The GOR method for predicting secondary structure in proteins. In: Fasman, GD, editor. *Prediction of Protein Structure and the Principles of Protein Conformation.* New York: Plenum Press, pp 417–65.

Gascuel O, Golmard JL (1988): A simple method for predicting the secondary structure of globular proteins: implications and accuracy. *CABIOS* 4:357–65.

Gerloff DL, Cannarozzi GM, Joachimiak M, Cohen FE, Schreiber D, Benner SA (1999): Evolutionary, mechanistic, and predictive analyses of the hydroxymethyldihydropterin pyrophosphokinase family of proteins. *Biochem Biophys Res Commun* 254:70–6.

Gerstein M, Levitt M (1997): A structural census of the current population of protein sequences. *Proc Natl Acad Sci USA* 94:11911–6.

Goffeau A, Slonimski P, Nakai K, Risler JL (1993): How many yeast genes code for membrane-spanning proteins? *Yeast* 9:691–702.

Gupta R, Jung E, Gooley AA, Williams KL, Brunak S, Hansen J (1999): Scanning the available Dictyostelium discoideum proteome for O-linked GlcNAc glycosylation sites using neural networks. *Glycobiology* 9:1009–22.

Hagler AT, Honig B (1978): On the formation of protein tertiary structure on a computer. *Proc Natl Acad Sci USA* 75:554–8.

Hansen J, Lund O, Tolstrup N, Gooley AA, Williams KL, Brunak S (1998): NetOglyc: prediction of mucin type O-glycosylation sites based on sequence context and surface accessibility. *Glycoconjugate J* 15:115–30.

Hargbo J, Elofsson A (1999): Hidden Markov models that use predicted secondary structures for fold recognition. *Proteins* 36:68–76.

Harris F, Wallace J, Phoenix DA (2000): Use of hydrophobic moment plot methodology to aid the identification of oblique orientated alpha-helices. *Molec Membr Biol* 17:201–7.

Heringa J (1999): Two strategies for sequence comparison: profile-preprocessed and secondary structure-induced multiple alignment. *Comput Chem* 23:341–64.

Hirokawa T, Boon-Chieng S, Mitaku S (1998): SOSUI: classification and secondary structure prediction system for membrane proteins. *Bioinformatics* 14:378–9.

Hofmann K, Stoffel W (1993): TMBASE—a database of membrane spanning protein segments. *Biol Chem Hoppe-Seyler* 374:166.

Holbrook SR, Muskal SM, Kim S-H (1990): Predicting surface exposure of amino acids from protein sequence. *Protein Eng* 3:659–65.

Hubbard TJP, Blundell TL (1987): Comparison of solvent-inaccessible cores of homologous proteins: definitions useful for protein modelling. *Protein Eng* 1:159–71.

Hubbard TJP, Park J (1995): Fold recognition and ab initio structure predictions using Hidden Markov models and $\beta$-strand pair potentials. *Proteins* 23:398–402.

Hubbard T, Tramontano A, Barton G, Jones D, Sippl M, Valencia A, Lesk A, Moult J, Rost B, Sander C, Schneider R, Lahm A, Leplae R, Buta C, Eisenstein M, Fjellström O, Floeckner H, Grossmann JG, Hansen J, Helmer-Citterich M, Joergensen FS, Marchler-Bauer A, Osuna J, Park J, Reinhardt A, Ribas de Pouplana L, Rojo-Dominguez A, Saudek V, Sinclair J, Sturrock S, Venclovas C, Vinals C (1996): Update on protein structure prediction: results of the 1995 IRBM workshop. *Folding & Design* 1:R55–R63.

Irving JA, Pike RN, Lesk AM, Whisstock JC (2000): Phylogeny of the serpin superfamily: implications of patterns of amino acid conservation for structure and function. *Genome Res* 10:1845–64.

Jackson RM, Russell RB (2000): The serine protease inhibitor canonical loop conformation: examples found in extracellular hydrolases, toxins, cytokines and viral proteins. *J Mol Biol* 296:325–34.

Jennings AJ, Edge CM, Sternberg MJ (2001): An approach to improving multiple alignments of protein sequences using predicted secondary structure. *Protein Eng* 14:227–31.

Jones DT (1998): Do transmembrane protein superfolds exist? *FEBS Lett* 423:281–5.

Jones DT (1999a): GenTHREADER: an efficient and reliable protein fold recognition method for genomic sequences. *J Mol Biol* 287:797–815.

Jones DT (1999b): Protein secondary structure prediction based on position-specific scoring matrices. *J Mol Biol* 292:195–202. [The alignments used by PHD are replaced by PSI-BLAST alignments. This replacement improves prediction accuracy significantly. However, possibly the most important aspect is the description of ways to run PSI-BLAST automatically without finding too many wrong hits.]

Jones DT, Taylor WR, Thornton JM (1994): A model recognition approach to the prediction of all-helical membrane protein structure and topology. *Biochemistry* 33:3038–49.

Jones DT, Tress M, Bryson K, Hadley C (1999): Successful recognition of protein folds using threading methods biased by sequence similarity and predicted secondary structure. *Proteins* 37:104–11.

Juan HF, Hung CC, Wang KT, Chiou SH (1999): Comparison of three classes of snake neurotoxins by homology modeling and computer simulation graphics. *Biochem Biophys Res Commun* 257:500–10.

Juretic D, Zucic D, Lucic B, Trinajstic N (1998): Preference functions for prediction of membrane-buried helices in integral membrane proteins. *Comput Chem* 22:279–94.

Kabsch W, Sander C (1983): Dictionary of protein secondary structure: pattern recognition of hydrogen bonded and geometrical features. *Biopolymers* 22:2577–637.

Karplus K, Barrett C, Hughey R (1998): Hidden Markov models for detecting remote protein homologies. *Bioinformatics* 14:846–56.

Karplus K, Barrett C, Cline M, Diekhans M, Grate L, Hughey R (1999): Predicting protein structure using only sequence information. *Proteins* IS Suppl 3:121–5.

Kelley LA, MacCallum RM, Sternberg MJ (2000): Enhanced genome annotation using structural profiles in the program 3D-PSSM. *J Mol Biol* 299:499–520.

Kessel A, Ben-Tal N (2002): Free energy determinants of peptide association with lipid bilayers. In: Simon S, and McIntosh T, editors. *Peptide-Lipid Interactions*. San Diego: Academic Press. Forthcoming.

Kirshenbaum K, Young M, Highsmith S (1999): Predicting allosteric switches in myosins. *Protein Sci* 8:1806–15.

Klein P, Kanehisa M, De Lisi C (1985): The detection and classification of membrane-spanning proteins. *Biochim Biophys Acta* 815:468–76.

Kleywegt G (2001): ProSAL: protein sequence analysis launcher. Swedish Structural Biology Network (http://alpha2.bmc.uu.se/~gerard/srf/prosal.html).

Koretke KK, Russell RB, Copley RR, Lupas AN (1999): Fold recognition using sequence and secondary structure information. *Proteins* 37:141–8.

Kyte J, Doolittle RF (1982): A simple method for displaying the hydropathic character of a protein. *J Mol Biol* 157:105–32.

Laval V, Chabannes M, Carriere M, Canut H, Barre A, Rouge P, Pont-Lezica R, Galaud J (1999): A family of Arabidopsis plasma membrane receptors presenting animal beta-integrin domains. *Biochim Biophys Acta* 1435:61–70.

Lee BK, Richards FM (1971): The interpretation of protein structures: estimation of static accessibility. *J Mol Biol* 55:379–400.

Lesk AM (1997): CASP-2: Report on ab initio predictions. *Proteins* IS Suppl. 151–66.

Lesk AM, Lo Conte L, Hubbard TJP (2001): Assessment of novel folds targets in CASP4: predictions of three-dimensional structures, secondary structures, and interresidue contacts. *Proteins*. Suppl.

Levin JM, Garnier J (1988): Improvements in a secondary structure prediction method based on a search for local sequence homologies and its use as a model building tool. *Biochim Biophys Acta* 955:283–95.

Levitt M, Warshel A (1975): Computer simulation of protein folding. *Nature* 253:694–8.

Li X, Pan XM (2001): New method for accurate prediction of solvent accessibility from protein sequence. *Proteins* 42:1–5.

Lichtarge O, Bourne HR, Cohen FE (1996): Evolutionarily conserved Galphabetagamma binding surfaces support a model of the G protein-receptor complex. *Proc Natl Acad Sci USA* 93:7507–11.

Lichtarge O, Yamamoto KR, Cohen FE (1997): Identification of functional surfaces of the zinc binding domains of intracellular receptors. *J Mol Biol* 274:325–37.

Lio P, Vannucci M (2000): Wavelet change-point prediction of transmembrane proteins. *Bioinformatics* 16:376–82.

Liu J, Rost B (2001): Comparing function and structure between entire proteomes. *Protein Sci* 10:1970–9. [Twenty-eight entirely sequenced genomes are compared based on predictions of coiled-coil proteins (COILS; Lupas, 1996), membrane helices (PHD; Rost, 1996), and functional classes (EUCLID; Tamames et al., 1998). In contrast to many other publications, the correlation between the complexity of an organism and its use of helical membrane proteins is not confirmed.]

Liu J, Rost B (2002): Target space for structural genomics revisited. *Bioinformatics* 18:922–33.

Liu J, Tan H, Rost B (2002): Loopy proteins appear conserved in evolution. *J Mol Biol* 322:53–61.

Liu LP, Deber CM (1999): Combining hydrophobicity and helicity: a novel approach to membrane protein structure prediction. *Bioorg Med Chem* 7:1–7.

Lomize AL, Pogozheva ID, Mosberg HI (1999): Prediction of protein structure: the problem of fold multiplicity. *Proteins* Suppl 3:199–203.

Lupas A (1996): Prediction and analysis of coiled-coil structures. *Methods Enzymol* 266:513–25.

Lupas A (1997): Predicting coiled-coil regions in proteins. *Curr Opin Struct Biol* 7:388–93. [Analysis of the performance of various programs predicting coiled-coil helices based on new structures.]

Lupas A, Van Dyke M, Stock J (1991): Predicting coiled coils from protein sequences. *Science* 252:1162–4.

Marcotte EM, Pellegrini M, Ng HL, Rice DW, Yeates TO, Eisenberg D (1999): Detecting protein function and protein–protein interactions from genome sequences. *Science* 285:751–3.

Marti-Renom MA, Madhusudhan MS, Fiser A, Rost B, Sali A (2002): Reliability of assessment of protein structure prediction methods at CASP. *Structure* 10:435–40.

Martin J, Hartl FU (1997): Chaperone-assisted protein folding. *Curr Opin Struct Biol* 7:41–52.

Maxfield FR, Scheraga HA (1979): Improvements in the prediction of protein topography by reduction of statistical errors. *Biochemistry* 18:697–704.

Möller S, Croning DR, Apweiler R (2001): Evaluation of methods for the prediction of membrane spanning regions. *Bioinformatics* 17:646–53.

Monge A, Friesner RA, Honig B (1994): An algorithm to generate low-resolution protein tertiary structures from knowledge of secondary structure. *Proc Natl Acad Sci USA* 91:5027–9.

Monne M, Hermansson M, von Heijne G (1999): A turn propensity scale for transmembrane helices. *J Mol Biol* 288:141–5.

Moult J, Pedersen JT, Judson R, Fidelis K (1995): A large-scale experiment to assess protein structure prediction methods. *Proteins* 23:ii–iv.

Moult J, Hubbard T, Bryant SH, Fidelis K, Pedersen JT (1997): Critical assessment of methods of protein structure prediction (CASP): Round II. *Proteins* Suppl 1:2–6.

Moult J, Hubbard T, Bryant SH, Fidelis K, Pedersen JT (1999): Critical assessment of methods of protein structure prediction (CASP): Round III. *Proteins* Suppl 3:2–6.

Mucchielli-Giorgi MH, Hazout S, Tuffery P (1999): PredAcc: prediction of solvent accessibility. *Bioinformatics* 15:176–7.

Mumenthaler C, Braun W (1995): Predicting the helix packing of globular proteins by self-correcting distance geometry. *Protein Sci* 4:863–71.

Naderi-Manesh H, Sadeghi M, Arab S, Moosavi-Movahedi AA (2001): Prediction of protein surface accessibility with information theory. *Proteins* 42:452–9.

Nakai K, Kanehisa M (1992): A knowledge base for predicting protein localization sites in eukaryotic cells. *Genomics* 14:897–911.

Neuwald AF, Liu JS, Lawrence CE (1995): Gibbs motif sampling: detection of bacterial outer membrane protein repeats. *Protein Sci* 4:1618–31.

Ng P, Henikoff J, Henikoff S (2000): PHAT: a transmembrane-specific substitution matrix. *Bioinformatics* 16:760–6.

Nielsen H, Engelbrecht J, Brunak S, von Heijne G (1997): Identification of prokaryotic and eukaryotic signal peptides and prediction of their cleavage sites. *Protein Eng* 10:1–6.

Nilges M, Brünger AT (1993): Successful prediction of coiled coil geometry of the GCN4 leucine zipper domain by simulated annealing: comparison to the X-ray. *Proteins* 15:133–46.

Nishikawa K, Ooi T (1982): Correlation of the amino acid composition of a protein to its structural and biological characteristics. *J Biochem* 91:1821–4.

Nishikawa K, Ooi T (1986): Amino acid sequence homology applied to the prediction of protein secondary structure, and joint prediction with existing methods. *Biochim Biophys Acta* 871:45–54.

O'Donoghue SI, Nilges M (1997): Tertiary structure prediction using mean-force potentials and internal energy functions: successful prediction for coiled-coil geometries. *Folding & Design* 2:S47–S52.

Ortiz AR, Kolinski A, Rotkiewicz P, Ilkowski B, Skolnick J (1999): Ab initio folding of proteins using restraints derived from evolutionary information. *Proteins* Suppl. 3:177–85.

Ota M, Kawabata T, Kinjo AR, Nishikawa K (1999): Cooperative approach for the protein fold recognition. *Proteins* 37:126–32.

Panchenko A, Marchler-Bauer A, Bryant SH (1999): Threading with explicit models for evolutionary conservation of structure and sequence. *Proteins* Suppl. 3:133–40.

Paquet JY, Vinals C, Wouters J, Letesson JJ, Depiereux E (2000): Topology prediction of Brucella abortus Omp2b and Omp2a porins after critical assessment of transmembrane beta strands prediction by several secondary structure prediction methods. *J Biomol Struct Dyn* 17:747–57.

Pasquier C, Promponas VJ, Palaios GA, Hamodrakas JS, Hamodrakas SJ (1999): A novel method for predicting transmembrane segments in proteins based on a statistical analysis of the SwissProt database: the PRED-TMR algorithm. *Protein Eng* 12:381–5.

Pazos F, Sanchez-Pulido L, Garcia-Ranea JA, Andrade MA, Atrian S, Valencia A (1997): Comparative analysis of different methods for the detection of specificity regions in protein families. In: Olsson B, Lundh D, Narayanan A, editors. *BCEC97: Bio-Computing and Emergent Computation*. Skövde, Sweden: World Scientific, pp 132–45.

Persson B, Argos P (1996): Topology prediction of membrane proteins. *Protein Sci* 5:363–71.

Persson B, Argos P (1997): Prediction of membrane protein topology utilizing multiple sequence alignments. *J Protein Chem* 16:453–7.

Phoenix DA, Stanworth A, Harris F (1998): The hydrophobic moment plot and its efficacy in the prediction and classification of membrane interactive proteins and peptides. *Membr Cell Biol* 12:101–10.

Pilpel Y, Ben-Tal N, Lancet D (1999): kPROT: a knowledge-based scale for the propensity of residue orientation in transmembrane segments. Application to membrane protein structure prediction. *J Mol Biol* 294:921–35.

Pollastri G, Przybylski D, Rost B, Baldi P (2002): Improving the prediction of protein secondary structure in three and eight classes using recurrent neural networks and profiles. *Proteins* 47:228–35.

Przybylski D, Rost B (2002): Alignments grow, secondary structure prediction improves. *Proteins* 46:195–205.

Przytycka T, Aurora R, Rose GD (1999): A protein taxonomy based on secondary structure. *Nat Struct Biol* 6:672–82. [Does a protein's secondary structure determine its 3D fold? This question is tested directly by analyzing proteins of known structure and constructing a taxonomy based solely on secondary structure. The taxonomy is generated automatically, and it takes the form of a tree in which proteins with similar secondary structure occupy neighboring leaves.]

Qian N, Sejnowski TJ (1988): Predicting the secondary structure of globular proteins using neural network models. *J Mol Biol* 202:865–84.

Richards FM (1977): Areas, volumes, packing, and protein structure. *Ann Rev Biophys Bioeng* 6:151–76.

Robson B (1976): Conformational properties of amino acid residues in globular proteins. *J Mol Biol* 107:327–56.

Romero P, Obradovic Z, Kissinger C, Villafranca JE, Garner E, Guilliot S, Dunker AK (1998): Thousands of proteins likely to have long disordered regions. *Pacific Symp Biocomput* 3:437–48.

Romero P, Obradovic Z, Li X, Garner EC, Brown CJ, Dunker AK (2001): Sequence complexity of disordered protein. *Proteins* 42:38–48.

Rost B (1995): TOPITS: threading one-dimensional predictions into three-dimensional structures. In: Rawlings C, Clark D, Altman R, Hunter L, Lengauer T, Wodak S, editors. *Third International Conference on Intelligent Systems for Molecular Biology.* Cambridge: AAAI Press, pp 314–21.

Rost B (1996): PHD: predicting one-dimensional protein structure by profile based neural networks. *Methods Enzymol* 266:525–39.

Rost B (1997): Protein structures sustain evolutionary drift. *Folding & Design* 2:S19–S24.

Rost B (1998): Marrying structure and genomics. *Structure* 6:259–63.

Rost B (1999a): Short yeast ORFs: expressed protein or not? CUBIC, Columbia University, Dept. of Biochemistry & Mol. Biophysics CUBIC preprint (http://cubic.bioc.columbia.edu/papers/1999_globe).

Rost B (1999b): Twilight zone of protein sequence alignments. *Protein Eng* 12:85–94.

Rost B (2000): *PredictProtein—internet prediction service.* Columbia University, New York, USA WWW document (http://cubic.bioc.columbia.edu/predictprotein).

Rost B (2001b): Protein secondary structure prediction continues to rise. *J Struct Biol* 134:204–18. [Brief walk through the recent highlights in the field of protein secondary structure prediction.]

Rost B, Sander C (1992): Exercising multi-layered networks on protein secondary structure. In: Benhar O, Brunak S, DelGiudice P, Grandolfo M, editors. *Neural Networks: From Biology to High Energy Physics.* Elba, Italy: International Journal of Neural Systems, pp 209–20.

Rost B, Sander C (1993a): Prediction of protein secondary structure at better than 70% accuracy. *J Mol Biol* 2232:584–99. [Original paper describing the first method that surpassed the threshold of 70% prediction accuracy through combining neural networks and evolutionary information.]

Rost B, Sander C, Schneider R (1993b): Progress in protein structure prediction? *TIBS* 18:120–3.

Rost B, Sander C (1994a): Combining evolutionary information and neural networks to predict protein secondary structure. *Proteins* 19:55–72.

Rost B, Sander C (1994b): Conservation and prediction of solvent accessibility in protein families. *Proteins* 20:216–26. [Analysis of the evolutionary conservation of solvent accessibility and description of an alignment-based neural network prediction method.]

Rost B, Casadio R, Fariselli P, Sander C (1995): Prediction of helical transmembrane segments at 95% accuracy. *Protein Sci* 4:521–33.

Rost B, Casadio R, Fariselli P (1996a): Topology prediction for helical transmembrane proteins at 86% accuracy. *Protein Sci* 5:1704–18.

Rost B, Sander C (1996b): Bridging the protein sequence-structure gap by structure predictions. *Annu Rev Biophys Biomol Struct* 25:113–136.

Rost B, O'Donoghue SI (1997): Sisyphus and prediction of protein structure. *CABIOS* 13:345–56.

Rost B, Sander C (2000): Third generation prediction of secondary structure. *Methods Mol Biol* 143:71–95.

Rost B, Eyrich V (2001a): EVA: large-scale analysis of secondary structure prediction. *Proteins* 45(5):S192–S199.

Rost B, Baldi P, Barton G, Cuff J, Eyrich V, Jones D, Karplus K, King R, Ouali M, Pollaastri G, Przybylski D (2001b): Simple jury predicts protein secondary structure best. *Proteins.* Forthcoming.

Russell RB, Copley RR, Barton GJ (1996): Protein fold recognition by mapping predicted secondary structures. *J Mol Biol* 259:349–65.

Rychlewski L (2000): META server. IIMCB Warsaw WWW document (www.bioinfo.pl/meta/) http://BioInfo.PL/LiveBench/.

Salamov AA, Solovyev VV (1997): Protein secondary structure prediction using local alignments. *J Mol Biol* 268:31–6.

Sali A (1998): 100,000 protein structures for the biologist. *Nat Struct Biol* 5:1029–32.

Samudrala R, Xia Y, Huang E, Levitt M (1999): Ab initio protein structure prediction using a combined hierarchical approach. *Proteins* Suppl 3:194–8.

Samudrala R, Huang ES, Koehl P, Levitt M (2000): Constructing side chains on near-native main chains for ab initio protein structure prediction. *Protein Eng* 13:453–7.

Saqi M (1995): An analysis of structural instances of low complexity sequence segments. *Protein Eng* 8:1069–73.

Schulz GE, Schirmer RH (1979): Prediction of secondary structure from the amino acid sequence. In: *Principles of Protein Structure*. Berlin: Springer-Verlag, pp 108–30.

Seshadri K, Garemyr R, Wallin E, von Heijne G, Elofsson A (1998): Architecture of beta-barrel membrane proteins: analysis of trimeric porins. *Protein Sci* 7:2026–32.

Seto MH, Liu HL, Zajchowski DA, Whitlow M (1999): Protein fold analysis of the B30.2-like domain. *Proteins* 35:235–49.

Shah PS, Bizik F, Dukor RK, Qasba PK (2000): Active site studies of bovine alpha1—>3-galactosyltransferase and its secondary structure prediction. *Biochim Biophys Acta* 1480:222–34.

Shapiro L, Harris T (2000): Finding function through structural genomics. *Curr Opin Biotech* 11:31–5.

Sipos L, von Heijne G (1993): Predicting the topology of eukaryotic membrane proteins. *Eur J Biochem* 213:1333–40.

Smith RF, Wiese BA, Wojzynski MK, Davison DB, Worley KC (1996): BCM Search Launcher—an integrated interface to molecular biology database search and analysis services available on the World Wide Web. *Genome Res* 6:454–62.

Sonnhammer ELL, von Heijne G, Krogh A (1998): A hidden Markov model for predicting transmembrane helices in protein sequences. In: *Sixth International Conference on Intelligent Systems for Molecular Biology* (ISMB98), pp 175–82. [Most advanced and seemingly most accurate method of predicting transmembrane helices through cyclic Hidden Markov models.]

Stawiski EW, Baucom AE, Lohr SC, Gregoret LM (2000): Predicting protein function from structure: unique structural features of proteases. *Proc Natl Acad Sci USA* 97:3954–8.

Sweet RM, Eisenberg D (1983): Correlation of sequence hydrophobicities measures similarity in three-dimensional protein structure. *J Mol Biol* 171:479–88.

Tamames J, Ouzounis C, Casari G, Sander C, Valencia A (1998): EUCLID: automatic classification of proteins in functional classes by their database annotations. *Bioinformatics* 14:542–3.

Tanford C (1978): The hydrophobic effect and the organization of living matter. *Science* 200:1012–8.

Tanford C (1980): *The Hydrophobic Effect: Formation of Micelles and Biological Membranes*. New York: John Wiley & Sons.

Taylor WR, Jones DT, Green NM (1994): A method for $\alpha$-helical integral membrane protein fold prediction. *Proteins* 18:281–94.

Thompson MJ, Goldstein RA (1996): Predicting solvent accessibility: higher accuracy using Bayesian statistics and optimized residue substitution classes. *Proteins* 25:38–47.

Thornton J (2001): Structural genomics takes off. *TIBS* 26:88–9.

Tusnady GE, Simon I (1998): Principles governing amino acid composition of integral membrane proteins: application to topology prediction. *J Mol Biol* 283:489–506.

Tusnady GE, Simon I (2001): Topology of membrane proteins. *J Chem Inf Comput Sci* 41:364–8.

van Gunsteren WF (1993): Molecular dynamics studies of proteins. *Curr Opin Struct Biol* 3:167–74.

von Heijne G (1986): The distribution of positively charged residues in bacterial inner membrane proteins correlates with the trans-membrane topology. *EMBO J* 5:3021–7.

von Heijne G (1989): Control of topology and mode of assembly of a polytopic membrane protein by positively charged residues. *Nature* 341:456–8.

von Heijne G (1992): Membrane protein structure prediction. *J Mol Biol* 225:487–94.

von Heijne G (1994): Membrane proteins: from sequence to structure. *Ann Rev Biophys Biomol Struct* 23:167–92.

von Heijne G (1996): Prediction of transmembrane protein topology. In: Sternberg MJE, editor. *Protein Structure Prediction*. Oxford: Oxford University Press, pp 101–10. [Review of methods predicting transmembrane helices based on hydrophobicity.]

Wako H, Blundell TL (1994): Use of amino acid environment-dependent substitution tables and conformational propensities in structure prediction from aligned sequences of homologous proteins I. Solvent accessibility classes. *J Mol Biol* 238:682–92.

Wallin E, von Heijne G (1998): Genome-wide analysis of integral membrane proteins from eubacterial, archaean, and eukaryotic organisms. *Protein Sci* 7:1029–38. [The authors analyze all membrane helix predictions for a number of entirely sequenced genomes. They conclude that more complex organisms use more helical membrane proteins than simpler organisms.]

Wang Z-X, Yuan Z (2000): How good is prediction of protein structural class by the component-coupled method? *Proteins* 38:165–75.

Wolf E, Kim PS, Berger B (1997): MultiCoil: a program for predicting two- and three-stranded coiled coils. *Protein Sci* 6:1179–89.

Wootton JC, Federhen S (1996): Analysis of compositionally biased regions in sequence databases. *Methods Enzymol* 266:554–71.

Wright PE, Dyson HJ (1999): Intrinsically unstructured proteins: re-assessing the protein structure-function paradigm. *J Mol Biol* 293:321–31.

Xu H, Aurora R, Rose GD, White RH (1999): Identifying two ancient enzymes in Archaea using predicted secondary structure alignment. *Nat Struct Biol* 6:750–4.

Yang AS, Honig B (2000): An integrated approach to the analysis and modeling of protein sequences and structures. II. On the relationship between sequence and structural similarity for proteins that are not obviously related in sequence. *J Mol Biol* 301:679–89.

Young M, Kirshenbaum K, Dill KA, Highsmith S (1999): Predicting conformational switches in proteins. *Protein Sci* 8:1752–64. [A data set of 16 protein sequences having functions that involve substantial backbone rearrangements are analyzed with respect to the ambivalence of predicted secondary structure. They find all segments involved in conformational switches to have ambivalent predictions, measured by the similarity in prediction probabilities for helix, sheet and loop, as reported by PHD.]

Zemla A, Venclovas C, Fidelis K (2001): Protein structure prediction center. Lawrence Livermore National Laboratory (http://PredictionCenter.llnl.gov/).

Zvelebil MJ, Barton GJ, Taylor WR, Sternberg MJE (1987): Prediction of protein secondary structure and active sites using alignment of homologous sequences. *J Mol Biol* 195:957–61.

# Section VIII

## THE FUTURE

# 29

# STRUCTURAL GENOMICS

## Stephen K. Burley and Jeffrey B. Bonanno

With access to sequences from the entire human genome plus those of the mouse, the fruit fly, the round worm, fungi, archaea, and a host of important bacterial pathogens, the relatively young discipline of structural biology is on the verge of a dramatic transformation. This wealth of genomic sequence information is the foundation for an important new "big science" initiative in biology. X-ray crystallographers, solution NMR spectroscopists, and computational biologists have embarked on a systematic program of high-throughput structure determination aimed at developing a comprehensive view of the protein structure universe. Structural genomics promises to yield a large number of experimental protein structures (tens of thousands) and an even larger number of calculated models of related proteins (millions). Well before this ambitious program is complete, the technological benefits of automated protein expression and purification, robotic crystallization, and NMR and X-ray data measurement/analysis should have a substantial impact on the pace of all structural biology research. Once development of the high-throughput structure determination pipeline is complete, the flood of publicly available structural information promises to accelerate discoveries in the biomedical sciences. In addition, it should revolutionize our understanding of protein evolution and the way we think about relationships between three-dimensional (3D) structure and macromolecular function. A comprehensive guide to structural genomics has been published in a Supplement to Volume 7 of *Nature Structure Biology* published in 2000.

## STRUCTURAL INFORMATION FOR EVERY PROTEIN SEQUENCE IN NATURE: THE PRIMARY GOAL

The benefits of combining 3D structural information with whole genome sequences are well documented. To paraphrase influential architects of the early 20th century, "function follows form" in biology. Virtually every one of the 15,500 plus experimentally

*Structural Bioinformatics*
Edited by Philip E. Bourne and Helge Weissig
Copyright © 2003 by Wiley-Liss, Inc.

determined protein structures in the Protein Data Bank (PDB; http://www.rcsb.org/pdb/; Chapter 9; Berman et al., 2000) has been of some benefit to biological and biochemical researchers. On the computational front, comparative protein structure modeling (also known as homology modeling, Chapter 25), protein fold assignment (Chapter 26), and *ab initio* protein structure prediction (Chapter 27) represent important bioinformatic tools, which bring mechanistic insights to biology, chemistry, and medicine (Chapters 19 and 23; Moult and Melamud, 2000; Skolnick, Fetrow, and Kolinski, 2000; Teichmann, Murzin, and Chothia, 2001). For a general review on protein structure prediction and structural genomics see Baker and Sali (2001).

## Protein Structure Space

The impact of combining experimental structure determinations with comparative protein structure modeling comes in no small part from the rather low complexity of protein fold space, which is quite unlike the Byzantine character of the estimated 40,000 human genes and more than 400,000 human gene products. Although there is still some uncertainty regarding the precise numerology, we now appreciate that the universe of compact globular protein folds or domains is small (Chapter 18; Zhang and DeLisi, 2001). Current estimates suggest that there are only 1000–5000 distinct, stable polypeptide chain folds in nature (Brenner, Chothia, and Hubbard, 1997; Liu and Rost, 2001). At present, we have experimental structures of less than 700 of these distinct protein folds (Orengo et al., 1999), with some popular folds such as the eightfold $\alpha\beta$ barrel of the triose phosphate isomerase type represented by more than 23 protein sequence families. In eukaryotes, most genes encode proteins that are comprised of multiple globular domains, with the average domain size $= 153 +/- 87$ residues (Orengo et al., 1999), giving larger proteins the appearance of beads on a string. Typically, a single bead is responsible for carrying out a specialized biochemical task.

## Evolutionary Considerations

A significant evolutionary change in gene sequence often manifests itself at the level of an individual protein functional unit or domain, which may be regarded as the focus of natural selection. Changes destabilizing the structure of a critical domain within an essential gene product cannot endure; whereas benign alterations in structure, many of which are found on domain surfaces or within linkers between domains, can persist. In some cases, gene duplication followed by mutation(s) of the gene can provide the host organism with a new biochemical capability that confers a selective advantage. For example, a mutation leading to a change in the surface properties of an enzyme active site (Chapters 20 and 21) might give rise to a new catalytic function that allows the organism to metabolize an environmental toxin or synthesize an essential substrate/cofactor/precursor (Ritz et al., 2001; Sheehan et al., 2001; Todd, Orengo, and Thornton, 2001). Beneficiaries of such genetic enhancements can proliferate at the expense of less fortunate neighbors, thereby ensuring that the newly created gene and its protein product are maintained on an evolutionary time scale.

## Homology Modeling

Computational biologists exploit the same features of protein evolution to leverage the work product of X-ray crystallographers and solution NMR spectroscopists. Each newly determined experimental structure typically provides useful structural information for many additional protein sequences (Chapter 25). In practice, however, homology modeling is currently restricted to protein sequences for which a nearby experimental template is available (amino acid sequence identity >30% over the entire length of the protein or domain). The coverage of the PDB currently limits the scope of homology modeling to about 50% of the open reading frames of the *Saccharomyces cerevisiae* genome. If one considers that only a portion of a given protein can be modeled in most cases, the situation looks decidedly worse (18% of all residues in yeast proteins; Sanchez and Sali, 1998). A recent study by Rost (Liu and Rost, 2001) of 29 fully sequenced genomes suggests that 20–40% of all proteins have a homologue with an experimentally determined structure, representing structural homology for 20–30% of all residues.

Structural biologists can contribute enormously to our understanding of biology and to biomedical research by providing full coverage of the protein structure universe (Hol, 2000). Determination of at least one experimental structure for every protein sequence family, defined at the level of 30% identity, would bring all globular proteins within the radius of convergence of current homology modeling tools. Analyses by Vitkup et al. (2001) have suggested that determination of as few as 16,000 carefully selected target structures will enable production of "accurate" homology models for 90% of all proteins found in nature.

Homology modeling can be distinguished from all other methods for analyzing relationships among protein sequences because it yields atomic coordinates suitable for direct comparisons with X-ray and solution NMR structures. Acceptable models can be divided into three accuracy classes, characterized with blind tests using known structures (Marti-Renom et al., 2000). Models based on >50% sequence identity with the template are comparable in accuracy to 3Å resolution X-ray structures or medium-resolution solution NMR structures (10 long-range restraints/residue). Models obtained with less similar templates (30–50% identity) typically have >85% of their $\alpha$-carbon atoms within 3.5Å of the correct position. When sequence identities fall below 30%, models with acceptable model scores (>0.7) may contain significant errors arising from ambiguities in the sequence alignment of the modeling candidate with the template, but can nevertheless be used for protein fold identification.

## Protein Data Bank

The PDB (http://www.rcsb.org/pdb/; Chapter 9; Berman et al., 2000) serves as *the central repository* for X-ray and solution NMR structures of proteins and other macromolecular complexes. Atomic coordinates for more than 15,500 experimentally determined structures of proteins and peptides can be freely downloaded from this single, publicly accessible archive, which is maintained by teams working at three centers (Rutgers University; Center for Advanced Research in Biotechnology; San Diego Supercomputer Center) with funding from U.S. government agencies. International access is enhanced by additional data deposition sites in Europe (http://pdb.ccdc.cam.ac.uk/pdb) and Japan (http://pdb.protein.osaka-u.ac.jp/pdb) and various PDB mirror sites located worldwide (see http://www.rcsb.org/pdb/mirrors.html).

## Homology Modeling Databases

As a general rule, the PDB does not include calculated models of protein structures. Instead, two major publicly accessible Web sites with search options offer access to the results of comparative protein structure modeling: MODBASE (http://pipe.rockefeller.edu/modbase-cgi/index.cgi) and SWISS-MODEL (http://www.expasy.org/swissmod/SM_3DCrunch_Search.html). In addition, researchers can submit protein sequences to MODWEB (http://pipe.rockefeller.edu/mwtest-cgi/main.cgi), SWISS-MODEL (http://www.expasy.ch/swissmod), FAMS (http://physchem.pharm.kitasatou.ac.jp/FAMS/fams.html), 3-D JIGSAW (http://www.bmm.icnet.uk/servers/3djigsaw/), CPHmodels (http://www.cbs.dtu.dk/services/CPHmodels/index.html), and the San Diego Supercomputing Center server (http://cl.sdsc.edu/hm.html) which are Web servers that return one or more homology models for the protein sequence in question. Structural biologists can also submit an atomic coordinate file to MODWEB and SWISS-MODEL and obtain homology models for all sequences that are within modeling distance of their structure. This feature will be particularly important for assessing the impact of each newly determined experimental structure produced by structural genomics centers.

## Strategic Considerations

Before describing how pilot programs in structural genomics are approaching the goal of providing structural information for every protein sequence found in nature, it is worth considering whether or not an organized, genome project scale effort is, in fact, necessary. Eventual success of existing experimental approaches was never in doubt. At issue is the timescale for substantial completion of the target list and the cost per structure determination.

Protein crystallographers and NMR spectroscopists currently deposit more than 2000 to 3000 structures per year into the PDB. An exhaustive analysis (Brenner, Chothia, and Hubbard, 1997) of all PDB submissions in 1994 revealed the following breakdown: 70% had essentially the same sequence as an existing PDB entry, 21% were closely related to an existing PDB entry, and 9% had no obvious homologue within the PDB (see at http://www.rcsb.org/pdb/holdings.html for up-to-date PDB growth statistics). Thus, even at the current rate of nearly 3000 PDB depositions per year, structural biology research as a whole will only make a modest annual impact on the estimated 16,000 protein structures required for the comprehensive homology modeling effort posited by Vitkup et al. (2001). No criticism of our colleagues is intended. Hypothesis-driven research in structural biology is directed toward understanding protein function, and not studying protein architecture per se.

The success of the Human Genome Project suggests that a systematic effort in high-throughput structural genomics may have a shorter timescale and lower cost per structure than "business as usual" structural biology. Although there are no hard numbers for the cost of each new structure determined in a structural biology laboratory, an estimate of $100,000 each has gained general acceptance. If this figure is accurate, completion of the minimal number of targets discussed by Vitkup et al. (2001) will cost no more than $1.6 billion. With efficiencies coming from economies of scale, the total cost of determining 16,000 structures should fall considerably.

The analogy with high-throughput sequencing is far from perfect. As a general rule, proteins do not behave like one another during the course of expression and purification, not to mention the vagaries inherent in crystallization and solution NMR spectroscopy. Some members of the structural biology community are, however, sufficiently encouraged by the impact of automation in genome sequencing to attempt the development of structural genomics pipelines.

## STRUCTURAL GENOMICS PILOT PROGRAMS

The feasibility of a large-scale, high-throughput program of structure determination is being explored by pilot studies underway in North America (Terwilliger, 2000) and South America (http://watson.fapesp.br/structur/menu.htm), Europe (Heinemann, 2000), and Asia (Yokoyama et al., 2000). Efforts involve both X-ray crystallography and solution NMR spectroscopy, with target lists derived from the genome sequences of archaea, eubacteria, and eukaryotes.

Nine pilot studies have been funded under the auspices of the National Institutes of Health (NIH) National Institute of General Medical Sciences (NIGMS) Protein Structure Initiative (PSI) (Norvell and Machalek, 2000). These NIH-funded groups have been charged with development and implementation of all high-throughput technologies required for going from gene sequences to disseminated protein structures. Some of the U.S. pilot studies are taking sequence family-based approaches with structure determination targets obtained from various bacterial and eukaryotic genomes. Others are targeting as many gene products as possible from particular model organisms or human pathogens (see Table 29.1). Among those groups studying protein families, some have elected to focus on targets involved in medically important biological processes such as signal transduction, antibiotic resistance, and human malignancies.

All nine NIGMS-funded centers maintain laboratory information management systems (LIMS) (Bertone et al., 2001), and publicly available Web sites that provide a full account of their targets and progress toward structure determination (see Table 29.1 for URLs). In addition, the U.S. government requires that this information be made available to the PDB, which is acting as a central repository for these data (http://targetdb.pdb.org/). As the structural genomics projects enter the production phase, the PDB will also capture interim results pertaining to protein expression, purification, and biophysical characterization. Selected reagents produced by the pilot studies, including expression plasmids/clones, purified proteins, and possibly crystals, will be archived and freely distributed by the NIGMS.

The methodological approach of one such NIGMS-funded pilot study, the New York Structural Genomics Research Consortium (NYSGRC, Table 29.1), is depicted schematically in Figure 29.1. The process involves modular automation of (1) target selection; (2) PCR amplification of the coding sequence from genomic or cDNA; (3) cloning the coding sequence into an appropriate expression vector; (4) expressing the protein at a sufficiently high level; (5) sequencing the cloned gene to verify that the coding sequence was correctly amplified; (6) confirming the identity of the expressed protein and characterizing it as a prelude to crystallographic studies; (7) obtaining the protein in sufficient amounts and purity; (8) defining suitable crystallization conditions; (9) X-ray diffraction measurements; (10) determining and refining the experimental structure; (11) calculating homology models using this new template; and (12) making functional inferences from the structure plus derived models and disseminating the findings. Failures are anticipated at every experimental step. Feedback loops provide for the possibility of both sample and process optimization.

Considerable efforts are now underway at a number of centers to automate protein production and crystallization, and useful reviews of the state of the art have been published by Edwards et al. (2000) and Stevens (2000). Once a statistically significant number of interim results are accumulated within the pilot study LIMS databases, we can look forward to a better understanding of correlations among amino acid sequence, resistance to proteolysis, globular domain structure, solubility, and crystallizability,

TABLE 29.1. Nine Pilot Studies Funded by National Institute of General Medical Sciences Protein Structure Initiative

| Consortium | Principle Investigation (PI) | Focus | URL |
|---|---|---|---|
| New York Structural Genomics Research Consortium | Stephen K. Burley The Rockefeller University[a] | Method: X-ray Targets: novel structural data; all Kingdoms of life with emphasis on medically relevant proteins | http://www.nysgrc.org |
| Structural Genomics of Pathogenic Protozoa Consortium | Wim G. J. Hol University of Washington[b] | Method: X-ray Targets: proteins from pathogens (Leishmania major, Trypanosoma brucei, Trypanosoma cruzi and Plasmodium falciparum) | http://depts.washington.edu/sgpp |
| The Midwest Center for Structural Genomics | Andrzej Joachimiak Argonne National Laboratory[c] | Method: X-ray Targets: novel structural data; all Kingdoms of life | http://www.mcsg.anl.gov |
| Berkeley Structural Genomics Center | Sung-Hou Kim Lawrence Berkeley National Lab[d] | Methods: X-ray, NMR Targets: whole organism proteomes (X-ray, Mycoplasma genitalium; NMR, Mycoplasma pneumoniae) | http://www.strgen.org |
| Center for Eukaryotic Structural Genomics | John L. Markley University of Wisconsin, Madison[e] | Methods: X-ray, NMR Targets: novel structural, functional data; eukaryotic proteins (model A. thaliana) | http:// www.uwstructuralgenomics.org |
| Northeast Structural Genomics Consortium | Gaetano Montelione Rutgers University[f] | Methods: X-ray, NMR Targets: novel structural data; eukaryotic proteins (e.g., S. cerevisiae, C. elegans, D. melanogaster, human) or practical prokaryotic homologues | http://www.nesg.org |

| Consortium | Director / Institution | Methods and Targets | URL |
|---|---|---|---|
| TB Structural Genomics Consortium | Thomas Terwilliger, Los Alamos National Lab[g] | Methods: X-ray, NMR<br>Targets: whole *M. tuberculosis* proteome with emphasis on functionally important proteins | http:// www.doe-mbi.ucla.edu/TB |
| The Southeast Collaboratory for Structural Genomics | Bi-Cheng Wang, University of Georgia[h] | Methods: X-ray, NMR<br>Targets: whole organism proteomes (*C. elegans*, *Pyrococcus furiosus*), human proteins | http:// 128.192.15.145/secsg |
| The Joint Center for Structural Genomics | Ian Wilson, The Scripps Research Institute[i] | Method: X-ray<br>Targets: novel structural data; proteins from *T. maritima* and *C. elegans* | http://www.jcsg.org |

[a] Albert Einstein College of Medicine, Brookhaven National Lab, Mount Sinai School of Medicine, Weill Medical College of Cornell University.

[b] Hauptman-Woodward MRI, Lawrence Berkeley Lab, Seattle Biomedical Research Institute. S.S R.L., University of Rochester.

[c] Northwestern University, University College London, UT Southwestern Medical Center at Dallas, University of Toronto, University of Virginia, Washington University School of Medicine.

[d] Stanford University, U.C. Berkeley, U.N.C., Chapel Hill.

[e] Hebrew University, Tokyo Metropolitan University, Medical College of Wisconsin, Washington State University.

[f] Columbia University, Cornell University, Genomic Sciences Center, Japan, Hauptman-Woodward MRI, Hebrew University, N.C.I. Frederick, Pacific Northwest National Lab, Robert Wood Johnson Medical School, S.U.N.Y. Buffalo, University of Toronto, Yale University.

[g] California State University, Fullerton, Case Western Reserve University, Florida State University, Lawrence Berkeley National Lab, Lawrence Livermore National Lab, N.Y.U., Pacific Northwest National Lab, Public Health Research Institute, Purdue University, St. Jude Children's Research Hospital, Stanford University, Texas A&M University, Texas Medical Center, University of Alberta, U.B.C., U.C., Berkeley, U.C.L.A., U.C.S.D., University of Colorado, Boulder, University of Kansas Medical Center, U.N.C., Chapel Hill, U. Penn, U. Texas Health Science Center, San Antonio, U. Texas, Austin, University of Virginia, et al.

[h] Georgia State University, Massachusetts General Hospital, Oklahoma Medical Research Foundation, Research Genetics, University of Alabama at Birmingham, University of Alabama at Huntsville, University of Oklahoma.

[i] G.N.F., S.D.S.C., S.S.R.L., Stanford University, U.C.S.D.

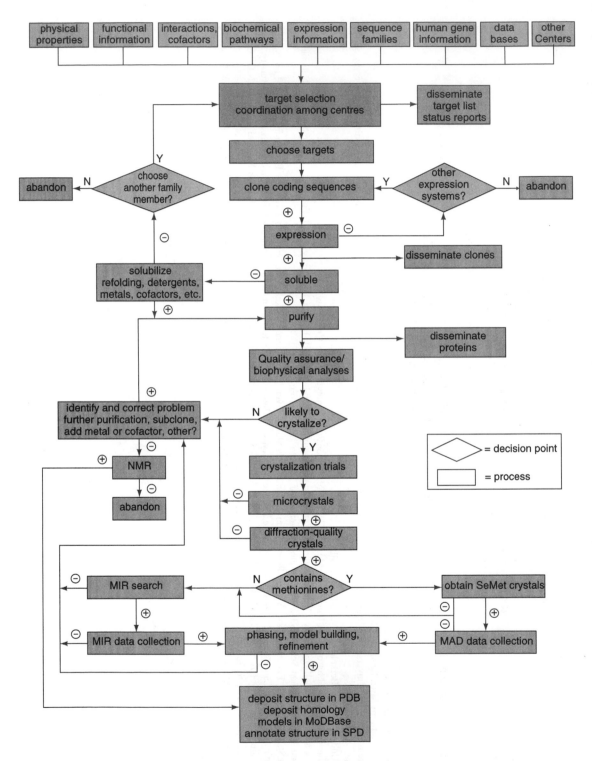

**Figure 29.1.** Flowchart depicting the processes involved in high-throughput structural genomics using X-ray crystallography. MIR denotes multiple isomorphous replacement, an alternative to MAD for structure determination. Reprinted with permission from Burley et al., (1999).

which should aid target selection and help increase the efficiency of structural genomics pipelines. In this context, it is imperative that negative information (i.e., susceptibility to proteases, conformational heterogeneity, poor solubility, and failure to crystallize) is captured with high fidelity.

Attempts to automate both X-ray diffraction experiments (Abola et al., 2000) and interpretation of experimental electron density maps (Lamzin and Perrakis, 2000; Terwilliger, 2001) have already shown considerable promise. Current estimates suggest that the total time required for data acquisition and structure determination/refinement could be reduced to no more than a few hours in many cases with little if any manual intervention (Walsh et al., 1999b; Walsh et al., 1999a). Similarly, developments in high field magnets and cryoprobe hardware, computational methods (automated resonance assignment [Moseley and Montelione, 1999; Duggan et al., 2001], TROSY [Pervushin et al., 1997], "direct methods" [Atkinson and Saudek, 2002]), chemical shift/sequence homology databases (Cornilescu, 1999; Meiler, Peti, and Griesinger, 2000), and protein biochemical labeling (Goto and Kay, 2000; Medek et al., 2000; Cowburn and Muir, 2001; Wider and Wuthrich, 1999) have advanced the pace of experimental structure determination via NMR spectroscopy.

A substantial number of reviews describing various aspects of structural genomics have been published over the last several years, including (Blundell and Mizuguchi, 2000; Brenner and Levitt, 2000; Burley et al., 1999; Chance et al., 2001; Erlandsen, Abola, and Stevens, 2000; Heinemann et al., 2000; Kim, 2000; Linial and Yona, 2000; Mittl and Grutter, 2001).

## NYSGRC CASE STUDY: YEAST MEVALONATE-5-DIPHOSPHATE DECARBOXYLASE

### Target Selection

Early NYSGRC structure determinations focused on proteins from the baker's yeast, *Saccharomyces cerevisiae*, an intensively studied eukaryotic organism with a fully sequenced genome containing numerous human gene homologues. Target selection was aided by the results of automated comparative protein structure modeling using the *S. cerevisiae* genome (Sanchez and Sali, 1998). In addition to providing homology models for all yeast proteins within "modeling distance" of an experimental structure already present in the PDB, the procedure identified yeast proteins for which no structural information was available. Initial targets were chosen with an emphasis on members of large protein families that would maximize the leverage of homology modeling.

The NYSGRC experience with mevalonate-5-diphosphate decarboxylase provides an informative case study for the development of a structural genomics pipeline. Mevalonate-5-diphosphate decarboxylase (MDD) is an enzyme from the sterol/isoprenoid biosynthesis pathway (Goldstein and Brown, 1990) (Figs. 29.2 and 29.3). MDD catalyzes the last of three sequential ATP-dependent reactions that convert mevalonate to isopentenyl diphosphate. A full account of the determination and bioinformatic analyses of the structure of MDD has been published by Bonanno et al. (2001) with Supplementary Materials available at http://www.pnas.org/.

### Sample Preparation and Characterization

The gene encoding yeast MDD (396 residues in length) was expressed in *Escherichia coli*, yielding a fusion protein with an N-terminal hexa-histidine tag

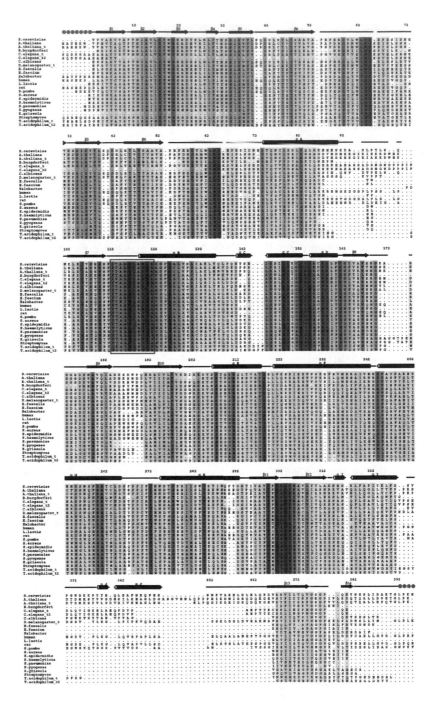

**Figure 29.2.** MDD sequence alignments. Proteins similar to *S. cerevisiae* MDD (E-value $<10^{-4}$) were identified using PSI-BLAST (Altschul et al., 1997), and aligned with CLUSTAL (Higgins, Bleasby, and Fuchs, 1992). Secondary structural elements of *S. cerevisiae* MDD are shown with cylinders ($\alpha$-helices) and arrows ($\beta$-strands). Grey dots denote poorly resolved residues in the final electron density map. Color coding denotes sequence conservation among MDDs (white → green ramp, 30 → 100% identity). Red box denotes the ATP-binding P-loop. Adapted from (Bonanno et al., 2001). Figure also appears in Color Figure section.

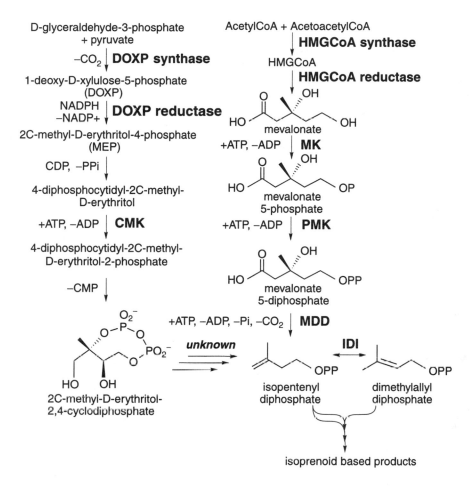

<u>Figure 29.3.</u> Pathways for biosynthesis of isopentenyl diphosphate. Isopentenyl diphosphate, the central intermediate in sterol/isoprenoid biosynthesis, is produced by two independent pathways, which have different evolutionary distributions (Eisenreich, Rohdich, and Bacher, 2001; Lange et al., 2000). Reprinted with permission from (Bonanno et al., 2001).

that facilitated purification to homogeneity using standard affinity and ion-exchange chromatographic methods.

Preliminary biophysical characterization of recombinant yeast MDD documented that the protein represented an ideal candidate for crystallization. Dynamic light scattering demonstrated that the purified protein is a highly soluble monodisperse dimer in aqueous solution (Ferre-D'Amare and Burley, 1995). Matrix-assisted laser desorption ionization mass spectrometry showed that the purified protein was neither post-translationally modified nor proteolyzed during expression and purification, and limited proteolysis revealed no evidence of conformational heterogeneity (Cohen, 1996).

Diffraction-quality crystals were obtained using incomplete factorial screens (Carter and Carter, 1979; Jancarik and Kim, 1991), which survey a large number of crystallization conditions (typically a few hundred) that have produced crystals of other proteins. Seleno-methionine substituted protein was produced and crystallized using similar procedures.

## X-Ray Structure Determination

X-ray diffraction data were recorded under cryogenic conditions at Beamline X25 at the National Synchrotron Light Source (Brookhaven National Laboratory). The structure of MDD was determined using the multiwavelength anomalous dispersion or MAD method (Hendrickson, 1991), from diffraction data recorded at three X-ray wavelengths in the vicinity of the selenium absorption edge. Following experimental electron density map fitting and structure refinement, structure quality/validation procedures (Chapter 14; Brünger et al., 1998; Laskowski et al., 1993; Vaguine, Richelle, and Wodak, 1999) were used to verify the accuracy of the refined atomic coordinates (publicly available at http://www.rcsb.org/pdb via PDB ID Code 1FI4).

## Structural Overview

MDD is a single $\alpha/\beta$ domain (Figure 29.4a) with a deep cleft. The structure consists of three antiparallel $\beta$-sheets and three sets of $\alpha$-helices. The crystallographic twofold

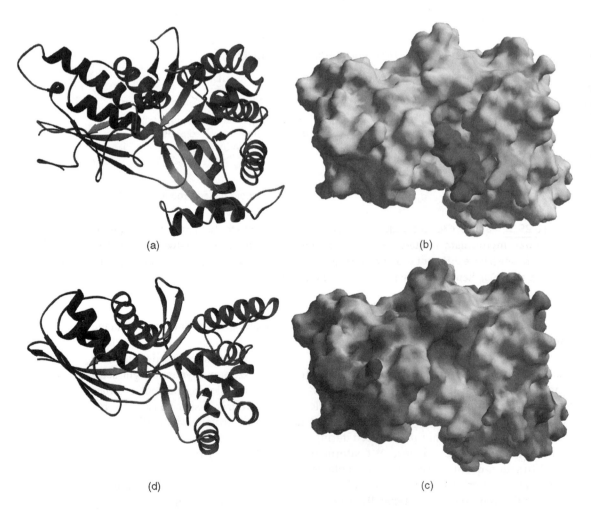

(a)

(b)

(d)

(c)

Figure 29.4.  *S. cerevisiae* MDD and *M. jannaschii* HSK. Ribbon drawings of MDD (a) and HSK (d) in the same orientation. Figure also appears in Color Figure section.

symmetry axis parallel to the $c$-axis generates a symmetric dimer. MDD sequence alignments (Figure 29.2) show that conserved segments surround the deep cleft (Figure 29.4b). A putative ATP-binding site or P-loop (outlined in red in Fig. 29.2 and colored red in Figs. 29.4a and 29.4b) lies within this surface concavity. Surface electrostatic calculations revealed a surface patch with positive electrostatic potential in the cleft (Fig. 29.4c), which represents an excellent candidate for binding the negatively charged substrate, mevalonate-5-diphosphate.

## Homology Modeling with MDD

Comparison of the structure of *S. cerevisiae* MDD with the PDB in November 2000 revealed no similar structures, as judged by the DALI protein structure similarity server (Holm and Sander, 1993a; Dietmann and Holm, 2001a). Put more succinctly, the structure of MDD represented a new protein fold. Homology modeling using MODPIPE (Sanchez et al., 2000) and the MDD template gave models for all known or putative MDDs plus various GHMP small-molecule kinases (Bork et al., 1993), including the galactokinases (GK), homoserine kinases (HSK), mevalonate kinases (MK; Fig. 29.3), and phosphomevalonate kinases (PMK; Fig. 29.3), plus diphosphocytidyl-2-C-methyl-D-erythritol kinases (CMK, an enzyme in the 1-deoxy-D-xylulose-5-phosphate pathway to isopentenyl diphosphate; Fig. 29.3), other enzymes, and some hypothetical proteins.

Subsequently, publication of a bona fide GHMP kinase structure (Zhou et al., 2000) (*Methanococcus jannaschii* HSK, PDB Code 1FWL) provided direct experimental confirmation that MDD is a member of the GHMP kinase superfamily. The only significant secondary structural differences between *S. cerevisiae* MDD and *M. jannaschii* HSK are two insertions within MDD (compare Figs. 29.4a and 29.4d).

## Mechanistic/Evolutionary Implications

A kinase-type mechanism of action for MDD is not unexpected, because MDD phosphorylates the substrate C-3 hydroxyl group, followed by elimination of both the added phosphate and carboxylate groups to give the product isopentenyl diphosphate, ADP, and $P_I$. Surprisingly, however, all three enzymes responsible for sequential conversion of mevalonate to isopentenyl diphosphate (MK, PMK, and MDD; Fig. 29.3) have the same fold.

It is almost certain that the genes encoding MDD, MK, and PMK arose from a common precursor, and active site mutations created three enzymes that bind structurally similar substrates and catalyze chemically similar reactions. The mevalonate-dependent pathway (Fig. 29.3) can be thought of as an example of retrograde evolution (Horowitz, 1945). It is also remarkable that CMK, which is part of the mevalonate independent pathway for sterol/isoprenoid biosynthesis in plastids and bacteria (Fig. 29.3), appears to have evolved from the same ancestral small molecule kinase.

## Homology Modeling with MDD and HSK

Using both experimental structures, MODPIPE produced models for 181 GHMP kinase superfamily members (113 with both templates, 60 with HSK only, and 8 with MDD only), including 22 MDDs, 31 GKs, 33 HSKs, 25 MKs, 9 PMKs, 25 CMKs, 7 archael shikimate kinases (Daugherty et al., 2001), and 8 D-glycero-D-manno-heptose 7-phosphate kinases (Kneidinger et al., 2001). In addition, 21 models fell into 3 sequence similarity groups, suggesting that additional enzyme activities are encompassed within the GHMP kinase superfamily.

## Implications for Target Selection

MDD was chosen for structure determination to provide a modeling template for a biomedically important family of enzymes. Greater coverage of protein sequence space was possible, because MDD proved to be a member of a protein superfamily encompassing both enzymes that catalyze chemically distinct reactions and proteins of unknown biochemical function. The availability of a second modeling template, HSK, gave yet wider modeling coverage of the GHMP kinase superfamily, because archaeal HSK and yeast MDD are only very distantly related to one another (16% identity for 272 structurally equivalent $\alpha$-carbon atomic pairs).

The modeling coverage of the GHMP kinase superfamily also provided some insights into the problem of target selection for structural genomics. The structure of *S. cerevisiae* MDD produced medium or high accuracy models for 80% of the modeled sequences thought to have MDD activity, but coverage of the HSK family provided by the *M. jannaschii* HSK template was not as broad (only 41% of HSK models are medium or high accuracy).

GHMP kinase superfamily members cluster into 19 discrete subfamilies (Fig. 29.5), encompassing one cluster of MDDs, three clusters of GKs, four clusters of HSKs, three clusters of MKs, one cluster of PMKs, one cluster of D-glycero-D-manno-heptose 7-phosphate kinases (DGMPKs), three clusters of CMKs, one cluster of archael shiki-mate kinases (aSKs), and two distinct clusters of hypothetical proteins. From this analysis, it was estimated that an additional 17 carefully chosen experimental struc-tures will be needed to produce medium or high accuracy models for ($\geq$80%) of

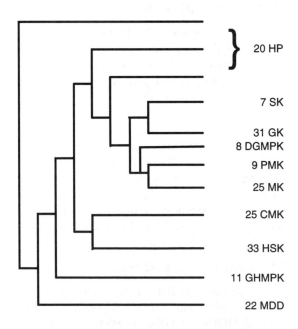

**Figure 29.5.** GHMP kinase superfamily clustering. Schematic dendrogram showing clusters within 181 GHMP kinase superfamily members.

each GHMP kinase subfamily. (The analyses of Vitkup et al. (2001) discussed earlier represents a generalization of this target selection procedure for all protein sequences in nature.)

The efficacy of this approach to target selection was explored by studying a representative PMK from *Streptococcus pneumoniae*. As expected, *S. pneumoniae* PMK is structurally similar to *S. cerevisiae* MDD and *M. jannaschii* HSK (Fig. 29.6) (Romanowski, Bonanno, and Burley, 2002). Homology modeling with the PMK structure yielded medium or high accuracy models for eight PMKs from eubacteria. Seven other models in the lower accuracy class were obtained for either bona fide or putative PMKs from four eubacteria, one archaebacterium, and two fungi (*S. cerevisiae* and *Schistosaccharomyces pombe*). It is not surprising that PMKs from higher eukaryotes were not modeled by MODPIPE, because these proteins constitute a distinct enzyme family that is not encompassed by the GHMP kinases.

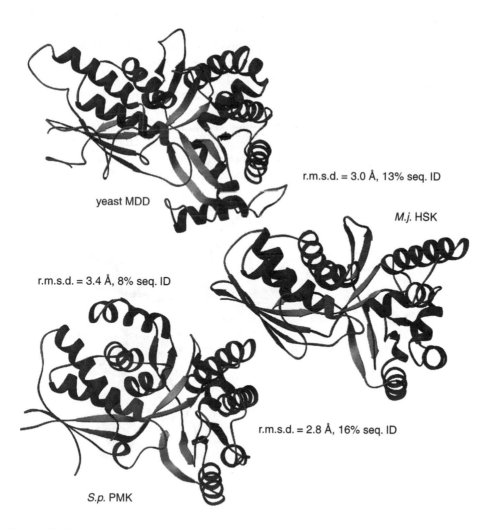

**Figure 29.6.** Structural comparison of three GHMP kinase superfamily members. Ribbon drawings of MDD, HSK, and PMK in the same orientation with root-mean-square deviations (rmsd's) and sequence identities.

The target selection exercise performed with the aid of models produced by the MDD and HSK templates was successful for the PMK subfamily. The structure of *S. pneumoniae* PMK produced much improved models (medium or high accuracy) for all of the PMKs that were previously modeled at low accuracy with the structures of MDD and HSK. In addition, the work on *S. pneumoniae* PMK provided a useful template for calculating models of other PMKs that were not within "modeling distance" of either HSK or MDD. These encouraging results underscore the value of a families-based approach to structural genomics and the importance of establishing rigorous target selection procedures before high-throughput production of proteins structures begins in earnest.

## Implications for Biomedical Research

Structures of GHMP kinase superfamily members and resulting homology models can guide experiments aimed at defining enzymatic function, cofactor requirements, and mechanism(s) of action, which should allow us to better understand how this family of structurally similar, yet functionally diverse enzymes evolved from a common ancestor.

The GHMP kinase models may also be of some medical relevance in understanding the structural bases of diseases associated with single nucleotide polymorphisms (SNPs) in coding regions. Figure 29.7 illustrates a homology model of human GK calculated with the archael HSK template. This schematic view shows the locations of disease-causing SNPs. Impairment of human GK function by missense mutations (Val32 → Met, Gly36 → Arg, His44 → Tyr, Gly346 → Ser, Gly349 → Ser) leads to galactosemia and cataract formation in newborns (Novelli and Reichardt, 2000), which can be reversed by restricting dietary galactose. The first three of these mutations map

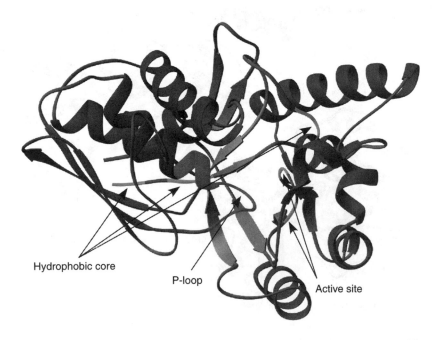

**Figure 29.7.** Understanding disease-causing SNPs with the homology model of human galactokinase. Ribbon drawing of GK in the same orientation as MDD and HSK in Figures 29.4a and 29.4d, respectively. Mutations due to disease-causing SNPs are labeled.

to the hydrophobic core of the protein, where they probably destabilize the structure of the enzyme, leading to complete loss of function. The remaining two mutations map to the conserved active site cleft (green in Fig 29.4b for MDD), where they almost certainly interfere with either galactose or ATP binding rendering the enzyme inactive.

## NYSGRC PROGRESS SUMMARY

The MDD case study provides an encouraging look at the potential impact of a program of high-throughput structure determination by X-ray crystallography and solution NMR spectroscopy. A review the overall progress of the NYSGRC during its first year of NIH funding also merits a brief discussion. As of August 31, 2001, the NYSGRC completed 27 X-ray crystal structures of recombinant proteins from eukaryotes, archaea, and eubacteria (Chance et al., 2001). Homology modeling with these 27 experimentally determined structures produced additional structural information for thousands of protein sequences. The NYSGRC structures and resulting models are publicly available via the PDB (http:/www.rcsb.org/pdb) and MODBASE (http://pipe.rockefeller.edu/modbase), respectively. These raw statistics show that the NYSGRC pilot study is off to an encouraging start. Other NIH-funded consortia reported similar outcomes for the first year of funding, and there is every reason to believe that the NIH-NIGMS PSI will succeed in demonstrating the feasibility of Structural Genomics. More important, perhaps, is the need to demonstrate the impact of the PSI on our knowledge of protein structure/sequence space.

With the results of quantitative structure–structure comparisons (DALI; Holm and Sander, 1993b; Dietmann et al., 2001b, SCOP; Chapter 12; Lo Conte et al., 2002; CATH; Chapter 13; Orengo et al., 1997; Orengo et al., 2002; and PRESAGE; Brenner, Barken, and Levitt, 1999) and automated homology modeling structures (Marti-Renom et al., 2000), the NYSGRC structures can be subdivided into the following five categories (the number of structures and the NYSGRC target identifier are shown in parentheses):

1. Structures such as MDD that represented new protein folds at the time of structure determination and thereby provide entirely new information about protein sequence/structure space (4/25, P008, P018, P100, T130).
2. Structures that are distantly related to previously known protein structures and thereby provide a considerable amount of new information about protein sequence superfamilies, despite the fact that they do not represent a new fold (10/25, P007, P097, P109, P111, P111a, T27, T127, T136, T139, T140).
3. Structures that are more closely related to previously known protein structures than class 2 proteins, and thereby provide incremental information about subfamilies within protein sequence superfamilies (6/25, P044a, P068, P102, T35, T45, T135).
4. Structures that are very closely related to previously known protein structures, and thereby provide little, if any, incremental information about protein sequence/structure space (5/25, P003, P048a, T129, T138, MHP).
5. Unclassified (2, P096, T132).

Of the first 27 NYSGRC structures, over half were distantly or entirely unrelated to known folds. Eighty percent of the structures satisfy the pragmatic target selection criterion based on the current limitations of homology modeling (i.e., <30% identical in amino acid sequence to a protein of known structure). The remaining 20% represent protein structure determination projects that were continued despite the fact that a closely related structure appeared in the PDB while work was in progress. As the NYSGRC moves beyond the pilot stage, redundant structure determination attempts will be abandoned once they have been superseded by PDB deposition of a closely related structure.

## LONG-TERM PROSPECTS AND CAVEATS

Although the early results from structural genomics pilot studies are encouraging, to say the least, there are some regions of protein structure space that will not succumb quickly to either solution NMR or X-ray methods. Membrane protein crystallization continues to represent a considerable technical challenge, but advances in robotic protein solubilization/purification and crystallization may ease these difficulties. Alternatively, the TROSY technology developed by Kurt Wuthrich and coworkers (Pervushin et al., 1997) may offer an approach to solution NMR spectroscopic studies of protein-detergent micelles.

There have also been predictions that many of the proteins normally found in macromolecular complexes can never be studied in isolation (Ban et al., 2000). This contention may be true, but misses the point of the genomewide philosophy of structural genomics. Somewhere in biology, the same fold will almost certainly be used in a context where it is not inextricably bound up in a large complex. The balance of protein sequence space is occupied by so-called low complexity regions, which may never adopt stable conformations or remain unstructured until they interact with their respective targets. Clearly, these cases are beyond the initial scope of structural genomics.

Finally, some discussion of what structural genomic programs are not designed to do is required. It would be most unfortunate if the perception that Structural Genomics Centers should become the only source of single domain protein structures were to arise. The technologies developed for structural genomics should accelerate the work of all structural biologists, but there is no expectation that these advances are going to put X-ray crystallographers and solution NMR spectroscopists out of business.

Structural genomics projects will not, at least in most cases, attempt to acquire data at atomic resolution or determine extremely accurate structures. Detailed studies of biochemical mechanism(s), protein-ligand complex structures, mutant proteins, and protein–protein interactions aimed at understanding biochemical function are also beyond the scope of structural genomics. Moreover, the families-based approach means that there will be no premium placed on obtaining a structure for the protein of interest from one particular organism. A homology model of a protein from an important pathogen calculated with a remotely related template may not suffice to understand virulence, and the structure of that particular virulence factor will have to be determined independent of any high-throughput effort. Similar considerations will almost certainly apply to human proteins that are targets of drug discovery efforts.

Structural genomics projects promise to deliver a very large number of experimental structures of protein domains, which will in turn yield a staggering number of homology models covering virtually all of protein sequence space. These data will be invaluable to chemists and biologists, providing a 3D view of the protein sequence universe and giving insights into biochemical and cellular function. With time, this enormous wealth of structures and homology models should permeate all of biomedical research, bringing more and more biologists into the structural "fold." In planning for this structured future, structural biologists should prepare themselves to satisfy an ever-increasing demand for X-ray or solution NMR structures of proteins and their complexes with small molecules, nucleic acids, carbohydrates, and their interaction partners.

## REFERENCES

Abola E, Kuhn P, Earnest T, Stevens RC (2000): Automation of X-ray crystallography. *Nat Struct Biol* 7:970–2.

Altschul SF, Madden TL, Schaffer AA, Zhang JZ, Miller W, Lipman DJ (1997): Gapped BLAST and PSI-BLAST: a new generation of protein database search programs. *Nucleic Acids Res* 25:3389–402. [The authors describe development of state-of-the-art sequence database searching tools.]

Atkinson RA, Saudek V (2002): The direct determination of protein structure by NMR without assignment. *FEBS Lett* 510:1–4.

Baker D, Sali A (2001): Protein structure prediction and structural genomics. *Science* 294:93–6. [Review of homology modeling and *ab initio* protein structure prediction.]

Ban N, Nissen P, Hansen J, Moore PB, Steitz TA (2000): The complete atomic structure of the large ribosomal subunit at 2.4 A resolution. *Science* 289:905–20.

Berman HM, Westbrook J, Feng Z, Gilliland G, Bhat TN, Weissig H, Shindyalov IN, Bourne PE (2000): The Protein Data Bank. *Nucleic Acids Res* 28:235–42. [The authors describe the contents and operation of the Protein Data Bank, which is the single repository for all 3D protein structure information.]

Bertone P, Kluger Y, Lan N, Zheng D, Christendat D, Yee A, Edwards AM, Arrowsmith CH, Montelione GT, Gerstein M (2001): SPINE: an integrated tracking database and data mining approach for identifying feasible targets in high-throughput structural proteomics. *Nucleic Acids Res* 29:2884–98.

Blundell TL, Mizuguchi K (2000): Structural genomics: an overview. *Prog Biophys Mol Biol* 73:289–95. [Review of structural genomics.]

Bonanno JB, Edo C, Eswar N, Pieper U, Romanowski MJ, Ilyin V, Gerchman SE, Kycia H, Studier FW, Sali A, Burley SK (2001): Structural genomics of enzymes involved in steroid/isoprenoid biosynthesis. *Proc Natl Acad Sci USA* 98:12896–901.

Bork P, Sander C, Valencia A (1993): Convergent evolution of similar enzymatic function on different protein folds: the hexokinase, ribokinase, and galactokinase families of sugar kinases. *Protein Sci* 2:31–40.

Brenner SE, Chothia C, Hubbard T (1997): Population statistics of protein structures: lessons from structural classifications. *Curr Opin Struct Biol* 7:369–76.

Brenner SE, Barken D, Levitt M (1999): The PRESAGE database for structural genomics. *Nucleic Acids Res* 27:251–3. [The authors describe the content and operation of the PRESAGE database for structural genomics.]

Brenner SE, Levitt M (2000): Expectations from structural genomics. *Protein Sci* 9:197–200. [Review of structural genomics.]

Brünger AT, Adams PD, Clore GM, DeLano WL, Gros P, Grosse-Kunstleve RW, Jiang JS, Kuszewski J, Nilges M, Pannu NS, Read RJ, Rice LM, Simonson T, Warren GL (1998): Crystallography and NMR system: a new software suite for macromolecular structure determination. *Acta Crystallogr* D54:905–21.

Burley SK, Almo SC, Bonanno JB, Capel M, Chance MR, Gaasterland T, Lin D, Sali A, Studier FW, Swaminathan S (1999): Structural genomics: beyond the human genome project. *Nat Genet* 23:151–7. [Review of structural genomics.]

Carter CW, Jr, Carter CW (1979): Protein crystallization using incomplete factorial experiments. *J Biol Chem* 254:12219–23.

Chance MR, Bresnick AR, Burley SK, Jiang J-S, Lima CD, Sali A, Almo SC, Bonanno JB, Buglino JA, Boulton S, Chen H, Eswar N, He G, Huang R, Ilyin V, McMahon B, Pieper U, Ray S, Vidal M, Wang L (2002): High throughput structural biology: a pipeline for providing structures for the biologist. *Protein Sci* 11:723–38.

Cohen SL (1996): Domain elucidation by mass spectrometry. *Structure* 4:1013–6.

Cornilescu G, Delagio Γ, Bax A (1999): Protein backbone angle restraints from searching a database for chemical shift and sequence homology. *J Biomol NMR* 13:289–302.

Cowburn D, Muir TW (2001): Segmental isotopic labeling using expressed protein ligation. *Methods Enzymol* 339:41–54.

Daugherty M, Vonstein V, Overbeek R, Osterman A (2001): Archaeal shikimate kinase, a new member of the GHMP-kinase family. *J Bacteriol* 183:292–300.

Dietmann S, Holm L (2001a): Identification of homology in protein structure classification. *Nat Struct Biol* 8:953–7.

Dietmann S, Park J, Notredame C, Heger A, Lappe M, Holm L (2001b): A fully automatic evolutionary classification of protein folds: Dali Domain Dictionary version 3. *Nucleic Acids Res* 29:55–7. [The authors describe operation of the DALI protein structure similarity search server.]

Duggan BM, Legge GB, Dyson HJ, Wright PE (2001): SANE (Structure Assisted NOE Evaluation): an automated model-based approach for NOE assignment. *J Biomol NMR* 19:321–9.

Edwards AM, Arrowsmith CH, Christendat D, Dharamsi A, Friesen JD, Greenblatt JF, Vedadi M (2000): Protein production: feeding the crystallographers and NMR spectroscopists. *Nat Struct Biol* 7:970–2.

Eisenreich W, Rohdich F, Bacher A (2001): Deoxyxylulose phosphate pathway to terpenoids. *Trends Plant Sci* 6:78–84.

Erlandsen H, Abola EE, Stevens RC (2000): Combining structural genomics and enzymology: completing the picture in metabolic pathways and enzyme active sites. *Curr Opin Struct Biol* 10:719–30. [Review of structural genomics focused on metabolic pathways and enzymes.]

Ferre-D'Amare AR, Burley SK (1995): Dynamic light scattering as a tool for evaluation crystallizability of macromolecules. In: Charles RMS, Carter W Jr, editors. *Methods in Enzymology*. San Diego: Academic Press, pp 157–66.

Gilson M, Sharp K, Honig B (1988): Calculating the electrostatic potential of molecules in solution: method and error assessment. *J Comput Chem* 9:327–35.

Goldstein JL, Brown MS (1990): Regulation of the mevalonate pathway. *Nature* 343:425–30.

Goto NK, Kay LE (2000): New developments in isotope labeling strategies for protein solution NMR spectroscopy. *Curr Opin Struct Biol* 10:585–92.

Heinemann U (2000): Structural genomics in Europe: Slow start, strong finish? *Nat Struct Biol* 7:940–2. [Review of structural genomics activities in Europe.]

Heinemann U, Frevert J, Hofmann K-P, Illing G, Maurer C, Oschkinat H, Saenger W (2000): An integrated approach to structural genomics. *Prog Biophys Mol Biol* 73:347–62. [Review of structural genomics.]

Hendrickson W (1991): Determination of macromolecular structures from anomalous diffraction of synchrotron radiation. *Science* 254:51–8.

Higgins DG, Bleasby AJ, Fuchs R (1992): CLUSTAL V: improved software for multiple sequence alignment. *Comput App Biosci* 8:189–91. [The authors describe state-of-the-art protein sequence comparison tools.]

Hol W (2000): Structural genomics for science and society. *Nat Struct Biol* 7:964–6. [Review of structural genomics.]

Holm L, Sander C (1993a): Families of structurally similar proteins, version 1.0. *J Mol Biol* 233:123–38.

Holm L, Sander C (1993b): Protein structure comparison by alignment of distance matrices. *J Mol Biol* 233:123–38.

Horowitz NH (1945): On the evolution of biochemical syntheses. *Proc Natl Acad Sci USA* 31:153–7.

Jancarik J, Kim S-H (1991): Sparse matrix sampling: a screening method for crystallization of proteins. *J Applied Crystallogr* 24:409–11.

Kim S-H (2000): Structural genomics of microbes: an objective. *Curr Opin Struct Biol* 10:380–3.

Kneidinger B, Graninger M, Puchberger M, Kosma P, Messner P (2001): Biosynthesis of nucleotide-activated D-glycero-D-manno-heptose. *J Biol Chem* 276:20935–44.

Lamzin VS, Perrakis A (2000): Current state of automated crystallographic analysis. *Nat Struct Biol* 7:978–81.

Lange BM, Rujan T, Martin W, Croteau R (2000): Isoprenoid biosynthesis: the evolution of two ancient and distinct pathways across genomes. *Proc Natl Acad Sci USA* 97:13172–7.

Laskowski RJ, MacArthur MW, Moss DS, Thornton JM (1993): PROCHECK: a program to check stereochemical quality of protein structures. *J Applied Crystallogr* 26:283–90.

Linial M, Yona G (2000): Methodologies for target selection in structural genomics. *Prog Biophys Mol Biol* 73:297–320.

Liu J, Rost B (2001): Comparing function and structure between entire proteomes. *Protein Sci* 10:1970–9.

Lo Conte L, Brenner SE, Hubbard TJ, Chothia C, Murzin AG (2002): SCOP database in 2002: refinements accommodate structural genomics. *Nucleic Acids Res* 30:264–7. [The authors describe the contents and operation of the SCOP database.]

Marti-Renom MA, Stuart AC, Fiser A, Sanchez R, Melo F, Sali A (2000): Comparative protein structure modeling of genes and genomes. *Ann Rev Biophys Biomol Struct* 29:291–325.

Medek A, Olejniczak ET, Meadows RP, Fesik SW (2000): An approach for high-throughput structure determination of proteins by NMR spectroscopy. *J Biomol NMR* 18:229–38.

Meiler J, Peti W, Griesinger C (2000): DipoCoup: a versatile program for 3D-structure homology comparison based on residual dipolar couplings and pseudocontact shifts. *J Biomol NMR* 17:283–94.

Mittl PRE, Grutter MG (2001): Structural genomics: opportunities and challenges. *Curr Opin Chem Biol* 5:402–8. [Review of structural genomics.]

Moseley HN, Montelione GT (1999): Automated analysis of NMR assignments and structures for proteins. *Curr Opin Struct Biol* 9:635–42.

Moult J, Melamud E (2000): From fold to function. *Curr Opin Struct Biol* 10:382–9. [Review of structural genomics.]

Norvell JC, Machalek AZ (2000): Structural genomics programs at the US National Institute of General Medical Sciences. *Nat Struct Biol* 7:931. [Description of the U.S. government-funded structural genomics initiatives.]

Novelli G, Reichardt JK (2000): Molecular basis of disorders of human galactose metabolism: past, present, and future. *Mol Genet Metab* 71:62–5.

Orengo CA, Michie AD, Jones S, Jones DT, Swindells MB, Thornton JM (1997): CATH—a hierarchic classification of protein domain structures. *Structure* 5:1093–108.

Orengo CA, Pearl FMG, Bray JE, Todd AE, Martin AC, LoConte L, Thornton JM (1999): The CATH Database provides insights into protein structure/function relationship. *Nucleic Acids Res* 27:275–9.

Orengo CA, Bray JE, Buchan DW, Harrison A, Lee D, Pearl FM, Sillitoe I, Todd AE, Thornton JM (2002): The CATH protein family database: a resource for structural and functional annotation of genomes. *Proteomics* 2:11–21. [The authors describe the contents and operation of the CATH database.]

Pervushin K, Riek R, Wider G, Wuthrich K (1997): Attenuated T2 relaxation by mutual cancellation of dipole–dipole coupling and chemical shift anisotropy indicates an avenue to NMR structures of very large biological macromolecules in solution. *Proc Natl Acad Sci USA* 94:12366–71.

Ritz D, Lim J, Reynolds CM, Poole LB, Beckwith J (2001): Conversion of a peroxiredoxin into a disulfide reductase by a triplet repeat expansion. *Science* 294:158–60.

Romanowski MR, Bonanno JB, Burley SK (2002): Crystal structure of Streptococcus pneumoniae phosphomevalonate kinase. *Proteins* 47:568–7.

Sanchez R, Sali A (1998): Large-scale protein structure modeling of the *Saccharomyces cerevisiae* genome. *Proc Natl Acad Sci USA* 95:13597–602.

Sanchez R, Pieper U, Mirkovic N, de Bakker PI, Wittenstein E, Sali A (2000): MODBASE, a database of annotated comparative protein structure models. *Nucleic Acids Res* 28:250–3.

Sheehan D, Meade G, Foley VM, Dowd CA (2001): Structure, function and evolution of glutathione transferases: implications for classification of non-mammalian members of an ancient enzyme superfamily. *Biochem J* 360:1–6.

Skolnick J, Fetrow JS, Kolinski A (2000): Structural genomics and its importance for gene functional analysis. *Nat Biotech* 18:283–7. [Review of structural genomics.]

Stevens RC (2000): High-throughput protein crystallization. *Curr Opin Struct Biol* 10:558–63.

Teichmann SA, Murzin AG, Chothia C (2001): Determination of protein function, evolution and interactions by structural genomics. *Curr Opin Struct Biol* 11:354–63. [Review of structural genomics.]

Terwilliger TC (2000): Structural genomics in North America. *Nat Struct Biol* 7:935–9. [Review of structural genomics focused on activities in North America.]

Terwilliger TC (2001): Maximum-likelihood density modification using pattern recognition of structural motifs. *Acta Crystallogr* D57:1755–62.

Todd AE, Orengo CA, Thornton JM (2001): Evolution of function in protein superfamilies, from a structural perspective. *J Mol Biol* 307:1113–43.

Vaguine AA, Richelle J, Wodak SJ (1999): SFCHECK: a unified set of procedures for evaluating the quality of macromolecular structure-factor data and their agreement with the atomic model. *Acta Crystallogr* D55:191–205.

Vitkup D, Melamud E, Moult J, Sander C (2001): Completeness in structural genomics. *Nat Struct Biol* 8:559–66. [Review of target selection in structural genomics.]

Walsh MA, Dementieva I, Evans G, Sanishvili R, Joachimiak A (1999a): Taking MAD to the extreme: ultrafast protein structure determination. *Acta Crystallogr* D55:1168–73.

Walsh MA, Evans G, Sanishvili R, Dementieva I, Joachimiak A (1999b): MAD data collection—current trends. *Acta Crystallogr* D55:1726–32.

Wider G, Wuthrich K (1999): NMR spectroscopy of large molecules and multimolecular assemblies in solution. *Curr Opin Struct Biol* 9:594–601.

Yokoyama S, Hirota H, Kigawa T, Yabuki T, Shirouzu M, Terada T, Ito Y, Matsuo Y, Kuroda Y, Nishimura Y, Kyogoku Y, Miki K, Masui R, Kuramitsi S (2000): Structural genomics projects in Japan. *Nat Struct Biol* 7:943–5.

Zhang C, DeLisi C (2001): Protein folds: molecular systematics in three dimensions. *Cell Molec Life Sci* 58:72–9.

Zhou T, Daugherty M, Grishin NV, Osterman AL, Zhang H (2000): Structure and mechanism of homoserine kinase: prototype for the GHMP kinase superfamily. *Structure* 8:1247–57.

# INDEX